Advances in Science, Technology & Innovation

IEREK Interdisciplinary Series for Sustainable Development

D1800011

Advances in Science, Technology & Innovation (ASTI) is a series of peer-reviewed books based on the best studies on emerging research that redefines existing disciplinary boundaries in science, technology and innovation (STI) in order to develop integrated concepts for sustainable development. The series is mainly based on the best research papers from various IEREK and other international conferences, and is intended to promote the creation and development of viable solutions for a sustainable future and a positive societal transformation with the help of integrated and innovative science-based approaches. Offering interdisciplinary coverage, the series presents innovative approaches and highlights how they can best support both the economic and sustainable development for the welfare of all societies. In particular, the series includes conceptual and empirical contributions from different interrelated fields of science, technology and innovation that focus on providing practical solutions to ensure food, water and energy security. It also presents new case studies offering concrete examples of how to resolve sustainable urbanization and environmental issues. The series is addressed to professionals in research and teaching, consultancies and industry, and government and international organizations. Published in collaboration with IEREK, the ASTI series will acquaint readers with essential new studies in STI for sustainable development.

More information about this series at http://www.springer.com/series/15883

Pravat Kumar Shit • Hamid Reza Pourghasemi •
Gouri Sankar Bhunia

Editors

Gully Erosion Studies from India and Surrounding Regions

Editors
Pravat Kumar Shit
Department of Geography
Raja N. L. Khan Women's College (Autonomous)
Medinipur, West Bengal, India

Hamid Reza Pourghasemi
Department of Natural Resources
and Environmental Engineering
College of Agriculture, Shiraz
University
Shiraz, Iran

Gouri Sankar Bhunia
Aarvee Associates Architects
Engineers & Consultants Pvt. Ltd
Hyderabad, India

ISSN 2522-8714   ISSN 2522-8722   (electronic)
Advances in Science, Technology & Innovation
IEREK Interdisciplinary Series for Sustainable Development
ISBN 978-3-030-23245-0   ISBN 978-3-030-23243-6   (eBook)
https://doi.org/10.1007/978-3-030-23243-6

This Springer imprint is published by the registered company Springer Nature Switzerland AG.
The registered company address is: Gewerbestrasse 11, 6330 Cham, Switzerland

Dedicated to beloved teachers and parents

Foreword

I am happy to learn that Springer-Nature Publishing Company is bringing out a book on "Gully Erosion Studies from India and Surrounding Regions" under the Advances in Science, Technology & Innovation (ASTI) series on Sustainable Development. The book is jointly edited by Pravat Kumar Shit, Hamid Reza Pourghasemi, and Gouri Sankar Bhunia who are eminent scholars and researchers in the field of geomorphology and advanced geospatial technology.

Gully erosion is the erosion process whereby runoff water accumulates and sediment production is varied in various temporal and spatial scales and under the different climatic conditions and land-use patterns. Soil is the most fundamental and basic resource that provides food, fodder, fuel, and fiber. It underpins food security and environmental quality. The essentiality of soil to human well-being is often not realized until the production of food drops is jeopardized when the soil is severely eroded or degraded to the level that it loses its inherent resilience.

India and its surrounding regions are experiencing major problems in land degradation and mismanagement in land-use practices. By contrast, mismanagement and also the sustainability of the current and future soil and land resource allocation are other concerns. Thus, it is important to use the newest technologies and tools to improve and properly develop sustainable management. In this context, the book has been very effectively organized into thematic sections, covering the fundamental concepts of the gully classification system, erosion processes, and prediction, modeling, and sustainable management strategies.

It is a cohesive effort of a number of authors, researchers, and experts in the field of geomorphology across the country and other parts of the world. The editors have done an exemplary job in collecting, compiling, and editing the papers in a book form. The quality of this interdisciplinary research can be realized by readers by going through chapters in this enriched volume. This book will be very much beneficial for geomorphologists, hydrologists, soil scientists, students, scientists, ecologists, and policymakers.

I extend my warm greetings to all those associated with the publication and congratulate Springer-Nature Publishing Company for launching this book.

Department of Geography, University of Allahabad   Savindra Singh
Allahabad, India

Preface

Gully erosion is the removal of soil along constricted channels via the accretion of surface runoff, which tends to yield more sediment loss than other types of soil erosion, such as overland drift or rilling. Gully erosion is physiographically prevalent and is recognized by numerous terms in various regions. Gullies cause widespread destruction to infrastructures and susceptible cultivated lands and adulterate water by confiscation and aggradation of soil elements and chemicals from manure. Gullies are everlasting erosional practices that grow in several areas of the biosphere, predominantly in arid and semiarid zones. Unscientific land-use practices and mismanagement of land use and strategies may accelerate gully development by head cutting, sidewall breakdown, piercing, floor corrosion, and other procedures, which bring about extensive land deprivation and probable mutilation to human erections and activities.

Gully erosion has worldwide effects on agricultural, financial, and sociopolitical circumstances; nevertheless, statistics concerning the amount of these influences have been principally unpredictable. This book emphasizes the reciprocal interactions of gully deprivation, management, and remediation. It highlights the instantaneous and long-lasting effects of gully degradation in association with the geospatial technology in arid and semiarid environments.

The collection of thirty-one chapters draws on the research of an international group of scholars and practitioners who work in college, universities, government sectors, private consultancies, and research centers. Their expertise is in the field of applied geomorphology, hydrology, sedimentology, ecology, and engineering. Their methods include intensive field investigations, laboratory experimentation, geospatial technology, and modeling of gully susceptibility mapping, monitoring, and management.

This book addresses the core subjects associated with morphology and development of rill-gully, restoration strategies of gully and ravine land, gully collapsing in lateritic belt, risk estimation of rill and gully erosion by random forest model, geomorphic threshold and SCS-CN-based runoff-sediment yield modeling, Bayesian weight of evidence, RUSLE, and SDR model, hydraulic flume experiment, SWAT model, plant roots—an experimental insight, MARS model, SVM machine learning algorithm, etc. Research results are presented in this book that forms the scientific basis for the extensive evaluation of gully erosion, control and greening for livelihood, and environmental sustainability. The book covers a wide range of vital topics in the areas of gully erosion and water erosion at lateritic uplands of India and its surrounding regions. Additionally, the book offers GIS-based advanced cartographic techniques for students to recognize both simple and classy concepts of applied geomorphology. It is an ardent challenge to efficiently guide the aspirants who are involved in research and development in applied geomorphology. Chapter-end references, which are recent and resourceful, shed light on topics mentioned in this book.

We thank all the authors who have meticulously completed their chapters at short notice and contributed in building this edifying and beneficial publication. We believe this book will be of great value to geographers, geologists, agricultural engineers, hydrologists, soil scientists, ecologists, research scholars, environmentalists, and policymakers.

Medinipur, West Bengal, India   Pravat Kumar Shit
                                                                          Hamid Reza Pourghasemi
                                                                               Gouri Sankar Bhunia

Acknowledgments

The preparation of this book has been guided by several geomorphologic pioneers. We are obliged to these experts for providing their time to evaluate the chapters required updation in this book. We are very much thankful to our respected teachers, Dr. Ramkrishna Maiti, Dr. Ashis Kumar Paul, Dr. Dilip Kr. Pal, Dr. Sunando Bandyopadhyay, Dr. Lakshminarayan Satpati, Dr. Malay Mukhopadhyay, Dr. Joyti Shankar Bandopadhay, Dr. Soumendu Chatterjee, Dr. Nilanjana Das Chatterjee, and Dr. Ratan Kumar Samanta, for their guidance, suggestions, encouragement, and immense support throughout the work.

We thank the anonymous reviewers for their constructive comments that led to substantial improvements to the quality of this manuscript.

We also thank Ranita and Debjani, whose love, encouragement, and support have motivated us to make this book a reality. Because this book was a number of years in the making, we want to thank our family and friends for their continued support.

Dr. Pravat Kumar Shit thanks Dr. Jayasree Laha, Principal, Raja N.L. Khan Women's College (Autonomous), Midnapore, for her administrative support to carry on this project. We also acknowledge the Department of Geography of this college for providing logistic support and infrastructure facilities.

Dr. Hamid Reza Pourghasemi thanks College of Agriculture, Shiraz University, and Watershed Management Society of Iran for supporting during the preparation of this book.

This work would not have been possible without constant inspiration from our students, knowledge from our teachers, enthusiasm from our colleagues and collaborators, and support from our family.

Finally, we also thank our publisher and its publishing editor Dr. Nabil Khélifi, the Middle East and North Africa, Springer Heidelberg, for their continuous support in the publication of this book.

Disclaimer

The authors of individual chapters are solely responsible for ideas, views, data, figures, and geographical boundaries presented in the respective chapters of this book, and these have not been endorsed, in any form, by the publisher, the editor, and the authors of forewords, preambles, or other chapters.

Contents

1　Spatial Extent, Formation Process, Reclaimability Classification System
and Restoration Strategies of Gully and Ravine Lands in India . . . .　1
Gopal Kumar, Partha Pratim Adhikary, and Ch. Jyotiprava Dash

2　Soil Disintegration Characteristics on Ephemeral Gully Collapsing
in Lateritic Belt of West Bengal, India . . . .　21
Pravat Kumar Shit and Partha Pratim Adhikary

3　Modeling of Gully Erosion Based on Random Forest Using GIS and R . . . . .　35
Amiya Gayen, Sk. Mafizul Haque, and Sunil Saha

4　Geomorphic Threshold and SCS-CN-Based Runoff and Sediment Yield
Modelling in the Gullies of Dwarka–Brahmani Interfluve, West Bengal,
India . . . .　45
Sandipan Ghosh and Sanat Kumar Guchhait

5　Assessing Gully Asymmetry Based on Cross-Sectional Morphology:
A Case of Gangani Badland of West Bengal, India . . . .　69
Aznarul Islam, Biplab Sarkar, Balai Chandra Das, and Suman Deb Barman

6　The Potential Gully Erosion Risk Mapping of River Dulung Basin,
West Bengal, India Using AHP Method . . . .　93
Kishor Dandapat, Rajkumar Hazari, Gouri Sankar Bhunia,
and Pravat Kumar Shit

7　Application of Field-Monitoring Techniques to Determine Soil Loss
by Gully Erosion in a Watershed in Deccan, India . . . .　109
Veena U. Joshi

8　Gully Erosion Susceptibility Mapping Based on Bayesian Weight
of Evidence . . . .　133
Pravat Kumar Shit, Gouri Sankar Bhunia, and Hamid Reza Pourghasemi

9　Understanding the Morphology and Development of a Rill-Gully:
An Empirical Study of Khoai Badland, West Bengal, India . . . .　147
Asish Saha, Manoranjan Ghosh, and Subodh Chandra Pal

10　Estimation of Erosion Susceptibility and Sediment Yield in Ephemeral
Channel Using RUSLE and SDR Model: Tropical Plateau Fringe Region,
India . . . .　163
Raj Kumar Bhattacharya, Nilanjana Das Chatterjee, and Kousik Das

11　Assessment of Potential Land Degradation in Akarsa Watershed,
West Bengal, Using GIS and Multi-influencing Factor Technique . . . .　187
Ujjal Senapati and Tapan Kumar Das

12   Using Ground-Based Photogrammetry for Fine-Scale Gully Morphology
     Studies: Some Examples..   207
     Priyank Pravin Patel, Rajarshi Dasgupta, and Sayoni Mondal

13   Effects of Grass on Runoff and Gully Bed Erosion: Concentrated Flow
     Experiment..   221
     Pravat Kumar Shit, Hamid Reza Pourghasemi, and Gouri Sankar Bhunia

14   Water Flow-Induced Gully Erosion in Himalayan Watershed Cum Plateau
     and Alluvial Plains..   235
     Anand Verdhen

15   Influence of Road-Stream Crossing on the Initiation of Gully: Case Study
     from the Terai Region of Eastern India....................................   251
     Suvendu Roy

16   Land Degradation Processes of Silabati River Basin, West Bengal, India:
     A Physical Perspective...   265
     Avijit Mahala

17   Assessment of Gully Erosion and Estimation of Sediment Yield
     in Siddheswari River Basin, Eastern India, Using SWAT Model..............   279
     Amit Bera, Bhabani Prasad Mukhopadhyay, and Swagata Biswas

18   Role of Plant Roots to Control Rill-Gully Erosion: Hydraulic Flume
     Experiment..   295
     Pravat Kumar Shit, Hamid Reza Pourghasemi, and Gouri Sankar Bhunia

19   Bamboo-Based Technology for Resource Conservation and Management
     of Gullied Lands in Central India...   307
     S. Kala, A. K. Singh, B. K. Rao, H. R. Meena, I. Rashmi, and R. K. Singh

20   Soil Erosion Protection on Hilly Regions Using Plant Roots:
     An Experimental Insight..   321
     R. Gobinath, G. P. Ganapathy, Isaac I. Akinwumi, E. Prasath, G. Raja,
     T. Prakash, and G. Shyamala

21   Planning, Designing and Construction of Series of Check Dams for Soil
     and Water Conservation in a Micro-watershed of Gujarat, India............   337
     Deodas Meshram, S. D. Gorantiwar, Saurabh Samadhan Wadne,
     and K. C. Arun Kumar

22   Impacts of Gully Erosion on River Water Quality and Fish Resources:
     A Case Study...   345
     Avijit Kar, Deep Sankar Chini, Manojit Bhattacharya, Basanta Kumar Das,
     and Bidhan Chandra Patra

23   Gully Erosion in I. R. Iran: Characteristics, Processes, Causes,
     and Land Use...   357
     Majid Soufi, Reza Bayat, and Amir Hossein Charkhabi

24   Factors Affecting Gully-Head Activity in a Hilly Area Under a Semiarid
     Climate in Iran..   369
     Narges Kariminejad, Mohsen Hosseinalizadeh, Hamid Reza Pourghasemi,
     Majid Ownegh, and Mauro Rossi

25   Topographic Threshold of Gully Erosion in Iran: A Case Study of Fars,
     Zanjan, Markazi and Golestan Provinces....................................   381
     Majid Soufi, Reza Bayat, Aliakbar Davudirad, Majid Zanjanijam,
     and Hossein Esaei

26 A Review on the Gully Erosion and Land Degradation in Iran 393
Mohsen Hosseinalizadeh, Mohammad Alinejad, Ali Mohammadian Behbahani,
Farhad Khormali, Narges Kariminejad, and Hamid Reza Pourghasemi

**27 Mapping and Preparing a Susceptibility Map of Gully Erosion Using
the MARS Model** 405
Mahdis Amiri and Hamid Reza Pourghasemi

**28 Gully Erosion Susceptibility Assessment Through the SVM Machine
Learning Algorithm (SVM-MLA)** 415
Hamid Reza Pourghasemi, Amiya Gayen, Sk. Mafizul Haque, and Shibiao Bai

**29 Data Mining Technique (Maximum Entropy Model) for Mapping Gully
Erosion Susceptibility in the Gorganrood Watershed, Iran** 427
Narges Javidan, Ataollah Kavian, Hamid Reza Pourghasemi,
Christian Conoscenti, and Zeinab Jafarian

**30 Land Degradation and Community Resilience in Rural Mountain Area
of Java, Indonesia** 449
Iwan Rudiarto, Isna Rahmawati, and Anang Wahyu Sejati

**31 Spatial Analysis and Prediction of Soil Erosion in a Complex Watershed
of Cameron Highlands, Malaysia** 461
Taofeeq Sholagberu Abdulkadir, Raza Ul Mustafa Muhammad,
Olayinka Gafar Okeola, Wan Yusof Khamaruzaman, Bashir Adelodun,
and Saheed Adeniyi Aremu

Index 479

Pravat Kumar Shit received his Ph.D. in Geography (Applied Geomorphology) from Vidyasagar University (India) in 2013, M.Sc. in Geography and Environment Management from Vidyasagar University in 2005, and PG Diploma in Remote Sensing & GIS from Sambalpur University in 2015. He is Assistant Professor in the Department of Geography, Raja N. L. Khan Women's College (Autonomous), Gope Palace, Midnapore, West Bengal, India. His main fields of research are soil erosion spatial modeling, badland geomorphology, gully morphology, water resources and natural resources mapping, and modeling and has published more than 45 international and national research articles in various renowned journals; also, he has published three books. His research work has been funded by the University Grants Commission (UGC), India, and Higher Education Science and Technology and Biotechnology, Government of West Bengal. He is Associate Editor and on the editorial boards of three international journals in geography and earth environmental sciences.

Hamid Reza Pourghasemi is an Associate Professor of Watershed Management Engineering in the College of Agriculture, Shiraz University, Iran. He has a B.Sc. in Watershed Management Engineering from the University of Gorgan (2004), Iran; an M.Sc. in Watershed Management Engineering from Tarbiat Modares University (2008), Iran; and a Ph. D. in Watershed Management Engineering from the same University (Feb 2014). His main research interests are GIS-based spatial modeling using machine learning/data mining techniques in different fields such as landslide, flood, gully erosion, forest fire, land subsidence, species distribution modeling, and groundwater/hydrology. Also, Hamid Reza works on multi-criteria decision-making methods in natural resources and environment.

He has published more than 90 peer-reviewed papers in high-quality journals, with three chapters in Springer. Also, he has published two books for Springer (https://www.springer.com/gp/book/9783319733821) and Elsevier (https://www.elsevier.com/books/spatial-modeling-in-gis-and-r-for-earth-and-environmental-science/pourghasemi/978-0-12-815226-3).

Gouri Sankar Bhunia received his Ph.D. from the University of Calcutta, India, in 2015. His Ph.D. dissertation work focused on environmental control measures of infectious disease (visceral leishmaniasis or kala-azar) using geospatial technology. His research interests include kala-azar disease transmission modeling, environmental modeling, risk assessment, data mining, and information retrieval using geospatial technology. He is Associate Editor and on the editorial boards of three international journals in health GIS and geosciences. He worked as a "Resource Scientist" in Bihar Remote Sensing Application Centre, Patna (Bihar, India). He is the recipient of the Senior Research Fellow (SRF) from the Rajendra Memorial Research Institute of Medical Sciences (ICMR, India) and has contributed to multiple research programs: kala-azar disease transmission modeling, development of customized GIS software for kala-azar "risk" and "non-risk" area, and entomological study.

Abbreviations

AGNPS	Agricultural Non-Point Source Pollution Model
AHP	Analytical Hierarchy Process
AMC	Antecedent Moisture Condition
ANSWERS	Areal Nonpoint Source Watershed Environment Response Simulation
AS	Anti-scourbility
ASTER	Advanced Spaceborne Thermal Emission and Reflection Radiometer
AUC	Area under the curve
BOD	Biological Oxygen Demand
BT	Brightness temperature
CEC	Cation Exchange Capacity
CF	Causative factors
CFSR	Climate Forecast System Reanalysis
COR	Calculation of correlation
CR	Consistency ratio
CWC	Central Water Commission
DEM	Digital Elevation Model
DN	Digital number
DPM	Dherua Paschim Medinipur
DT	Decision tree
EGEM	Ephemeral Gully Erosion Model
EPIC	Erosion Productivity Impact Calculator
FR	Frequency ratio
GCP	Ground Control Point
GE	Gully erosion
GEIM	Gully erosion inventory mapping
GESM	Gully erosion susceptibility mapping
GG	Gangani Garhbeta
GIS	Geographical Information System
GSI	Geological Survey of India
HRU	Hydrological Response Units
HSPF	Hydrologic Simulation Program-Fortran
IDW	Inverse Distance Weighted
ILP	Iranian Loess Plateau
IMD	India Meteorological Department
IWD	Irrigation and Waterways Department
Kc	Soil anti-disintegration index
LL	Liquid Limit
LMT	Logistic model tree
LS	Slope Length
LST	Land surface temperature
LULC	Land Use and Land Cover

MAE	Mean absolute error
MARS	Multivariate Adaptive Regression Splines
ME	Maximum entropy
MIF	Multi-Influencing Factor
MLA	Machine Learning Algorithm
NCEP	National Centre for Environmental Prediction
NDVI	Normalized Difference Vegetation Index
OSM	Open Street Map
PCI	Principal component image
PL	Plastic Limit
PLDZ	Potential Land Degradation Zone
PMSE	Potential Mean Soil Erosion Rate
PSD	Soil Particle Size Distribution
PV	Proportion vegetation
RD	Root Density
RI	Resilience Index
RLD	Root Length Density
RM	Rangamiti Medinipur
RMSE	Root mean square error
ROC	Receiver Operating Characteristic
RS	Remote Sensing
RSAD	Root Surface Area Density
RSC	Road-Stream Crossing
RSD	Relative Soil Detachments
RSP	Relative slope position
RUSLE	Revised Universal Soil Loss Equation
SCS-CN	Soil Conservation Service-Curve Number
SDR	Sediment Delivery Ratio
Sfm-MVS	Structure-from-motion together with multi-view stereo
SOI	Survey of India
SOM	Soil Organic Matter
SPI	Stream Power Index
SVM	Support vector machine
SWAT	Soil and Water Assessment Tool
SY	Sediment Yield
TDS	Total Dissolve Solid
TDS	Total Dissolved Salt
TM	Thematic Mapper
TT	Topographic threshold
TWI	Topographical wetness index
USLE	Universal Soil Loss Equation
UTM	Universal Transverse Mercator
WEPP	Water Erosion Prediction Project
WI	Wetness Index
WLC	Weighted Linear Combination
WoE	Weight of Evidence

Spatial Extent, Formation Process, Reclaimability Classification System and Restoration Strategies of Gully and Ravine Lands in India

1

Gopal Kumar, Partha Pratim Adhikary, and Ch. Jyotiprava Dash

Abstract

Land degradation has been a major global issue due to its adverse effect on food security, environment and ecology. Among different degraded lands, gullied and ravine lands are very important and remained a highly researchable topic. Gullies are continuous depression on the sloping land surface as a result of soil displacement caused by overland water flow and aided by gravity force, whereas ravines are most extreme form of erosion with intricate network of various forms of gullies, having high drainage density and multidirectional slopes. In India, ravines are mostly found in four states such as Uttar Pradesh, Madhya Pradesh, Gujarat and Rajasthan. During 1976, total ravine land in India was 3.67 million ha, which has been reduced to about 60% at present, and the treatable area including peripheral land is likely to be as high as 1.5 times the actual ravine.

High-intensity rainfall, loose, friable soil devoid of organic carbon and vegetation, faulty agricultural practices, removal of vegetation and overgrazing of lands along with upliftment of central highlands, Aravalli range, Bundelkhand and Chhota Nagpur plateau against lowering of Himalayan base are some of the major factors which are responsible for formation and extension of ravine. As these lands are socio-economically very important, they need reclamation. The main objectives of ravine reclamation are to arrest degradation process, promote ecological restoration, positive on-site and off-site hydrological influences and to establish socio-economic balance with a defined benefit-sharing mechanism. Land shaping, levelling and bench terracing along with riser stabilization are recommended for reclamation and productive utilization of narrow ravine systems with depth up to 3 m. Marginal bund of 1.5 m^2 cross section with 0.1–0.2% grades can be constructed at the periphery of agricultural land to regulate entry of runoff into ravine lands. Similarly, for ravine with gullies deeper than 3 m, reclamation process involves stabilization of gully heads, gully bed and side slopes, establishment of protective vegetation with economic importance and encouraging socio-ecological harmony for sustenance of protective measures. Perennial vegetation and wildlife along with eco-tourism can be considered as the best option for most degraded ravines. As the ravine area is ecologically sensitive complex system, its reclamation and sustainable development can be achieved through scientific planning at micro level and socio-eco-friendly policies.

Keywords

Chambal valley · Yamuna valley · Gully · Mahi · Ravine · Rehabilitation · India

1.1 Introduction

Land degradation has been a major global issue during the twentieth century and still remains high on the international agenda in the twenty-first century. The threat of food insecurity and deterioration of environmental quality have kept alive the issue of land degradation in the present century. Land degradation can be considered in terms of the loss of actual or potential productivity or utility of the land; it is the decline in land quality or reduction in its productivity (Eswaran et al. 2001). When the productivity does not match with land quality, the issue of land degradation arises. Principal processes of land degradation include erosion by water and wind, physical deterioration and chemical degradation (Adhikary et al. 2018). Important among physical processes are a decline in soil structure leading to crusting,

G. Kumar
ICAR—Indian Institute of Soil and Water Conservation, Dehradun, Uttarakhand, India

P. P. Adhikary (✉) · C. J. Dash
ICAR—Indian Institute of Soil and Water Conservation, Research Centre, Koraput, Odisha, India
e-mail: partha.adhikary@icar.gov.in

© Springer Nature Switzerland AG 2020
P. K. Shit et al. (eds.), *Gully Erosion Studies from India and Surrounding Regions*, Advances in Science, Technology & Innovation,
https://doi.org/10.1007/978-3-030-23243-6_1

compaction, erosion, desertification, environmental pollution and unsustainable use of natural resources (Singh et al. 2004). The chemical processes responsible for land degradation are acidification and alkalization, salinization, nutrient leaching, decrease in action retention capacity and fertility depletion. The statistics of variously degraded lands in the world are given in Fig. 1.1.

Among the different degraded lands, gullied and ravine lands are very important and remained a highly researchable topic. Gully erosion has been described in a large variety of landscapes throughout the world. In simple terms, gullies may be described as continuous depression on the sloping land surface as a result of soil displacement caused by overland water flow and aided by gravity force. They may extend in length from a few meters at the initiation stage and up to hundreds of meters if the erosion is not checked (Kumar et al. 2018). The commonly accepted definition of gullies is that they are larger than rills, which cannot be ploughed or easily crossed but smaller than streams, creeks or river channels. The most commonly described gullies are the 'hillslope gullies', which are present in the upland portions of catchments. Gully erosion and the associated soil loss have caused major environmental disasters worldwide. Many urban and rural communities have been severely affected, while the sustainability of the total landscape has been threatened. Human and animal population, physical infrastructure, agricultural lands and socio-economic system of the land/areas are adversely exposed to multifaceted hazards. In many developing countries, several villages and communities have been displaced and virtually disappeared as a result of the scourges of gully erosion. Gully erosion, which generally starts after sheet erosion if remained unchecked for some time, can render large areas useless.

Ravines are a result of formation of the gullies within unconsolidated, relatively loosely bound material such as soft sediments, which can make the land totally unproductive and ruin the livelihood of the people residing nearby (Chaturvedi et al. 2014).

Gullied and ravine lands are the most degraded and vulnerable eco-system which pose a threat to the livelihood of the peoples residing along with those. Apart from the deterioration of the physical, chemical and biological quality of the lands, gullied lands are the hot spot of the major soil erosion as huge amount of soils are being eroded from these lands, which increases the risk of flooding and sedimentation (Kumar et al. 2018). The subsistence farming practised by the resource-poor farmers in the gullied and ravine lands also contributed to the progressive degradation of this fragile eco-system. Erratic and short-duration high-intense rainfall, loosely bound deep alluvial soils, undulating landscape, faulty agricultural practice, illegal cutting of trees and bushes and overgrazing are some of the factors responsible for the formation of gully and ravine land (Rao et al. 2015).

With increased pressure on the land, attention has been shifted on the so far neglected lands like gullied and ravine lands. Being at very low production baseline, these extremely degraded lands offer an opportunity to improve farmer's income and livelihood. Ravine is an extreme form of land degradation developed through soil erosion. Most of the Indian workers believe it as work of concentrated flow of runoff which is accelerated by improper land management and poorly consolidated earth materials. There are several hypotheses for ravine formation including neo-tectonic upliftment and intensification of rainfall. With new information and shreds of evidence available, these hypotheses were revisited. The ravine land is not only problematic at the place

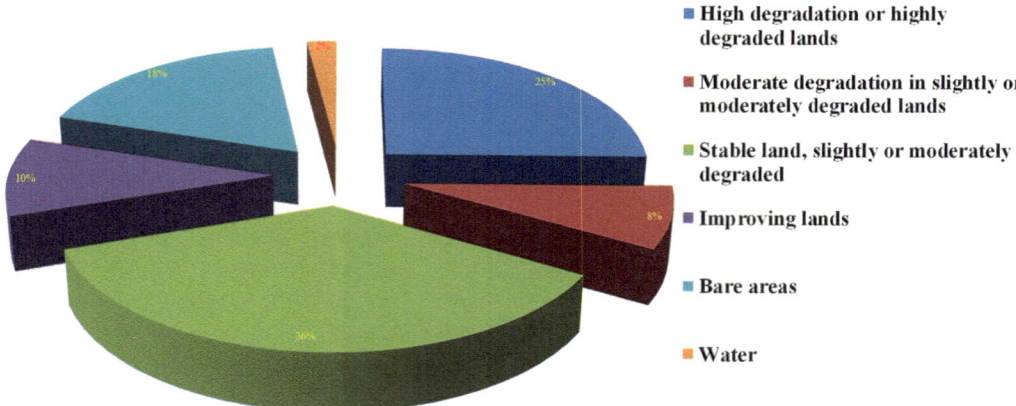

Fig. 1.1 Extent of land degradation in the world (Source: http://www.fao.org/fileadmin/templates/solaw/files/thematic_reports/SOLAW_thematic_report_3_land_degradation.pdf)

of existence but also poses a continuous threat to adjoining tableland. Proper assessment of ravine problem is essential for developing management plan. Therefore, in this article, an attempt has been made to assess the spatial extent of gullied and ravine lands in India. Their formation processes were discussed and reclaimability classification systems were also analysed. As these lands are very potential but are hidden, the various possibilities to restore the gullied and ravenous lands to their ultimate potential were also discussed.

1.2　Spatial Extent of Gully and Ravines in India

India accounts for a meagre 2.4% of the world surface area of 135.79 million km^2, covering 3,287,263 km^2, extending from the snow-covered Himalayan heights in the North to the tropical rain forests of the South. Extensive land degradation in the form of deep gullies has occurred along some of the major river systems of the country in various states. The largest is the Yamuna–Chambal ravine zone. The Chambal ravines flank the River Chambal in a 10-km-wide belt, which extends southwards from the Yamuna confluence to 480 km up to the town of Kota in Rajasthan. Ravines also affect basins of several Chambal tributaries, for example Mej, Morel, Kalisindh, etc. In Gujarat, ravine belt is spread over the southern bank of the Tapti, banks of the Narmada, Watrak, Sabarmati and Mahi basins. Besides these river basins, ravines are also found in Jharkhand (Chhota Nagpur), Bihar and Mahanadi and upper Sone Valley (Fig. 1.2).

The first authentic and reliable assessment of ravine lands in India was published by National Commission on Agriculture (NCA 1976) in which 3.67 million ha of ravine has been reported in India. The states of Uttar Pradesh, Madhya Pradesh, Rajasthan and Gujarat were reported to be the major ravine states, comprising about 75% (2.76 million ha) of the total ravine area of the country. NCA estimate was not on the basis of systematic ground survey and hence the committee suggested further survey and mapping using aerial photographs. Subsequent periodical assessments by National Remote Sensing Centre (NRSC) in 2000, 2003, 2005 and 2008–2009 have shown a sharp reduction in the ravine area of the country. In the backdrop of renewed public interest for reclaiming ravine lands and lack of availability of realistic estimate for the extent of the problem, the Indian Council of Agricultural Research–Indian Institute of Soil and Water Conservation (ICAR-IISWC) regional centres at Kota (Rajasthan) and Vasad (Gujarat) initiated a ravine area delineation project in 2014 with visual delineation approach using

high-resolution LISS-IV and Google Earth imageries. Total ravine area delineated in four states of Uttar Pradesh, Madhya Pradesh, Rajasthan and Gujarat is 1.036 million ha. Figure 1.3 shows the spatial extent of the ravine and gullied lands of four states of India, that is Gujarat, Rajasthan, Uttar Pradesh and Madhya Pradesh. The current ravine area assessment indicated about 60% reduction in the total ravine area since 1976 in the four major ravine states. Table 1.1 compares the spatial extent of gullied and ravine lands estimated by different agencies.

Development and dissemination of technologies for reclamation and productive utilization of ravine lands through field demonstration and capacity-building programmes conducted by ICAR-IISWC regional centres located at Kota (Rajasthan), Vasad (Gujarat) and Agra (Uttar Pradesh) has helped reclamation of about 1.7 million ha of ravine land in four major ravine states of UP, MP, Rajasthan and Gujarat as indicated by nearly 62.5% reduction in ravine land since 1976 in a recent estimate by ICAR-IISWC (Kumar et al. 2018). This reclamation largely comprises shallow ravines that too not to the level of potential production. As per remote sensing-based estimate by ICAR-IISWC, un-reclaimed ravine (rugged land only) spread in the states of Uttar Pradesh, Rajasthan, Gujarat and Madhya Pradesh was 1.036 million ha. The treatable area including peripheral land is likely to be as high as 1.5 times the actual ravine. The ravine reclamation packages developed and recommended by the ICAR-IISWC can transform these wastelands into productive land by encouraging sediment entrapment and water harvesting.

Despite extreme form of degradation, these lands offer potential for productive utilization and economic upliftment of the ravine dwellers. The production potential estimated of ravine areas in Uttar Pradesh, Madhya Pradesh and Rajasthan was estimated long back (1960s) to be 3 million tonnes of food grains annually. With the advent of new tools, cultivation practices and technologies generated for reclamation and productive utilization, this figure can conveniently be proportionately higher per unit land. In addition, fruits, fodder, fuel, timber and raw material for industrial can be produced. By not reclaiming these ravines, the revenue loss was estimated by Planning Commission to be about Rs. 157 crores a year. The reclamation strategy for a ravine land largely depends on degree of terrain deformation, soil quality, accessibility to water and other resources. The target of ravine reclamation is to arrest degradation process, promote ecological restoration, positive on-site and off-site hydrological influences and to establish socio-economic balance with a defined benefit-sharing mechanism.

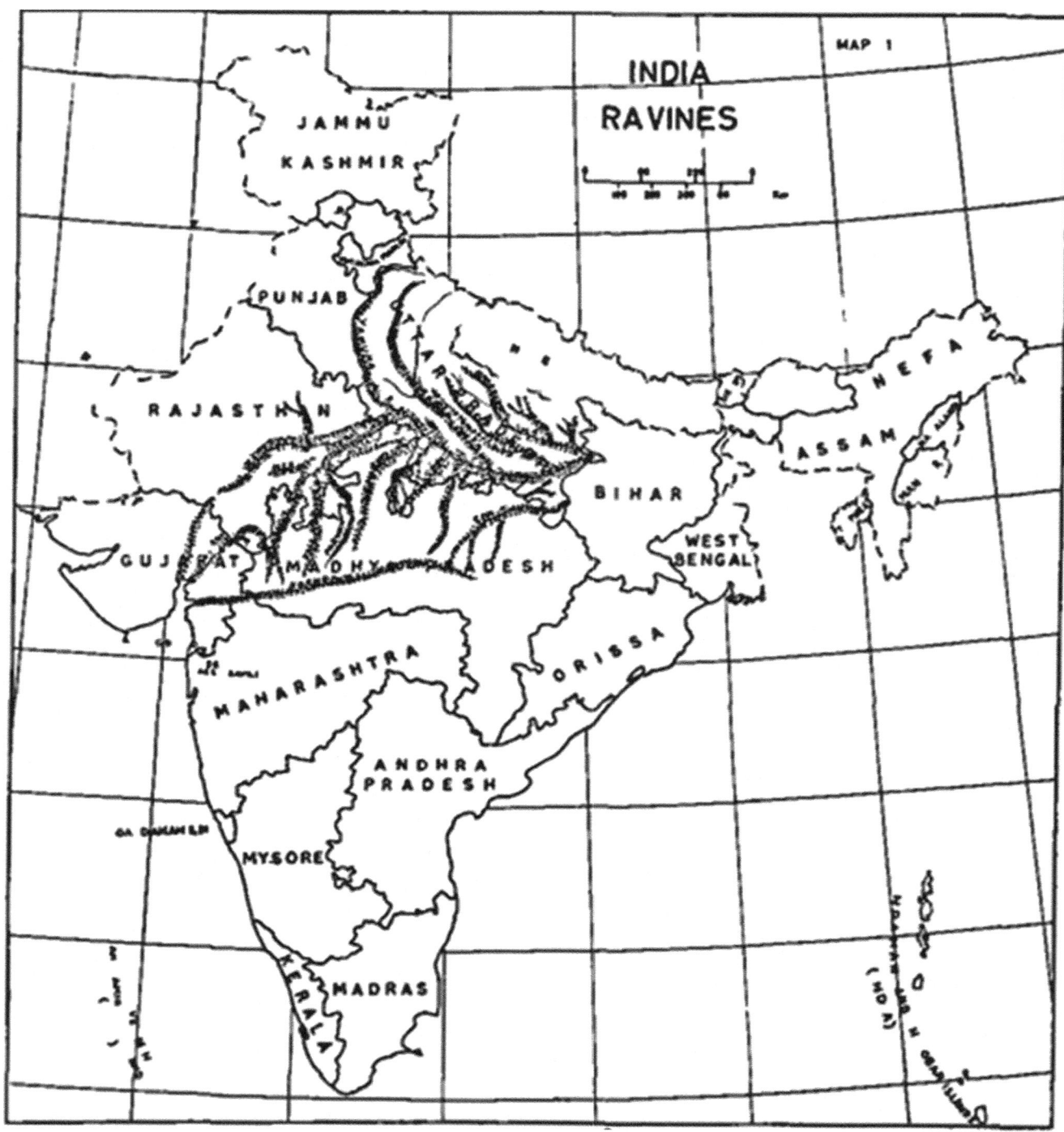

Fig. 1.2 Geographical position of major gullied and ravine lands in India (Source: Sharma 1980)

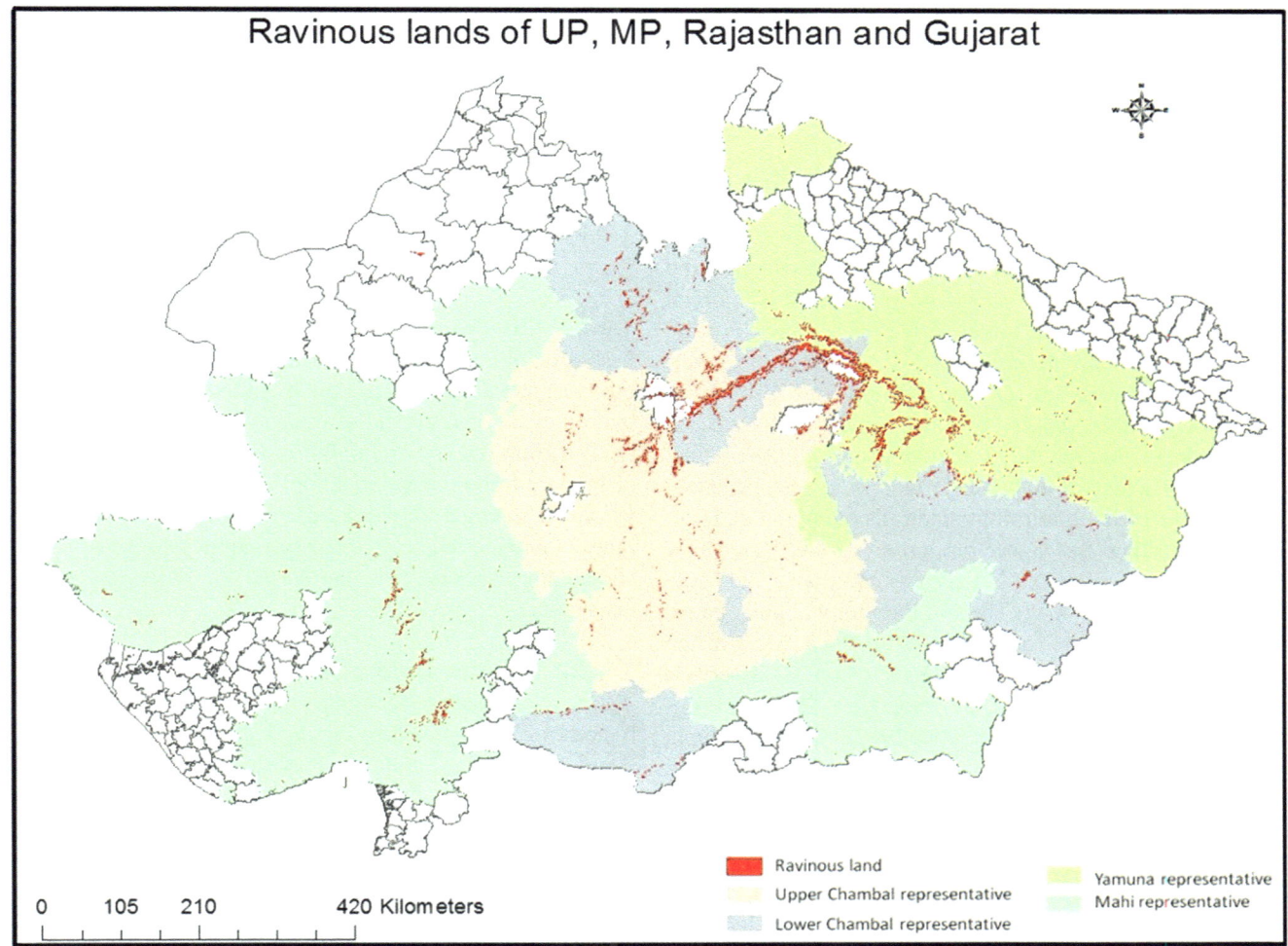

Fig. 1.3 Spatial distribution of gullied ravinous lands of Uttar Pradesh, Madhya Pradesh, Rajasthan and Gujarat (Source: Kumar et al. 2018)

Table 1.1 Estimate of spatial extent of ravine area (lakh hectare) in different states of India by different sources

S. no.	State	NCA (1976)	Chaturvedi et al. (2014)	NRSC (2000)	NRSC (2008)	Kumar et al. (2018)
1	Uttar Pradesh	12.30	12.30	3.25	1.199	3.40
2	Madhya Pradesh[a]	6.83	6.83	5.274	1.453	3.12
3	Rajasthan	4.52	4.52	6.6	1.525	2.74
4	Gujarat	4.00	4.00	0.39	0.339	1.101
5	Maharashtra	0.20	0.20			
6	Punjab	1.20	1.20			
7	Bihar[b]	6.00	6.00			
8	Tamil Nadu	0.60	0.60			
9	West Bengal	1.04	1.04			
	Odisha	–	0.11			
	Others	–	0.19			
Total		36.69	39.8			

[a]Area included: Chhattisgarh
[b]Area included: Jharkhand

1.3 Gully and Ravine Formation Process and Hypothesis

1.3.1 Gully and Ravine Formation Process

Soil material displacement by water initially results in loss of top-soil, but if left unattended may lead to terrain deformation through splash, sheet and rill erosion. Gully erosion is an advanced stage of rill erosion where rills get widened and deepened enough that it cannot be obliterated by normal tillage operations. In the initial active phase, gully erosion may surpass rill and sheet erosion; however, with time, it gets stabilized and causes less erosion than sheet or rill erosion. Gullies occur when runoff flows over land concentrate and cut a channel through the soil. Continuous undercut and resulting collapsing of gully head triggers the upslope extension of most of the gullies but sidewalls slumping and collapsing contribute the higher proportion of soil loss. Gully may start from any depression such as cart tracks and cattle trails if neglected for long. The soil instability at gully banks leads to sloughing and cave in the bank slope. Most of the Indian workers considered the gully erosion, as described above, responsible for ravine formation. Also land-use-induced concentrated runoff (Gupta 1973; Ali 1974; Gupta and Prajapati 1983), poor management of runoff (Prajapati et al. 1982), poor land management, or high-intensity rainfall (Babu et al. 1978) were considered responsible. However, some geologists (Sharma 1968, 1976; Ahmed 1973) put alternate explanation and additional geological factor, including tectonic uplift. With varied perception, an attempt is made to discuss gully under two categories.

Category I: Gully which is formed by concentrated flow through several stages of erosion and is an advance phase of rill erosion. This type of gully is observed in moderate-to-high land slope area, including hills. The rate of gully extension primarily depends on runoff producing characteristics of catchment including alignment, size and shape of the gully, soil characteristic, bank and bed slopes.

Category II: Gully which is formed by progressing slope failure or other mechanism but not essentially by concentrated flow and cannot be considered as an advanced stage of sheet or rill erosion as there may not be an appreciable volume of runoff reaching the gully. Gullies of ravine systems in alluvial soils along Chambal, Yamuna, Mahi, Sabarmati and other rivers are in this category. These gullies are developed on land with gentle slope but high elevation difference (almost vertical drops along the riverbank) between ground level (higher) and main drainage system (riverbed).

Land with loose structured soil, no or poor vegetation and poor soil organic matter ends up to a gully easily when runoff is concentrated (Table 1.2). Rarely any distinction was made in these gullies despite different mechanisms operating. Most often, Category I has been focused on and reported.

1.3.2 Hypothesis of Gully and Ravine Formation

1.3.2.1 Climate and Land-Use Theory

It is difficult to ascertain when the land deterioration started but most popular hypothesis put by Indian workers including hydrologists considers indiscriminate use of land as the main reason. High-intensity rainfall, loose and friable soil devoid of organic carbon and vegetation, erratic, faulty agricultural practices, removal of vegetation, overgrazing of lands and other biotic interference might have aggravated the situation, and continue to be the main factor of gully and ravine extension.

1.3.2.2 Tectonic Upliftment Theory

Another explanation for ravine formation comes from the hypothesis of upliftment of central highlands, Aravalli range, Bundelkhand and Chhota Nagpur plateau against lowering of Himalayan base, thus lowering base level of rivers (Ahmed 1973; Sharma 1976). Steepening of stream gradient due to tectonic wrap, deep incision led to high elevation difference between riverbed and adjoining tableland, which might be the reason for regressive slope failure and ravine formation. River backflow during flood also adds up to wide extent erosion by wet slip and removal of eroded materials. Uniform skyline of Aravalli ranges and presence of hard formations like quartz conglomerate on hill slopes and phyllites and schist forming the lowlands in Chambal–Yamuna ravine area are some of the supporting shreds of evidence to upliftment theory.

1.3.2.3 Aggradation and Degradation Theory

Intensification of monsoon rain in the past is also believed as the reason for ravine formation. The polycyclic nature of river floodplain in which floodplains go through aggradational and degradational phases associated with the change in monsoon intensity and ravine is the symptom of degradational phases has been hypothesized (Gibling et al. 2005). The Chambal ravine is considered a late Pleistocene–Holocene degradational landscape. In the aggradational phase, the large amount of sediment gets deposited by the river in the flood plain. This periodic deposition aggrades the floodplain. However, this condition changes, as during longer

Table 1.2 Distinction between two categories of gullies

Category I (high slope area)	Category II (flat alluvium)
Gully joins higher order, almost perpendicularly	Gully joins higher order from multiple directions
No or less crisscrossing or networking.	Networks of gullies are main features
Runoff is main cutting agent	Role of runoff is more in removing eroded materials instead of active cutting
Human factors including land-use changes and runoff mismanagement are important	Natural factors dominate; however, human activities can accelerate the process
Tunnelling/piping is less observed as mechanism for head extension	Tunnelling/piping is frequently associated to head extension and also along bank slope
Factors affecting: • Improper land use, forest cutting, shifting cultivation, hill cutting for road and rail construction, livestock and vehicle trails, nature of soils • Rainfall intensity is more important than durations • Catchment characteristics: Topography, slope, shape and size of catchment have influence on runoff rate and quantity which ultimately effect gully erosion • Runoff management including diversion and safe disposal through spillways are important	Factors affecting: • Elevation difference with very steep slope, left unattended without conservation measures Unconsolidated soil with poor cementing materials or very fine materials (black soils), lack of vegetation • Rainfall: High duration (wet period) is relatively more important than high intensity • Catchment is usually small, and thus catchment characteristics are not a prime factor • Slope stabilization is more important; however, at sites of substantial runoff, spillways are required
Generally observed in high annual rainfall usually more than 1000 mm	Generally observed in poor-to-moderate annual rainfall (300–1200 mm)
Have deeper channels down slope until the point where deposition or local base level limits extent	No such clear distinction, but ravines are generally deeper close to stream
Loose geological material, poor vegetation, poor soil organic matter and lack of conservation measures are common to both the categories	

wet periods and increased rain intensity, river discharge increases and sediments are not deposited locally, but are carried out of the system to the sea. The rivers cut its own deposits and the channels become deep and remain detached from its floodplain. In this situation, the flood plain not only gets starved of sediment, but the elevation difference created between tableland and riverbed due to deep incision leaves the area vulnerable to wet as well as dry slips and slope failure. In this phase, the floodplain degrades due to bank erosion along the main channel which ingresses gradually or sometimes abruptly into tableland; once inside the tableland, this slope failure moves in multiple directions creating a network of the gully which forms ravine. The runoff generated over adjoining tableland (which depends on runoff-generating properties of catchment) catchment of gully also aggravates the degradation process. Loose and friable soil type, poor organic carbon content, poor vegetation and biotic interference add to the pace of gully head extension.

1.3.2.4 Oceanic Upwelling Theory

Based on the sedimentological and stratigraphic analysis of facies and dating of sediment, it is also suggested that ravine formation coincided with the intensification of the south-west Indian monsoon at the end of the last glacial maximum around 15,000 years ago (Gibling et al. 2005). This degradation process probably got amplified in recent past due to intense human interference including, removal of vegetation and improper land use and management.

The older sediment exposed under the incised main channel of the river and the ravines leds to the evidence of earlier degradational and aggradational episodes coincide with fluctuations in monsoon intensity. The coeval Arabian Sea cores containing variations in pollen abundance which records variation in terrestrial vegetation and planktonic foraminiferal abundance also indicate oceanic upwellings related to monsoonal circulation.

Ahmed (1973) opined that though the gully erosion and ravine formation are not prominently related to sea-level regression, tectonic upliftment during early Holocene may be responsible for quick ravine formation that almost ceased during middle Holocene mainly because of sea-level rise that resulted in the second terrace along riverbank (up to 6 m deep). The marine deposition confined near the riverbank and within old cliffs may be considered supporting evidence. Incision of the lower terrace can be attributed to the phase of tectonic upliftment that probably continues even today. Ravine erosion perhaps extended to the older marine terrace (second surface).

1.3.2.5 Concave Riverbank Elevation Theory

Western theory relates gully and ravine formation with climate, mainly rainfall intensification, which may be valid for gully in high-slope hills (Category I) but found very little logic in alluvial ravine of India, as high concentration of ravine is found in poor rainfall (500–600 mm annual rainfall) areas. Indian soil conservation workers have preferred explanations couched in land-use terms along with high

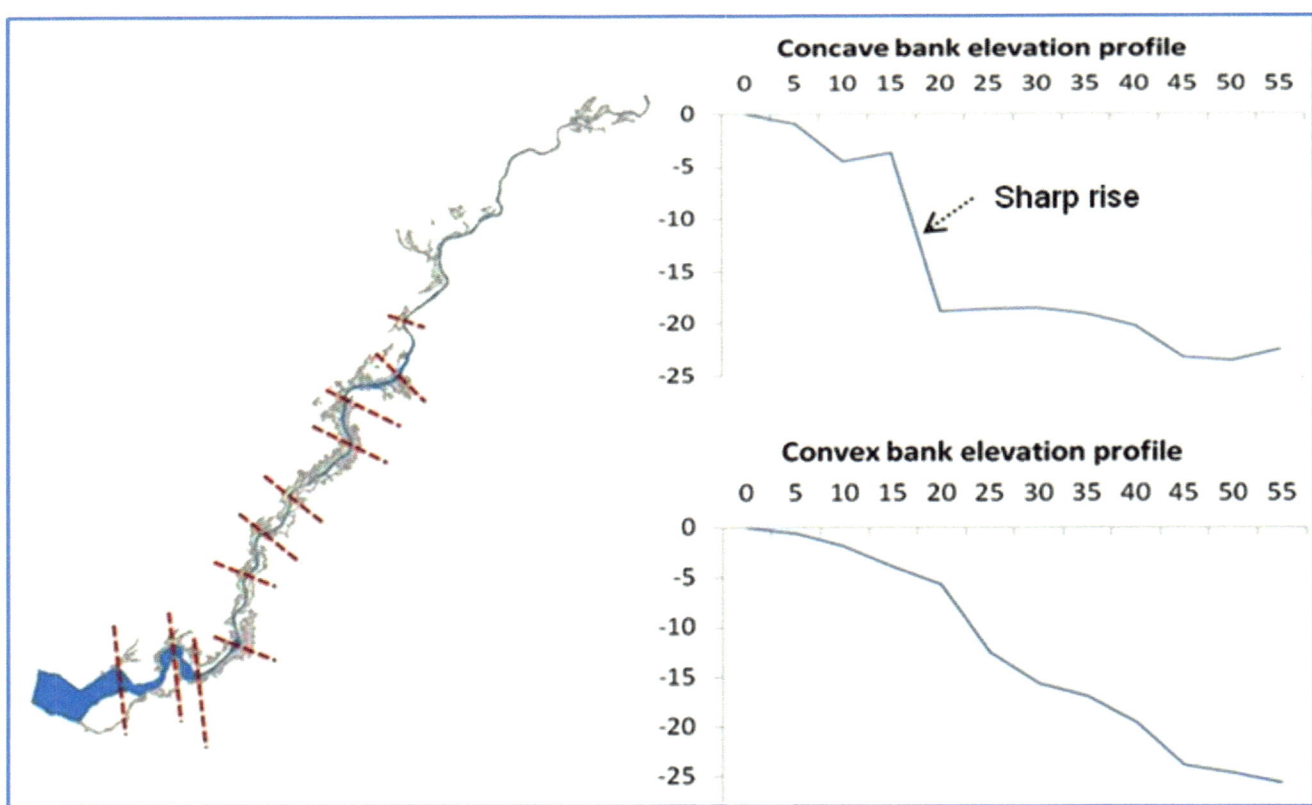

Fig. 1.4 Association of concave bank with ravine formation and elevation profile and concave and convex bank of Mahi ravine

intensity and concentration of rainfall during monsoon. High erodibility of the deep, alluvial soil was also pointed as responsible for ravine formation. Forest thinning and indiscriminate use that led to disturbed hydrological balance have been the other reasons found in Indian studies.

In many of the hypothesis, it is common that sharp elevation change, probably far more than the angle of internal friction, is responsible for ravine formation. Placement of most of the ravine along the concave bank where the sharp slope is seen indicates that damage would have started from riverbank at many places preferably from the concave bank where stream bank erosion is prominent (Kumar et al. 2018). About 83% of ravine patches about River Mahi was found associated with the concave bank. Elevation profile of this concave bank reveals a sharp elevation rise along the bank (Fig. 1.4). Slope failure under wet as well as dry conditions and removal of fallen loose materials during the wet season either by runoff or by rainfall seem to be the most plausible consideration for headward progress of instability. Multiple factors can be attached to landward extension of the gully. When slope/bank instability coincided with the drainage line, ravine ingression would have progressed very fast. Unlike gullies in high-slope area which is an advanced form of rill erosion and initiates over the land, these gullies have probably started from the riverbank. Headward progress is

common to both categories of the ravine. Once inside the land, instability propagates in different direction resulting in new branches that often crisscross each other and form a network of the gully. Concave bank stream is always at the risk of gully initiation and subsequent ravine formation.

1.3.3 Other Mechanisms of Gully Head Extension

Discontinuous gullies which are features of ravine begin as an abrupt head cut, as usually associated with the concentration of flow in soil pipes. Gully extension also takes place by means of tension cracks and tunnelling.

1.3.3.1 Tension Cracks
Tensional cracks decrease the stability of the gully head and gully wall by reducing cohesion. Tension cracks developed around 20–60 cm upslope from the gully head under the combined influence of desiccation and tension stress, which is often observed in lateritic soils of eastern India. The pore water pressure increases multifold when these cracks get filled with runoff water, which results in failure or toppling (Collison 2001; Bull and Kirkby 2002). The presence of tension cracks, therefore, indicates that gully is active.

Tension cracks development is also influenced by the undercutting of gully walls and gully heads.

1.3.3.2 Tunnelling

Tunnelling is an important mechanism for headword and lateral gully expansion. Tunnelling is more frequently observed in dispersible soils. It is also observed in black soils where enlarged cracks develop into tunnels which carry a suspension of soil and water. Tunnel collapse causes a rapid extension of the gully (Fig. 1.5). Tunnel usually develops below the course of surface flow. In dispersible or loose soil materials, it may develop below flatter runoff course or on a gentle slope as it provides greater opportunities for infiltration and subsurface flow. Grain size distribution plays a very important role in tunnelling. Failure of burrow pits created by animals also promotes tunnelling and gully extension.

1.3.3.3 Suffusion/Internal Erosion/Internal Instability

Fine grains and soluble materials can pass through coarse soil matrix due to seepage flow. This process is called suffusion or, internal erosion or, internal instability. When surface water moves into the soil along cracks or channels or through burrows pits, the cavity of tree roots or through natural pore system and moves along the subsurface, it carries finer materials. Dispersive clays are the first to be removed. With enlargement, more water enters and erodes the soil. Sinkholes are formed due to the partial collapse of the tunnel roof.

Water seepage at foot of a slope and fine sediment fans at outlet of tunnels are the indication of the tunnelling. Tunnelling is mainly observed in non-cohesive and dispersive soils in which fine particles are in abundance. The geometric criterion is though considered to avoid suffusion in embankment and bunds, and the same holds good regarding tunnelling. Kenney and Lau (1985) proposed F–H diagram by transforming the ordinary grain size distribution curve. F is the mass percentage of grains with diameters less than a particular grain diameter d and H is the mass percentage of grains with diameters between d and $4d$. For poorly graded soils $H/F \geq 1.0$ for $F \leq 0.3$ and for well-graded soils $H/F \geq 1.0$ for $F \leq 0.2$ are considered stable (Kenney and Lau 1986). Kezdi (1979) proposed splitting up the grain size distribution of a soil into two distributions of the fine and coarse parts, and assessing the stability by Terzaghi's well-known filter criterion applied to the two distributions: $d_{c,15} \leq 4\,d_{85,f}$ with $d_{c,15}$ = grain diameter for which 15% of the grains by weight of the coarse soil are smaller and $d_{85,f}$ = grain diameter for which 35% of the grains by weight of the fine soil are smaller. Recent earth fill over compact materials is prone to quick tunnelling and collapse.

In some places, tunnels are visible and may be of 10–50 cm equivalent diameter before it collapses, but often may not be visible until collapse because as a result of suffusion, it is made up of many micro tunnels with the intact matrix. The non-visible suffusion and collapse are mainly observed in the coarse and relatively graded materials. For

Fig. 1.5 Tunnelling as observed on adjacent tableland near ravine and on bank slope

poorly graded (in terms of particle size) soil, suffusion is not an issue; however, hydraulic failure may occur under upward seepage flow. In fine-textured soil, widening of cracks may serve as a channel.

1.4 Gully Reclaimability Classification System

Realizing the potential of gully and ravine reclamation and use for the productive purpose to address the socio-economic and environmental concern, the Government of India took several initiatives in these parts. This includes the setting of the ravine research centres at Vasad (Gujarat), Kota (Rajasthan) and Agra (Uttar Pradesh) and during the 1950s, establishing the Ravine Reclamation Board and National Ravine Reclamation Policy during 1967–1970, and launching of several centrally and externally funded ravine reclamation programmes during 1970–1995. Development and dissemination of technologies for reclamation and productive utilization of ravine lands through field demonstration and capacity-building programmes conducted by ICAR-IISWC regional centres located at Kota (Rajasthan), Vasad (Gujarat) and Agra (Uttar Pradesh) has helped reclamation of about 1.7 million ha of ravine land in four major ravine states of UP, MP, Rajasthan and Gujarat as indicated by nearly 62.5% reduction in ravine land since 1976 in a recent estimate by Kumar et al. (2018). In a recent remote sensing-based survey, ravine area (rugged land) in Gujarat, Rajasthan, Madhya Pradesh and Uttar Pradesh has been estimated to be 1.036 million ha. The treatable area including peripheral land is likely to be as high as 1.5 times the actual ravine. The ravine reclamation packages developed and recommended by the ICAR-IISWC can transform these wastelands into productive land by encouraging sediment entrapment and water harvesting.

Extensive areas under medium-deep and deep ravines continue to remain unattended despite exorbitantly escalated land value and available economically viable and ecologically sustainable ravine reclamation technology developed and demonstrated by ICAR-IISWC over a period of about 60 years. Understanding the problem is prerequisite for selecting proper invention and approach. As alluvial ravines are made of network of gullies, the treatment plan is generally gully centric. Several gully classifications have been used in the past.

As per the gully classification scheme developed at ICAR-IISWC, classes 1–6 lands indicate progressively increased terrain deformation with corresponding increase in erosion hazards and land-use restrictions. The target of ravine reclamation is to arrest the degradation process, promote ecological restoration, positive on-site and off-site hydrological influences and to establish socio-economic balance with a defined benefit-sharing mechanism.

Ravine restoration starts with protection of peripheral land of ravine. Safe disposal of runoff from marginal/peripheral land by means of peripheral bund and spillways is crucial to check gully head extension into marginal land. Contour bund and field bunds in adjoining arable land for in situ water harvesting also help reduce runoff load, and thus help to protect gully extension. Land levelling and periodic slope smoothening promote in situ soil and water conservation, ease of cultivation operation, higher water and nutrient-use efficiency. Contour tillage, deep summer ploughing, vegetative barriers, intercropping, mulching, etc. are helpful to promote in situ water conservation.

For narrow ravine systems with depth up to 3 m deep, land shaping, levelling and bench terracing along with riser stabilization are important measures for reclamation and productive utilization. Marginal bund of 1.5 m^2 cross section (bottom width, 2.5 m; top width, 0.5 m; height 1.0 m) with 0.1–0.2% grades is recommended; lesser section (0.6 m^2) may be used towards ridge. For ravine with gullies deeper than 3 m, it is difficult to reclaim through levelling. Reclamation process for deeper gully involves stabilization of gully heads, gully bed and side slopes, establishment of protective vegetation with economic importance and encouraging socio-ecological harmony for sustenance of protective measures.

A comprehensive gully classification scheme evolved at ICAR-IISWC (Table 1.3) classifies the gully network into size categories based on gully dimensions, soil characteristics, water availability and climate (Kumar et al. 2018). With progressively increased terrain deformation from class 1 to class 6, there is a corresponding increase in erosion hazards and land-use restrictions.

Marginal lands at the periphery of ravine system with occasional presence of shallow gullies are classified as Gully Reclaimability Class I (GRC-I). These lands require minor levelling work on gently sloping sides and bed for reclamation. Favourable soil texture and deep soil depth facilitate cultivation of most crops possible and also alleviate levelling and maintenance cost of these lands. Narrower and deeper gullies make the reclamation process expensive and challenging. Accordingly, land-use restrictions are imposed as land degrades from GRC-I to GRC-VI. Generally, GRC-I to -IV are recommended for cultivation with increased level of management and restriction. GRC-V has similar topographic feature as of GRC-I to -IV, but this cannot be reclaimed for agriculture or horticulture as the gully is prone to seasonal backflows from a nearby river or it has developed water logging, salinity due to irrigation or any other adverse factor like arid climate. Such lands may be put under perennial vegetation like suitable fuel and fodder

Table 1.3 Gully classification scheme developed at ICAR-IISWC (Kumar et al. 2018)

Class	Texture	Bed width (m)	Gully depth (m)	Side slope (%)	Climate
1	sicl, cl, l, sl, sil, scl	>18 or <6, (W$_3$) (W$_1$)	<1.5 (gd$_1$)	<5% (S$_1$)	Humid climate with well-distributed rainfall or having perennial irrigation
2	sicl, cl. sl, sil, scl.	6–18 (W$_2$)	1.5–3.0 (gd$_2$)	5–10% (S$_2$)	Humid climate with occasional dry spells or dry-cum-wet irrigation
3	sc, sic, c,ls	>18 (W$_3$)	3–6 (gd$_3$)	10–15% (S$_1$)	Sub-humid climate or situated at tail end of an irrigation system where water is occasionally not adequately available
4	Any texture excluding sand and with gravel	Any width	6–9(gd$_4$)	10–15% (S$_3$)	Semiarid climate (without irrigation) with enough rainfall to sustain hardy horticultural plants
	Reclaimable classes decrease in priority for reclamation for agriculture from Class 1 to 3. Class 4 may be put under horticulture				
5	Reclaimable classes 1–4 having the following hazards:				
	Water logging/salinity has developed due to irrigation system				
	Back flow and flooding from a nearby stream or river. Hence, the gully need not be reclaimed for agriculture. It may be put under any suitable grass or tree species				
6	Any texture including gravelly, etc.	Any width	>9 (gd$_5$)	15% irregular (S$_4$)	Semiarid with long dry spells and arid
	Gully humps may also be included in this class				

trees and grasses. Grass cutting for stall feeding may be allowed. Fuel trees may be exploited on rotational basis. GRC-VI lands present limitations severe than in GRC-IV in any one or more soil, topographic or climatic, and are best suited for perennial vegetation and wildlife. Eco-tourism can also be a possible economic activity suited to these lands.

Gully varies in size along a regular side that can be expressed in terms of depth, bed width and side slope. Tejwani (1974) has developed a simple system of gully classification based on depth and width and side slope of gully. The classification scheme by different workers is given in Table 1.4.

The classification system developed by Bali and Karale (1977) has more comprehensive information including soil and climate that can be utilized for ravine reclamation planning, and is presented in Table 1.5.

Due to advent of new earth-moving machinery, increased access to ravine area, cost escalation, and more options of technology, the cost–benefit aspect of ravine reclamation has been drastically changed. The shallow ravine (depth < 3 m) was probably decided as the manual workability for levelling in absence of heavy tools but now depth up to 8–10 m is being levelled by means of JCB and other machinery. New portable construction materials, low-cost micro-irrigation setup, portable light pump and machinery has paved the way for developing ravine as specific fruit or medicinal belt; therefore, for reclamation purpose under current scenario, new classification scheme is highly desired.

1.5 Technology for Gully and Ravine Area Restoration

An effective ravine restoration plan with sustainable benefits needs to target the following three objectives:

- To arrest the degradation process and initiate ecological restoration
- Establish a sustainable socio-ecological balance
- Promote positive on-site and off-site hydrological, ecological and climatic influences

Steps involved in the planning process are as follows:

- Acquiring high-quality topographic maps and cadastral maps of the area.
- Participatory appraisal of the problem, resources and possible solutions integrating local wisdom.
- Selection of appropriate land-use systems and befitting soil and water conservation measure following sound scientific principles.
- Skill development and capacity development of beneficiary groups for establishing self-governed and self-sustaining socio-ecological system.

The gully reclaimability classification system facilitates bio-physical characterization of ravine area. However, there

Table 1.4 Classification scheme of gully given by different workers for different states of India (Source: Hajgh 1984)

Gully type	Gujarat (Tejwani 1974)	Madhya Pradesh (Gupta and Prajapati 1983)	Rajasthan (Bhulyan 1967)	Chambal river basin (Seth et al. 1969)
Very small	G1: Gullies up to 1 m deep and 18 m bed width	D1: Shallow ravines up to 1.5 m deep	G1: Gullies up to 1 m deep (18% overall)	D0: Gullies less than 1.5 m deep and 3 m wide (10% overall)
				D1: Gullies less than 1.5 m deep and more than 3 m wide (60% overall)
Small	G2: Gullies up to 3 m deep and more than 18 m bed width	D2: Medium ravines up to 1.5–5 m deep	G2: Gullies up to 1–5 m deep (39% overall)	D2: Gullies between 1.5 and 5 m deep, more than 3 m wide (24% overall)
Medium	G3: Gullies 3–9 m deep and not less than 18 m bed width, side slope 8–15%	D3: Deep ravines, 5–10 m deep	G3: Gullies more than 5 m deep (43% overall)	D3: Trenches 5–10 m deep (6% overall)
Large	G4: (a) Gullies 3–9 m deep and less than 18 m bed width	D4: Very deep ravines, 10 m deep		
	(b) Ravines more than 9 m deep, side slopes steep to vertical and gullied			

Table 1.5 Comprehensive classification scheme of gully given by Bali and Karale (Source: Hajgh 1984 and Ravine Reclamation Group by Bali and Karale 1977)

Class	RSG	Description
Class-I	Very good	Ravinous land that can be reclaimed readily with minor reclamation operation of levelling and scraping. These lands include shallow ravines (G1) with widths up to 30 m or more, simple and compound bed slopes of up to 5%, lack extremes of texture and do not have within a depth of 1.5 m a calcareous layer, hard pan, bed rock, salinity or alkalinity. Groundwater level is below 2.5 m.
Class-II	Good	Land that can be reclaimed readily with minor reclamation operations but requiring more dozing work than class-I. The depth and width specifications for this group are the same as class-I, but bed slopes range between 5 and 10%. They have a low frequency of gullies. They also lack extremes of texture, hard pans, caliche, salinity and alkalinity within the depth of 1–1.5 m.
Class-III	Moderate	Land that can be reclaimed with medium intensity reclamation measures. These lands have moderately deep ravines with narrow to moderate width, and medium frequency with compound bed slopes of 10–15%. Soil textures are moderately fine and fine, often demanding heavy draft. Soils have slight to moderate salinity and/or alkalinity and possess calcium carbonate, hard pan or bed rock within 0.5–1 m from the surface.
Class-IV	Poor	Includes medium to deep gullies of narrow width. There are severe limitations due to complex and strong bed slopes, a high frequency of gullies, extremes of textures, salinity and alkalinity, and the occurrence of hard layers, bed rocks or pans within 0.5 m from the surface. These lands are very costly to reclaim or may not be suitable for agriculture after reclamation. Reclamation of such lands for horticultural uses is recommended.
Class-V	Unsuitable	There are severe limitations for reclamation of these lands for agriculture or horticulture, and the very high costs means reclamation is not desirable, and such lands would be best developed for forestry and grassland.

are some additional concerns such as land tenancy, resource availability, community preferences and marketing opportunities for preparing a land-use plan. Soil and water conservation structures are specific to terrain conditions and intended land use (Table 1.6).

1.5.1 Management of Shallow Ravines and Marginal Lands

Soil erosion threatens millions of hectares of productive lands along the periphery of ravine systems. Packages of practices have been developed, evaluated and recommended for Chambal, Yamuna and Mahi ravine systems. A choice of appropriate conservation measure would depend on technical feasibility, economic viability, intended land use, community preferences and long-term sustainability. After completing

basic resource survey and physical characterization of ravine area, the treatment plan for arable lands is planned in the following steps:

- Avert the process of gully head extension into tablelands by channelizing runoff from the marginal land and providing safe disposal into the ravine system. This is achieved by constructing marginal bunds and spillways.
- Levelling and slope smoothening of arable lands to promote in situ rainwater conservation
- Installation of contour or field bunds to facilitate intra-plot runoff harvesting
- Bench terracing and land shaping to promote inter-plot runoff harvesting and crop diversification

For regulated entry of runoff into ravine lands, marginal bund of 1.5 m^2 cross section (bottom width, 2.5 m; top width,

Table 1.6 Suitability of land treatment options under ravine area development project

Site characteristics	Land tenancy	Preferred land use	Recommended land treatment
Zone I: Levelled (<1% slope) marginal lands along the ravine area and vulnerable to gully head extension if left unattended	Private	Field crops	Conservation agronomy, contour or graded bunding, peripheral bunds with masonry spillways
	Community lands	Silvi-pasture or horti-pasture	Peripheral bunds with masonry spillways, in situ moisture conservation measures with mini bunds, silvi-pastoral system
	Forest land	Silvi-pasture or fuel–fodder plantations	Peripheral bunds with masonry spillways, afforestation with moisture conservation measures for overall habitat improvement and support for fuel and fodder need
Zone II: Flat or mildly sloping lands or shallow and small gullies located between Zone I and deep gully system (Ravine reclaimability class I, II and III)	Private	Field crops or agri-horticulture systems	Gully head stabilization, land levelling to form inward sloping terraces with proper outlets. Conservation bench terracing, gully plugs
	Community lands	Silvi-pasture or horti-pasture	Gully head stabilization, half-moon terracing or staggered trenching. Silvi-pastoral systems, gully plugs
	Forest lands	Silvi-pasture	Gully head stabilization, half-moon terracing or staggered trenching, afforestation and other moisture conservation measures for overall habitat improvement and support for fuel and fodder needs, gully plugs
		Fuel–fodder plantations	
Zone III: Severely degraded medium deep to very deep ravines which cannot be economically reclaimed for cultivation (Ravine Reclaimability classes IV and VI)	Private lands	Agri-horticulture and horticulture systems	Narrow horticulture terraces with half-moon terracing or staggered trenching and side slope stabilization measures, gully bed stabilization, micro-irrigation system
	Community lands	Silvi-pasture or horti-pasture	Silvi-pastoral system with half-moon terracing or staggered trenching and side slope stabilization measures, gully bed stabilization measures
	Forest lands	Silvi-pasture, bamboo plantations	Afforestation and vegetation modification interventions with suitable moisture conservation measures for overall habitat improvement and fuel and fodder needs. Gully bed stabilization measures
Zone IV: This black flow zone is situated between Zone III and river stream (Ravine Reclaimability class V)	Private	Rabi cropping	Levelling, bio-engineering measures for erosion control
	Community land	Water harvesting for life-saving irrigations, bamboo plantations, wildlife sanctuary, fishery	Stream bank and side slope stabilization measures, Earthen CDs for embankment-type WHS, gully bed stabilization measures
	Forest land		

Table 1.7 Types and suitability of spillways as gully head control structure

Type of spillway	Site suitability		Advantages	Disadvantages
	Overfall (m)	Peak runoff (m^3/s)		
Drop spillway	<3	>2.5	Relatively stable and safe, easy to construct and maintain	Requires stable grade on either side, unsuitable for detention storage, high initial cost
Chute spillway	3–6	<2.5	Relatively low cost	Susceptible to failure due to seepage or rodents activity
Pipe spillway	>3	Any	Suited for upstream detention storage, road culverts, economical, stable and safe	Sensitive to clogging by debris, requires careful field execution to avoid channelling along pipe

0.5 m; height 1.0 m) with 0.1–0.2% grades should be constructed at the periphery of agriculture land. The cross section of bund may be reduced to 0.6 m^2 towards the ridge. The bund should be constructed at a distance of twice the depth of active gully heads. Runoff diverted by peripheral bunds is then safely discharged into gully network through appropriate type of spillways for the given vertical fall at gully head and gully catchment area (Table 1.7).

Levelling and terracing of multidirectional slopes are recommended on the upstream side of peripheral bunds. Suitable conservation measures are adopted for promoting in situ moisture conservation such as deep summer ploughing, contour tillage, intercropping, vegetative barriers, contour or graded bunding with contour furrows in inter-bunded areas, etc.

1.5.2 Rehabilitation of Medium and Deep Ravines

A ravine area with gullies deeper than 3 m is difficult to level and therefore requires a different approach for its reclamation and productive utilization. Medium-deep gully refers to 3–6 m gully depth, while a gully deeper than 6 m is classified as deep gully. The reclamation process for such lands involves the following steps:

- Stabilization of gully heads
- Stabilization of gully bed and side slopes
- Establishment of vegetation having protective and economic value
- Establish socio-ecological harmony for ensuring self-sustenance of protective measures

1.5.2.1 Gully Head Stabilization

Preventing the gully head extension through appropriate head stabilization measures is much easier and less expensive than reclaiming deeper section of ravine lands. Therefore, planning regulated and safe disposal of runoff from gully catchment to ravine system is the key component of any gully control programme which is supported by catchment treatment for reducing and diverting of surface runoff by the structural and non-structural measures. Three types of spillways are generally recommended based on the expected peak runoff volumes and fall at the gully head, which is the elevation difference between the land above the gully head and gully bed (Table 1.7). The straight drop spillways are adapted to control the gully head with less than 3 m fall. The low straight drop is also advisable for small gully head cuts (1–2 m). The chute spillways are preferred for gully heads with 3–6 m fall and large drainage area. The pipe spillway is usually constructed in conjunction with earthen embankment with no provision for permanent water storage behind the structure and very little detention storage above the spillway inlet. For identifying suitable site and type of spillway for gully head stabilization, the potential sites need to be surveyed with depth and width measurements. Stability of gully head banks should be studied carefully for proper anchoring of the headwall extensions into banks. The spillway is designed for expected peak discharge volumes from the catchment.

1.5.2.2 Stabilization of Gully Bed and Side Slopes

The second step in rehabilitation process of medium-deep gullies is to stabilize gully bed and side slope by means of a series of check dams. The type of check dam, its size and placement are scientifically planned to optimize cost and conservation efficiency of these structures (Table 1.8). The selection of the check dams depends on the size of the gully catchment area, the gradient and the length of the gully channel. A gully network treated with a series of check dams eventually gets transformed from silt-producing site to silt entrapment site. Gradually, the terrain roughness is diminished giving rise to levelled and fertile gully beds. During initial years, these structures act as embankment-type water-harvesting structures and promote groundwater recharge.

Bio-engineering measures are planned for stabilization of steep side slope of deep gullies and riverbanks. For stabilizing gully side slopes, 1:1 slope-easing treatment is desirable to establish a good cover. *Dichanthium annulatum* was found very suitable for sodding the eased slopes (Rao et al. 2015). This type of work is expensive and is recommended where protection of costly construction like roads, buildings, etc. is required. The gully stabilization can also be achieved slowly if the area is protected against grazing.

1.5.2.3 Vegetation Establishment

The ravine area vegetation in India is in highly degraded condition with very low density of xerophytic vegetation. Excessive erosion and biotic pressure give little scope for natural regeneration of trees and grass. Any growing tree would be heavily grazed or lopped for fuel. Sustainable restoration of ravine lands is possible only if these lands are vegetated with ecologically efficient and economically viable perennials. A successful vegetation establishment programme would include the following steps:

- Regulating biotic interferences
- Identifying suitable tree and grass species and land-use system
- Establishing silvopastoral systems in community lands
- Habitat improvement in forest lands

Empowering local beneficiaries to develop a social mechanism or social fencing is an effective and very well-tested tool for regulating biotic pressure on the planted vegetation. It offers an opportunity for local stakeholders to formulate a strategy to control grazing and protect the planted vegetation with the aim of utilizing the developed resources in a socially and ecologically acceptable way. The inbuilt mechanism of conflict resolution due to intimate involvement of society in protection of planted vegetation with the aim of equitable sharing of usufructs ensures that vegetation is protected

Table 1.8 Different types of check dams for gully bed stabilization

| Check dam type | Site suitability | | | | Advantage and disadvantage |
	Stream order	Catchment size (ha)	Peak runoff (m³/s)	Slope (%)	
LBCD	First	<2	<0.25	<1	Low cost, easy to plan and install, reduce the runoff velocity and soil loss, improved moisture availability and vegetation establishment, silt deposited behind structures leads to stabilization of gully beds
Earthen gully plug	First	<2	<0.25	<1	
Bori bunds	First and second	<2	<0.25	<3	*Limitations*: These are temporary structures with a lifespan of 3–5 years
Earthen CDs	Second	2–20	0.25–2.5	<3	Very effective in medium deep gully, retain large quantity of runoff and silt, improve groundwater availability, promote vegetation/regeneration, helps in drought mitigation. Longer life span
Masonry CDs	Third and higher order	>20	>2.5	Any	*Limitation*: High initial cost, requires good technical know-how for planning and execution
Composite CDs	Third and higher order	>20	>2.5	Any	
Bio-engineering measures	First	<2	<0.25	<1	Low cost and easy to install, effective. Reduce the runoff velocity, help in reduction of soil loss, help in filtering of the soil moving with the water, improved moisture availability, leads to better survival and growth of vegetation
					Limitation: Temporary conservation measures, lifespan is less than 5 years

without erecting physical fence system. Involving community-based management institutions is often considered as a critical precondition for effective implementation for management of forest through people's participation and protection strengthened by empowering the local communities with adequate power and responsibilities and equitable benefit sharing. In highly wildlife-infested areas, live fencing is also a simple and economical way of protection. The hardy and thorny species like *Agave americana*, *Agave sisilana*, *Caesalpinia bondac*, *Caesalpinia decapetala*, *Jatropha curcus* and *Euphorbia* spp. in conjugation with ditch-cum-wire fences may be erected to provide protection to vegetation during initial years of establishment and resurrection. Figure 1.6 shows the native as well as planted vegetation in a ravenous land.

The community land may preferably be utilized for silvipasture while other revenue land can be utilized for bamboo, MPTs plantation or silvi-pasture. Severely degraded government land can also be made productive through appropriate MPTs plantation and assisted natural regeneration. Animal production systems are major livelihood source in ravine areas. Shifting of focus to cash crops from traditional cereal and millet crops in past few decades has also reduced the availability of fodder from cropped lands. There is need to augment supply of fodder for improving productivity of animal production system. Silvi-pasture systems provide opportunities for farmers to utilize marginal and waste lands to increase productivity on a unit of land by producing both livestock and timber products. Trees can be planted in pastures or existing woodlots can be thinned using timber

stand improvement techniques to create environment that will support both forage and tree production. Suitable fodder species for ravine region are *Faidherbia albida*, *Leucaena leucocephala*, *Acacia nilotica*, *Azadirachta indica*, *Gliricidia sepium*, *Prosopis cineraria*, *Salvadora oleoides*, *Tecomella undulata*, *Balanites aegyptica*, *Cassia siamea*, *Inga dulce*, *Anogeissus pendula*, *Pongamia pinnata*, *Dalbergia sissoo*, *Albizzia lebbeck*, *Grewia species*, *Melia azedarach*, *Erythrina indica*, *Holoptelea integrifolia*, *Ailanthus excelsa* and *Zizyphus mauriatina*. Recommended grass species are *Cenchrus ciliaris* and *Dicanthum annulatum* (Table 1.9).

Habitat improvement strategy is planned for forest lands. To improve the composition of existing ravine forest vegetation, the following steps are recommended:

- Singling of bushy trees and removal of unwanted vegetation
- Augmented natural regeneration
- Introduction of valuable mixed species by enrichment planting

The dominant vegetation in ravine areas is *Prosopis juliflora* which has limited utility. Singling of bushy trees provides open spaces for planting useful species while improving the quality of existing vegetation. For developing single strong bole, one strong shoot is retained and all other shoots are removed. The main healthy stem or shoot develops into a leading and strong bole in due course of time. The tree with clear bole has potential value for use as timber, pole or other construction material. The natural regeneration of

Fig. 1.6 Well-established native and planted vegetation in a ravine of India

desirable species is usually low in the degraded forests due to non-availability of seed-bearing mother trees and adverse edaphic conditions. For improving the density of the native valuable species, those species have to be raised either by direct seeding or enrichment planting. Usually, 1-year-old seedlings are planted with optimum soil working technique to supplement the natural regeneration and re-vegetation establishment.

The supportive moisture conservation measures like staggered contour trenching (SCT) or half-moon terracing enhance the success and survival of native plants on moisture-stressed multidirectional slopes of ravine systems. Compartmental bunding is a suitable moisture conservation practice for levelled ravine humps. Staggered trenching is a practice of excavating trenches across the land slope directly below one another in alternate rows that breaks slope length and increases opportunity time for retention of runoff. Staggered contour trenching density of 417 trench/ha at 75% runoff trapping potential has been found to be the most protective, productive and economical conservation measure for the medium and deep ravines. The technology conserved on an average 85% runoff and reduced 9.1 ton/ha/year soil losses over no trench.

1.5.2.4 Bamboo-Based Production System

Bamboo has high soil conservation potential. It acts as a barrier and filter strip for overland flow, promoting soil deposition and slowing down water flow speed along riverbanks and in deforested areas. *Dendrocalamus strictus* which covers about 45% of the area under bamboo in India has been demonstrated to be very useful species for economic utilization of ravines. It is estimated that out of 3 million ha area of ravines, approximately 1 million ha area of deep ravines is potentially suitable for planting of *D. strictus*. With an average annual harvest of 4000 culms of bamboo/ha, the benefit–cost ratio of *D. strictus* plantation in ravines is lucrative 1.98 with an economic return of 19.3% over a 20-year period. The intangible benefits of carbon sequestration and prevention of soil erosion are the added benefits. Figure 1.7 shows the profuse bamboo growth in gullied and ravinous land of India.

1.5.3 Performance Evaluation of Ravine Reclamation Works

Ecological and socio-economic benefits of ravine reclamation were assessed in a watershed of Rajasthan which was treated during 1998–2003 with limited budgetary provisions under Integrated Wasteland Development Programme. The direct and indirect benefits accrued from major watershed interventions are listed in Table 1.10. Despite abnormal monsoon pattern and severe drought conditions experienced during the project period, the overall B:C ratio worked out at the

Table 1.9 Criteria for selection of land use and species in ravines

S. no.	Topographic location	Soil type	Resource availability	Land use	Tree/shrub species	Land ownership
1.	Marginal lands along ravines	Heavy	Supplemental irrigation available	AHS	Amla, guava, lemon, bel, anar, ber, karonda, lasoda, tenti	Private and community lands
2.	Marginal lands along ravines/reclaimed shallow ravines	Heavy	Supplemental irrigation available for crops	AFS boundary plantation	Desi babul, khejra, subabul, neem, siris, rohan, ratanjot	Private/community/revenue forest lands
3.	Marginal lands along ravines	Light	Supplemental irrigation available	AHS	Amla, guava, lemon, belpatra, ber, karonda, lasoda, sharifa	Private and community lands
4.	Marginal lands along ravines/reclaimed shallow ravines	Light	Supplemental irrigation available for crops	AHS boundary plantation	Shisham, desi babul, khejari, khejidi, subabul, bakain, semal, ardu, neem, siris, rohan, ratanjot, pilu and Marwari teak	Private/community/revenue forest lands
5.	Medium and deep ravines / Ravine top gently sloping	Heavy	Rainfed lifesaving irrigation available	HPS	Amla, lemon, belpatra, anar, ber, lasoda, tenti + grasses, (karad, dhaman, rodes)	Private and community lands
6.	Medium and deep ravines / Ravine top gently sloping	Light	Rainfed lifesaving irrigation available	HPS	Amla, lemon, bel, ber, lasoda, drumstick + grasses (karad, dhaman, black dhaman, rodes), stylo (legume)	Private and community lands
7.	Medium and deep ravines slopes <33%	Heavy	Rainfed	MPTS/Grass production	Neem, churel, ratanjot + grasses	Private/community/revenue lands
8.	Medium and deep ravines slopes <33%	Light	Rainfed	SPS	Neem, churel, bakain + grasses	Private/community/revenue lands
9.	Ravine slopes >33%	Heavy or light	Rainfed	Grass production	Karad, dhaman, black dhaman, rodes	Community/revenue lands
10.	Medium and deep ravines – Beds	Heavy	Rainfed	MPTS/bamboo cultivation	Bamboo, cassia, desibabool, khejidi, subabool, neem + grasses	Private/community/revenue lands
11.	Medium and deep ravines-Beds	Light	Rainfed	MPTS/bamboo cultivation	Bamboo, shisham, desi babool, khejiri, cassia, subabool, bakain, ardu, neem + grasses	Private/community/revenue lands
12.	Waterlogged ravine beds	–	–	MPTS/Grass production	Tamarix spp, para grass	Wastelands/revenue lands

MPTS multi-purpose trees and shrubs, *AFS* agroforestry systems, *AHS* agri-horticulture systems, *SPS* Silvi-pastoral systems, *HPS* horti-pastoral systems

project completion was 1.54:1 (Singh et al. 2004). Reclamation of gullied land and increased soil moisture were the most apparent and immediate benefits experienced. With improved cropping practices, these ameliorative effects can be enhanced to quickly recover the cost of conservation measures. Bunding and levelling require high initial investment in ravine reclamation plan, which can be recovered in following 3–5 years period through increased crop productivity (Singh et al. 2005). Strengthening existing water-harvesting structure and developing additional village pond effectively reduced the drought effect.

Erosion control measures with and without vegetative support reduced soil loss by 68 and 55% compared to untreated area. About 57,000 ton of soil was retained in upstream side of 36 check dams having catchment of 507 ha. Therefore, as cumulative effect, silt retained behind gully control structure was equivalent to arresting 112.5 ton/ha of soil. These structures reclaimed 9.12 ha of the severely gullied land and 24.6 ha of the moderately degraded land. Structures in upper reach retained greater soil volumes compared to lower reach structures. It is estimated that the runoff from watershed was about 30% during the initial years which reduced to less than 10% in response to conservation measure. Similarly, in Yamuna ravines at Agra district, a 5-year (2010–2014) field study evaluated impact of minor land levelling and peripheral bunding on crop yields (Singh

Fig. 1.7 Profuse bamboo growth in gullied and ravine land

Table 1.10 Economic returns from different interventions in a watershed of Rajasthan (Singh et al. 2016)

Intervention	Total cost (Rs) for 6-year period	Area covered (ha)	Direct benefits	Indirect benefits	Estimates value of benefits (Rs/year)
Bunding and levelling	169,611	90.3	Increased crop yield (89% increase in mustard observed)	Improvement in land quality	328,500
				Ground water recharge	
Peripheral bund with spillways	888,172	381.9	Increased crop yield through increased moisture regime in the vicinity of structures (24.6 ha)	Improvement in land quality	89,446
				Ground water recharge	
Gully control structures	342,727	131.6	Increased area under cultivation (9.12 ha)	Improvement in land quality	70,350
			Increased fuel and fodder production[2]	Ground water recharge	
Water harvesting structures	576,873	21.65[a]	Increased water availability for irrigation and drinking for cattle	Ground water recharge	135,846
Crop demonstrations	19,466	17.6	Increased crop yields	Sustainability in agriculture production	24,262
Horticulture plantations	7200	1.3	Increased fruit production from fifth year onwards	Sustainability, stability and diversification in agriculture production	72,200
Afforestation works	15,298	38.5	Increased fuel and fodder production[b]	Erosion control	NIL
				Microclimate improvement	

[a]Cultivated area estimated to receive lifesaving irrigation
[b]The benefit could not be realized during project period due to severe biotic interference and persistent drought

et al. 2016). Grain yields of pearl millet, green gram and sesame improved by 90, 164 and 179% over respective yields received from untreated area.

1.6 Road Ahead

Located along the major rivers and their tributaries, the ravine area is ecologically sensitive riparian zones which are highly responsive to management. Recognizing far-reaching and multiplicity of benefits of reclaiming these lands, several initiatives have been taken in the past at various levels. There is a need to do collaborative project regarding reclamation of ravine lands for productive utilization involving delineation of ravine area in four states of UP, MP, Rajasthan and Gujarat; for ravines area development and rehabilitation. The ravine areas development is also needed to include the climate change impact mitigation activity, and therefore, international collaboration is also needed.

To facilitate scientific micro-level planning and ensure sustainable development of these fragile and complex ecosystems, the following collateral research, developmental and policy initiatives are recommended:

- Establishment of benchmark model ravine development projects in representative agro-ecozones for field evaluation and demonstration of reclamation technology.
- Establish centralized repository of high-resolution digital maps of ravine regions.
- Identify expert groups and initiate collaborative research for developing DSS and planning tools required for scientific planning.
- Up-scaling and dissemination of ravine technologies through modelling applications, software and web technologies.
- Realistic policy gap assessment by a dedicated agency especially related to tenure provision, illegal land occupancy and land-use policy.
- Establishing a national-level central unit to expedite launching of ravine reclamation programme, address policy gaps and coordinate research & development activities.

References

Adhikary, P. P., Hombegowda, H. C., Barman, D., & Madhu, M. (2018). Soil and onsite nutrient conservation potential of aromatic grasses at field scale under ashifting cultivated, degraded catchment in Eastern Ghats, India. *International Journal of Sediment Research*, 33(3), 340–350.

Ahmed, E. (1973). Soil Erosion in India, Asia Publishing House, Mumbai, ISBN 10:0210222476, 99p.

Ali, M. (1974). Halt of menace of slopy land. *Indian Farmers Digest*, 7 (9), 31–34.

Babu, R., Tejwani, K.G. Agarawal, M.C. and Bhushan, L.S. (1978). Distribution of erosion index and isoerodent map of India. *Indian Journal of Soil Conservation*, 6 (1), 1–12.

Bali, Y.P. and Karale, R.L. (1977). Reclaimability classification of ravines for agriculture, *Soil Conservation Digest*, 5(2), 40–47.

Bhulyan, S. (1967). Survey of ravine lands in Rajasthan, In: Eleventh Silvicultural Conference, Proceedings. Forest Research Institute, Dehra Dun, U.P. (cited in GUPTA and PRAJAPATI, 1983, 249).

Bull, L.J. and Kirkby, M.J. (2002). Dryland Rivers- Hydrology and Geomorphology of the semi arid channels. John Wiley & Sons, Chichester, UK. ISBN: 978-0-471-49123-1. 398p.

Chaturvedi, O.P., Kaushal, R., Tomar, J.M.S., Prandiyal, A.K., and Panwar, P. (2014). Agroforestry for Wasteland Rehabilitation: Mined, Ravine, and Degraded Watershed Areas. In: J. C. Dagar et al. (eds.), *Agroforestry Systems in India: Livelihood Security & Ecosystem Services* Advances in Agroforestry 10, DOI: 10.1007/978-81-322-1662-9_8.

Collison, A.J.S. (2001). The cycle of instability: stress release and fissure flow as controls on gully head retreat. *Hydrological processes*. 15, 3–12.

Eswaran, H., Lal, R., Reich, P.F. (2001). Land degradation. An overview conference on land degradation and desertification. KhonKaen, Thailand: Oxford Press, New Delhi, India.

Gibling, M.R. Tandan, S.K., Sinha, R. and Jain, M. (2005). Discontinuity bounded alluvial sequences of southern Gangetic plain, India: Aggradation and Degradation in response to monsoonal strength, *Journal of Sediment Research*, 75(3), 369–385.

Gupta, R.K. (1973). Publication Corner: annotated bibliography on ravine land, *Soil Conservation Digest*, 2(1), 88–94.

Gupta, R.K. and Prajapati, M.C. (1983). Reclamation and use of ravine lands. *Desert Resources and Technology* (Jodhpur), 1, 221–262.

Hajgh, J.M. (1984). Ravine erosion and reclamation in India, *Geoforum*, 15(4), 543–561.

Kenney, T.C. and Lau, D. (1985). Internal stability of granular filter, *Canadian Geotechnical Journal*, 22 (2), 215–225.

Kenney, T.C. and Lau, D. (1986). Internal stability of granular filters: reply, *Canadian Geotechnical Journal*, 23 (4), 420–243.

Kezdi, A. (1979). Soil Physics, selected topics Elsevier Scientific Publishing Co. Amsterdam, 160p.

Kumar, G., Sena, D.R., Rao, B.K., Pande, V.C., Kurothe, R.S., Bhatnagar, P.R. and Mishra, P.K. (2018). Ravines of Gujarat-Delineation and Characterization. Technical Bulletin No. TB-04/V/E-2018. ICAR- Indian Institute of Soil and Water Conservation, Research Centre, Vasad-388 306, District Anand, Gujarat, India. 33p.

N.C.A. (1976). Report of the National Commission on Agriculture, Part 5: Resource Development, Ministry of Agriculture and Irrigation, Government of India, New Delhi, pp. 107–322.

N.R.S.C. (2000). Wasteland Atlas of India-2000. Ministry of Rural Development and NRSC Publ. NRSC, Hyderabad.

N.R.S.C. (2008). Wasteland Atlas of India-2008. Ministry of Rural Development and NRSC Publ. NRSC, Hyderabad.

Prajapati, M.C., Joshi, P., Rathore, B.L. and Dubey, L.N. (1982). Surface water management for grass and tree land development in ravinous watershed- a case study. *Indian Journal of Soil Conservation*, 10(2&3),30–38.

Rao, B.K., Mishra, P.K., Kurothe, R.S., Pande, V.C. and Kumar, G. (2015). Effectiveness of Dichanthiumannulatum in watercources for reducing sediment delivery from agricultural watersheds. *Clean-Soil Air Water*, 43(5), 710–716.

Seth, S.P., Bhatnagar, R.K. and Chauhan, S.S. (1969). Reclaimability classification and nature of ravines of Chambal Command Areas. *Journal of Soil and Water Conservation, India*, 17(3/4), 39–44.

Sharma, H.S. (1980). Ravine Erosion in India, Concept. New Delhi.

Sharma, H.S. 1976. Morphology of Ravines of the Morel Basin (Rajasthan) India. *International Geography*, 219–23.

Sharma, H.S. 1968. Genesis of Ravines of the Lower Chambal Valley, India. Selected Papers, *21ˢᵗInternational Geographical Union Congress*, 1 (1968), 114–18.

Singh, A.K., Sharma, K.K., Dubey, S.K. and Chandra, S. (2016). Peripheral bund: Controlling measures for ravine formation in arable land. Technical bulletin No. T-71/ A-02 published by ICAR-IISWC, Dehradun (Uttrakhand).

Singh, K.D., Prasad, S.N., Singh, R.K., Ali, S., Prasad, A., Singh, S.V., Parandiyal, A.K. and Ashok Kumar. (2004). Participatory watershed management for sustainable development in Badakhera watershed, Bundi (Raj.). CSWCRTI, Research Centre Kota. *pp.* 1–76.

Singh, R.K., Prasad, S.N., Ali, Shakir, Kumar, A., Singh, K.D., Prasad, A., Singh, S.V. and Parandiyal, A.K. (2005). On-farm evaluation of conservation measures to performance of rainfed crops in semi-arid region. *Indian Journal of Soil Conservation*, 33(2), 141–143.

Tejwani, K.G. (1974). Classification and reclamation of gullied lands. *Journal of Soil and Water Conservation in India*, 24, 29–40.

Gopal Kumar is presently working as Senior Scientist at ICAR-Indian Institute of Soil and Water Conservation (IISWC), Dehradun, Uttarakhand. He has experience of more than 15 years of research, training, and extension in the field of soil and water conservation, watershed management, ravine land management, geostatistics, remote sensing, and GIS. He is the recipient of various awards from Indian Council of Agricultural Research, New Delhi, Indian Association of Soil and Water Conservationists, Dehradun, etc. Dr. Kumar has more than 100 publications to his credit which include research papers in journals of national and international repute, books, book chapters, technical and extension bulletins, popular articles, technology brochures, e-publications, etc.

Partha Pratim Adhikary is presently working as Senior Scientist at ICAR-Indian Institute of Soil and Water Conservation (IISWC), Research Centre, Sunabeda, Koraput, Odisha. He has experience of more than 12 years of research, training, and extension in the field of soil and water conservation, watershed management, geostatistics, pedotransfer function, surface and groundwater modeling, remote sensing, and GIS. He is the recipient of various awards from Indian Council of Agricultural Research, New Delhi, Indian Association of Soil and Water Conservationists, Dehradun, SADHNA, Solan, ICAR-IISWC, Dehradun, CCSHAU, Hisar, etc. Dr. Adhikary has more than 100 publications to his credit which include research papers in journals of national and international repute, books, book chapters, technical and extension bulletins, popular articles, technology brochures, e-publications, etc.

Ch. Jyotiprava Dash is presently working as Scientist at ICAR: Indian Institute of Soil and Water Conservation (IISWC), Research Centre, Sunabeda, Koraput, Odisha. She has experience of more than 8 years of research, training, and extension in the field of soil and water conservation, watershed management, geostatistics, surface and groundwater modeling, remote sensing, and GIS. She is the recipient of various awards from the Indian Council of Agricultural Research, New Delhi, Soil Conservation Society of India, New Delhi, Indian Association of Soil and Water Conservationists, Dehradun, SADHNA, Solan, etc. Dr. Dash has more than 50 publications to her credit which include research papers in journals of national and international repute, book chapters, technical and extension bulletins, popular articles, e-publications, etc.

Pravat Kumar Shit and Partha Pratim Adhikary

Abstract

Gully erosion has become a menace to agricultural and other development in the world. In the tropical and subtropical region of India, it is of very importance because of its advancement due to collapsing of head and gully wall. Quantification of different soil physico-chemical properties and soil disintegration characteristics within different weathering profiles (surface layer, red soil layer, sandy soil layer and detritus layer) and its relationships with different soil physico-chemical properties is necessary to understand the mechanism of the forming process and development of the collapsing gully. In this study, three collapsing gullies under red soil region of subtropical India were analysed for their physico-chemical properties and their relationships with the disintegration ability of the gully. The anti-disintegration ability of the different weathering profiles with two different moisture conditions (the natural state soil moisture condition and the air-dried condition) was determined by the anti-disintegration index (K_c) and measured by the submerging test. The results showed that surface soil layers are high in finer soil particles and organic matter; and the sandy soil layer and the detritus soil layer are rich with coarser soil particles. The anti-disintegration coefficient gradually decreases with the increase in soil depth. The anti-disintegration coefficient decreases sharply with the increase of soil moisture. Therefore, sandy soil layer and detritus layers are vulnerable to disintegration due to the effect of external factors compared to the upper two layers. The anti-disintegration coefficient is positively correlated with clay and soil organic matter. Therefore, in soils with high clay and organic matter content smaller gully can be noticed, which is a common fact, and thus been established in this study with the help of anti-disintegration coefficient.

Keywords

Gully erosion · Bulk density · Disintegration index · Particle size distribution · Soil organic matter

2.1 Introduction

Gully erosion is a common type of soil erosion found extensively all over the world. Over the last few decades, extensive research on the formation of gullies and the control factors to increase the gully head was done. Researchers even tried to model the collapsing of gully head under different scenario. Yet there remain many questions unanswered with respect to better understanding of the collapsing of gully head. The collapsing of gully head is very common in tropical and subtropical region of the world, and in the red soil region, it is most common.

Tropical and subtropical gullies are most extensive by gully collapsing due to soil disintegration in red soil region in India. Accelerated rate of soil erosion through gully is a serious endemic long-term environmental problem in the humid tropical, subtropical, and temperate region (Singh and Dubey 2002). The collapsing is a main factor of gully erosion in lateritic red soil during the rainy season (Shit et al. 2015). According to field observation, in lateritic collapsing, gully walls are subdivided into five gully profiles (surface layer, red soil layer, sandy soil layer, detritus layer and bed rook) based on degrees of weathering, soil structure, grain size distribution, soil colour, soil moisture contents, etc. However, the soil disintegration property is one of the major gully erosion characteristics. Hence, to understand the soil disintegration property and find out the mechanism of forming processes and development of gully collapsing in lateritic soil is necessary.

P. K. Shit (✉)
Department of Geography, Raja N. L. Khan Women's College (Autonomous), Medinipur, West Bengal, India

P. P. Adhikary
Central Soil and Water Conservation Research and Training Institute, Research Centre, Koraput, Odisha, India

P. K. Shit et al. (eds.), *Gully Erosion Studies from India and Surrounding Regions*, Advances in Science, Technology & Innovation,
https://doi.org/10.1007/978-3-030-23243-6_2

The soil disintegration property refers to the phenomenon that the soil is separated after being absorbed in water. With the infiltration of water and saturation of soil, the cementation between the soil particles is weakened and the soil structure becomes unstable, and therefore soil disintegration occurs (Lan 2013; Xia et al. 2018). Zhang (2009) reported that soil disintegration property is an irreversibly physical process. Many researchers studied the influencing factors of gully erosion, process and rate of gully head retreated, but few studies are there on soil disintegration processes in gully collapsing (Zhang and Tang 2013; Xia et al. 2015, 2016). The soil disintegration rate is one of the evaluation indices of soil erosion, which can be used to reveal some of the causes of the collapsing gully erosion (Lado et al. 2004; Liu et al. 2016; Xia et al. 2016, 2018).

Soil disintegration has very strong relation with soil moisture, soil texture and other soil properties. But these relationships will vary on the types of soil and other climatic conditions where the soil generated. Moreover, it is very difficult to quantify soil disintegration. One widely accepted expression is soil disintegration index (K_c). But its relationships with different soil properties in a wide range of red soils are yet to be ascertained properly. Therefore, in this chapter, we analysed the relationship between soil anti-disintegration index (K_c), particle size distribution (PSD) and soil organic matter (SOM) based on experimental study on three collapsing gullies in different weathering profiles.

2.2 Materials and Methods

2.2.1 Site Description

Three collapsing gullies region in Paschim Medinipur district in different locations in lateritic tract were selected according to topographic variation and climatic condition (Fig. 2.1). Gangani, Garbheta (GG) is situated on the right bank of river Shilai with the location of 22°51′30.04″N latitude and 87°20′40.23″E longitude and average altitude of 64 m above mean sea level. Rangamati, Medinipur (RM) is located on the left bank of river Kangsaboti with 22°24′44.59″N latitude and 87°17′50.26″E longitude. The altitude of RM site is 45 m above mean sea level. Dherua, Paschim Medinipur (DPM) site is also located on the left bank of river Kangsaboti. The location of this site is 22°29′51.49″N 87°5′40.00″E with the altitude of 49 m above mean sea level. A detailed description of the gully sites is presented in Table 2.1. In this region, typical monsoon climate prevails where mean annual temperature varied from 20 °C to 28 °C and annual precipitation varied from 1350 mm to 1500 mm with high variation (70–75%) in July to October (Shit et al. 2013). From

geomorphological point of view, this region is a part of the Chhotanagpur plateau margin with extremely dissected, discontinued landforms and pleistocene lateritic characterized. The formation of Pali (~1000 m) is portrayed by pebbly to coarse-grained micaceous sandstones, medium- to fine-grained sandstones, and red- and green-coloured mudstones in the study area (Dey et al. 2009). Niyogi (1970) and Paul (2002) have recognized this region as a part of paleo-coastal zone of Bengal basin.

2.2.2 Soil Sampling

Soils were sampled in three collapsing gullies from different weathering profiles. The soil samples were collected in three gully profiles (i.e., surface layer, red soil layer, sand soil layer, detritus layer and bedrock). A gully profile depends on degrees of weathering, soil structure, grain size distribution, soil colour and moisture contents (Luk et al. 1997). A detailed description of the weathering profile is presented in Table 2.2. According to the height of the gully profiles, we collected 3, 10, 9 and 2 soil samples, respectively, from each layers, that is surface layer or top soil layer, red soil layer, sandy soil layer and detritus layer. Sixty small soil blocks (diameter 3–5 cm) were obtained from each soil layer under wet and dry condition. During the soil sampling at each layer, about 1 kg soil sample were obtained by mean of quartering and transported to the laboratory for measurement of SOM and soil particle size distribution. Using cutting ring, five soil samples at each layer were obtained to estimate soil bulk density. The detailed information of soil sampling sites and soil sampling depth is given in Table 2.3.

2.2.3 Soil Analysis

The soil samples were air-dried and then sieved through a 2 mm mesh size to remove vegetation roots and stone for laboratory analysis. The soil bulk density was analysed using cutting ring method (Adhikary et al. 2017). SOM was measured using Walkey and Black method (Nelson and Sommers 1996).The soil particle size distribution (PSD) was analysed by percentage of gravel (1.0–2.0 mm), coarse sand (0.25–1.0 mm), find sand (0.05–0.25 mm), silt (0.002–0.05 mm) and clay (<0.002 mm). The clay and silt fractions were separated by using pipette method and gravel and silt fractions were determined by wet sieving method (Gee and Bauder 1986). The water stability of soil aggregates test was used for the soil anti-disintegration index (K_c) of soil blocks (Fig. 2.2). The soil anti-disintegration index (K_c) is calculated according to the following formula (Liu et al. 2009):

Fig. 2.1 Location of the weathering collapsing gullies profile: (**a**) Gangani, Garbheta (GG); (**b**) photo of the typical collapsing gully in Gangani; (**c**) Rangamati, Medinipur (RM); (**d**) photo of the typical collapsing gully in Rangamati; (**e**) Dherua, Paschim Medinipur (DPM); (**f**) photo of the typical collapsing gully in Dherua

$$K_c = \frac{(a_1 \times 5 + a_2 \times 15 + a_3 \times 25 + \ldots a_{10} \times 95 + a_\infty \times 100)}{N},$$

N is the total number of soil blocks (diameter 3–4 cm). The total immersion experiment time was 10 min. In the experiment, 50 soil blocks of 3–4 cm diameter, collected from different weathering profiles of the three collapsing gullies, were randomly selected and were placed uniformly on a sieve with aperture mesh size 2 mm (Fig. 2.2a), then water was added and lowered the screen gradually into water (Fig. 2.2b). The time was recorded to count the number of disintegration within a certain time. The a_1, a_2, ... a_{10}

Table 2.1 Description of the gully cluster sites from where soil samples were collected

Location	Collapsing gully code	Geographical location	Altitude (m)	Height of the collapsing gully wall (m)	Vegetation cover (%)	Vegetation community
Gangani, Garbheta	GG	22°51′30.04″N 87°20′40.23″E	64	8–12	18	*Cashew tree, Eragrostis cynosuroides* (Poaceae), *Andropogon aciculate, Panicum maxima (Poaceae)*
Rangamati, Medinipur	RM	22°24′44.59″N 87°17′50.26″E	45	4–8	22	*Eragrostis cynosuroides* (Poaceae), *Saccharum munja* (Poaceae); *Miscanthus sinensis, Andropogon aciculate* (Poaceae)
Dherua, Paschim Medinipur	DPM	22°29′51.49″N 87°5′40.00″E	49	1–3	27	*Eragrostis curvula*, weird grass and Eucalyptus tree

Table 2.2 Description of weathering profiles (modified after Luk et al. 1997)

Weathering profiles	Description
Surface layer/top soil layer	Clayed sand with granular structure. Brownish red colour with minor organic matter content. Shallow vegetation roots and some desiccation cracks
Red soil layer	Reddish brown layer with high content of Al_2O_3 and Fe_2O_3. Intense kaolinization and oxidation. Massive structure. Soft lay mixed with residual quartz grains. Dominant desiccation cracks with acidic soil.
Sandy soil layer	Light red material with low Fe_2O_3 content. Molded weathering, clayey material with dominant desiccation cracks. Easily broken down under hand pressure
Detritus layer	Layer of light yellowish colour. Crystalline structure of bedrock presented. Easily broken down under hand pressure
Bedrock	Original rock

Table 2.3 Description of sampling profile of the collapsing gully sites

Soil layer	Gangani, Garbheta Soil layer code	Sampling height (m)	Rangamati, Medinipur Soil layer code	Sampling height (m)	Dherua, Paschim Medinipur Soil layer code	Sampling height (m)
Top soil layer	GG1	0–0.2	RM1	0–0.2	DPM1	0–0.2
Red soil layer	GG2	0.5	RM2	0.5	DPM2	0.5
	GG3	0.9	RM3	1.1	DPM3	0.9
	GG4	1.2	RM4	1.3	DPM4	1.6
	GG5	1.8				
Sandy soil layer	GG6	2.5				
	GG7	3.8	RM5	2.1	DPM5	2.3
	GG8	4.7	RM6	3.6		
	GG9	6.6	RM7	4.4		
	GG10	8.2				
Detritus layer	GG11	10.4	RM8	6.2		

express the accumulative number of completed disintegration soil blocks at the immersion times of 1 min, 2 min, 3 min, ... 0.10 min. a_∞ is the number of small blocks which remain not totally disintegrated after an immersion time of 10 min. The determined water stability of the total immersion time 10 min can represent 0, 10, 20, 30, ..., 0.100. The coefficient of disintegration of 5, 15, 25, ... 0.100 represent the water stability of soil aggregates. We calculated the disintegration coefficient value of the first 1 min as 5 = (0 + 10)/2. If aggregate collapse during the immersion time of 1–2 min, its coefficient value is 15 = (10 + 20)/2. If the soil block never disintegrates during the immersion time of 10 min, its coefficient is defined as 100. We calculate K_c of soil blocks by computing the average value. The coefficient of disintegration is an index to represent the anti-disintegration ability of soil block in a period of time. The increase in K_c value indicates the more difficulty to collapse.

2.2.4 Statistical Analysis

All the data were analysed in MS excel. Regression analysis was used to examine the relationship between soil particle size distribution, SOM and soil anti-disintegration index. A one-way analysis of variance (ANOVA) was used to examine the differences of soil properties among the each layer at 0.05 level of significant. The Fisher's least squares difference (LSD) test was used to compare means of soil variable. All

Fig. 2.2 Different procedure of static-water measurement for anti-disintegration of soil: (**a**) preparation, (**b**) absorbing water, (**c**) immersion and (**d**) collapse

figures were drawn using Origin 6.1 software and all results are illustrated as the mean value.

2.3   Results and Discussion

2.3.1   Soil Moisture Content of the Collapsing Gullies

The average soil moisture content under air-dry condition and natural conditions was analysed and shown in Table 2.4. The soil moisture content at natural state at GG gully was always higher than RM gully and DPM gully. Among the three gully sites, the DPM gully sites always showed lower soil moisture than the other two sites. The ranking of profile soil moisture content was in the order of GG > RM > DPM. The higher clay content at the GG gully sites is responsible for higher soil moisture content at that site.

2.3.2   Soil Organic Matter (SOM) and Bulk Density

Soil organic matter is an important parameter affecting the formation of soil structure and also stabilizes the soil profile by soil aggregates. In our experimental result, it was found that with the increase in soil depth, SOM contents sharply decreased ($p < 0.005$) for all the three collapsing gully profiles (Table 2.5). The relatively lower SOM contents in the sandy soil and detritus soil layer with respect to top soil layer and red soil layer was observed. The highest SOM

Table 2.4 Description of average moisture content (kg kg^{-1}) of different collapsing gullies

Collapsing gully ID	Air-dried condition	Natural state condition
GG1	0.046	0.208
GG2	0.041	0.187
GG3	0.034	0.195
GG4	0.047	0.231
GG5	0.052	0.209
GG6	0.045	0.228
GG7	0.039	0.245
GG8	0.033	0.222
GG9	0.037	0.242
GG10	0.029	0.251
GG11	0.041	0.203
RM1	0.038	0.189
RM2	0.044	0.257
RM3	0.038	0.249
RM4	0.031	0.247
RM5	0.034	0.194
RM6	0.028	0.199
RM7	0.032	0.208
RM8	0.047	0.191
DPM1	0.082	0.173
DPM2	0.046	0.193
DPM3	0.078	0.231
DPM4	0.054	0.183
DPM5	0.047	0.145

contents was observed at GG1 (28.23 g kg^{-1}), followed by RM1 (22.39 g kg^{-1}) and DPM1 (32.65 g kg^{-1}). The bulk density was lowest among the top soil surface layer of the collapsing gully sites, that is 1.27, 1.32 and 1.31 g cm^{-3} for GG, RM and DPM, respectively. With increase in soil depth, bulk density in the soil layer gradually shows increasing trend ($p < 0.005$). So, the higher SOM contents in surface soil are caring to the increase of anti-disintegration of soil particles, which is of great significance to reduce erosion of soil. Comparatively higher rainfall in the GG sites was responsible for higher plant density there. High plant density resulted higher amount leaf fall in these sites and increased the

SOM content of the surface soils. The higher root growth and activity in these sites also indirectly helped to move the SOM to the lower layers and decomposition of plant roots also helped positively to accumulate some amount of organic matter in the soil profile.

2.3.3 Soil Particle Size Distribution (PSD)

Soil particle size distribution is one of the most important physical parameters in soil system (Hillel 1980). Soil properties are mainly affected by soil particle size distribution due to the horizontal and vertical movement groundwater, retention of water, air pressure, drying-wetting and solutes of soil particles. The highest clay contents were observed as 36.59, 32.44 and 39.16% in GG, RM and DPM sites, respectively. Silt contents ranged between 22.62 and 37.69% of GG, 26.58 and 35.71% of RM and 26.18 and 38.25% of DPM site, respectively. The average fine sand contents (1.0–0.25 mm) were greater than the contents of coarse sand (2.0–1.0 mm) of all the three collapsing gullies, but, top surface layer soil is different except DPM gully site. The average percentages of diffident particle-size distributions of collapsing gullies are shown in Table 2.6. The results shows that the finer soil particles gradually declined and coarse soil particles increased from the surface layer to bottom layer. The surface layer of GG, RM and DPM collapsing gullies were evidenced higher clay (<0.002 mm) contents of 33.44%, 32.44% and 27.79%, respectively. In red soil layer (lateritic horizon) results indicated that the highest percentage of sand and clay contents of the three collapsing gullies. This occurrence can be attributed to the different degree of weathering and laterization processes.

2.3.4 Soil Anti-Disintegration Index (K_c)

Soil disintegration represents the vulnerability of the soils due to one or many erosive agents. Soil anti-disintegration

Table 2.5 Average of soil properties for different weathering gully profiles

Collapsing gully profile	Weathering layer	Bulk density (g cm^{-3})	SOM (g kg^{-1})
GG	Top soil layer	1.27	28.23
	Red soil layer	1.33	6.28
	Sandy soil layer	1.37	1.12
	Detritus layer	1.41	0.70
RM	Top soil layer	1.32	22.39
	Red soil layer	1.36	4.36
	Sandy soil layer	1.40	1.08
	Detritus layer	1.43	0.56
DPM	Top soil layer	1.31	32.65
	Red soil layer	1.34	6.82
	Sandy soil layer	1.37	1.67

Table 2.6 Average percentages of diffident particle-size distributions of collapsing gullies

Soil layer code	Mass percentage particle-size distributions (mm)				
	Coarse sand	Fine sand	Very Fine sand	Silt	Clay
	2.0–1.0	1.0–0.25	0.25–0.05	0.05–0.002	<0.002
GG1	19.32	14.25	10.37	22.62	33.44
GG2	8.99	9.17	16.84	28.41	36.59
GG3	8.21	9.01	17.33	31.42	34.03
GG4	9.84	10.17	21.15	34.59	24.25
GG5	8.84	13.64	26.79	34.12	16.61
GG6	9.81	9.19	32.86	35.91	12.23
GG7	10.13	15.25	25.37	37.69	11.56
GG8	13.13	14.22	25.3	33.68	13.67
GG9	9.91	10.11	33.16	36.04	10.78
GG10	12.02	18.68	25.44	34.05	9.81
GG11	14.64	17.03	23.12	29.69	15.52
RM1	18.35	10.24	12.39	26.58	32.44
RM2	23.14	8.25	10.47	28.89	29.25
RM3	22.32	10.46	8.07	31.55	27.6
RM4	14.41	9.86	10.11	34.04	31.58
RM5	22.16	9.68	7.82	35.71	24.63
RM6	23.05	12.44	8.41	32.45	23.65
RM7	20.08	10.68	15.04	34.38	19.82
RM8	9.14	14.38	13.67	29.90	32.91
DPM1	8.78	13.67	11.51	38.25	27.79
DPM2	5.61	10.22	16.87	28.14	39.16
DPM3	4.23	15.37	25.21	28.33	26.86
DPM4	3.08	14.48	22.72	26.18	33.54
DPM5	2.02	22.18	23.81	37.36	14.63

index is the quantitative representation of the stability of any soil against dispersive or erosive agents. Figure 2.3 represents the variations of soil anti-disintegration values (K_c) of different soil layers of the three collapsing gullies of Gangani, Garbheta (GG); Rangamati, Medinipur (RM); and Dherua, Paschim Medinipur (DPM) sites. We analysed the K_c values under two moisture conditions, that is air-dried condition and natural state condition. The K_c values ranged from 7.38 to 92.70 at GG, 7.11 to 84.68 at RM, and 20.30 to 70.30 at DPM sites, respectively, in the natural state condition. In air-dried condition, the Kc values varied from 3.93 to 24.75 in GG, 5.67 to 32.55 in RM and 7.30 to 12.70 in DPM sites, respectively. It was observed that the K_c values decrease with increase in the soil depth (surface layer to detritus layer) for all the soil profiles but with various rates. The K_c values in natural-state soil moisture condition are much higher than that in the air-dried condition of all collapsing gully sites (Fig. 2.3), with no significant different between the two moisture conditions in the detritus layer. These results reveal that addition of water from outside to any soil layer of collapsing gully will trigger the disintegration process of gully head or wall easily. Figure 2.4 demonstrated this fact in the laboratory where addition of water-facilitated disintegration and collapse. This collapse is of two types, gradual

collapse and explosive collapse, based on soil materials and minerals. Meanwhile, the interesting part of this study is the surface layer and red soil layer under natural state soil moisture condition; high Kc values indicated that soils in these two layers are difficult to disintegrate during dry season, but during wet season, these layers are vulnerable to disintegrate due to the underlying washing and collapsing of soil layers. Figure 2.5 demonstrates the collapsing of gully wall due to removal of underlying support because of disintegration. Huge chunks of top soils are being collapsed and the surface layers are being exposed due to this process and thereby weaken the underlying layers. Thus, maintain this cycle continuously unless some external soil and water conservation measures are imposed.

2.3.5 Relationship Between the Soil Anti-disintegration (K_c) and Particles Size Distribution (PSD) and SOM

In this study, linear and nonlinear regression analyses were performed to determine the degree of relationships between K_c values and PSD contents and SOM (Fig. 2.6; Table 2.7). Nonlinear regression analyses showed that the SOM contents

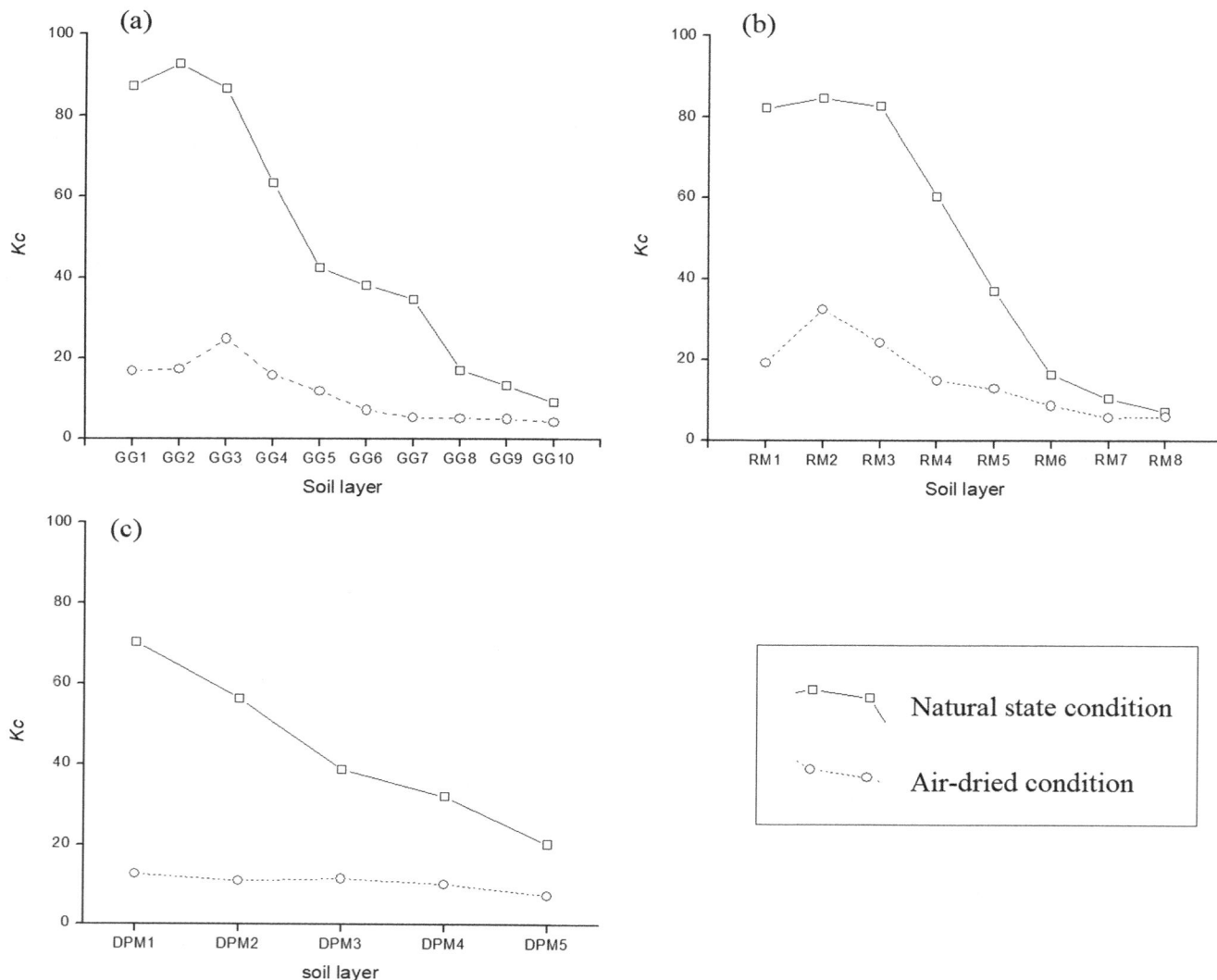

Fig. 2.3 Variations in condition of soil anti-disintegration (Kc) values of the different soil layers of the three collapsing gullies. (**a**) Gangani, Garbheta (GG); (**b**) Rangamati, Medinipur (RM); and (**c**) Dherua, Paschim Medinipur (DPM)

have a significant and positive logarithmic relationship under two different moisture conditions with R^2 values of 0.673 ($p < 0.001$) and 0.361 ($p < 0.001$) in natural state soil moisture condition and air-dried condition, respectively (Table 2.7). The results of regression analysis indicated that K_c values have a week negative correlation with fine sand, very fine sand, and silt under both the conditions. The linear regression analyses showed that the K_c value and coarse sand contents were found to have low positive relationship ($R^2 = 0.026$ for natural state condition and $R^2 = 0.078$ for air-dried condition at $P < 0.001$). The clay content and SOM played a very significant role for aggregate stability. The results indicated that clay contents and Kc values had very significant and positive relationship with $R^2 = 0.415$ and $R^2 = 0.387$ ($P < 0.001$) in nature and air-dried conditions, respectively (Table 2.7).

2.3.6 Analysis of Soil Disintegration Characteristics of Different Soil Layers of Collapsing Gullies

Gully developments have been found to occur in three phases: (1) failure of gully head and gully banks, (2) transport of the debris by stream-flow and (3) degradation of the channel. The soil disintegration rate increases with the soil depth. The stabilization of the gully wall or bank depends on various factors, such as laterization processes and degree of weathering processes. The cation exchange capacity and concentration of Fe, Ca and Al, oxides, and hydroxides of Fe and Al play an important role in soil structure stability of collapsing gullies bank. Figure 2.7 shows the typical structure of soil profile of a collapsing gully site. During rainy season (monsoon period: July–September), gully bank used

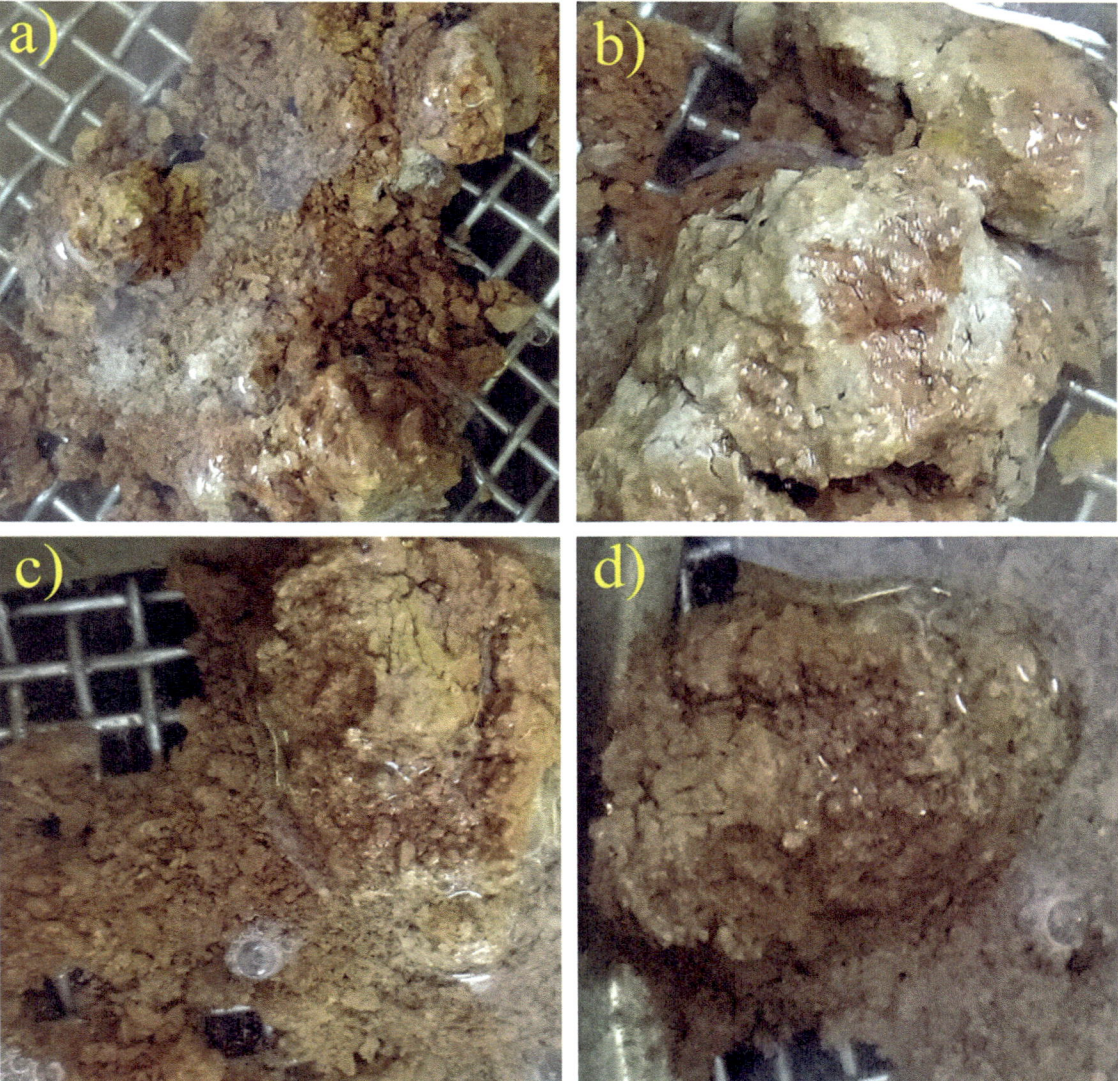

Fig. 2.4 Images of different collapse patterns; (**a**) and (**b**) gradual collapse; (**c**) and (**d**) explosive collapse

to collapse every year due to washing out of sandy soil layer and cementing materials (silt and clay). Because the coarse particles in the sandy soil layer and detritus layer are higher than that of the surface layer and the red soil layer, these are easily removed by the flowing water. In this study, sandy soil layer, the cohesive force between the soil particles declined due to low concentration of Fe and Al oxides, clay content and SOM, which led to unstable structure of soil. When water or rainfall touches the sandy layer, the hydrated films surrounding the soil particles weaken the cohesion force and, subsequently, the soil disintegrates. In surface soil layer and red soil layer, the higher concentrations of SOM, Fe, Al and clay contents make these layers more stable compared to the other layers. Therefore, the results indicated that the sandy soil layer and detritus layer are much easily

disintegrated than the surface layer and red soil layer when rainfall occurs.

2.4   Conclusion

The finer soil particles are higher in the surface soil layer, whereas the coarse particles are higher in the sandy soil layer and the detritus layer of collapsing gully. The surface soil layer and red soil layers are more stable than sandy soil layer and detritus layer in the red and lateritic zones of India. With the increase of soil moisture, the disintegration rate increases, and at air-dry condition, the rate is very low. A positive correlation between soil anti-disintegration constant and finer soil particles and cementing agents like SOM was

Fig. 2.5 Gullies wall collapsing due to removal of underlying support; (**a**) GG and (**b**) RM sites

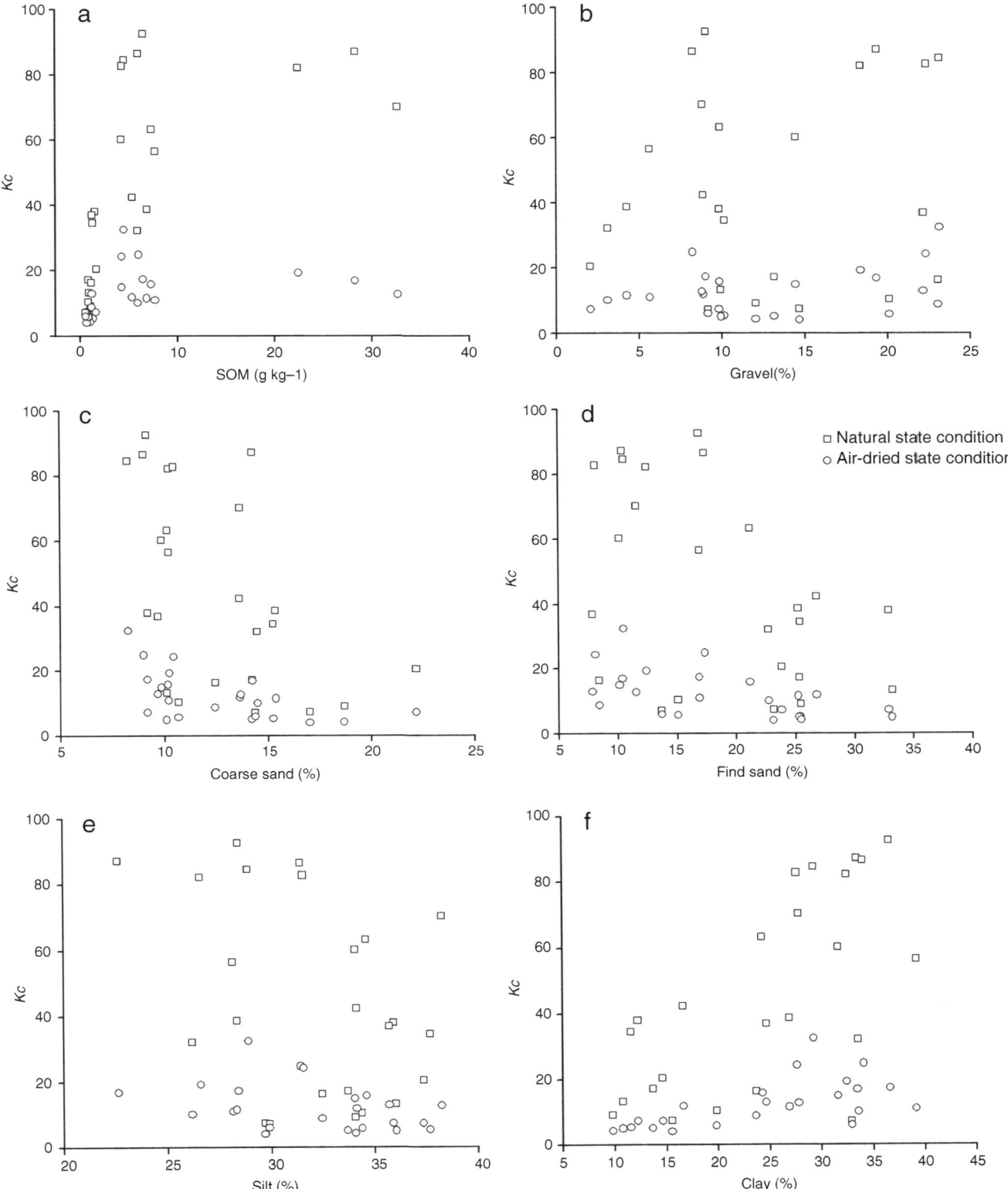

Fig. 2.6 Correlation between soil anti-disintegration index (Kc) and soil particle-size distribution and SOM with two different moisture conditions

Table 2.7 Best-fit regression models between anti-disintegration coefficient and other soil physical parameters

Particle-size distribution	Natural state condition		Air-dried condition	
	Regression	R^2	Regression	R^2
Coarse sand	$y = 0.750x + 35.98$	0.026	$y = 3.264\ln(x) + 4.494$	0.078
Fine sand	$y = -63.2\ln(x) + 203.5$	0.304	$y = -16.9\ln(x) + 54.47$	0.352
Very fine sand	$y = -1.783x + 78.39$	0.218	$y = -0.522x + 21.89$	0.303
Silt	$y = -88.2\ln(x) + 350.6$	0.155	$y = -0.684x + 34.19$	0.143
Clay	$y = 44.19\ln(x) - 91.88$	0.415	$y = 10.61\ln(x) - 20.76$	0.387
SOM	$y = 20.35\ln(x) + 21.97$	0.673	$y = 3.706\ln(x) + 7.955$	0.361

Fig. 2.7 Soil structure of gully collapsing site (as adopted after Xia et al. 2018)

observed. Thus, gully collapsing can be prevented by adopting such land use which promotes increase of soil organic matter and other cementing agents.

References

Dey S, Ghosh S, Debbarma C and Sarker P (2009) Some observation of regional evidences of Tertiary–Quaternary geo-dymanics in a paleo-coastal of Bengal basin, India. Russian Geol Geophys 50(11): 884–894

Gee GW, Bauder JW. Particle-size analysis (1986) In: Klute A, et, al. (Ed.), Methods of Soil Analysis Part 1, Physical and Mineralogical Methods, 2nd ed.; ASA, Inc, Madison, WI, pp 383–411.

Hillel D (1980) Fundamentals of soil physics. Academic Press, New York

Lado M, Benhur M, Shainberg I. (2004) Soil wetting and texture effects on aggregate stability, seal formation, and erosion. Soil Sci. Soc. Am. J.; 68: 1992–1999.

Lan ZX. (2013) Experimental study on disintegration behavior of granite residual soil. Doctoral dissertation, South China Univ Techno, Guangzhou, China. (in Chinese)

Liu J., Shi B., Jiang H., Bae S., Huang H. (2009) Improvement of water-stability of clay aggregates admixed with aqueous polymer soil stabilizers. Catena, 77: 175–179

Liu XL, Qiu JA, Zhang DL. (2016) Analysis of soil wetting mechanism and influencing factors on the headwall of collapsing and erosional gully. J. Soil Water Conserv. 30: 80–84. (in Chinese)

Luk SH, Yao QY, Gao JQ, Zhang JQ, He YG, Huang SM. (1997) Environmental analysis of soil erosion in Guangdong Province: a Deqing case study. Catena. 29: 97–113.

Nelson DW, Sommers LE (1996) Total carbon, organic carbon, and organic matter. In: Sparks DL, Page AL, etc. (eds), Methods of Soil Analysis, Part 3. Chemical methods. Wisconsin, WI, USA: Soil Science Society of America Book Series, 5, 961–1010

Niyogi D (1970) Geological background of beach erosion of Digha, West Bengal. Bull Geol Mining Metall Soc India 43:1–36

Paul AK (2002) Coastal Geomorphology and Environment. ACB Publications, Kolkata.

PP Adhikary, HC Hombegowda, D Barman, P Jakhar, M Madhu (2017) Soil erosion control and carbon sequestration in shifting cultivated degraded highlands of eastern India: performance of two contour hedgerow systems, Agroforestry Systems 91 (4), 757–771.

Shit PK, Bhunia G, Maiti R (2013) Assessment of Factors Affecting Ephemeral Gully Development in Badland Topography: a Case Study at Garbheta Badland (Pashchim Medinipur. Int J Geosci 4 (2):461–470. doi:https://doi.org/10.4236/ijg.2013.42043

Shit PK, Paira R, Bhunia GS, Maiti R (2015) Modeling of potential gully erosion hazard using geo-spatial technology at Garbheta block, West Bengal in India. Model. Earth Syst. Environ. 1:2, DOI https://doi.org/10.1007/s40808-015-0001-x

Singh S. and Dubey A. (2002) Gully erosion and management methods and applications (A Field Manual). New Academic Publishers, Delhi, 1–248 pp.

Xia D, Zhao B, Liu D, Deng Y, Cheng H, Yan Y, et al. (2018) Effect of soil moisture on soil disintegration characteristics of different weathering profiles of collapsing gully in the hilly granitic region, South China. PLoS ONE 13(12): e0209427. https://doi.org/10.1371/journal.pone.0209427

Xia D., Deng Y., Wang S., Ding S., Cai C. (2015) Fractal features of soil particle-size distribution of different weathering profiles of the collapsing gullies in the hilly granitic region, south China. Nat Hazards, 79:455–478.

Xia, D., Ding, S. W., Long, L., Deng, Y. S., Wang, Q. X., Wang, S. L., and Cai, C. F.(2016) Effects of collapsing gully erosion on soil qualities of farm fields in the hilly granitic region of south China, J. Integr. Agr., 15, 2873–2885.

Zhang S, Tang HM. (2013) Experimental study of disintegration mechanism for unsaturated granite residual soil. Rock Soil Mech.; 34: 1668–1674. (in Chinese)

Zhang S. (2009) A study on disintegration behavior of granite residual soil in Guangzhou. Doctoral dissertation, China Univ Geosci, Wuhan, China. (in Chinese)

Pravat Kumar Shit received his Ph.D. in Geography (Applied Geomorphology) from Vidyasagar University (India) in 2013, M.Sc. in Geography and Environment Management from Vidyasagar University in 2005, and PG Diploma in Remote Sensing & GIS from Sambalpur University in 2015. He is Assistant Professor in the Department of Geography, Raja N. L. Khan Women's College (Autonomous), Gope Palace, Midnapore, West Bengal, India. His main fields of research are soil erosion spatial modeling, badland geomorphology, gully morphology, water resources and natural resources mapping, and modeling and has published more than 45 international and national research articles in various renowned journals; also, he has published three books. His research work has been funded by the University Grants Commission (UGC), India, and Higher Education Science and Technology and Biotechnology, Government of West Bengal. He is Associate Editor and on the editorial boards of three international journals in geography and earth environmental sciences.

Partha Pratim Adhikary is presently working as Senior Scientist at ICAR-Indian Institute of Soil and Water Conservation (IISWC), Research Centre, Sunabeda, Koraput, Odisha. He has experience of more than 12 years of research, training, and extension in the field of soil and water conservation, watershed management, geostatistics, pedotransfer function, surface and groundwater modeling, remote sensing, and GIS. He is the recipient of various awards from Indian Council of Agricultural Research, New Delhi, Indian Association of Soil and Water Conservationists, Dehradun, SADHNA, Solan, ICAR-IISWC, Dehradun, CCSHAU, Hisar, etc. Dr. Adhikary has more than 100 publications to his credit which include research papers in journals of national and international repute, books, book chapters, technical and extension bulletins, popular articles, technology brochures, e-publications, etc.

Modeling of Gully Erosion Based on Random Forest Using GIS and R

Amiya Gayen, Sk. Mafizul Haque, and Sunil Saha

Abstract

Generally, gully erosion and its areal extension is a natural process fully controlled by external forces and shaped by internal settings. It adversely impacts soil productivity, eco-system function, and quality of environment as it affects land and water quality. For the development of sustainable land utilization strategy, it is initially required to develop an effective management process. This study aims to develop a gully erosion potentiality map using a well-known machine learning algorithm, that is, random forest (RF) model in the River Bakulla basin area, Jharkhand, India. In this work, 12 gully erosion predisposing factors (i.e., altitude, plan curvature, slope length, land use, soil types, slope gradient, topographical wetness index, distance from river, drainage density, distance from road, and distance from lineament) were selected based on available data and literature review. Finally, the gully erosion susceptibility map (GESM) generated by the RF model and the output was validated by employing the unused gully locations with ROC curve. The predicted results reveal that RF model has high prediction accuracy; the AUC value was 91% at end of the analysis. RF-generated GESM can be a very useful tool for the management action and land improvement measures in initial stages of gully development, to protect the development of gully erosion.

Keywords

Random forest · Area under the curve · Gully erosion susceptibility · Machine learning model

A. Gayen (✉) · S. M. Haque
Department of Geography, University of Calcutta, Kolkata, West Bengal, India

S. Saha
Department of Geography, University of Gour Banga, Malda, West Bengal, India

3.1 Introduction

Gully erosion and its interrelated land degradation process are most important in rapid erosion areas of the river basin. River basins are dynamic in nature that can change watershed boundary by the headward erosions and anthropogenic activities like unplanned agriculture, grazing activity, infrastructural development, and deforestation. Generally, gully erosion is denoted as the most intense land degradation over short periods of time (Torri and Borselli 2003). Today, majority of the country population is suffering from huge soil degradation problems that have affected agriculture productivity and acreage of agriculture by degrading nutrients from topsoil (Hoyos 2005; Arekhi et al. 2012; Gayen and Saha 2018). The soil erosion is a severe geo-environonmental problem in various parts of India: the north-eastern states, Himalayan ranges, Western Ghats, and Jharkhand together constitute 45% (130 Mha) of the total geographical area, which is affected by serious soil erosion through gullies, ravines, and shifting agriculture (Narayan and Babu 1983).

In a basin area, gully erosion extends the channels larger than rill by which 10–94% erosion occurs in respect to total erosion (Poesen et al. 2003). To minimize these problems, there is a need to find out the magnitude and spatial distribution of gully erosion. Many studies have been done by researchers on gully erosion susceptibility mapping using RF model (Chen et al. 2018). Earlier gully erosion mapping was based on the direct field survey, which was more expensive and time-consuming process (Poesen et al 2003). And the outputs were mainly micro-scale mapping. In recent times, remote sensing, GIS, along with other earth observatory techniques are extensively applied to map the GESM in a large area with a short period of time and as a low cost-effective measure (Zabihi et al. 2018).

In the previous research work, GIS-based model has been successfully applied to identify the GESM with high prediction rate (Conoscenti et al. 2014; Pourghasemi et al. 2017; Gayen and Pourghasemi 2019). Since the last three decades,

© Springer Nature Switzerland AG 2020
P. K. Shit et al. (eds.), *Gully Erosion Studies from India and Surrounding Regions*, Advances in Science, Technology & Innovation, https://doi.org/10.1007/978-3-030-23243-6_3

combined studies of RS and GIS techniques have been employed in different aspects for GESM (Conoscenti et al. 2008; Conforti et al. 2010; Buttafuoco et al. 2011; Gayen and Saha 2017; Zabihi et al. 2018). In addition, several statistical techniques have been employed along with RS and GIS techniques for the gully erosion potential mapping such as frequency ratio (Zabihi et al. 2018), weights of evidence (Dube et al. 2014), bi-variate statistical models (Rahmati et al. 2016), and probabilistic approach (Svoray et al. 2012), but they were special-purpose inventory.

The main objective of this present study is to determine the gully erosion susceptibility zone and its mapping in the River Bakulla basin, Jharkhand, India, using GIS-based modeling. The outcomes of this study will help decision-makers, government agencies, and the private sector for sustainable land-use management practices in the study area.

3.2 Material and Methods

The present study has employed different thematic layers, later which have transformed into the spatial database in ArcGIS environment. Then, the RF model has been used to analyze the relationship between gully locations and gully erosion-influencing factors; finally, gully erosion potential map has been prepared using the RF model. The accuracy of the models has been examined using the 'area under the curve' (AUC) tool.

3.2.1 Study Area

River Bakulla basin is a fourth-order sub-watershed of the River Ajay (Fig. 3.1). It lies between 24°16′23″N and 24°20′11″N and 86°30′35″E and 86°37′18″E. The elongated river basin has an area of about 37 sq. km. Geologically, the upper portion of this watershed is dominated by the Panchet formation with sandstones, granite gneiss, and clays. This study area is a part of Chotanagpur Plateau.

Gully erosion Inventory Map has been analyzed very carefully for potential mapping to explore the relationship between gully point locations and conditioning factors of gully erosion. In the inventory map of the study area, different thematic maps have been prepared (Fig. 3.1) using satellite image and published maps. A total of 543 gully locations have been identified in the study area from topographical maps and extensive field study has been carried out in all locations. Random partition algorithm has also been deployed to separate the gully points for training and validation purpose, where 380 (70%) gullies have been preferred for training and the other 163 (30%) for validation of dataset.

3.2.2 Affecting Factors Related to Gully Erosion

It is essential to select gully erosion effective factors for creating a gully erosion susceptibility mapping. Based on

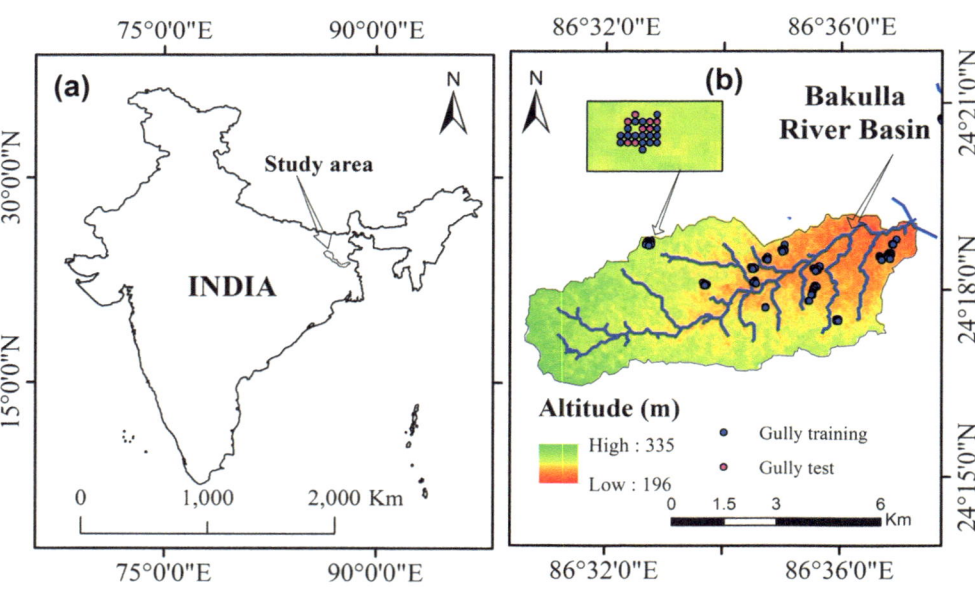

Fig. 3.1 Study area with gully training and gully test dataset (Source: Prepared by authors)

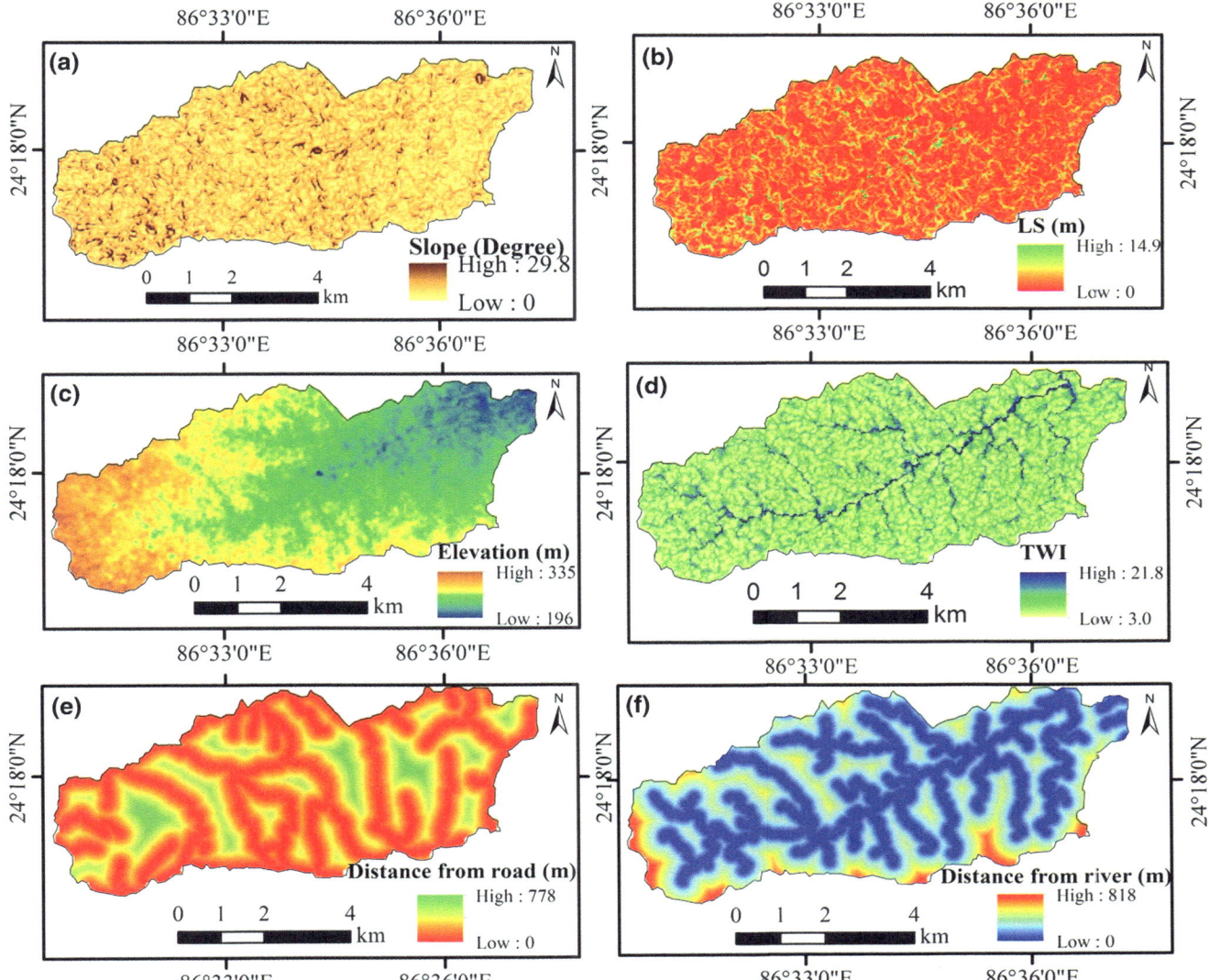

Fig. 3.2 Conditioning factor maps of gully erosion and its variables: (**a**) slope, (**b**) LS, (**c**) elevation, (**d**) TWI, (**e**) distance from roads, (**f**) distance from river, (**g**) land-use land cover, (**h**) plan curvature, (**i**) soil types, (**j**) slope aspect, (**k**) distance from lineaments, and (**l**) drainage density (Source: Prepared by authors)

the previous studies (Svoray et al. 2012; Rahmati et al. 2016; Zabihi et al. 2018) and field examination, 12 effective factors of gully erosion have been employed for GESM in the study area. These factors are elevation, slope, aspect, slope length (LS), plan curvature, distance from the river, distance from the lineament, soil, land use, drainage density, distance from road, and Topographical Wetness Index (TWI). All the thematic layers were converted into the spatial database using ArcGIS 10.2 software.

A 30 m resolution SRTM digital elevation model (DEM) was employed to prepare the elevation, slope, slope length, aspect, and plan curvature maps. Landsat 8 OLI data were used to create the land use and distance from lineament

maps. Drainage map has been developed from the topographical maps in 1:50,000 scale.

The 'elevation' parameter along with the local slope is considered by many researchers (Hongchun et al. 2014; Gomez Gutierrez et al. 2009) for gully erosion-affecting factor. The elevation of this study area varies from 196 m to 335 m from the mean sea level (Fig. 3.2c). Slope played an essential role in formation of gully development. The extracted slope angle in the study region ranges from 0° to 29° (Fig. 3.2a). The gully erosion rate varies with a varying slope angle of the region. The direction of slope is one of the most important controlling factors for the GESM (Rahmati et al. 2016; Pourghasemi et al. 2017). It defines which

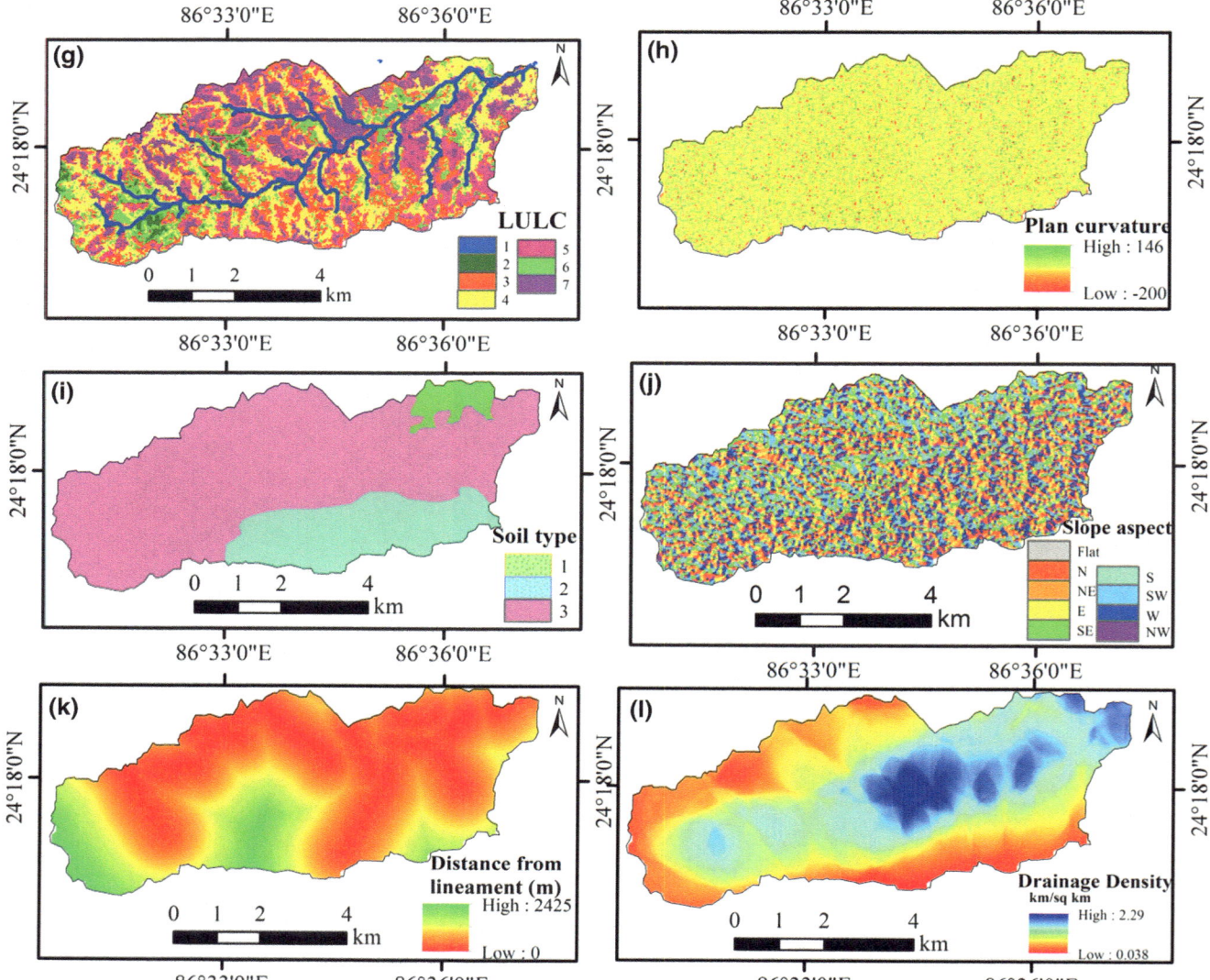

Fig. 3.2 (continued)

directional land segments will be best associated with the exposure to sunlight, winds, lineament, and rainfall. Aspect map was created in geospatial tool to correlate the gully erosion at different aspects of the slope (Fig. 3.2j). Slope length (*LS*) defines the rational expression of length (*L*) and gradient of slope (*S*) of the topography that influences the gully erosion (Gayen and Pourghasemi 2019). The result of *LS* has been calculated with the help of the following equation (Moore and Burch 1986):

$$LS = (\text{fa} \times \text{cell size}/22.13)^{0.4} \times (\sin\theta/0.0896)^{1.3} \quad (3.1)$$

where fa is defined as flow accumulation and θ represents the slope in degrees.

Curvature is a key factor that influences the surface and subsurface hydrology (Regmi et al. 2016). Curvature is the rate of change of slope gradient where profile curvature is parallel to the maximum slope in a particular direction (ArcGIS for Desktop). Plan curvature defines the maximum slope in a perpendicular direction. It has described the convergence and divergence of water flow in the earth surface. Negative value describes the concave slope in the surface which shows the convergence of water flow and positive values illustrate the convex slope of the surface that indicates the divergence of water flow in the region (ArcGIS). TWI defines the effect of topography on the moisture condition and its location related to soil situation of the area. TWI is calculated based on the following equation (Moore et al. 1991):

$$\text{TWI} = \text{In}(A/\tan\beta) \qquad (3.2)$$

where A is defined as the flow accumulation and β is the slope angle at the point.

Distance from the rivers also contributes in this study for the assessment of the gully erosion potentiality of the study area. In general, less distance contributes more gully erosion potential of an area because it works as a driver of eroded materials. Surface lineaments represent weak zones, which are of high permeability, and low in strength, which influence the slope stability and increase the gully erosion. These are linear or curvilinear patterns on the earth surface and identified from the satellite imagery (Magesh et al. 2012; Rahmati et al. 2016). Lineaments are extracted from the superimposed shaded relief maps at an interval of 45° azimuth angle. High lineament indicates more gully erosion potentiality in the area (Magesh et al. 2012). Euclidean distance method has been applied to examine the relationship between gully occurrences and distance from river, roads, and lineaments. Soil plays a significant role in infiltration, recharge, runoff, and soil moisture of the gully erosion. Soil map from National Bureau of Soil Survey and Land Use Planning was used to reclassify (Fig. 3.2) the soil into three categories: silt clay loam (marked as 3), silt loamy soil (marked as 2), and clay loamy soil (marked as 1). Land use influences the surface components such as infiltration, runoff, recharge, and soil moisture (Balamurugan et al. 2017). Unsupervised classification and iso-cluster algorithm were also used to produce the land-use map from the Landsat 8 OLI image. The study area has been classified into seven major land-use classes: water bodies (marked as 1), dense forest (marked as 2), built-up area (marked as 3), agriculture land (marked as 4), barren land (marked as 5), shrub forest (marked as 6), and fallow land (marked as 7). The accuracy of the land-use classification has been calculated as 86%, using Kappa index algorithm.

3.2.3 Methods

In this work, three machine learning model RFs were employed for gully erosion potential mapping for the study area. The algorithms of the models were compatible with R (R Development Core Team 2005) software, version 3.2.4. To run the model successfully, 380 gully points have been adopted using random partition algorithm in ArcGIS environment. The raster values of 12 gully erosion-affecting factors of each gully location have been imported to R software then; the models have been simultaneously adopted according to the algorithm of the models. The final output values of the models have transformed into a spatial dataset

for GESM classification using the natural break statistics in ArcGIS tool.

3.2.3.1 Random Forest

Random forest (RF) is a popular supervised machine learning technique for both classification and regression tasks (Breiman 2001; Youssef et al. 2016; Naghibi et al. 2017). For classification and regression of the datasets, a decision tree is created to get the output of the class and to obtain the dependent variable, respectively (Kim et al. 2018). This method consists of multiple decision trees and merges them together to explain the spatial relationship between gully occurrence and gully erosion-affecting factors (Gayen and Pourghasemi 2019). RF classification used the resampling technique by randomly changing the predictive variables to increase the diversity in each tree (Youssef et al. 2016; Naghibi et al. 2017). The predictive variable is expressed by log $2(M + 1)$, where M denotes the number of inputs to the algorithm (Kim et al. 2018). In this model, the mean-square error was calculated by Kim et al. (2018)

$$\varepsilon = \left(V_{\text{observed}} - V_{\text{response}}\right)^2 \qquad (3.3)$$

where ε represent the mean-square error of the algorithm, V_{observed} is the observed data of the variable, and V_{response} is the result of the variable.

RF algorithm was used for a number of trees and predictive variables to regulate the split at each node (Naghibi et al. 2017). The average prediction of the tree is computed as

$$S = \frac{1}{K} \sum K^{\text{th}} v^{\text{response}} \qquad (3.4)$$

where S denotes any forest prediction and K represents the individual trees in the model. Using these guidelines, the RF model has been run into R software.

3.2.3.2 Validation of Gully Erosion Potential Map

Validation is a fundamental step in mode-building approach for the scientific significance of the research (Chen et al. 2019; Naghibi et al. 2016). In this research, one popular and widely accepted index namely area under the receiver operating characteristic curve i.e., ROC has been implemented for evaluation of the model. ROC is a graphical plot which determines the performance of the model in a diagnostic test (Egan 1975; Golkarian et al. 2018). The curve plots the 'sensitivity' of the model (predicts the erosional feature as 'gully') on Y-axis and '1-specificity' (which predicts the feature as 'non-gully') on X-axis (Youssef et al. 2016; Golkarian et al. 2018). Model prediction for occurrence and nonoccurrence of gully locations was evaluated using the

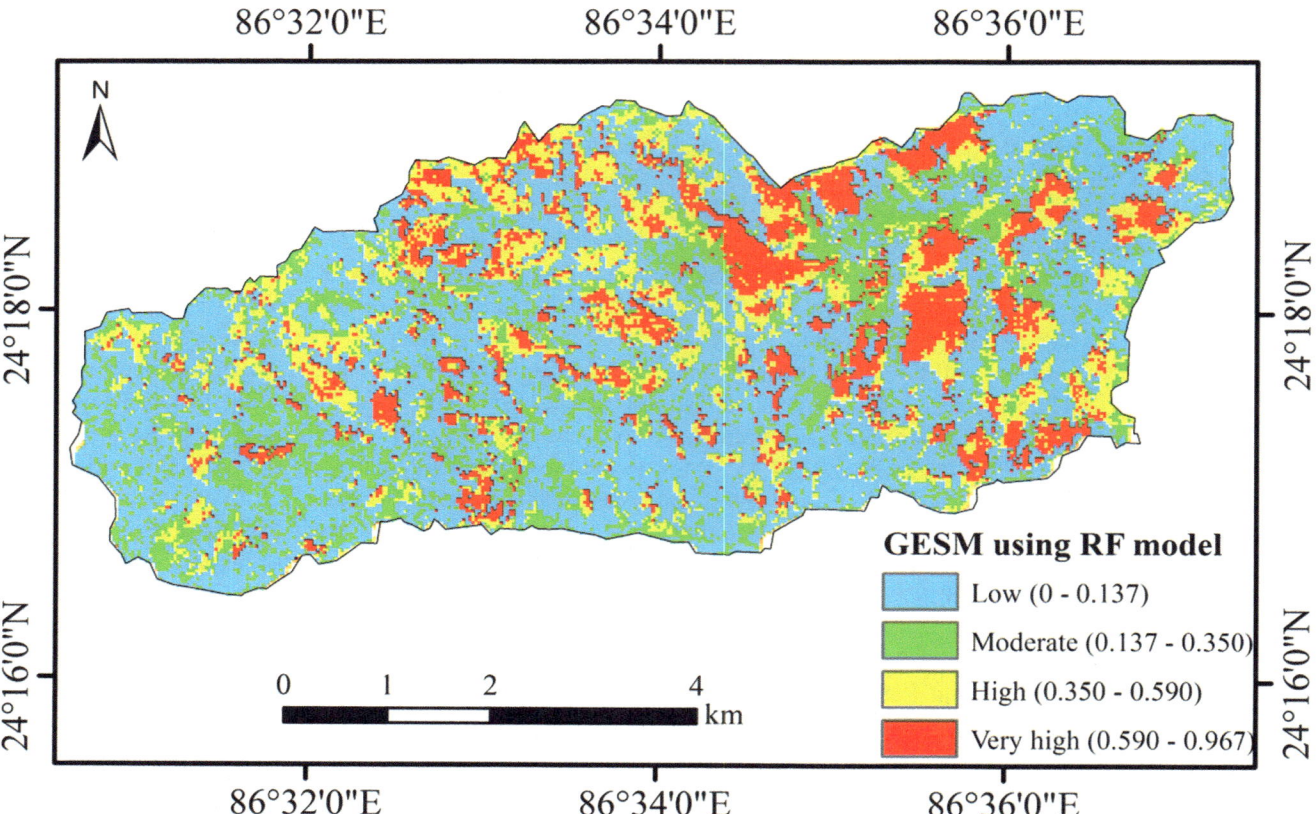

Fig. 3.3 Model output on gully erosion susceptibility zones and its spatial distribution (Source: Prepared by authors)

area under the ROC curve. AUC (area under curve) represents the value in between 0.5 and 1, and the higher value represents a better performance of the model (Youssef et al. 2016; Naghibi et al. 2016; Golkarian et al. 2018; Chen et al. 2018, 2019). To determine the most influencing factors, the final result has been examined by mean decrease in Gini assessment.

3.3   Results

3.3.1   Gully Erosion Potential Zone Models (GEPM)

Gully erosion potential zone map was prepared using the RF machine learning algorithms (Fig. 3.3). The model result was classified into four categories: very high, high, moderate, and low, based on natural break method. The higher value represents the very high gully erosion potential zones and vice versa. The GEPM, prepared from the RF model, shows the values from 0.0 to 0.967. According to the RF model, the high and very high gully erosion potential zone covered

13.19% and 8.04% of the study area, respectively, whereas, low and moderate gully erosion potential zones made up 21.39% and 57.38% of the total area (Table 3.1).

3.3.2   Validation of Machine Learning Model

The validation of the models is very important for the assessment of gully erosion potential zone mapping. ROC curve has been employed to validate the results of RF model. In many studies, ROC curve was used for quantitative validation of the models with high prediction rate (Golkarian et al. 2018; Chen et al. 2018). Based on ROC result, the validation of RF model is 91% (Fig. 3.4) with field verification (Fig. 3.5). RF model provides better result due to its interaction ability between effective factors and nonlinearities (Catani et al. 2013). This model also provides good predictive results in other fields of research like ecology, wildlife, spring potential assessment, and landslides mapping (Naghibi et al. 2015). Moreover, the RF model is promising and sufficient to be advised as a method to prepare groundwater susceptibility map at regional scale.

Table 3.1 Geo-environmental characteristics of GESM zones in the study area

GESM zone	Index value	Area (in %)	Leading process	Geo-environmental settings	Affected area (locality wise)
Low	0–0.137	57.38	Surface cracking, sheet erosion (in low elevated area), initiation of channels (in high area)	Soil formed in pocket, exposed bedrock, patches of vegetation and bushes, mainly found in high elevated area, moderate to gentle slope	Madhupur railway station area, Parwaria, Salia, Dulampur, Chapri, Singha, Domohani, etc.
Moderate	0.137–0.350	21.39	Sheet erosion and initiation of rills	Area of break of slope, less vegetated cover, red soil, and laterite dominancy	Durgapur, Parwaria, Lalpur, and its surroundings
High	0.350–0.590	13.19	Land subsidence, head-ward erosion, topple, etc.	Moderate to steep slope, area of leaching and percolation, presence of less vegetation, fragmented forest patches, seasonal cultivation	Dhab, Jhumka, Raghunathpur, Paharpur
Very high	0.590–0.967	8.04	Collapsing of gully wall, channel development, toe erosion	Area of high runoff efficiency, flow-dominated area, degraded forest	Bakulia, Dalha, Dubra, Mordih

Source: Composed by authors based on model output and field observation, 2016–2019

Fig. 3.4 Model validation result of GESM using validation data set (Source: Prepared by authors)

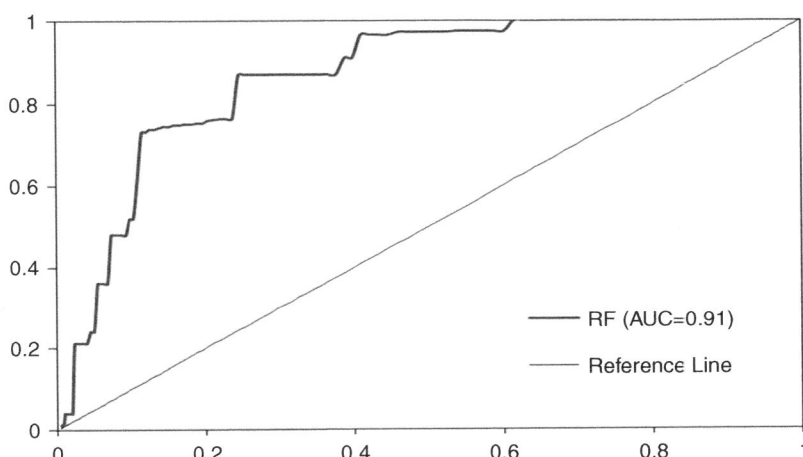

3.3.3 Important Effective Factors for GESM

The importance of the 12 effective factors of gully erosion has been illustrated using mean decrease Gini. In this context, drainage density, distance from lineament, LULC, distance from road, and soil types have the highest importance, while LS, plan curvature, and slope aspect have the moderate importance (Fig. 3.6).

In earlier studies carried out by Gayen and Saha (2017) and Gayen et al. (2019a, b) regarding the gully erosion susceptibility mapping, results reveal that the most important factors are the soil types, lineament, and LULC. This work also provided the same results in the abovementioned study area. The importance of drainage density, LULC, and soil types has matched with the result of mean decrease Gini. However, the properties of the study area and methods have decided the important effective factors for GESM in this research work.

3.4 Conclusion

The increasing nature of gully erosion-affected area decreases the productivity of land, and thus the gully erosion environment creates a concern for suitable GESM, especially in Jharkhand, a state of India. For accurate assessment of gully erosion potential zone mapping, different methods have been employed in various parts of the world. In this research, RS and GIS based machine learning algorithms have been applied to demarcate the suitable gully erosion potential zones in Bakulla watershed. Based on literature review and field knowledge, 12 gully erosion-effective factors such as elevation, slope, aspect, LS, plan curvature, TWI, distance from river, distance from lineament, distance from roads, drainage density, soil, and land use were overlaid with gully occurrence in GIS platform and integrated the result with RF models. According to RF model results, very high,

Fig. 3.5 Field measurement of gully erosion

Fig. 3.6 Comparative assessment of influencing factors in GESM

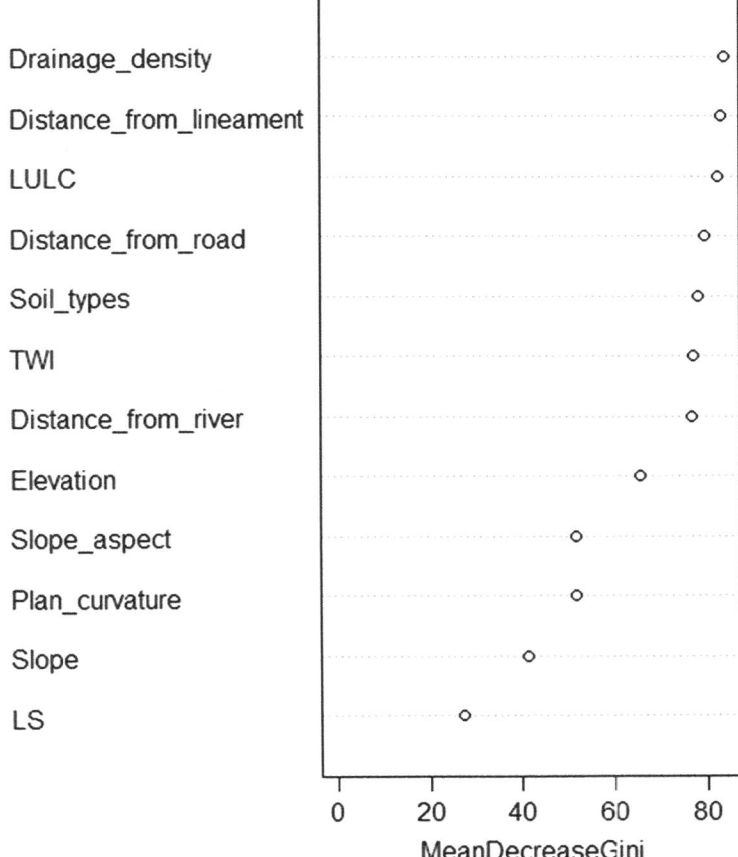

high, moderate, and low GESZ covered 8.04%, 13.19%, 21.39%, and 57.38%, respectively, in the study area. The validation of prediction of the models is portrayed in this article with the help of the ROC curve. The validation results of the RF model are highly efficient evaluation in the study place. The three most important factors like drainage density followed by distance from lineaments and LULC are the most influential for the high gully erosion potential zone. The obtained results of the present study can be useful for government, planner, decision-makers, and private agencies for land resource management, land-use planning, and environmental protection of the Bakulla watershed in Jharkhand, India.

References

Arekhi, S., Niazi, Y., Kalteh, A.M., 2012. Soil erosion and sediment yield modelling using RS and GIS techniques: a case study, Iran. Arab J Geosci 5, 285–296.

Balamurugan, G., Seshan, K., Bera, S., 2017. Frequency ratio model for groundwater potential mapping and its sustainable management in cold desert, India. Journal of King Saud University Science, 29 (3), 333–347.

Breiman, L., 2001. Random forests. Machine learning, 45(1), 5–32.

Buttafuoco, G., Conforti, M., Aucelli, P.P., Robustelli, G., Scarciglia, F., 2011. Assessing spatial uncertainty in mapping soil erodibility

factor using geostatistical stochastic simulation. Environmental Earth Sciences 66 (4), 1111–1125.

Catani, F., Lagomarsino, D., Segoni, S. Tofani, V., 2013. Landslide susceptibility estimation by random forests technique: sensitivity and scaling issues. Natural Hazards and Earth System Sciences, 13 (11), 2815–2831.

Chen, W., Chen, W., Li, H., Hou, E., Wang, S., Wang, G., Peng, T., 2018. GIS-based groundwater potential analysis using novel ensemble weights-of-evidence with logistic regression and functional tree models. Science of the Total Environment, 634, 853–867.

Chen, W., Panahi, M., Tsangaratos, P., Shahabi, H., Ilia, I., Panahi, S., Ahmad, B.B., 2019. Applying population-based evolutionary algorithms and a neuro-fuzzy system for modeling landslide susceptibility. Catena 172, 212–231.

Conforti, M., Aucelli, P.P., Robustelli, G., Scarciglia, F., 2010. Geomorphology and GIS analysis for mapping gully erosion susceptibility in the Turbolo stream catchment (Northern Calabria, Italy). Natural Hazards 56 (3), 881–898.

Conoscenti, C., Angileri, S., Cappadonia, C., Rotigliano, E., Agnesi, V., Ma̋rker, M., 2014. Gully erosion susceptibility assessment by means of GIS-based logistic regression: a case of Sicily (Italy). Geomorphology 204 (1), 399–411.

Conoscenti, C., Maggio, C.D., Rotigliano, E., 2008. Soil erosion susceptibility assessment and validation using a geostatistical multivariate approach: a test in Southern Sicily. Natural Hazards 46 (3), 287–305.

Dube, F., Nhapi, I., Murwira, A., Gumindoga, W., Goldin, J., Mashauri, D.A., 2014. Potential of weight of evidence modelling for gully erosion hazard assessment in Mbire District—Zimbabwe. Phys. Chem. Earth 67, 145–152.

Egan, J.P. Signal detection theory and ROC-analysis. New York: Academic Press. 1975.

Gayen, A., Pourghasemi, H.R., 2019. Spatial Modeling of Gully Erosion: A New Ensemble of CART and GLM Data-Mining Algorithms. Spatial Modeling in GIS and R for Earth and Environmental Science. pp. 653–669.

Gayen, A., Saha, S., 2017. Application of weights-of-evidence (WoE) and evidential belief function (EBF) models for the delineation of soil erosion vulnerable zones: a study on Pathro river basin, Jharkhand, India. Modeling Earth Systems and Environment 3 (3), 1123–1139.

Gayen, A., Saha, S., 2018. Deforestation probable area predicted by logistic regression in Pathro river basin: a tributary of Ajay River. Spatial Information Research 26 (1), 1–9.

Gayen, A., Saha, S., Pourghasemi, H.R., 2019a. Soil erosion Assessment using RUSLE model and its Validation by FR probability model. Geocarto International. doi: https://doi.org/10.1080/10106049.2019.1581272

Gayen, A., Pourghasemi, H.R., Saha, S., Keesstra, S. Bai, S., 2019b. Gully erosion susceptibility assessment and management of hazard-prone areas in India using different machine learning algorithms. Science of the Total Environment. 668, 124–138

Golkarian, A., Naghibi, S.A., Kalantar, B. Pradhan, B., 2018. Groundwater potential mapping using C5.0, random forest, and multivariate adaptive regression spline models in GIS. Environmental monitoring and assessment, 190(3), 149

Gomez Gutierrez, A., Schnabel, S., Felicisimo, A.M., 2009. Modelling the occurrence of gullies in rangelands of southwest Spain. Earth Surf. Process. Landforms 34, 1894–1902.

Hongchun, Z.H.U., Guoan, T., Kejian, Q., Haiying, L., 2014. Extraction and analysis of gully head of loess plateau in china based on digital elevation model. Chin Geographical Sci. 24 (3), 328–338.

Hoyos, N., 2005. Spatial modeling of soil erosion potential in a tropical watershed of the Colombian Andes. Catena 63, 85–108.

Kim, J.C., Lee, S., Jung, H.S. Lee, S., 2018. Landslide susceptibility mapping using random forest and boosted tree models in Pyeong-Chang, Korea. Geocarto international, 33(9), 1000–1015.

Magesh, N.S., Chandrasekar, N. Soundranayagam, J.P., 2012. Delineation of groundwater potential zones in Theni district, Tamil Nadu, using remote sensing, GIS and MIF techniques. Geoscience Frontiers, 3(2), 189–196.

Moore, I.D., Burch, G.J., 1986. Physical basis of the length-slope factor in the universal soil loss equation 1. Soil Science Society of America Journal, 50(5), 1294–1298.

Moore, I.D., Grayson, R.B. and Ladson, A.R., 1991. Digital terrain modelling: a review of hydrological, geomorphological, and biological applications. Hydrological processes, 5(1), 3–30.

Naghibi, S.A., Pourghasemi, H.R., Razaei, A., 2015. Groundwater qanat potential mapping using frequency ratio and Shannon's entropy models in the Moghan watershed, Iran. Earth Sci. Information 8, 171–186.

Naghibi, S.A., Pourghasemi, H.R. Dixon, B., 2016. GIS-based groundwater potential mapping using boosted regression tree, classification and regression tree, and random forest machine learning models in Iran. Environmental monitoring and assessment, 188(1), 44.

Naghibi, S.A., Ahmadi, K. Daneshi, A., 2017. Application of support vector machine, random forest, and genetic algorithm optimized random forest models in groundwater potential mapping. Water Resources Management, 31(9), 2761–2775.

Narayan, D.V.V., Babu, R., 1983. Estimation of soil erosion in India. J. Irrig. Drain. Engng. ASCE 109(4), 419-435.

Poesen, J., Nachtergaele, J., Verstraeten, G., Valentin, C., 2003. Gully erosion and environmental change: importance and research needs. Catena 50 (2–4), 91–133

Pourghasemi, H.R., Yousefi, S., Kornejady, A., Cerdà, A., 2017. Performance assessment of individual and ensemble data-mining techniques for gully erosion modeling. Science of The Total Environment 609, 764–775.

R Development Core Team, 2005. R: a language and environment for statistical computing. R Foundation for Statistical Computing, Vienna, Austria, www.Rproject.org

Rahmati, O., Haghizadeh, A., Pourghasemi, H.R., Noormohamadi, F., 2016. Gully erosion susceptibility mapping: the role of GIS-based bivariate statistical models and their comparison. Natural Hazards 82 (2), 1231–1258.

Regmi, N.R., Giardino, J.R., McDonald, E.V. and Vitek, J.D., 2016. A review of mass movement processes and risk in the critical zone of Earth. In Developments in Earth Surface Processes, 19, 319–362.

Svoray, T., Michailov, E., Cohen, A., Rokah, L., Sturm, A., 2012. Predicting gully initiation: comparing data mining techniques, analytical hierarchy processes and the topographic threshold. Earth surf. Process. Landforms 37, 607–619.

Torri, D., Borselli, L., 2003. Equation for high-rate gully erosion. Catena 50 (2–4), 449–467.

Youssef, A.M., Pourghasemi, H.R., Pourtaghi, Z. S., Al-Katheeri, M. M., 2016. Landslide susceptibility mapping using random forest, boosted regression tree, classification and regression tree, and general linear models and comparison of their performance at Wadi Tayyah Basin, Asir Region, Saudi Arabia. Landslides, 13(5), 839–856.

Zabihi, M., Mirchooli, F., Motevalli, A., Darvishan, A.K., Pourghasemi, H.R., Zakeri, M.A., Sadighi, F., 2018. Spatial modelling of gully erosion in Mazandaran Province, northern Iran. Catena 161, 1–13.

Amiya Gayen is a Ph.D. student, Department of Geography, University of Calcutta, Kolkata, India. He obtained his M.Sc. in Geography from Presidency University, Kolkata. His research field is soil geography, natural hazard, forest health assessment, and environmental geography including GIS and R techniques. Simultaneously, he is working as an Assistant Professor under RUSA in the Department of Geography, Midnapore College (Autonomous), Midnapore, West Bengal.

Sk. Mafizul Haque was awarded M.A. in geography and environment management from Vidyasagar University, Midnapore, in the year of 2004. He was an Assistant Professor in the Department of Geography, Aliah University, Kolkata, from 2012 to 2015. After he was awarded the Ph.D. degree in 2014 from the University of Calcutta, he joined the Department of Geography, University of Calcutta, as an Assistant Professor in 2015. His area of inclination is change science—mainly the application of geospatial technology on urban landscape, micro-climate, habitat analysis, urban environment, and palaeo-climate study. Dr. Haque has published more than 17 research papers in reputed national and international journals. He was engaged in the training program of SIPRD, Kalyani, Government of West Bengal, during March–June, 2015. Currently, he is coordinator of IIRS outreach program at the Department of Geography, University of Calcutta.

Sunil Saha working as Assistant Professor in the Department of Geography of University of Gour Banga, Malda, West Bengal. He is research interests are toward understanding fluvial landforms, surface and sub-surface hydrological conditions, and their impacts on land uses regarding both time and space. Currently, some articles on groundwater modeling, gully erosion, and soil erosion have been published from some reputed journals.

Geomorphic Threshold and SCS-CN-Based Runoff and Sediment Yield Modelling in the Gullies of Dwarka–Brahmani Interfluve, West Bengal, India

Sandipan Ghosh and Sanat Kumar Guchhait

Abstract

Gully erosion signifies instability in the landscape, and it is regarded as a threshold phenomenon under certain conditions in the landscape, relating to flow erosivity and surface resistance. The main cause of gully formation is too much water at a certain location of slope—a threshold condition that may be brought about by external factors or internal factors. Intense rainfall is the primary trigger, but the local conditions such as slope morphometry, land-use and soil characteristics control the triggering of gully erosion. The catchment size above a stream or gully head and the land-use characteristics determine the volume of overland flow but it is different from the concentrated-flow area by its position on the slope. In the present study, the models of geomorphic threshold and the M–D Envelope have emphasized on the role of overland flow (as surface runoff) for initiation of gullies in the laterite slope of Dwarka–Brahmani Interfluve, West Bengal (Neogene–Early Pleistocene lateritic terrain located in between the Rajmahal basalt traps and Bengal basin). The upstream laterite slopes above gully heads are negatively correlated ($r = -0.55$) with upstream drainage areas which are used as surrogate for the volume of runoff yield in the gullies of the study area. The calculated empirical trend line ($S = 17.419\ A^{-0.2517}$, with R^2 of 0.52) represents an approximation to critical slope–area threshold relationship for gully incision in this region. Then, the Soil Conservation Service-Curve Number (SCS-CN; now called the National Resource Conservation Service, NRCS-CN) method is used for quick and accurate estimation of surface runoff in any storm event in the ungauged watersheds of gullies. Experimenting in three sample watersheds of gullies, it is found that on the basis of rainfall range of 42–137.2 mm the gullies can yield runoff of 40.02–118.0 mm in excess moisture condition of monsoon. The differences of runoff in the catchments for the same rainfall event are the direct effects of land use–land cover derived from CN values. In the gully catchments, the estimated runoff is increased above 22% from Antecedent Moisture Condition (AMC) II to AMC III and in most gullies it ranges from 22.33% to 85.73%. It is understood that if the rainfall amount is increased day by day, the runoff coefficient (R_c) is also increased consecutively and it will be high runoff event which is the sign of high vulnerability of flow erosion on the bare laterite slope. The SCS-CN analyses reflect that in prolong rainfall event of tropical depression more than 86% of rainfall can be transformed into direct runoff, as R_c is reaching up to 0.86. Alongside, it gives more hydraulic energy to gully initiation and gully head migration in AMC III condition. Based on the daily rainfall-runoff modelling and annual potential erosion rate, the estimated sediment yield varies from 5.0 to 13.45 t ha^{-1} in the gullies. It is learnt that in prolong rainstorm event or in cyclonic rainfall, with increase of rainfall and moisture content of surface, the transport of eroded materials is increased and substantially, the sediment yield of gully catchment is also escalated.

Keywords

Gully erosion · Geomorphic threshold · Runoff · Sediment yield · Laterite

S. Ghosh (✉)
Department of Geography, Chandrapur College, Barddhaman, West Bengal, India

S. K. Guchhait
Department of Geography, The University of Burdwan, Barddhaman, West Bengal, India

4.1 Introduction

Land degradation and desertification issues are now milestone pillars of many international environmental and development agendas and soil erosion is now identified as key process of land degradation (Lal 1992). In different parts of

© Springer Nature Switzerland AG 2020
P. K. Shit et al. (eds.), *Gully Erosion Studies from India and Surrounding Regions*, Advances in Science, Technology & Innovation, https://doi.org/10.1007/978-3-030-23243-6_4

the world, the most intensified soil erosion is the gully erosion which is an extreme form of soil erosion and land degradation, affecting multiple soil and land functions through interconnected networks of narrow channels over the slope (Singh and Dubey 2002). Many governmental organizations of India have estimated the extent of land degradation in India and the value varies from 53.28 to 173.64 M ha. Land degradation due to soil erosion is a momentous hazard in India and gully erosion already engulfs about 3.975 million ha of land in India (Yadav and Bhushan 2002; Pathak et al. 2005; Singh et al. 2015). It is estimated that soil erosion takes place at the rate of 1.35 t ha^{-1} year^{-1} in India, and about 29% of total eroded soil is slot permanently to sea and 10% is deposited in the reservoirs (Narayana and Babu 1983; Sharda et al. 2010; Sharda and Dogra 2013). Singh et al. (1992) estimated that soil erosion took place at a rate of exceeding 40 t ha^{-1} year^{-1} in the ravines and badlands of India. In the humid subtropical region of India, soil erosion (about 15 million t year^{-1}) leads to low crop productivity and an annual loss of 13.4 million tonnes in the production of crops due to water erosion equivalent to about $2.51 billion (Sharda et al. 2010; Sharda and Dogra 2013). Understanding that emerging issue of soil erosion it is utmost necessity to identify the key factors of erosion and their estimation.

The research reveals that after the rainsplash detachment the second erosive factor is overland flow in catchment scale. Intense rainfall is the primary trigger, but the local conditions such as slope morphometry, land-use and soil characteristics control the triggering of gully erosion (Rossi et al. 2015). It is found that after a critical distance from the water divide, the gully head is formed because the depth of the overland flow increases with distance and it cumulatively increases shear stress on the surface to allow incision at a certain part of slope (Fig. 4.1) (Morgan 2005). After determination of thresholds, the estimation of runoff is crucial for understanding the energy factor of erosion. The location of gully head reflects the critical hydraulic condition where flow erosivity overcomes the erodibility (Toy et al. 2013). This critical condition can be explained and determined by the concept of threshold in a wide range of conditions. Before taking any steps or strategies to protect erosion, it is of the utmost necessity to recognize the dominant factors and processes of erosion in a particular land. Therefore, this chapter deals with, firstly, the topographic thresholds of gully initiation on the lateritic terrain and, secondly, the estimation of overland flow (as surface runoff) which is key factor for hydraulic threshold. In this study, the prime objective is setup to identify and estimate the geomorphic threshold of gully erosion and runoff flow erosivity in the lateritic badlands of Dwarka–Brahmani Interfluve (western part of the Bengal basin, India) through hydrogeomorphic quantitative techniques.

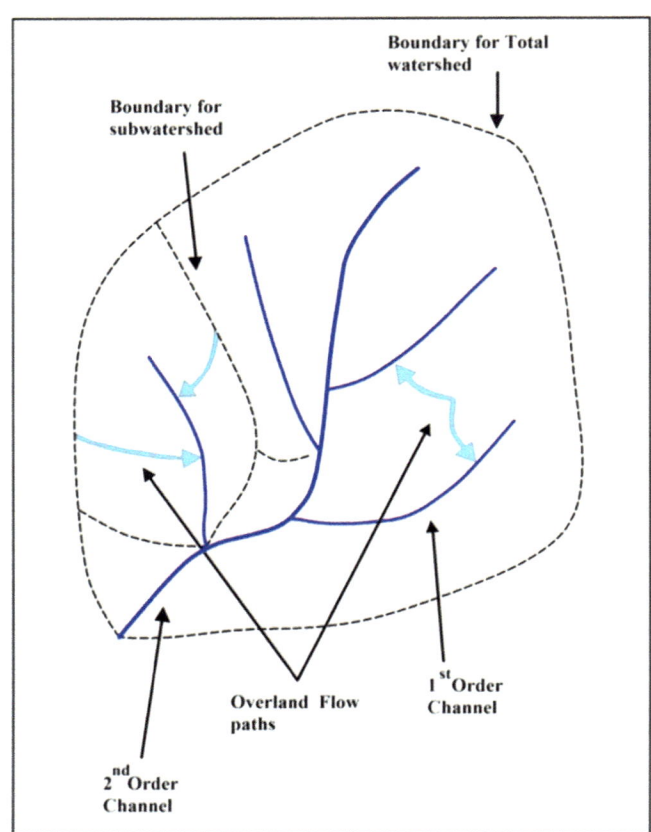

Fig. 4.1 Nested hierarchy of gullies and overland flow paths in a watershed

4.2 Materials and Methods

To carry on the proposed research operations, a full, well-equipped research design (Fig. 4.2) is always needed, which is flexible, appropriate, efficient, economical and scientific. Soil erosion study has an interdisciplinary outlook which incorporates geology, geomorphology, pedology and hydrology under the shade of one umbrella—pedogeomorphology. So, the framework of research design should be strong enough to make it viable for long term, adopting sound methodology of aforesaid disciplines.

4.2.1 Study Area

The selected study area of Dwarka–Brahmani interfluve (about 176 km^2, confined by 24°08′N to 24°14′N and 87°38′E to 87°44′E latitude and longitude respectively), covering Shikaripara block (Dumka, Jharkhand), Rampurhat I and Nalhati I blocks (Birbhum, West Bengal). The geomorphic unit of study area is recognized as an interfluve in

Fig. 4.2 Flow chart of methodology used in the research

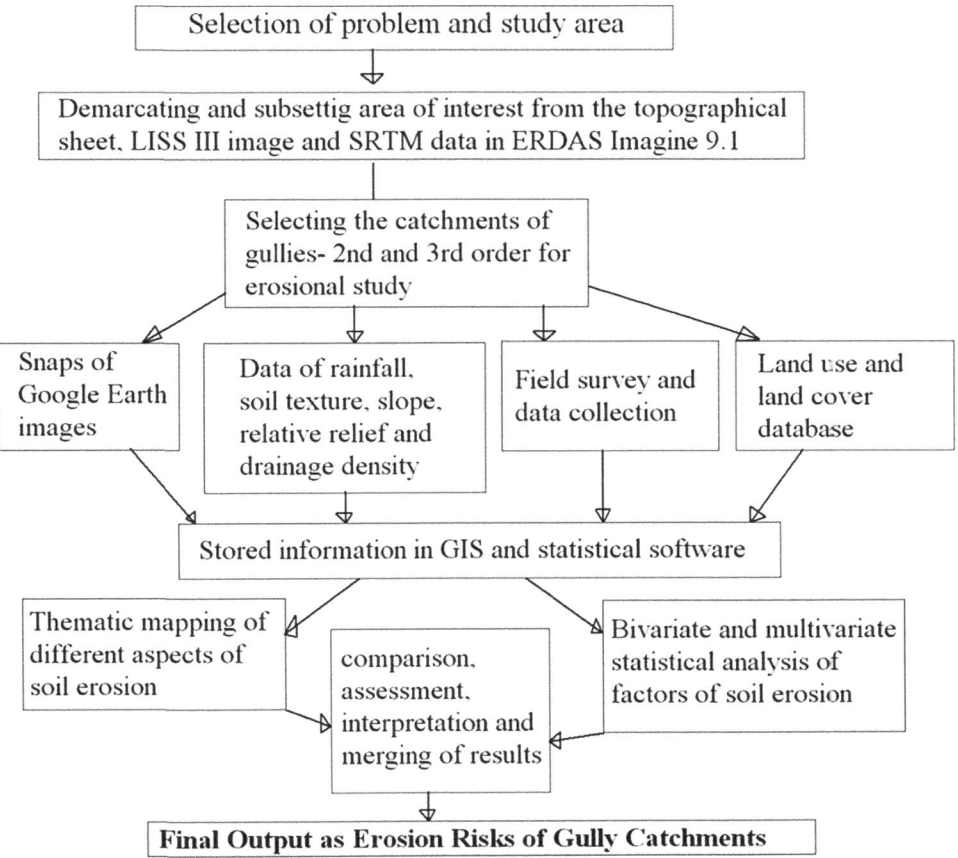

Selection of problem and study area

Demarcating and subsettig area of interest from the topographical sheet, LISS III image and SRTM data in ERDAS Imagine 9.1

Selecting the catchments of gullies- 2nd and 3rd order for erosional study

Snaps of Google Earth images

Data of rainfall, soil texture, slope, relative relief and drainage density

Field survey and data collection

Land use and land cover database

Stored information in GIS and statistical software

Thematic mapping of different aspects of soil erosion

comparison, assessment, interpretation and merging of results

Bivariate and multivariate statistical analysis of factors of soil erosion

Final Output as Erosion Risks of Gully Catchments

between Brahmani (north) and Dwarka (south) rivers (confined by 24°20′N to 23°40′N, and 87°26′E to 88°21′E) (Fig. 4.3). River Dwarka basin (2978 km²) is a sub-basin of River Mayurakshi basin and River Brahmani (1139 km²) is a sub-basin of River Dwarka. Both the rivers have originated from the Rajmahal Hills of Dumka District, Jharkhand, and River Dwarka flows through Birbhum and Murshidabad districts of West Bengal, where it joins with Mayurakshi to form River Babla that finally outfalls into River Bhagirathi. River Brahmani meets Dwarka near Nabagram, Murshidabad. Geomorphologically, the interfluve of Dwarka–Brahmani is associated with plateau proper and plateau fringe of Chotanagpur, having laterite exposures and basaltic hills, and also the northern part of the Rarh Plain (Biswas 1987). Most of the peninsular rivers flow from west to east direction, guided by the general basement slope towards the Ganga–Brahmaputra delta. Geologically, the present research work deals with the contiguous Early–Late Pleistocene unit between the Rajmahal basalt traps (RBT) (Early Cretaceous origin) and the Bengal basin (Late Pleistocene to recent origin) which exhibits shallow Quaternary alluvium deposits and palaeogenesis of the deep-weathering profiles under intense tropical wet–dry

palaeoclimate on the basaltic surface to form hard ferruginous crust, that is ferricrete.

Observing the slope morphology of gully catchments, it is found that the slope steepness above gully head (i.e. convex part) varies from 5° to 13° in the study area and high degree of slope favours flow convergence and rill/gully initiation. The elevation zones of study area vary from 20 m to 100 m from mean sea level (Fig. 4.4). The most of gullies are developed in between the elevation range of 80–100 m. In and around the study area, the soil series of Bhatina, Raspur and Jhinjharpur (Sarkar et al. 2007) have been developed in the present geo-climatic setting. Generally, the thin solum is loamy-skeletal and hypothermic in nature developing on the barren lateritic wastelands with sparse bushy vegetation and grass. These soil series has weak fine crumb and granular structure (slightly hard, friable and slightly sticky), 2–5 mm size of manganese nodules, >2 mm size of ferruginous nodules with goethite cortex, 30–40% gravels and pebbles, excessive drained surface and pH of 5.4–5.7. The climate of this region has been identified as sub-humid and subtropical monsoon type, receiving mean annual rainfall of 1300–1437 mm. The peak monsoon and cyclonic rainfall intensity of 21.51 mm h^{-1} (minimum) to 25.51 mm h^{-1}

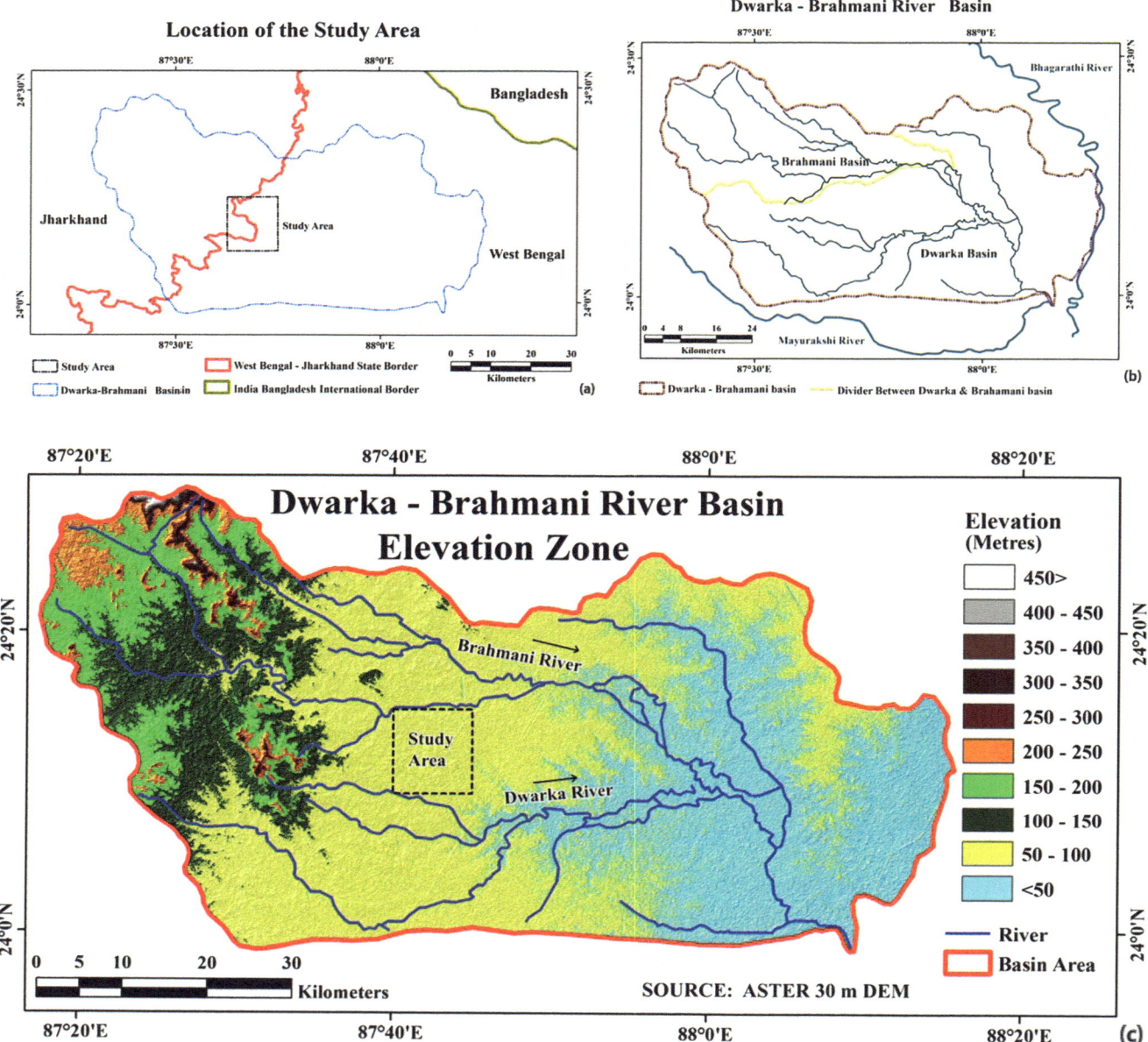

Fig. 4.3 Location map of study area and elevation zonation of Dwarka–Brahmani river basin

(maximum) is the most powerful climate factor to develop this lateritic badlands. The natural vegetation of the study area belongs to the tropical moist and dry deciduous type with few evergreen types. Though once upon a time the most of the region was covered under thick forest, mainly Sal (*Shorea robusta*); due to encroachment of stone crushers, mining and agriculture, the forests are fragmented and vanished from some places.

4.2.2 Secondary Data Collection

The key sources of main secondary data are regional soil report, geology report and other physical environmental report published by NBSS and LUP, Census of India, district gazetteer, official websites of IMD Pune and Kolkata, Irrigation and Waterways Dept. of Govt. of West Bengal, Geological Survey of India (GSI), related e-books and e-journals.

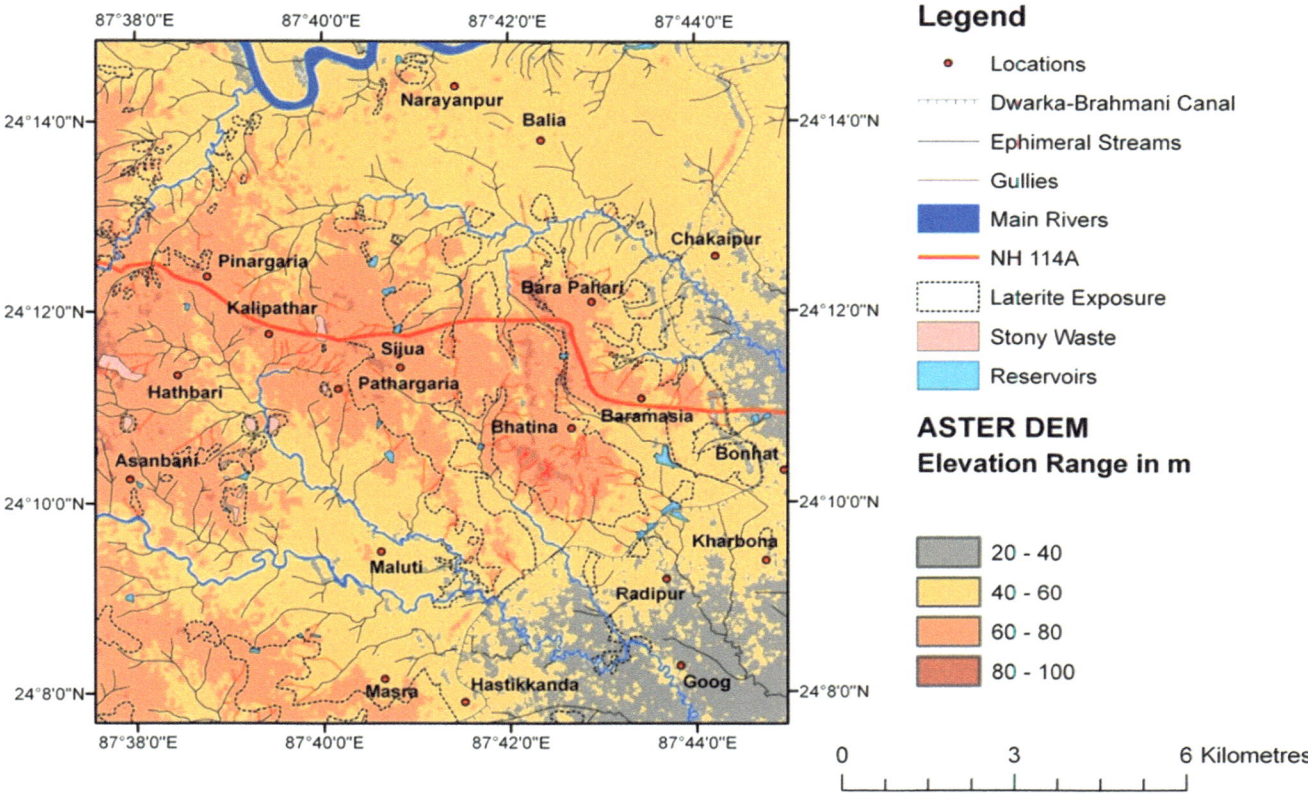

Fig. 4.4 Geomorphic map of study area to show the locations of gullies in different elevation zones

The topographical sheets of Survey of India (72 P/12/NE, R.F. 1:25,000 and 72 P/12, R.F. 1:50,000), District Resource Map of Geological Survey of India, District Planning Map of National Atlas Thematic Mapping Organization (NATMO) and block map of Census of India are most important sources of spatial information. Landsat TM and ETM+ (30 m resolution) images are downloaded from the website of Global Land Cover Facility (GLCF) and Shuttle Radar Topography Mission (SRTM; 90 m resolution) and Advanced Spaceborne Thermal Emission and Reflection Radiometer (ASTER; 30 m resolution) elevation data are downloaded from the websites of GLCF and Consortium for Spatial Information (CGIAR-CSI). The spatial information is stored in Geographic Information System (GIS) and the thematic maps are prepared using GIS software (ArcGIS 9.2, Erdas Image 9.1 and MapInfo Professional 11.5). The different statistical analysis (e.g. linear and curvilinear regression, correlation, principal component analysis, cluster analysis and multiple regressions) is done in Microsoft Excel 2007 and SPSS 14.0 software.

4.2.3   Field Research Design

4.2.3.1 Geomorphic Analysis

For in-depth research, first of all, it should be focused on the spatial scale or spatial unit of study. The drainage basin is universally considered as fundamental unit of geomorphic study. It is planned to select different orders (Strahler 1964) of gully catchments where the distinctiveness of land use–land cover, different profiles of laterites and soil erosion processes are easily analysed (Fig. 4.3). The spatial scale to study erosion threshold is here plot scale (10–100 m^2) and field scale (100–10,000 m^2). In terms of identifying the geomorphic thresholds in gully initiation, the present experimental work includes the 118 gully heads (both valley-floor and valley-side gullies) (Fig. 4.5). Using Garmin GPS (Global Positioning System) receiver 76csX (horizontal accuracy of ±3 m) it has positioned exact locations of slope facets, rills, gull heads and eroded features including up-to-date spatial information. Sprinter 150 m of Leica Geosystem (height accuracy ±1.5 mm, distance accuracy ±1 mm) was

Fig. 4.5 Spatial extent of gullies and sample locations of gully heads in the areas of (**a**) Maluti (24°09′45″N, 87°41′14″E) and (**b**) Bhatina (24°10′25″N, 87°42′33″E) (Google Earth imagery date: 13/01/2014), and filed photographs showing (**c**) collection of sediment at the base of gully head at Maluti, Jharkhand, (**d**) barren lateritic upstream landscape of gully-head catchment at Bhatina, West Bengal, and (**e**) downstream dissection of laterites by deeply incised gully and expansion of gully heads at Bhatina, West Bengal (Ghosh and Gucchait 2017)

Fig. 4.6 Cross and long-profile survey using Leica Sprinter 150 m at (**a**) V-shaped gully of catchment 1 and (**b**) wide U-shaped gully of catchment 3 (arrow signifies the direction of flow)

used to measure the angle of slope facets (Fig. 4.6). Alongside, in few cases (due to obstacles) from ASTER DEM (Digital Elevation Model), the slope length and angle (usually from gully headcut to water divide) are measured to judge the length of surface flow (responsible for gully erosion). Drainage area is calculated from the flow direction and flow accumulation algorithm of Arc GIS 9.3 using drainage lines (digitized from toposheets) and DEM.

4.2.3.2 Hydrologic Analysis

The curve number method requires a watershed/catchment scale and per day rainfall data of a permanent rain-gauge station, but the study area does not have any such station of India Meteorological Department (IMD), Central Water Commission (CWC) and Irrigation and Waterways Department (IWD). For that reason, we have chosen three rain-gauge stations of IWD (under the Government of West Bengal) adjacent to the study area (located in between 12 and 25 km) and these stations are placed in Nalhati (24°17′25″N, 87°49′44″E), Rampurhat (24°10′13″N, 87°46′50″E) and Mollarpur (24°04′35″N, 87°42′36″E) of Birbhum District (Table 4.1 and Fig. 4.7).

From the central IWD office of Rampurhat (the nearest gauged station of the field site), the data (2014–2016) of daily, monthly and yearly rainfall of three stations are collected and the tabulated for this investigation. As our field-based erosion study was conducted from January 2016 to January 2017, the daily rainfall database of year 2016 was chosen for the SCS-CN method. The extreme rainfall events greater than 40 mm day^{-1} are chosen to estimate runoff in a catchment. In this study, three sample catchments of permanent gullies are selected as gully catchment 1, gully catchment 2 and gully catchment 3 (Fig. 4.8). The runoff of each

Table 4.1 Summary of daily rainfall database, 2016 in the rain-gauge stations of Nalhati, Rampurhat and Mollarpur to show maximum rainfall greater than 40 mm day^{-1}

	Sl. no.	Date	Rainfall (mm)
Nalhati	1	13-06-2016	58.8
	2	25-06-2016	61.4
	3	06-07-2016	68.2
	4	16-07-2016	125
	5	20-07-2016	60.6
	6	22-07-2016	58.0
	7	05-08-2016	42.0
	8	11-08-2016	43.8
	9	26-08-2016	43.8
	10	09-10-2016	57.6
Rampurhat	10	13-06-2016	107.4
	11	03-06-2016	78.2
	12	22-06-2016	67.6
	13	30-06-2016	57.6
	14	08-07-2016	61.4
	15	16-07-2016	85.8
	16	30-07-2016	57.6
Mollarpur	17	13-06-2016	137.2
	18	24-06-2016	56.4
	19	03-07-2016	46.2
	20	06-07-2016	80.2
	21	16-07-2016	130.2
	22	17-07-2016	130.2
	23	21-07-2016	56.4
	24	11-07-2016	40.3
	25	05-08-2016	57.4
	26	06-08-2016	83.1
	27	30-07-2016	45.2

Source: Central Office, IWD, Govt. of West Bengal, Rampurhat Division of Birbhum district

Fig. 4.7 Location of IWD rain-gauge stations at Nalhati, Rampurhat and Mollarpur and their proximity to study area (map source: Google Earth Imagery 2016)

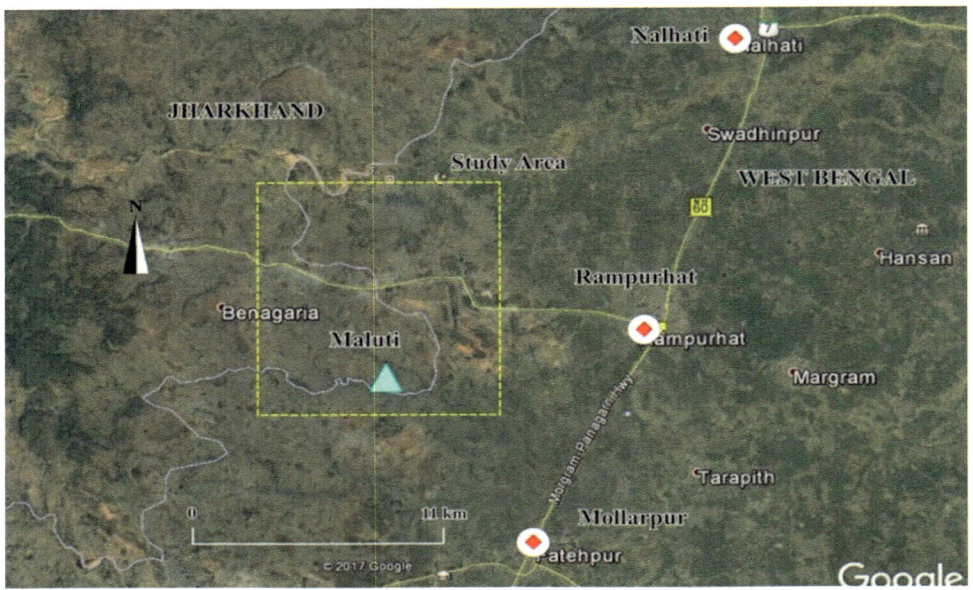

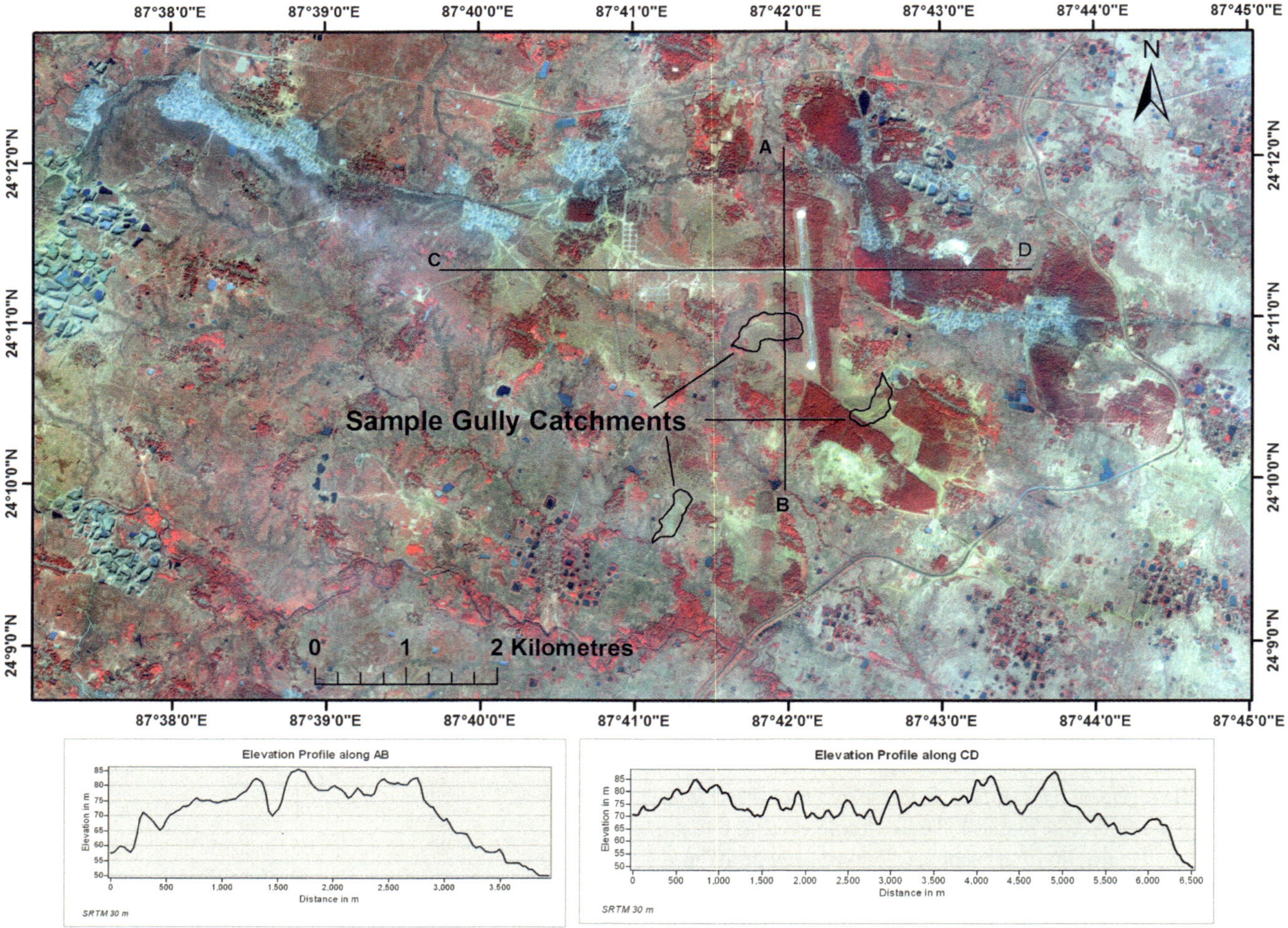

Fig. 4.8 Standard False colour composite IRS LISS IV image (Dec 2015) of a portion of study area showing location of sample gully catchments and ASTER DEM elevation profile

event (using database of three stations) is calculated in the AMC II and AMC II conditions of each gully catchment, because we have also used the data of peak monsoon season (July–October), that is high moisture saturation condition of soils. The mapping of gully catchment has been done using GPS field survey and Google Earth imagery. The land use and land cover map is prepared using IRS P6 LISS IV satellite image (2015) and Erdas Imagine 10.1 version, along with GPS ground truth verification.

4.2.4 Quantitative Models and Techniques

4.2.4.1 Geomorphic Threshold Model

Geomorphic thresholds can be defined in terms of ratios, the numerator and denominators of which describe opposing tendencies and which may be simple or complex depending on the needs of the investigator or the complexity of the real world (Bull 1980; Coates and Vitek 1980). The part of the system under consideration may be considered to be at a threshold or equilibrium condition when the ratio is equal to 1.0. When the derived value exceeds 1.0, a threshold has been reached. A threshold that described changes in dominant hydro-geomorphic processes is the hillslope runoff threshold (Bull 1980):

Factors that promote runoff/Factors that promote infiltration = 1.0.

Probably, at first Horton (1945) explained the mechanism of channel initiation on a hillslope where after a critical distance from the divide the overland flow exceeds threshold condition to incise a channel (Fig. 4.9). In western Colorado, Patton and Schumm (1975) have reported on a relation between drainage area and valley floor slope above which incision of the valley floor is likely to take place.

In general, thresholds for gully head position in the landscape traditionally take into consideration local slope angle (as topographic variable) and gully head drainage area (as runoff variable) (Rossi et al. 2015). There are wide ranges of threshold conditions or values (viz., thresholds of hydraulic, rainfall, topography, lithology and land use–land cover control, etc.) which are responsible for the initiation of gullies in different environments (Horton 1945; Patton and Schumm 1975; Begin and Schumm 1984; Ebisemiju 1989; Vandaele et al. 1996; Vandekerckhove et al. 1998; Moeyersons 2003; Morgan and Mngomezulu 2003; Poesen et al. 2003; Montgomery and Dietrich 2004; Valentin et al. 2005; Samni et al. 2009; Dong et al. 2013; Torri and Poesen 2014; Ghosh and Gucchait 2017).

The relation between critical valley slope and drainage basin area ($S = aA^{-b}$, where a = coefficient and b = exponent

of relative area) is used as a predictive model to locate those areas of instability within alluvial valleys where gullies will form.

$$S = aA^{-b}.$$

The idea of taking critical slope as threshold reveals that gully incision demands a minimum runoff discharge in the function of slope (Moeyersons 2003). A threshold line is drawn through the lower limit of scatter points and this line represents, for a given area, a critical value for valley slope above which entrenchment of the laterite should occur. This relationship can be written as $SA^b > T$ (where T = threshold value, i.e. area^b), defining the limit of threshold value to start gully initiation (Morgan and Mngomezulu 2003; Torri and Poesen 2014). A theoretical division of the landscape into process regimes in terms of log S (X axis) and log A (Y axis) signifies different geomorphic thresholds to gully erosion and the resultant critical threshold line is demarcated as Montgomery–Dietrich (M–D) envelope, through A–S threshold (Montgomery and Dietrich 1988, 1992; Moeyersons 2003; Samni et al. 2009).

4.2.4.2 SCS-CN Method and Sediment Yield Model

The SCS-CN method is well explained and elaborated by Chow et al. (1988), Mishra and Singh (2003), Mishra et al. (2006), Bhunya et al. (2014), Gajbhiye et al. (2014) and Srivastava and Imtiyaz (2016). The SCS-CN rainfall-runoff model is based on the water balance equation and two fundamental hypotheses. The first hypothesis equates the ratio of the actual amount of direct surface runoff (Q) to the total rainfall (P) (or maximum potential surface runoff) to the ratio of the amount of actual infiltration (F) to the amount of the potential maximum retention (S) (Mishra et al. 2006). The second hypothesis relates the initial abstraction (I_a) to the potential maximum retention (Mishra et al. 2006). Thus, SCS-CN method consists of the following equation:

(a) Water balance equation:
$P = I_a + F + Q$

(b) Proportional equality hypothesis:
$Q/P - I_a = F/S$.
The fundamental hypothesis is primarily a proportionality concept. Apparently, it explained as $Q \rightarrow (P - I_a)$, $F \rightarrow S$. This proportionality enables portioning (or dividing) $(P - I_a)$ into two-surface water (Q) and subsurface water (F) for given watershed characteristic or S (Mishra and Singh 2003). The extent of runoff contribution of a storage element depends on its capacity or, alternatively, the magnitude of S. Therefore, the whole watershed should contribute to runoff, if S is

Fig. 4.9 Horton's model of channel initiation—a simple model of threshold condition (modified from Horton 1945)

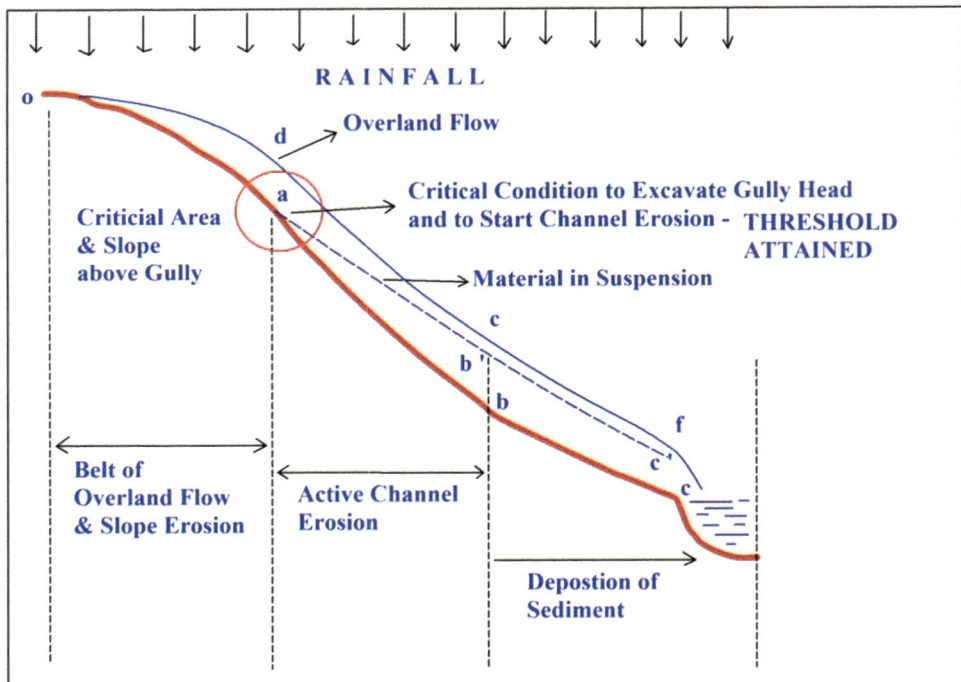

taken to be a definite quantity (Mishra and Singh 2003). Parameter S of the SCS-CN method depends on the soil type, land use, hydrologic condition and antecedent moisture condition (AMC) (Mishra and Singh 2003).

(c) $I_a - S$ hypothesis:

$I_a = \lambda S$,

where, P = total rainfall, I_a = initial abstraction, F = Cumulative infiltration excluding I_a, Q = direct runoff, S = potential maximum retention of infiltration, and λ = abstraction coefficient.

The second hypothesis of the SCS-CN method reflects that initial abstraction is linearly related to the potential maximum retention (Mishra and Singh 2003).

Thus, the direct runoff (Q) can be derived from the above equations, and presented as follows (Mishra and Singh 2003):

$$Q = (P - I_a)^2 / (P - I_a + S).$$

The equation is valid for $P \geq I_a$; $Q = 0$, otherwise. According to Mishra and Singh (2003), for $\lambda = 0.2$, the equation can be written as

$$Q = (P - 0.2S)^2 / (P + 0.8S).$$

In practice, S is derived from a mapping equation expressed in terms of the curve number (CN) (Table 4.2) (Mishra et al. 2006):

$$S = 1000/\mathrm{CN} - 10 \quad (\text{where } S \text{ in inch})$$

The equation is transformed to SI units.

$$S = 25,400/\mathrm{CN} - 254 \quad (\text{where } S \text{ in cm}).$$

The current practice is to derive hydrologic information from a rainfall-runoff model and utilize it in the computation of potential erosion using USLE for determining the sediment yield. Both the SCS-CN method and the USLE method share a common characteristic, in that they account for watershed characteristic, albeit differently. It is therefore conjectured that by coupling these two methods one can compute the sediment yield from the knowledge of rainfall, soil type, land use and antecedent soil moisture condition. The methodology of SCS-CN and USLE coupling (i.e. rainfall-runoff erosion modelling) is adapted from the works of Mishra and Singh (2003), Mishra et al. (2006), Bhunya et al. (2014), Gajbhiye et al. (2014) and, Srivastava and Imtiyaz (2016).

The USLE estimates the potential soil erosion (sheet and rill erosion) from upland areas and it is expressed as follows (Wischmeier and Smith 1978):

$$A = R.K.LS.C.P,$$

where A is the annual potential soil loss per unit area (t ha^{-1} year^{-1}), R is the rainfall and runoff factor

Table 4.2 Runoff curve numbers for hydrologic cover complexes and land uses (AMC II and $I_a = 0.2\,S$) (Mishra and Singh 2003)

Sl. no.	Land-use description	Hydrologic condition	Hydrologic soil group			
			A	B	C	D
1	Bare soil/fallow	–	77	86	91	94
2	Crop residue cover	Poor	76	85	90	93
		Good	74	83	88	90
3	Pasture or range land	Poor	68	79	86	89
		Fair	49	69	79	84
		Good	39	61	74	80
4	Forest land	Poor	45	66	77	83
		Fair	36	60	73	79
		Good	25	55	70	77
5	Meadow-continuous grass	Good	30	58	71	78
6	Herbaceous mixture of grass, weeds and low-growing brush	Poor	–	80	87	93
		Fair	–	71	81	89
		Good	–	62	74	85

(MJ mm ha^{-1} year^{-1}), K is the soil erodibility factor, L is the slope-length factor, S is the slope-steepness factor, C is the cover and management factor and P is the support practice factor.

The sediment yield (S_y) is determined from the USLE computed potential erosion (A) using the sediment delivery ratio (DR). DR is dimensionless ratio of the sediment yield to the total potential erosion in the watershed (Mishra et al. 2006).

$$DR = S_y/A.$$

The SCS-CN and USLE coupling is based on three hypotheses: (1) the runoff coefficient is equal to the degree of saturation, (2) the potential maximum retention can be expressed in terms of the USLE parameters and (3) the sediment delivery ratio is equal to the runoff coefficient (Mishra and Singh 2003; Mishra et al. 2006).

- Hypothesis: $C = S_r$.
  For $I_a = 0$ (i.e. immediate ponding situation), the SCS-CN proportionality hypothesis ($Q/P - I_a = F/S$) equates the runoff factor C ($C = Q/P$) to the degree of saturation (S_r).
  $S_r = F/S = V_w/V_v$,
  where V_v is the void space and V_w is the space occupied by the infiltrated moisture.
- Hypothesis: physical significance of S.
  The potential maximum retention (S) depends on all factors affecting the potential erosion of the watersheds; the higher the potential erosion from a watershed, the higher will be the value of S and vice versa.
  $S = [n(1 - S_{ro})/(1 - n)p_s]\ R.\ K.\ LS.\ C.\ P$,
  where n is the soil porosity, S_{ro} is the initial degree of saturation and p_s is the density of solids.

- Hypothesis: DR = C.
  Similar to the SCS-CN proportional equality (or $C = S_r$) concept, it is possible to extend it for sediment yield as:
  $C = S_r = DR$ (all variables range from 0 to 1).
  The above equation of hypothesis DR = C can be expanded using the usual definitions and $I_a = 0$ as:
  $Q/P = F/S = P/P + s = S_y/A$,
  $S_y = CA$.

It implies that sediment yield is directly proportional to the potential maximum erosion A and the runoff factor C (proportionality constant). Alternatively,

$$S_y = AP/P + S.$$

As $S \to 0$ (or CN $\to 100$), $S_y \to A$ since $Q \to P$. Similarly as $S \to \alpha$ (or CN $\to 0$), $S_Y \to 0$, since $Q \to 0$. It confirms that higher the rainfall, higher will be the sediment erosion and its transport and hence higher the sediment yield, and vice versa. It also shows that direct surface runoff primarily drives sediment yield.

The initial abstraction, I_a can be incorporated in the above equation as:

$$S_y = (P - I_a)A/P - I_a + S.$$

Taking $I_a = 0.2S$ which is a standard practice, the equation can be recast as:

$$S_y = (P - 0.2S)A/P + 0.8S,$$

which suggests that sediment yield reduces with the increasing initial abstraction and vice versa.

4.2.4.3 Statistical Test of Model and Model Validation

To judge the slope–area relation (i.e. statistically fit or not), we have performed two statistical techniques, viz., (1) Student's *t* test of correlation coefficient (*r*) and (2) significance test of standard error of *b* (S_E) (Sarkar 2013).

$$\text{Student's } t = r\sqrt{(N-2)}/\sqrt{(1-r^2)}$$

where *r* is Pearson product moment correlation coefficient, *N* is total number of sample and *N* – 2 is the degree of freedom.

$$S_E = b\sqrt{(1-r^2)/N}$$

where the confidence limit of calculated S_E of *b* is ($b \pm 1.96$ S_E).

We have tried to establish the *S–A* non-linear relationship (i.e. influence of intrinsic threshold) as a model to analyse the initiation criteria of gullies. The performance of this model is validated by the value of model efficiency coefficient (MEC) which was developed by Nash and Sutcliffe (1970) and this equation is applied successfully by Morgan and Duzant (2008) and Cao et al. (2013) in soil erosion research.

$$\text{MEC} = 1 - \Sigma(Q_{obs} - Q_{pred})^2/(Q_{obs} - Q'_{obs})^2.$$

In the above equation, Q_{obs} is measured value, Q_{pred} is calculated value and Q'_{obs} is mean of measured value.

4.3 Results

4.3.1 Estimating Geomorphic Threshold of Gully Erosion

Based on the data of slopes (*S*) and drainage areas (*A*) of 118 gully-head catchments an empirical power regression is adopted which can be used as geomorphic intrinsic threshold for gully initiation on this lateritic terrain. The upstream slopes above gully heads are negatively correlated (*r* = −0.55) with upstream drainage areas which are used as surrogate for the volume of runoff yield in the study area. A significant line is fitted through the lower-most scatter points for the study sites which are incised to form gully heads. This empirical straight line ($S = 17.419 \ A^{-0.2517}$, with R^2 of 0.52) represents an approximation to critical slope–area threshold relationship for gully incision (Fig. 4.10). Any site (may be un-trenched or trenched by gullies) lying above this critical line is much prone to gully erosion on this terrain of laterites. It is derived that mean critical threshold slope for the initiation of gullies is 2.34°.

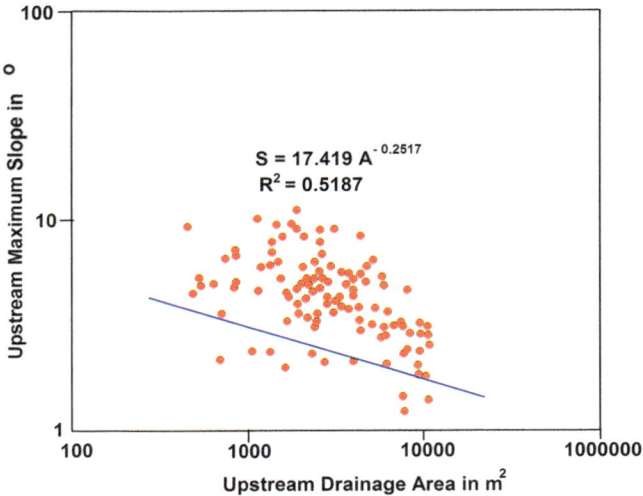

Fig. 4.10 Establishing critical slope–area threshold relation ($S = 17.419 \ A - 0.2517$) for the gullies of lateritic terrain on the basis of intrinsic thresholds *S* (in degree) and *A* (m^2) (Ghosh and Gucchait 2017)

The high value of *a* (i.e. 17.419) signifies the initiation of gullies by high volume of overland flow and landsliding at micro scale in the study sites (Morgan and Mngomezulu 2003). Most importantly the constant *b* is variously interpreted as relative area exponent or relative shear stress indicator (Begin and Schumm 1984; Morgan and Mngomezulu 2003). The negative value of *b* (i.e. −0.2517) and in general consideration *b* > 0.2 is considered to identify the dominancy of overland flow erosion over sub-surface processes in the study area (Vandaele et al. 1996; Vandekerckhove et al. 1998; Morgan and Mngomezulu 2003; Samni et al. 2009; Dong et al. 2013).

The null hypothesis (H_O) is that there is no significant correlation between the two variables. For 116 degree of freedom (*N* – 2) the tabulated *t* value is 3.29 in 0.01 significance level (two-tailed) but our calculated *t* value (7.09) much greater than tabulated *t*. Thus H_O is rejected and alternative hypothesis is accepted, which favours a significant inter-relation between *S* and *A* in the geomorphic system of gully erosion. The calculated confidence limit of calculated S_E of *b* (0.271–0.232) does not include zero (i.e. zero gradient). It signifies that the power regression ($S = 17.419$ $A^{-0.2517}$) is certainly significant at 5% level.

Through inserting the values of drainage area (Q_{obs}) in the equation of $S = 17.419 \ A^{-0.2517}$ the predicated slope values (Q_{pred}) of each gully are calculated. The mean slope of sample gullies (Q'_{obs}) is 4.6°. EC is estimated in the case of slope prediction and its value is greater than 0.63 (greater than 0.5) which is generally interpreted to denote that this model performs satisfactorily (Morgan and Duzant 2008). Therefore, this model is validated in the study area.

Fig. 4.11 The diagram showing S (in °)–A (in m²) scatter plot in M–D Envelope (i.e. red curve) to depict erosion dominant gullies in the study area (Ghosh and Gucchait 2017)

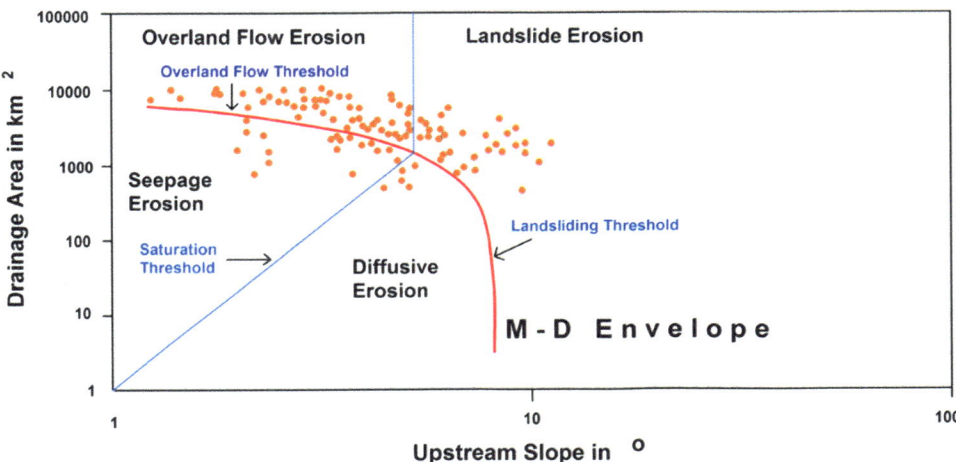

Table 4.3 Distribution of gully heads in respect of dominant erosion process using M–D envelope

Dominant gully erosion process	Percentage of gully heads	Slope range	Area range (m²)
1. Overland flow erosion	52.51	1.2–5.2°	2129.1–10,513.9
2. Seepage erosion	15.25	2.2–4.6°	685.5–3843.7
3. Landslide erosion	27.96	5.2–9.5°	457.1–5702.5
4. Diffusive erosion	4.28	4.4–5.3°	483.2–879.9

Source: Ghosh and Gucchait (2017)

4.3.2 M–D Envelope and Dominancy of Erosion Processes

The trend line of A–S empirical relationship and regression slope (b value) can determine relative importance of overland flow erosion, subsurface flow erosion, diffusive erosion and mass movement or landslide erosion (Fig. 4.11). Here, on the basis of slope (X axis) and drainage area (Y axis) we have classified the gully heads to determine erosion dominancy which is clearly depicted through a threshold line, that is called Montgomery–Dietrich (M–D) envelope. The estimated M–D envelope distinguishes mass movement dominated gullies from hydraulic erosion dominated gullies. In this study area, 52.51% of gullies are affected by overland flow erosion (S—1.2° to 5.2° and A—2129.1 to 10,513.9 m²) while 27.96% belongs to landslide erosion (S—5.2° to 9.5° and A—457.1 to 5702.5 m²). Only 15.25% of gullies (S—2.2 to 4.6° and A—685.5 to 3843.7 m²) are affected by tunnel erosion or seepage erosion (Table 4.3).

4.3.3 Catchment-Wise Runoff Yield

In gully catchment 1 (basin area of 109,250 m²), the principal land use–land cover is identified as natural vegetation (25.35%), grassland (37.67%) and bare laterite land (36.98%) (Fig. 4.12a). Based on the areal coverage of these land uses, CN II is calculated in AMC II condition using Table 4.4 and then CN III is calculated in AMC III condition

(as the rainfall occur in monsoon months). Thereafter, S is calculated in AMC II and III condition using CN values. The values of weighted CN II and CN III are 85.88 and 93.45 respectively (Table 4.4). The values of S of AMC II and III condition are 41.75 mm and 17.80 mm respectively. It is understood that for AMC III condition the CN value is increased and S is decreased, because in AMC III condition the soils hold maximum moisture previously due to heavy rainfall of monsoon and the situation promotes more runoff than storage in a rainfall event. From the database, we have selected the rainfall events of greater than 40 mm and these events vary from 42 mm to 137.2 mm of rainfall in 2016. The calculation of runoff (Q_c) during those events reflects that Q_c ranges from 15.03 mm to 97.41 mm in AMC II condition, and it varies from 26.30 to 118.03 mm in AMC III condition. The runoff coefficient (R_c) is increased from low to high rainfall events. R_c varies from 0.35 to 0.71 in AMC II condition and it increases from 0.62 to 0.86 in AMC III condition respectively. Therefore, it is realized that AMC III condition promotes more runoff in the same rainfall event and this condition increases the vulnerability of laterite slope in escalating flow erosion. Applying the linear regression method, it is found that in both AMC (Table 4.5) conditions the Q_c is positively related to rainfall (P) and the regression line has high slope (b value) (i.e. 0.851–0.958)—(1) in AMC III condition, $Q_c = 0.958\ P - 14.39$ ($R^2 = 0.999$) and (2) in AMC II condition, $Q_c = 0.851\ P - 21.87$ ($R^2 = 0.997$) (Fig. 4.13).

In gully catchment 2 (basin area of 118,325 m²), the areal coverage of natural vegetation, grassland and bare laterite soil

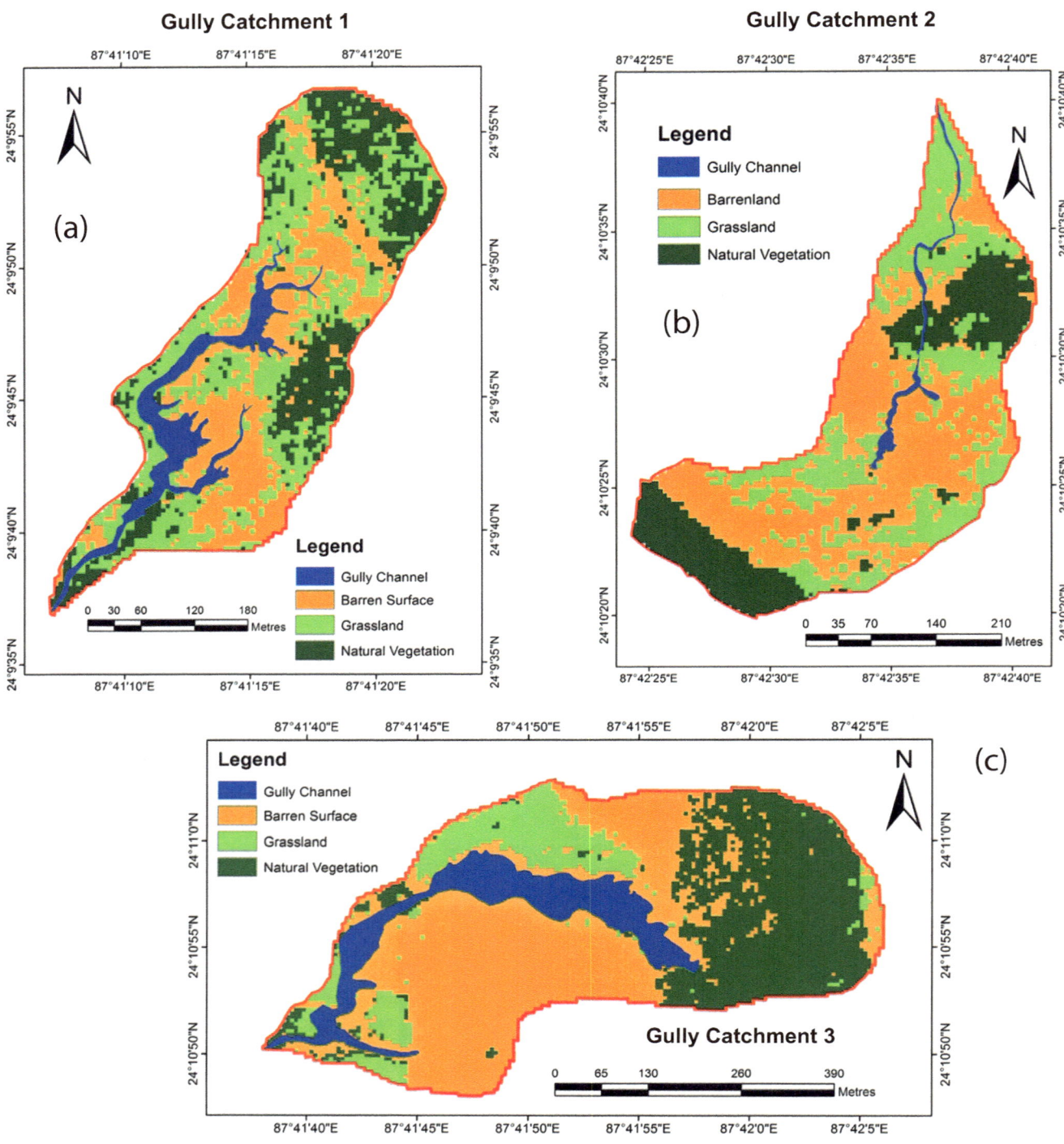

Fig. 4.12 Supervised image classification of IRS P6 LISS IV image (2015) and thematic mapping of land use–land cover in gully catchment 1, 2 and 3

are 21.88%, 30.65% and 41.47% respectively (Fig. 4.12b). The weighted CN values are 85.52 in AMC II condition and 93.26 in AMC III condition respectively (Table 4.5). The derived S varies from 42.97 mm (AMC II) to 18.35 mm (AMC III). In the same rainfall events, Q_c ranges from 14.62 mm to 97.41 mm in AMC II condition and it varies

from 25.94 mm to 118.03 mm in AMC III condition (Table 4.6 in Chapter "Assessing gully asymmetry based on cross-sectional morphology: A case of Gangani Badland of West Bengal, India"). Therefore, R_c of AMC II is increased from 0.34 to 0.71 and it increased from 0.61 to 0.86 in AMC III condition. Similarly the Q_c is positively related to rainfall

Table 4.4 Estimated CN values of AMC II and III condition in the sample gully catchment 1, 2 and 3 on the basis of existing land use–land cover

HSG group	LULC	CN II	Area (m^2)	Product of CN II × area	CN II weighted	CN III AMC III	S (mm) II	S (mm) III
Gully catchment 1								
C	Natural vegetation	73	27,700	2,022,100	85.88	93.45	41.72	17.79
B	Grassland	86	41,150	3,538,900				
B	Bare surface	91	40,400	3,680,040				
Gully catchment 2								
C	Natural vegetation	73	25,900	1,890,700	85.52	93.26	42.97	18.35
B	Grassland	86	36,275	3,119,650				
B	Bare surface	91	56,150	5,109,650				
Gully catchment 3								
C	Natural vegetation	73	64,500	4,708,500	84.9644	92.97	44.92	19.18
B	Grassland	86	28,600	2,459,600				
B	Bare surface	91	122,950	11,188,450				

Note: HSG hydrologic soil group, *LULC* land use–land cover, *CN* curve number, *S* maximum surface storage

(P), having high slope (b value) of regression—(1) in AMC II, $Q_c = 0.895P - 22.07$ ($R^2 = 0.997$), and (2) in AMC III, $Q_c = 0.956P - 14.67$ ($R^2 = 0.999$).

In gully catchment 3 (basin area of 216,050 m^2), the total areal coverage of natural vegetation, grassland and bare laterite land are 29.85%, 13.23% and 56.94% respectively (Fig. 4.12c). CN value of AMC II is 84.94 and the value of AMC III is 92.97 (Table 4.4). The maximum storage, S, varies from 44.92 mm (AMC II) to 19.18 mm (AMC III). In the same rainfall events of 2016, the calculated runoff, Q_c, ranges in between 13.99 mm and 95.02 mm in AMC II condition, and it ranges in between 25.41 mm and 116.07 mm in AMC III condition (Table 4.5). R_c of AMC II varies from 0.33 to 0.69 and it increased from 0.60 to 0.85 in AMC III condition.

Now, it is required to compare the SCS-CN runoff results of gully catchments in both AMC condition. The differences of runoff in the catchments for a same rainfall event are the direct effects of land use–land cover derived from CN values. In AMC II condition, the weighted CN II is maximum in the gully catchment 1 (85.88) and lowest in the gully catchment 3 (84.94). In AMC III condition, the weighted CN III is maximum in the gully catchment 1 (93.45) and lowest in the gully catchment 3 (92.97). The increase of runoff from AMC II to AMC III condition is calculated here for each catchment. In the gully catchments, the estimated runoff is increased above 22% from AMC II to AMC III condition—(1) in gully catchment 1 from –22.33% to 78.63%, (2) in gully catchment 2 from –22.97% to 81.28%, and (3) in gully catchment 3 from –24.04% to 85.73% respectively. Therefore, high CN values mean the high runoff potential in the catchment. So, Gully catchment 1 has runoff potential and it depicts high density of deep gullies in the field. It is understood that if the rainfall amount is increased day by day, the R_c is also increased consecutively and it will be high runoff event which is the sign of high vulnerability of flow erosion on the bare slope. The SCS-CN analyses reflect that in

prolong rainfall event of tropical depression more than 86% of rainfall can be transformed into direct runoff, as R_c is reaching up to 0.86. Alongside, it gives more hydraulic energy to gully initiation and gully head migration in AMC III condition.

The runoff response of each rainfall event is turned to be maximum in gully catchment 1 compared to gully catchment 2. To understand the response, five daily rainfall events are arbitrarily selected (year—2016) of 42 mm, 58.8 mm, 80.2 mm, 107.4 mm and 137.2 mm respectively. The results of runoff show that hydraulic response of gully catchment 1 and 2 is more or less same but it slightly differs in the case of gully catchment 3 due to differences of land use–land cover area. In each rainfall event, the runoff values of catchment 3 are less than other tow catchments. For example, the runoff yield of rainfall event of 80.2 mm is near about 45.49 mm in the gully catchment 1 and 2 in AMC III condition, but it is 43.70 mm in the catchment 3. Similarly, for the rainfall event of 137.2 mm, the runoff of gully catchment 1 and 2 is near about 97.40 mm, whereas it is 95.01 mm in the catchment 3. In short, it can be said that as the CN values of AMC II and AMC III condition are quite high (because of low areal coverage of natural vegetation and grassland), the runoff potential and erosion potential of the catchments will remain high in the extreme rainfall event or any torrential rainfall event, if any conservation measure or the transformation of existing land use–land cover practice is not taken.

4.3.4 SCS-CN Model Validation

Now, the question is how far the results of calculated direct runoff are taken into account to this region and how far the SCS-CN method has applicability and validity. To justify the above questions, the previously used Model Efficiency Coefficient (MEC) is applied in this case using ten observed

Table 4.5 SCS-CN based daily runoff of AMC II and III condition and runoff coefficient of daily event

Rainfall (mm)	Gully catchment 1				Gully catchment 2				Gully catchment 3			
	Q_c (mm) AMC II	Q_c (mm) AMC III	R_c AMC II	R_c AMC III	Q_c (mm) AMC II	Q_c (mm) AMC III	R_c AMC II	R_c AMC III	Q_c (mm) AMC II	Q_c (mm) AMC III	R_c AMC II	R_c AMC III
58.8	27.64	41.82	0.47	0.71	27.07	41.39	0.46	0.7	26.21	40.77	0.446	0.693
61.4	29.72	44.27	0.48	0.72	29.14	43.84	0.47	0.71	28.25	43.2	0.46	0.704
68.2	35.3	50.73	0.52	0.74	34.66	50.28	0.51	0.74	33.7	49.62	0.494	0.728
125	85.99	106	0.69	0.85	85.09	105.5	0.68	0.84	83.69	104.7	0.67	0.837
58	27	41.07	0.47	0.71	28.5	43.09	0.47	0.71	27.62	42.45	0.456	0.701
42	15.04	26.3	0.36	0.63	26.44	40.64	0.46	0.7	25.59	40.02	0.441	0.69
43.8	16.3	27.93	0.37	0.64	14.62	25.94	0.35	0.62	14	25.41	0.333	0.605
43.8	16.3	27.93	0.37	0.64	15.87	27.56	0.36	0.63	15.21	27.02	0.347	0.617
57.6	26.69	40.69	0.46	0.71	15.87	27.56	0.36	0.63	15.21	27.02	0.347	0.617
107.4	69.75	88.72	0.65	0.83	26.13	40.27	0.45	0.7	25.29	39.65	0.439	0.688
78.2	43.77	60.32	0.56	0.77	69.75	88.72	0.65	0.83	67.62	87.44	0.63	0.814
67.6	34.8	50.16	0.51	0.74	43.77	60.32	0.56	0.77	42	59.16	0.537	0.756
57.6	26.69	40.69	0.46	0.71	34.8	50.16	0.51	0.74	33.21	49.05	0.491	0.726
61.4	29.72	44.27	0.48	0.72	26.69	40.69	0.46	0.71	25.29	39.65	0.439	0.688
57.6	26.69	40.69	0.46	0.71	29.72	44.27	0.48	0.72	28.25	43.2	0.46	0.704
137.2	97.41	118	0.71	0.86	50.38	67.67	0.59	0.79	48.51	66.46	0.565	0.775
56.4	25.74	39.56	0.46	0.7	26.69	40.69	0.46	0.71	25.29	39.65	0.439	0.688
46.2	18.02	30.11	0.39	0.65	97.41	118	0.71	0.86	95.02	116.7	0.693	0.85
80.2	45.49	62.25	0.57	0.78	25.74	39.56	0.46	0.7	24.37	38.54	0.432	0.683
130.2	90.84	111.1	0.7	0.85	18.02	30.11	0.39	0.65	16.87	29.18	0.365	0.632
130.2	90.84	111.1	0.7	0.85	45.49	62.25	0.57	0.78	43.7	61.07	0.545	0.762
56.4	25.74	39.56	0.46	0.7	90.84	111.1	0.7	0.85	88.51	109.8	0.68	0.843
40.3	13.87	24.77	0.34	0.61	90.84	111.1	0.7	0.85	88.51	109.8	0.68	0.843
57.4	26.53	40.5	0.46	0.71	25.74	39.56	0.46	0.7	24.37	38.54	0.432	0.683
83.1	48.01	65.05	0.58	0.78	13.87	24.77	0.34	0.61	12.87	23.91	0.319	0.593
45.2	17.3	29.2	0.38	0.65	26.53	40.5	0.46	0.71	25.13	39.47	0.438	0.688

Note: Q_c = SCS-CN calculated runoff, R_c = Runoff coefficient

Fig. 4.13 Interrelation and linear regression between rainfall and runoff in the gully catchment 1, 2 and 3 for AMC II and AMC III condition

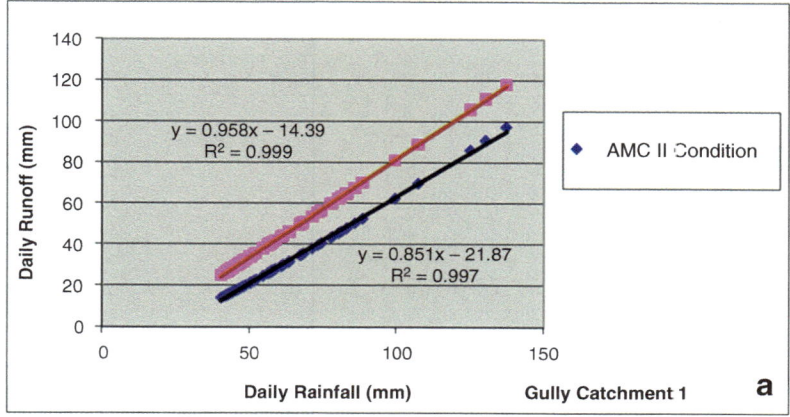

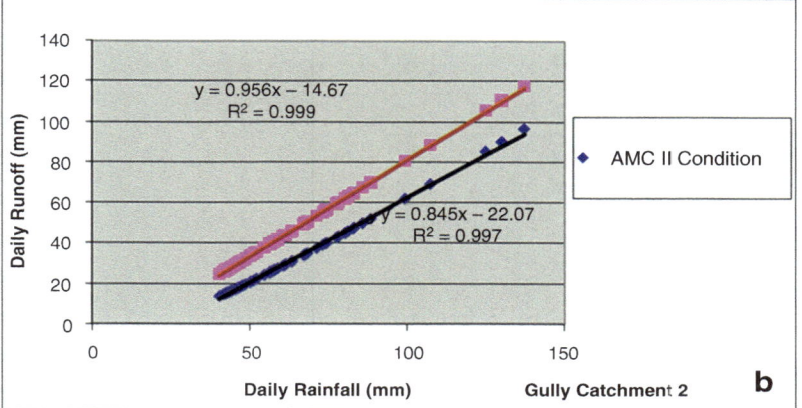

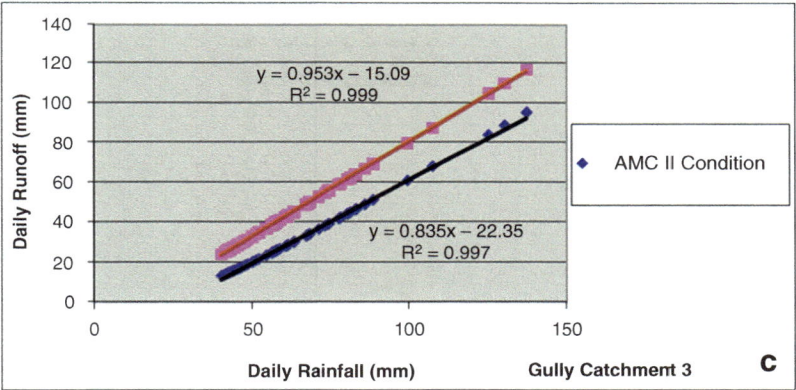

rainfall–runoff events of different times which are collected from the central office of IWD, Rampurhat Division, Birbhum. The amount of rainfall varies from 109 mm to 488 mm. The MEC value is derived using observed runoff (Q_{obs}), mean of observed runoff (Q'_{obs}) and predicted runoff (Q_{pred}).

$$MEC = 1 - \Sigma \left(Q_{obs} - Q_{pred}\right)^2 / \left(Q_{obs} - Q'_{obs}\right)^2.$$

It is known that if the MEC value is 0.5 or more, then the model performs satisfactorily (Morgan 2005). We have performed the MEC in each catchment. It is found that the SCS-CN runoff responses (AMC III condition) of these catchments are (Table 4.6): (1) 89.67–465.91 mm in gully catchment 1, (2) 89.65–465.91 mm, and (3) 88.36–464.81 mm. The observed runoff ranges in between 107 mm and 409 mm. In all cases, there are positive relation between Q_{obs} and Q_{pred}, and the relative error ranges in between 15.77 and –18.16 (Fig. 4.14). The MEC values of three catchments are 0.889, 0.886 and 0.887 respectively (greater than 0.5). It means that the SCS-N model predicts the runoff very accurately in the study area. So, having good

Table 4.6 Comparison between observed and predicted runoff in the catchments

Date of event	Total rainfall (mm)	Observed runoff, Q_{obs} (mm)	Calculated runoff, Q_{pred} (mm)			Relative error		
			G1	G2	G3	G1	G2	G3
10/12/1991	231	223	210.051	209.75	208.53	−5.8	−5.94	−6.48
16/09/1992	287	279	265.539	265.44	264.18	−4.82	−4.86	−5.31
20/08/1994	244	268	222.791	222.67	221.44	−16.87	−16.91	−17.37
27/09/1995	148	116	127.836	134.3	126.5	10.2	15.77	9.05
28/09/1995	156	129	135.708	135.5	134.37	5.2	5.04	4.16
18/09/2000	247	259	225.768	226.5	224.42	−12.83	−12.55	−13.35
19/09/2000	488	409	465.948	465.91	464.81	13.92	13.91	13.64
20/09/2000	403	355	382.896	381.06	379.76	7.86	7.34	6.97
21/09/2000	238	262	216.838	214.41	215.49	−17.24	−18.16	−17.75
21/07/2007	109	107	89.677	90.17	88.36	−16.19	−15.73	−17.42

Note: *G1* gully catchment 1, *G2* gully catchment 2, *G3* gully catchment 3

Fig. 4.14 Comparison and interrelation between observed and calculated SCS-CN runoff in certain storm events through linear regression, showing positive trend with good similarity

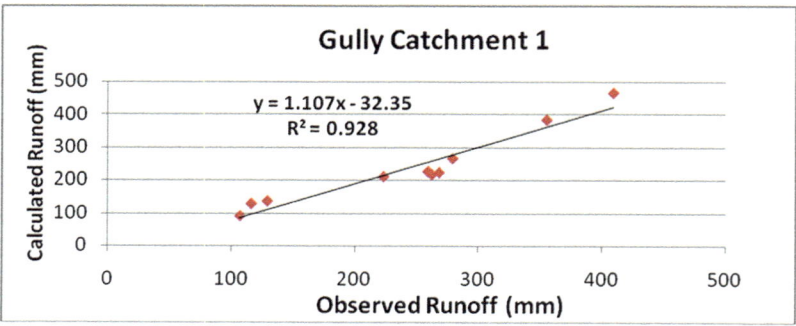

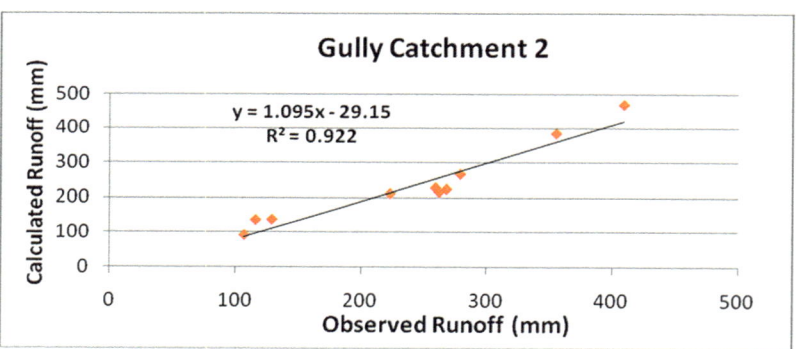

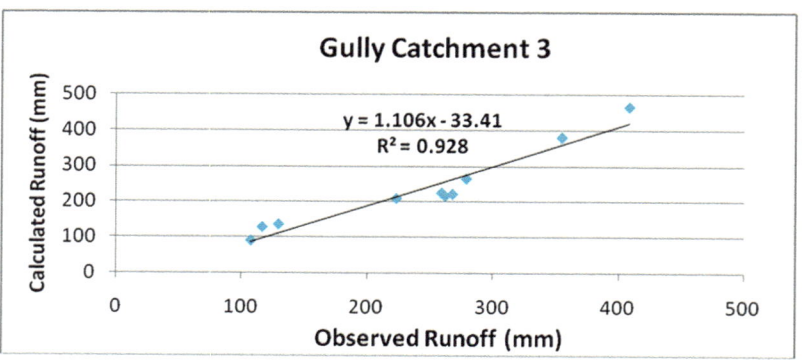

performance of the model we can apply the SCS-CN method in any un-gauged catchment and our derived runoff results of gully catchment 1, 2 and 3 are much acceptable.

4.3.5 SCS-CN-Based Sediment Yield

The sediment yield of a catchment is obtained from the measurements of the quantity of sediment leaving a catchment along the river over time (Morgan 2005). The sediment yield of a drainage basin (t km^{-2} $year^{-1}$) is the result of the processes of erosion, transportation and deposition operating in that basin and its magnitude reflects the sediment delivery ratio. Erosion is distinguished from the sediment yield in that the former represents the potential erosion that is taken equal to the sum of sheet and channel erosion and the sediment yield refers to the sediment measured in the receiving water body in a given time period (Mishra et al. 2006). Estimating the sediment yield, it can be understood that how much of eroded materials are removed from the catchment during the rainstorm events.

In gully catchment 1, through sample study of five slope elements the USLE estimated annual potential erosion rate (A) varies from 9.49 t ha^{-1} $year^{-1}$ to 19.22 t ha^{-1} $year^{-1}$, with an average of 15.25 t ha^{-1} $year^{-1}$. In this catchment based on the selected rainfall event of 2016, the daily runoff ranges from 15.03 mm to 97.40 mm in antecedent moisture condition II, i.e. normal condition (AMC II). In AMC III condition (i.e. high moisture content of soil during monsoon), it varies from 26.30 mm to 118.03 mm. Based on the daily rainfall-runoff modelling and A value, the estimated S_y of gully catchment 1 varies from 6.8 to 11.51 t ha^{-1} in AMC II condition, whereas due to increase in runoff in AMC III condition, S_y varies from 10.42 to 13.45 t ha^{-1}. Similarly, in gully catchment 2, the mean annual potential soil erosion rate is near about 11.46 t ha^{-1} $year^{-1}$. Then the S_y of gully catchment 2 varies from 5.0 to 8.58 t ha^{-1} in AMC II condition and in AMC III condition, S_y varies from 7.74 to 10.07 t ha^{-1}. In gully catchment 3, the mean annual potential soil erosion rate is 14.87 t ha^{-1} $year^{-1}$. In all three catchments, the S_y (Y) is increasing with logarithmic increase of daily runoff (X) ($Y_c = a + b\log X$) in both AMC II and AMC III condition, having coefficient of determination of 0.99 (Fig. 4.15).

Therefore, in prolonged rainstorm event or in cyclonic rainfall, with increase of rainfall and increase in moisture content of surface, in high risk class of catchments the transport of eroded materials is increased and substantially, the sediment yield from the catchment is also escalated. Therefore, if the effective rainfall and runoff are checked by the vegetation cover and the infiltration capacity of the lateritic surface is increased, the sediment yield of a catchment can be minimized.

4.4 Discussion

4.4.1 Hydraulic and Topographic Threshold

From the above analysis, it is confirmed that prolong cyclonic rainfall generated overland flow has significant role in the gully catchments of lateritic terrain. So the processes of overland flow erosion are needed to explain the nature of gully initiation and sediment yield. This study reveals that under the influence of hydraulic threshold of overland flow, the instability of gully erosion system is finally triggered by the intrinsic threshold (slope and drainage area) which already exists within the system. Slope is the major determinant to flow accumulation and flow erosion in a hillslope scale and watershed scale. It is understood from the threshold model that slope and basin size are acted as intrinsic geomorphic threshold in the erosion system to run it as positive feedback mechanisms. The profile of slope or stream profile has a stream–gradient index (Hack 1973) which is believed to reflect the power of flow or competence. The condition of higher the slope steepness and larger the basin area generates huge overland flow in the prolong rainstorm event and that mass of runoff water, when accumulates to form sufficient depth, trigger the incision on the bare slope to initiate gully.

Our present research is mainly concentrated on a certain section of gully system which the area between gully headcut and water divide. It is observed that after a critical distance from the water divide, the gully head is formed because the depth of overland flow increases with distance (slope gives it kinetic energy) and it cumulatively increases shear stress on the surface to allow incision on a concave part of the slope (Charlton 2008). On that incision point or headcut point, the flowing energy (i.e. critical shear stress of flow) is much greater than the resistance of laterite surface. In the catchment of a gully the factors of surface roughness, land use, soil permeability, rainfall intensity and effective rainfall, altogether have generated overland flow (as surface runoff) which becomes high energy erosive agent due to high slope steepness and long slope length in the gullies of laterite terrain. Usually under existing geo-climatic conditions when the bare slope of deposited laterites (secondary laterites) crosses the threshold limit of steepness, the overland flow (guided by basin size) becomes erosive to incise on that slope.

4.4.2 Role of Flow Erosivity in Gullies

Gullies are thought to form when a break in the vegetation covers and exposure of weak zone of secondary laterite surface (i.e. hollows, seepage lines or cracks) allows erosional hollows to form, in which overland flow water accumulates (Charlton 2008). If sufficient flow concentration occurs, on incipient gully head or headcut, forms and it

Fig. 4.15 Established positive logarithmic trend between sediment yield (*y*) and daily runoff (*x*) (AMC II and III conditions) in (**a**) gully catchment 1, (**b**) gully catchment 2 and (**c**) gully catchment 3

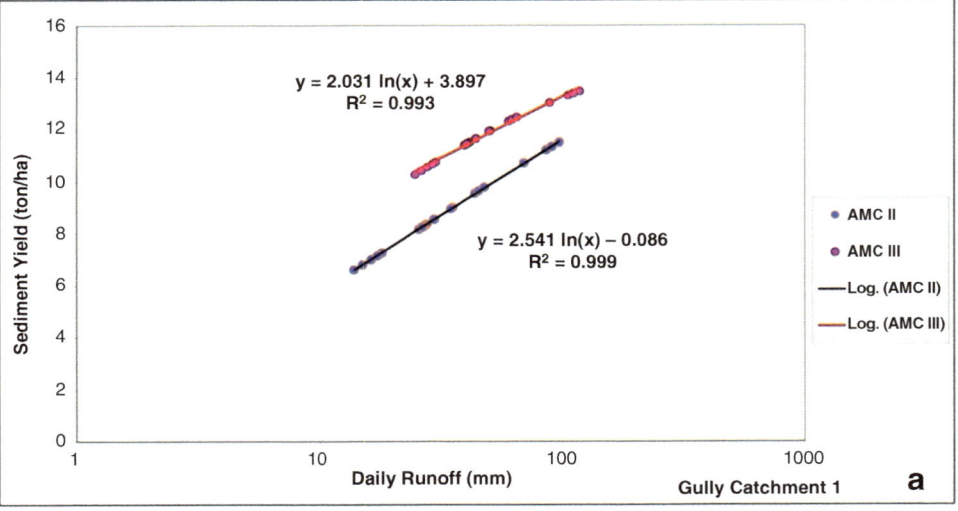

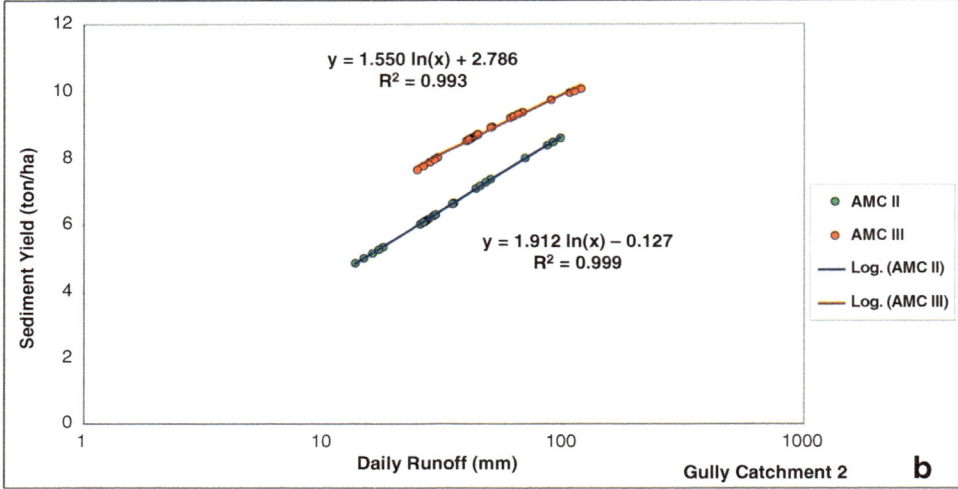

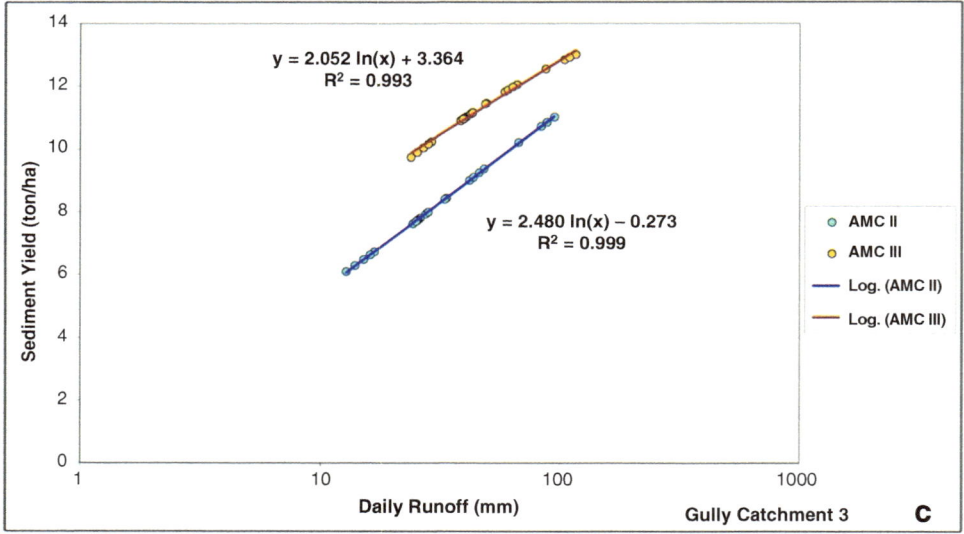

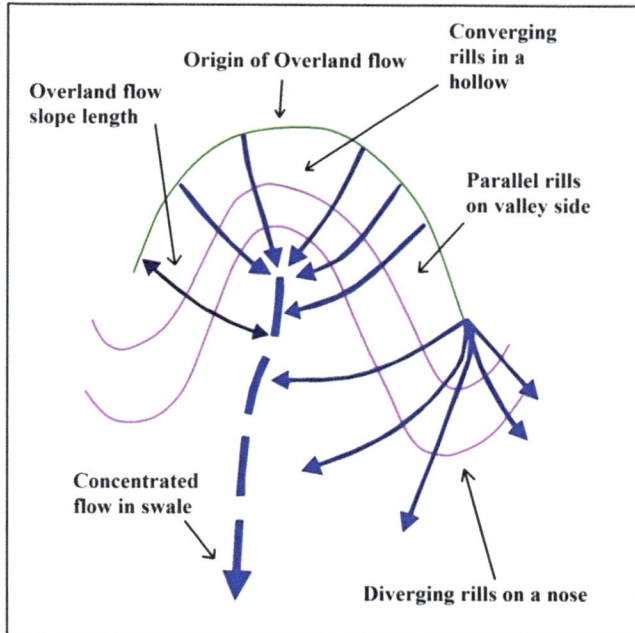

Fig. 4.16 Overland flow paths and concentrated flow areas in a hill-slope of gully (Toy et al. 2013)

further migrates upstream. The main cause of gully initiation on a certain slope is the accumulation of too much water or convergence of rills (i.e. concentrated flow areas) and it exceeds the threshold condition (Morgan 2005). Area between a channel and the divide that delineates the channel's drainage area is the overland flow portion of the landscape (Fig. 4.16). Concentrated flow areas start at the end of over-land flow path, that is slope length between divide and channel. Ephemeral gullies occur in concentrated flow areas in the swales of the landscape. Just as inter-rill areas deliver

runoff and sediment to rills areas, overland flow areas deliver runoff and sediment to the concentrated flow areas.

An ephemeral gully begins at a finite point in a hollow on a converging landscape where overland flow merges into a single, definable channel. One definition for the beginning of an ephemeral gully is here the inter-rill slope length becomes zero because of the convergence of the rill areas on the converging landscape. The permanent incised gullies form in concentrated flow areas by headcuts advancing upstream and these expand by lateral erosion (Fig. 4.17). The flow rate increases along the gully as runoff from adjacent overland flow areas enters the gully, and the transport capacity of eroded sediment first increases and then decreases along the gully to make it a graded profile.

Flow shear stress (τ) is the key erosivity variable to understand the erosive force of overland flow on the surface of laterites. Through inserting appropriate values of S and A for each sample site in the slope–area threshold relation ($\tau = SA^{0.2517}/17.419$, neglecting negative sign of b), we have estimated relative shear stress (ratio scale of threshold) of 118 slope elements as gradational threshold which is a geomorphic indicator of energy state expression of the gully system (Table 4.7). The logic of S–A model is put here to draw that threshold line, taking into consideration of lower most points. It is found that these slopes have critical situation regarding gully erosion prospect, because those points are located high above the trend line. So these slopes on laterites are needed special attention and prevention to avoid initial rill and gully formation. These vulnerable slopes vary in length from 72.2 to 221.6 m and in angle from 5.1° to 13.6°. The only safety factor of these sites is that the lateritic terrain is covered widely by bushes, herbaceous grasses, few tropical deciduous trees (mainly *Sal*) and *Acacia* plantation.

Fig. 4.17 (**a**) Deeply dissected laterites by gullies and (**b**) development of badlands on laterites in gully catchment 3 (arrow signifies the direction of flow)

Table 4.7 Summary of parametric values of selected variables of gully erosion (i.e. major determinants of geomorphic thresholds) in study area

Slope length (L)		Slope gradient (S)		Drainage area (A)		Overland flow eroding force range (F)	Upstream overland flow range (Q)	Detachment of lateritic surface by overland flow (H)	Transport capacity by overland flow (G)
Range	Mean	Range	Mean	Range	Mean				
24.63–200.3 m	71.72 m	1.39–12.58°	4.6°	457.08–10513.9 m²	4112.8 m²	0.58–5.32 N m⁻²	560.58–693.45 mm	2.33–14.6 kg m⁻² year⁻¹	8.8–72.3 kg m⁻² year⁻¹

4.5 Conclusion

From the above study, it is learnt that the hillslope profiles of study area have two crucial intrinsic factors—(1) elevation difference or vertical fall (influencing potential energy) and (2) gradient or steepness (influencing kinetic energy). The character of laterite hillslope directs the location of channel initiation and growth of gully because slope is a prime intrinsic factor in the erosion system, catalysing the flow energy down the slope. Area above the gully head provides the surface of overland flow and the present position of gully headcut reflects the location of flow convergence under the influence of ground cover and gradient. The average slope in between crest and head varies from 5°58′ to 11°06′ in the selected hillslope segments, but in the initial convex part the slope varies much from 14°35′ to 38°06′ which provides more kinetic energy to overland flow.

SCS-CN method reveals that in a rainfall amount of 42 mm to 137.2 mm in 2016, the estimated runoff of gully catchments varies from 13.99 mm to 97.41 mm in AMC II condition, and it varies from 25.41 mm to 118.03 mm in AMC III condition. In short, it can be said that as the CN values of AMC II and AMC III condition are quite high (because of low areal coverage of natural vegetation and grassland), the runoff potential and erosion potential of the catchments will remain high in the extreme rainfall event or any torrential rainfall event, if any conservation measure or the transformation of existing land use–land cover practice is not taken.

Soil erosion continually shapes and reshapes the land, and it has crafted the lateritic terrain as ravine through dense network of gullies. This is natural or geological erosion but in Anthropocene, various human actions such as deforestation, overgrazing, unscientific land uses, over tilling and shifting cultivation have accelerated soil erosion beyond the tolerance limit which can vary from 2 to 11 t ha⁻¹ year⁻¹. The main erosion processes are rainsplash erosion, inter-rill and rill erosion, gully erosion and tunnel erosion, but human activities (e.g. deforestation, agriculture and grazing) have aggravated these processes and the erosion risk and vulnerability of land degradation are increased in the Rarh region.

References

Begin ZB, Schumm SA (1984) Gradational thresholds and landform singularity; significance for quaternary morphology, Geological Society of American Bulletin 56 (3): 267–274.

Bhunya PK, Jain SK, Singh PK, Mishra SK (2014) A simple conceptual model of sediment yield. Water Resource Management 24: 1697–1716.

Biswas A (1987). Laterites and lateritoids of Bengal. In: Datye VS, Diddee J, Jog SR (eds.), Exploration in the Tropics, University of Pune, Pune, pp. 157–167.

Bull WB (1980) Geomorphic thresholds as defined by ratios. In: Coates DR, Vitek JD (eds), Threshold in Geomorphology, George Allen & Unwin, pp. 259–263.

Cao L, Zhang K, Dai H, Liang Y (2013) Modelling inter-rill erosion on unpaved roads in the Loess Plateau of China. Land Degradation and Development 26 (8): 825–832.

Charlton R (2008) Fundamentals of Fluvial Geomorphology. Routledge, New York.

Chow VT, Maidment DR, Mays LW (1988) Applied Hydrology. McGrawHill Book Company, New York.

Coates DR, Vitek JD (1980) Perspectives on geomorphic thresholds. In: Coates DR, Vitek JD (eds), Threshold in Geomorphology, George Allen & Unwin, pp. 3–23.

Dong Y, Xiong D, Su Z, Li J, Yang D, Zhai J, Lu X, Liu G, Shi L (2013) Critical topographic threshold of gully erosion in Yuanmou dry – hot valley in southwestern China. Physical Geography 34 (1): 50–59.

Ebisemiju FS (1989) Thresholds of gully erosion in a lateritic terrain, Guyana. Singapore Journal of Tropical Geography, 10 (2): 136–143.

Gajbhiye S, Mishra SK, Pandey A (2014) Relationship between SCS-CN and sediment yield. Applied Water Science, 4: 363–370.

Ghosh S, Gucchait SK (2017) Estimation of geomorphic threshold to permanent gullies of lateritic terrain in Birbhum, West Bengal, India. Current Science, 113 (3): 478–485

Hack JT (1973) Stream-profile analysis and stream-gradient index. Journal of Research of the U.S. Geological Survey, 1 (4): 421–429.

Horton RE (1945) Erosional development of streams and their drainage basins; hydrophysical approach to quantitative morphology. Geological Society of America Bulletin, 56, 275–370.

Lal R (1992) Restoring land degraded by gully erosion in the Tropics. In: Lal R, Stewart BA (eds), Soil Restoration. Springer – Verlag, New York, pp. 123–152.

Mishra SK, Singh VP (2003) Soil Conservation Service Curve Number (SCS-CN) Methodology. Kluwer Academic Publishers, Dordrecht.

Mishra SK, Tyagi JV, Singh VP, Singh R (2006) SCS-CN-based modelling of sediment yield. Journal of Hydrology, 324: 301–322.

Moeyersons J (2003) The topographic thresholds of hillslope incisions in southwestern Rwanda. Catena 50: 381–400.

Montgomery DR, Dietrich WE (1988) Where do channels begin. Nature 336 (6196): 232–234.

Montgomery DR, Dietrich WE (1992) Channel initiation and the problem of landscape scale. Science 255: 826–830.

Montgomery DR, Dietrich WE (2004) Landscape dissection and drainage area – slope thresholds. In: Kirkby MJ (ed), Process Models and Theoretical Geomorphology. John Wiley & Sons, New York, pp. 221–246.

Morgan RPC (2005) Soil Erosion and Conservation. Blackwell Publishing, Oxford.

Morgan RPC, Duzant JH (2008) Modified MMF model for evaluating effects of crops and vegetation cover on soil erosion. Earth Surface Processes and Landforms 32: 245–253.

Morgan RPC, Mngomezulu D (2003) Threshold conditions for initiation of valley-side gullies in the Middle Veld of Swaziland. Catena 50: 401–414.

Narayana, DVV, Babu R (1983). Estimation of soil erosion in India. Journal of Irrigation Drainage Engineering, 109 (4): 419–434, 1983.

Nash JE, Sutcliffe JV (1970) River flow forecasting through conceptual models I Discussion and Principles. Journal of Hydrology 10: 282–290.

Pathak P, Wani SP,Sudi R (2005). Gully Control in SAT Watersheds. Retrieved from Global Theme on Agroecosystems Report No.15 website:http://www.icrisat.org/journal/agroecosystem/v2i1/v2i1gully.pdf

Patton PC, Schumm SA (1975) Gully erosion, north-western Colorado: a threshold phenomenon. Geology 3: 88–90.

Poesen J, Nachtergaele J, Verstraeten G, Valentin C (2003) Gully erosion and environmental change: importance and research needs. Catena 50: 91–133.

Rossi M, Torri D, Santi E (2015) Bias in topographic thresholds for gully heads. Natural Hazards, 79: S51–S69.

Samni AN, Ahmadi H, Jafari M, Boggs G, Ghoddousi J, Malekian A (2009) Geomorphic threshold conditions for gully erosion in southwestern Iran (Boushehe – Samal watershed). Journal of Asian Earth Sciences 35 (2): 180–189.

Sarkar A (2013) Quantitative Geography: Techniques and Presentations. Orient Blackswan, New Delhi.

Sarkar D, Nayak DC, Dutta D, Gajbhiye KS (2007). Optimizing Land Use of Birbhum District (West Bengal) Soil Resource Assessment. NBSS & LUP, NBSS Publ.130, Nagpur.

Sharda VN, Dogra P (2013) Assessment of productivity and monetary losses due to water erosion in rainfed crops across different states of India for prioritization and conservation planning. Agricultural Research,2 (4): 382–392.

Sharda VN, Dogra P, Prakash C (2010) Assessment of production losses due to water erosion in rainfed areas of India.Journal of Soil and Water Conservation, 65 (2): 79–91.

Singh AK, Kala S, Dubey SK, Pande VC, Rao BK, Sharma KK, Mahapatra KP (2015). Technology for rehabilitation of Yamuna ravines – cost-effective practices to conserve natural resources through bamboo plantation. Current Science, 108 (8): 1527–1533.

Singh G, Babu R, Narain P, Bhushan LS, Abrol IP (1992). Soil erosion rates in India.Journal of Soil and Water Conservation, 47(1): 97–99.

Singh S, Dubey A (2002). Gully Erosion and Management: Methods and Application. New Academic Publishers, New Delhi

Srivastava RK, Imtiyaz M (2016) Testing of coupled SCS curve number model for estimating runoff and sediment yield for eleven watersheds. Journal Geological Society of India, 88: 527–636.

Strahler AN (1964). Quantitative Geomorphology of Drainage Basins and Channel Networks. In: Chow VT (ed.), Handbook of Applied Hydrology, McGraw Hill Book Company, New York, pp. 4–39 – 4-75.

Torri D, Poesen J (2014) A review of topographic threshold conditions for gully head development in different environments. Earth-Science Reviews, 130: 73–85.

Toy TJ, Foster GR, Renard KG (2013) Soil Erosion: Processes, Prediction, Measurement and Control. John Wiley & Sons, New York.

Valentin C, Poesen J, Li Y (2005) Gully erosion: impacts, factors and control. Catena 63: 132–153.

Vandaele K, Poesen J, Govers G, Wesemael B (1996) Geomorphic threshold conditions for ephemeral gully incision. Geomorphology 16 (2): 161–173.

Vandekerckhove L, Poesen J, Wijdnes DO, Nachtergaele J, Kosmas C, Roxo MJ, Figueiredo TD (1998) Thresholds for gully initiation and sedimentation in Mediterranean Europe. Earth Surface Processes and Landforms 25 (11): 1201–1220.

Wischmeier WH, Smith DD (1978). Predicting Rainfall Erosion Losses – A Guide to Conservation Planning. USDA, Agricultural Handbook No. 537, pp. 1–51.

Yadav RC, Bhushan LS (2002) Conservation of gullies in susceptible riparian areas of alluvial soil regions. Land Degradation and Development, 13 (3): 201–219.

Sandipan Ghosh holds the postgraduate degrees of M.Sc., M.Phil., and Ph.D. from the University of Burdwan, West Bengal, India. He has published 4 books and more than 40 international and national research articles in various geography and geosciences journals. His principal research field includes fluvial geomorphology, regolith geology, and quaternary geomorphology. Currently, he is working as an Assistant

Professor in the Department of Geography, Chandrapur College (Bardhaman, West Bengal).

Sanat Kumar Guchhait is an Applied Geographer with postgraduate and Ph.D. degrees from the University of Burdwan, West Bengal, India. He has published 3 books and more than 30 research articles in various international and national journals of geography. His principal research field includes social geography, environmental geography, and applied geomorphology. Seven research scholars have completed their Ph.D. under his supervision. Currently, he is working as a Professor in the Department of Geography, the University of Burdwan (Bardhaman, West Bengal).

Aznarul Islam, Biplab Sarkar, Balai Chandra Das, and Suman Deb Barman

Abstract

Asymmetric channels are more common than symmetric due to the innate nature of movement of water in sinuous path. The extensive field investigation executed on 62 cross sections of different orders of four major gully systems (first order, 20; second order, 13; third order, 11; fourth order, 10; fifth order, 8) in Gangani badland reveals that lower-order and higher-order gullies portray lesser asymmetry compared to the intermediate orders which is reflected by the three areal asymmetry indices ($A*$-first order, 0.19; third order, 0.24; and fifth order, 0.14; A_1-first order, 0.44; third order, 0.56; and fifth order, 0.29; A_2-first order, 0.26; third order, 0.31; and fifth order, 0.13) and four morphometric indices (shape index-first order, 0.3; third order, 0.3; and fifth order, 0.41; width-depth ratio-first order, 10.21; third order, 12.63; and fifth order, 7.47; erosiveness-first order, 0.49; third order, 0.48; and fifth order, 0.52; and concavity-first order, 0.50; third order, 0.49; and fifth order, 0.59). This order-asymmetry relationship is addressed in terms of morphological and sedimentological combination of various slope segments (convex, free face, rectilinear and concave). Finally, we have proposed a simple model which portrays that lower- and higher-order gullies are approximating the geometrical shape (V and U) and intermediate-order gullies are having more irregular shape due to free wandering of the concentrated flow of water as a result of the interplay between erodibility and erosivity factors in complex response system. Therefore, this study would be helpful to understand the mechanism of asymmetric behaviour of gullies.

A. Islam (✉) · B. Sarkar
Department of Geography, Aliah University, Kolkata, India

B. C. Das
Department of Geography, Krishnagar Government College, Nadia, West Bengal, India

S. D. Barman
The University of Burdwan, Bardhaman, West Bengal, India

Keywords

Gully asymmetry · Cross-sectional morphology · Sedimentological combination · Slope segments · Order-asymmetry relation · Gangani badland

5.1 Introduction

Gullies are identified as steep incised channels with ephemeral stream flow and rapid head growth often with a sharp headcut (Radoane et al. 1995; Woodburn 1949; Zachar 1982; Bradford and Piest 1980; Motoc 1963; Nordstorm 1988). The previous works indicate that gullies are studied with special interest in fluvial geomorphology as an important part of drainage network (Hu et al. 2009; Ding et al. 2017; Vandaele et al. 1996; Seginer 1966; Rowntree 2010; James et al. 2007; Harvey 1974; Cheng et al. 2007; Das 2014). They are commonly developed in the hilly region or other regions having moderate to high degree of slope. Land degradation through the process of gullying is recognized as an important environmental threat (Li et al. 2016; Boughton 1989; Yang et al. 2012; Mashi et al. 2015; Zgłobicki et al. 2015; Ionita et al. 2015a, b; Maerker et al. 2015; Momm et al. 2015; Guerra et al. 2018). Therefore, to prevent the adverse effects of gully erosion, adequate knowledge is required regarding controlling factors, developmental processes and forms, measuring magnitude (distribution and density) and estimating the rate and amount of soil erosion through gullying (Sidorchuk 1999). In many scholastic works, models are found to be used for predicting the future rate of gully advancement (Samani et al. 2018). The United States Department of Agriculture (USDA) (1966) developed Soil Conservation Services (SCS)-I and SCS-II models for simulating rate of headcut retreat based on factors like geology, topography, land use and volume of runoff. In addition, models like Ephemeral Gully Erosion Model (EGEM) of Woodward (1999), Gully Thermoerosion and Erosion model (GULTEM) of Sidorchuk (2001) and Channel-Hillslope

© Springer Nature Switzerland AG 2020

P. K. Shit et al. (eds.), *Gully Erosion Studies from India and Surrounding Regions*, Advances in Science, Technology & Innovation, https://doi.org/10.1007/978-3-030-23243-6_5

Integrated Landscape Development Model (CHILD) of Tucker et al. (2001) were developed for predicting the rate of headcut retreat. More importantly, gully cross sections help in computing the rate and eroded volume of gully erosion, understanding the dynamics of fluid and finally revealing the changing pattern of cross-sectional morphology downslope. Each parameter of gully cross section (width, depth, slope segments, slope materials, etc.) is intrinsically linked with others and the deviation of each parameter from normal distribution is commonly observed within and in between gully orders. Deng et al. (2015) extensively studied the size and proportional parameters of gully cross section. The former includes width (upper, quarter, bottom and average), depth (maximum and average) and area. The latter includes the ratio of width-depth and maximum-average depth. Finally, they extracted key factors influencing the morphology of gully cross section by principal component analysis (PCA). Areal asymmetry of channel cross section is important in portraying fluid and morphological dynamics (Leopold and Wolman 1960; Knighton 1981). Generally, asymmetric channels are more common than symmetric. It is common due to the innate nature of the movement of water in sinuous path. Channel asymmetry has been extensively studied by Knighton (1981) who developed three indices ($A*$, A_1 and A_2) on areal asymmetry of the channel. Post-Knightonian development through the works of Das and Islam (2015a, b, 2016a) has furthered areal asymmetry indices and developed bed asymmetry indices. Gullies are also asymmetric channel (mostly dry) developed as a response of intrinsic and extrinsic thresholds in the system (Kar and Bandyopadhyay 1974). With time, gully matures in its dimension, and consequently the degree of asymmetry changes from first order to next order. Therefore, it is important to quantify the asymmetric behaviour of gully cross sections and to quest for the principal factors responsible for asymmetry. However, Garhbeta is not adequately explored in this direction. Most of the previous works focussed on the evolution of badland (Chakraborty 1969; Ghosh and Majumder 1981), slope morphometry (Sen 2008), headward erosion and growth of gullies (Das and Bandyopadhyay 1995), development of drainage network on badland (Bandyopadhyay 1988) and formation of nodular laterite (Niyogi and Mallick 1973). Similarly, fluvial forms and processes of this area were also adequately addressed by Sen et al. (2000, 2002). Besides, Garhbeta as a geomorphic site was introduced by Ray and Biswas (1989). The management of gully erosion was also studied by the researchers with equal importance (Sen et al. 2000, 2002). Keeping these views in mind, an endeavour is taken to address the following objectives:

1. To quantify the channel asymmetry and trace out the trend existing among different orders of gully cross sections.
2. To find out the key factors responsible for gully asymmetry.

3. To establish an empirical regularity in the order-asymmetry relation.

5.2 Study Area

Garhbeta badland is widely known as the *Ganganir Danga*, i.e., the 'land of fire' (Sen 2008; Shit et al. 2015). This area is located in the Silai river basin, a basin extending from 86° 38′ 37 E to 87° 46′ 46″ E and 22° 23′ 49″ N to 23° 14′ 21″ N covering an area of about 4188 km². However, Garhbeta is a very small spatial unit in Paschim Medinipur district, West Bengal, extending from 22° 51′ 18″ N to 22° 51′ 30″ N and 87° 20′ 20″ E to 87° 20′ 28″ E (Fig. 5.1) covering an area of only about 3.5 km² (Shit et al. 2014). Geomorphologically, it fringes Chhotonagpur Plateau and is marked by an escarpment of about 25 m height above the right bank of the Silai river, while the left bank of the river is a featureless plain (Fig. 5.2a). Geochronologically, it has been evolving under subaerial denudation processes since Pleistocene (Bandyopadhyay 1988), and presently it is highly dissected by the network of ravines and gullies of different orders (Fig. 5.2b). Therefore, it is often called as the grand canyon of West Bengal (Sen 2008). The hot humid climate under the stable platform has produced spectacular duricrust (ferricrete surface) over a vast span of the area (Fig. 5.2c). Besides, at the upper stretches of the gullies the loose iron nodule dominates the surface (Fig. 5.2d). Thus, active laterization characterizes the area.

5.3 Database and Methodology

5.3.1 Database

The present study has been carried out based on the data collected from the extensive primary fieldworks during January 2017 to April 2017. We measured 62 cross sections (CS) of four gullies (designated as A, B, C and D) located at Garhbeta badland taking 20 CS from first order (Gully A, 5; B, 6; C, 5; D, 4), 13 CS from second order (Gully A, 3; B, 3; C, 5; D, 2), 11 CS from third order (Gully A, 4; B, 2; C, 2; D, 3), 10 CS from the fourth order (Gully A, 2; B, 5; C, 1; D, 2) and 8 CS from the fifth order (Gully A, 4; C, 2; D, 2) (Fig. 5.3). Linear dimension of gully cross section is measured in terms of width (total width, width of the half depth, width of the quarter depth) and depth (maximum depth, average depth, depth of the half-left side, depth of the half right side) (Fig. 5.4). In addition, for quantifying channel asymmetry, cross-sectional area (left, right and total) was calculated using the trapezoidal method (Das and Islam 2016b). The magnitude of slope of each segment (convex, free face, rectilinear and basal) on both sides was measured

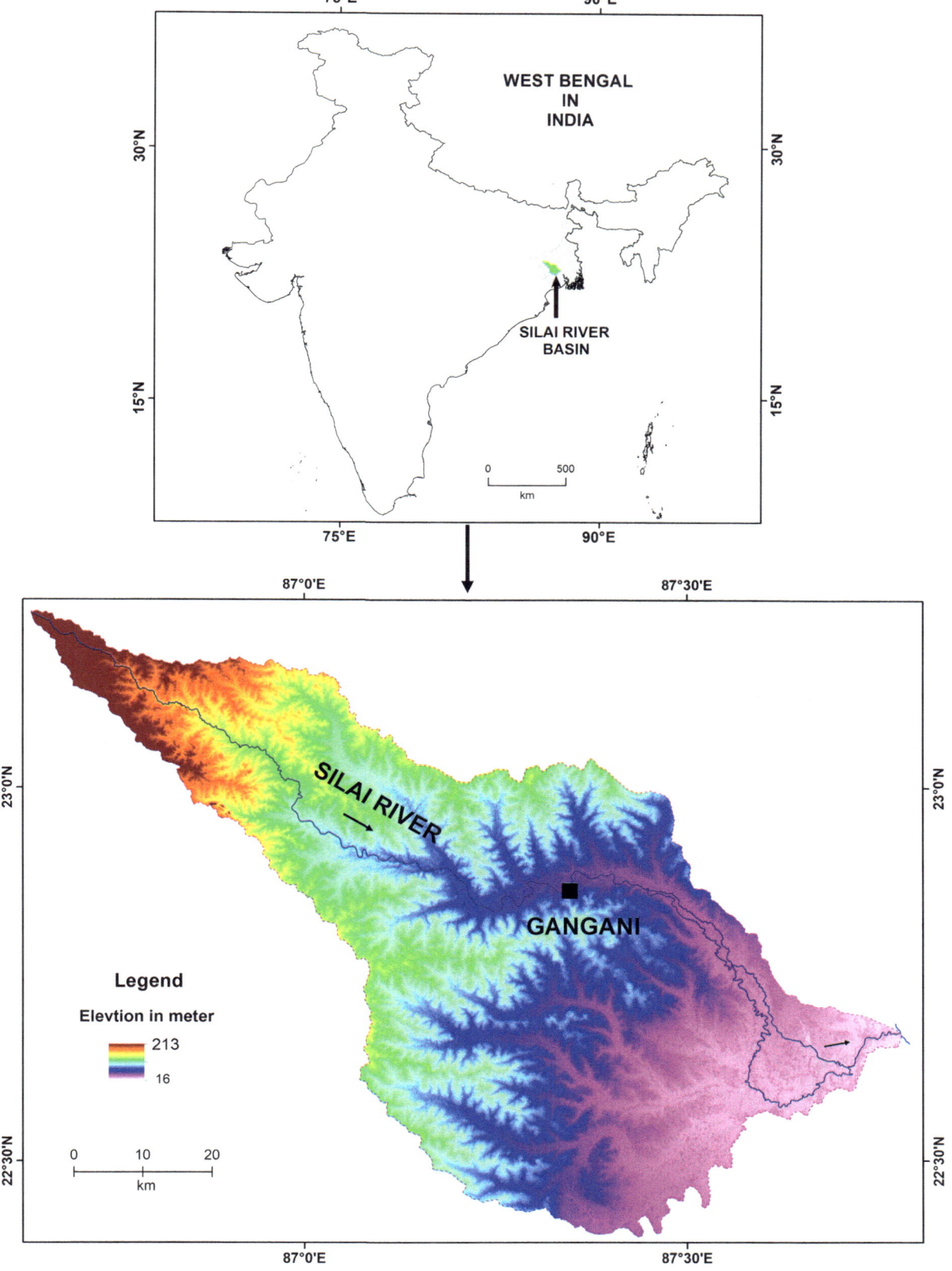

Fig. 5.1 Location of the study area

Fig. 5.2 Geomorphic nature of the study area. (**a**) Escarpment on the right side of the Silai river. (**b**) Highly dissected topography at Garhbeta. (**c**) Duricrust (ferricrete) surface. (**d**) Iron nodule strewn surface at the upper stretches of the gully (Source: Field Photograph 2017)

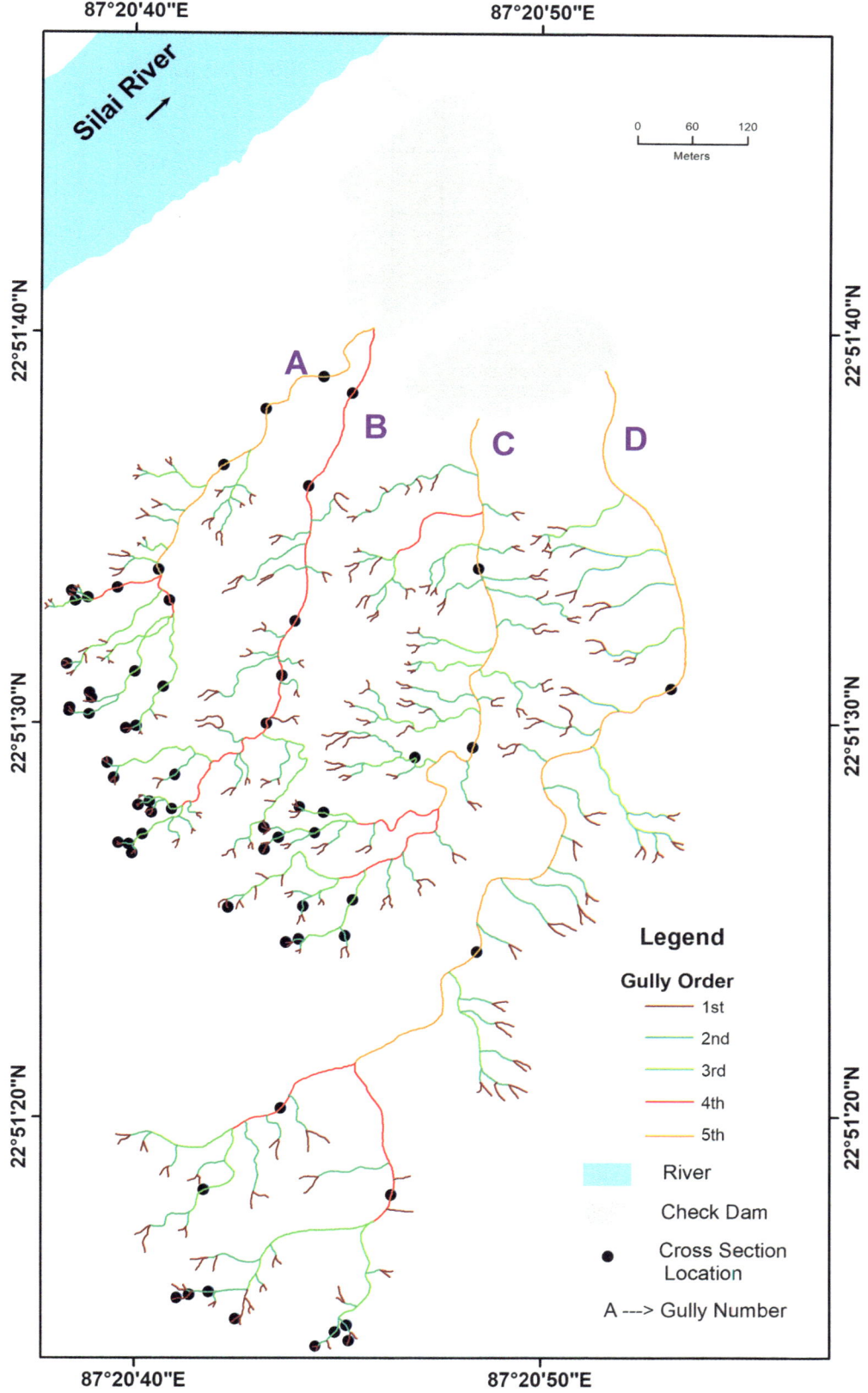

Fig. 5.3 Location of the gully cross sections

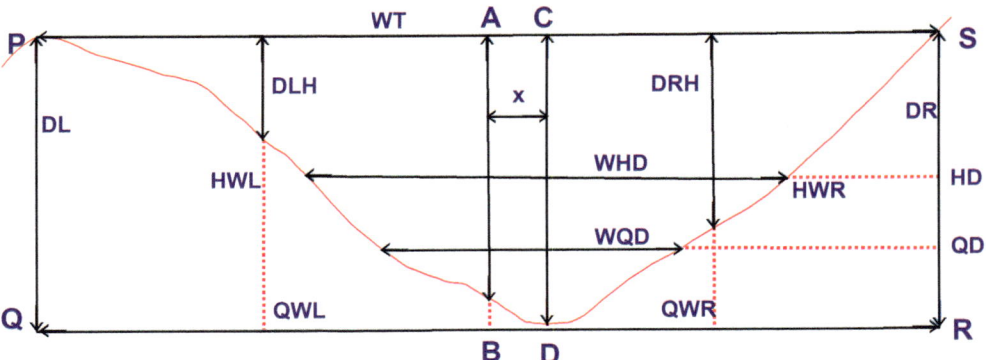

Fig. 5.4 Parameters of the cross-sectional morphology of the gully (Note: WT for total width, DR depth of the right side, DL depth of the left side, DLH depth of the half left side, DRH depth of the half right side, HWL half left width, HWR half right width, QWL quarter left width, QWR quarter right width, HD half depth, QD quarter depth, WQD width with reference to QD, WHD width with reference to HD, PQRS represents the bounding limit of the gully, AB channel centreline, CD line of maximum depth (d_{max}), x distance between the centreline and d_{max} (Source: Based on Deng et al. 2015 and Knighton 1981)

with the help of a clinometer (Fig. 5.5a). Besides, for analysing the material composition of slopes, soil samples were collected and median particle size (D_{50}) of each sample is computed (Fig. 5.5c).

5.3.2 Methodology

5.3.2.1 Areal and Slope Asymmetry Indices

Natural water channel is more asymmetric than symmetric in nature throughout its course. Investigation of the nature and pattern of asymmetry of natural water channel is essential for better understanding hydraulic behaviour and channel morphological dynamics. Knighton (1981) formulated the channel asymmetry indices—$A*$, A_1 and A_2. The $A*$ measures the degree of asymmetry calculating the actual difference in area between left and right part of channel demarcated by centreline. Moreover, in order to develop the index more precisely, he considered vertical asymmetry and formulated A_1 and A_2. Besides, the asymmetry of slope is also important. Thus, the algorithms for computing the asymmetry indices are as follows:

$$A* = \frac{A_r - A_l}{A} \tag{5.1}$$

$$A_1 = \frac{2x \times D_{max}}{A} \tag{5.2}$$

$$A_2 = 2x \frac{D_{max} - D}{A} \tag{5.3}$$

$$A_s = \frac{S_r - S_l}{S_r + S_l} \tag{5.4}$$

where A_r is the cross-sectional area of the right of the channel centreline; A_l, the cross-sectional area of the left of the channel centreline; $A = (A_r + A_l)$, the total cross-sectional area of the channel; D_{max}, the highest depth; D, the average depth; x, the distance between channel centreline and D_{max}; S_r, the slope of the right bank; and S_l, the slope of the left bank. The value of all indices being zero represents no asymmetry present in cross section. Moreover, the ranges of them are −1 and +1 representing extreme left and right asymmetry, respectively, for $A*$ and As. Similarly, higher values of A_1 and A_2 (>1) indicate higher asymmetry.

5.3.2.2 Form Indices

We have employed four indices to find out the morphometric dimensions of the gullies. Besides the one general measure width to depth ratio (*W/D* ratio), other three indices—shape index (SI), erosiveness index (EI) and concavity index (CI)—were computed following the algorithms of Deng et al. (2015) as follows:

$$SI = \frac{WQD}{WT} \tag{5.5}$$

$$EI = \frac{Al + Ar}{(WT * (\max(DR, \; Dl))} \tag{5.6}$$

$$CI = \frac{Al + Ar}{(WT * (\max(DR, Dl))} - Al - Ar \tag{5.7}$$

where WQD is the width of one-fourth depth; WT, total width; Al and Ar, area of the left side and area of the right side, respectively; and DR and DL, depth of the right side and depth of the left side respectively.

It may be noted that the SI < 0.4, 0.4 < SI < 0.6, and SI > 0.6 indicate V-shaped, intermediate shape and U-shaped types of cross section. Similarly, regarding EI and CI zero indicates active in the developmental process while 1 for stable level of development.

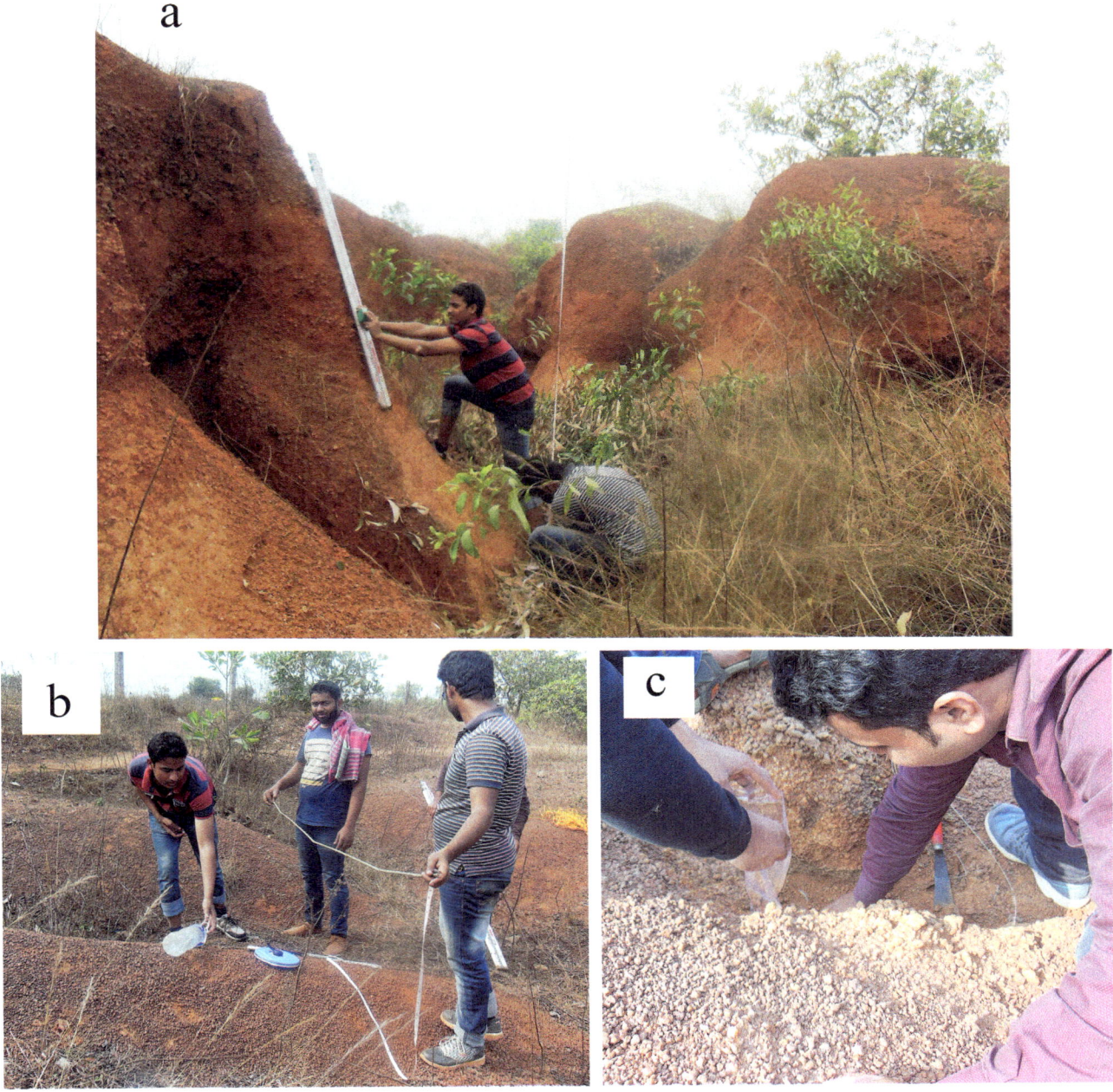

Fig. 5.5 Measurements of gully. (**a**) Determination of slope by clinometer. (**b**) Cross-sectional profiling on the lower-order gully. (**c**) Collection of sediment sample for particle size analysis (Source: Field Photograph 2017)

5.3.2.3 Principal Component Analysis

In multivariate statistical analysis, PCA technique is widely used to reduce the dimensionality of correlated variables into uncorrelated variables explaining the same amount of variance in reduced variables (principal components). In order to find out the underlying key factors regulating the gully cross-sectional morphology at every gully order, PCA has been run using SPSS 14 software based on the mathematical algorithm of Hotelling (1933) as follows:

$$Z_{ij} = \alpha_{i1} X_{1j} + \alpha_{i2} X_{2j} + \alpha_{i3} X_{3j} + \cdots \alpha_{im} X_{mj} \quad (5.8)$$

where Z is the component score; α, component loading; X, measured value of variables; i, component number; j, sample number; m, total number of variables.

5.3.2.4 Residual Analysis

Residual (R) measures the difference between observed distribution (y) and expected distribution (yc). It is generally calculated as follows:

$$R = y - yc \text{ and } yc = a + bx \qquad (5.9)$$

where 'a' is the y intercept with x and b indicates the slope of the line.

5.3.2.5 Particle Size Analysis

The median particle size (D_{50}) was determined by using manual sieve. Four sieves of different sizes (0.15, 0.60, 1.18, and 2.36 mm) were used to segregate the grain size. Then the segregated sediment was weighted by electronic balance. Weight of sample of each size class was used as frequency to find out median particle size (D_{50}).

5.4   Results

5.4.1   Linear Perspective on Cross-Sectional Morphology

5.4.1.1 Depth Analysis

Depth has been found to increase with gully order. The average depth of the first-order gully is 0.43 m, while it is 2.33 m for the fifth-order gullies. Similarly, QD and HD are also increasing with order (Table 5.1). However, this empirical regularity of gradual increase of the depth with order is not followed everywhere. For example, regarding DRH third-order gullies have recorded a lower average of about 0.83 m compared to second-order gullies (1.04 m). Besides, the range and the fluctuation in the depth parameters are also highly variable between the gullies (Table 5.1). The boxplot (Fig. 5.6a) portrays that interquartile ranges (IQR) are highest for third order (IQR = 1.01 m) while median (Med) is highest for fifth order (Med 2.18 m) regarding average depth (D). Besides, we have observed the nature of IQR and Med variable from one order to the next, e.g., IQR = 0.31, Med = 0.29 for first order; IQR = 0.84, Med = 0.36 for second order; IQR = 0.73, Med = 1.16 for fourth order; and IQR = 0.87 for fifth order. Similarly, regarding DLH (Fig. 5.6b), the highest IQR (1.65) is found in third order and that of Med in fifth order (4.0), while the lowest IQR (0.30) and Med (0.40) are recorded for first order. Moreover, the IQR for second (0.87) and fourth (0.35) order are below the average IQR (0.95) while fifth order (1.57) exceeds the average. Besides, for Med a raising trend was observed among gully orders (0.5, 0.8, 1.5 and 4.0 for second, third, fourth and fifth orders, respectively). Regarding DRH (Fig. 5.6c), IQR has followed the rising trend (0.49, 0.5, 0.9 and 0.925 for first, third, fourth and fifth orders, respectively,

with exception for second order (9.25)). Similarly, Med has followed the rising trend (0.30, 0.40, 0.60, 1.4, and 2.2 for first, second, third, fourth and fifth orders, respectively).

5.4.1.2 Width Analysis

Width of the gullies has been analysed with reference to total width (WT), width of the half depth (WHD) and width of the one-fourth depth (WQD). In general, gully width increases with gully order like steam order. In the study area, there are gullies up to fifth order. Regarding WT and WHD, this trend has been followed strictly. The first order is having the minimum width (2.81 m), while the fifth order is having the maximum (about 19 m) in the case of WT. Similarly, in the case of WHD, minimum is observed for the first order (1.32 m) and maximum (~14 m) for the fifth order. However, for WQD this gradually increasing trend has not been observed, and a lower figure has been registered by the third-order gully. Besides these average statistics, the range and the deviations in the data are highly variable from one gully order to the next (Table 5.1). The boxplot (Fig. 5.6d) portrays that fifth-order gullies have outcompeted the rest orders in terms of both the IQR (23.95) and Med (16.0). Besides, that of rests are observed as IQR = 0.97, Med = 2.85 for first order; IQR = 9.20, Med = 7.20 for second order; IQR = 10.35, Med = 6.75 for third order; and IQR = 7.37, Med = 13.5 for fourth order. Similarly regarding WHD (Fig. 5.6e), a uniform trend is found for both the IQR and Med, i.e., they are rising with increasing order (IQR 0.68, 5.22, 6 and 9.45 for first, second, fourth and fifth orders and Med 1.3, 3.5, 9.0, 12.0 for first, second, fourth and fifth orders with the exception for third order (IQR 5.02 and Med 2.2)). Regarding WQD (Fig. 5.6f), the highest IQR (15.29) and Med (10.0) are found in fourth order while the lowest IQR (0.36) and Med (0.90) for the first order. Moreover, the IQR for second (2.27) and third (4.45) order are below the average (5.66) where fifth order (5.93) exceeds the average. Similarly, the Med for second (1.42) and third (1.10) order are well below the average (4.18) with the exception of fifth order (7.50).

5.4.1.3 Width-Depth Ratio

Width-depth ratio reflects the nature of channel shape. With increasing gully order the *W/D* ratio increases to a certain order and then decreases. For example, first order is having the ratio of about 11 and second order is having about 18, while the fifth order is having the ratio of only about 9. Thus, it portrays an inverted U-shaped pattern. We observed higher deviations in the *W/D* ratio for intermediate order of gullies as reflected through the higher standard deviation (SD) compared to average (Table 5.1). Thus, it appears that first and fifth-order gullies are more efficient channel forms compared to the intermediate orders as higher *W/D* ratio induces lesser channel efficiency (Das 2015). The

Table 5.1 Descriptive statistics of the width and depth

Parameters	First order	Second order	Third order	Fourth order	Fifth order
Average depth (D)	0.43 ± 0.27 (0.05–0.92)	0.66 ± 0.69 (0.05–2.52)	0.85 ± 0.52 (0.17–1.86)	1.21 ± 0.53 (0.46–1.95)	2.33 ± 0.72 (1.48–3.45)
Depth of half left side (DLH)	0.47 ± 0.34 (0.07–1.7)	0.74 ± 0.86 (0.05–2.5)	1.11 ± 0.81 (0.2–2.20)	1.60 ± 0.71 (0.36–3)	3.59 ± 1.10 (0.03–5.60)
Depth of half right side (DRH)	0.59 ± 0.69 (0.07–3.1)	1.04 ± 1.31 (0.08–4.5)	0.83 ± 0.84 (0.28–3.3)	1.59 ± 0.80 (0.6–3.0)	3.03 ± 1.39 (1.75–5.60)
Total width (WT)	2.81 ± 0.78 (1.2–4)	7.3 ± 5.45 (1–17)	8.64 ± 6.38 (2.1–23)	14.52 ± 4.43 (9.2–20.5)	18.63 ± 12.80 (1.75–40.0)
Width of the half depth (WHD)	1.32 ± 0.50 (0.37–2.3)	3.44 ± 2.56 (0.6–8.0)	4.25 ± 4.27 (1.2–15)	8.69 ± 3.38 (3.2–14)	13.86 ± 6.64 (3.75–25.0)
Width of the one-fourth depth (WQD)	0.88 ± 0.34 (0.35–1.5)	1.87 ± 2.56 (0.6–8)	2.48 ± 2.71 (0.4–8)	5.10 ± 2.55 (0.8–10.5)	9.0 ± 4.11 (3.75–15)
Width-depth ratio (*W/D*)	10.21 ± 8.65 (1.33–40)	17.52 ± 13.46 (3.03–50.76)	12.63 ± 10.74 (4.45 41.81)	14.22 ± 8.32 (7.93–33.69)	7.47 ± 4.25 (1.16–11.76)
Shape index (SI)	0.30 ± 0.12 (0.05–0.5)	0.24 ± 0.13 (0.03–0.5)	0.30 ± 0.16 (0.05–0.6)	0.34 ± 0.31 (0.08–0.58)	0.41 ± 0.15 (0.25–0.57)

Note: Computed by the authors 2017 (Values within parentheses indicate range of the observation, and values preceding and following the ± denote arithmetic mean and standard deviation, respectively)

boxplot (Fig. 5.6g) portrays that interquartile ranges (IQR) are highest in second order (IQR = 17.93, Med = 14.73) while that of rests are observed as IQR = 6.12, Med = 8.73 for first order; IQR = 9.53, Med = 9.79 for third order; IQR = 12.72, Med = 11.53 for fourth order; and IQR = 9.07, Med = 9.32 for fifth order. An interesting observation is that asymmetry has increased with increased *W/D* ratio. For example, the *W/D* ratio for first- and fifth-order gullies is smaller in comparison to second-, third- and fourth-order gullies, whereas $A*$, A_1 and A_2 are smaller in first- and fifth-order gullies.

5.4.1.4 Shape Index

Based on the observation of 62 gullies of different orders, it appears that 75% of the first-order gullies ($n = 20$), about 85% of the second-order gullies ($n = 13$) and about 82% of the third order ($n = 11$) are 'V'-shaped, i.e., having shape index below 0.4 while the rest has emerged in the intermediate class (SI = 0.4–0.6). However, only 60% of the fourth gullies ($n = 10$) and 50% of fifth-order gullies ($n = 8$) have appeared as 'V'-shaped and the rest in the intermediate class. No gullies are found to be falling in the 'U'-shaped categories. But one trend is clear that fourth- and fifth-order gullies compared to others are higher in the intermediate class and having a tendency to come under 'U'-shaped signifying more hydraulic actions over the multi-slope segments. The boxplot (Fig. 5.6h) portrays that the highest IQR (0.23) is observed for fifth order and the highest Med (0.38) for fourth order. Besides, the rests are observed as IQR = 0.19, Med = 0.33 for first order; IQR = 0.20, Med = 0.25 for second order; IQR = 0.19, Med = 0.28 for third order; IQR = 0.14 for fourth order; and Med = 0.25 for the fifth order.

5.4.2 Areal Perspective on Cross-Sectional Morphology

5.4.2.1 Cross-Sectional Area

The cross-sectional (CS) area is found to be increasing gradually with the gully order. For example, the first-order gully portrays the average CS area of about 1.4 m^2 which has increased steadily throughout the intermediate order to reach a maximum of about 70 m^2 at the fifth-order gully (Table 5.2). This gradual increase is obvious because of the gully widening and deepening actions in response to excessive volume of water downstream (Fig. 5.7a–e). Concerning total area, both the Med and IQR are gradually rising with increasing gully. The order-wise distribution (from first to fifth) of Med is 1.07, 4.53, 6.19, 10.51 and 57.54 and IQR is 1.11, 14.78, 17.53, 18.26 and 55.12 (Fig. 5.8a).

5.4.2.2 Areal Asymmetry ($A*$, A_1, A_2)

Assessing gully asymmetry is best understood measuring the areal dimensions of the gullies. The results indicate that there is a distinct trend in the asymmetry pattern. The initial gully order (first order) and highest gully order (fifth order) appear more symmetric compared to the intermediate orders (Fig. 5.7a–e). Regarding $A*$ index the cross-sectional asymmetry has been computed to be around 0.15 for the initial and the terminal orders, while it is quite high (~0.25) for the second and third orders. Similarly, the range and standard deviation are also high for the intermediate orders (Table 5.2). Based on the absolute values of $A*$, the distributional pattern of the cross sections indicates that out of 62 cross sections, 29 cross sections are falling in the left asymmetric category while the 28 in the right asymmetry category and the 5 are perfectly symmetric. The boxplot

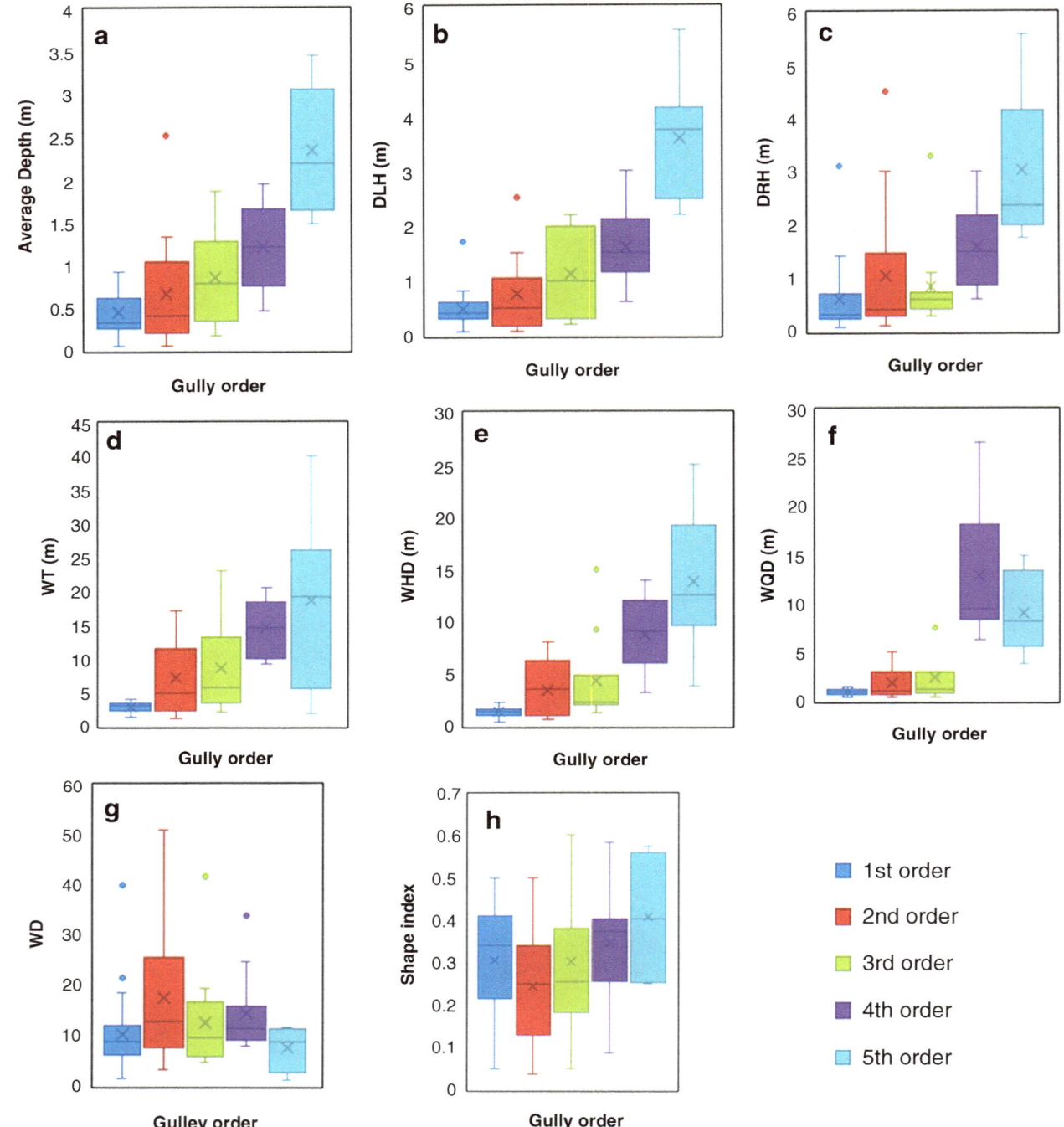

Fig. 5.6 Width and depth parameters. (**a**) Average depth, (**b**) DLH, (**c**) DRH, (**d**) WT, (**e**) WHD, (**f**) WQD, (**g**) *W/D*, (**h**) shape index

(Fig. 5.8a) portrays that interquartile ranges (IQR) are highest in third order (IQR = 0.49, median (Med) = 0.11) while that of rests are observed as IQR = 0.32, Med = −0.06 for first order; IQR = 0.31, Med = 0.01 for second order; IQR = 0.28, Med = 0.00 for fourth order; and IQR = 0.29, Med = −0.03 for fifth order. Regarding A_1 index having a special consideration on thalweg line and channel centreline, similar results

were echoed—the highest average asymmetry was observed for the third-order gullies (~0.60) and lowest for the fifth-order gullies (~0.30). The distributional pattern of the cross sections indicates that out of 62 cross sections, 52 cross sections are falling in the right asymmetric category while the rest 10 cross sections are perfectly asymmetric. The boxplot (Fig. 5.8b) portrays that interquartile ranges (IQR)

Table 5.2 Descriptive statistics of the areal asymmetry

Parameters	First order	Second order	Third order	Fourth order	Fifth order
Total area (m^2)	1.32 ± 0.78 (0.1–2.98)	9.33 ± 12.51 (0.08–42)	10.37 ± 10.88 (0.57–35.16)	22.24 ± 17.14 (5.54–48)	67.27 ± 42.15 (25.25–154.75)
$A*$	0.19 ± 0.15 (0.00–0.54)	0.21 ± 0.16 (0.00–0.43)	0.24 ± 0.13 (0.07–0.50)	0.12 ± 0.09 (0.00–0.26)	0.14 ± 0.12 (0.00–0.31)
A_1	0.44 ± 0.40 (0.00–1.22)	0.37 ± 0.31 (0.00–0.95)	0.56 ± 0.28 (0.20–0.99)	0.32 ± 0.36 (0.00–0.95)	0.29 ± 0.32 (0.00–0.73)
A_2	0.26 ± 0.25 (0.00–0.86)	0.22 ± 0.19 (0.00–0.60)	0.31 ± 0.16 (0.13–0.60)	0.16 ± 0.17 (0.00–0.47)	0.13 ± 0.15 (0.00–0.33)
Concavity (CI)	0.50 ± 0.09 (0.30–0.62)	0.50 ± 0.11 (0.33–0.67)	0.49 ± 0.15 (0.17–0.68)	0.52 ± 0.14 (0.22–0.79)	0.59 ± 0.13 (0.29–0.73)
Erosiveness (EI)	0.49 ± 0.07 (0.31–0.65)	0.52 ± 0.15 (0.36–0.96)	0.48 ± 0.15 (0.18–0.80)	0.51 ± 0.12 (0.21–0.69)	0.52 ± 0.22 (0.002–0.69)

Note: Computed by the authors 2017 (Values within parentheses indicate range of the observation, and values preceding and following the ± denote arithmetic mean and standard deviation respectively)

are highest in fourth order (IQR = 0.75, Med = 0.19) while that of rests are observed as IQR = 0.56, Med = 0.36 for first order; IQR = 0.46, Med = 0.35 for second order; IQR = 0.58, Med = 0.49 for third order; and IQR = 0.33, Med = 0.02 for fifth order. Similarly regarding A_2 index the highest asymmetry was recorded by the third-order gullies (0.31) and lowest for the fifth order (0.13). The distributional pattern of the cross sections indicates that out of 62 cross sections, 52 cross sections are falling in the right symmetric category while the rest 10 cross sections are perfectly asymmetric. The boxplot (Fig. 5.8c) depicts that fourth-order records the highest IQR (0.38) while rests are observed as IQR = 0.37 for first order; IQR = 0.26 for second order; IQR = 0.31 for third order; and IQR = 0.30 for fifth order. Similarly the highest Med (0.27) is observed for third order, while the rests are observed as Med = 0.18 for first order, Med = 0.21 for second order, Med = 0.11 for fourth order, and Med = 0.04 for fifth order.

5.4.2.3 Concavity Index and Erosiveness

To estimate the degree of erosion in the cross-sectional morphology, concavity (CI) and erosiveness (EI) have been computed which reflects that the higher gully orders have more concavity and erosiveness. This is due to the development of the gullies with time. In the study area, we have found that highest CI (~0.60) and EI (~0.55) are recorded by the fifth-order gullies while the lowest concavity (0.49) and erosiveness (0.48) are observed for the third-order gully. The boxplot for concavity (Fig. 5.8e) portrays that IQR is highest for third order (IQR = 0.34) while highest Med (0.64) is noted for fifth order. Besides, we have observed IQR = 0.09, Med = 0.52 for first order; IQR = 0.16, Med = 0.51 for second order; Med (0.48) for the third order; IQR = 0.14, Med = 0.50 for fourth order; and IQR = 0.14 for fifth order.

The boxplot for EI (Fig. 5.8f) portrays that fifth-order gullies depict the highest IQR (0.20) and the highest Med (0.61) as well. Besides, IQR (0.05), Med (0.50) have been observed for the first order; IQR(0.13), Med = (0.50) for

second order; IQR(0.16), Med (0.51) for third order; and IQR (0.11), Med (0.51) for fourth order.

5.4.3 Slope Asymmetry

The law of unequal slope of Gilbert (1877) states that differential erosion by the subaerial denudation processes produces largely asymmetric ridge. Following Wood (1942), we have measured the basic four elements of slope (convex, free face, rectilinear and concave) on both the left and right bank of each cross section of different orders. The slope asymmetry depicts that convex and rectilinear have more asymmetry compared to the free face and basal component (Fig. 5.9). However, there exists inter-order variation in each slope segment. For example, rectilinear has in general more asymmetry (about 0.15) but lesser (about 0.05) for the lower-order gullies (Fig. 5.9).

It is worth noting that though inter-order variation exists for different slope segments, each fundamental slope segment is unique and hence demands treatment separately. Field measurements in the study area suggest that slope magnitude for convex segment is oscillating around 16° on the left bank (LB) and 17° on the right bank (RB). Similarly, concave also portrays 12° for LB and 11° for RB. However, the free face exceeds 75° on LB and 70° on RB. Besides, rectilinear is found to be moving around 30° on LB and 31° RB (Table 5.3).

To detect the asymmetry in each slope segment, we have performed residual analysis that depicts the divergence between the LB and RB. The highest average residual is recorded by the rectilinear and convex segments, while the lowest is recorded by the free face and basal (Fig. 5.10). This states that the cross sections consisting of basal segments would be more symmetrical compared to that of the convex. Rectilinear is a common element found in almost every cross section. That's why all cross sections tend to be asymmetrical

Fig. 5.7 Nature of gully asymmetry of different orders. (**a**) First-order gully, (**b**) second-order gully, (**c**) third-order gully, (**d**) fourth-order gully and (**e**) fifth-order gully (Source: Field Photograph 2017)

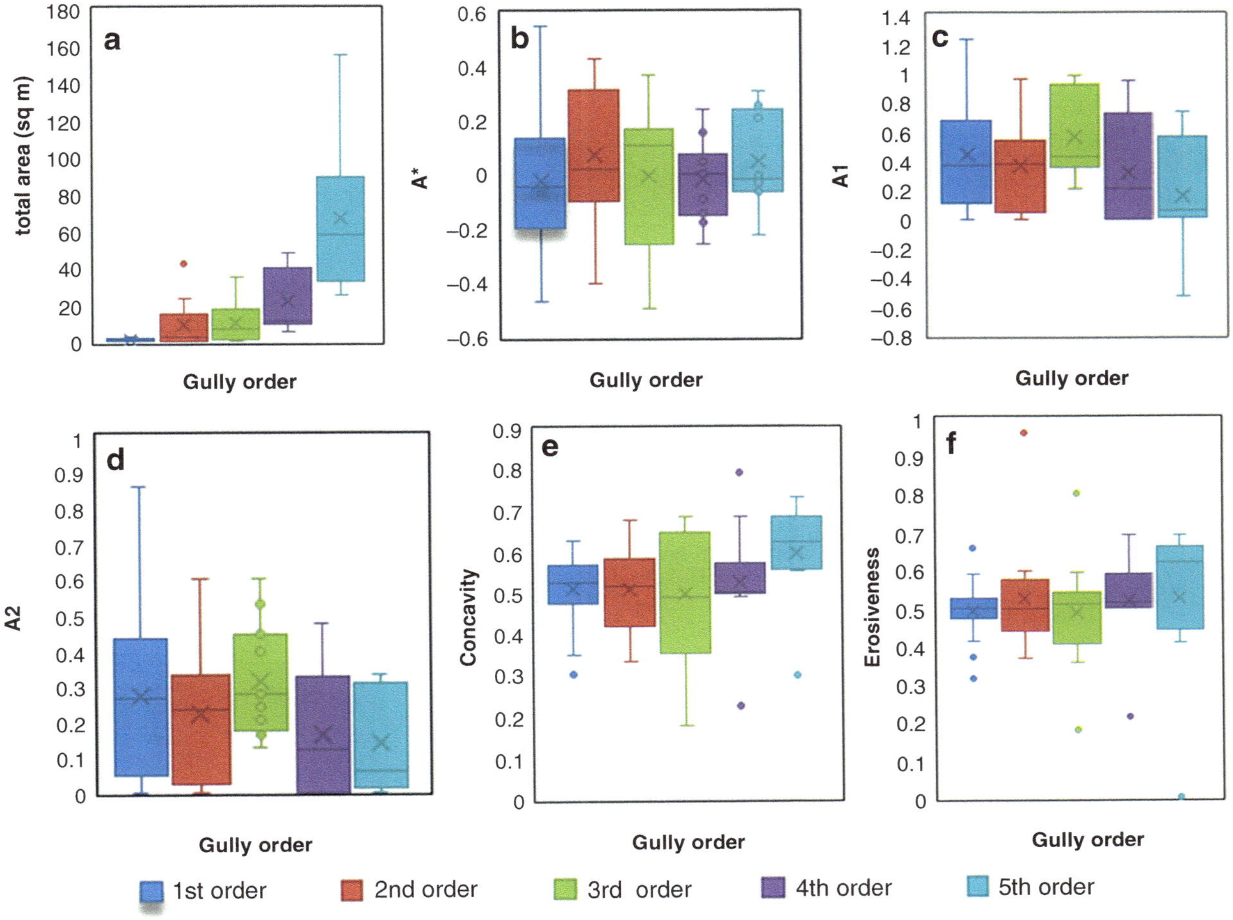

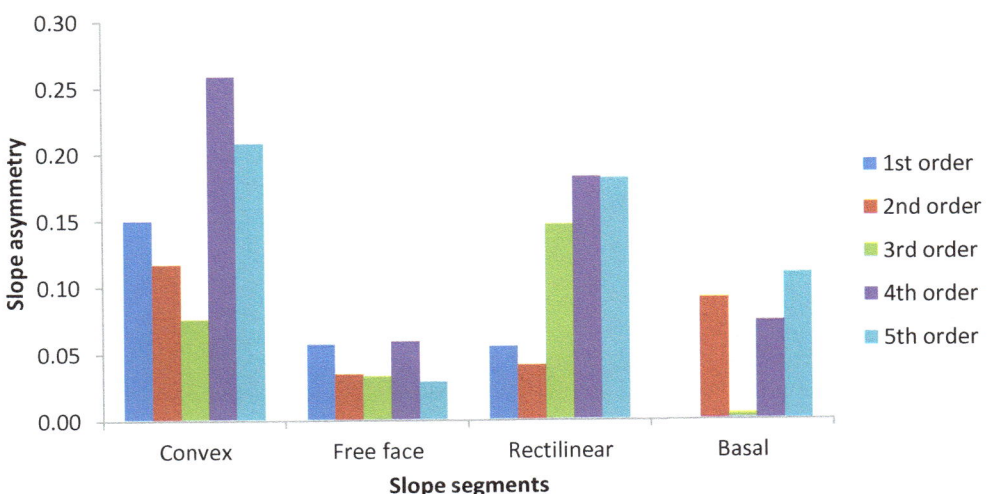

Fig. 5.8 Areal asymmetry of the gully. (**a**) Total area, (**b**) A∗ index, (**c**) A_1 index, (**d**) A_2 index, (**e**) concavity, (**f**) erosiveness

Fig. 5.9 Segment-wise slope asymmetry (mod value) of different orders (Note: Convex 'n' = 2 for each first, second and third order, n = 4 for fourth order, n = 3 for fifth order; free face 'n' = 1 for first order, n = 2 for each second and third order, n = 4 for fourth order, n = 3 for fifth order; rectilinear 'n' = 4 for each five order; basal 'n' = 2 for each second and third order, n = 4 for fourth order, n = 3 for fifth order)

Table 5.3 Slope magnitude in various slope segments

Slope	Convex		Free face		Rectilinear		Concave	
	LB	RB	LB	RB	LB	RB	LB	RB
Average	15.87°	16.07°	75.10°	70.30°	29.74°	30.71°	11.33°	10.79°
CV (%)	31.99	68.33	19.82	22.36	27.08	46.66	71.12	59.82
n	13	13	12	12	20	20	12	12

Note: Computed by the authors 2017

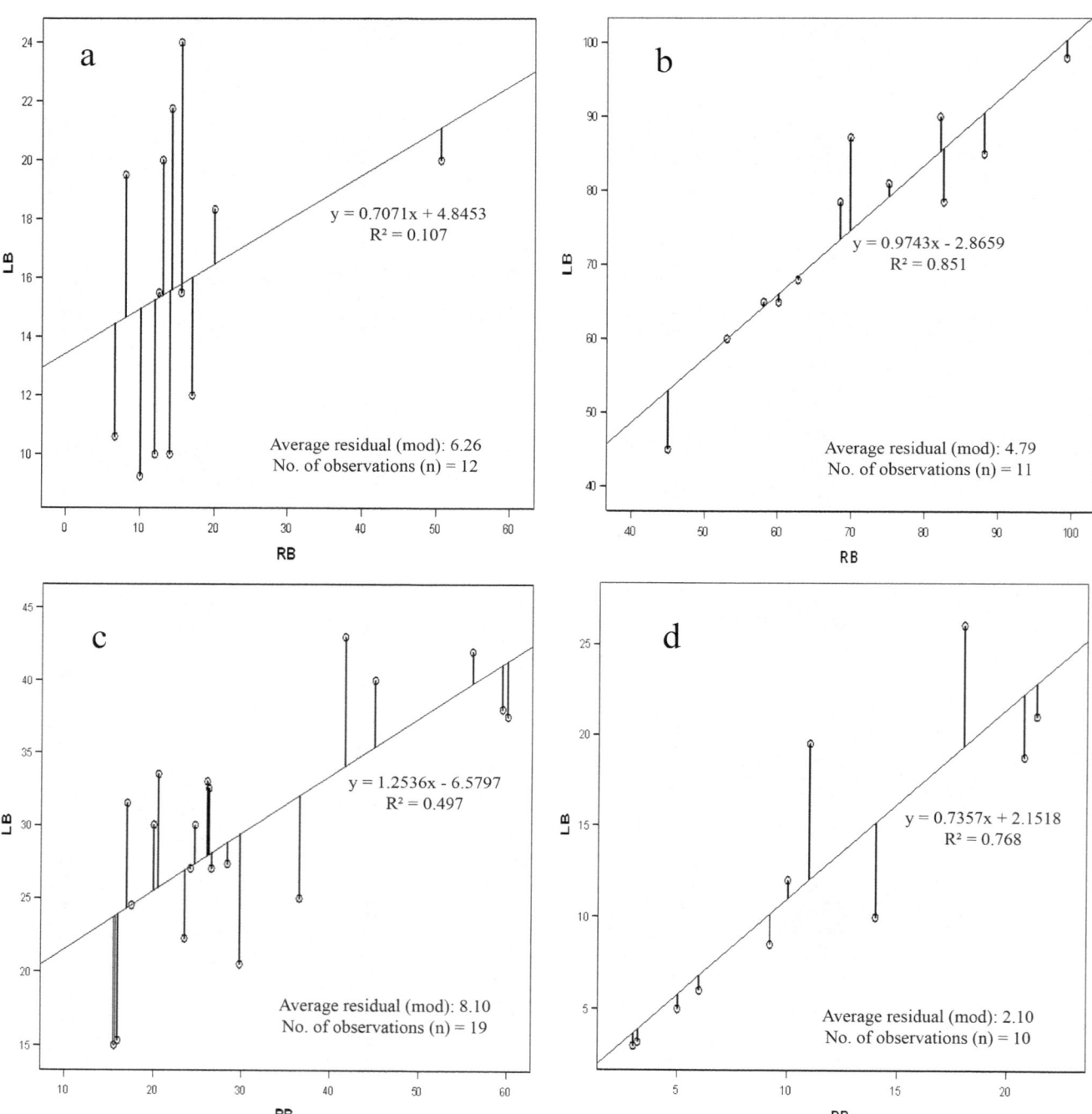

Fig. 5.10 Residual between left bank (LB) and right bank (RB) for the different gully slope segments. (**a**) Convex, (**b**) free face, (**c**) rectilinear, (**d**) concave

Table 5.4 Factor loadings and eigenvalues of the main components of the gullies

Basic variables	First order PC-I loadings	First order PC-II loadings	Second order PC-I loadings	Second order PC-II loadings	Third order PC-I loadings	Third order PC-II loadings	Fourth order PC-I loadings	Fourth order PC-II loadings	Fifth order PC-I loadings	Fifth order PC-II loadings
D	0.978	−0.107	0.918	0.354	0.250	0.891	0.951	0.236	0.924	0.324
DLH	0.507	−0.332	0.885	0.326	−0.004	0.790	0.793	0.160	0.076	0.968
DRH	0.739	−0.321	0.965	0.148	0.274	0.786	0.791	0.433	0.782	0.286
WT	−0.107	0.673	0.418	0.892	0.947	0.109	0.416	0.888	0.928	−0.112
WHD	0.042	0.820	0.235	0.959	0.965	0.152	−0.030	0.969	0.974	0.063
WQD	−0.176	0.760	0.258	0.941	0.909	0.333	−0.858	0.429	0.815	−0.016
A	0.892	0.329	0.799	0.578	0.644	0.721	0.789	0.594	0.901	0.227
Eigenvalue	2.599 (37.131%)	2.036 (29.083%)	3.493 (49.900%)	3.190 (45.566%)	3.207 (45.810%)	2.701 (38.581%)	3.689 (52.699%)	2.532 (36.175%)	4.758 (67.964%)	1.193 (17.042%)

Note: Numerals within parentheses indicate the loadings/eigenvalues after varimax rotation with Kaiser normalization

to some extent. However, the higher magnitude depends on the presence of other segments, mainly convex. Thus, the fifth-order gully dominated by the huge concave and free face portrays less asymmetry compared to the rest. So it can be mentioned that the sequence of slope segments greatly determines the asymmetrical behaviour of the gullies.

5.4.4 Identification of Elements Influencing Cross-Sectional Morphology of Gullies

To identify the major elements of cross-sectional morphology of gully across different orders, we run a PCA taking seven basic parameters—D, DLH, DRH, WT, WHD, WQD and A. Gullies of the first order, second order, and fourth order have appeared similar in respect of factor-component relation. Four major factors—D, DLH, DRH and A—having factor loadings >0.5 control the system in positive dimension. The D emerged as the principal component in the PC-I for the first order and fourth order while DRH for the second order. But the factor loadings of the third-order gullies have behaved quite differently. In the case of WT, WHD, WQD and A are found to be dominating the system with higher factor loadings (>0.6). It is to note that in general with increasing order of gully, eigenvalues in PC-I have been increased substantially. The eigenvalues in PC-I for first, second, third, fourth and fifth are 37, 49, 45, 52% and 67%, respectively (Table 5.4). For fifth-order gullies, all the six factors except DLH have factor loading >0.75, thereby resulting in very high eigenvalues that indicate a strong factorial relationship. However, in PC-II there is no systematic factorial relation across the gullies—WT for first order has appeared as the main component. The two-dimensional component plot (Fig. 5.11) after varimax rotation for the different gully orders portrays that the centroid or the distribution is located at positive-positive (++) quadrant. In general, we have observed that the width-related parameters are located opposite the depth parameter

with the location of 'A' transverse to the width-depth axis. So the clustering of variables is observed for WHD, WQD, and WT and another for DLH, DRH, and D especially observed for second- and third-order gullies. However, the first-order gullies only have portrayed a minor cluster between WT and WQD. In the case of first-, second- and fourth-order gullies, width parameters are pointing towards north-west corner, while depth parameters are tipping towards south-west corner. However, third-order gullies portray the reverse. One general trend that appears in this plot is that with increasing gully order, factor loadings have been increasing for most of the variables. So the gradual transformation from a distal component location to the nearest location establishes the systematic relationship between the variables for higher gully order.

5.4.5 Association Among Gully Morphometric Indices

To glean out the perspective of association between different morphometric indices, another PCA is undertaken considering three areal asymmetry indices ($A*$, A_1 and A_2) and four shape indices (W/D ratio, SI, CI and EI). Regarding factor loadings, first-order and fifth-order gullies appear similar. In the PC-I after varimax rotation, four indices ($A*$, SI, CI and EI) having loadings >0.6 appear to be dominating the system for both the first and fifth orders (Table 5.5). The only difference between these two systems lies in that $A*$ tends to dominate the system with negative loadings for first order while positive for the fifth order (Fig. 5.12a–e). For the second-and third-order gullies, areal asymmetry indices ($A*$, A_1 and A_2) were the main components to control the system, while the fourth-order gullies are portraying the system controlled by the shape indices (SI, CI and EI). Besides, we have observed that the eigenvalue also has increased with increasing gully order, thereby implying the higher strength of association among the indices (Fig. 5.12).

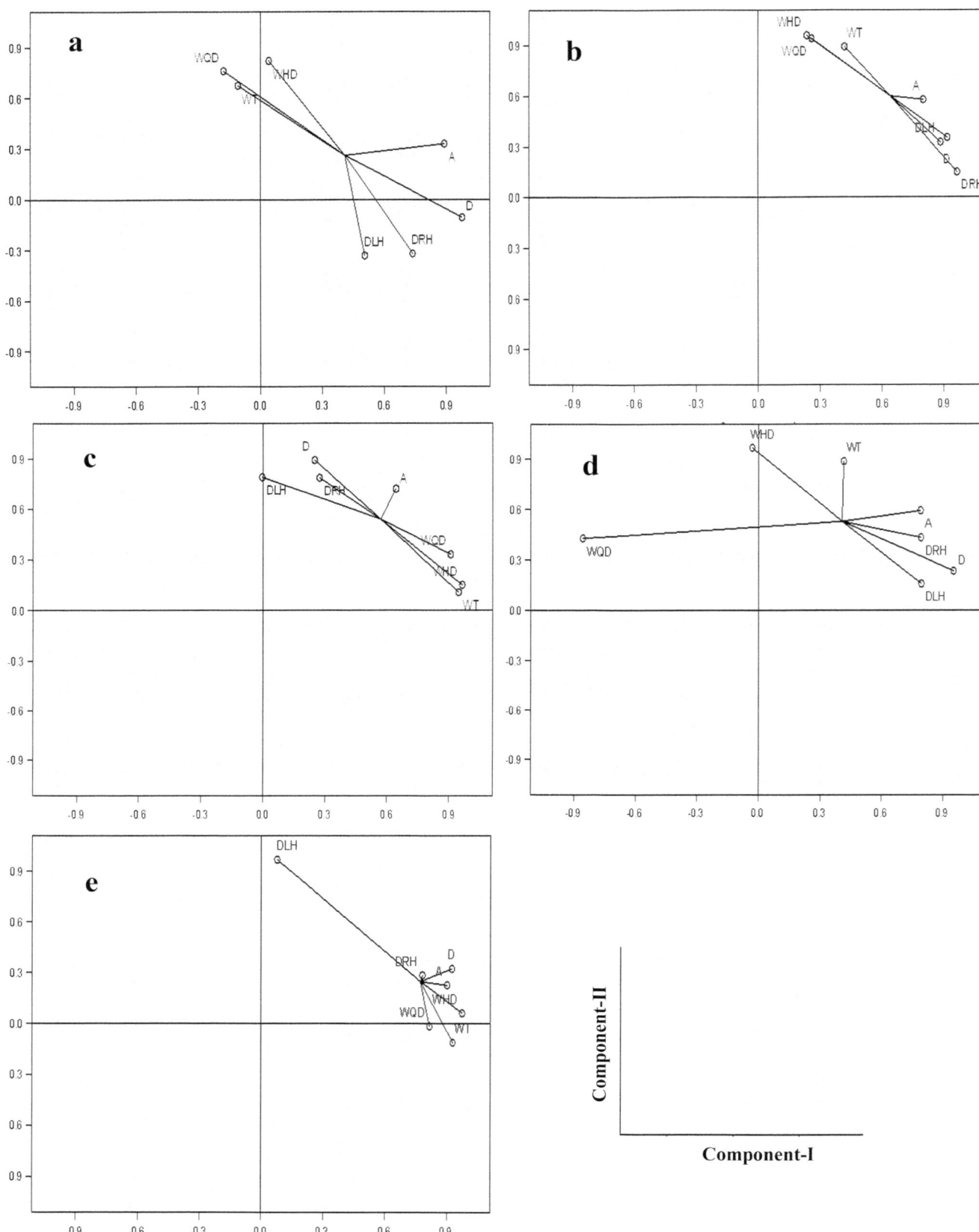

Fig. 5.11 Component plot on rotated space. (**a**) First-order gully, (**b**) second-order gully, (**c**) third-order gully, (**d**) fourth-order gully, (**e**) fifth-order gully

Table 5.5 Factor loadings and eigenvalues of various morphometric indices

Ratio variables	First order		Second order		Third order		Fourth order		Fifth order	
	PC-I loadings	PC-II loadings	PC-I loadings	PC-II loadings	PC-I loadings	PC-II loadings	PC-I loadings	PC-II loadings	PC-I loadings	PC-II loadings
$A*$	−0.664	0.170	0.877	0.251	0.480	−0.437	−0.029	0.723	−0.825	0.339
A_1	0.013	0.982	0.931	−0.140	0.914	0.325	0.307	0.857	0.679	0.484
A_2	−0.079	0.980	0.932	−0.171	0.923	0.274	0.165	0.945	−0.112	0.924
W/D	0.235	−0.430	−0.043	0.762	−0.681	0.151	0.091	0.658	0.887	−0.128
SI	0.599	0.063	−0.380	−0.512	−0.206	0.020	0.950	0.015	0.720	0.467
CI	0.952	−0.131	−0.541	0.776	−0.052	0.993	0.976	0.099	0.841	−0.277
EI	0.809	−0.225	0.000	0.907	0.209	0.868	0.920	0.336	0.710	0.091
Eigen value	2.423 (34.613%)	2.211 (31.582%)	2.946 (42.079%)	2.380 (33.993%)	2.470 (35.286%)	2.135 (30.497%)	2.831 (40.449%)	2.706 (38.652%)	3.354 (47.916%)	1.841 (26.295%)

Note: Computed by the authors 2017

5.5 Discussions

5.5.1 Identifying Inter- and Intra-Order Variation in Gully Morphology

To detect the differences in cross-sectional morphology, ANOVA has been carried out for seven areal and morphometric indices ($A*$, A_1, A_2, W/D ratio, SI, CI and EI) for five different orders of gully. Regarding $A*$, A_1, A_2, W/D ratio, CI and EI, the first-order gullies (n=20) portray no significant differences between the cross-sectional morphology as depicted by the higher P value (> 0.05) and acceptance of null hypothesis at 0.05 significance level for 3 degrees of freedom (Table 5.6). However, there is a difference in SI as portrayed by the higher F value ($F = 3.243$, F critical $= 3.239$) and lesser P value (0.05) at 0.05 significance level at 3 degrees of freedom. Regarding $A*$, A_2, W/D ratio, SI and EI portray the similar picture that of the first-order gully, i.e., null hypothesis is accepted for this case, thereby signalling no significant differences between 13 CS of second-order gullies in their morphological parameters. However, A_1 and CI depict that there is significant difference among the CS of the second order as indicated by the lower value of P (<0.05). Besides, 10 CS of third order, 10 CS of fourth order and 8 CS of fifth-order gullies appear similar in their morphology as inferred from the F test (null hypothesis) and P value (>0.05) for the all seven morphometric indices (Table 5.6).

Besides, in an attempt to identify the variation of CS morphology between lower order (first order), intermediate order (second and third orders together) and higher order (fourth and fifth orders together), another set of ANOVA is performed. Regarding $A*$ and SI we have observed significant difference among lower, intermediate and higher gully orders as indicated by the confirmation of alternative hypothesis at 90% confidence level against 2 degrees of freedom (df). Besides, A_1, A_2, W/D ratio and CI though not portraying significant difference among the different orders of gullies are marginal implying a tendency to become dissimilar. However, in terms of EI there is absolutely no difference among different orders of gully (Table 5.7).

5.5.2 Compositions of Slope Segments, Processes and Gully Asymmetry

The material composition of each slope segment plays an instrumental role in slope morphology. One of the most interesting observations is the sequential distribution of bed and bank materials' size. The convex portrays the heavily specialized distribution of material dominated by fine gravel (>2.36 mm) and coarse sand (0.6–2.36 mm). There is clear gradation of material from convex to free face and rectilinear slope elements. The material is trending to the more uniform in distribution instead of the heavily specialized one. This trend is clearly observed for the basal components where there is relatively equal proportion of fine sand (<0.6 mm), coarse sand and fine gravel (Fig. 5.13).

In other words, the largest particle size is found at the top and smallest on the base of the gully slope (Table 5.8).

The average D_{50} at summital convex slope element is the largest (D_{50}:5.66 mm) followed by 5.42 mm at free face, 4.86 mm at rectilinear slope element and 4.42 mm at concave slope element at base, respectively. It is apparent that these lithological arrangements have significant bearing on gully asymmetry. First-order gullies are associated with summital convex and rectilinear slope elements. The summital convexity of Gangani gullies is shaped by combination of processes (Fig. 5.14) like decreasing overland flow towards crest (Gilbert 1909); upward increase of rock strengths (Schumm and Mosley 1973); rapid degradation of shoulder in comparison to upper portion of cliff (Wood and King,); progressive steepening of slope away from crest through the process of disposal of increased waste by creep (Schumm and Mosley 1973); increased weathering at the shoulders (Ahmad 1985); creep at crest and rain wash at slope (Savigear 1952).

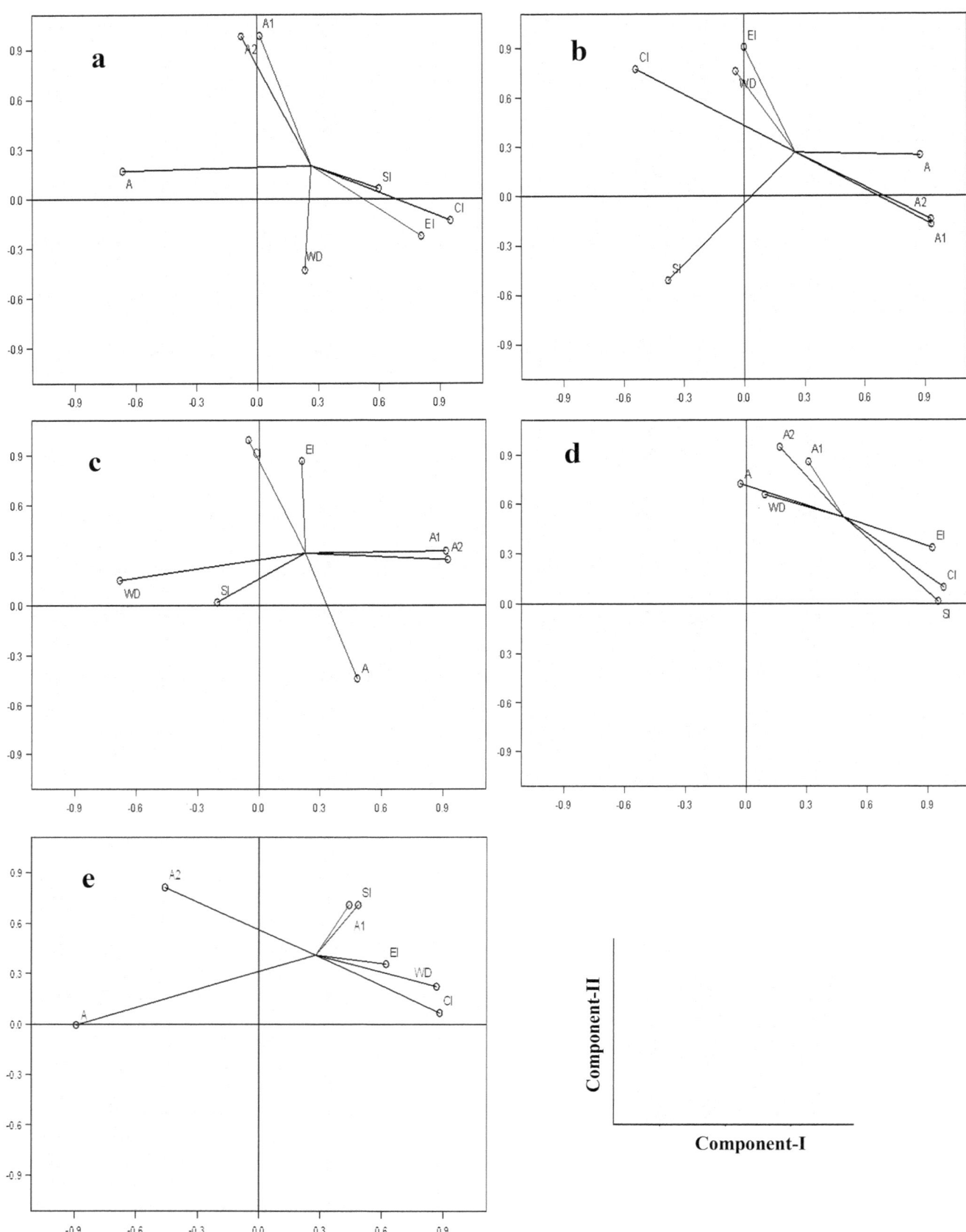

Fig. 5.12 Component plot on rotated matrix. (**a**) First-order gully, (**b**) second-order gully, (**c**) third-order gully, (**d**) fourth-order gully, (**e**) fifth-order gully

Table 5.6 ANOVA for intra-order differentiation

Order		df	Significance level	F	P-value	F critical	Remarks
First	$A*$	3	0.05	0.206	0.891	3.239	Null
	A_1	3	0.05	2.692	0.081	3.239	Null
	A_2	3	0.05	3.184	0.052	3.239	Null
	W/D	3	0.05	0.038	0.990	3.239	Null
	SI	3	0.05	3.243	0.050	3.239	Alternative
	CI	3	0.05	0.088	0.966	3.239	Null
	EI	3	0.05	0.362	0.781	3.239	Null
Second	$A*$	3	0.05	0.786	0.531	3.863	Null
	A_1	3	0.05	5.510	0.048	5.409	Alternative
	A_2	3	0.05	0.990	0.440	3.863	Null
	W/D	3	0.05	0.410	0.750	3.863	Null
	SI	3	0.05	0.309	0.818	3.863	Null
	CI	3	0.05	5.165	0.023	3.863	Alternative
	EI	3	0.05	3.052	0.085	3.863	Null
Third	$A*$	3	0.05	0.271	0.844	4.347	Null
	A_1	3	0.05	0.691	0.586	4.347	Null
	A_2	3	0.05	0.812	0.527	4.347	Null
	W/D	3	0.05	1.384	0.325	4.347	Null
	SI	3	0.05	1.402	0.320	4.347	Null
	CI	3	0.05	0.266	0.848	4.347	Null
	EI	3	0.05	0.916	0.481	4.347	Null
Fourth	$A*$	3	0.05	0.203	0.891	4.757	Null
	A_1	3	0.05	0.382	0.770	4.757	Null
	A_2	3	0.05	0.369	0.779	4.757	Null
	W/D	3	0.05	0.817	0.530	4.757	Null
	SI	3	0.05	1.195	0.388	4.757	Null
	CI	3	0.05	1.460	0.316	4.757	Null
	EI	3	0.05	0.936	0.479	4.757	Null
Fifth	$A*$	2	0.05	1.783	0.260	5.786	Null
	A_1	2	0.05	0.344	0.725	5.786	Null
	A_2	2	0.05	0.454	0.659	5.786	Null
	W/D	2	0.05	0.618	0.576	5.786	Null
	SI	2	0.05	2.091	0.219	5.786	Null
	CI	2	0.05	1.315	0.348	5.786	Null
	EI	2	0.05	0.802	0.499	5.786	Null

Note: Computed by the authors 2017

Table 5.7 ANOVA for inter-order differentiation

Morphometric indices	df	Significance level	F	P-value	F critical	Remarks
$A*$	2	0.1	2.449	0.095	2.395	Alternative
A_1	2	0.1	2.086	0.133	2.395	Null
A_2	2	0.1	2.029	0.141	2.395	Null
W/D	2	0.1	1.612	0.208	2.395	Null
SI	2	0.1	2.839	0.066	2.395	Alternative
CI	2	0.1	1.696	0.192	2.395	Null
EI	2	0.1	0.206	0.815	2.395	Null

Note: Computed by the authors 2017 (ANOVA for $A*$ index is undertaken on mod value)

As overland flow remains far below the threshold point to initiate erosion, free face element is least affected by the erosivity of flowing water. It is subjected to slipping and sliding (King 1957). Unless channels are dominated by processes of sinuous and asymmetric flow of water, the gully shape maintains its symmetry. First- and second-order gullies are excavated in the coarsest (D_{50} : 5.66mm) materials and due to insufficient flow lengths, erosivity of concentrated

Fig. 5.13 Ternary plot of sediment size class. (**a**) Convex (left bank), (**b**) convex (right bank), (**c**) free face (left bank), (**d**) free face (right bank), (**e**) rectilinear (left bank), (**f**) rectilinear (right bank), (**g**) concave (left bank), (**h**) concave (right bank)

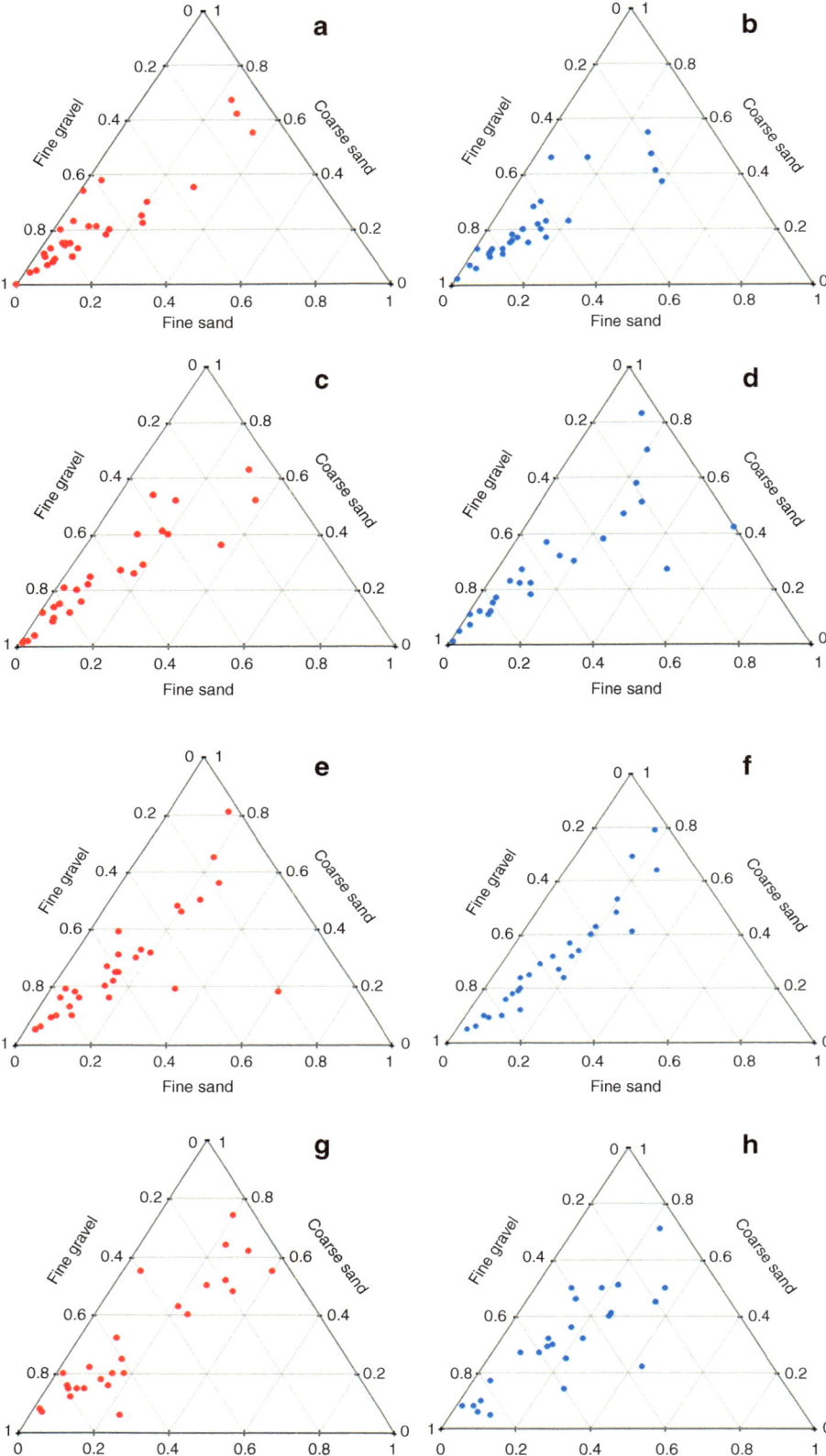

Table 5.8 Average D_{50} of various slope segments

Slope element	Average D_{50} (mm)
Convex	5.66
Free face	5.42
Rectilinear	4.86
Basal concave	4.42

flow was insignificant there. As a result, gullies are much more symmetric ($A*$, A_1, A_2) in nature with summital convex and free face element (Fig. 5.7a, b).

It is noted that particle size (D_{50}) of substrate materials decreases downwards from the surface. The third- and fourth-gullies have excavated channels to a sufficient depth and receive sufficient volume of flowing water that supersedes the resistance of relatively finer materials (D_{50}: 5.42 to 4.86 mm). So, erosive action of flowing water dominates over processes of weathering and mass-wasting which were principal processes in first- and second-order gullies. As a result, easily adjustable substrate materials get readily shaped by sinuous and asymmetric flow of water. Although summital convex and free face elements of third- and fourth-order gullies are identical in nature on both banks, due to asymmetric nature of flowing water, lengths and slope angles of rectilinear elements from opposite banks became different (Fig. 5.7c, d) resulting in higher gully asymmetry ($A*$, A_1, A_2).

Fig. 5.14 Essence of gully asymmetry lies in processes operating on slope elements. Weathering and mass-wasting on similar substrate result in similar slopes in upper slope elements. Rectilinear and basal elements are much more affected by the action of flowing water which produces different slope angles and lengths resulting in gully asymmetry

Morphological Combination   Sedimentological Combination

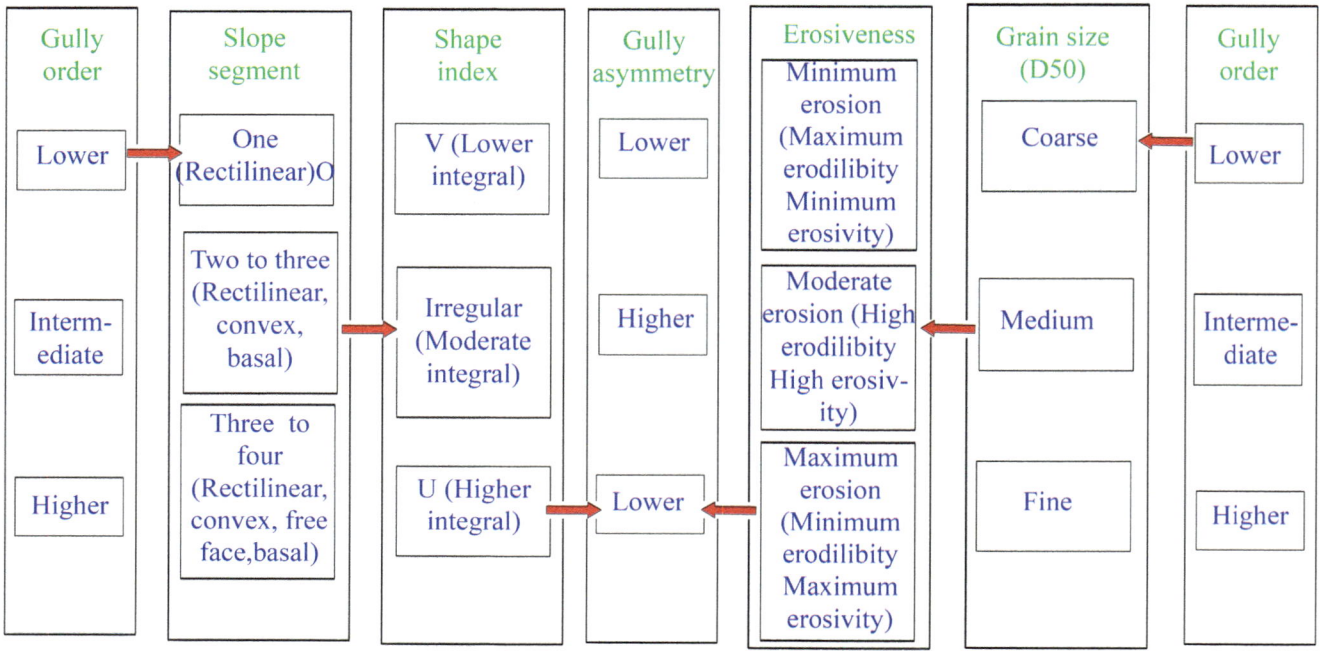

Fig. 5.15 Order-asymmetry model of gully development

Volume of flowing water in shaping form is the highest in fifth-order gullies in comparison to first-, second-, third- and fourth-order gullies. Yet, finer and cohesive particle of substrate (D_{50}: 4.42 mm) of concave slope element with higher resistance against erosion is not readily susceptible to water action. Moreover, wide flat channel bed occasionally covered with grasses (finer particle and higher depth facilitate soils to retain sufficient water to support grasses in non-perennial channel bed) forces moving sheet of water to flow in laminar fashion which is turbulent/helical in bare channels of third- and fourth-order gullies. As a result, similar concave element and wide flatbed produce more symmetric gullies (Fig. 5.7e).

5.5.3 Modelling Asymmetric Behaviour of Gully

In nature though there is uncertainty and variability in order-asymmetry relation, it may be deduced that the first-order and fifth-order gullies emerge similar from the perspective of cross-sectional morphology and asymmetry behaviour. Hence, simplifying the complexity of real-world gully, we have proposed a simple model on gully order-asymmetry that depicts the relationship exquisitely (Fig. 5.15).

It may be due to the morphological and sedimentological factors. The morphological perspective takes the consideration of the shape of the gullies. In general, lower-order gully

is 'V'-shaped, while higher order becomes 'U'-shaped and middle order is located in between. It is established that if a gully is perfectly 'V'-shaped, it reflects symmetric behaviour because of minimum erosion while 'U'-shaped gullies having maximum erosion also portray the same. However, intermediate orders portray differential erosion and higher asymmetry. In general, the free face and concave segment maintains the symmetry, while the convex and rectilinear makes slope profile asymmetric. This is reflected from the higher residual of slope angle between the opposite banks for rectilinear (8.10 (mod)) and convex (6.26 (mod)) and lower residual for free face (4.79 (mod)) and concave (2.10 (mod)). However, lower order convex and rectilinear elements maintain a relative symmetry by dint of which the lower order CS also become symmetric. The fifth-order gullies dominated by convex, free face and concave elements are symmetric compared to the gullies of intermediate orders dominated by convex and rectilinear elements. On the other hand, the sedimentological perspective also states that lower-order gully has coarser D_{50} but minimum erosion due to minimum erosivity while higher-order gully has fine cemented materials offering resistance to erosion by channelized flow of water, thereby producing lesser asymmetry. However, the intermediate gully materials move freely due to moderate erodibility and high erosivity. In other words, they are susceptible to asymmetric processes (sinuous and helical flow) of water action (with higher erosivity) which erode relatively easily adjustable sediments from bed and lower part of banks giving them asymmetric shape.

5.6 Conclusion

Gully asymmetry is a fundamental essence of nature triggered due to crossover of the threshold of erodibility and erosivity sequence and complex geomorphic response. In the present study, we found that virtually all the 62 CS have emerged as asymmetrical. The areal asymmetry index indicates that lower-order and higher-order gullies are more symmetrical ($A*$ (0.15), A_1 (0.35), A_2 (0.19)) while the intermediate-order gullies depict a relatively higher asymmetry ($A*$ (0.20), A_1 (0.4), A_2 (0.25)). The shape index similarly portrays that lower-order gullies are perfectly 'V'-shaped (0.5) and higher having a tendency towards 'U'-shaped (0.56). It is apparent that this basic nature is due to the morphological and sedimentological differences in various segments of the slope on either side of the CS. Thus, the interplay between erodibility and erosivity factors in general and weathering, creep and water action in particular made first- and fifth-order gullies less asymmetric in comparison to intermediate gullies. Asymmetry in forms and processes in gully geomorphology is of keen interest for knowing all about gully asymmetry which may armament geomorphology providing answer to where, how, why and how much related to gullying and land degradation and basin management. However, to reach that end of generalization, more studies from diversified regions are yet to be done.

Acknowledgements The authors acknowledge the active participation of the postgraduate special paper (Advanced Geomorphology) students (session: 2015–2017) of the Department of Geography, Aliah University, India, in the fieldwork.

References

Ahmad E (1985). Geomorphology, Kalyani Publishers, Ludhiana, p.38–78

Bandyopadhyay, S. (1988). Drainage evolution in badland terrain at Gangani in Medinipur District, West Bengal. *Geographical Review of India, 50*(3), 10–20

Bradford, J. M., & Piest, R. F. (1980). Erosional development of valley-bottom gullies in the Upper Midwestern United States. In D. R. Coates, & J. D. Vitek, Thresholds in Geomorphology (pp. 75–101)

Boughton, W. C. (1989). A review of the USDA SCS curve number method. *Australian Journal of Soil Research, 27*(3), 511–523

Chakraborty, S. C. (1969). Physiographic evolution of Bengal basin. In A. B. Chatterjee, A. Gupta, & M. K. Mukhopadhyay, *West Bengal.* Calcutta

Cheng H., Zou X., Wu Y., Zhang C., Zheng Q., Jiang Z. (2007). Morphology parameters of ephemeral gully in characteristics hillslopes on the Loess Plateau of China. Soil and Tillage Research 94 (1):4–14

Das, D. (2014). Identification of erosion prone areas by morphometric analysis using GIS. *J. Inst. Eng. India Ser. A, 95*(1), 61–74

Das, BC (2015). Modeling of most efficient channel form: A quantitative approach. *Model. Earth Syst. Environ.* 1:15. https://doi.org/10.1007/s40808-015-0013-6

Das, K., & Bandyopadhyay, S. (1995). Badland development over laterite duricrust. In S. R. Jog, *Indian geomorphology* (Vol. 1, pp. 31–42). New Delhi: Rawat Publication

Das, BC, Islam, A (2015a). Channel asymmetry of an Ox-Bow Lake: A different perspective. *International Journal of Ecosystem*, 5 (3A):69–74 https://doi.org/10.5923/c.ije.201501.10

Das, BC, Islam, A (2015b). Formulation of channel bed-asymmetry indices and their application to the River Jalangi, India. *Asian Profile*, 43 (5), 457–463

Das, BC, Islam, A (2016a). Analysis of channel asymmetry: A different perspective (pp. 33–42) In Das, BC et al. (eds.), *Neo-thinking on Ganges-Brahmaputra Basin geomorphology*, Springer Geography, https://doi.org/10.1007/978-3-319-26443-1_3

Das, BC, Islam, A (2016b). An enquiry into fitting natural channel shape to geometric shape: A study on River Jalangi, India (pp. 65–80) In Das, BC et al. (eds.), *Neo-thinking on Ganges-Brahmaputra Basin Geomorphology*, Springer Geography, https://doi.org/10.1007/978-3-319-26443-1_5

Deng, Q. et al. (2015). Characterizing the morphology of gully cross-sections based on PCA: A case of yuanmou dry-hot valley. *Geomorphology, 228*, 703–713

Ding, L. et al. (2017). Morphology and controlling factors of the longitudinal profile of gullies in the yuanmou dry-hot valley. *Journal Of Mountain Science, 14*(4), 674–693

Ghosh, R. N., & Majumder, S. (1981). Neogene- quaternary sequence of Kasai Basin. *Neogene-quaternary boundary field con. India*, (pp. 63–73.)

Gilbert, GK (1877). Report on the geology of Henry Mountains, department of the interior. U. S. geographical and geological survey of the Rocky Mountain Region

Gilbert, GK (1909). The convexity of hilltops, Journal of Geology Vol. 17, No. 4, pp. 344–350

Guerra, A. J. et al. (2018). Gully erosion and land degradation in Brazil: A case study from São Luís municipality, Maranhão State. In J. Dagar, & A. Singh, Ravine Lands: Greening for livelihood and environmental security. Singapore: Springer.

Harvey, A. M. (1974). Gully erosion and sediment yield in the Howgill Fells, Westmorland. In K. J. Gregory, & D. E. Walling, Fluvial processes in instrumented watersheds (Vol. 6, pp. 45–58). Institute of British Geographers Special Publication

Hotelling, H. (1933). Analysis of a complex of statistical variables into principal components. *Journal of Educational Psychology, 24*, 417–441 & 498–520

Hu G., Wu Y., Liu B., Zhang Y., You Z., Yu Z. (2009). The characteristics of gully erosion over rolling hilly black soil areas of Northeast China. Journal of Geographical Sciences 19 (3):309–320

Ionita, I. et al. (2015a). Gully development in Eastern Romania: A case study from Falciu Hills. *Nat Hazards, 79*, S113–S138

Ionita, I. et al. (2015b). Gully erosion as a natural and human-induced hazard. *Nat Hazards, 79*, S1–S5

James, LA., Watson DG., Hansen WF. (2007). Using LiDAR data to map gullies and headwater streams under forest canopy: South Carolina, USA. CATENA 71 (1):132–144

Kar, A., & Bandyopadhyay, M. K. (1974). Mechanisms of rills: An investigation in microgeomorphology,. *Geographical Review of India, 36*(3), 204–215

King, LC (1957). The uniformitarian nature of hill slopes, Trans. Edin. Geol. Soc. Vol. 17, Part. 1, p. 81–102

Knighton, A. D. (1981). Asymmetry of river channel cross-sections: Part 1. quantitative indices. *Earth Surf. Process. Landf., 6*, 581–588

Li, H. et al. (2016). Effects of topography and land use change on gully development in typical mollisol region of Northeast China. *Chinese Geographical Science, 26*(6), 779–788

Leopold, L. B., & Wolman, M. G. (1960). River meanders. Geological Society of America Bulletin, 71(6), 769–793

Maerker, M. et al. (2015). A simple DEM assessment procedure for gully system analysis in the Lake Manyara area, Northern Tanzania. *Nat Hazards, 79*, S235–S253

Mashi, S. A. et al. (2015). Causes and consequences of gully erosion: Perspectives of the local people in Dangara Area, Nigeria. *Environ Dev Sustain, 17*, 1431–1450

Momm, H. et al. (2015). GIS technology for spatiotemporal measurements of gully channel width evolution. *Nat Hazards, 79*, S97–S112

Motoc, M. (1963). Soil erosion on agricultural land and its control. Ed. Agro-silvica, 318

Niyogi, D., & Mallick, S. (1973). Quaternary laterite of West Bengal: it's geomorphology, stratigraphy and genesis. *Quat J Geol Min Metal Soc India, 45*(4), 155–174

Nordstorm, K. (1988). Gully erosion in the Lesotho Lowlands. A geomorphological study of the interactions between intrinsic and extrinsic variables. PhD Thesis Uppsala: Uppsala University

Radoane, M. et al. (1995). Gully distribution and development in Moldavia, Romania. *Catena, 24*, 127–146

Ray, R., & Biswas, A. (1989). Environment and Tourism(mimeo). Conf. Nat. Geog. Assoc. , (pp. 60, 65–66, 230–231, 241). India, New Delhi

Rowntree, K. (2010). An assessment of the potential impact of alien invasive vegetation on the geomorphology of river channels in South Africa. Southern African Journal of Aquatic Sciences 17 (1–2):28–43

Samani, N. A et al. (2018). Assessment of the sustainability of the territories affected by gully head advancements through aerial photography and modeling estimations: A case study on samal watershed, Iran. *Sustainability, 10*(8), 2909. https://doi.org/10.3390/su10082909

Savigear RAG (1952). Some observations on slope development in south wales transactions and papers (Institute of British Geographers), hyperlink "https://www.jstor.org/publisher/black" Wiley on behalf of hyperlink "https://www.jstor.org/publisher/rgs" The royal geographical society (with the Institute of British Geographers), https://doi.org/10.2307/621019

Schumm SA and Mosley MP (1973). Slope morphology, benchmark papers in geology, Dowden, Hutchison & Ross, Inc, Stroudsburg, Pennsylvania

Seginer, I. (1966). Gully development and sediment yield. Journal of Hydrology 4:236–253

Sen, J. (2008). Geomorphology of Garhbeta Badlands: West Medinipur District, West Bengal. (Unpublished Doctoral Dissertation). University of Calcutta, Kolkata, India

Sen, J.et al (2000). Geomorphological processes & landforms: A study on Garhbeta Badlands West Bengal, India. *International Symposium on Gully erosion under global Change*, (p. 116). Leuven, Belgium

Sen, J. et al. (2002). Application of geotextile for soil erosion control in the gully eroded lateritic wastelands of West Bengal (Abstr.). *State Science Congress 2002*

Shit, P.K et al. (2014). Morphology and development of selected Badlands in South Bengal (India). *Indian Journal of Geography & Environment*. 13:161–171

Shit, P. K. et al. (2015). Modeling of potential gully erosion hazard using geo-spatial technology at Garbheta Block, West Bengal in India. *Modeling Earth Systems and Environment, 1*(2)

Sidorchuk, A. (1999). Dynamic and static models of gully erosion. *Catena, 37*, 401–414

Sidorchuk, A. (2001). GULTEM–The model to predict gully thermoerosion and erosion (Theoretical framework). In D. E. Stott, R. H. Mohtar, & G. C. Steinhardt, *Sustaining the global farm* (pp. 966–972). Purdue University and the USDA-ARS National Soil Erosion Research Laboratory

Tucker, G. et al. (2001). The channel-hillslope integrated landscape development model (CHILD). In R. S. Harmon, & W. W. Doe, *Landscape erosion and evolution modeling* (pp. 349–388). Boston, MA: Springer

United States Department of Agriculture (USDA). (1966). Procedure for Determining Rates of Land Damage, Land Depreciation and Volume of Sediment Produced by Gully Erosion. Washington DC, USA: USDA Soil Conservation Service Technical Release No. 32

Vandaele K., Poesen J., Govers G., van Wesemael B. (1996). Geomorphic threshold conditions for ephemeral gully incision. Geomorphology 16 (2):161–173

Wood, A. (1942). *The development of hillside slopes*. Proceedings of the Geologists Association, 63, 128–40

Woodburn, R. (1949). Science studies a gully. Soil Conservation, 15(1), 11–13

Woodward, D. E. (1999). Method to predict cropland ephemeral gully erosion. *Catena, 37*, 393–399

Yang, T. et al. (2012). DEM-based numerical modelling of runoff and soil erosion processes in the hilly–gully loess regions. *Stoch Environ Res Risk Assess, 26*, 581–597

Zachar, D. (1982). Soil Erosion. Amsterdam, Oxford and New York: Elsevier Scientific Publishing Company

Zgłobicki, W. et al. (2015). Gully erosion as a natural hazard: The educational role of geotourism. *Nat Hazards, 79*, S159–S181

Aznarul Islam is an Assistant Professor in the Department of Geography, Aliah University, Kolkata, India. He did Master of Science in Geography from Kalyani University, India, and M.Phil. and Ph.D. in Geography from the University of Burdwan, India. He has already published more than 25 research papers in different national and international journals and edited volumes and conference proceedings. He is an editorial board member of five international journals and also acting as the reviewer of eight international journals. He is an editor of *Neo-Thinking on Ganges Brahmaputra Basin Geomorphology*, Springer International Publishing, and *Quaternary Geomorphology in India - Case Studies from the Lower Ganga Basin*, Springer International Publishing. His principal area of research includes geomorphology of Bengal basin especially river bank erosion, channel migration, flood, anthropogeomorphology and channel decaying.

Biplab Sarkar is a Research Scholar in the Department of Geography, Aliah University, working on hydraulic processes and channel forms of alluvial rivers of Bengal Delta. In addition, he is sincerely carrying out his works with keen interest on hydrochemistry of polluted river and its adverse consequences on aquatic ecology, irrigation system and sustainable agriculture. Hitherto, he has published three research papers in different journals.

Balai Chandra Das is an Assistant Professor at the Department of Geography, Krishnagar Government College, India. He holds a postgraduate degree in geography from the University of Burdwan and a Ph.D. in Geography from the University of Calcutta and has published more than 30 research articles in reputed international journals. Dr. Das has served as an editorial board member for two international journals and as a reviewer for five more. He is the main editor of two books *Neo-Thinking on Ganges Brahmaputra Basin Geomorphology* and *Quaternary Geomorphology in India - Case Studies from the Lower Ganga Basin*, Springer International Publishing.

Suman Deb Barman is an Independent Researcher with Master of Science degree in Geography from the University of Burdwan, India. He has already published two research papers in national and international journals. He has presented six research papers in various national and international seminars and conferences. His main research interest encompasses fluvial Geomorphology, neo-tectonics and remote sensing.

Kishor Dandapat, Rajkumar Hazari, Gouri Sankar Bhunia, and Pravat Kumar Shit

Abstract

Rill and gully erosion are an important morphological feature of lateritic terrain in Jhargram District, a part of rolling topography of Chhotonagpur Plateau. The present work took the River Dulung basin's rill and gully erosion as a study and used analytical hierarchy process (AHP) to extract the anatomical line of rill and gullies. This chapter analyzes the probable risk patches of rill and gullies along the River Dulung basin. To evaluate the risk zone of rill and gully erosion, eight biophysical variables were selected. The analytical hierarchy process (AHP) and weighted linear combination (WLC) were considered to functionalize the conceptual model within a geographic information system (GIS) framework. Results revealed that 2.74 km^2 of the study area falls into the very high-risk zone; 201.18 km^2 area comes under high-risk zone; 570.04 km^2 area falls into moderate risk zone; 111.63 km^2 area comes under the low-risk zone and 0.226 km^2 area falls into the very low-risk zone.

Keywords

Gully erosion · GIS · Analytical hierarchy process (AHP) · Weighted linear combination (WLC) · Risk estimation

K. Dandapat
Department of Geography, Seva Bharati Mahavidyalaya, Kapgari, Jhargram District, West Bengal, India

R. Hazari
Onze Technologies (India) Private Limited, Bengaluru, Karnataka, India

G. S. Bhunia
Aarvee Associates Architects, Engineers & Consultants Pvt. Ltd, Hyderabad, India

P. K. Shit (✉)
Department of Geography, Raja N. L. Khan Women's College (Autonomous), Medinipur, West Bengal, India

6.1 Introduction

An understanding of the erosion process of rill erosion is not only important for the prevention of soil erosion in slope lands, but it is also very significant to prepare the prediction models of soil erosion. Rill erosion has been defined as a series of little channels or rills up to 30 cm deep that can be abolished by cultivation (Cerdan et al. 2002). The top soils' and nutrients' loss caused by rill erosion will reduce soil productivity and the off-site deposition of sediments can bring sedimentation and water quality deterioration in stream and reservoirs. Continuous and concentrated flow is one of the main sources of soil detachment energy in rills; while raindrops play more significant roles for interrill erosion (Bradford et al. 1987; Govers et al. 2007). Different researches have been conducted including qualitative and quantitative approaches on rill ignition, rill networks, and rill formation since last decades (Bryan and Poesen 1989; Wang and Shangguan 2012). Several approaches on field experiments, laboratory experiments, digital photogrammetry (Gessesse et al. 2010), and satellite remote sensing (Vrieling 2006) have been improved to analyze and predict the sediment detachment and transport in rills (Hessel and Jetten 2007; Wirtz et al. 2012).

Erosion of soil is an important environmental issue in underdeveloped country like India. In India, one of the most important land degradation factors is soil erosion (Vrieling 2006). In India, the rate of soil erosion is 16 tons per hectare per annum, which is more than three times the acceptable norm. More than 60% of the world's population depend on agriculture (FAO 2017), whereas in India more than 62% of the Indian population are from rural belt and depend on agriculture (Indian Census 2011). Generally, four types of soil erosions have been recognized: sheet erosion, rill erosion, gully erosion, and in-stream erosion (Merritt et al. 2003). The erosion of the flow has been defined as a series of small channels or furrows up to 30 cm deep that can be removed by cultivation (Cerdan et al. 2002). The loss of soil and nutrients in the upper part caused by the erosion of the embankment will reduce soil productivity and deposition of sediments outside the site can cause sedimentation and deterioration of water quality in watercourses and reservoirs.

In Jhargram District, most of the people live in rural area and their main source of income comes from agricultural activities. Top soil erosion has reduced the agricultural production, which has become a prime factor in demarcating the local economic development. More than 90% of the people living in the River Dulung basin depend on agriculture. Therefore, rill and gully erosions have become a very serious concern for the people of the area of the River Dulung basin that has affected the livelihood pattern of the people.

6.2 Materials and Methods

6.2.1 Study Area

The basin of River Dulung lies between 86°40′0″E to 87°10′0″E longitude and 22°10′0″N to 22°40′0″N latitude on the fringe area of eastern Chhotonagpur Plateau, enclosing an area of 1176.38 km^2 (Fig. 6.1). It originates from the height of 110–130 m, namely Dulung Diha (22°37′45″N and 86°43′53″E) in Jhargram District of West Bengal State, flows in south-east direction within Jhargram District, and falls into River Subarnarekha near Kuthighat (22°09′27″N and 87°05′31″E) in Sankrail Block of Jhargram District. River Dulung has three headstreams: Katchua Khal (22°37′14″N and 86°42′11″E), Dakai Khal (22°34′58″N and 86°42′08″E), and the Dulung itself. The major tributaries of River Dulung are Deb Nala, Kupan Nala, Gandharpi Nala, Shimana Nala, Ruti Khal, Palpala Khal, Bansi Khal, and the Champa Khal. The total length of River Dulung in Jhargram District is 110 km. The major rill and gully erosion of the Dulung basin was found up to Fekoghat from the source region of Jhargram District.

6.2.2 Methodology

6.2.2.1 Database

Biophysical variables data associated to rill and gully erosion were collected from the various sources. The geomorphological map, geological map, and soil map were collected from Geological Survey of India, Kolkata (Table 6.1). The watershed boundary, slope characteristics, and drainage density were extracted from Cartosat 1 DEM dataset (date of acquisition 29.04.2015). Land-use/land-cover and lineaments data were obtained from the Landsat 8-Operational Land Imager (OLI) and Landsat 5 Thematic Mapper (TM) data, respectively.

6.2.2.2 Layer Creation and Digital Analysis

Selection of parameters in the study has been conducted through literature evaluation, field investigation, and expert's results. In this study, eight geo-environmental parameters were considered to map the rill and gully erosion. These representations can be based on the different parameters including biophysical parameters (e.g., slope, soil, geology, geomorphology, rainfall, drainage density, lineament, and land use/land cover). The Survey of India (SOI) topographical maps (F45B, F45C, F45H, F45I) were digitally scanned in TIFF format and imported into ArcGIS software. Using the geo-referencing option, the SOI topographical sheet was registered through the Universal Transverse Mercator (UTM) projection system based on the World Geodetic System (WGS) 84 datum. After that, the study area was extracted from the projected scene using boundary as a mask. Subsequently, the secondary map (e.g., geology, soil, and geomorphology) was geo-referenced through map-to-map rectification process. After completion of digitization of all thematic layers, the coverages were edited to remove topological errors. Finally, digital database on each of the aspects was prepared from the geo-referenced map. The slope map was produced from the Advanced Spaceborne Thermal Emission and Reflection Radiometer (ASTER) Digital Elevation Model (DEM) with 30 m pixel resolution. The lineament map of the study area was extracted from the Landsat OLI data through linear image enhancement process. Land use/land-cover (LULC) map has been generated based on the maximum likelihood algorithm of supervised image classification from Thematic Mapper imagery with a spatial resolution of 30 m.

6.2.2.3 Portfolio of Analytical Hierarchy Process (AHP)

There are several techniques available to calculate the vulnerability to natural hazards and can be classified as inductive or deductive (Yoon 2012). Likewise, the variables used to derive the vulnerability of an area depend on several factors, such as the scale of analysis, the extent of the study area, and the data vulnerability (Fekete et al. 2010). The analytical hierarchy process is based on several criteria that prioritize the criteria identified by the different groups of people involved in the decision-making process to achieve the best decision (Dandapat and Panda 2017). The hierarchical decision-making model is the design phase of AHP, in which the upper level shows the general objective of the decision. Indicators can be criteria at the highest level of the model, and each is subdivided into subcriteria. Each pair of criteria or subcriteria of elements is compared in terms of their relative importance through a system of 9 points from 1 (if two indicators contribute equally to the objective) to 9 (when one indicator is strongly favored over another to

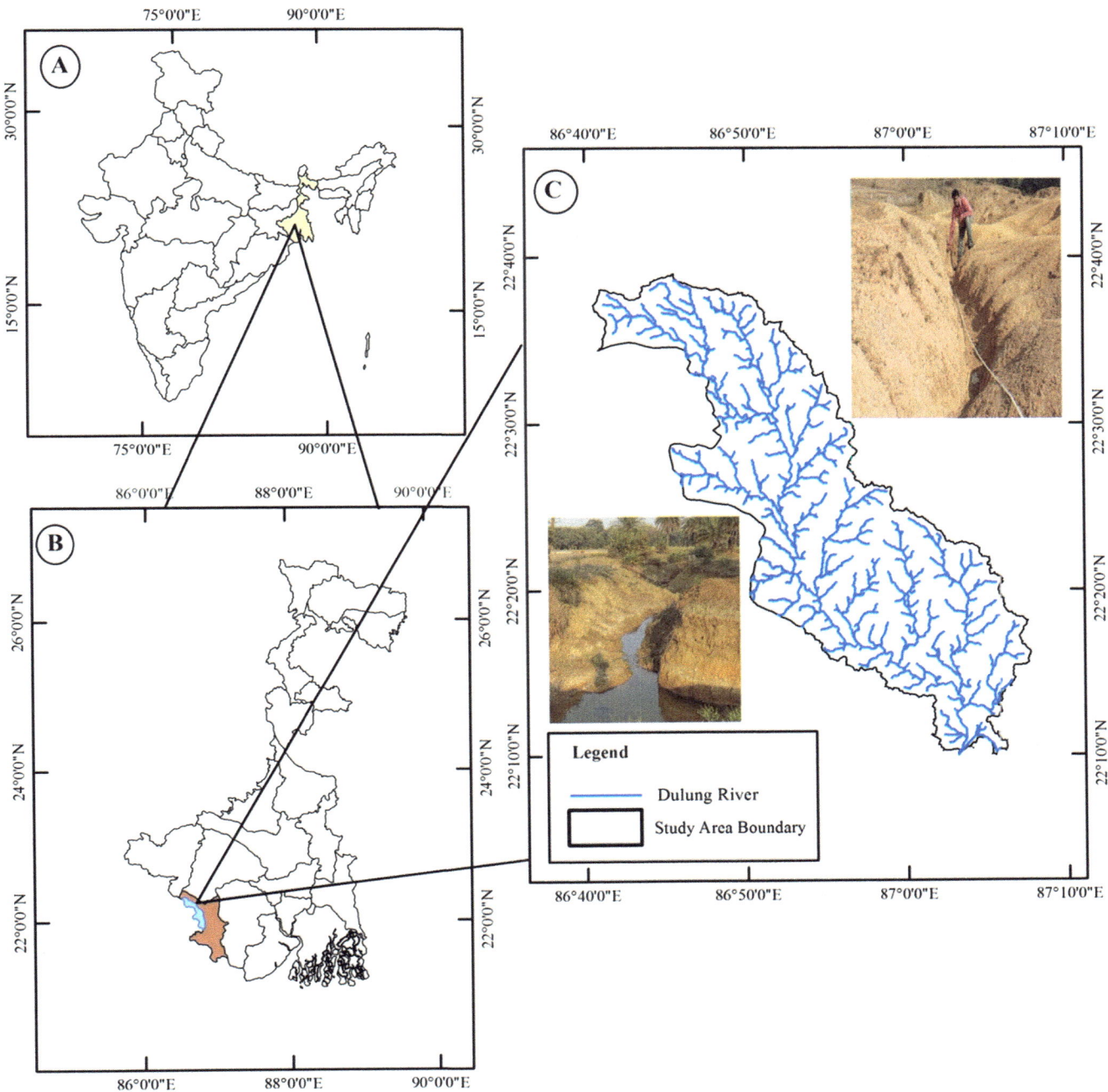

Fig. 6.1 Geographical location of the study area. (**a**) State of West Bengal in India, (**b**) Jhargram district with Dulung basin in West Bengal, (**c**) Dulung basin with their tributaries and partial view of rill and gully erosion

Table 6.1 Details of the data used in the present research work

Sl. no	Type	Format	Scale/resolution	Source
1	Survey of India	Analog	1:50,000	SOI, Kolkata
2	Geology	Analog	1:2,50,000	GSI, Kolkata
3	Geomorphology and soil	Analog	1:2,50,000	GSI, Kolkata
4	Cartosat 1 DEM	Digital	2.5 m	BHUVAN
5	Landsat 5 TM	Digital	30 m	USGS
6	Landsat 8 OLI	Digital	30 m	USGS

Table 6.2 Semantic scale of the AHP method

Comparative importance	Definition	Description
1	Equal importance	Two indicators equally influence the decision of the parents
3	Weak importance	One factor is moderately influential relative to the other
5	Essential or strong importance	One factor is strongly favored compared to the other
7	Demonstrated importance	One decision factor has a significant influence on another
9	Absolute importance	The evidence that favors a decision-making factor with respect to the other is the highest order of affirmation
2,4,6,8	Intermediate	When a commitment is required, the values are used between two adjacent judgments
Reciprocals	If A_i is the critical value with respect to j, then A_j has the reciprocal value with respect to A_i	A reasonable assumption

Source: Ramanathan (2001)

reach the "target") (Table 6.2) and forms a comparison matrix, that is, pairwise comparison matrix (Table 6.3). A score of "1" denotes equal importance; score of "3" refers to a weak preference, while the scores of "5" and "7" represent an evident and strong preference. Even numbers (i.e., 2, 4, 6, and 8) were used when a trade-off between odd numbers was needed (Dandapat and Panda 2017).

The local priority (weight) for a criterion is calculated from a pairwise comparison matrix by normalizing the points in the columns (dividing a cell value by the sum of a column) and averaging the normalized points in the criteria row. The consistency of comparisons is evaluated by calculating a consistency ratio (CR). If the CR is equal to or less than 0.1, the comparisons are considered coherent, otherwise they would be reviewed. The CR is defined by the following equation:

$$CR = \text{Coherence index} / \text{Random index}$$

The random index (RI) refers to a mutual matrix randomly generated by the 9-point scale and can be obtained by referring to the table of RI Satty (1980), which provides a function of "n" in relation (Table 6.4). The consistency index (CI) is defined as:

$$CI = (\lambda_{max} - n)/(n - 1)$$

In this analysis, λ_{max} was the largest eigenvalue derived from the comparison matrix and "n" was the number of criteria. Once the consistency was validated, the final priority of the criteria at the highest level of the hierarchy model was

obtained by adding the local policy priorities to their lower level. In addition to AHP, WLC was also used in this study. Usually, the condition of the values of the features associated to each other is allied to the assortment of the decision-maker. Because of its simplicity, the analysis of decisions of multiple criteria in terms of WLC is a technique frequently used in a GIS (Malczewski 2006). In the GIS analysis, each indicator is treated as a data layer. A WLC was done by multiplying the indicators for the corresponding weights and adding up all the weighted levels (Fig. 6.2; Table 6.5).

6.3 Results

6.3.1 Factor Analysis of Rill and Gully Erosion

6.3.1.1 Slope Analysis
In the study area, most of the high elevated zones or high sloping zones fall into northern, upper part of north-western, and north-eastern portion, which are expected to be the very high risk zone (VHRZ) of rill and gully erosion. Therefore, the highest weight (0.470) was given to the category of greater than 20% slope and the lowest weight (0.05) was given to the category of less than 3% slope (Table 6.6; Fig. 6.3).

6.3.1.2 Soil Analysis
The soil characteristics data comprise of six categories, namely, fine aericochraqualfs (FAO), fine loamy aericochraqualfs (FLAO), fine loamy typic paleustalfs (FLP), fine loamy ultipaleustalfs (FLUP), fine loamy typic ustifluvents (CLU), and coarse loamy typic haplustalfs (CLH). Fine aeric and fine loamy aericochraqualfs are most commonly associated with rill and gully erosion (see Fig. 6.3). So, the highest weight (0.33) was given to the fine aericochraqualfs (FAO) and second highest weight (0.32) given to the fine loamy aericochraqualfs (FLAO) in the analytical hierarchy process (AHP) (Table 6.6; Fig. 6.3).

Table 6.3 Example of a pairwise comparison matrix

Biophysical variables	Topography	Rainfall	Drainage density
Topography	1	5	6
Rainfall	1/5	1	5
Drainage density	1/6	1/5	1

Table 6.4 Indices of random average consistency for different "*n*"

n	1	2	3	4	5	6	7	8	9
RI	0.0	0.0	0.58	0.90	1.12	1.24	1.32	1.41	1.45

"*n*" represents the number of criteria

6.3.1.3 Geological Characteristics and Association with the Rill and Gully Erosion

The area is covered mostly by quaternary sediments, except in the north-western part, where older rocks are exposed (Fig. 6.3). The older rocks of the area belonging to the Paleoproterozoic age are represented by—(a) Singbhum Group consisting of mica schist, phyllite, garnet-staurolite, schist, and quartzite and (b) younger intrusive belonging to Mesoproterozoic age, consisting of Kuilapal granite and quartz-tourmaline rocks. The Cainozoic gravel bed, namely, tertiary gravel bed constitutes gravels and pebbles of quartz, which are occasionally embedded in the laterites. The Cainozoic laterites in the area are observed at many places of the Dulung basin from north to south, representing a hard crust at the top, followed by a layer of nodular lateritic mass that grades down through a saprolite zone to an unconsolidated parent material. There are numerous exposures of laterite in the area giving rise to bi-or-tri profile sequence indicating "in situ" nature. The oldest quaternary deposits exposed in the area comprise lalgarh formation of Early Pleistocene age consisting of fragments of quartz, phyllite, granite pebbles, and gravels occasionally laterized. The Sijua formation constitutes the sediments of older alluvium, comprising hard clay and silt, impregnated with caliche concentrations. The overlying sediments of Panskura formation constitute older flood plain deposits consisting of sand, silt, and clay of different flood regimes. The singbhum group

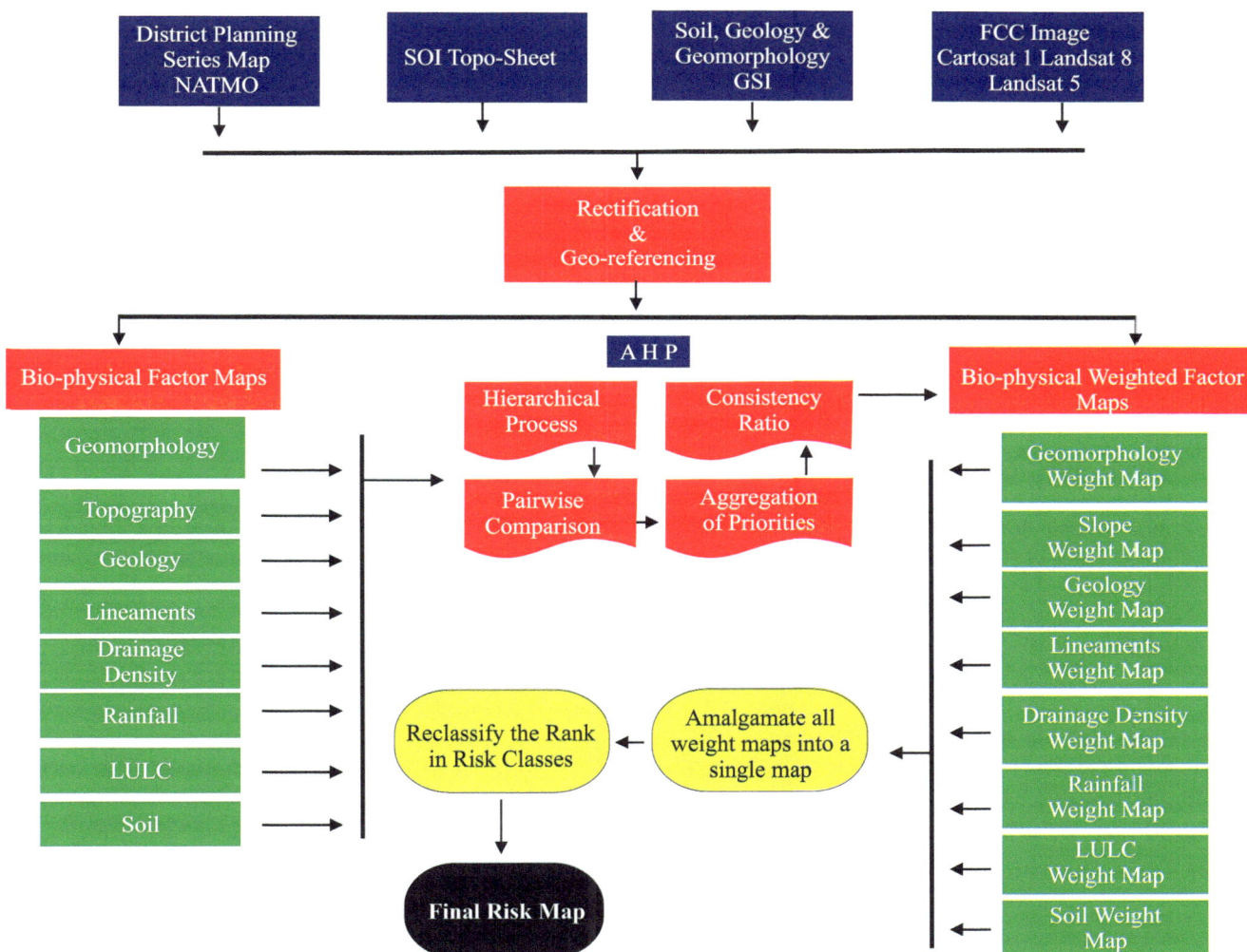

Fig. 6.2 Methodology flowchart of the proposed study

Table 6.5 Pairwise comparison matrixes for biophysical indicators

Parameter	Slope	Soil	Geology	Geomorphology	Drainage density	Rainfall	Lineaments	LULC	Overall weight
Slope	1	2	3	4	5	6	7	8	0.33
Soil	1/2	1	2	3	4	5	6	7	0.23
Geology	1/3	1/2	1	2	3	4	5	6	0.16
Geomorphology	1/4	1/3	1/2	1	2	3	4	5	0.11
Drainage density	1/5	1/4	1/3	1/2	1	2	3	4	0.07
Rainfall	1/6	1/5	1/4	1/3	1/2	1	2	3	0.05
Lineaments	1/7	1/6	1/5	1/4	1/3	1/2	1	2	0.04
LULC	1/8	1/7	1/6	1/5	1/4	1/3	1/2	1	0.03
CR = 0.06									

and younger intrusive were given the highest and second highest weight (0.33 and 0.32, respectively) (Table 6.6). Because of high sloping value and undulating topography, they fall into the VHRZ category of rill and gully erosion, although cainozoic laterite is a very good exposure of rill and gully erosion because of unconsolidated parent material. The cainozoic laterite was given the third highest weight (0.29) (Table 6.6) because of low sloping value; otherwise, it is a very good land for high-risk zone of rill and gully erosion.

6.3.1.4 Geomorphological Characteristics and Association Rill and Gully Erosion

Geomorphologically, the study area can be divided into five categories: deep buried pediments, moderately buried pediments with lateritic capping, pediments, denudational terrace and rocky outcrops, flood plain deposits, and valley fill deposits (Fig. 6.3). Because of the above characteristics, deep buried pediments category of geomorphology is getting the highest weight (0.47) and valley fill deposits plain is getting the lowest weight (0.03) because of absolute low-lying land (Table 6.6).

6.3.1.5 Drainage Density and Association Rill and Gully Erosion

Drainage density also an important indicator of soil erosion as well as rill and gully erosion, which is the total length of streams per unit area of the watershed and it is dependent on the lithology, permeability, vegetation cover, etc. In the study, area of highest drainage density is found along the channel (350–520) (Fig. 6.4). So, that is the cause of 350–520 category getting relatively highest weight (0.497) (Table 6.6). The following formula has been used to estimate the drainage density (D) using the following equation:

$$D = \sum L/A$$

where $\sum L$ is the total length of the hydrographic network (km) and A is the hydrographic basin area (km^2).

6.3.1.6 Annual Average Rainfall and Association Rill and Gully Erosion

The ground is eroded by the action of water whether it is in the form of raindrops or by the surface flow action. The runoff water moves rapidly over the surface of the soil, cutting well-defined groove structures in the form of fingers, which appear as thin channels or streams. For the study of rainfall impact on rill and gully erosion, one day maximum rainfall (mm) has been taken. High intensity of rainfall greater than 120 mm h^{-1} plays a vital role in creating the rill and gully erosion (Lal 1992). Therefore, the range of average annual rainfall greater than 400 mm was given higher importance at all levels of the decision hierarchy. Higher rainfall was allocated high weight (0.470) and lower rainfall (<100 mm) was allocated low weight (0.05) (Fig. 6.4; Table 6.6).

6.3.1.7 Spatial Characteristics of Lineaments and Association Rill and Gully Erosion

Lineaments had a positive role in rill and gully erosion, as they create the weak zone of earth surface like fractures, discontinuities, and shear zones. The spatial distribution of lineaments in the study area is illustrated in Fig. 6.4. The weight was given to the category of Lineament 0.04 among all eight variables of the study (Table 6.6).

6.3.1.8 Characteristics of LULC and Its Bearing on Rill and Gully Erosion

Land-use and land-cover (LULC) characteristics of the study area have been categorized into eight classes (Fig. 6.4). LULC plays an important role for the stability of surface slope. In this study, LULC factors, mainly lateritic/barren land, were given the highest weight (0.330) (Table 6.6). Also, the cultivated land (0.230), agricultural fellow land (0.160), and waste land/fellow land (0.110) have positive impact on rill and gully erosion. On the other hand, river, open forest, and dense forest represented negative impact on rill and gully erosion (Table 6.6).

Table 6.6 Weighted linear combination of biophysical variables associated to rill and gully erosion

Variables	1 Criteria	Weight	2 Criteria	Weight
Biophysical variables	Slope	0.33	Less than 3%	0.05
			3–6%	0.06
			6–10%	0.13
			10–20%	0.25
			Above 20%	0.47
	Soil	0.23	FAO	0.33
			FLAO	0.32
			FLP	0.15
			FLUP	0.11
			CLU	0.05
			CLH	0.04
	Geology	0.16	Present dry flood	0.02
			Panskura formation	0.03
			Lalgarh formation	0.04
			Sijua formation	0.05
			Tertiary gravel bed	0.07
			Laterite formation	0.23
			Younger intrusive	0.25
			Singbhum formation	0.31
	Geomorphology	0.11	Deep buried pediments	0.47
			Moderately buried pediments with lateritic capping	0.21
			Pediments	0.17
			Denudational terrace and rocky outcrops	0.09
			Flood plain deposits	0.04
			Valley fill deposits	0.03
	Drainage density (km^2)	0.07	0.5–86	0.036
			87–160	0.069
			170–240	0.136
			250–340	0.262
			350–520	0.497
	Rainfall (mm)	0.05	Less than 100	0.05
			100–200	0.06
			200–300	0.13
			300–400	0.25
			Above 400	0.47
	Lineaments	0.04	–	–
	LULC	0.03	Dense forest	0.03
			Open forest	0.04
			Scrub/degraded forest	0.05
			River and water body	0.07
			Waste land/fellow land	0.11
			Agricultural fellow land	0.16
			Cultivated land	0.23
			Laterite/barren land	0.33

6.3.2 Pairwise Comparison Matrix of the Biophysical Variables Associated to Rill and Gully Erosion

Analytical hierarchy process (AHP) weight of each of the variables associated with the rill and gully is illustrated in Table 6.5. All the weight was shown in the same trend in exponential curve. The evaluation values for each class of every influencing feature, which had an impact on rill and gully susceptibility, had been intended for the Expert choice software. The software computes the heft of each feature and based on this, the weighting aspects were ranked. Regarding

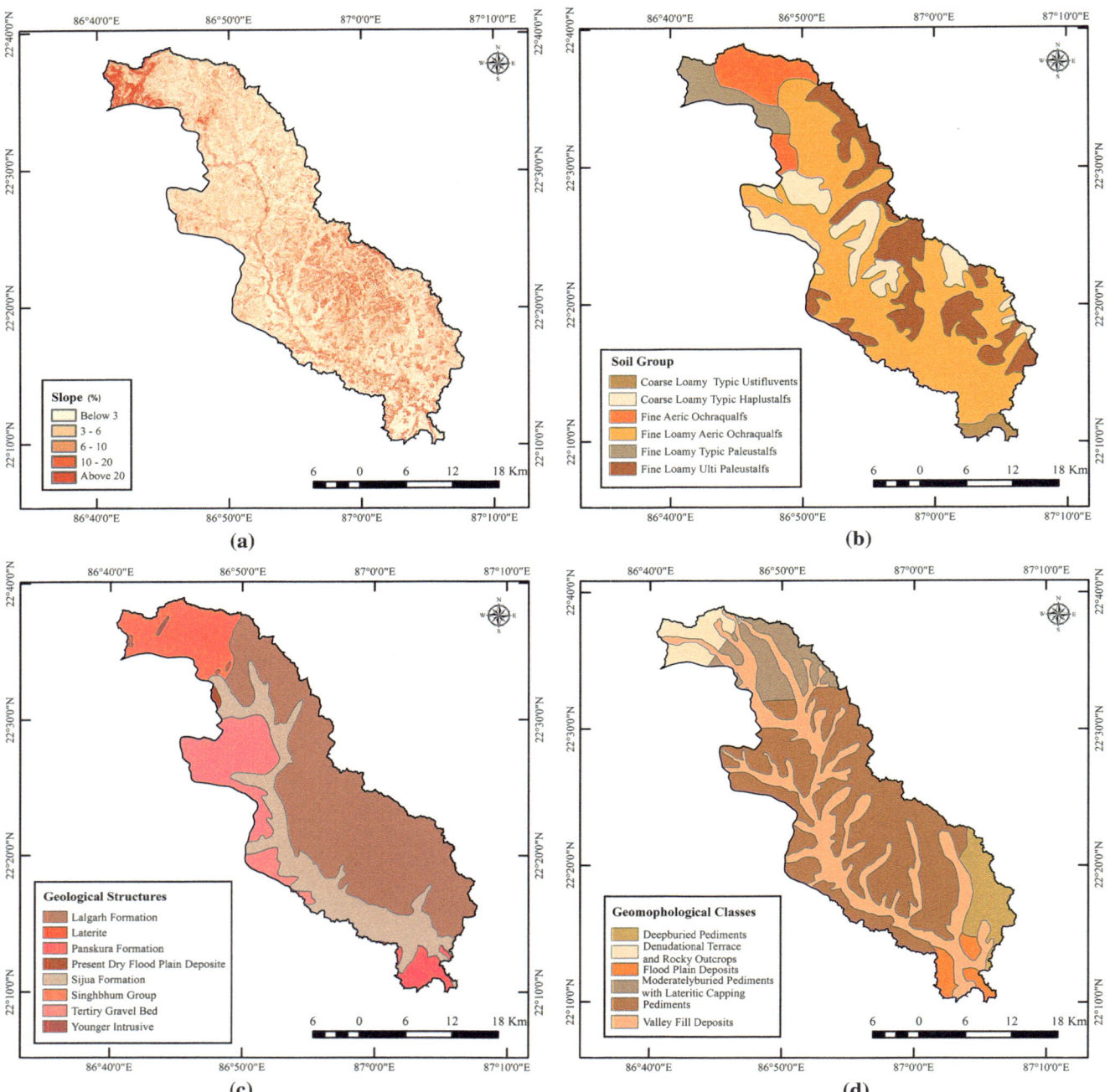

Fig. 6.3 Spatial characteristics of slope, soil, geology, and geomorphology along the Dulung River Basin

the importance of the criteria (AHP) at the upper level (Level 1), slope was considered the most important criterion and given the highest weight (0.330). The second most-important criterion was soil (weight 0.230), because slope stability and instability mostly depend on compactness of soil characteristics.

To evaluate the rill and gully erosion, it is necessary to delineate the erosion susceptible areas in relation to the degree of predisposition. Therefore, after preparation of

thematic layers of each parameter and conveying the weights of the discrete factor and their subclass, spatial combination of all thematic layers was completed using the raster calculator tool of ArcGIS software. For giving weights, spatial occurrence of every parameter in the study area has been observed carefully. A CR value of less than 0.1 indicates an acceptable level of consistency in pairwise comparison to recognize the weights applied. Table 6.5 shows the CR value of the variables, justifying the good level of

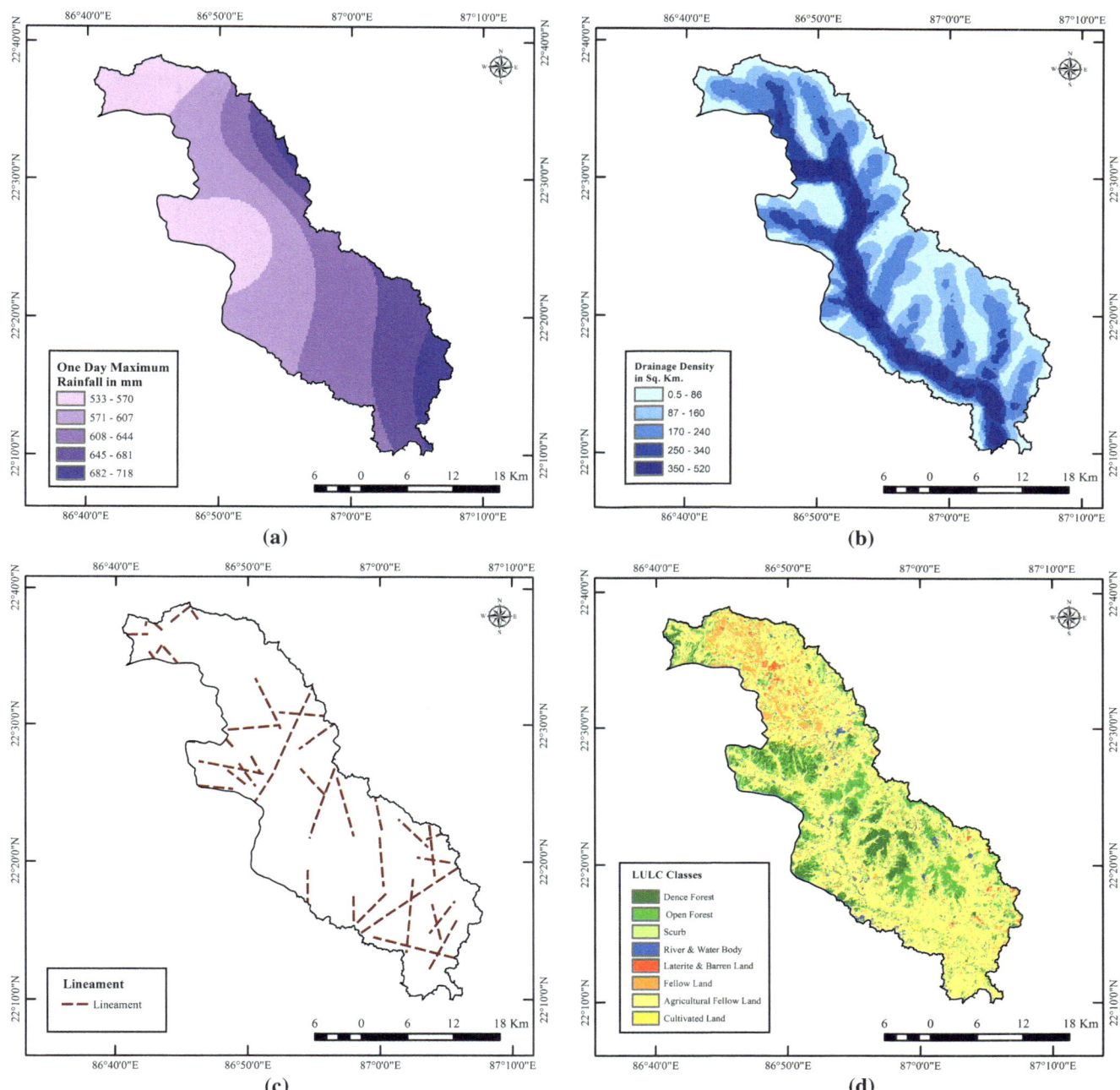

Fig. 6.4 Spatial characteristics of rainfall, drainage density, lineaments, and land use/land cover of the River Dulung basin

consistency in pairwise comparison. Finally, based on the weights of the influential factors, a weighted linear combination (WLC) model was espoused for receiving the rill and gully erosion susceptibility zone of River Dulung basin in remote sensing and GIS environment (Fig. 6.5). The attained consequence of this criterion is improved into percentage. All these parameter weights are united to estimate rill and gully liability and categorize zones in relative vulnerability classes.

Slope with AHP weight (0.33), soil with AHP weight (0.23), geology with AHP weight (0.16), geomorphology with AHP weight (0.11), drainage density with AHP weight (0.07), rainfall with AHP weight (0.05), lineament with AHP weight (0.04), and LULC with AHP weight (0.03) were found to be the important parameters in occurrence of rill and gully susceptibility in the study area. The thematic layers have been classified as per the standard classification schemes

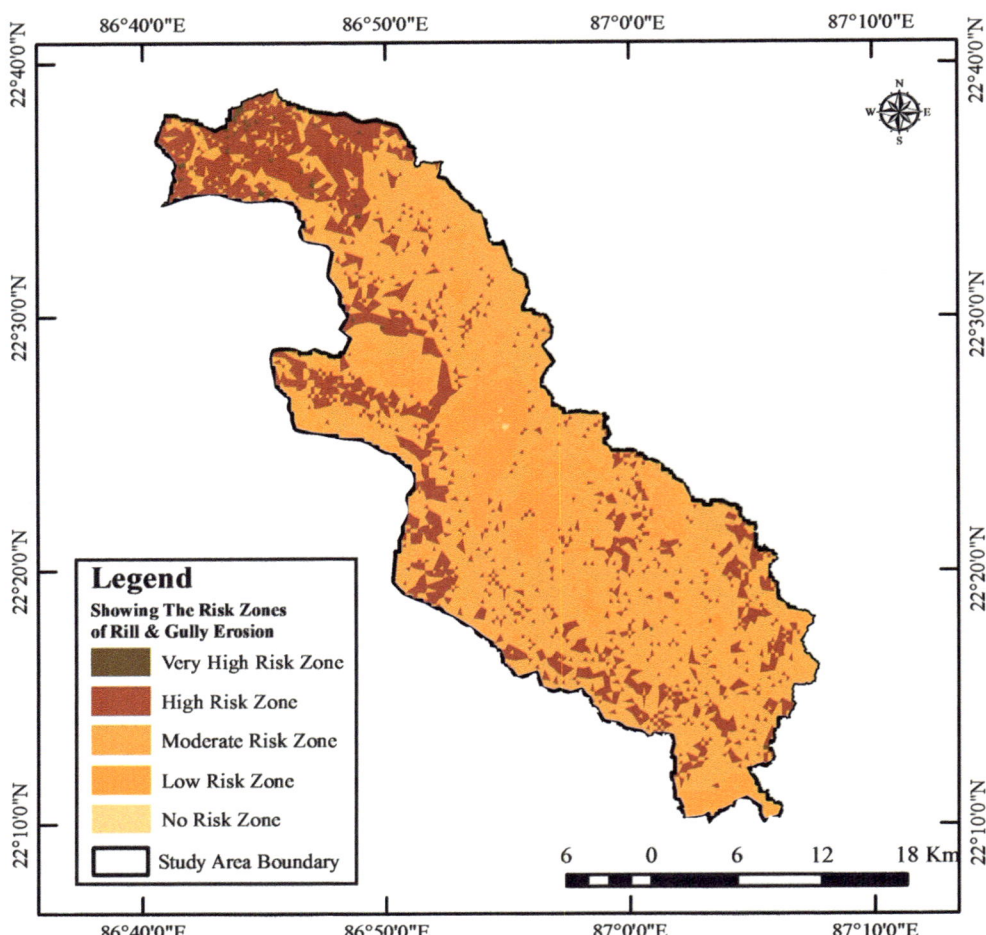

Fig. 6.5 Risk zonation map of rill and gully erosion at the Dulung River Basin

adopted by the expert's opinion and literature survey. The different thematic units in each layer have been ranked as per their priority in rill and gully erosion.

The rill and gully erosion risk zone map of the study area has been prepared and illustrated in Fig. 6.6 that represents the division and preferably subdivision of land surface into numerous regions consistent with the grades of real erosion instigated by the rill and gully. The probable rill and gully erosion map was prepared by assimilating the consequence of numerous eliciting factors.

The thematic map has been classified into five categories as follows: (i) very high-risk zone, (ii) high-risk zone, (iii) medium risk zone, (iv) low-risk zone, and (v) very low-risk zone. The results also showed that 0.31% (2.74 km²) of the study area comes under the very high-risk zone for rill and gully erosion, 64.35% (570.04 km²) of the study area comes under the medium risk zone for rill and gully erosion, followed by high-risk zone (22.71%, 201.18 km²) and low-risk zone (12.60%, 111.63 km²) (Table 6.7).

6.3.3 Validation of Map

Validating the predictive results of a model is very significant in a research investigation. The rill and gully erosion hazard map has been validated with the survey points collected from the field through the handheld Geographical Positioning System (GPS). There are totally 10 points which were collected (Fig. 6.6a, b) and overlay on the probable rill and gully eroded area on River Dulung basin. The result of the analysis showed positive relation between predictive model and ground reality (Fig. 6.7).

6.4 Discussion

Mechanisms for rill and gully erosion form and progression are still poorly tacit. In our study area, rill and gully development is concomitant with both natural landscapes and land-management practices. The development of rill and gully

Fig. 6.6 (**a**) Field visualization and field measurement with measuring tape and GPS. (**b**) Field visualization and field measurement with measuring tape and GPS

g

h

i

j

Fig. 6.6 (continued)

networks in the River Dulung basin seemed to be measured by numerous aspects that comparatively produce large runoff volume. The rill and gully erosion area map of the River Dulung basin has been prepared using AHP and WOA methods, in which eight thematic aspects, namely, slope, soil, geology, geomorphology, rainfall, drainage density, lineament, and land use/land cover, were addressed. The

thematic layers were integrated into GIS platform based on their respective ranks and finally the probable risk zone of rill and gully erosion was derived. The ranking of the themes is done in an unbiased manner. Higher values indicated the higher susceptibility and the lower values indicated the low susceptibility to rill and gully erosion.

Earlier study suggested that the shape of slopes plays a very effective role in the formation of rill patterns, rill distribution, rill density, sediment yield, and runoff production (Rieke-Zapp and Nearing 2005). Many investigations have been revealed that if the slope gradients have been increased, then the amount of rill and gully erosion has also increased (Liu et al. 1994; Wang 1998; Berger et al. 2010). In our study, more than 20% of slope was considered as very high risk for the rill and gully erosion. Generally, high slope leads to quick detachment of soil and indicates highly to severe

Table 6.7 Areal distribution of risk zone area of rill and gully erosion in Dulung River Basin

Risk zone category	Area (km^2)	Percent
Very high risk zone	2.74	0.31
High risk zone	201.18	22.71
Moderate risk zone	570.04	64.35
Low risk zone	111.63	12.60
Very low risk zone	0.23	0.03

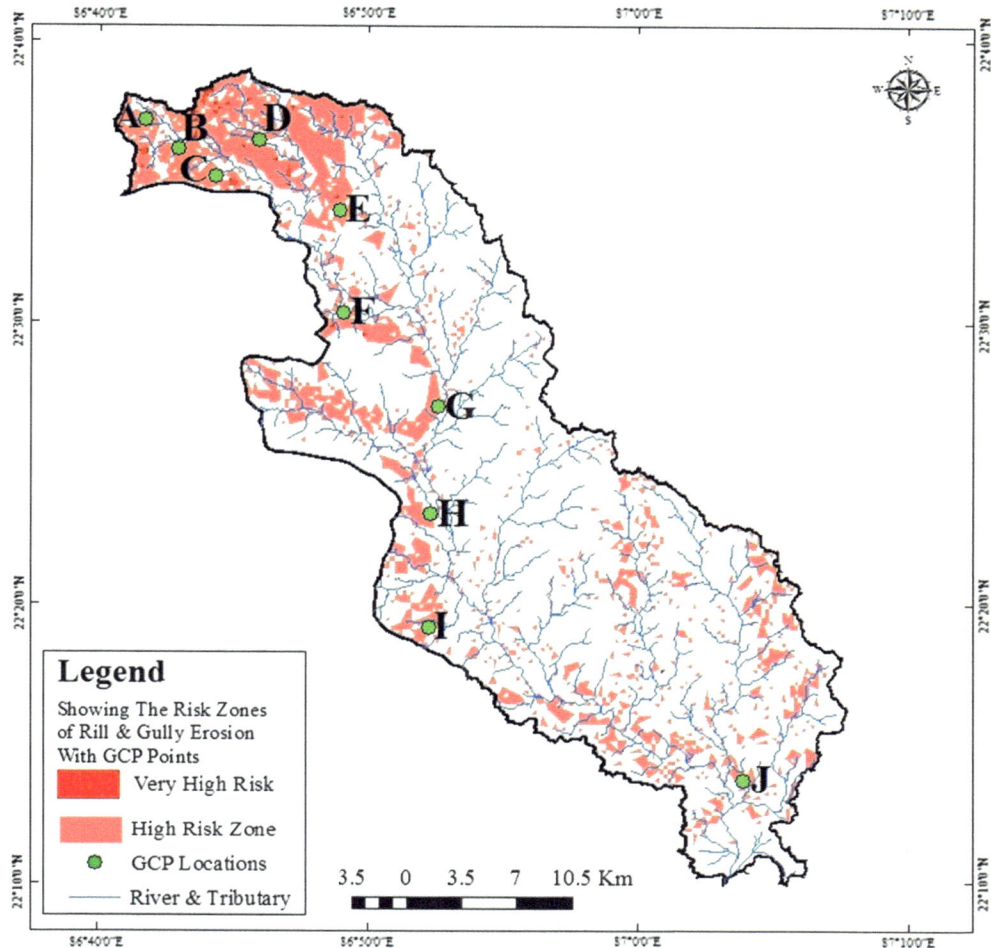

Fig. 6.7 Ground control point verification map with VHRZ and HRZ

erosion susceptible areas. Carefully elaborated the "geomorphic facies" are important tools for planning and management of natural resources and geohazards (Ownegh 1996). Generally, "a Pediment is a broad, gentle sloping rock—floored erosional surface, developed due to the process of denudation by the subaerial agents like running water" (Fairbridge 1968). When the Pediment is covered under a thick weathered mantle, it is termed as a buried pediment Bandyopadhyay (1998). In case of open and fallow, LULC are more erosive. The presence of plant cover decreases the intensity of rill and gully erosion because it is the decreasing erosion factor of surface runoff (Shit et al. 2015). Barren and sparsely vegetated areas are exaggerated by previous erosion and greater unsteadiness than forest (Dai et al. 2001; Cevik and Topal 2003). On the other hand, modification in LULC, declining vegetation cover, and overgrazing are peripheral issues in rill and gully erosion in the study area. The area with the high intensity of rainfall helps to find the ways of water flow on bare surface and weakened because of the loss of cohesiveness that may later form the rill and gullies.

Moreover, the rainfall characteristics aid in headcut erosion of permanent gully in the study area. Overall, due to barrenness, triviality, and little water-holding capability of lateritic soil and cementation of ferruginous concretions, there is partial hoard of soil moisture and inadequate development of trees excluding bushes and shrubs, scrub, and thin grasslands. These aspects convert the land as degraded barren waste and dismembered badlands. The additional significant noticeable issues of long-term extreme soil loss include high rainfall, deforestation due to morum quarrying, livestock grazing played an important role of rill, and gully development in the study area.

Weighted linear combination (WLC) is a mixture between qualitative and quantitative technique (Ghosh and Bhattacharya 2012). This practice is a current method that is tailored in numerous GIS and is appropriate for the lithe amalgamation of maps. Therefore, the slabs of scores and the map encumbrances can be attuned based on the expert's decision in the field. For instance, the very low and low gully eroded areas coincide with less than 6% slope,

CLU/CLH soil category, alluvial plain/tertiary gravel bed, denudational terrace and rocky outcrops/flood plain deposits/valley fill deposits, low drainage density, low rainfall, and forest/surface water bodies/fallow land type of land-use category. Correspondingly, the high-risk zones are associated with the higher slope, heavy rainfall, bare land surface, high drainage density, and lateritic formation region.

6.5 Conclusion

The rill and gully network along with its geometric topographies differs each year, perhaps as a consequence of scour and rill process. Most importantly, when the sloping value >20% rises, fine aericochraqualfs soil, singbhum formation geological type, deep buried pediments' geomorphological characteristics, drainage density 350–520 categories, above 400 mm annual average rainfall, fracture zone of earth surface, namely lineaments and lateritic or barren land category of land-use land-cover category are amalgamated than very high risk zone (VHRZ) rill and gully erosion have been formed. The VHRZ of rill and gully erosion are found at the upper part of the basin, not all the places but some patches have been found that covered the 2.74 km² area. The high-risk zone (HRZ) has been formed when the factors of sloping value 10–20% rise, fine loamy aericochraqualfs soil, younger intrusive geological type, moderately buried pediments with lateritic capping, drainage density 250–340 category, 300–400 mm one day maximum rainfall, and cultivated land category of land-use land-cover category have been amalgamated that covered the maximum risk zone of 201.18 km² area. This study also revealed that during the monsoon period from June to September, rill and gully erosion are maximum with 10–20% and more than 20% rise sloping gradient. These types of risk zone determination are very helpful to the planners or decision-makers to protect the soil erosion and enrich the landscape condition that may be suitable for the agriculture practices.

References

Bandyopadhyay S (1998) Drainage Evolution in Badland Terrain at Gangani in Medinipur District, West Bengal. Geogr Rev India 50 (3): 10–20

Berger C, Schulze M, Rieke-Zapp D et al., (2010) Rill development and soil erosion: A laboratory study of slope and rainfall intensity. Earth Surface Processes and Landforms, 35(12): 1456–1467. doi: https://doi.org/10.1002/esp.1989

Bradford J M, Ferris J E, Remley P E (1987) Interrill soil erosion processes: I. Effect of surface sealing on infiltration, runoff, and soil splash detachment. Soil Science Society of America Journal, 51(6): 1566–1571

Bryan R B, Poesen J (1989) Laboratory experiment on the influence of slope length on runoff, percolation and rill development. Earth

Surface Processes and Landforms, 14(3): 211–231. doi: https://doi.org/10.1002/esp.3290140304

Cerdan C, Le Bissonnais Y, Couturier A et al., (2002) Rill erosion on cultivated hillslopes during two extreme rainfall events in Normandy, France. Soil & Tillage Research, 67(1): 99–108. doi: https://doi.org/10.1016/S0167-1987(02)00045-4

Cevik E, Topal T (2003) GIS-based landslide proneness mapping for a problematic segment of the natural gas pipeline, Hendlok (Turkey). Environ Geol 44:949–962

Dai FC, Lee CF et al (2001) Assessment of landslide proneness on the natural terrain of Lantau Island, Hang Kong. Environment Geol 40:381–391

Dandapat K; Panda G.K (2017) Flood vulnerability analysis and risk assessment using analytical hierarchy process. Model. Earth Syst. Environ. 3:1627. doi:https://doi.org/10.1007/s40808-017-0388-7

Fairbridge, R.W (1968) -The Encyclopedia of Geomorphology. Reinhold, New York, 1295 pp.

FAO (2017) The future of food and agriculture – Trends and challenges. Rome. 180 pp. http://www.fao.org/3/a-i6583e.pdf

Fekete A, Damm M, Birkmann J (2010) Scales as a challenge for vulnerability assessment. Nat Hazard 55(3):729–747

Gessesse G D, Fuchs H, Mansberger R et al., (2010) Assessment of erosion, deposition and rill development on irregular soil surfaces using close range digital photogrammetry. The Photogrammetric Record, 15(131): 299–318.

Ghosh, S, Bhattacharya, K (2012). Multivariate erosion risk assessment of lateritic badlands of Birbhum (West Bengal, India): A case study. J. Earth Syst. Sci. 121(6): 1441–1454

Govers G, Giménez R, Oost K V (2007) Rill erosion: Exploring the relationship between experiments, modelling and field observations. Earth-Science Reviews, 84(3–4): 87–102. doi: https://doi.org/10.1016/j.carscirev.2007.06.001

Hessel R, Jetten V (2007) Suitability of transport equations in modelling soil erosion for a small Loess Plateau catchment.

Indian Census (2011) http://censusindia.gov.in/

Lal, R (1992) Restoring land degradation by Gully Erosion in the tropics. Adv. Soil Sci., 17:123–152

Liu B Y, Nearing M A, Rise L M (1994) Slope gradient effects on soil loss for steep slopes. Transactions of the ASAE, 37(6): 1835–1840.

Malczewski J (2006) GIS-based multi-criteria decision analysis: a survey of the literature. Int J Geogr Inf Sci 20(7):703–726

Merritt W S, Letcher R A, Jakeman A J. (2003) A review of erosion and sediment transport models. Environmental Modelling & Software, 18 (8–9): 761–799. doi: https://doi.org/10.1016/S1364-8152(03)00078-1

Ownegh, M. (1996). The role of geomorphology in soil erosion and land degradation assessment. Proc. Inter. Conf. on Land Degradation, 10–14 June 1996, Adana, Turkey, p 31–32

Ramanathan R (2001) A note on the use of the analytic hierarchy process for environmental impact assessment, Journal of Environmental Management, 63(1), 27–35.

Rieke-Zapp D H, Nearing M A (2005) Slope shape effects on erosion: A laboratory study. Soil Science Society of American Journal, 69(5): 1463–1471. doi: https://doi.org/10.2136/sssaj2005.0015

Satty, T. (1980) The Analytical Hierarchy Process. McGraw Hill, New York.

Shit PK, Paira R, Bhunia G, Maiti R (2015) Modeling of potential gully erosion hazard using geo-spatial technology at Garbheta block, West Bengal in India. Model. Earth Syst. Environ. (2015) 1:2 DOI https://doi.org/10.1007/s40808-015-0001-x

Vrieling A (2006) Satellite remote sensing for water erosion assessment: A review. Catena, 65(1): 2–18. doi: https://doi.org/10.1016/j.catena.2005.10.005

Wang G (1998) Summary of rill erosion study. Soil and Water Conservation in China, (8): 23–26. (in Chinese)

Wang K, Shangguan Z (2012) Simulating the vegetation-producing process in small watersheds in the Loess Plateau of China. Journal of Arid Land, 4(3): 300–309. doi: https://doi.org/10.3724/SP.J.1227.2012.00300

Wirtz S, Seeger M, Ries J B (2012) Field experiments for understanding and quantification of rill erosion processes. Catena, 91(s1): 21–34. doi: https://doi.org/10.1016/j.catena.2010.12.002

Yoon DK (2012) Assessment of social vulnerability to natural disasters: a comparative study. Nat Hazard 63(2):823–843

Kishor Dandapat is currently working as a government-approved Lecturer at the Seva Bharati Mahavidyalaya, Jhargram District, under Vidyasagar University, Paschim Medinipur, West Bengal, and he received his Doctoral Research Degree (Ph.D.) from Utkal University, Bhubaneswar, India. His major research areas are natural hazard and disaster management. His research interest lies in the area of fluvial geomorphology, hydrology, and also man and natural hazard.

Rajkumar Hazari is currently working as MLE at WAPCOS Limited, Jaipur, Rajasthan. He obtained his M.Sc. in remote sensing and GIS from Vidyasagar University, West Bengal, India. His major research interests are drought monitoring, watershed management, monitoring landslide zones, rural-urban planning, and coastal zone management.

Gouri Sankar Bhunia received his Ph.D. from the University of Calcutta, India, in 2015. His Ph.D. dissertation work focused on environmental control measures of infectious disease (visceral leishmaniasis or kala-azar) using geospatial technology. His research interests include kala-azar disease transmission modeling, environmental modeling, risk assessment, data mining, and information retrieval using geospatial technology. He is Associate Editor and on the editorial boards of three international journals in health GIS and geosciences. He worked as a "Resource Scientist" in Bihar Remote Sensing Application Centre, Patna (Bihar, India). He is the recipient of the Senior Research Fellow (SRF) from the Rajendra Memorial Research Institute of Medical Sciences (ICMR, India) and has contributed to multiple research programs: kala-azar disease transmission modeling, development of customized GIS software for kala-azar "risk" and "non-risk" area, and entomological study.

Pravat Kumar Shit received his Ph.D. in Geography (Applied Geomorphology) from Vidyasagar University (India) in 2013, M.Sc. in Geography and Environment Management from Vidyasagar University in 2005, and PG Diploma in Remote Sensing & GIS from Sambalpur University in 2015. He is Assistant Professor in the Department of Geography, Raja N. L. Khan Women's College (Autonomous), Gope Palace, Midnapore, West Bengal, India. His main fields of research are soil erosion spatial modeling, badland geomorphology, gully morphology, water resources and natural resources mapping, and modeling and has published more than 45 international and national research articles in various renowned journals; also, he has published three books. His research work has been funded by the University Grants Commission (UGC), India, and Higher Education Science and Technology and Biotechnology, Government of West Bengal. He is Associate Editor and on the editorial boards of three international journals in geography and earth environmental sciences.

Application of Field-Monitoring Techniques to Determine Soil Loss by Gully Erosion in a Watershed in Deccan, India

Veena U. Joshi

Abstract

The study is a compilation of three field-monitoring techniques applied in a watershed in Deccan Traps, India, to measure soil erosion through gullies and rills. More than 4 million hectares of land in India is severely affected by rill and gully erosion. It is necessary to investigate the gully formation process and the problem of gully erosion that will help in proper planning of the landuse in this region. The study area falls in a semi-arid tract of Maharashtra where the landscape is densely dissected by gullies to form badlands. These badlands are greatly disturbed and reclaimed for agriculture. A microprofilometer was fabricated to monitor gully heads and five profiles were monitored every 6 months for 3 years that displayed rapid expansion of gully network annually. Within the same basin, two badland catchments were selected and erosion pins were inserted into the ground. The exposed pinheads were measured after every monsoon for 2 years and volume loss were calculated from these basins for two time periods that yielded the soil loss of 0.76 kg/m^2/year from sample basin 1 and 1.79 kg/m^2/year for sample basin 2 during the monitoring period. The third experiment was rainfall simulation under controlled condition on six experimental plots within the same basin. The average soil loss after 1 h experiment was 0.8 km/m^2 which is a very high value for such sediment-starved terrain. A glance at the results obtained from the three techniques reveal that gully network expansion and soil erosion are happening in a big way in the region. Such field-monitoring techniques are helpful not only in understanding the actual soil loss scenario in the field, but such data are valuable because nowadays, soil loss estimates are done using remote-sensing data in GIS environment, and such field generated data provides validity to studies conducted using RS and GIS techniques.

Keywords

Profilometer · Rainfall simulator · Erosion-pin · Gully erosion · Badlands

7.1 Introduction

Soil erosion and land degradation are a problem worldwide, though the degree of severity differs from one country to another. The problem became more intensified in the last few decades, more so in India, due to the increasing population pressure (Joshi and Nagare 2009). Rill and gully erosion are consuming several million hectares of agricultural lands in India in an unprecedented rate (Singh and Dubey 1998). The removal of vegetative cover and unplanned land-use practices have been the main reasons behind the acceleration in the erosion and have transformed large areas into ravine lands (Singh and Agnithoti 1987). Badlands are intensely dissected lands, mostly fluvial in origin, characterized by high drainage densities, v-shaped valleys and short steep slopes where vegetation is absent or sparse and is of no economic use, especially agriculture (Bryan and Yair 1982). Badlands though develop in a wide range of materials and climate are usually found in arid and semi-arid environment and develop on unconsolidated and poorly cemented material known as marls. Badlands can be formed naturally on the sites where lithology is favourable, vegetation is sparse and climate is semi-arid with prolonged dry season, but many of the badlands are also induced by human activities (Wells and Andreamiheja 1995; Bryan and Yair 1982). Even the natural badlands have become more actively undergoing erosion now due to the human disturbances of the fragile system that exists there.

Badlands are dynamic landscape where erosion rate and sediment yield are high. Review of the literature reported from all over the world (Bryan and Yair 1982; Wells and Andreamiheja 1995; Gallart et al. 2013; Nadal-Romero et al.

V. U. Joshi (✉)
Department of Geography, Savitribai Phule Pune University, Pune, India

© Springer Nature Switzerland AG 2020
P. K. Shit et al. (eds.), *Gully Erosion Studies from India and Surrounding Regions*, Advances in Science, Technology & Innovation
https://doi.org/10.1007/978-3-030-23243-6_7

2018) suggests that though the badlands are dynamic system, they maintain a stability if undisturbed. Over the years in general and last two decades in particular, badlands have been extensively reclaimed wherever accessible, and it holds true for the badlands that are found along Pravara River and two of its tributaries in Deccan Traps Region, Maharashtra, India. River Pravara is a tributary of Godavari River in Maharashtra. Deccan Volcanic Province exhibits rocky landscape and there is dearth of sediments and soils are very thin everywhere. These valuable soils are lost heavily every year through rill and gully erosion. Evidence reveal that the degradation of land and the rate of erosion have increased many folds due to the human interference in the landscape in the last few decades. India is an agricultural country and land is the most precious resource. Hence it is a matter of serious issue for concern, and there is a need for a thorough investigation to minimize the problem effectively.

The chapter presents a compilation of three different field-monitoring techniques applied to a badland topography to estimate rill and gully erosion in a watershed in the Deccan Trap Region in Maharashtra.

Investigators from across the globe have carried out several studies on soil loss estimation over the years using different techniques. Watson and Evans (1991) compared estimates of soil erosion made in the field with that from the photographs. A mesh-bag method was used by Hsieh (1992) for field measurement of soil erosion. Sirvent et al. (1997) calculated soil loss from a field using profilometer, collectors and erosion pins. Heng et al. (2011) conducted flume scale experiments to model the dynamics of soil erosion. Bowyer-Bower and Burt (1989), Bhardwaj and Singh (1992), Cerda et al. (1997) etc. estimated sediment yield from experimental plots using rainfall simulators. Nowadays, most of such types of studies to monitor gully erosion are done using high-resolution images such as LiDAR or laser scanner images or Unmanned Aerial Vehicle or UAVs (Tucker et al. 2006; Evans and Lindsay 2010; Wang et al. 2016 etc.). The best result is obtained when field-monitoring and RS data are combined to calculate gully erosion (Slimane et al. 2018).

The outcome of such studies is to find suitable land-use practices to minimize further land degradation in such sensitive areas. Such outcomes have been reported earlier for some watersheds in the Deccan Traps Region of Maharashtra (Joshi 2014; Joshi and Tambe 2010).

7.2   Study Area

The study has been conducted in a watershed in Pravara Basin in Maharashtra, India. This area is a part of the Deccan Trap Region where semi-arid conditions prevail. Figure 7.1 demonstrates the location of the study area. Rivers mostly flow through rocks and do not develop well-defined floodplains. This study area falls under Godavari Basin, which is an exception in Maharashtra where thick alluvial deposits are found along the riverbanks at several locations and within the same basin, there are pediment slopes with thick sediment cover exceeding 10 m in thickness. These deposits are severely dissected by gullies and form badlands at many places. There are long stretches of badlands along the banks of the Pravara River and its tributaries. Apart from that, few pediment foot slopes also demonstrate similar badland features. Joshi and Nagare (2013) produced the evidence of neotectonism as the initiating factor of these badlands. Further acceleration of the gullying process in the region is evident that are believed to be the response of the new land-use practices introduced recently in these watersheds (Joshi and Nagare 2009). It is one of the most important mesoscale sub-watersheds of Godavari Basin, having an area of 2930 km^2 and badlands occupying approximately 18% of the area. These badlands are levelled extensively for agriculture now. The reclamation became more intensified post year 2000 when irrigation facilities became available to the farmers in these areas. Due to the rugged nature and inhospitability of the terrain, the badlands were left alone from any activity until year 2000 when government took the initiative to construct dams and weirs across the main Pravara River and its tributaries for irrigation. It has boosted agriculture in the area like never before and soon the plain fertile regions became exhausted and farmers started levelling these badlands. Landuse took a sharp turn following this period. Since then land levelling and expanding agriculture has become a common practice in the area, which has induced accelerated erosion in the area.

The region experiences monsoonal climate. Rain-shadow effect of the Western Ghat induces low average annual rainfall and high rate of evapotranspiration. The region is usually hot and dry in summer with mild winter, and precipitation is confined during the four monsoon months from June to September with a mean annual rainfall range between 470 and 500 mm.

Although the amount of rainfall is less, there are few rainstorms every year with fairly high intensity exceeding 50 mm/h (Joshi and Nagare 2013). This is responsible for initiating new rills and gullies every monsoon in these disturbed badlands. Winter is dry and invigorating, that lasts from November to mid-February. The average temperature of the region is 25 °C. The pre-monsoon period of March till June is hot throughout the region, with May experiencing highest temperature of over 40 °C on many days. Southwest monsoon arrives late June/early July and last till September, sometimes till the beginning of October. Vegetation cover is sparse and dominantly includes thorny acacias, which are typical of semi-arid regions. By and large, silt and sand predominate the sediment and clay content is less. Field observation of the area reveals various renewed evidences

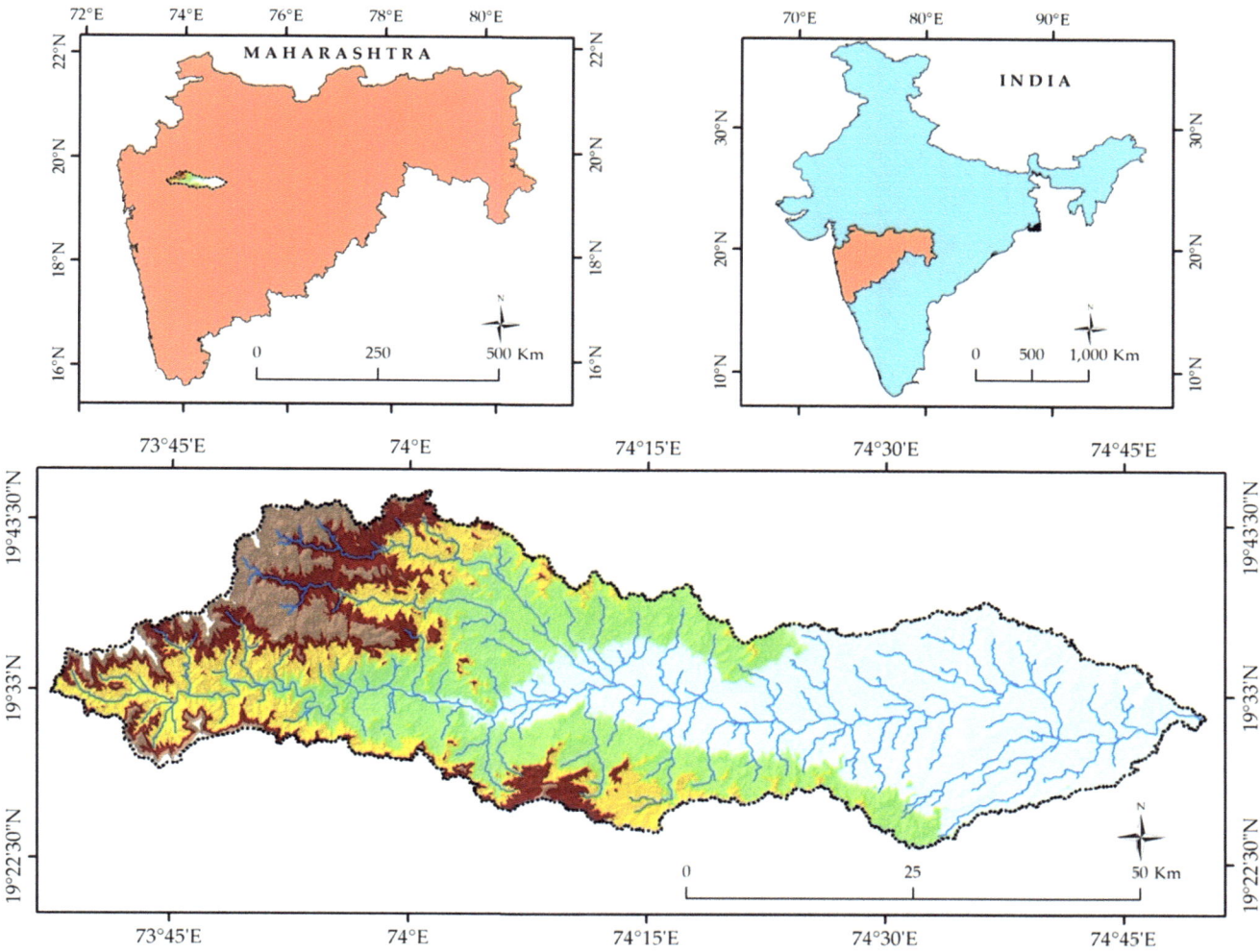

Fig. 7.1 Location map of the study area

of land degradation and gully erosion. So far there had been rarely any sediment yield and erosion rate data for the area under review. Hence the author monitored gully erosion rates in a few catchments within the basin for a period of 3 years using a self-fabricated microprofilometer, self-fabricated rainfall simulator and erosion-pin techniques. The chapter is the compilation of the three techniques applied in the area.

7.3 Monitoring and Assessment

7.3.1 General Sedimentological Properties Assessment

Sedimentological properties play an important role in the development of badlands as well as further degradation of them. Certain soils are more erodible than others and they crumble quickly under natural processes as well as anthropogenic activities. Fifteen samples were collected from the field to detect erodibility of these sediments by using a few indices that are relevant. Granulometric analysis was conducted for the samples by using X-ray-based sedigraph. Silt and sand dominate the sample population though the entire range is from gravel to clay.

7.3.1.1 Results of the Sedimentological Analyses

Figure 7.2 displays the textural characteristics of these sediments. It is evident that sand and silt constitute 3/4 of the entire sample size fractions. Particle size statistics were determined adopting Folk and Ward (1957) technique and found that all the samples are very poorly sorted (Table 7.1). There is wide spectrum in mean phi range as can be seen in the Table. Except for a few, most of the samples demonstrate positive skewness indicating the predominance of the finer fractions. Samples reveal low-peaked platykurtic to strongly peaked leptokurtic. All these undoubtedly suggest variation

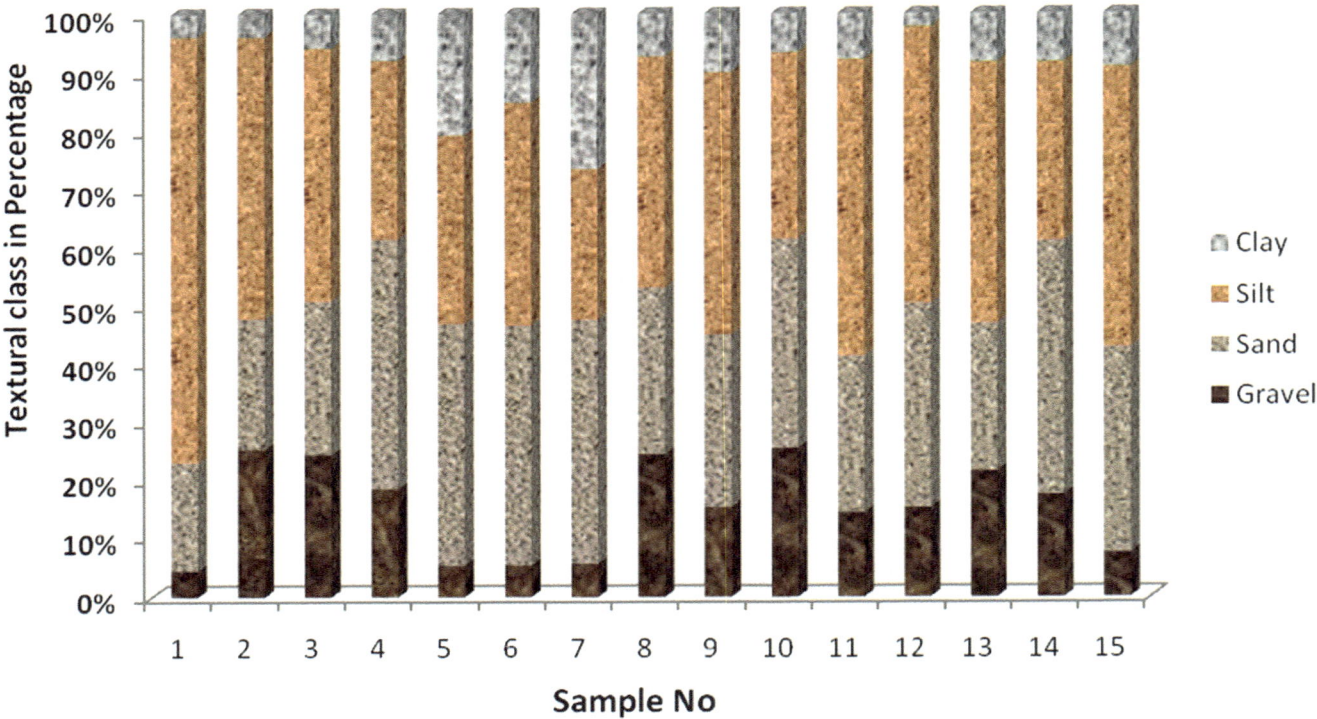

Fig. 7.2 Textural characteristics of the sediments

in the energy level during the deposition of these sediments. The predominance of silt in the sample population implies that these are more erodible soils.

Clay is an important sediment property that controls the susceptibility of them to outside forces, especially water. Smectite clays have high shrinking and swelling properties, thus promotes loosening of the soil catena and induce erosion. The clay fractions were subjected to examination by X-ray diffraction (XRD) of oriented K/Ca saturated samples

Table 7.1 Sample statistics of the sediments

Sample no.	Mean Phi size	Sorting	Skewness	Kurtosis
1	−0.29	1.77	0.33	1.06
2	−0.21	1.36	0.12	1.01
3	0.54	2.26	0.22	1.47
4	1.13	3.01	0.51	0.65
5	0.38	2.68	0.64	1.19
6	3.96	4.12	0.68	0.54
7	2.33	2.69	−0.38	0.77
8	−0.63	1.29	0.20	0.72
9	−0.17	2.09	0.46	1.31
10	1.25	3.03	0.57	0.88
11	−0.96	1.42	0.56	1.93
12	1.75	3.03	0.29	0.72
13	3.08	3.23	0.32	0.64
14	1.08	2.75	0.44	0.80
15	−0.71	1.67	0.61	1.07

using a Philips Diffractometer and Ni-filtered Cu Kα radiation. Figure 7.3 displays the result of five samples taken depthwise from a vertical sediment profile exposed along a gully wall. Samples reveal abundance of expandable smectite clays with traces of chlorite and kaolinite. Presence of smectite is very distinct in all the samples and this is a major cause of slacking in the sediments. Ten undisturbed samples were collected and tested in the laboratory for their physical properties. Results indicate that field dry density values range from 1.11 to 1.72 g/cm^3 suggesting low density and permeability ranges between 1.10×10^{-3} and 1.25×10^{-6}, indicating high permeability of soils.

Soil erodibility plays a vital role in understanding the development of badlands and rate of erosion. Some soil yield easily to water erosion while some show fair resistance based on their properties. The textural, mineralogical and few physical properties indicate that the soil is susceptible to erosion. Further actual K values have been calculated for these samples using soil erodibility nomograph to quantify their erodibility status (Wischmeier et al. 1971). The result of the calculation has been presented in Table 7.2. Results indicate that actual K values are not very high except for just a few samples. Review of the literatures reveals that soil can be erodible if exchangeable sodium percentage is high and if the soil contains smectite clays. Hence few chemical properties that are relevant have been studied and the exchangeable sodium percentage has been calculated.

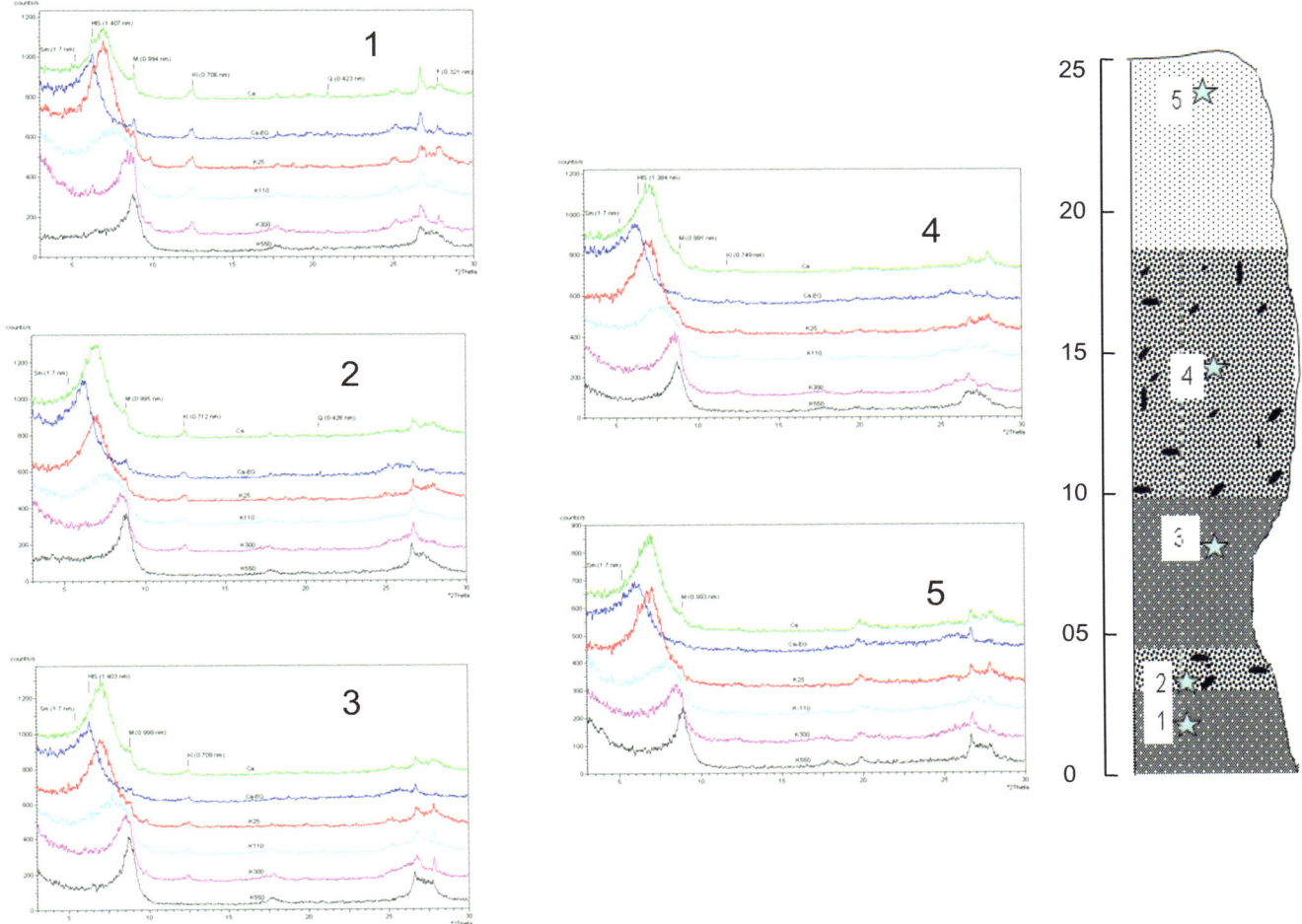

Fig. 7.3 Clay mineralogy of the sediment samples taken at different depths

Table 7.2 Erodibility

Sample no.	Sfv%	S%	Organic matter	Structure	Permeability	K value
1	6.45	52.58	0.06	Coarser granular	Rapid	0.03
2	3.64	79.15	0.06	Blocky massive	Rapid	0.05
3	51.15	58.67	0.12	Coarser granular	Slow to moderate	0.62
4	12.49	46.9	0.24	Coarser granular	Moderate	0.09
5	37.02	57.13	0.12	Coarser granular	Slow to moderate	0.32
6	31.23	48.89	0.6	Coarser granular	Moderate	0.19
7	21.65	62.34	0.21	Coarse massive	Moderate	0.18
8	31.49	65.05	0.06	Coarse massive	Moderate	0.28
9	13.15	51.17	0.18	Coarser granular	Moderate	0.09
10	27.28	53.74	0.06	Coarser granular	Moderate	0.17
11	17.94	56.22	0.54	Coarser granular	Moderate	0.1
12	14.12	71.19	0.09	Blocky massive	Moderate	0.17
13	26.14	63.84	0.15	Coarse massive	Moderate	0.22
14	18.11	64.03	0.12	Coarse massive	Moderate	0.16
15	4.01	66.2	0.06	Coarse massive	Rapid	0.07

K values of the sediment samples
Calculated using the soil erodibility nomograph (Wischmeier et al. 1971)
Sfv silt + VeryFine sand, *S* sand

Table 7.3 Chemical properties of the sediment and exchangeable sodium percentage (ESP)

Sample no.	pH	Org. carbon (%)	CaCO$_3$ (%)	Exch. Ca (meL^{-1})	Exch. Mg (meL^{-1})	Exch. sodium (meL^{-1})	Exch. potassium (meL^{-1})	Cation Exc. Cap. (me/100 g soil)	ESP
1	9.20	0.06	2.25	9.00	1.20	6.50	0.10	9.00	7.22
2	8.98	0.06	7.00	10.60	0.20	4.50	0.10	10.50	4.29
3	8.31	0.12	10.75	9.80	4.00	4.50	0.12	16.00	2.81
4	8.86	0.24	32.75	16.40	3.80	9.00	0.12	17.00	5.29
5	8.93	0.12	20.25	8.80	3.20	4.50	0.14	13.50	3.33
6	8.30	0.60	21.75	16.80	1.40	7.50	0.16	14.50	5.17
7	7.95	0.21	26.75	16.60	3.40	8.50	0.12	11.50	7.39
8	8.43	0.06	8.25	15.80	3.00	4.50	0.08	6.50	6.92
9	8.40	0.18	12.00	15.40	2.60	5.50	0.08	10.00	5.50
10	8.78	0.06	16.00	12.80	3.20	7.00	0.14	11.00	6.36
11	8.10	0.54	16.25	18.40	0.20	9.00	0.14	12.50	7.20
12	7.80	0.09	18.50	11.40	9.00	5.50	0.16	13.00	4.23
13	8.30	0.15	5.75	11.00	4.20	6.50	0.18	14.50	4.48
14	7.80	0.12	14.50	14.40	4.40	8.00	0.22	16.00	5.00
15	8.10	0.06	14.50	8.80	5.80	4.50	0.18	13.00	3.46
16	7.58	0.18	14.00	8.50	2.10	4.00	0.38	14.50	2.76
17	7.70	0.12	13.75	10.40	7.60	6.50	0.12	12.00	5.42
18	7.69	0.18	35.75	11.60	7.00	5.50	0.26	13.50	4.07

Fig. 7.4 Exchangeable sodium percentage against depth

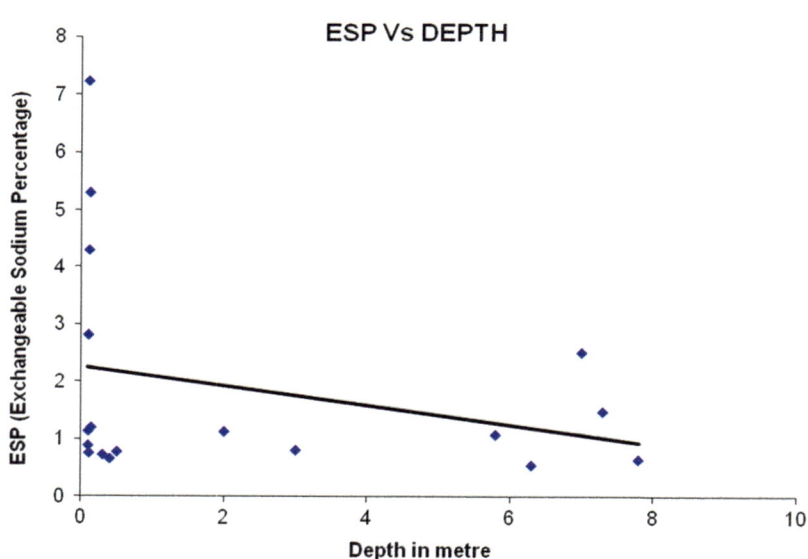

Sodium is known to disintegrate soil aggregates and induce dispersion of clay particles. Geochemical properties of these samples were obtained, and exchangeable sodium percentage (ESP) was calculated and produced in Table 7.3. An additional 3 surface samples were added to the analysis. ESP ranges from 2.5% to 7.5%. Fifteen percent is considered threshold of the ESP for erosion, but in smectite-rich soils, the threshold goes much below. Figure 7.4 indicates the relationship between ESP and depth. Many of the values cluster around 3 and few surface samples go up to 7. A weak negative relationship can be seen in the figure indicating that surface samples are more sodic than the ones

at the depth. This has a significant implication in badland erosion. The combination of the results of these analyses suggests moderately to fairly erodible sediments.

7.3.2 Microprofilometer Technique

The area chosen for study is a badland and observation of the field over the years indicates that these badland gullies expand their network through head ward as well as linear erosion every year. Evidences are clearly visible all along the badland stretch after each monsoon season. New gullies appear on the reclaimed fields and the natural gullies change their cross-sectional profiles. The first experiment was to understand the mode of gully expansion using a microprofilometer.

Before the high-resolution DEMs such as LiDAR and Terrestrial Laser Scanners became available, microprofilometer technique was used to monitor gully heads in many areas of the world. Until today, results obtained from microprofilometer technique are widely used. Microprofilometer was common in the late 1960s/1970s and popular till the 1990s. Many of such profilometers were used for agricultural engineering, though they were used for academic purpose also. Newton (1968) designed a device to measure ripple marks underwater. Curtis and Cole (1972) constructed one microtopographic profile gauge for agricultural engineering. Mosley (1975) designed and used a profilometer for the accurate survey of small-scale slopes. Leatherman (1987) developed a microtopography profiler (MTP) to measure the amount of direct sand displacement resulting from off-road vehicles (ORVs) travelling along the beach. Microerosionmeters were designed for different environment and used by Robinson (1976), Trudgill et al. (1981) and Toy (1983). Benito et al. (1992), Sirvent et al. (1997) and Desir and Marín (2007) used microprofilometer to measure erosion rates in badlands. Malinov and Ilieva (2017) employed a microtopographic profile gauge to monitor erosion on gullies developed along the erosion concentration lines caused by vehicle wheels on unmetalled roads. Microtopographic profile gauge was used by Karimov et al. (2015) to monitor the development of ephemeral gullies on agricultural lands. A microprofilometer was fabricated to measure gully expansion and erosion rate for the present study and used to detect annual changes in the cross-sectional profiles of five badland gully heads in the study area.

7.3.2.1 Fabrication of the Microprofilometer

The fabricated microprofilometer is a wooden frame of 1.13 m/0.93 m where steel rods have been inserted against the frame at the interval of every 2 cm (Fig. 7.5). The wooden frame is painted white to resist the weather and a graph is drawn on it and calibrated. The structure has two legs that have flat bottoms designed to be placed with balance on fixed erosion pins. The steel rods can slide through the holes made at the bottom of the frame. When this instrument is placed on gullies, we remove the lock at the bottom of the frame and the rods slide down and rest on the ground below, thus creating a graph of the cross-section on the board. Once placed on gully profiles, all that is required is a perfect front shot picture by a camera. XY of the points are calculated by a programme generated in PASCAL for every cross-section.

Adula and Mahalungi are two tributaries of Pravara River. Along the banks for these rivers, extensive badlands have been formed on alluvium. Two ephemeral gullies at Adula and three at Mahalungi were selected for monitoring their

Fig. 7.5 Microprofilometer fabricated for the study

expansion after every monsoon. The monitoring started in 2007 and continued till the end of 2009. Two readings were taken each year, one before and one after the monsoon. The monitoring profiles were close to the source of the gullies where the erosion is maximum. Two pins were inserted to the ground across the cross-section where the instrument was to be placed. Since the area is characterized by rolling badland topography, maintaining horizontality was not achieved every time, but these pins fixed the spots and helped in maintaining uniformity at every time of measurement.

The first monitoring was on May 29, 2007, because that is the driest time of the year in the area, and the second reading was taken in November after complete cessation of the monsoon. This continued till the end of 2009. All the five sections are the first-order gullies of the badland system with general slope ranging from 25° to 35°. The monitored cross-sections of these gullies are presented in Fig. 7.6. Each time the same camera was used to click the photos and care was taken to take the photo with constant angle and height. Of the two sites, Mahalungi is more disturbed than Adula. Badland

reclamation is a lot more at Mahalungi than the later. All the five cross-sections were carefully selected away from the reclaimed areas where no animal or human would disturb the profiles. Dates of the surveys were kept closest to each other every season.

7.3.2.2 Results of the Microprofilometer Monitoring

The result of monitoring after 2 years reveals that gullies are expanding the cross-sections rapidly. Figure 7.7 shows the changes in the cross-sections during the monitoring period and Tables 7.4 and 7.5 demonstrate the actual cross-section as well as changes after every survey. It is clear that considerable changes have occurred in all the profiles within the monitoring period. Some values are negative indicating reduction in the area and implying deposition. This suggests that erosion of the bed and banks of the gullies was rapid but sediments were transferred downstream at different rain episodes, hence temporarily held storage on the beds to be flushed out in the next rain. Most of the profiles showed

Profile No 1

Profile No 2

Profile No 5

Profile No 3

Profile No 4

Fig. 7.6 Figure depicts the monitoring five cross-profiles

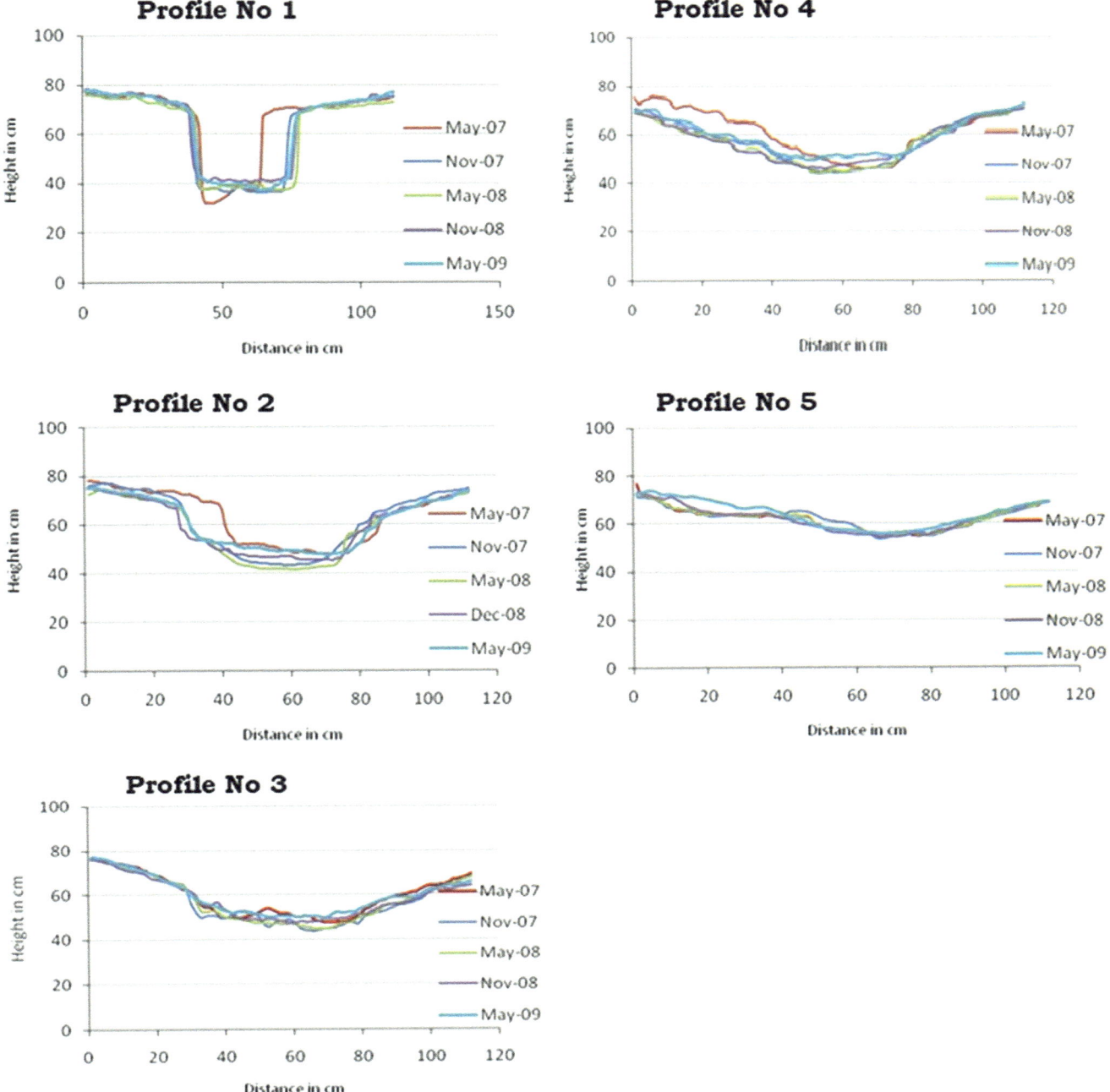

Fig. 7.7 Figure shows changes in the cross-sectional areas of the gullies during the monitoring period

Table 7.4 Cross-sectional area (cm^2) from the day of monitoring till the end of monitoring

Cross-section no.	May 2007	Nov 2007	May 2008	Nov 2008	May 2009
1	1175.12	1556.42	1482.73	1597.03	1582.50
2	1369.45	1677.42	1670.42	1714.01	1523.30
3	1536.67	1711.18	1610.67	1358.11	1351.40
4	1463.93	1478.21	1489.32	1466.32	1355.40
5	1213.77	851.92	1008.74	980.34	754.50

Table 7.5 Changes in the cross-sectional area (cm^2) during the monitoring period

Cross-section no.	May 2007 Nov 2007	Nov 2007 May 2008	May 2008 Nov 2008	Nov 2008 May 2009	May 2007 May 2009
1	353.3	150.3	−244.1	13.4	273
2	224	278.6	−107.6	−123.5	271.5
3	247.3	−10.9	−62.2	−174.6	−0.4
4	395.1	117.5	−34.2	−267.9	210.5
5	−53.8	61.6	−28.4	−199.4	−220

remarkable increase in the cross-sectional area. Most of the erosion occurred during the early monsoon. Rate of erosion declined as vegetation started growing over the badland slopes. After December, grasses turned dry and ground became bare again until monsoon arrived in June. As the ground dried, bank failure was common and the sediments were slumped at the rill bottoms. By and large, greater and more intense rainfall events promoted erosion because such events generated run-off capable of transporting the sediments deposited at the bed and also progressively developed new rills and gullies. The seasonal behaviours in sediment erosion and transportation have been reported by many researchers earlier (Schumm 1964; Hodges and Bryan 1982; Benito et al. 1993). It can be clearly observed in the present study area also, such as that the region experiences long dry season from October till mid-June. Rate of erosion retards during the dry period but sediments slump and deposit to the gully bottoms. Occasional light showers are unable to move these sediments and are held storage at the rill bottoms. Zanchi (1988) reported that seasonal variations in rain produce variations in monthly erodibility in such landscapes. Changes in the moisture content and little variation in the physical properties of the soil throughout the year result in variability in erodibility in the present study area also. There is a cyclic pattern of erosion and deposition following the seasonality of rainfall. Within the Godavari Basin, the entire badland area occupies 2930 km^2. We can imagine the depth of the problem that all the gullies expand the network at this rate in the entire region under review.

7.3.3 Erosion-Pin Technique

It is evident from the results of the profilometer technique that every year, considerable erosion is experienced in the area under investigation. The results revealed only the cross-sectional expansion but did not give volume of soil erosion. To evaluate total soil loss, erosion-pin technique was employed in the same two sites, selecting two catchments, one from each site. Before the experiment, the technical bulletin of Haigh (1997) was carefully studied. The erosion-pin method to calculate soil loss was also employed by other investigators (Hessel and Asch 2003; Ghimire et al. 2013

etc.). The first catchment was at Adula (Fig. 7.8) and the second was at Mahalungi site (Fig. 7.9). Firstly, the basins were surveyed using a theodolite in order to select spots for pin installation. The approximate area of Adula catchments is 1600 m^2 and that of Mahalungi catchment is 2100 m^2. DEMs were generated in Arc-GIS for the basins from Shuttle radar topographic mission (SRTM) DEM, and the erosion-pin locations were determined on the map for both the basins. Fifty erosion pins were installed in each basin. Each pin was of 50 cm length and they were driven into the spots marked on the DEM with only 1 cm of the rod exposed above the ground. For easy location, the pins were painted white. The pins were installed at the same time when profilometer monitoring began in 2007. After 1 year, the exposed parts of the pins were located and the height was measured. Out of the 50 pins, only 35 pins could be located because many were buried and could not be found. The exercise was repeated again in 2009 and this time all 35 pins were located. The data was used to calculate net loss and gain in Arc-GIS. Table 7.6 indicates the results of the experiment.

7.3.3.1 Results of the Microprofilometer Monitoring

Results of the experiment indicate that between May 2007 and May 2008, the calculated sediment yield from Adula Basin was 0.87 kg/m^2 and that from Mahalungi Basin was 1.71 kg/m^2. In the following year between May 2008 and May 2009, it was 0.65 kg/m^2 for the first basin and 1.87 kg/m^2 for the second basin. The average of these 2 years indicates an annual sediment yield of 0.76 $kg/m^2/year$ for Adula Basin and 1.79 $kg/m^2/year$ for Mahalungi Basin. The difference in the sediment yield between the two basins could be attributed to many factors. Both the catchments are very similar in terms of sediment properties and terrain characteristics. Mahalungi is more populated than Adula and currently it is undergoing massive transformation by various activities, especially agriculture. Badlands are levelled and practised agriculture and such remodelling of the slopes increase susceptibility to erosion. Adula is not much reclaimed hence the natural slopes have more stability than the former. The tolerance limit of soil loss for Deccan Plateaus has been presented by Mandal and Sharda (2011) as between 0.25 and 1.25 $kg/m^2/year$, and comparing the soil

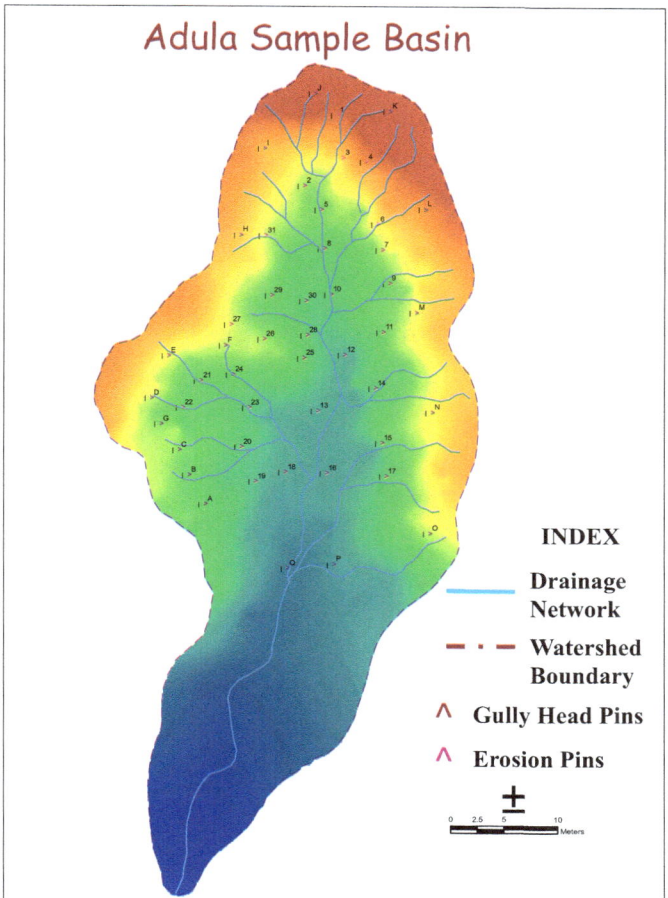

Fig. 7.8 The field pictures and the DEM of Adula Basin. The location of the erosion pins have been depicted in the DEM

loss values of both the basins with the tolerance limit of the soil for Deccan Region indicates that the values are above the threshold of tolerance limit.

7.3.4 The Rainfall Simulator Experiment

A rainfall simulator has been designed to conduct experiment to measure infiltration, run-off and sediment yield under controlled condition for the study area. Though rainfall simulation experiments are used worldwide, such studies are not common in the area under review. Few studies have been conducted in India using a rainfall simulator (Bhardwaj and Singh 1992; Singh et al. 1999). Rawat et al. (1992), Rawat and Rawat (1994) and Rawat and Rai (1997) have performed rainfall simulation to understand infiltration, discharge, run-off and sediment yield from different watersheds. Outside India, such studies are reported often (Young and Burwell 1972; Tricker 1979; Dunne et al. 1980; Bryan 1981; Pall et al. 1983; Roth et al. 1985; Exeter 1990; Hignett et al. 1995; Singh et al. 1999; Wright et al. 2002; Sukhanovskii 2007; Sangüesa et al. 2010; Haffzullah et al.

2017; Wang and Lai 2018 and others). Almost all these simulators were designed and fabricated to suit local conditions and different requirements. The present rainfall simulator also has been designed for the present study and many factors were taken into considerations while designing it. The design of this simulator is almost identical with the one used by Dunne et al. (1980), but few modifications have been made to suit the local condition. It is an auto-simulator but has been adjusted in such a way that in difficult terrain, it could be operated manually. However, for the present study, all the simulations had been performed under auto set-up.

7.3.4.1 Design of the Simulator

The rainfall simulator which has been designed for the present work consists of a bolted steel frame of 2×2 m in dimension. The height of the simulator was 3 m but the legs were made adjustable in length to facilitate a firm footing along an uneven terrain. Four custom made nozzles were attached to a wheeled trolley in the central line of the frame. The nozzles were 1.2 mm in diameter and the wheels moved freely and automatically along the track. The ends of the track were protected with foam rubber. The trolley was programmed to slide 1.9 m

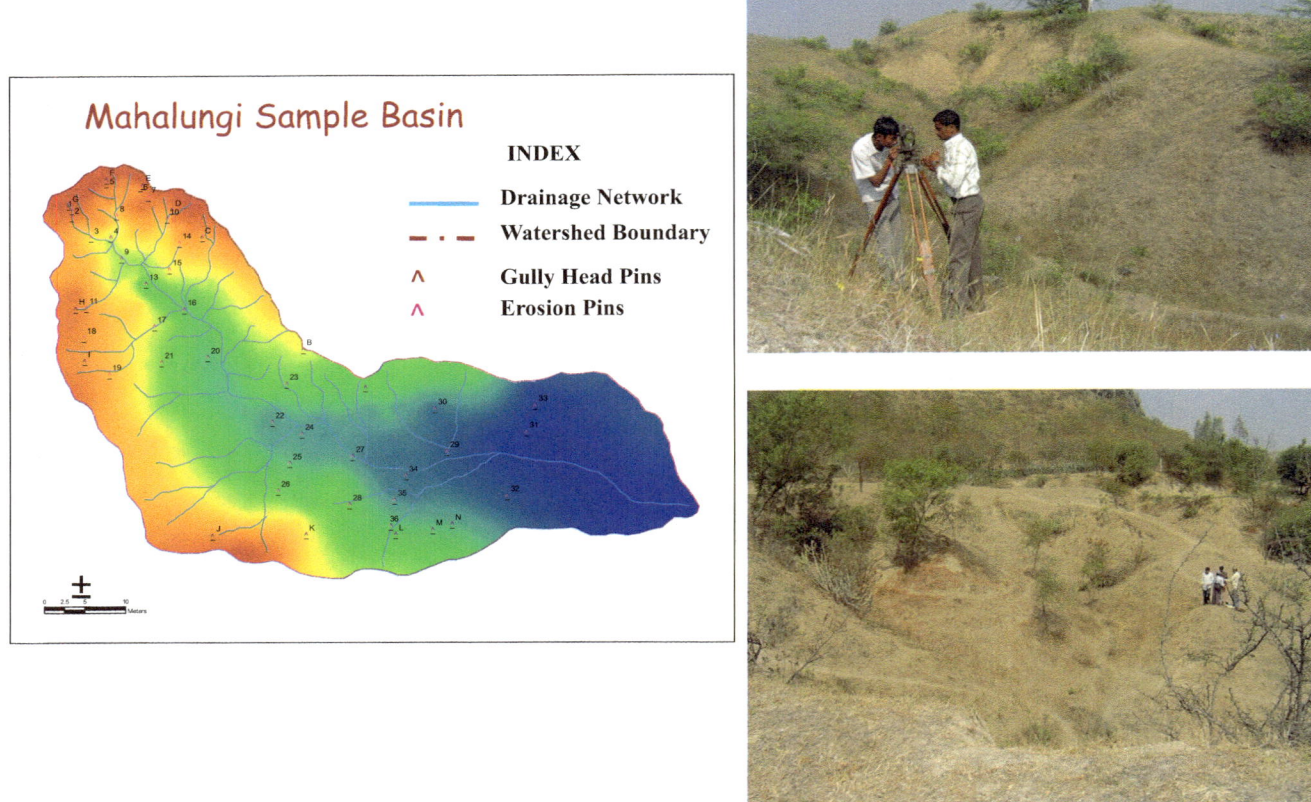

Fig. 7.9 The field pictures and the DEM of Mahalungi Basin. The location of the erosion pins has been depicted in the DEM

Table 7.6 Erosion-pin monitoring result

Sample basins	Basin area (m^2)	Eroded volume (m^3)	Deposited volume (m^3)	Bulk density (g/cm^2)	Eroded volume (tonnes)	Deposited volume (tonnes)	Net loss volume (tonnes)	Sediment yield (kg/m^2)
Adula catchment 2007–2008	1670	5.12	4.12	1.45	7.43	5.97	1.45	0.87
Mahalungi catchment 2007–2008	2100	8.59	6.01	1.45	12.45	8.71	3.74	1.71
Adula catchment 2008–2009	1670	2.98	2.23	1.45	4.33	3.24	1.09	0.65
Mahalungi catchment 2008–2009	2100	4.85	2.04	1.45	7.04	2.96	4.08	1.87
Adula catchment 2007–2009	1670	8.11	6.35	1.45	11.75	9.21	2.54	1.52
Mahalungi catchment 2007–2009	2100	13.44	8.05	1.45	19.49	11.67	7.82	3.58

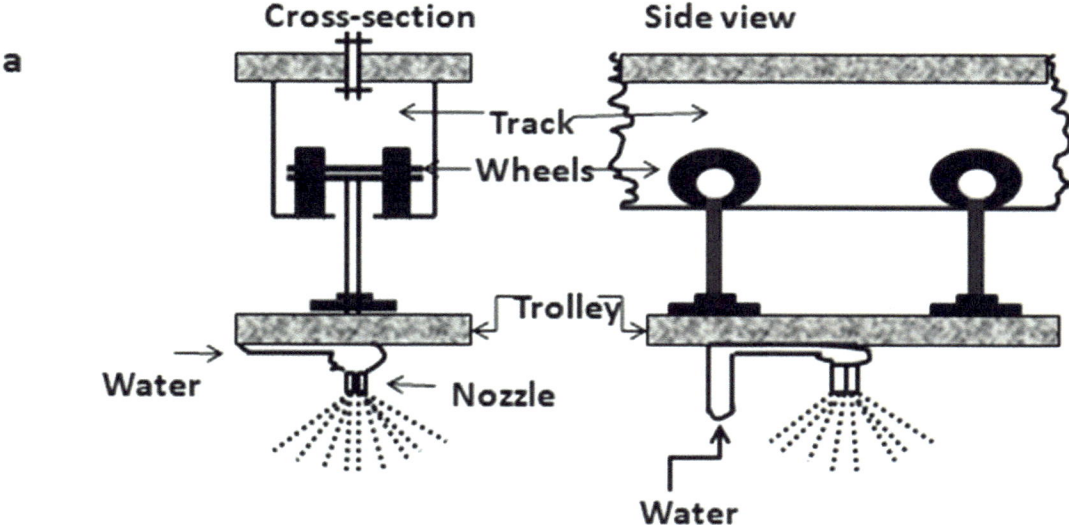

End and side diagrams of the suspension of a spray nozzle from a trolley which moves on wheels in an overhead track running the length of the rainfall simulator

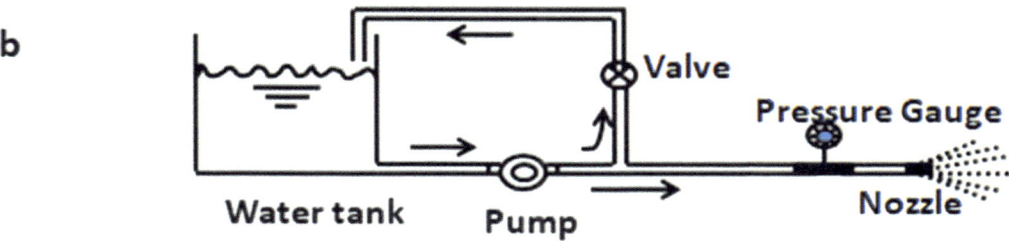

Schematic diagram of the system used for supplying water to the rainfall simulator. A recirculating loop has been included to allow fine adjustment of pressure in the nozzle

Fig. 7.10 (**a**) It is a schematic sketch of the end and side diagrams which shows the suspension of a spray nozzle from a trolley that is moving on wheels in an overhead track running the full length of the rainfall simulator. (**b**) It is a schematic sketch of the fabric that has been used to supply water to the simulator. (Designed after Dunne et al. 1980)

length in 3 s along the track. This speed had been so decided to ensure uniform spray of rain all over the plot during 1 h of experiment. A removable canvas sheet had been used as a part of the design to shield the plot during the experiment on windy days. A schematic diagram of the instrument and different parts of the systems have been displayed in Fig. 7.10. The design is parallel to the one used by Dunne et al. (1980).

Firstly the amount of water that would be required during 1 h experiment was carefully calculated and based on the estimate; a water tank of 500 l capacity was used for the experiment. An HP diesel pump was used to supply water from the tank to the nozzles. One pipe of the pump was connected to the water tank while another bifurcation of the pipe was attached to the nozzles. Pressure was controlled by a valve which was fixed to the pipe. This allowed adjustment of

the water flow rate during the experiment. Trial experiments were conducted many times before the actual use of the instrument and many adjustments were made following each trial. Final calibration was complete after several trial experiments and instrument was ready for real experiment.

At the lower end of the plots, a collector flume was placed firmly to receive the run-off and sediment. The collector was designed exactly the same as Dunne et al. (1980)'s without any modification. Figure 7.11 shows the collector sheet metal trough and the dimension specifications.

The metal trough was adjusted to match the slope of the plot and care was taken not to let it spill out during the experiment. At the end of the plot, a 2 m long, 25 cm deep and 22 cm wide trench was dug where the metal trough was firmly inserted. When the roof of the metal trough was laid horizontal, the floor

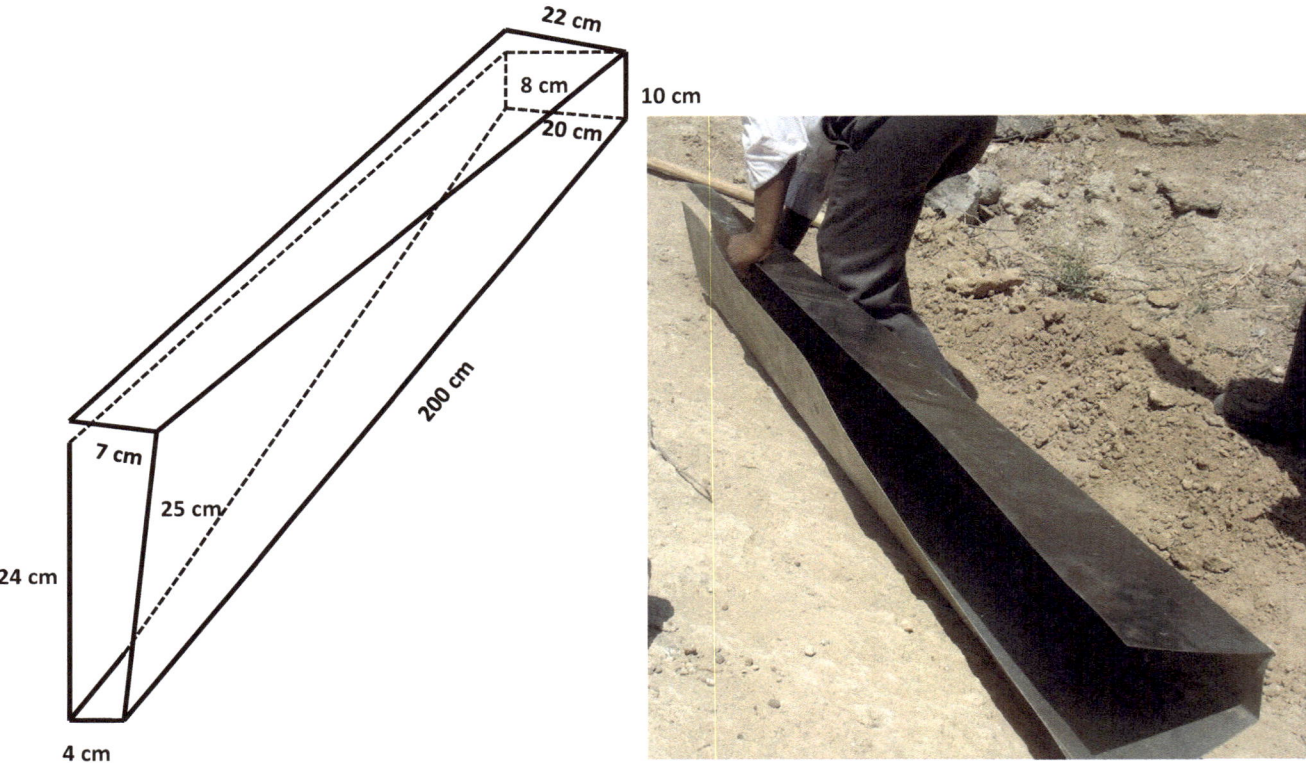

Fig. 7.11 Run-off collector, cut from a piece of sheet metal bent to form a trough with a slope of 10%

had a gradient of 10%, to ensure sufficient delivery of all sediment discharge effectively to the outlet. After the complete installation of the collector, all the gaps between the soil and metal trough were filled with plaster of Paris.

Within the watershed under review, experiments were conducted regularly for a period of 2 years on different plots. Figure 7.12 depicts the simulator at different experimental plots. In the present chapter, the results of six plots that are akin to badlands have been discussed.

7.3.4.2 Results of the Rainfall Simulation Experiment

Figure 7.13 presents the results of the infiltration rates during 1 h experiment for all the six plots. Within the first few minutes of the experiment, rate of infiltration increased rapidly till the middle of the experiment for plot no 1, 3 and 4 and slowly started declining. Pattern is lightly different for plot 2, 5 and 6 that showed a sharp decline within just a few minutes of the experiment. This may be due to the variation in the clay content of the sediments of different plots. As the experiment continued, the differences became subtle till the end of the experiment.

Run-off began slowly soon after the experiment for all the plots and increased rapidly till the middle of the experiment and became stable. The rate of run-off for the plots during the

experiments can be seen in Fig. 7.14. The pattern is the exact reverse of the infiltration as expected that prevailed till the end.

Sediment yield during the experiment as depicted in Fig. 7.15 that displays different patterns for all the plots. Plots 1 and 3 show gradual increase as the experiment continued, with the highest value around 20 g/m^2 for plot no. 1. Plot no. 4 shows drastic difference both in pattern and value of the soil loss. Within the first 5 min, sediment discharge shot up to 300 g/m^2 and continued till the end. Minor fluctuations can be seen in plot 5 and 6 in the sediment discharge pattern.

The variation in the infiltration, run-off and final soil loss of the six plots are attributed to presence/absence of grasses, properties of the sediments and slope. All the six plots represent the variation in the badland topography in the study area, such as slope and ground cover. The relationships between infiltration, run-off and sediment loss against time have been individually obtained and presented in Figs. 7.16, 7.17 and 7.18. The regression lines confirm what was described in the previous paragraphs. Infiltration and run-off generation time were influenced by several factors differently for each plot. Grass cover played an important role, since their ability to hold water promoted delay in the run-off until the soil was drenched till the root depth. Where clay was high on the bare

Fig. 7.12 Rainfall simulation experiment on different plots

plots, the swelling smectite clays created a thin crust after the initial flushing of the loose earth, hence retarded infiltration after a few minutes of the experiment. Slope also played an important role especially on the bare plots. Ploughed and loosened fields rapidly lost sediments within the first few minutes of the experiments and continued the pattern till the end but the plots covered with crop residue protected the surface from raindrop impacts and the results showed negative correlation with erosion. Bradford and Huang (1994) in a similar experiment in Central Illinois presented the same findings that whether it was conventional till or no-till plot, crop residue changed the infiltration, run-off and soil loss significantly. The plots yielded different results based on the slope and ground cover. The objective was to understand a general scenario of soil loss from the badland under review and hence the average has been calculated that resulted in 0.8 kg/m²/year.

7.4 Discussion and Conclusion

The main aim of the chapter was to report the application of three field experimental techniques applied for the first time in a watershed in the Pravara Region, Deccan, India and their significance in soil loss estimates. Microprofilometer technique, erosion-pin technique and rainfall simulation under controlled conditions were applied in a badland watershed along Pravara River, Maharashtra. The region contains

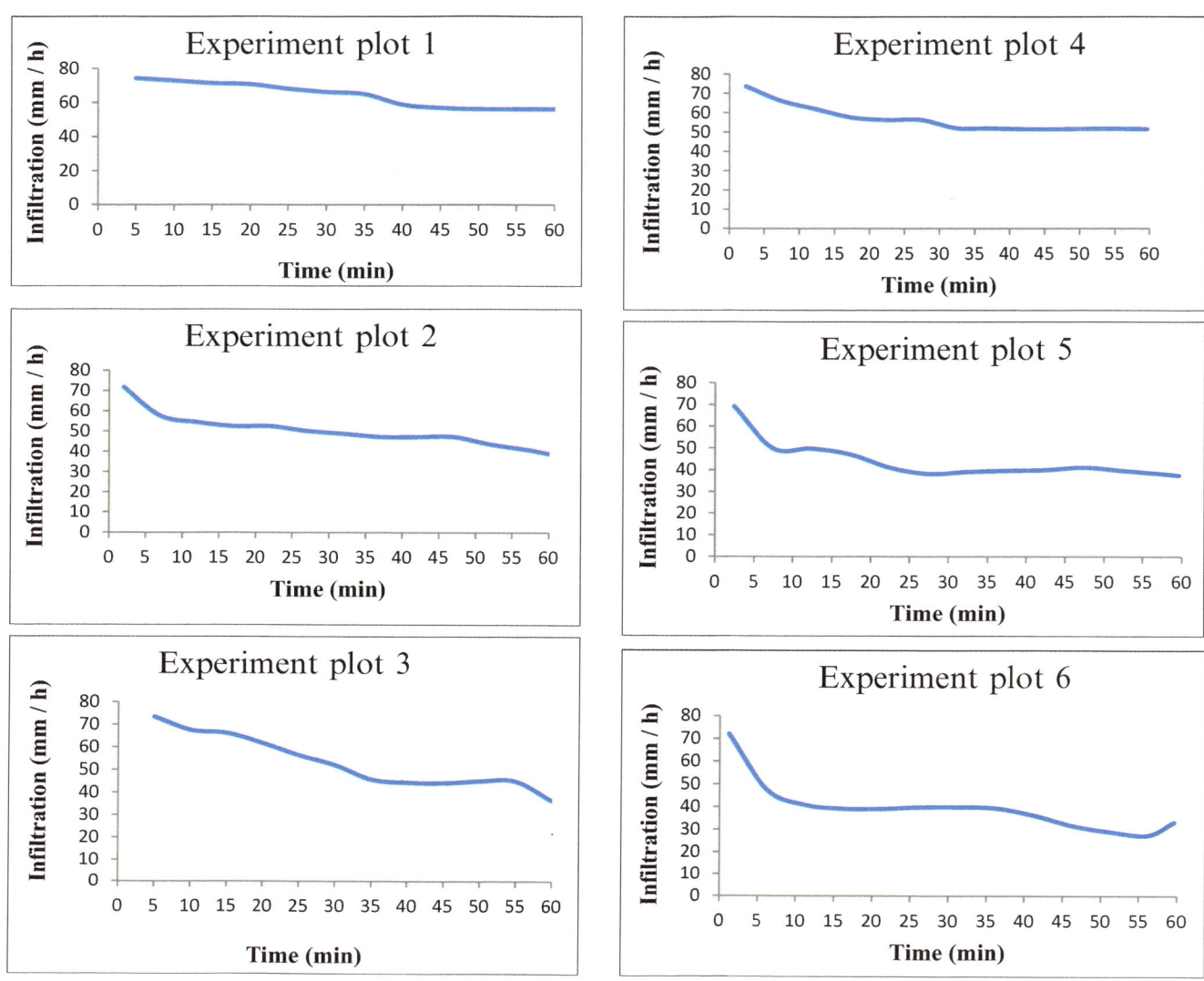

Fig. 7.13 Infiltration during the experiment for the six experimental plots

smectite swelling clays and significantly high exchangeable sodium percentage that makes them fairly erodible soil. The region experiences a semi-arid climate with prolonged dry season that is favourable for the formation of badland topography. Monitoring with the help of microprofilometer technique reveals that the area is experiencing rapid expansion in gully network. All the five sections monitored showed remarkable changes in the cross-sectional area within 2 years. The region is characterized by badland topography with dense fabric of gully network. All the gullies in the area are expanding at that rate continuously. It also has been observed that erosion and deposition are episodic. However, the net yield is positive by the end of the monitoring period.

The result is in agreement with cyclic erosion-deposition presented by Sirvent et al. (1997) as a response to seasonal rains in badland area.

Erosion-pin method resulted in annual sediment loss from the two basins as 0.76 kg/m^2/year from Adula and 1.79 kg/m^2/year from Mahalungi. Though these two watersheds do not represent the entire soil loss from the badland under review, it is clear that from each small badland catchment, enormous volume of sediment is eroded every year by the gullies. Differences in the catchments also were revealed due to the human interference of these slopes. The tolerance limit of the soil loss for Deccan Peninsular region has been reported in an earlier study by Mandal and Sharda (2011)

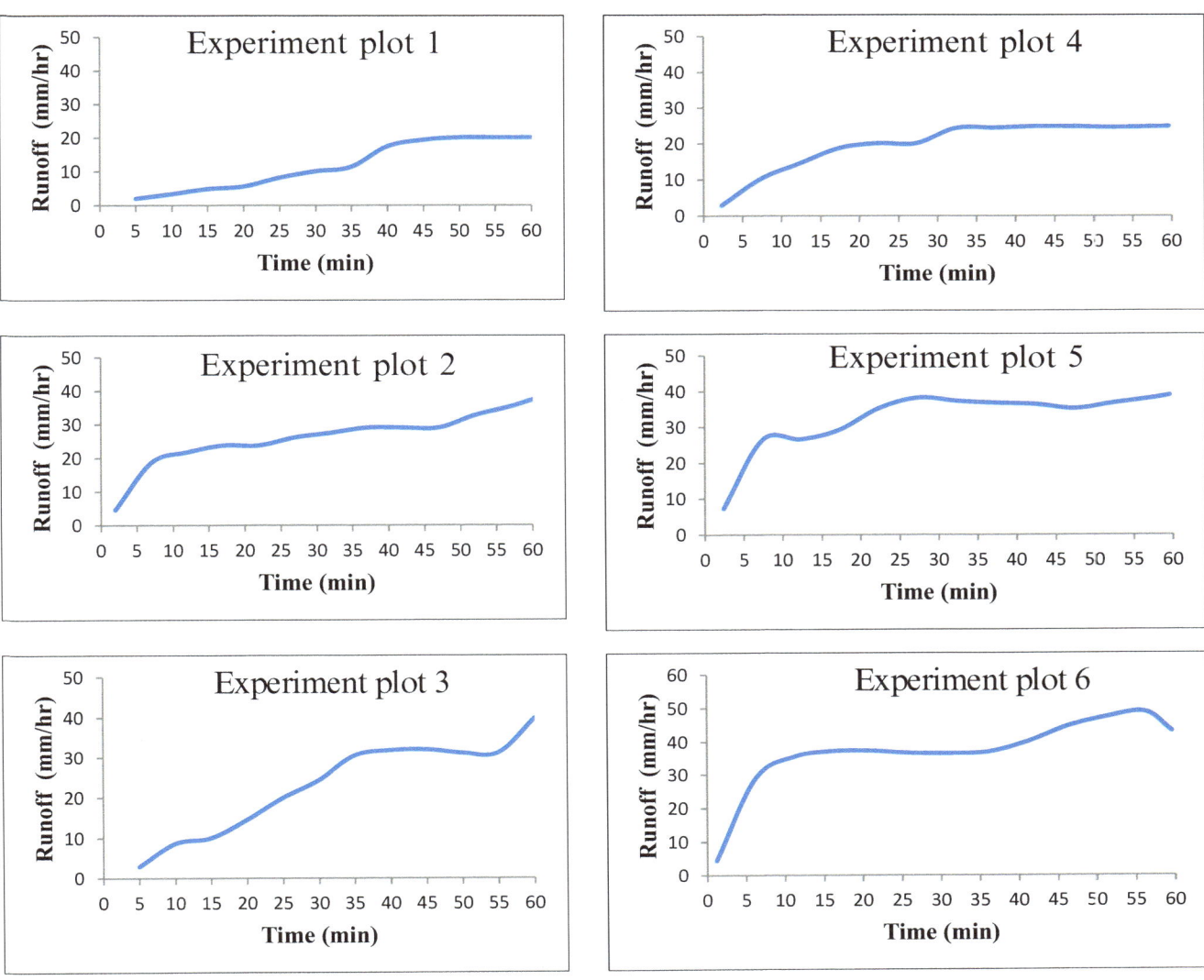

Fig. 7.14 Run-off during the experiment for the six experimental plots

that ranges between 0.25 and 1.25 kg/m²/year. The soil loss values indicated by these two basins after all the three experiments indicate high limit already. If the current landuse is continued and no conservation practice is adopted, the region will bear serious soil loss in the following years. There is a cyclic pattern of erosion and deposition following the seasonality of rainfall. Similar findings have been presented by Johnston et al. (1980) and Sala (1988) in their studies.

Determining sediment yield from different representative plots under controlled rainfall experiment demonstrated that 1 h of rainstorm during the early monsoon rain can erode huge volume of sediments from these gullied watersheds. Indirectly the experiment has revealed that within the same region, different ground cover and terrain characteristics have differently influenced sediment yield, thus suggesting possible choices of a more feasible way to utilize these badlands. Due to the increasing population pressure all over the

country, it is difficult to protect and conserve the natural geomorphic terrain any more now. Reclamation for the purpose of agriculture, housing and other commercial activities will go on. What is necessary is to bring an optimum balance to these land-use practices. It can be possible to achieve this with thorough investigation of such sensitive environments and the field-monitoring techniques have proved to be advantageous.

With the increasing use of RS and GIS and continued availability of high-resolution DEMs, such types of studies are mostly conducted nowadays using remote-sensing and GIS techniques. But ground validation is required for such investigations. Hence this chapter is a contribution to the understanding of the erosion scenario in the real ground condition. Improvement is necessary in these techniques by increasing frequency in data collection and expanding the duration of monitoring period for better results and for meaningful planning purposes in future.

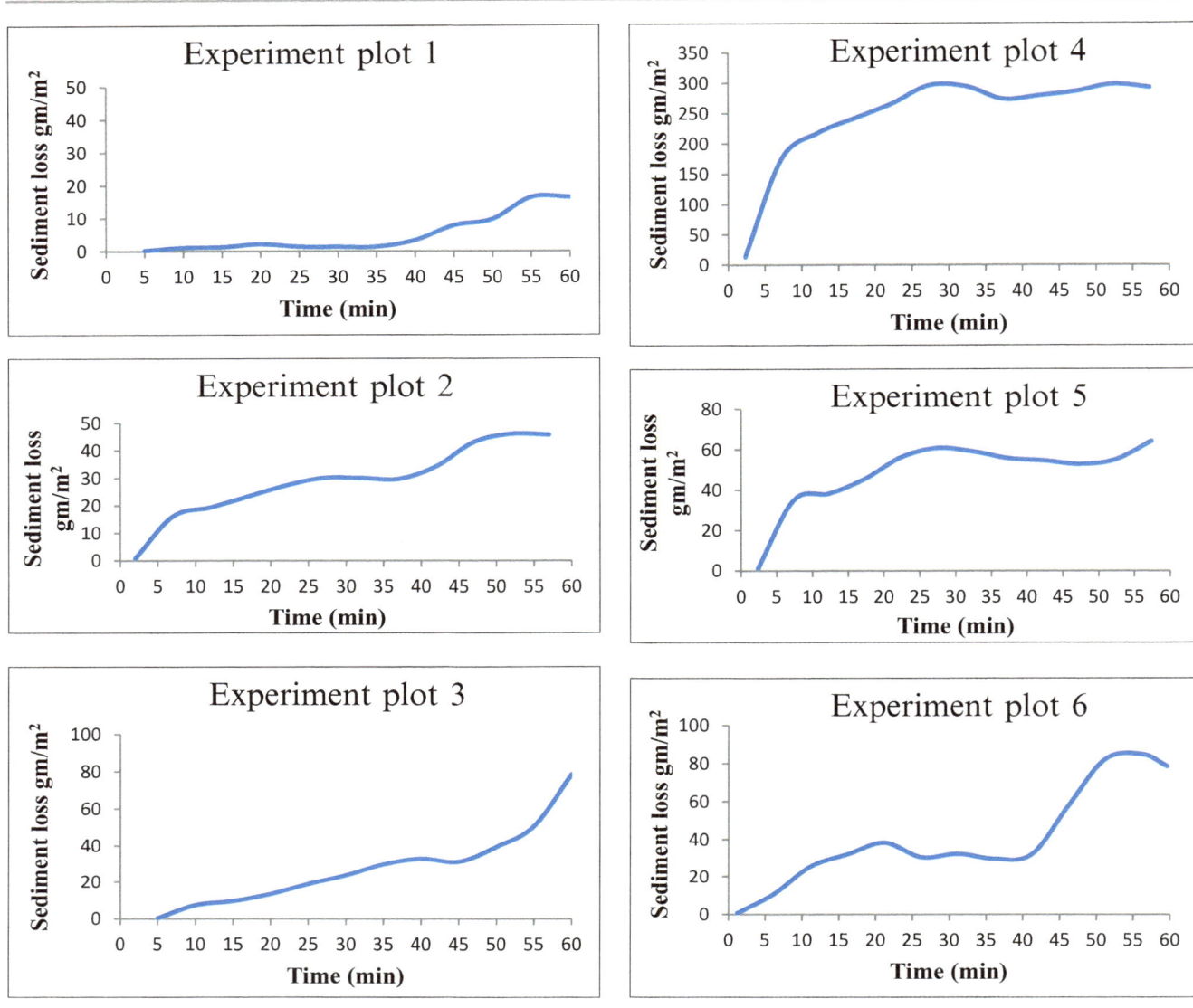

Fig. 7.15 Sediment loss during the experiment for the six experimental plots

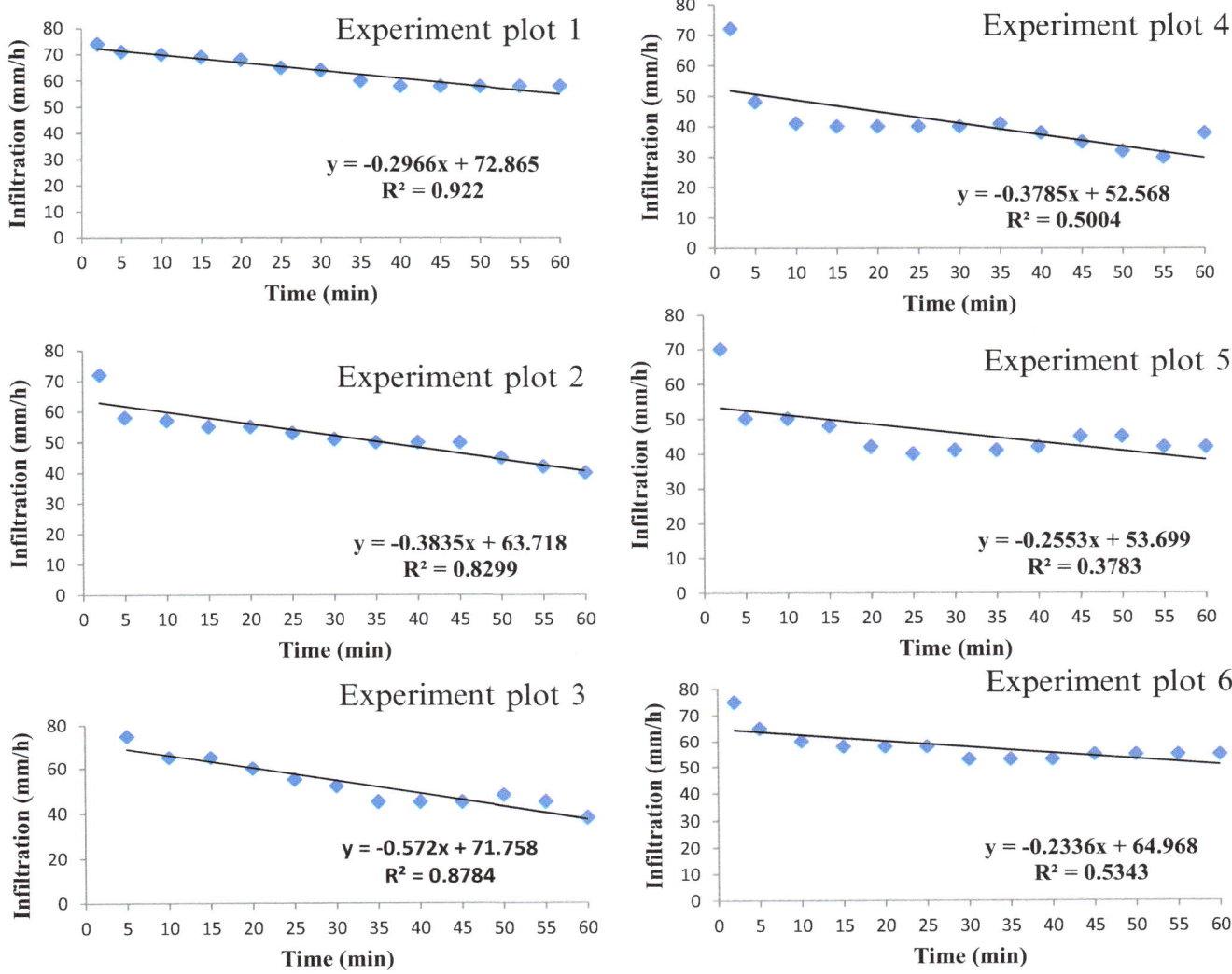

Fig. 7.16 Relationship between infiltration and time during the experiment

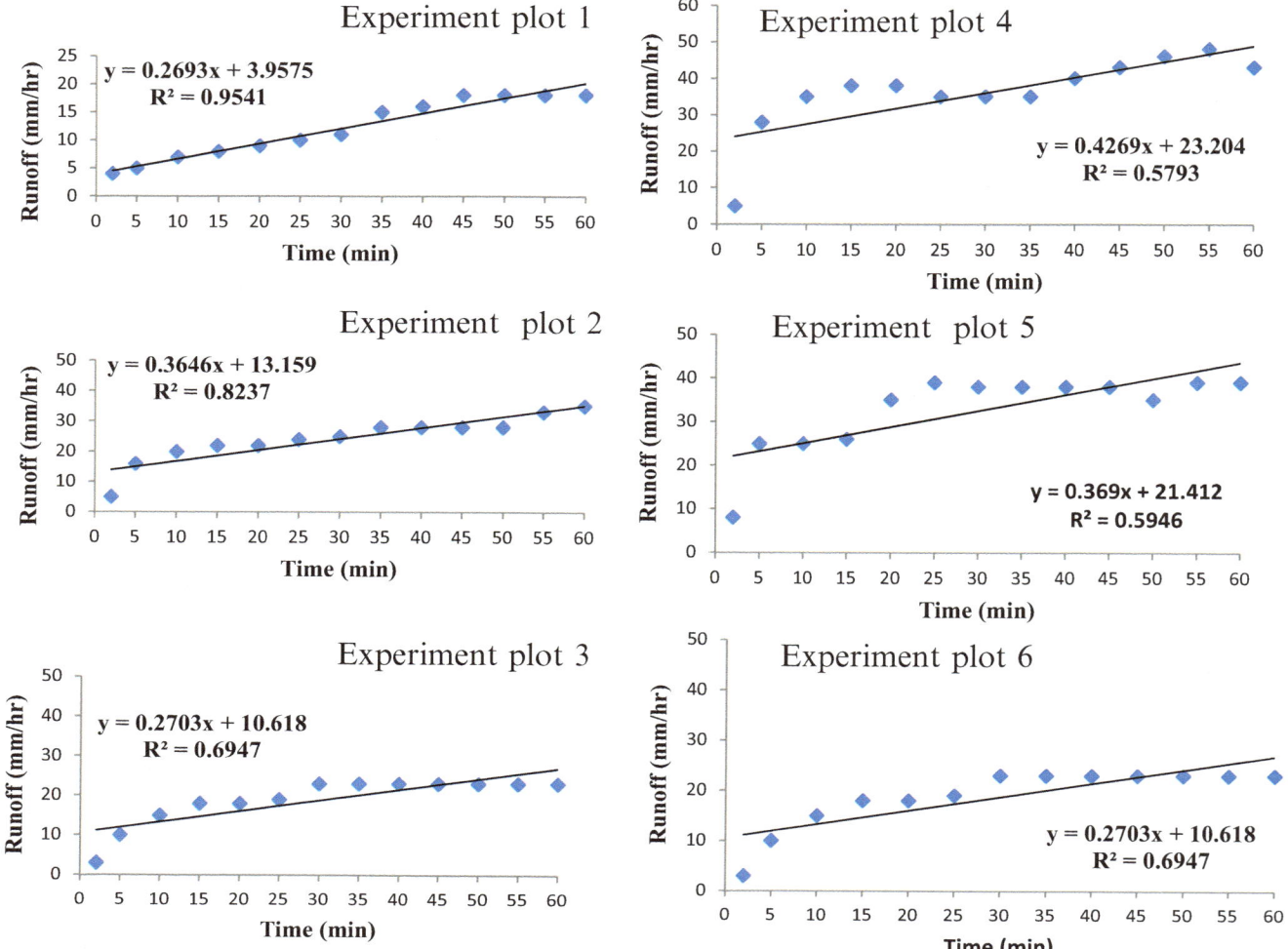

Fig. 7.17 Relationship between run-off and time during the experiment

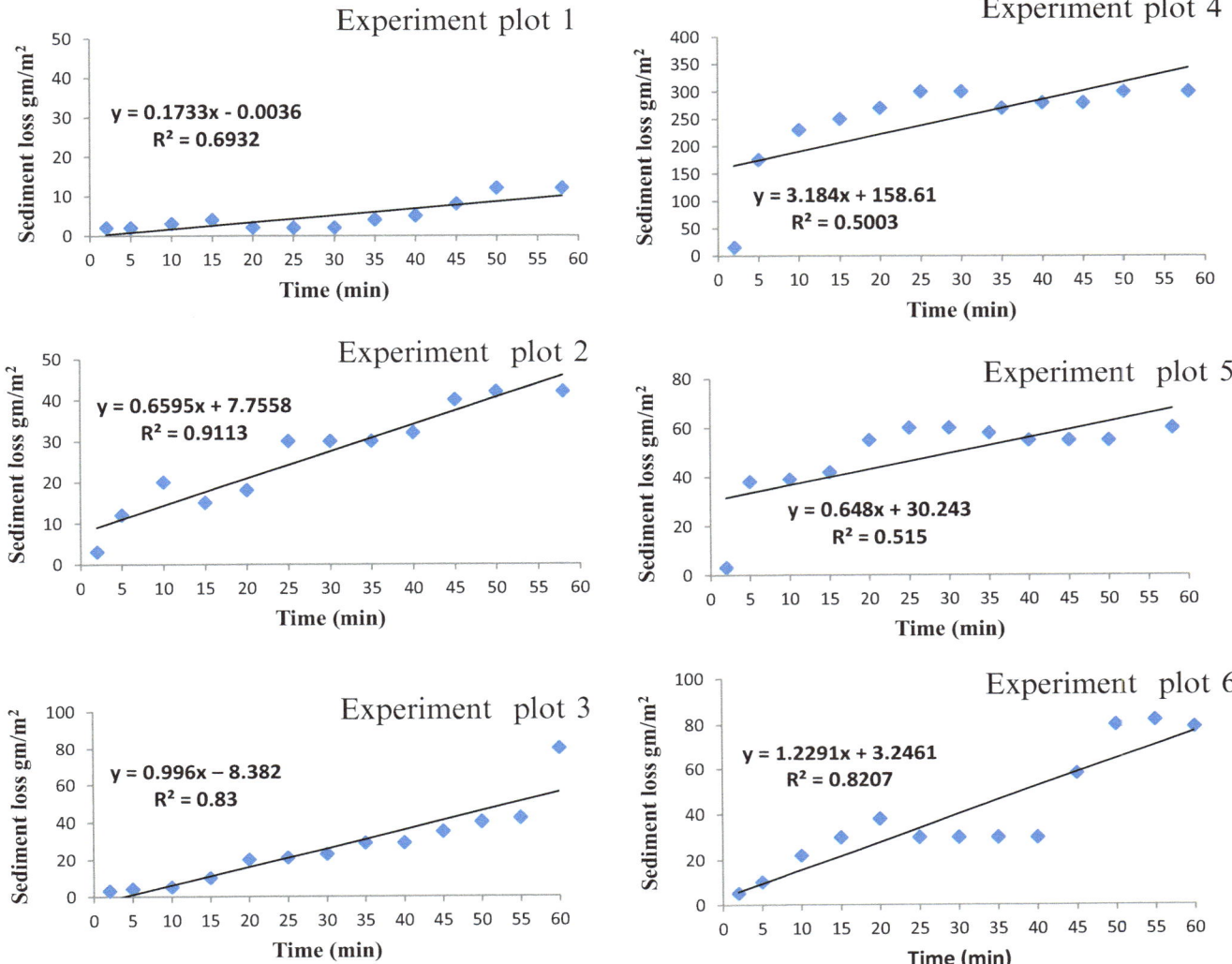

Fig. 7.18 Relationship between sediment loss and time during the experiment

References

Benito G, Gutierrez M, Sancho C et al (1992). Erosion rates in Badlands areas of the Central Ebro Basin (NE Spain). Catena 19: 269–286

Benito G, Gutirrez M. and Sancho C et al (1993) The influence of physico-chemical properties on erosion processes in badland areas, Ebro Basin, NE-Spain. Zeist. Geom. 37: 199–214

Bhardwaj A and Singh R (1992) Development of a portable rainfall simulator infiltrometer for infiltration, runoff and erosion studies. Agri Wat Man. 22: 235–248

Bowyer-Bower TAS and Burt TP (1989) Rainfall simulators for investigating soil response to rainfall; Soil Technol. 2: 1–16

Bradford JM and Huang CH (1994) Interill soil erosion as affected tillage and residue cover. Soil and tillage research. 31(4) 353–361

Bryan RB (1981) Soil erosion under simulated rainfall in the field and laboratory: variability of erosion under controlled conditions. Erosion and Sediment Transport Measurement (Proceedings of the Florence Symposium, June 1981). 1AHS Publ. no. 133

Bryan R and Yair A (1982) Badland geomorphology and piping. In Geo Books University Press Cambridge, 1–11

Cerda A, Ibanez S and Calvo A et al (1997) Design and operation of a small and portable rainfall simulator for rugged terrain; Soil Technol. 11: 163–170

Curtis WR and Cole WD (1972) Micro-topographic profile gauge. Agri Engg. 53: 17

Desir G and Marín C (2007) Factors Controlling the Erosion Rates in a Semi-arid Zone (Bardenas Reales, NE Spain). Catena 71: 31–40

Dunne T, Dietrich WE and Brunengo MJ et al (1980) Simple, portable equipment for erosion experiments under artificial rainfall; J. Agric. Engg. Res. 25: 161–168

Evans M and Lindsay J (2010) High resolution quantification of gully erosion in upland peatlands at the landscape scale. Earth Surf Proc Land. 35, 876–886 (2010)

Exeter NL (1990) Design and operation of a rainfall simulator for field studies of runoff and soil erosion; Soil Technol. 3: 385–397

Folk RL and Ward WC (1957) Brazos river bar: A study in the significance of the grain size parameters. Jour. Sed. Pet. 27: 3–26

Gallart F, Marignani M, Pérez-Gallego N, Santi E and Maccherini S et al (2013) Thirty years of studies on badlands, from physical to vegetational approaches. A succinct review, Catena, 106: 4–11

Ghimire SK, Higaki D and Bhattarai TP et al (2013) Estimation of Soil Erosion Rates and Eroded Sediment in a Degraded Catchment of the Siwalik Hills, Nepal. Land 2: 370–391

Haffzullah A, Ebru E and Gokmen T et al (2017) Empirical Sediment transport models based on indoor rainfall simulator and erosion flume experimental data. Land Degra and Dev. 28: 1320–1328

Haigh M (1997) Use of Erosion Pins in the Study of Slope Evolution, A technical report at: https://www.researchgate.net/publication/281004697

Heng BCP, Sander GC, Armstrong A, Quinton JN and Chandler JH et al (2011) Modelling the dynamics of soil erosion and size-selective sediment transport over non uniform topography in flume-scale experiments. Wat Reso Res. 47(2): W02513

Hessel R and Asch TV (2003) Modelling gully erosion for a small catchment on the Chinese Loess Plateau. Catena. 54: 131–146

Hignett CT, Gusli S, Cass A and Besz W et al (1995) An automated laboratory rainfall simulation system with controlled rainfall intensity, raindrop energy and soil drainage; Soil Technol. 8: 31–42

Hodges WK and Bryan R (1982) Influence of material behavior on runoff initiation in the Dinosaur Badlands, Canada. In: R. Bryan and A. Yair (Eds.), Badland Geomorphology and Piping. Geobooks, Norwich, pp. 13–46

Hsieh YP (1992) A mesh-bag method for field measurement of soil erosion.-J of Soil and Wat Cons. 47(6), pp. 495–499

Johnston HT, Elsawy EM and Cochrane SR et al (1980) A study of the infiltration characteristics of undisturbed soil under simulated rainfall, Earth Surf Proc and Land, 5: 159–174

Joshi VU and Nagare V (2009) Land Use and Land Cover Change Detection along the Pravara River Basin in Maharashtra, using Remote Sensing and GIS techniques. Acta Geodaetica ET Geophysica Hungarica, AGD Landscape & Environment 3(2): 71–86

Joshi VU and Nagare V (2013) Badland formation along the Pravara River, Western Deccan, India, Can tectonism be the cause? Zeist Geom. 57(3): 349–370

Joshi VU and Tambe DT (2010) Estimation of Infiltration Rate, Run-off and Sediment Yield under simulated rainfall experiments in Upper Pravara Basin India, Effect of slope angle and ground cover, Journal of Earth System Science, 119 (6): 763–773

Joshi VU (2014) Soil loss estimation based on profilometer and erosion pin techniques along the badlands of Pravara Basin, Maharashtra. Journal of Geological Society of India, 83(6): 613–624

Karimov V, Sheshukov A, Barnes P et al (2015) Impact of Precipitation and Runoff on Ephemal Gully Development in Cultivated Croplands. Sediment Dynamics from the Summit to the Sea, December 11–14, 2014, USA

Leatherman SP (1987). Field Measurement of Microtopography. J of Coast Res. 3(2): 233–235

Malinov I and Ilieva D (2017) Gully and Rill Erosion in the National Park "Central Balkan". Bulg J of Agri Sc. 23(2): 238–241

Mandal D and Sharda VN (2011) Assessment of permissible soil loss in India employing a quantitative bio-physical model. Current Sc. 100 (3): 383–390

Mosley MP (1975) A device for the accurate survey of small scale slopes. British Geomorph Res Group. Tech. Bull. 17: 11–6

Nadal-Romero E, Martinez-Murillo J and Kuhn N (eds) (2018) Badlands Dynamics in a Context of Global Change, Elsevier

Newton RS (1968) Anew device for measuring ripple mark profiles underwater. Marine Geology. 6: 73–75

Pall R, Dickinson WT, Reals D, and McGirr R et al (1983) Development And Calibration of a Rainfall Simulator. Cana Agri Eng. 25(2): 181–187

Rawat JS and Rai SP (1997) Pattern and intensity of erosion in the environmentally stressed Khulgad watershed, Kumaun Himalaya. J. Geol. Soc. India. 50: 331–338

Rawat JS and Rawat MS (1994) Accelerated erosion and denudation in the Nana Kosi watershed, Central Himalaya, India. Part I: Sediment Load. Mt. Res Dev. 14: 25–38

Rawat JS, Haigh MM and Rawat MS et al (1992) Hydrological response of a Himalayan pine forest micro-watershed; Preliminary results; Proc. Int. Symp. on Hydrology of Mountainous Areas, Simla 28–30 235–258

Robinson LA (1976) The micro-erosion meter technique in a littoral environment. Marine Geology. 22, M51-M58

Roth CH, Meyer B and Frede HG et al (1985) A portable rainfall simulator for studying factors affecting runoff, infiltration and soil loss. Catena. 12: 79–85

Sala M (1988) Slope runoff and sediment production in two Mediterranean mountain environments. Catena Supplementary. 12: 13–29

Sangüesa C, Arumí J, Pizarro R, and Link O et al (2010) A rainfall simulator for the in situ study of superficial runoff and soil erosion. Chilean Journal of Agricultural Research 70(1): 178–182

Schumm SA (1964) Seasonal variations of erosion rates and processes on hillslopes in western Colorado. Zeitsc Geom Supplementary. 5: 215–238

Singh S and Agnithoti SP (1987) Rill and gully erosion in the subhumid tropical riverine environment of Teonthar Tahsil, M.P, India. Geografiska Annaler, 69A-1: 227–236

Singh S and Dubey A (1998) Rate of erosion in the hierarchical orders of natural and cultivated gully basins of Deoghat area, Allahabad District, India. Ind. J. of Geomorphology, 3(1): 75–94

Singh R, Panigrahy N and Philip G et al (1999) Modified rainfall simulator infiltrometer for infiltration, runoff and erosion studies. Agri Wat Managt. 41: 167–175

Sirvent J, Desir G. Gutierrez M. Sancho C and Benito G (1997) Erosion rates in badland areas recorded by collectors, erosion pins and profilometer techniques (Ebro Basin, NE-Spain) Geomorphology. 18: 61–75

Slimane AB, Raclot D, Rebaid H, Bissonnais YL, Planchon O and Bouksila F et al (2018) Combining field monitoring and aerial imagery to evaluate the role of gully erosion in a Mediterranean catchment (Tunisia). Catena. 170: 73–83

Sukhanovskii YP (2007) Modification of a Rainfall Simulation Procedure at Runoff Plots for Soil Erosion Investigation. Eur Soil Sc. 40 (2): 195–202

Toy TJ (1983) A linear erosion/elevation measuring instrument (LE MI). Earth Surf Proc and Land. 8: 313–3

Tricker AS (1979) The design of a portable rainfall simulator infiltrometer; J. Hydrol. 41: 143–147

Trudgill ST, High CJ and Hanna FK (1981) Improvements to the micro erosion meter. British Geom Res Group Tech Bull. 29: 3–17

Tucker GE, Arnold L, Rafael L, Flores BH, Istanbulluoglu E and Sólyom P et al (2006) Headwater channel dynamics in semiarid rangelands, Colorado high plains, USA, Geol Soc of Am Bull. 118 (7/8): 959–974

Wang YC and Lai CC (2018) Evaluating the Erosion Process from a Single-Stripe Laser-Scanned Topography: A Laboratory Case Study. Water. 10, 956

Wang R, Zhang S, Pu L, Yang J, Yang C, Chen J, Guan C, Wang Q, Chen D, Fu B and Sang X et al (2016) Erosion Mapping and Monitoring at Multiple Scales Based on Multi-Source Remote Sensing Data of the Sancha River Catchment, Northeast China. ISPRS Int. J. Geo-Inf. 5: 200

Watson A and Evans R (1991) A comparison of estimates of soil erosion made in the field and from photographs. Soil and Tillage Res. 19: 17–27

Wischmeier, W.H., Johnson, C.B. and Cross, B.V. (1971): A soil erodibility nomograph for farmland and construction sites, J. Soil and Wat. Conserv. 26, 189–193

Wells N. A and Andreamiheja B (1995) The initiation and growth of gullies, Madagascar- Are humans to be blamed? Geomorphology, 8: 1–46

Wright JA, Smith J, Gundry SW and Glasbey CA et al (2002) A spatial rainfall simulator for crop production modeling in southern Africa; Math. Comput. Model. 35: 1459–1466

Young RA and Burwell RE (1972) Prediction of Runoff and Erosion from Natural Rainfall Using a Rainfall Simulator. Soil Sci. Soc. Amer. Proc. 36: 827–830

Zanchi C (1988) Soil loss and seasonal variation of erodibility in two soils with different texture in the Mugello valley in Central Italy. Catena Supplementary. 12: 167–173

Veena U. Joshi is currently a Professor in the Department of Geography, Savitribai Phule Pune University. She did her M.A., B.Ed., M.Phil. and Ph.D. from the same university. She has been conducting research on gully formation and dynamics in the Deccan Traps Region of Maharashtra for the last 35 years and published many articles in both national and international journals and chapters on Badland special volumes. She has been awarded Fulbright Nehru Senior Research Fellowship, Fulbright Nehru Environmental Leadership Fellowship for USA, ERASMUS MUNDUS Fellowship for Belgium, Endeavour Award for Australia, and JSPS (Japan Society for the Promotion of Science) for Tokyo to conduct collaborative research with international scientists. She is member of several scientific committees including the Gully Erosion Research group. She has completed two international collaborative projects funded by the European Union with nine partner countries. She is instrumental in the student's exchange programs with international universities and home university. She has established soil monitoring lab in the institute to promote soil erosion research. She is actively engaged in gully erosion research and working on the development of an indigenous soil erosion model for India.

Gully Erosion Susceptibility Mapping Based on Bayesian Weight of Evidence

Pravat Kumar Shit, Gouri Sankar Bhunia, and Hamid Reza Pourghasemi

Abstract

Identifying gully erosion susceptibility in cultivated region is important for the manager and decision makers. The present study demonstrated the application of the weight of evidence (WoE) model (a Bayesian probability model) for gully erosion susceptibility mapping using geographic information system (GIS) and remote sensing (RS) tools in the southwestern part of West Bengal, India. Eight gully erosion conditioning geo-environmental factors were considered for the susceptibility analysis, such as lithology, geomorphology, soil type, land use, slope, slope length (LS), stream power index (SPI), and wetness index (WI). Tests of conditional independence were performed for the selection of eight gully conditioning factors. Finally, gully erosion susceptibility map was prepared using the ratings of each gully conditioning factor. The resultant susceptibility map was validated using the area under the curve (AUC) method. The results indicated that the WoE model had an AUC value of 67.8%. Therefore, the WoE model is useful in gully erosion susceptibility mapping and helps decision makers in land-use planning.

Keywords

Gully erosion · Weight of evidence model · Geo-environmental factors · GIS · India

8.1 Introduction

Gully erosion is one of the most important problems of the soil erosion processes in terms of soil loss and sediment production over the world and has negative impacts on soil features, crops, and water resources (Valentin et al. 2005; Whitford et al. 2010; Lucà et al. 2011). During field surveys, time constraints and difficult terrain allow only partial or local measurements of gully erosion. Gullies may remain unobserved when visually obstructed by vegetation, and recording their dimensions can be quite challenging when gullies are large or when they expand over vast undulating landscapes (Frankl et al. 2013). Moreover, field observations only provide limited information on the historical importance of gully erosion. Therefore, several studies explored the potential of remote-sensing products to facilitate research on gully erosion (Marzolff et al. 2011; Frankl et al. 2013; Khadse et al. 2015). Remote sensing (RS) and geographic information system (GIS) integrated with erosion forecast models do not only approximate soil loss but also offer the spatial distribution of the erosion (Okalp 2005; Shit et al. 2015; Khadse et al. 2015). Particularly, generating precise erosion risk maps in GIS platform is extremely significant to establish the areas with high erosion risks (Mitasova et al. 1996; Lucà et al. 2011) and to expand plenty erosion deterrence techniques (Vrieling et al. 2002).

However, natural events such as floods, earthquakes and avalanches, landslides, gully erosion, and soil erosion are often difficult to predict, because they are uncertain with potentially detrimental consequences. Governments and research institutions worldwide have attempted for years to assess natural and man-made hazards in order to predict risk (Pradhan et al. 2010); but a few studies have investigated the prediction of gully erosion and its susceptibility mapping (Conforti et al. 2011; Conoscenti et al. 2013; Dube et al. 2014; Shit et al. 2015) using the weight of evidence (WoE) model.

P. K. Shit (✉)
Department of Geography, Raja N.L. Khan Women's College (Autonomous), Medinipur, West Bengal, India

G. S. Bhunia
Aarvee Associates Architects, Engineers & Consultants Pvt. Ltd, Hyderabad, India

H. R. Pourghasemi
Department of Natural Resources and Environmental Engineering, College of Agriculture, Shiraz University, Shiraz, Iran

© Springer Nature Switzerland AG 2020
P. K. Shit et al. (eds.), *Gully Erosion Studies from India and Surrounding Regions*, Advances in Science, Technology & Innovation, https://doi.org/10.1007/978-3-030-23243-6_8

The WoE is based on a bivariate statistical approach and was originally developed for mineral potential assessment (Agterberg et al. 1990). Several authors have applied the WoE model for mineral potential mapping (Venkataraman et al. 2000; Asadi and Hale 2001) and landslide susceptibility mapping (Barbieri and Cambuli 2009; Regmi et al. 2010) using GIS and RS. The method has also been used for habitat quality assessment (Romero-Calcerradaa and Luqueb 2006). So, the aim of the current study is modeling gully erosion using a bivariate and probabilistic model named the WoE in southwestern part of West Bengal, India.

8.2 Study Area

The study area (Garbheta block) is located in the middle reach of Silai River basin, lying in the northern part of Paschim Medinipur District of West Bengal. It lies between 22° 47′ 12″ to 22° 56′ 27″ N latitude and 87° 13′ 17″ to 87° 23′ 29″ E longitude (Fig. 8.1). The study area is about 35.20 km^2 with an average elevation of 65 m above sea level. Gully erosion is a very conspicuous feature throughout the region and affecting different lithologies and soil types. The climate is of tropical type with hot and dry summers, whereas precipitation is concentrated in monsoon periods from July to September (Sen et al. 2004).

Recently, Shit et al. (2013, 2014, 2015) identified slope failure along gully walls and water erosion processes as the main denudational processes acting in the Garbheta block. Clayey, steep, and weathered slopes experience mainly rotational slip and topping, whereas sheet and rill erosion and gully incision particularly affect areas without vegetation cover, in cultivated fields and pasture lands. In the study area, many permanent large gullies consist of deep incisions with vertical sidewalls and depth often more than 20 m (Fig. 8.2a). In the study area, where sands and conglomerates outcrop, gully channels are narrow and usually V-shaped, with almost bare side slopes indicating an active stage of dissection (Fig. 8.2b). During heavy rainfall events, many gullies also undergo headcut, and valley-side retreat processes lead to their lengthening and widening (Fig. 8.2c). The most evident and spectacular landforms related to gully erosion in the study area are represented by calanchi landforms, mainly developed into clay deposits (Fig. 8.2d); they locally develop from slopes affected by landslide scars and promote based on the concentration of running water (Morgan and Mngomezulu 2003).

8.3 Materials and Methods

8.3.1 Gully Inventory Mapping

The gully inventory map at Garbheta block of Paschim Medinipur, West Bengal, was prepared using spot and Google Earth images and validated with extensive field observations (Fig. 8.3). For this aim, the "add polygon" function in Google Earth was used to digitize gullies in the selected areas. The gully polygons were saved as KML (Keyhole Markup Language) files. In the next step, the KML to Shape file function in ArcGIS 10.2.2 was used to convert the digitized gullies to Shape file (".shp file format") (Frankl et al. 2013). Also, some gully polygons were digitized from the spot images. Extensive field surveys were done for the identified gullies for recording of gully morphological features, i.e., gully length, width, and depth.

8.3.2 Gully Erosion Conditioning Geo-Environmental Factors

Gully erosion occurrence and behavior of this phenomenon depend on several factors including climate, topography, lithology, soil characteristics, and land use. The choice of a number of predisposing factors is needed to assess a gully erosion susceptibility map (Shit et al. 2015). In the present study, eight factors namely lithology, geomorphology, soil type, land use, slope, slope length (LS), stream power index, (SPI), and wetness index (WI) were considered to map the gully erosion susceptibility zone using the WoE Bayesian model. For this aim, soil map was extracted from the soil database with a scale of 1:50,000 (NBSS & LUP, National Bureau of Soil Survey and Land Use Planning, India). Lithology was extracted from the geological database with a scale of 1:50,000 (GSI, Geological Survey of India). The land-use data were classified from the Landsat Thematic Mapper (TM) and Indian Remote Sensing (IRS) 1C LISS-III imagery at 30 m × 30 m resolution. A digital elevation model (DEM) was prepared using the topographic database of The Advanced Spaceborne Thermal Emission and Reflection Radiometer (ASTER). All the morphometric characteristics relevant to gully erosion analysis were calculated from the DEM. The images were resampled to a pixel size of 2.5 m. Table 8.1 represented the predisposing factors that were calculated following described methods (Shit et al. 2015; Moore and Burch 1986; Moore et al. 1991). All the data

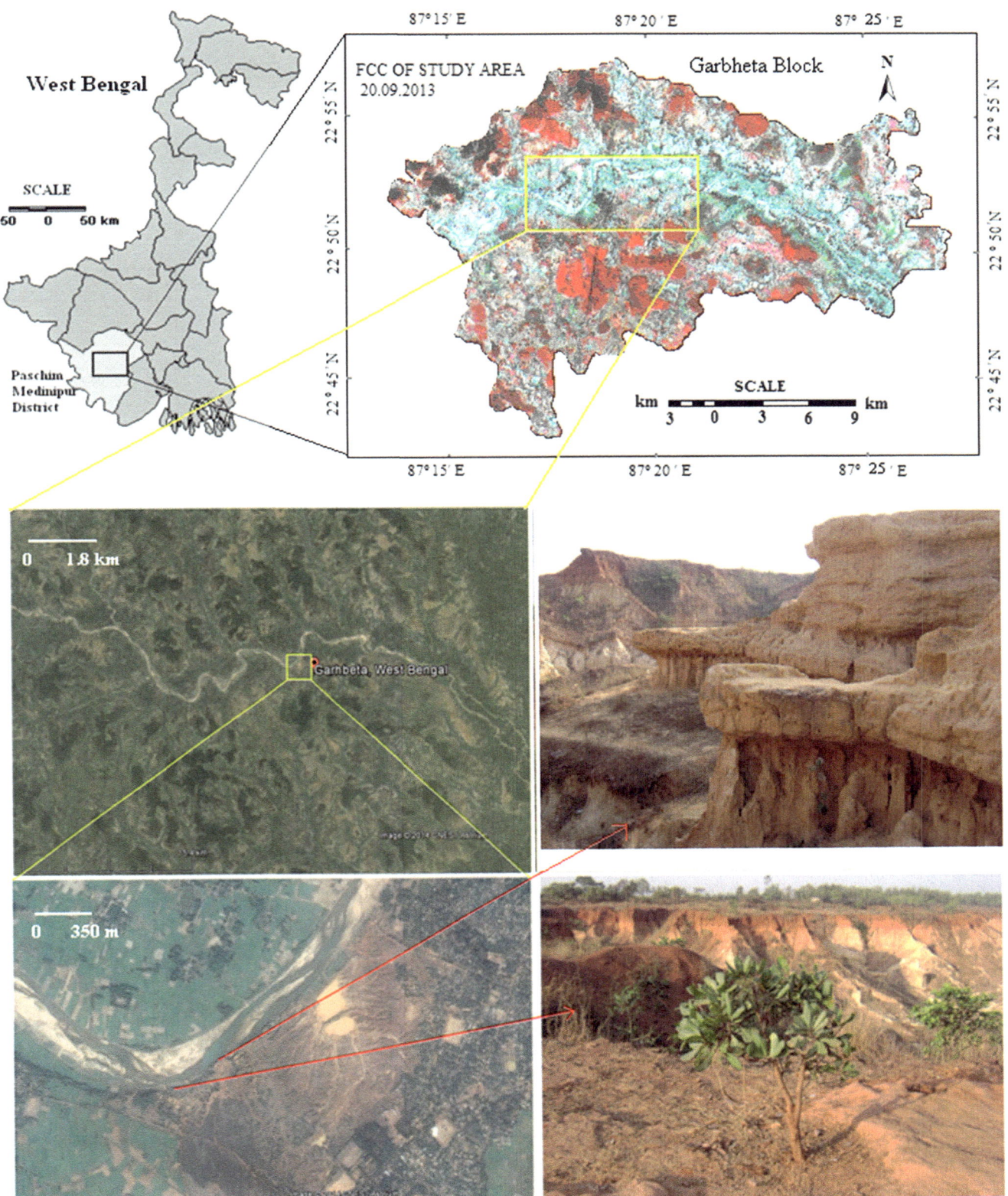

Fig. 8.1 Location of the study area: Garbheta block, Paschim Medinipur (West Bengal) (Source: Google Earth and Landsat Images, 2014), and gully erosion at Garbheta, Paschim Medinipur

Fig. 8.2 Overview of gully erosional landforms: (**a**) Affecting permanent gully on the left side of Silai River. (**b**) "U" shape gully valley. (**c**) Gully wall erosion. (**d**) Pillar formation by sand stone. (**e**) Gully heads retreat during rainy season, 2014. (**f**) Subsidence landform on clayey slope close to the valley floor

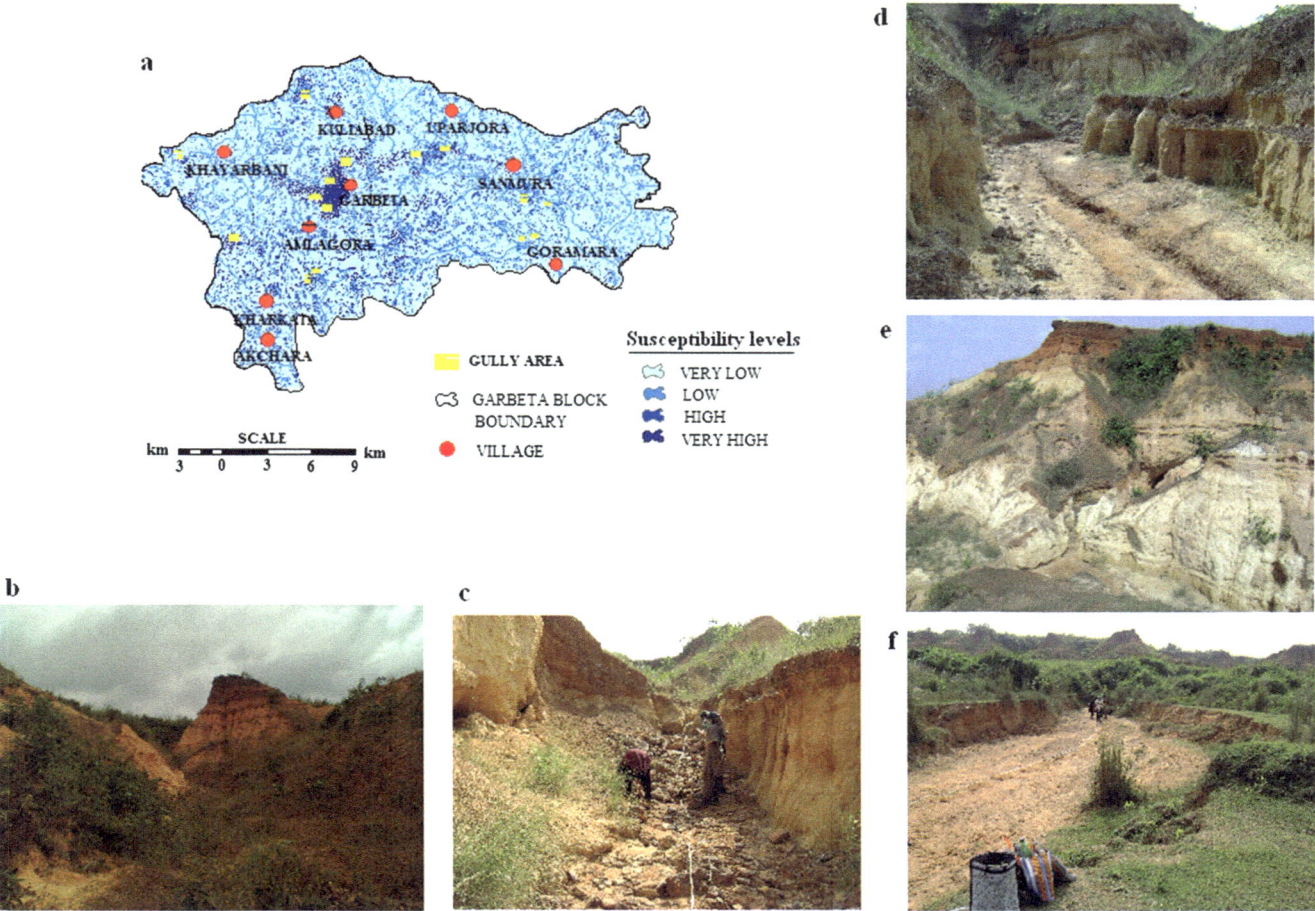

Fig. 8.3 Gully erosion inventory map (**a**) and pictures of some of the gullies identified in the study area (**b–f**)

Table 8.1 Methods of calculation of different geo-environmental factors for gully erosion

Predisposing factors	Methods	References
Land use	Supervised image classification using the maximum likelihood classifier (MLC) algorithm	Shit et al. (2015)
Slope length (LS)	LS = $(fa \times \text{cellsize}/22.13)^{0.4} \times (\sin \sigma/0.0896)^{1.3}$	Moore and Burch (1986)
Wetness index (TWI)	WI = $\ln [As/\tan (\beta)]$	Moore et al. (1991)
Stream power index (SPI)	SPI = $As \times \tan \sigma$	Moore et al. (1991)

were saved in a single platform in GIS environment. In the present study, weight value for a parameter class was delineated as the natural logarithm of the gullies density class divided by the area of gullies density over the entire study area.

Finally, the weighted overlay index method was used to calculate the susceptibility value to delineate gully eroded area and characterize the Badlands units. Thematic maps were reclassified in GIS environment (Fig. 8.4). All the layers are overlaid by applying the "Raster Calculator" tool in the "Spatial Analyst" extension of ArcGIS v9.2 in order to calculate the gully erosion susceptible zone in the study area. The gully dominated ranges of values have been categorized

into five susceptibility classes based on the natural breaks (Shit et al. 2015).

8.3.3 Weight of Evidence Model (WoE) for Preparing Gully Erosion Map

In the present study, according to Eqs. 8.1–8.5 of the Bayesian probability model, known as the weight of evidence model (WoE), was applied for gully erosion susceptibility mapping. The WoE model was used for assessing the relations between the spatial distribution of the areas affected by gullies and the spatial distribution of the predisposing

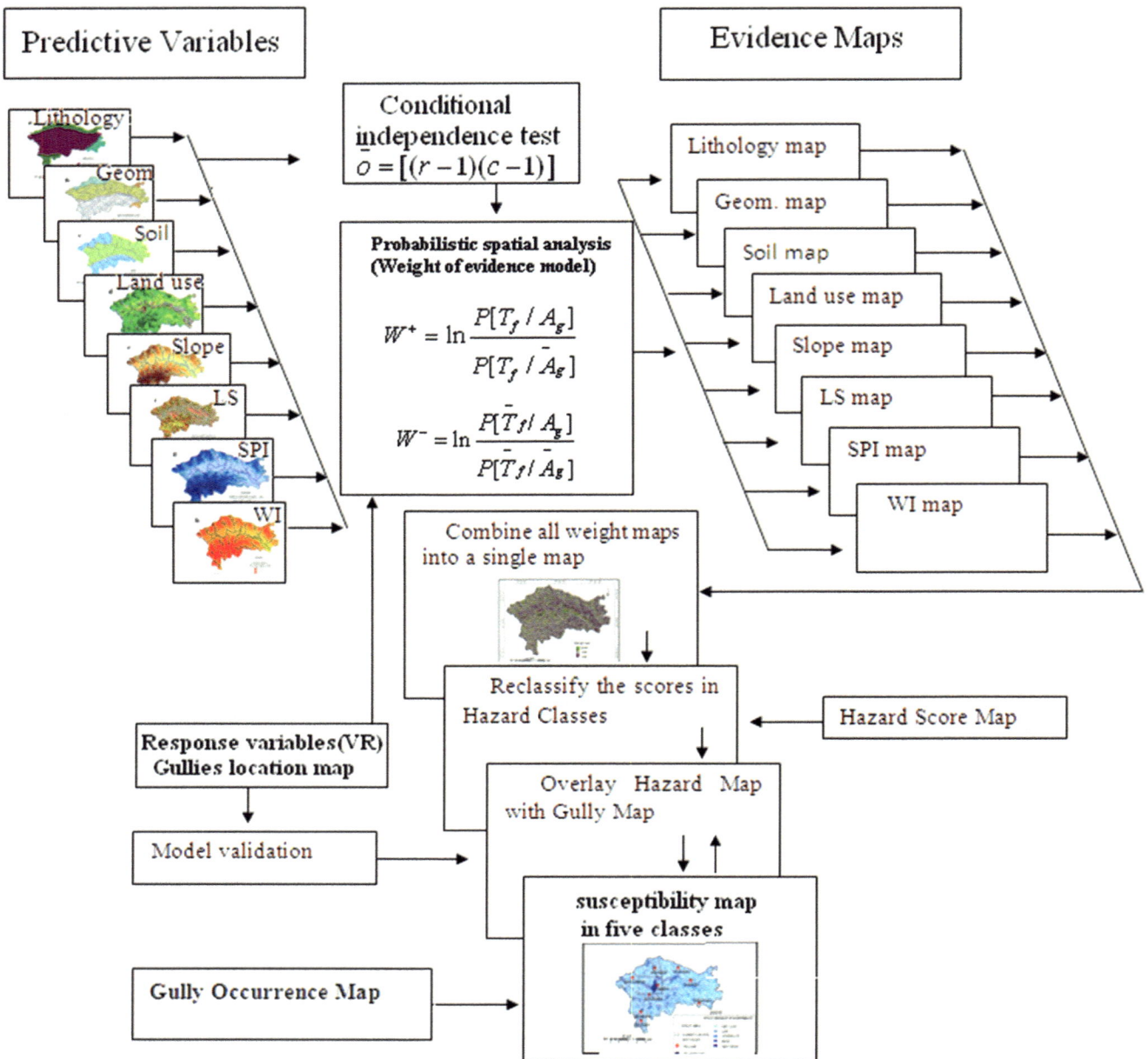

Fig. 8.4 Flow chart of the methodology (WoE model) modified from van Western (2003)

factors (Bonham-Carter 1994; Barbieri and Cambuli 2009; Dube et al. 2014).

This method is based on calculation of positive and negative weights to define degree of spatial association between gully occurrence and each explanatory variables class. The positive weights (W^+) and negative weights (W^-) are consigned to each pixel of the factor maps and sequentially the contrast weight (C_w) was calculated as suggested by van Western (2003). The possible combinations of each factor are represented in Table 8.2, of which the frequency is expressed as number of pixels, calculated within a GIS platform (van Western 2003). The weights are calculated as the following:

$$W^+ = \ln \frac{P\left[T_f/A_g\right]}{P\left[T_f/\overline{A}_g\right]} = \ln \left(\frac{\frac{T_f \cap A_g}{A_g}}{\frac{T_f \cap \overline{A}_g}{\overline{A}_g}}\right)$$

$$= \ln \left(\frac{\frac{\text{Gully\_area\_in\_class}}{\text{Total\_gully\_area}}}{\frac{\text{Stable\_area\_in\_class}}{\text{Total\_stable\_area}}}\right) = \ln \left(\frac{\frac{N\text{pix}_1}{N\text{pix}_1+N\text{pix}_2}}{\frac{N\text{pix}_3}{N\text{pix}_3+N\text{pix}_4}}\right) \quad (8.1)$$

$$W^- = \ln \frac{P\left[\overline{T}_f/A_g\right]}{P\left[\overline{T}_f/\overline{A}_g\right]} = \ln \left(\frac{\frac{\overline{T}_f \cap A_g}{A_g}}{\frac{\overline{T}_f \cap \overline{A}_g}{\overline{A}_g}} \right)$$

$$= \ln \left(\frac{\frac{Total\_gully\_area\_in\_other\_class}{Total\_gully\_area}}{\frac{Total\_stable\_area\_in\_other\_class}{Total\_stable\_area}} \right)$$

$$= \ln \left(\frac{\frac{N\mathrm{pix}_2}{N\mathrm{pix}_1 + N\mathrm{pix}_2}}{\frac{N\mathrm{pix}_4}{N\mathrm{pix}_3 + N\mathrm{pix}_4}} \right) \quad (8.2)$$

where $N\mathrm{pix}$ = number of pixels, W^+ = positive weight, and W^- = negative weight.

In order to calculate the weights using Eqs. (8.1) and (8.2), the following columns were created in Microsoft Office Excel:

N_{map} = total number of pixels in the map
n_{gully} = number of pixels with gullies in the map
n_{class} = number of pixels in the class
$n_{\mathrm{gullyclass}}$ = number of pixels with gullies in the class

The values needed for the weight formulas are:

N_{pix1} = ngullyclass
N_{pix2} = ngully-ngullyclass
N_{pix3} = nclass-ngullyclass
N_{pix4} = nmap-ngully-nclass + ngullyclass

The positive weight (W^+) is directly proportional to the influence that the n class of the i parameter has on gully development. For each factor, W^+ is used to signpost the significance of the occurrence of the factor for the gullies, and if W^+ is negative, it is not auspicious. The negative weight (W^-) is used to evaluate the importance of the absence of the factor for the occurrence of gullies. Weights with extreme values determine susceptibility mapping of gullies using these factors. The weight of the factors around zero designates no influence on gully occurrence.

The contrast weight (C_{w}) was calculated using Eq. (8.3):

$$C_{\mathrm{w}} = W^+ - W^- \quad (8.3)$$

where: W^+ = positive contrast of each factor class, W^- = negative contrast of each factor class.

Since the factor maps used are not binary rather they are multiclass, this means C_{w} calculation will involve adding the negative weights of other classes to W^+ (Van-Western 2003). The studentized contrast (C_{s}, which is the contrast divided by its standard deviation) was calculated, and the values range

Table 8.2 Contingency table showing four possible combinations of a gully conditioning factor and a gully map (van Western 2003)

		F—potential gully conditioning factor		
		Present	Absent	Diagnosed total
G—Gully	Present	$N\mathrm{pix}_1$	$N\mathrm{pix}_2$	$N\mathrm{pix}_1 + N\mathrm{pix}_2$
	Absent	$N\mathrm{pix}_3$	$N\mathrm{pix}_4$	$N\mathrm{pix}_3 + N\mathrm{pix}_4$

from $1.96 < C_{\mathrm{s}} < -1.96$ indicating the hypothesis that $C = 0$ cannot be rejected at 0.05 confidence level (Mihalasky and Moyer 2004). The standardized value of C, calculated as the ratio of C to its standard deviation, C_{s}, serves as a guide to the significance of the spatial association, and acts as a measure of the relative certainty of the posterior probability (Bonham-Carter 1994). The standard deviation of C is calculated as

$$C_{\mathrm{s}} = \sqrt{S^2(W^+) + S^2(W^-)} \quad (8.4)$$

where $S^2(W^+)$ and $S^2(W^-)$ are variance of W^+ and W^- and C_{s} is the standard deviation of the contrast.

The association between a factor and the contrast value was assessed using the Spearman's Rank Correlation Coefficient for assessing correlation between ranked variables. The statistical analysis has been performed using SPSS v12.0 statistical software. Before summing up the weight map, the factors were tested for conditional independence using the chi-square test. The contrast of each factor class or type was summed to calculate the total weight map (TWmap) as shown in Eq. (8.5).

$$\mathrm{TWmap} = \sum C_{\mathrm{w}} \quad (8.5)$$

where C_{w} is contrast of each factors class.

8.3.4 Accuracy Assessment

The WoE model validation was executed via validation set of randomly selected 30 locations in the field and eroding gully activity which had been surveyed in the previous season. Field survey through GPS coordinates of the newly formed gullies were converted into a gully point map and overlaid to the gully erosion susceptibility map. The accuracy of prediction of the model was determined from the area under the curve (AUC) of cumulative percentage gully area against cumulative percentage of the study area (Regmi et al. 2010; Dube et al. 2014).

8.4 Results and Discussion

8.4.1 Role of Land Use/Land Cover on Gully Occurrence

In general, weights of each factor according to relationship between each predisposing factor and gully erosion locations are given in Table 8.3. Land-use factor plays an important role in gully erosion occurrence, mainly on barren land and wastelands (C_s = 16.383 and 17.858, respectively). A significant (C_s > 1.96) association was observed between gully occurrence and land cover types (barren land, wasteland, agricultural land, and forest areas). The presence (W^+) of the different land-use types is more important than the absence (W^-) of the land-use types, with all the W^+ having a high magnitude than the W^-; for example, the class surface water bodies has a high W^+ (−2.7705) but a very low W^- (0.0049) (Table 8.3). Results also illustrate a high positive contrast value for barren land (C_w = 3.9265) and negative value for dense forest (C_w = −1.3742). Dube et al. (2014) described that removal of forests for the creation of settlements promotes gully development, and thus increases in settlements could promote further gully erosion. These findings are in agreement with the knowledge that forested areas experience less erosion in the form of gullies than barren land or wastelands (Harvey 1992; Zheng 2006). However, no significant association was observed between rivers and gully occurrence (C_s = −7.361). Scrub land areas have a negative association (C_w = −0.2079) with gully occurrence, indicating very less influence on gully occurrence. Therefore, Garbheta block is gully erosion-prone area with a registered decrease in the forest cover.

8.4.2 Role of Lithology and Soil Type on Gully Occurrence

The results of analyses also showed fragments of pebbles, boulder, and gravels have positive roles on gully erosion processes (C_w = 7.636 and C_s = 6.254) (Fig. 8.5). Consequently, the influence of the absence of pebbles, boulder, and gravels was pronounced than the influence of the presence of the factor (W^- = −0.5739 and W^+ = 0.1897). A negative association was observed for unconsolidated sands, silts, clay, and fine medium sands (C_w = −0.4604, −0.2845), respectively (Table 8.3). Results implied that gully erosion occurrence is positively associated with the soil type. The two main soil types are observed in Garbheta block, namely, lateritic soil and older alluvial soil. The influence of soil type on gully occurrence can be observed in Table 8.3. In the case of lateritic soils, the contrast value was higher (C_w = 0.2429).

A negative association was observed for older alluvial soil (C_w = −0.2429, C_s = −2.302), respectively.

This implies that gully erosion occurrence is positively associated with the lateritic soil type. Lateritic soils experience relatively lower infiltration rates during a heavy rainstorm; when the rain falling on the surface of this soil far exceeds infiltration, it gives rise to runoff and results in losses of soil erosion. Older alluvial soils have high clay content and thus are resistant to soil erosion.

8.4.3 Role of Slope Degree on Gully Occurrence

The factor slope gradient has a significant positive correlation with gully occurrence (ρ = 0.7178 (p < 0.05). All slope classes were significantly associated with gully occurrence (C_s > 1.96) except for 5–10 (C_s = −6.954) and 10–15 degrees, respectively (C_s = −5.782). In total, 15–30-degree slopes were found strongly associated with gullies in the area (C_s = 4.622) and >30-degree slope gradient indicates strong positive association with gully occurrence (C_s = 12.201). The influence of the absence of slope more than 30 degrees (W^- = −0.2783) was more pronounced than the influence of the presence of the factor (W^+ = 1.2410), and the contrast value was higher (C_w = 1.5193).

8.4.4 Role of Wetness Index (WI) on Gully Occurrence

Table 8.2 shows that there was a positive significant association between gully occurrence and WI (ρ = 0.9632, p < 0.05). A negative association was observed with WI values less than 2.0; however, positive association was observed with a TWI values >2.0 (C_w = 4.3448). These results are similar to the findings of Moore et al. (1991) and Pathak et al. (2005) that topographic nonuniformity within small catchments is a major factor controlling the spatial variability of soil water and the location and development of gullies. This can be explained by the fact that at low WI, there is less runoff water generated, and as the WI increases, it promotes gully development. This is because WI sets susceptibility to gully of the catchment area in relation to the slope gradient.

8.4.5 Role of Slope Length on Gully Occurrence

Table 8.3 shows that there was a significant positive association between gully occurrence and the LS factor (ρ = 0.9674, p < 0.05). At LS of less than 2.50, there was a negative association (C_w = −0.75). Higher positive association

Table 8.3 Weight of evidence analysis showing factors, factor classes, weights, and spearman's rho (p)

Factors	Class	W^+	W^-	Constant (C_W)	Studentized contrast (C_s)	Correlation Coefficient and p value
Soil	Lateritic soil	0.0885	−0.1544	0.2429	2.501	
	Older alluvial soil	−0.1544	0.0885	−0.2429	−2.302	N/A
Geomorphology	Upland plains	0.0060	−0.0043	0.0103	0.214	
	Paradeltaic fan surfaces	−2.0892	0.0164	−2.1056	−11.308	N/A
	Duricrusts	−1.8929	0.0116	−1.9045	−12.305	
	Pediments and pediplans	−2.2898	0.0960	−2.3858	−14.003	
	Flood plains	0.2300	−0.2351	0.4651	4.301	
Lithology	Unconsolidated sands, silts, and clay	−0.3636	0.0968	−0.4604	−4.105	N/A
	Fine and medium sands	−0.2168	0.0677	−0.2845	−1.348	
	Fragments of pebbles, boulder, and gravels	0.1897	−0.5739	0.7636	6.254	
Land use and land cover	Barren land	3.6533	−0.2732	3.9265	16.383	
	Agricultural land	0.0219	−0.0086	0.0305	0.351	
	Scrub land	−0.1873	0.0206	−0.2079	−2.170	N/A
	Dense forest	−1.1075	0.2667	−1.3742	−11.391	
	Open forest	−0.8511	0.1901	−1.0412	−10.785	
	Wasteland	2.1309	−0.1575	2.2884	17.858	
	River	−2.7705	0.0049	−2.7754	−7.361	
Slope (degree)	<5	0.5624	0.1674	0.3950	3.590	
	5–10	−1.2613	0.1368	−1.3981	−6.954	
	10–15	−1.1442	0.1353	−1.2795	−5.782	0.7178 ($p < 0.05$)
	15–20	0.3936	−0.0923	0.4859	4.622	
	20–30	0.3461	−0.1312	0.4773	4.753	
	>30	1.2410	−0.2783	1.5193	12.201	
Stream power index (SPI)	<0.50	−1.4909	0.0430	−1.5339	−6.240	
	0.50–1.00	−1.3981	0.0643	−1.4624	−8.332	0.8874 ($p < 0.05$)
	1.00–1.50	−1.3082	0.2400	−1.5482	−9.689	
	1.50–2.00	−1.1870	0.5425	−1.7295	−10.221	
	2.00–2.50	1.0674	−0.1778	1.2452	6.892	
	>2.50	4.3317	−0.7079	5.0396	15.614	
Slope length (LS)	< 0.07	−1.9903	0.6822	−2.6725	−6.583	
	0.07–0.81	−0.9514	0.1111	−1.0625	−5.772	0.9674 ($p < 0.05$)
	0.81–2.50	−0.6820	0.0686	−0.7506	−3.251	
	2.50–5.00	1.1658	−0.2688	1.4346	9.980	
	>5.00	1.7581	−0.5894	2.3475	11.242	
Topographic wetness index (TWI)	0.0–0.5	−2.6793	0.2703	−2.9496	−13.875	
	0.5–1.0	−1.8764	0.1030	−1.9794	−8.343	
	1.0–1.5	−0.8104	0.1177	−0.9281	−8.124	
	1.5–2.0	−0.1165	0.0830	−0.1995	−2.202	0.9632 ($p < 0.05$)
	2.0–2.5	3.6803	−0.6645	4.3448	18.315	

($C_w = 2.34$) was observed for LS between 5.0 and 7.5. The results can be explained by the fact that LS accounts for the effect of topography on erosion. Gullies could occur in almost every slope aspect according to WoE analysis and slopes facing north seem to be more susceptible to their development. With respect to the slope gradient, concave zones (hollows and valley heads) appear to be more prone to gully development. Under this condition, in fact, the concentration of flow downslope and the possible saturation of

soil due to surface and subsurface water enhance gully erosion (Lucà et al. 2011).

8.4.6 Role of Stream Power Index (SPI) on Gully Occurrence

The values of SPI factor were categorized into six classes based on geometric interval (Table 8.3). The SPI showed

Fig. 8.5 Gully headcut and valley-side retreat processes at Ganganir Danga, right side of Silai River (at Garbheta)

significant positive correlation with gully occurrence ($\rho = 0.8874$, $p < 0.01$). There was a significant association of all the factor classes and gully occurrence ($C \neq 0$) except of SPI classes of <2.00 ($C_s = -10.221$). The results indicated that gully occurrence increases with an increase of SPI and determines the areas of concentrated runoff. The results also corroborated with the previous findings (Morgan 1995; Dube et al. 2014).

8.4.7 Gully Susceptibility Map

The probability of gully occurrence for all the factors map was computed by performing the WoE weight analyses. The chi-square test indicated that the factors slope gradient, LS, SPI, and WI were conditionally independent from each other, so; all of these factors are useful for the final map using the WoE model. Figure 8.6 shows the total weight map (TWmap) which is a result of summing the weight contrast (C_w) values presented in Table 8.3. The total weighted map (TWmap) was converted into four classes (Fig. 8.7). The

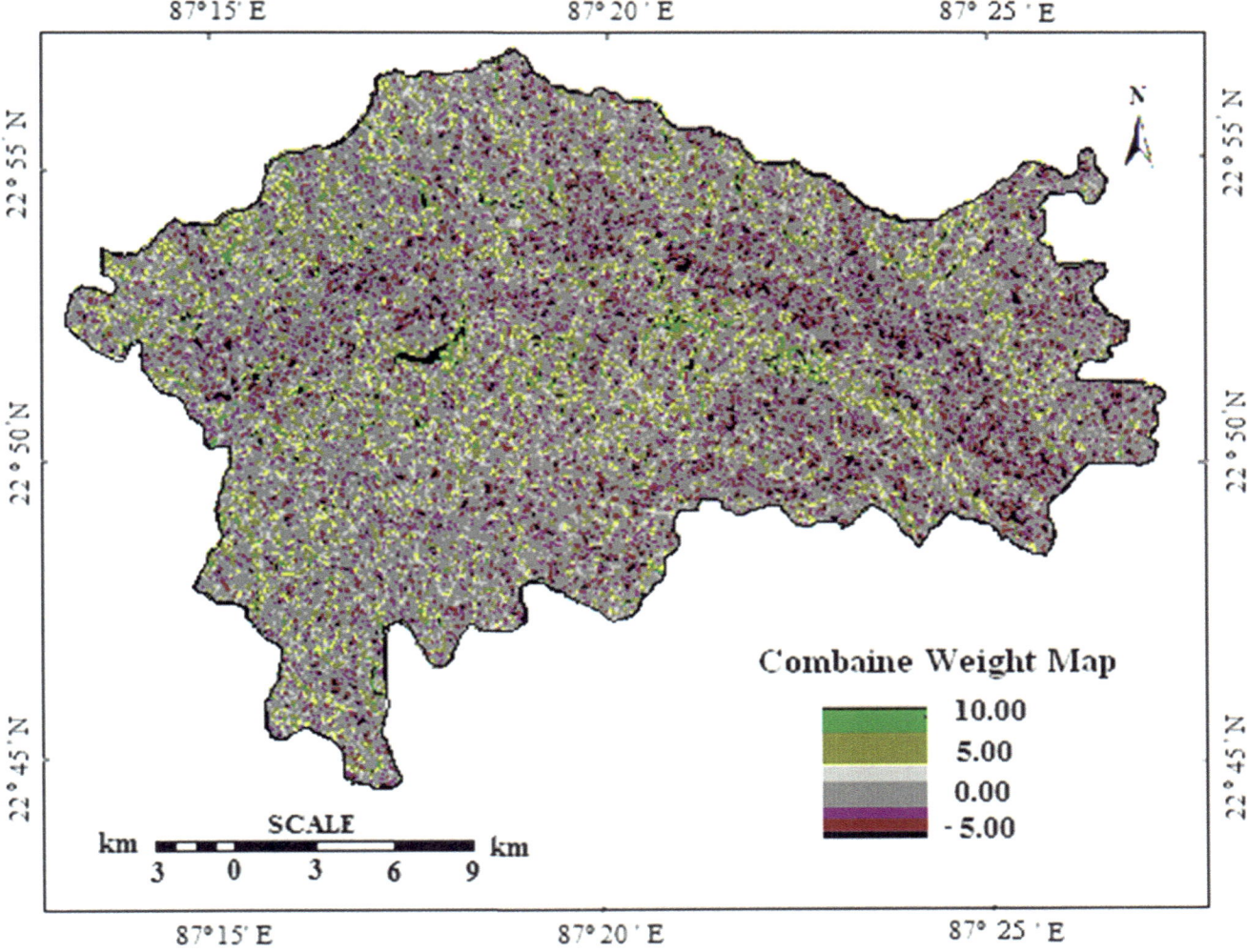

Fig. 8.6 Total weight map (TWmap) of the study area

classes were based on the change in gradient in the frequency distribution curve of the cumulated weight with a change in gradient signifying a change within class. The very high susceptibility zone has weight values ranging from 8.0 to 10.0, the high susceptibility has a weight values ranging from 5.00 to 8.0, the low susceptibility has weight value ranging from 0.0 to 5.0, and the very low susceptibility has a weight value of less than 0.

High gully erosion-susceptible zones were observed in the northeast and in the central part of the study area. Riverine zone is characterized by almost undifferentiated high susceptibility conditions (Fig. 8.7). On the other hand, very low susceptibility values are associated with the alluvial plain and

the sector near the eastern part of the water divide. A large number of quantity of cells with null density values guides to dilemma in delineating the susceptibility levels along with an equal area principle and usually lessen the prophetic concert of the model.

8.4.8 Validation of Gully Erosion Susceptibility Map

The fit of the susceptibility map with the spatial distribution of gullies was evaluated using randomly selected recent gully area in the field. For the recent gullies identified in the field,

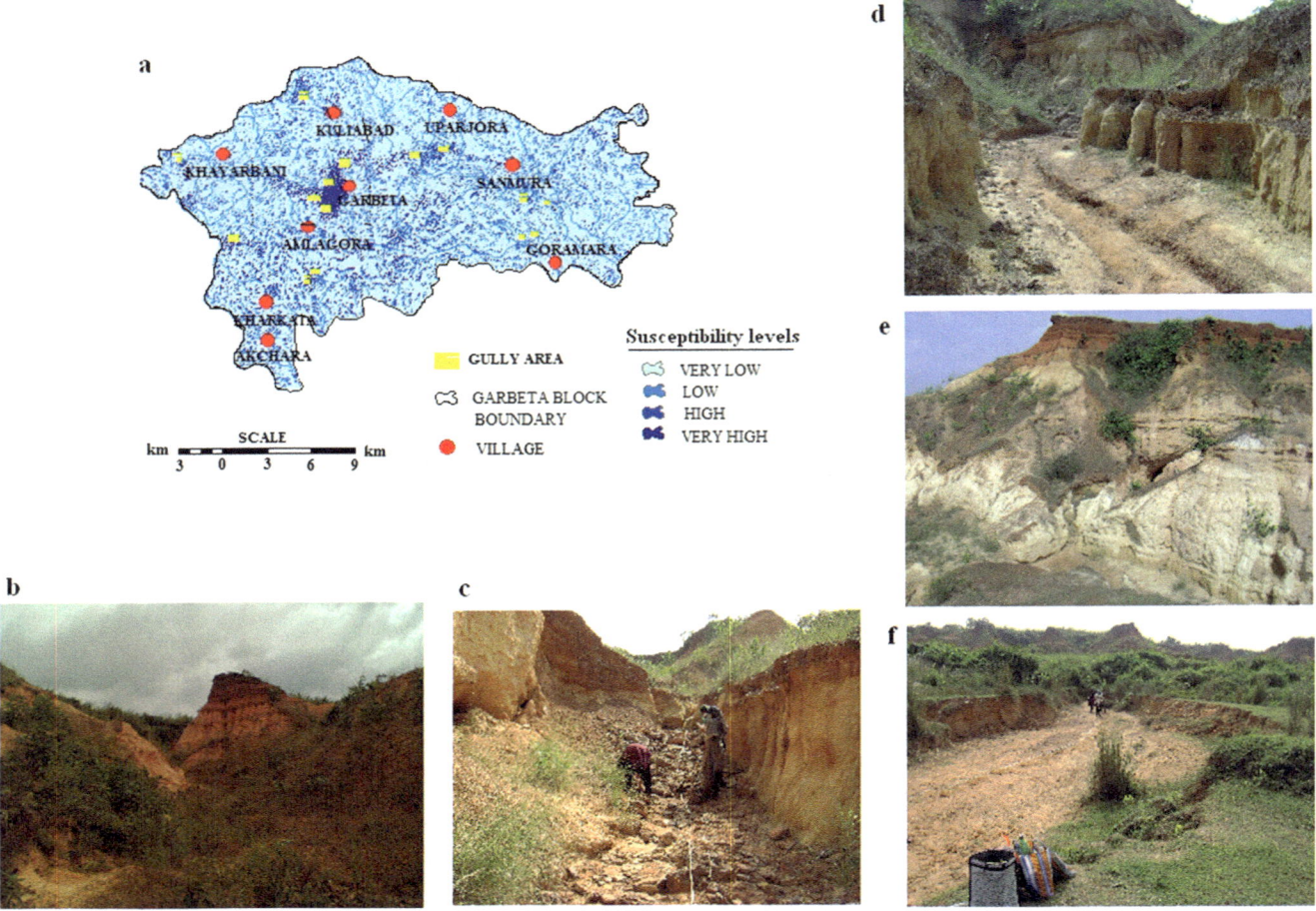

Fig. 8.7 Gully erosion susceptibility map (**a**) and pictures of some of the mapped recently gullies actives (**b**–**f**)

about 35% of the gully points were in the very high suscepti-bility class, 30% were in the high susceptibility, 35% were in the low susceptibility class, and no gully was observed in the very low susceptibility class. It was found that the ROC-AUC curve has an accuracy of 67.8% (Fig. 8.8).

8.5 Conclusion

The recent study reveals a correlation between gully occurrences and geo-environmental factors such as soil, lithology, geomorphology, land use, slope, SPI, and TWI. The relationships between gullies and the selected geo-environmental factors have been assessed using the

WoE Bayesian model. The lithology, SPI, TWI, slope, and land use appear to play an important role in controlling gully erosion; in particular, land surfaces shaped in wasteland and barren land and/or fragments of pebbles, boulder, and gravels were shown to be the most prone areas to gully erosion. Geologically, the region is situated between Indian Shield to the west and north and Indo-Burma Ranges to the east. The expansion of the region was ruined with extensive Rajmahal volcanism (Late Jurassic–Early Cretaceous) and enclosed the Gondwana sediments because of the occurrence of Kerguelen hot spot. Consequently, the region is differentiated by shal-low depth of Archean and Gondwana rocks and western shelf zone, portrayed by less slope of basement toward the east and southeast, and allows the tracks of tropical strong wet-dry

Fig. 8.8 Illustration of cumulative frequency diagram showing cumulative gully occurrence (%; y-axis) occurring in gully susceptibility index rank (%; x-axis)

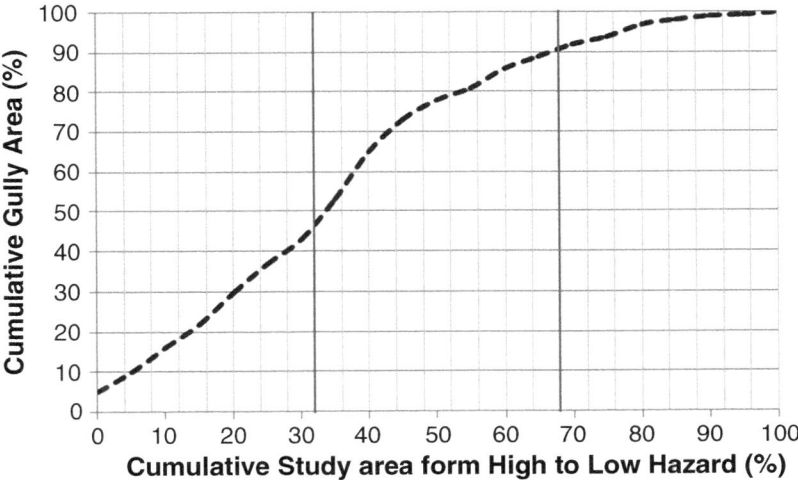

morphogenetic landform, i.e., lateritic terrain with Paleogene sediments. The present study shows that remote-sensing techniques aided by GIS can provide a useful tool for studying potential gully areas. However, the accuracy of the final result depends on the various parameters that are included in the dataset. The present study suffers from several limitations like small area and several other parameters like hydrological (infiltration, drainage) and soil erosion process (i.e., tillage erosion, land leveling, etc.) that strongly influenced the gully erosion process; these are needed and required for better accuracy. For the method to be applied in general, more studies are required to establish the efficiency and competence of gully erosion control programs at different temporal and spatial scales.

Acknowledgments The authors thank Mr. Kartic Rishi and Mr. Nitaynanda Sar for their help in the field data collection. Ms. Rumpa Paira is sincerely acknowledged for her technical support during the preparation of this manuscript.

References

Agterberg FP, Bonham-Carter GF, Wright DF (1990) Statistical Pattern Integration for Mineral Exploration, In GaalG., Merriam, D.F. (eds) Computer Applications in Resource Estimation Prediction and Assessment for Metals and Petroleum, Pergamon, Oxford, 1–21.

Asadi HH, Hale M (2001) A Predictive GIS Model for Mapping Potential Gold and Base Metal Mineralization in Takab Area, Iran, Computers and Geosciences, 27: 901–912.

Barbieri G, Cambuli P (2009) The Weight of Evidence Statistical Method in Landslide Susceptibility Mapping of the Rio Pardu Valley (Sardinia, Italy), In Proceedings of the 18th World IMACS/MODSIM Congress, held in Cairns, Australia, 13-17 July 2009, 2658-2664.

Bonham-Carter GF (1994) Geographic Information Systems for Geoscientists: Modelling with GIS, Pergamon Press, Oxford, UK.

Conforti M, Aucelli PPC, Robustelli G, Scarciglia F (2011) Geomorphology and GIS analysis for mapping gully erosion susceptibility in the Turbolo stream catchment (Northern Calabria, Italy). Nat Hazards, 56:881–898.

Conoscenti C, Angileri S, Cappadonia C, Rotigliano E, Agnesi V, Märker M (2013) Gully erosion susceptibility assessment by means of GIS-based logistic regression: A case of Sicily (Italy), doi:https://doi.org/10.1016/j.geomorph.2013.08.021

Dube F, Nhapi I, Murwira A, Gumindoga W, Goldin J, Mashauri DA (2014) Potential of weight of evidence modelling for gully erosion hazard assessment in Mbire District—Zimbabwe. Phys Chem Earth67–69:145–152

Frankl A, Zwertvaegher A, Poesen J, Nyssen J (2013) Transferring Google Earth observations to GIS-software: example from gully erosion study. Int J Digital Earth6(2):196-201. doi:https://doi.org/10.1080/17538947.2012.744777.

Harvey AM (1992) Process Interactions, Temporal Scales and the Development of Hillslope Gully Systems: Howgill Fells, Northwest England. Geomorphology, 5:323–344.

Khadse GK, Vijay R, Labhasetwar PK (2015) Prioritization of catchments based on soil erosion using remote sensing and GIS, Environ Monit Assess,187: 333, DOI https://doi.org/10.1007/s10661-015-4545-z

Lucà F, Conforti M, Robustelli G (2011) Comparison of GIS-based gullying susceptibility mapping using bivariate and multivariate statistics: Northern Calabria, South Italy, Geomorphology134: 297–308

Marzolff I, Ries JB, Poesen J (2011) Short-term versus medium-term monitoring for detecting gully-erosion variability in a Mediterranean environment. Earth Surf Process. Landforms36:1604–1623

Mihalasky MJ, Moyer LA (2004). Spatial databases of the Humboldt Basin mineral resource assessment, northern Nevada, U.S. Geological Survey Open-File Report 2004-1245, U.S. Department of the Interior, U.S. Geological Survey. Available at: https://pubs.usgs.gov/of/2004/1245/of2004-1245.pdf

Mitasova H, Hofierka J, Zlocha M, Iverson RL (1996) Modeling topographic potential for erosion and deposition using GIS, International Journal of Geographical Information Science10(5): 629-641.

Moore ID, Burch GJ (1986) Physical basis of the length-slope factor in the universal soil loss equation. Soil Sci Soc Am J50:1294–1298.

Moore ID, Grayson RB, Ladson AR (1991) Digital terrain modelling: a review of hydrological, geomorphological, and biological applications. Hydrol Process5:3–30

Morgan RPC (1995) Soil erosion and conservation. Longman, Group Ltd, ESSEX

Morgan RPC, Mngomezulu D (2003) Threshold conditions for initiation of valley-side gullies in the Middle veld of Swaziland. Catena50:401–414

Okalp K (2005) Soil erosion risk mapping using geographic information systems: a case study onKocadere Creek watershed, Izmir. MSc Thesis. Department of Geodetic and Geographic Information Technologies, Natural and Applied Sciences of Middle East Technical University. Ankara, Turkey, pp 20-21

Pathak P, WaniSP, SudiR (2005) Gully Control in SAT Watersheds. Global Theme on Agroecosystems, International Crops Research Institute for the Semi-Arid Tropics, p28.

Pradhan B,OH HJ, Buchroithner M (2010) Weights-of-evidence model applied to landslide susceptibility mapping in a tropical hilly area, Geomatics, Natural Hazards and Risk1(3): 199–223

Regmi NR, Giardino JR, Vitek JD (2010) Modelling susceptibility to landslidesusing the weight of evidence approach: Western Colorado USA. Geomorphology115:172–187

Romero-Calcerradaa R, Luqueb S (2006) Habitat Quality Assessment Using Weights-Of-Evidence Based GIS Modeling: The Case of *Picoides tridactylus* as Species Indicator of the Biodiversity Value of The Finnish Forest, Ecological Modelling196: 62–76.

Sen J, Sen S, Bandyopadhyay S (2004) Geomorphological investigation of badlands- a case study at Garbheta, West Medinipur District, West Bengal, India. In Singh S, Sharma HS, De SK (eds) Geomorphology and environment. ACB Publication, Kolkata, pp 204–234

Shit PK, Bhunia G, Maiti R (2013) Assessment of Factors Affecting Ephemeral Gully Development in Badland Topography: a Case Study at Garbheta Badland (PashchimMedinipur). Int J Geosci4 (2):461–470. doi:https://doi.org/10.4236/ijg.2013.42043

Shit PK, Bhunia G, Maiti R (2014) Morphology and development of selected Badlands in South Bengal (India). Indian J Geogr Environ13:161–171

Shit PK, Paira R, Bhunia GS, Maiti R (2015) Modeling of potential gully erosion hazard using geo-spatial technology at Garbheta block, West Bengal in India. Model. Earth Syst. Environ.1:2, DOI https://doi.org/10.1007/s40808-015-0001-x

Valentin C, Poesen J, Li Y (2005) Gully erosion: impacts, factors and control. Catena63:132–153. doi: https://doi.org/10.1016/j.catena.2005.06.001

Van-Western CJ (2003) Use Of Weights Of Evidence Modelling for Landslide Susceptibility Mapping Lecture Notes, International Institute for Geoinformation Science and Earth Observation (ITC), Enschede, The Netherlands.

Venkataraman G, Madhavan BB, Ratha DS, Antony JP, Goyal RS, Banglani S, Roy SS (2000) Spatial Modeling for Base-Metal Mineral Exploration Through Integration of Geological Data Sets, Natural Resources Research, 9: 27–42.

Vrieling A, Sterk G, Beaulieu N (2002) Erosion risk mapping; a methodological case study in the Colombian Eastern Plains. Journal of Soil and Water Conservation, 57(3): 158-163.

Whitford JA, Newham LTH, Vigiak O, Melland AR, Roberts AM (2010) Rapid assessment of gully sidewall erosion rates in data-poor catchments: a case study in Australia. Geomorphology118: 330–338.

Zheng F (2006) Effect of Vegetation Changes on Soil Erosion on the Loess Plateau, Pedosphere16 (4):420-427.

Pravat Kumar Shit received his Ph.D. in Geography (Applied Geomorphology) from Vidyasagar University (India) in 2013, M.Sc. in Geography and Environment Management from Vidyasagar University in 2005, and PG Diploma in Remote Sensing & GIS from Sambalpur University in 2015. He is Assistant Professor in the Department of Geography, Raja N. L. Khan Women's College (Autonomous), Gope Palace, Midnapore, West Bengal, India. His main fields of research are soil erosion spatial modeling, badland geomorphology, gully morphology, water resources and natural resources mapping, and modeling and has published more than 45 international and national research articles in various renowned journals; also, he has published three books. His research work has been funded by the University Grants Commission (UGC), India, and Higher Education Science and Technology and Biotechnology, Government of West Bengal. He is Associate Editor and on the editorial boards of three international journals in geography and earth environmental sciences.

Gouri Sankar Bhunia received his Ph.D. from the University of Calcutta, India, in 2015. His Ph.D. dissertation work focused on environmental control measures of infectious disease (visceral leishmaniasis or kala-azar) using geospatial technology. His research interests include kala-azar disease transmission modeling, environmental modeling, risk assessment, data mining, and information retrieval using geospatial technology. He is Associate Editor and on the editorial boards of three international journals in health GIS and geosciences. He worked as a "Resource Scientist" in Bihar Remote Sensing Application Centre, Patna (Bihar, India). He is the recipient of the Senior Research Fellow (SRF) from the Rajendra Memorial Research Institute of Medical Sciences (ICMR, India) and has contributed to multiple research programs: kala-azar disease transmission modeling, development of customized GIS software for kala-azar "risk" and "non-risk" area, and entomological study.

Hamid Reza Pourghasemi is an Associate Professor of Watershed Management Engineering in the College of Agriculture, Shiraz University, Iran. He has a B.Sc. in Watershed Management Engineering from the University of Gorgan (2004), Iran; an M.Sc. in Watershed Management Engineering from Tarbiat Modares University (2008), Iran; and a Ph.D. in Watershed Management Engineering from the same University (Feb 2014). His main research interests are GIS-based spatial modeling using machine learning/data mining techniques in different fields such as landslide, flood, gully erosion, forest fire, land subsidence, species distribution modeling, and groundwater/hydrology. Also, Hamid Reza works on multi-criteria decision-making methods in natural resources and environment.

He has published more than 90 peer-reviewed papers in high-quality journals, with three chapters in Springer. Also, he has published two books for Springer (https://www.springer.com/gp/book/9783319733821) and Elsevier (https://www.elsevier.com/books/spatial-modeling-in-gis-and-r-for-earth-and-environmental-science/pourghasemi/978-0-12-815226-3).

Understanding the Morphology and Development of a Rill-Gully: An Empirical Study of Khoai Badland, West Bengal, India

9

Asish Saha, Manoranjan Ghosh, and Subodh Chandra Pal

Abstract

The lateritic region of the Birbhum District of West Bengal is part of the low-level unconsolidated erosional deposits from the eastern Chotanagpur plateau. Topographically, the region is the part of the 'Rarh Plain' of western West Bengal. A localized badland, namely 'Khoai,' has developed in the west–south to north–east direction on the bank of the River Kopai in this lateritic region. The aim of the present study is to understand the slope, channel profile, and development processes of the rill-gully of Khoai badland topography. Therefore, both quantitative analysis and field investigation have been carried out to fulfill these objectives. To understand the nature of the gullies profiles, a least squares linear regression model as well as Hack's Stream Length–Gradient Index (SL) has been used in this study. The Soil Conservation Service-Curve Number (SCS-CN) method has been applied for the computation of the rainfall–runoff relationship of the study area. The rate of sediment transportation was calculated on the basis of the J.R. Williams Sediment Delivery Ratio (SDR). It was found that the existing badland topography in this region has been developed mainly by the climatogenetic processes of water erosion. The various water erosion processes, such as rain splash erosion, sheet erosion, and inter-rill erosion, have been very active over a long period. The laterites of this region have been dissected and shaped into numerous rills and gullies by the aforementioned erosion and weathering processes over time. It was also observed that the dominance of lower-order gullies indicates a high rate of soil erosion. Furthermore, it was found that a huge volume of sediment has been transported by surface runoff in this region. It was estimated that the region experiences a high rate of SDR (0.87–1.01).

Keywords

Water erosion · Stream length–gradient index · SCS-curve number · Sediment delivery ratio

9.1 Introduction

Badland is a term generally applied to those landscapes that are barren land with nearly absent or sparse vegetation, intensely dissected, mostly useless for agriculture, and mainly developed by water erosion (Joshi 2014; Aown and Kar 2016). Badland topography also includes that dissected rugged topography produced by the combined action of piping or tunnel erosion with fluvial processes (Joshi 2014; Pal and Shit 2017). The distinct characteristic of badland topography is that it has fluvial eroded channels variously termed as 'gully,' 'ravine,' 'wadi,' or 'dongas' (Stone 1967; Pal and Chakrabortty 2018). A gully is a relatively permanent narrow channel bound on the Earth's surface by the action of water, especially a miniature valley resulting from falling rain (Fairbridge 1968; Charlton 2008). Morphologically, gullies are different from stream channels, being narrow and deep, with almost vertical sidewalls. The development of rills and gullies entails the action of raindrops falling on the soil surface, refit of overland flow, shortening of runoff lag time, and surface runoff (Charlton 2008; Shit et al. 2014). The nature of badland topography anywhere in the world is always dynamic because of its morpho-climatic processes. Because of this dynamic nature, the rate of erosion and sediment yield fluctuate from one region to another. In this study, the sediment transportation rate has been taken into consideration on the basis of the sediment delivery ratio that was carried out by previous researchers (Williams 1977; Walling 1983; Ferro and Minacapalli 1995). The longitudinal

A. Saha · S. C. Pal (✉)
Department of Geography, The University of Burdwan, Bardhaman, West Bengal, India

M. Ghosh
Rural Development Centre, Indian Institute of Technology Kharagpur, Kharagpur, India

© Springer Nature Switzerland AG 2020
P. K. Shit et al. (eds.), *Gully Erosion Studies from India and Surrounding Regions*, Advances in Science, Technology & Innovation, https://doi.org/10.1007/978-3-030-23243-6_9

profiles are very important in understanding the nature and the degree of adjustment of a gully system to landform evolution. The longitudinal profile of a gully is obtained from the relationship between height and distance downstream of a gully. Generally, a longitudinal profile indicates the control of the gradient on channel behavior (Hack 1957). The least squares linear regression method has been widely used for analysis of longitudinal profiles. The long profile of a channel provides knowledge about the geomorphological and evolutionary history of a basin area (Magar and Magar 2016). The stream length–gradient index proposed by Hack (1973) is widely used to identify the factors that control the channel network such as the occurrence of resistant rocks, tributary confluence, and human interference (Magar and Magar 2016). Hack's (1973) stream length-gradient index (SL index) attempted to determine whether a channel is in morphological equilibrium on the basis of the relationship between channel slope and channel length. The Soil Conservation Service-Curve Number (SCS-CN) technique is one of the simplest methods developed by the U.S. Bureau of Agriculture National Resources Conservation Service (NRSC) (formerly known as Soil Conservation Service) for rainfall–runoff modeling. Remote sensing and GIS have a significant role for calculation of curve number (CN), which is essential for runoff estimation (Siddi et al. 2018). By using this SCS-CN method, we can find the amount of rainfall–runoff of a particular area. The amount of rainfall–runoff determines the rate of soil erosion, and soil erosion is the main cause of sedimentation of a particular region. This sedimentation process gradually degrades the land surface, which ultimately takes the form of badland topography.

It has been found that more than 50% of the geographical area of India has been affected deleteriously by land degradation and soil erosion (Sehgal and Abrol 1994). The substantial cause of soil degradation through rill and gully erosion shows the severe endemic problems of our society, including economic activities; it also leads toward underdevelopment (Jha and Gupta 2003; Shit et al. 2014). Geographers and geomorphologists have studied the various dimension of this subject area, such as gully morphology and morphometry, development, and formation, in different parts of the country. The badland topography of the state of West Bengal in the districts of West Medinipur, Bankura, Puruliya, Birbhum, and parts of Bardhaman has been primarily studied by many scholars (Bhattacharya 1957; Biswas 1987; Laha 2011; Shit and Maiti 2012; Bera and Bandyopadhyay 2013; Shit et al. 2013, 2014). The Khoai badland of Birbhum District has been studied by several major investigators (Basu 1972; Bandhopadhyay 1987; Mukhopadhyay 1992; Das and Bandyopadhyay 1996; Jha and Kapat 2009, 2011).

The objective of the present study is to understand the development process of badland topography, particularly the development and morphology of the rill-gully in the Khoai

badland of Birbhum District. In addition to morphology, the study has also attempted to examine the channel character, both cross section and long, runoff, with the sediment delivery ratio (SDR).

9.2 Materials and Methods

9.2.1 Study Area

The western part of West Bengal, popularly known as the 'Rarh plain of Bengal,' spreads with a few patches of badlands of low-level Pleistocene laterite. In this region, low levels of laterite and varying textures of lateritic soils are the outcome of morpho-climatic processes of subtropical and subhumid types of monsoonal climate. These laterites are considered to have originated over the unconsolidated secondary deposits from the eastern part of Chotanagpur Plateau in the late Pleistocene epoch and to have been reworked later by subaerial weathering (Chakraborty 1970; Niyogi et al. 1970; Basu 1972; Bandhopadhyay 1987, 1988; Das and Bandyopadhyay 1995; Dey et al. 2009; Ghosh and Guchhait 2012). This unconsolidated lateritic soil is more prone to gully erosion and piping or tunnel erosion (Sarkar et al. 2007). The region of this low-level laterite is dissected and shaped by numerous rills, gullies, and ravines, giving rise to the development of badland topography (Ghosh and Guchhait 2012; Joshi 2014). Such lateritic badland topography is commonly affected by the intense process of water-induced soil erosion, that is, rain splash erosion, sheet erosion, inter-rill erosion, rilling, and gullying erosion (Nadal-Romero and Regues 2010; Shit et al. 2013).

The present study was accomplished at the Khoai badland of Birbhum District in the state of West Bengal, India. Geographically, the Khoai badland extends from $87°38'34.94''E$ to $87°40'23.44''E$ and $23°40'08.71''N$ to $23°41'25.17''N$, with a total area of 1.85 km^2 (Fig. 9.1). The lateritic region of Birbhum District is part of the northern slope of River Ajay–Kopai interfluvial land and a localized low-level Pleistocene lateritic badland developed on the bank of the Kopai South Main Canal (KSMC) of the Mayurakshi River Valley Project (Das and Bandyopadhyay 1996). It is also about 2 km northwest of the popular tourist destination, Santiniketan.[1]

The general elevation of the study area varies from 49 m in the southwestern part to 67 m in the northeastern part; the slope varies from low to moderate ($0.00°$–$3.14°$) (Fig. 9.2). The entire region belongs to the subtropical monsoon climatic subdivision with a mean annual temperature around 26 °C and mean annual rainfall about 1463 mm. The soil

---

[1] Santiniketan is a popular tourist spot of the state of West Bengal. The place is well known for the natural forest, bird sanctuary, and red soil as well as the distinct rural and cultural landscape.

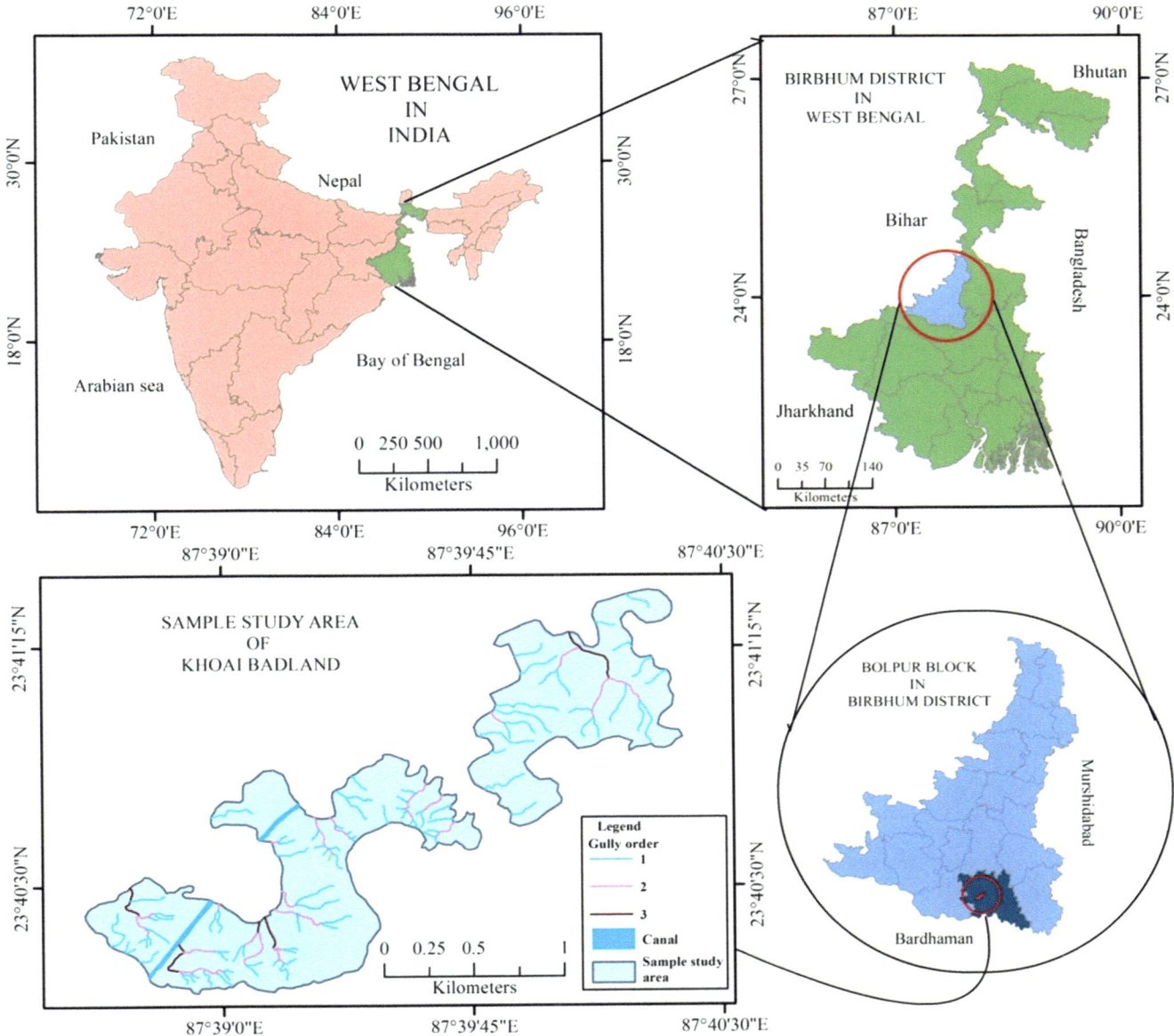

Fig. 9.1 Location of the study area and major rills and gullies

texture of this area is mainly fine cracked soils, occurring on level to nearly level low-lying alluvial plains with a clayey surface (Sarkar et al. 2007).

9.2.2 Methodology

9.2.2.1 Collection of Data

For conducting the present research, both primary and secondary data sources were used. To study the gully profile (cross section and longitudinal), a primary field survey was conducted in the summer of 2018 using tape, staff, clinometer, and Garmin GPS. Secondary data were collected from the

Geological Quadrangle Map (Geological Survey of India 1948), Survey of India topographical map of 1972 (73 M/ 10, on 1:50,000 scale), Google Earth imagery (2018), soil texture map from National Bureau of Soil Survey and Land Use Planning (NBSS and LUP) (Kolkata 2010), meteorological data from India Meteorological Department (IMD), Shuttle Radar Topographic Mission (SRTM) Digital Elevation Model (DEM) of 30 m, and LISS III satellite data from ISRO's Geo-portal and U.S. Department of Agriculture Technical Release-55 (1986). The land use–land cover (LULC) map of the study area was delineated with the help of IRS LISS III satellite data using an unsupervised classification method. Table 9.1 shows details about the different

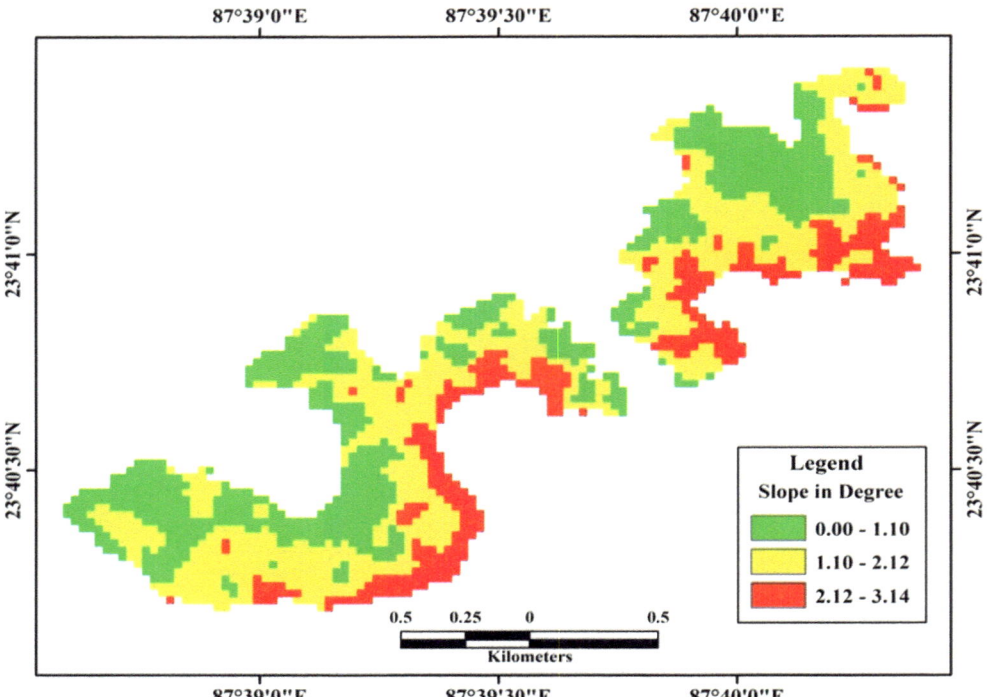

Fig. 9.2 Slope map of the study area

Table 9.1 Details of the data sources used in the study

Data	Source	Year
Topographical map (73 M/10)	Survey of India	1972
Geological quadrangle map	Geological Survey of India	1948
Soil map	National Bureau of Soil Survey and Land Use Planning	2010
Climatic data	India Meteorological Department	2008–2012
NRCS Technical Release 55	United State Department of Agriculture	1986
Google Earth imagery	earth.google.com	2018
LISS III (23.5 m)	www.isro.gov.in	2015
SRTM DEM (30 m)	USGS Earth Explorer	2018

data sources for the respective years. For analyzing the field survey and secondary data, Microsoft Excel 2007, Origin Pro 8.0 was used; and for generating the different maps of the study area, ArcGIS 10.3 and ERDAS IMAGINE 2014 software were also used in the present study.

9.2.2.2 Gully Morphology and Profile Character

For analysis of morphology and profile character (cross section and longitudinal) of the gullies, width (m), depth (m), and slope (degree, °) were measured during the field survey.

Longitudinal profile analyses of gullies were carried out by using the least squares linear regression method with respect to the elevation and distance data. To understand the nature of the gully profile (whether graded or not), the long profile of the selected gully was plotted on semilogarithmic graph paper. Hack's (1973) stream length–gradient index (SL) was used to identify the morphometric anomalies of different gullies along the longitudinal profiles. Seeber and Gornitz (1983) computed the ratio of $SL_{section}$ to SL_{total} to understand the morphometric anomalies. If the result of SL ratio is less than 2, the longitudinal profile is said to be graded. The Hack (1973) method consists of the applications of the following equations:

$$SL = (\Delta H/\Delta L)L \qquad (9.1)$$

where SL = stream length–gradient index, ΔH = difference between the highest and lowest point of a given section of the channel, ΔL = horizontal distance of the given section of the channel, and L = total length of the channel from the source to the furthest point of the channel section.

SL_{total} of the channel was calculated using the following equations:

$$SL_{total} = \Delta H/\ln L \qquad (9.2)$$

where ln = natural log.

9.2.2.3 Curve Number Method (CN) and Sediment Delivery Ratio (SDR)

Accurate estimation of the rainfall–runoff relationship and rate of sediment yield are the most important tasks for the management of a basin area (Gajbhiye et al. 2014; Gajbhiye 2015). Several methods are available to estimate the runoff from rainfall, the Soil Conservation Service Curve Number (SCS-CN) method being the most applicable method as reported in the literature. The SCS-CN has been used to determine the rate of surface runoff. The CN method is an index that represents the runoff potential for different hydrological soil groups combining LULC and soil texture (Ahmed et al. 2015; Roy and Saha 2017). On the basis of the minimum rate of infiltration, USDA NRCS has developed a soil classification system of four groups: A, B, C, and D. In the present study area, as fine soil texture is dominant, the C group, HSG soil, was used for choosing the CN value. However, direct runoff of a particular area depends on soil texture, land use, intensity, and duration of rainfall. Therefore, estimation of the rainfall–runoff condition is essential because there is a relationship between runoff and soil erosion. There is a positive correlation between rainfall runoff and sedimentation processes: a high rate of rainfall–runoff means a high rate of sedimentation and vice versa. The conventional methods of rainfall–runoff estimation are more time consuming, with the possibility of inaccuracy in the results. Remote sensing and GIS techniques have been used to enhance the accuracy of the estimation of rainfall–runoff with a wider perspective (Das et al. 1997; Martz and Garbrecht 1999). On the other hand, as we know that the different factors of SCS-CN model are geographic in nature, therefore application of remote sensing and GIS technology is very convenient for analyzing the sediment transport and rainfall–runoff relationships of a particular basin area (Gajbhiye 2015). The equation used for estimation of direct runoff by the CN method of an area is as follows:

$$Q = (P - 0.2S)^2 / (P + 0.8S) \qquad (9.3)$$

where Q = peak rate of runoff, P = rainfall in millimeters (mm), and S = maximum potential storage, calculated by the following equation:

$$S = (25,400 / \text{CN}) - 254 \qquad (9.4)$$

where CN = curve number.

The equation of the weighted curve number is

$$\text{CN}_w = \sum (\text{CN}_i \times A_i) / A \qquad (9.5)$$

where CN_w = weighted curve number, CN_i = curve number from 1 to any number N, A_i = area with curve number CN_i, and A = total area.

Sediment delivered from water erosion causes large-scale land degradation (Onyando et al. 2005). However, different factors including basin area, regional gradient, climatic condition, and land use–land cover (LULC) contribute to the process of sediment transportation. There is no one appropriate method to estimate sediment delivery ratio. The USDA Soil Conservation Service has published a National Engineering Handbook, in which it is mentioned that the extent of a drainage area is correspondingly related to its sediment delivery ratio (USDA SCS 1972). Sediment with coarser texture is more likely to be deposited compared to fine-textured sediment. Thus, the coarser-textured sediment has a relatively lower sediment delivery ratio (SDR) than the fine-textured sediment (Ouyang and Bartholic 1997). It is well known that areas with a steep slope have a higher sediment delivery ratio than areas with a flat level plain. In general, to estimate SDR, the size of the study area should be defined. Therefore, the SDR method proposed by Williams (1977) has been used to estimate the sediment delivery ratio of the study region because Williams' SDR is the functional result including drainage area, the relief–length ratio, and weighted SCS curve number (Ouyang and Bartholic 1997). The equation of Williams' method of SDR is as follow:

$$\begin{aligned} \text{SDR} = 1.366 \\ \times 10^{-11} \, (\text{DA})^{-0.0998} \, (\text{ZL})^{0.3629} \, (\text{CN})^{0.5444} \end{aligned} \qquad (9.6)$$

where DA = the drainage area in km^2, ZL = the relief–length ratio in m/km, and CN = the long-term average SCS curve number (SCS-CN).

To estimate the SDR, the entire study area was divided into three sample areas. The SDR was calculated using Eq. (9.6). Sedimentation largely depends on such factors as soil texture, land use–land cover, and runoff produced by rainfall. Depending on soil moisture conditions, the rate of rainfall–runoff varies. When the soil is in a dry condition, the rate of rainfall–runoff is lower because the maximum portion of rainfall percolates through the soil. On the other hand, in moist soil conditions, maximum rainfall–runoff takes place because only a small portion of the rainfall percolates and the remaining large portion of rainfall flows as overland flow (runoff). During the maximum runoff period, soil erosion takes place in a devastating way, and sediment starts to flow from upslope to downslope segments.

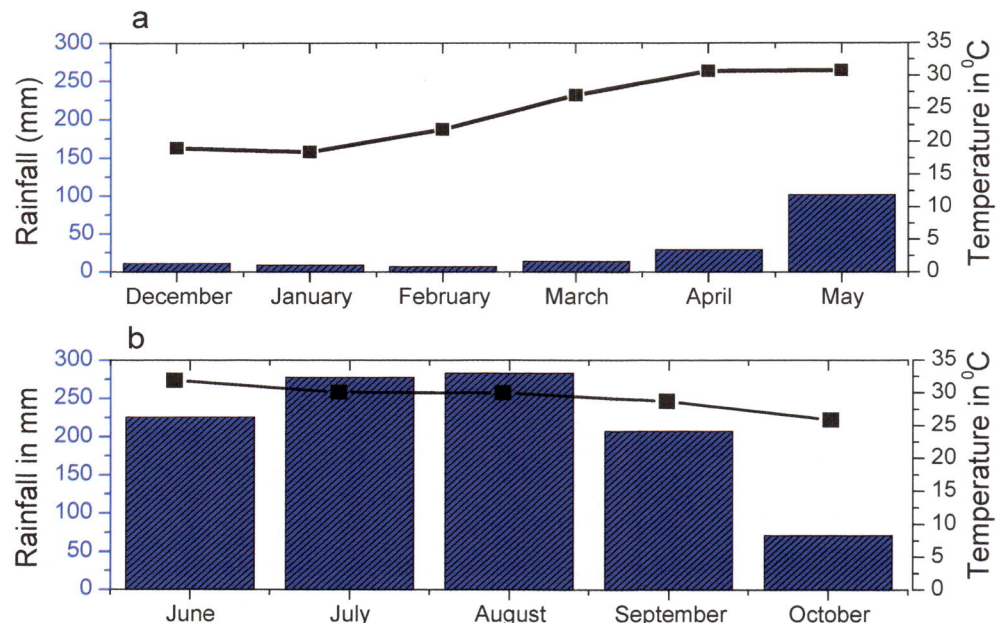

Fig. 9.3 Average monthly temperature and rainfall for upper surface build-up stage (**a**) and active surface erosion stage (**b**), 2008–2012

9.3 Results and Discussion

9.3.1 Processes of Gully Formation and Badland Development

Climate has always had a vital role in the soil erosion process of any region in the world (Jha and Gupta 2003). As was mentioned in the study area section, the climate of the present study area has been dominated by the subtropical monsoon type; hence, it is ideal for laterite development (Thomas 1974; Battaglia et al. 2011). Data for 5 years (2008–2012) of the monthly average temperature of the study area have shown that a more or less high temperature prevails throughout the year. The 5-year monthly average rainfall has shown that this region experiences mostly dry seasons, except for the monsoon months (June–October), in a year. However, the surface soil erosion processes have been actively going on, mainly in two seasonal stages. The upper surface build-up process involves the period including the winter season (December–February) and summer season (March–May). Active surface erosion dominated in the months of June–October (Sen et al. 2004) (Fig. 9.3a, b). In the upper-surface build-up stage, the dry season is responsible for lateritic surface loss from mechanical weathering. These loose particles wash out at the onset of 'Kalbaisakhi'[2] thunderstorms (these occur during mid-March–April). In the active erosion stage

(mid-June–October), rain splash erosion and overland flow start, which detach and transport soil particles downslope.

The kinetic energy,[3] duration, and intensity of rainfall always have a direct influence on the water-induced soil erosion process. The various kinds of water erosion processes, such as rainsplash erosion, sheet erosion, rill erosion, inter-rill erosion, and gully erosion, are highly responsible for the badland topography formation in Khoai of the Birbhum District of West Bengal. Subsurface piping or tunnels, gully bank collapse, and exposure of tree roots from sheet erosion are noticeable phenomena in the present study area (Fig. 9.4a, b). Direct fall of raindrops (raindrops that do not touch plants) on the bare soil surface affects the soil particles, and the soil becomes loose and splashes in all direction from the impact of kinetic energy. This is the first stage of water-induced soil erosion, known as splash erosion or raindrop erosion. In the second stage, after the soil is saturated, excess water moves as overland flow, which is usually known as sheet erosion. During the sheet erosion stage, maximum sedimentation takes place, moving from upslope toward downslope. Some basic signs of sheet erosion found in the study area are exposed tree roots, gravels, and stone (hard laterite). During the rainy season (June–October), numerous discontinuous small rills are formed in the area. These small rills are the evidence of lateritic undulations. Finally, the gully developed

---

[2] Kalbaisakhi normally originate over the Chotanagpur plateau and are carried eastward by the westerly wind. The term 'Kalbaisakhi' literally

means the black storms or a mass of dark clouds of the month of *Vaiasakha*.

[3] Kinetic energy is determined by the mass of the object and its velocity (Charlton 2008).

(a)

(b)

Fig. 9.4 Field photograph shows tunnel or piping erosion on gully sidewall (**a**) and exposed tree roots caused by sheet erosion (**b**)

as a result of water moving in rills, which concentrate to form larger and deeper channels.

9.3.2 Morphology and Profile Character of Selected Gullies

The first step in the morphometric analysis is stream ordering, which is a linear property of a particular drainage basin area. Based on Strahler's law of stream order, the study area has been identified as a third-order gully. The highest bifurcation ratio (R_b = 4.8) was found between second- and third-order gullies, whereas the mean bifurcation ratio (R_b = 4.3) of the entire study area was estimated. The relationship between gully order and the number of gullies (Fig. 9.5) neatly indicated the high rate of active soil erosion from the dominance of lower-order gullies. On the other hand, drainage density was first introduced by Horton (1932) to indicate the closeness of spacing of streams. Drainage density is a measure of the total stream length to the total basin area (Strahler 1964). A small area with a higher gully density has a higher sediment delivery ratio, whereas a large area with a low gully density has a low SDR. In hard, resistant rocks, the density will be low whereas in the soft and easily eroded type of rocks, density will be high (Sen 1993). High gully density implies a finer soil texture and greater gully frequency, which was observed in the present study area. The gully density map (Fig. 9.6) of the study area reveals that areas with high gully density have impermeable soil, which is a highly erosion-prone zone. Areas with low gully density have permeable, unconsolidated subsurface soil, which is more prone to subsurface erosion such as tunnels or piping.

In the present study, two cross sections, namely, P-Q and R-S, and four long profiles of the major gullies, namely, A, B, C, and D, show detailed morphometric analysis (Fig. 9.7). From the cross profiles of P-Q and R-S (Fig. 9.8), it was found that gully sidewall types are vertical as well as sloping. In the Khoai badland, it was observed that the vertical gully sidewall developed mainly as the result of bank collapse and the sloping gully sidewall has developed from sheet erosion and small rill formation.

The longitudinal profiles of selected gullies (Fig. 9.9) have drawn on the basis of distance versus elevation. The least squares linear regression analysis of the long profile of four selected gullies has shown a remarkably low concavity (Fig. 9.9a–d); this means that the progressive decrease in channel slope or gradient downstream is responsible for the concave-up shape of gully long profile. The concave-up shape reflects the downstream increase in discharge and decrease in sediment size (Leopold et al. 1964). It has been observed that all the gully profiles are smooth, which indicated the presence of the same lithology in the Khoai badland. However, long profiles of the gully A and C showed slight changes at the confluence of tributary gullies.

The semi-logarithmic plots of long profile give the best idea about the profile characters of a particular gully, because the straight line of semi-logarithmic graph paper represents the graded character of a particular gully. It is well known that gullies generally tend to a concave-up profile in their downstream segment (Leopold et al. 1964), which is the result of huge sediment contributions from upstream tributaries. If the gully segments are above the graded line (red line of Fig. 9.10), then the gully represents the above-grade condition. If the gully segments are below the

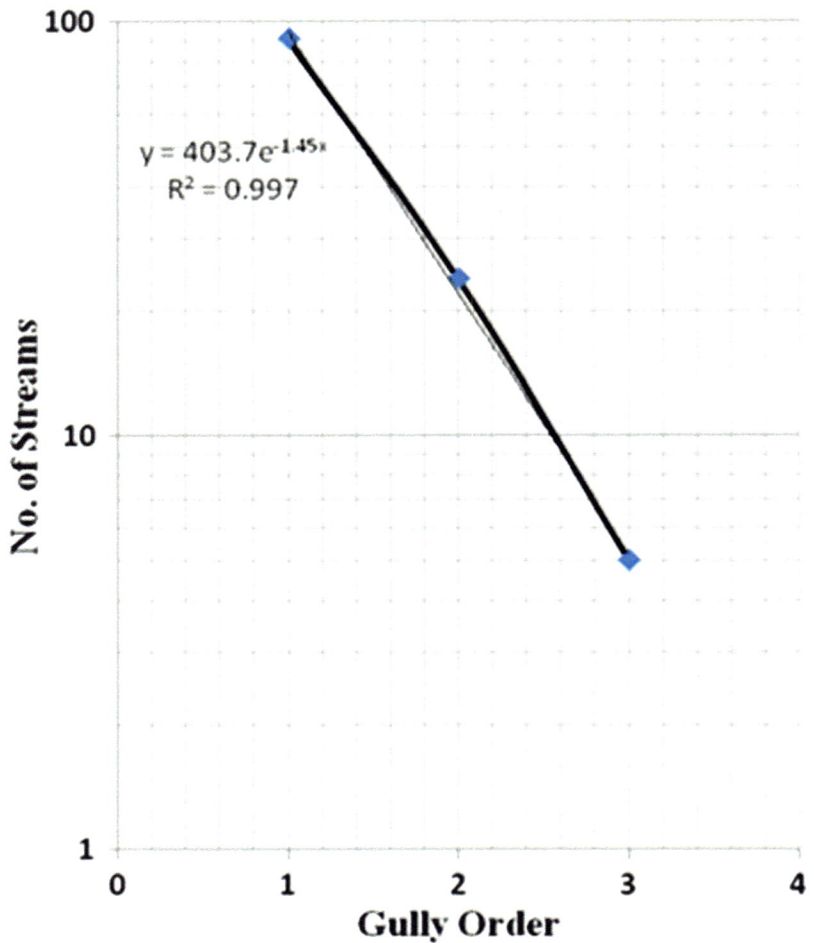

Fig. 9.5 Relationship between gully order and number of gullies

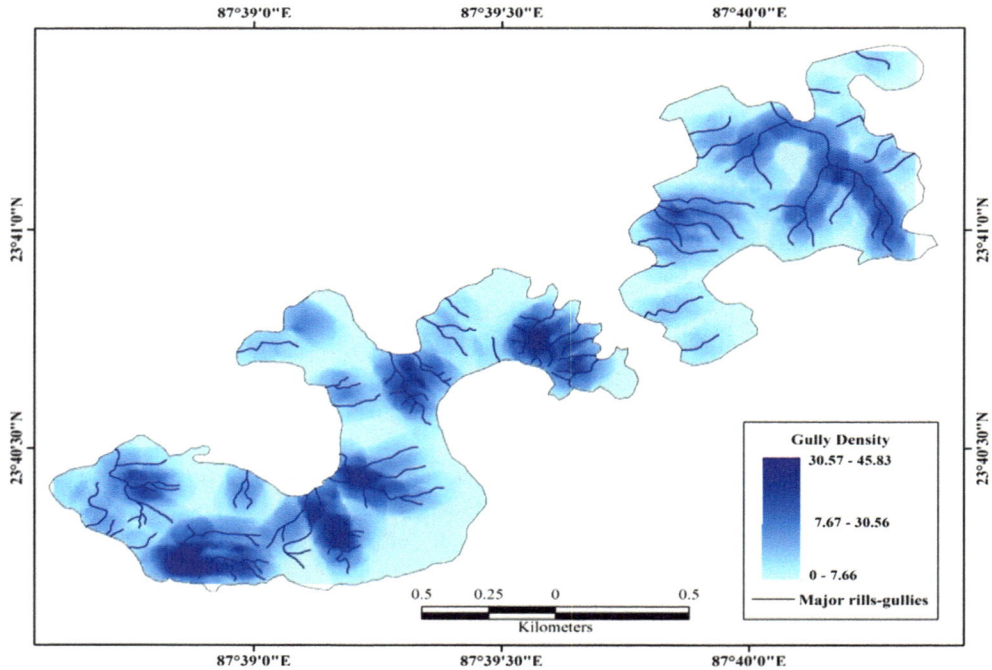

Fig. 9.6 Gully density map of the study area

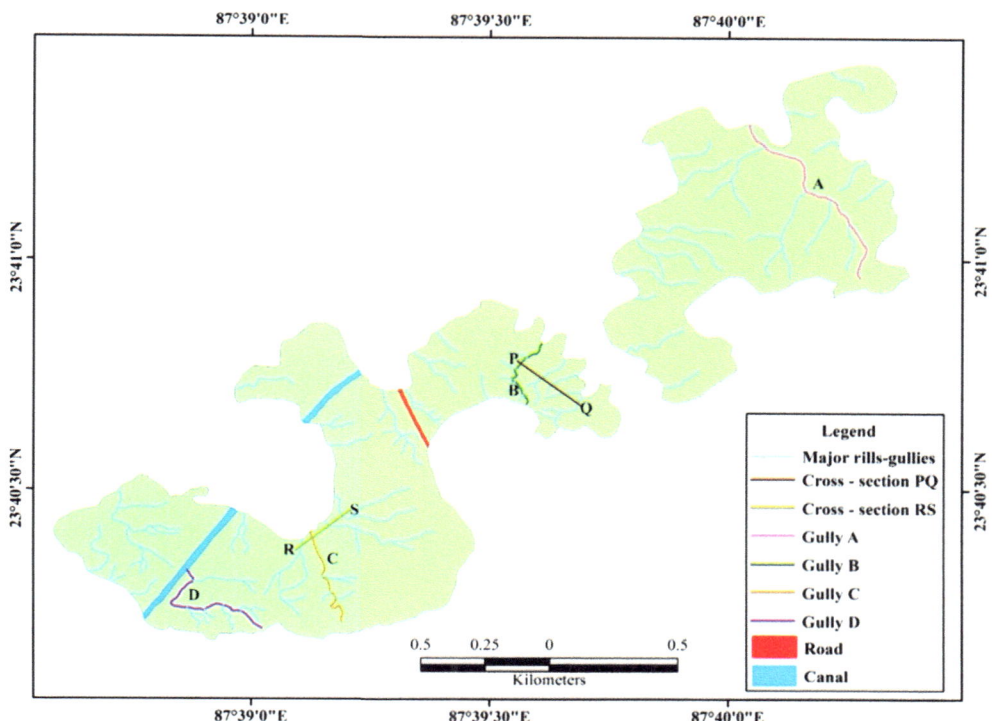

Fig. 9.7 Map showing site for cross profile and selected gullies for long profile analysis

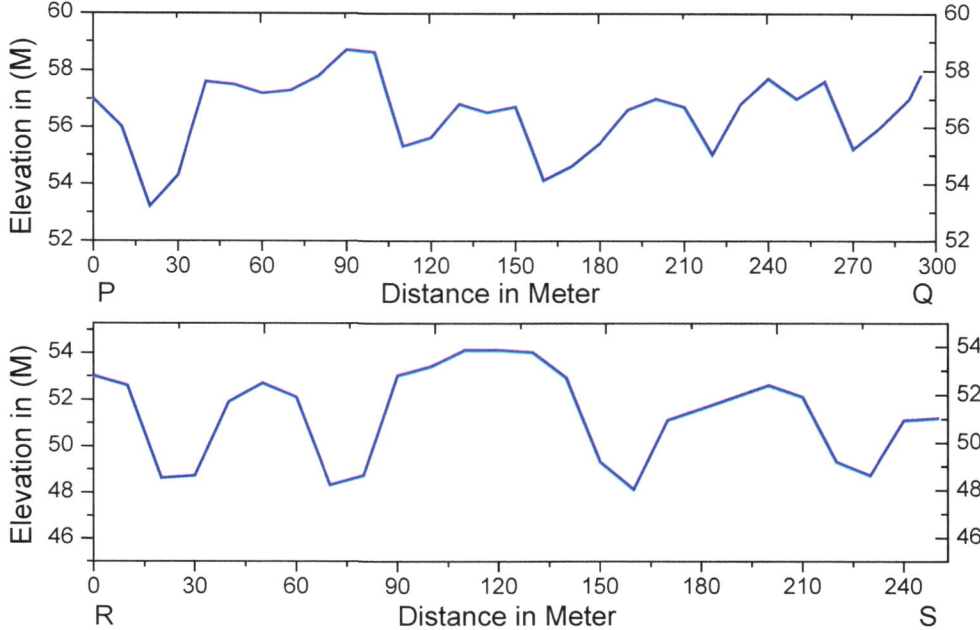

Fig. 9.8 The selected cross-profile PQ and RS showing the nature of the gully side-wall

graded line, then it represents a below-grade condition, which means the gully may not be able to transport bed-loads toward downstream. The source point of all the selected gullies has been taken as 0.1 m because zero (0) cannot be plotted on a semi-logarithmic scale. The semi-logarithmic long profiles of all these gullies are shown in (Fig.9.10a–d). From semi-logarithmic long profiles, it has been observed that all the gullies of the study area are in the above-grade condition. Thus, all these gullies may able to transport bed-loads downstream.

(a)

(b)

(c)

(d)

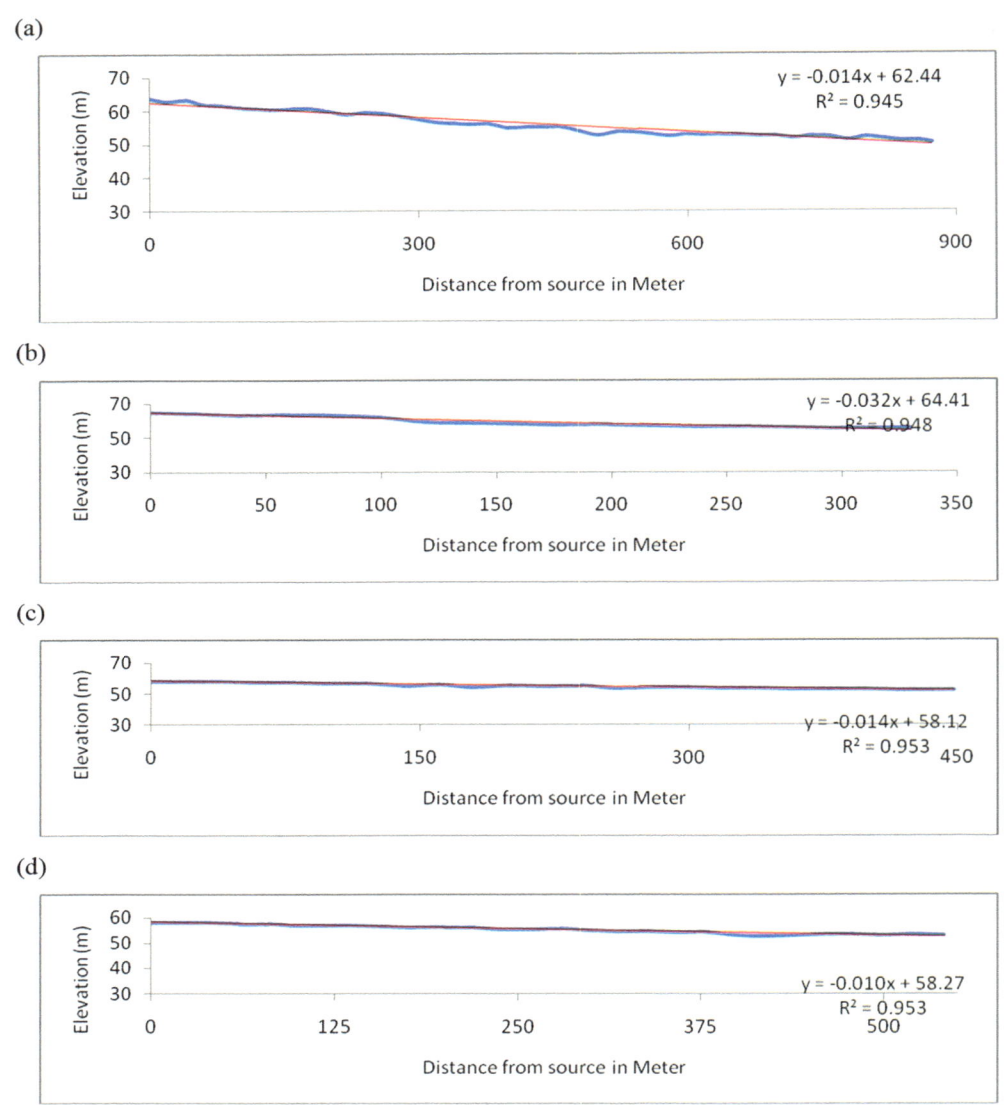

Fig. 9.9 Longitudinal profiles of selected gully A (**a**), gully B (**b**), gully C (**c**) and gully D (**d**)

The different types of properties, such as highest elevation (source), lowest elevation, altitudinal difference between the highest and lowest, and the total length of each gully, show the gully SL index, total SL index, and SL ratio (Table 9.2) of each gully. From the calculated result of SL ratio (the ratio of $SL_{section}$ to SL_{total}), in every reach of the gullies the value is less than two; hence, the longitudinal profile of all these gullies is said to be graded.

9.3.3 Estimation of Runoff and Sediment Delivery Ratio

9.3.3.1 LULC
The amount of rainfall–runoff largely depends on the surface cover types, that is, land use–land cover (LULC), whereas the pattern of rainfall–runoff changes with the changes of LULC

of the particular area (Guzha et al. 2018). In the present study, five LULC classes (Fig. 9.11) have been identified: canal, water body, road, open fallow land, and vegetation. In addition, we have deleted the built-up area class because this region has no such built-up area.

9.3.3.2 Hydrological Soil Group (HSG) Condition
On the basis of the minimum rate of infiltration, USDA NRCS has developed a soil classification system of four groups: A, B, C, and D. Runoff curve numbers for antecedent soil moisture condition (AMC-II) for hydrological soil cover types are shown in Table 9.3.

9.3.3.3 Weighted Area Curve Number
The calculated value of weighted CN of the study area is 83.17 (taking the rounded number of CN = 83) (Table 9.4).

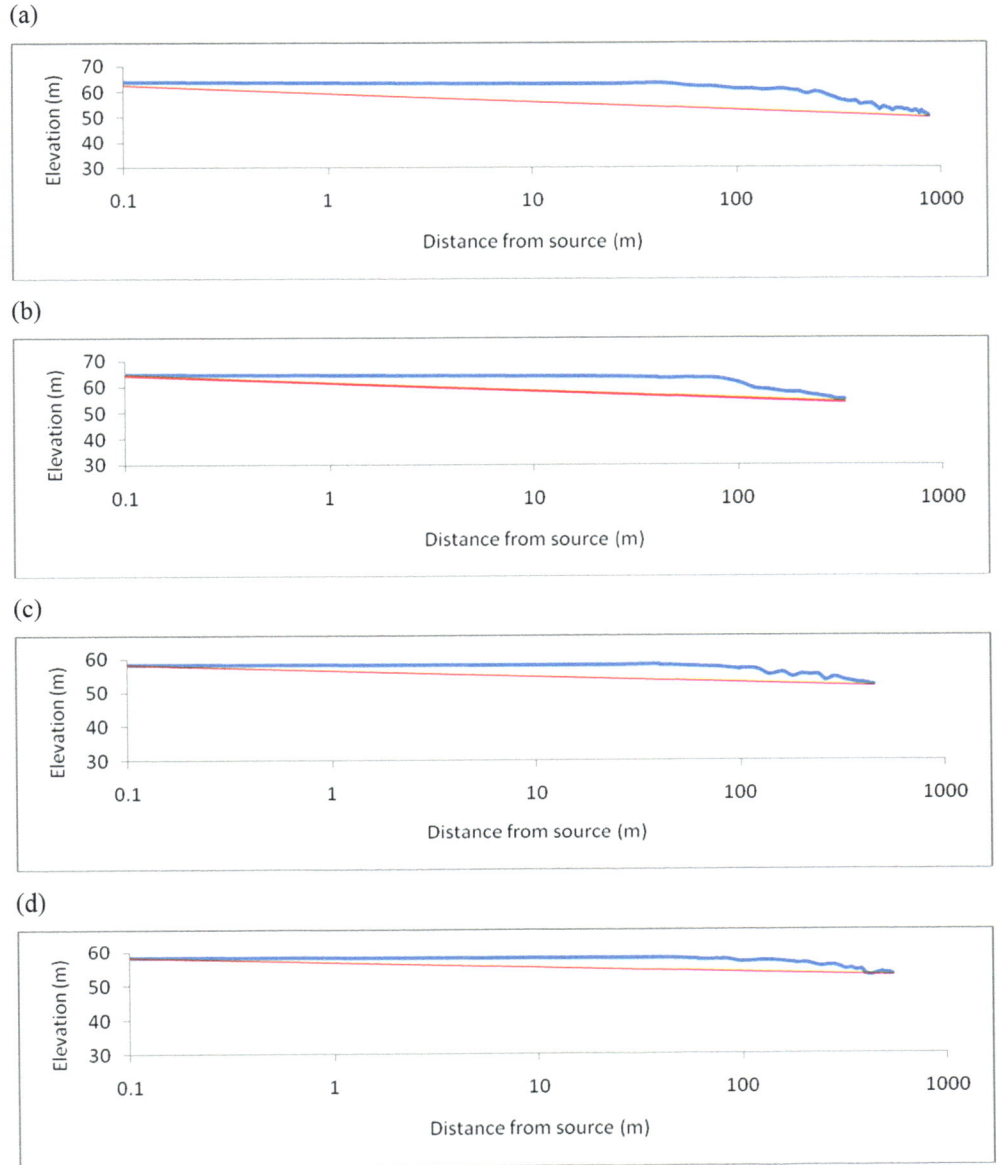

Fig. 9.10 Semi-logarithmic long profiles of selected gully A (**a**), gully B (**b**), gully C (**c**), and gully D (**d**). The straight red line indicates the graded profile

9.3.3.4 Runoff Estimation

In the present study, it has been estimated from Eq. (9.4) that the value of maximum potential storage (S) is 52.02, where S is related to the soil and cover condition of a particular area (USDA SCS 1972); it also means potential infiltration after runoff begins. In the present study, S is slightly high because of the presence of permeable unconsolidated subsurface soil: this is the main cause of numerous tunnels or pipings in the present study area. From Eq. (9.3), the average runoff (Q) for the period of 5 years (2008–2012) is 1186.392 mm, which indicates that runoff rate is much higher in comparison to the amount of rainfall in the study area. Thus, maximum runoff

causes maximum sedimentation from upslope to downslope segments, and the land is gradually degraded day by day.

9.3.3.5 Sediment Delivery Ratio (SDR)

The proportion of sediment yield and erosion rate of a particular area is called the sediment delivery ratio (Ghosh and Bhattacharya 2012). As we mentioned, to estimate the SDR, the entire study area was divided into three sample areas (Fig. 9.12a). Then, from the spatial distribution of the SDR map (Fig. 9.12b), we found that the central portion of the study area generated maximum sediment yield, and sedimentation gradually decreased from the central area to the north–

Table 9.2 Gully reach-wise stream length (SL) index and SL ratio

Gully	Source elevation ($h1$) (m)	Lowest elevation ($h2$) (m)	Elevation difference (ΔH) (m)	Total gully length (m)	Gully reach	SL index SL$_{Channel}$	SL$_{Total}$	SL ratio SL$_{Channel}$/ SL$_{Total}$
A	63.8	50.3	13.5	873	1	3.4	4.59	0.74
					2	3.59		0.78
					3	6.79		1.48
B	64.8	54.8	10	330	1	2.45	3.97	0.62
					2	3.45		0.86
					3	5.71		1.43
C	58.2	52.1	6.1	448	1	1.55	2.3	0.67
					2	3.09		1.34
					3	2.56		1.11
D	58.3	53.1	5.2	541	1	0.62	1.9	0.33
					2	3.33		1.75
					3	3.03		1.59

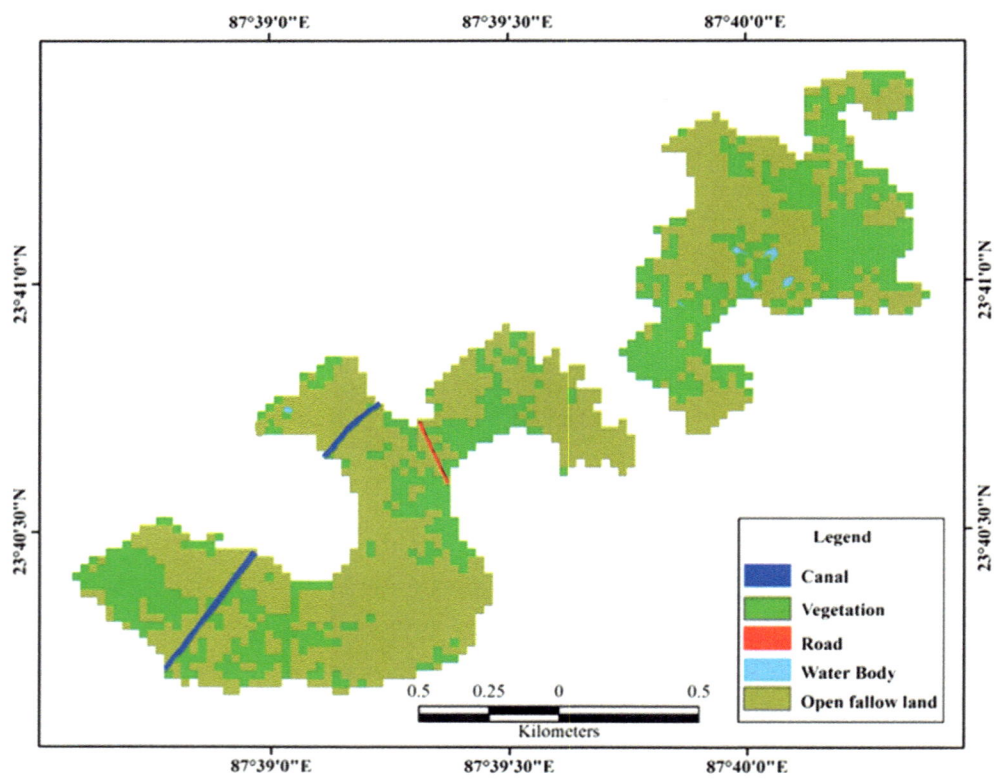

Fig. 9.11 LULC map of the study area

Table 9.3 Runoff curve number (AMC II) for hydrological soil group (USDA TR-55, 1986)

Sample no.	Land use	Cover description	Hydrological soil group (C)
1	Canal	–	100
2	Vegetation	Good	70
3	Road	Paved	98
4	Water body	–	100
5	Open fallow land	Bare soil	91

east and south–west direction. In addition (Table 9.5), it was observed that the value of SDR varies from 0.87 to 1.01 in the present study area. The maximum value of SDR (1.01) was found in sample area B; SDR of 0.87 and 0.91 was found in sample areas A and C, respectively. However, the study area is in the high SDR zone; hence, the rate of soil erosion is also very high. Therefore, some crucial measures against the existing erosion process, including check-dams and eucalyptus plantation, have been observed during the field survey (Fig. 9.13a, b). These measures are important to reduce the surface soil erosion.

Table 9.4 Calculation of weighted curve number (WCN)

Sample no.	Class	Area (km²)	Percentage of area	CN	Percent (%) of area × CN	WCN
1	Canal	0.0136	0.735	100	73.5	83.17
2	Vegetation	0.7002	37.85	70	2649.5	
3	Road	0.006	0.324	98	31.752	
4	Water body	0.0057	0.308	100	30.8	
5	Open fallow land	1.1245	60.78	91	5530.98	
Total		1.85	100			

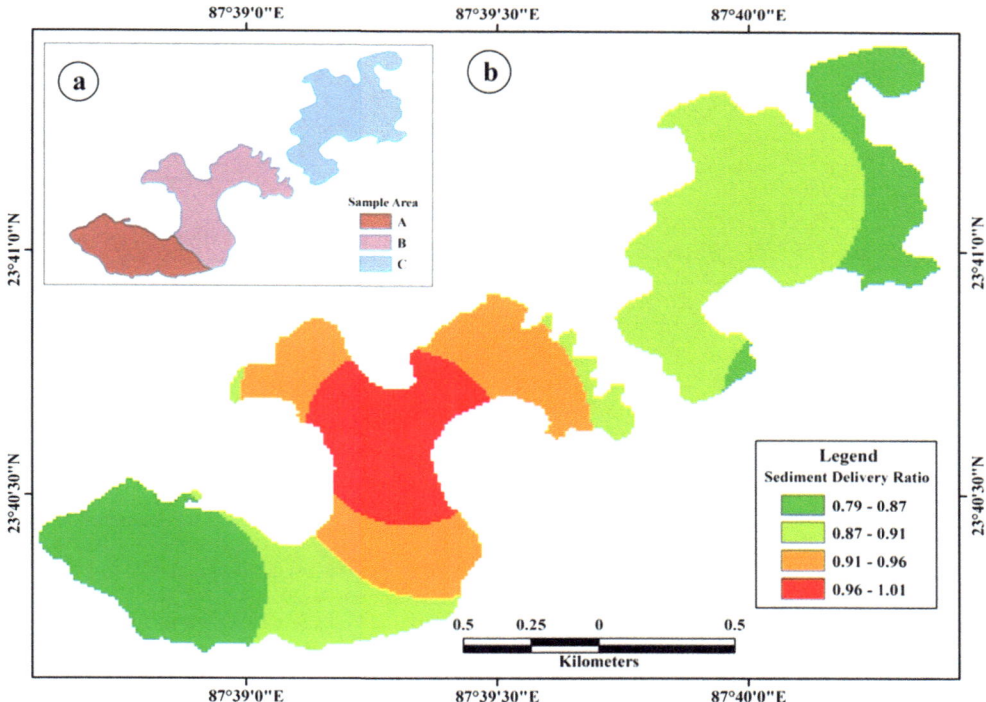

Fig. 9.12 Sample areas (**a**) and spatial distribution (**b**) of SDR map. (After Williams 1977)

Table 9.5 Calculation of sediment delivery ratio (SDR)

Sample area	Maximum basin length (km)	Relief (m)	Relief–length ratio (m/km)	Weighted curve number	Drainage area (km²)	Delivery ratio
A	1.22	10.5	8.606557377	83	0.62	0.87676
B	1.14	13.6	11.92982456	83	0.51	1.00648
C	1.4	14.1	10.07142857	83	0.72	0.91447

9.4 Conclusion

The nature of rill-gully development and its associated morphology is a very complex process, and it is not the same in all environments, because it depends on the climate of that particular region. In the present empirical study of the Khoai badland, it was shown that the area is highly dissected with well-developed rills and gullies resulting from the subtropical monsoon climate character. The long profiles of all these gullies are in above-grade condition. Therefore, the rate of sediment transportation is also very high. The computed value of high rainfall–runoff from the SCS-CN method also indicated high sediment delivery ratio (0.87–1.01), because there is a positive relationship between runoff and sediment delivery ratio. Thus, land degradation is a serious problem in the lateritic badland of the Khoai of Birbhum District of the state of West Bengal. To avoid land degradation of the Khoai badland region, different measures have been taken. From the field survey, it has been observed that to minimize the rate of

(a)   (b)

Fig. 9.13 Field photograph showing small stone check-dam (**a**) and extensive afforestation program (**b**) in the study area

soil detachment and soil transportation, small stone check-dams and afforestation programs have been undertaken by the government in this area. In addition, it has observed that the present badland topography of Khoai acts as a 'geomorphosite,' which is an emerging tool for geo-tourism, rather than as cropland.

References

Ahmed I, Verma V, Verma M (2015) Application of curve number method for estimation of runoff potential in GIS environment. In: 2nd International Conference on Geological and Civil Engineering, Singapore 80(4):16–20

Aown A, Kar N (2016) Lateritic badland of Sinhati, Bankura, West Bengal: a geomorphic investigation. In: Das BC, Ghosh S, Islam A, Ismail M (eds) Neo-thinking on Ganges-Brahmaputra Basin geomorphology. Springer, Berlin pp 19–31

Bandhopadhyay S (1987) Man-initiated gullying and slope formation in a lateritic terrain at Santiniketan West Bengal. Geogr Rev India 49 (4):21–26

Bandyopadhyay S (1988) Drainage evolution in a badland terrain at Gangani in Mednipur district, West Bengal. Geogr Rev India 50 (3):10–20

Basu SR (1972) On the formation of a shoal on the concave bank of lateritic river Kopai, West Bengal. Geogr Rev India 34(3):287–297

Battaglia S, Leoni L, Rapetti F et al (2011) Dynamic evolution of badlands in the Rogilo basin (Tuscany, Italy). Catena 86(1):14–23

Bera K, Bandyopadhyay J (2013) Prioritization of watershed using morphometric analysis through geoinformatics technology: a case study of Dungra water sub-basin, West Bengal, India. Int J Adv Remote Sens GIS 2(1):1–8

Bhattacharya JC (1957) Erosion studies in the lateritic areas of West Bengal. J Indian Soc Soil Sci 5:103–108

Biswas A (1987) Laterites and lateritiods of Bengal. In: Datye VS, Diddee J, Jog SR, Patil C (eds) Exploration in the tropics. Indian Institute of Tropical Meteorology (IITM), Pune, pp 137–146

Chakraborty SC (1970) Some consideration on the evolution of physiography of Bengal. In: Chatterjee AB, Gupta A, Mukhopadhyay (eds) West Bengal. Geographical Institute, Presidency College, Calcutta, pp 16–29

Charlton R (2008) Fundamentals of fluvial geomorpholgy. Routledge, New York

Das K, Bandyopadhyay S (1995) Badland development over laterite duricrust. In: Jog SR (ed) Indian geomorphology 1. Rawat Publications, New Delhi, pp 31–42

Das K, Bandyopadhyay S (1996) Badland development lateritic terrain: Santiniketan, West Bengal. In: Jog SR (ed) National Geographer 31 (1, 2):87–103

Das S, Behera S, Kar A et al (1997) Hydrogeomorphological mapping in ground water exploration using remotely sensed data– a case study in Keonjhar district, Orissa. J Indian Soc Remote Sens 25 (4):247–259

Dey S, Ghosh S, Debbarman C et al (2009) Some regional indicators of the tertiary-quaternary geodynamics in the palaeocostal part of Bengal Basin (India). Russ Geol Geophys 50(10):884–894

Fairbridge RW (ed) (1968) Encyclopedia of geomorphology. Reinhold, New York

Ferro V, Minacapalli M (1995) Sediment delivery processes at basin scale. Hydrol Sci J 40(6):703–717

Gajbhiye S (2015) Estimation of surface runoff using remote sensing and geographic information system. Int J Serv Sci Technol 8 (4):113–122

Gajbhiye S, Mishra S, Pandey A (2014) Relationship between SCS-CN and sediment yield. Appl Water Sci 4(4):363–370

Ghosh S, Bhattacharya K (2012) Multivariate erosion risk assessment of lateritic badlands of Birbhum (West Bengal, India): a case study. J Earth Syst Sci 121(6):1441–1454

Ghosh S, Guchhait S (2012) Soil loss estimation through USLE and MMF methods in the lateritic tracts of eastern plateau fringe of Rajmahal traps, India. Ethiop J Environ Stud Manag 5(4):529–541

Google earth pro V 7.3.2.5776. (July 18, 2018) Khoai badland, West Bengal, India. 23°40'24.89"N, 87°39'08.23"E, Eye alt 12680 feet. DigitalGlobe 2018. http://www.earth.google.com

GSI (1948) Published Geological quadrangle map, Kolkata, India

Guzha AC, Rufino MC, Okoth S, Jacobs S, Nobrega RLB (2018) Impact of land use and land cover change on surface runoff, discharge and low flows: evidence from east Africa. J Hydrol Reg Stud 15:49–67

Hack JT (1957) Studies of longitudinal stream profiles in Virginia and Maryland. US Geol Surv Prof Papers 294:45–97

Hack JT (1973) Stream profile analysis and stream-gradient index J Res US Geol Surv 1(4):421–429

Horton RE (1932) Drainage basin characteristics. EOS Trans Am Geophys Union 13(1):350–361

Jha V, Gupta K (2003) Land degradation in tropical lands: a case study. In: Jha VC (ed) Land degradation and desertification. Rawat Publications, New Delhi, pp 279–290

Jha V, Kapat S (2009) Rill and gully erosion risk of lateritic terrain in south-western Birbhum district, West Bengal, India. Soc Nat 21 (2):141–158

Jha V, Kapat S (2011) Degraded lateritic soils cape and uses in Birbhum district, West Bengal, India. Soc Nat 23(3):545–558

Joshi V (2014) Soil loss estimation by field measurements in the badlands along Pravara River (Western India). J Geol Soc India 83:613–624

Laha M (2011) Spatio-social impact of miniwatershed project at Bhalki, Bardhaman district, West Bengal. Wesleyan J Res 14(1):155–174

Leopold LB, Wolman MG, Miller JP (1964) Fluvial process in geomorphology. WH Freeman, San Francisco

Magar P, Magar N (2016) Application of Hack's stream gradient index (SL index) to longitudinal profiles of the river flowing across Satpura-Purna plain, Western Vidarbha, Maharashtra. J Indian Geomorphol 4:65–72

Martz LW, Garbrecht J (1999) An outlet breaching algorithm for the treatment of closed depressions in a raster DEM. Comput Geosci 25 (7):835–844

Mukhopadhyay S (1992) Soil erosion in Kopai basin, Birbhum. J Land Syst Ecol Stud 15(2):22–23

Nadal-Romero E, Regues D (2010) Geomorphological dynamics of sub-humid mountain badland areas: weathering, hydrological and suspended sediment transport processes. A case study in the Araguas catchment (Central Pyrenees) and implications for altered hydroclimatic regimes. Prog Phys Geogr Earth Environ 34(2):123–150

NBSS & LUP (2010) Published soil texture map of West Bengal, ICAR-Kolkata, West Bengal, India

Niyogi D, Mallick S, Sarkar S (1970) A preliminary study of laterites of West Bengal. India. In: Chatterjee SP, Das S, Gupta P (eds) Selected papers in physical geography vol 1. 21st International Geographical Congress, Calcutta, National Committee for Geography, pp 443–449

Onyando JO, Kisoyan P, Chemelil MC (2005) Estimation of potential soil erosion for River Perkerra catchment in Kenya. Water Resour Manag 19:133–143

Ouyang D, Bartholic J (1997) Predicting sediment delivery ratio in Saginaw Bay watershed. In: Orlando FL (ed) Proceedings of the 22nd National Association Environmental Professionals Conference, USA, pp 19–23

Pal SC, Chakrabortty R (2018) Modeling of water induced surface soil erosion and the potential risk zone prediction in a sub-tropical watershed of Eastern India. Model Earth Syst Environ 125

Pal SC, Shit M (2017) Application of RUSLE model for soil loss estimation of Jaipanda watershed, West Bengal. Spatial Inform Res 25(3):399–409

Roy J, Saha S (2017) Measuring the spatial pattern of surface runoff using SCS-CN method of Hinglo river basin: RS-GIS approach. Int Res J Earth Sci 5(8)1–7

Sarkar D, Nayak D, Dutta D et al (2007) Optimizing land use of Birbhum district (West Bengal) soil resource assessment. NBSS Publ. 130. NBSS and LUP Nagpur

Seeber L, Gornitz V (1983) River profiles along the Himalayan arc as indicators of active tectonics. Tectonophysics 92:335–367

Sehgal J, Abrol IP (1994) Soil degradation in India: status and impact. Oxford/IBH, New Delhi

Sen J, Sen S, Bandyopadhyay S (2004) Geomorphological investigation of badlands: a case study of Garhbeta, West Medinipur district, West Bengal, India. In: Singh S, Sharma HS, De SK (eds) Geomorphology and environment. ACB, Kolkata, pp 204–234

Sen P K (1993) Geomorphological analysis of drainage basins. Burdwan University Press, Burdwan, India

Shit P, Bhunia G, Maiti R (2013) Assessment of factors affecting ephemeral gully development in badland topography: a case study at Garhbeta badland (Paschim Medinipur). Int J Geosci 4 (2):461–470

Shit P, Bhunia G, Maiti R (2014) Morphology and development of selected badlands in South Bengal (India). Indian J Geogr Environ 13:161–171

Shit PK, Maiti RK (2012) Mechanism of gully-head retreat: a study at Ganganir Danga, Paschim Medinipur, West Bengal. Ethiop J Environ Stud Manag 5(4):417–431

Siddi R, Sudarsana G, Rajasekhar M (2018) Estimation of rainfall-runoff using SCS-CN method with RS GIS techniques for Mandavi basin in YSR Kadapa district of Andhra Pradesh, India. Hydrospatial Anal 2(1):1–15

Stone RO (1967) A desert glossary. Earth Sci Rev 211–268

Strahler AN (1964) Quantitative geomorphology of drainage basins and channel networks. In: Chow VT (ed) Handbook of applied hydrology McGraw-Hill, New York, pp 739–476

Thomas MF (1974) Tropical geomorphology. Macmillan, London, pp 49–82

USDA (1972) Sediment sources, yields, and delivery ratios. National Engineering Handbook, Section 3, Sedimentation. USDA, Washington, DC

USDA (1986) Natural Resources Conservation Service. Conservation Engineering Division, Technical Release 55. USDA, Washington, DC

Walling DE (1983) The sediment delivery problem. J Hydrol 65:209–237

Williams JR (1977) Sediment routing for agricultural watersheds. Water Resour Bull 11(5):965–974

Asish Saha obtained his both Bachelor and Master degree in Geography from The University of Burdwan. He has also completed Post Graduate Diploma in Geoinformatics from Department of Science and Technology, Government of West Bengal with collaboration of Maulana Abul Kalam Azad University of Technology. He has cleared UGC-NET JRF and WBCSC SET in geography. Presently, he is an independent scholar and highly passionate about physical geography. Climate change, fluvial geomorphology, and application of Remote Sensing-GIS in applied geography are core areas of his research interest.

Manoranjan Ghosh is Doctoral Fellow (UGC-SRF) at Indian Institute of Technology Kharagpur. He has received an undergraduate degree in Geography from University of North Bengal in 2012 and postgraduate degree in Geography from The University of Burdwan in 2015. Presently, he is doing his doctoral research on climate change vulnerability in Sub-Himalayan region using mixed research methodology and GIS technology. Climate change adaptation, forestry, and rural regional planning are also core areas of his research interest.

Subodh Chandra Pal is currently working as an Assistant Professor in the Department of Geography, The University of Burdwan, Bardhaman, West Bengal, India. He has completed his Doctoral research in the fields of Hydrogeomorphology from Visva-Bharati, Santiniketan, India. His fields of expertise are Fluvial Geomorphology, Pedo-geomorphology, Land Degradation, Soil Erosion, Gully Erosion, Landslide Vulnerability, and Climate Change.

Estimation of Erosion Susceptibility and Sediment Yield in Ephemeral Channel Using RUSLE and SDR Model: Tropical Plateau Fringe Region, India

Raj Kumar Bhattacharya, Nilanjana Das Chatterjee, and Kousik Das

Abstract

Soil erosion susceptibility and huge sedimentation are the major problems in plateau fringe basin under tropical monsoon climate. Ephemeral channel plays a dominant role in a linkage between erosion and sediment deposition in the plateau fringe basin. The objective of this study is to estimate the spatial distribution of potential mean soil erosion (PMSE) rate and gross volume of sediment yield (SY) along with sediment delivery ratio (SDR) in ephemeral outlets at sub-basin scale in Kangsabati basin, situated under Chota Nagpur plateau, India, using coupling models of Revised Universal Soil Loss Equation (RUSLE) and SDR. The models estimated that there is a maximum PMSE (226 t ha^{-1} year^{-1}) occurrence in upper basin along with high mean SY (32 t ha^{-1} year^{-1}), whereas lower basin has PMSE (74 t ha^{-1} year^{-1}) with low SY deposition (13.23 t ha^{-1} year^{-1}). According to sub-basin scale, most of the SY correspondence with PMSE occurs in Khatra-1 (10.22 t ha^{-1} year^{-1}) and Lalgarh-4 sub-basin (7.2 t ha^{-1} year^{-1}) out of 27 sub-basins. The result indicates that PMSE is positively significantly correlated with SY ($R^2 = 0.975$) following RUSLE parameters like rainfall erosivity, soil erodibility, slope factor, land cover management, and conservation practice, whereas the role of SDR on SY is determined by basin area and travel time. In spite of model parameters, first stream number ($R^2 = 0.895$) and basin length ($R^2 = 0.88$) positively signify on SY through enhancing the erosion parameters. Gradient ratio positively signifies on SDR ($R^2 = 0.674$) and also determines the role of SDR on SY which controlled the travel time and basin parameter. Therefore, several geo-environmental parameters govern the sediment deposition in plateau fringe basin.

Keywords

Kangsabati basin · Gully erosion · Potential mean soil erosion (PMSE) · Sediment yield (SY)

10.1 Introduction

In the tropical region, soil erosion is a very common phenomena occurring in two common ways, i.e., water and wind action and increased soil degradation as well as huge sedimentation on the river bed or dam and then reduction in their storage capacity (Bagherzadeh and Daneshvar 2013; Samad et al. 2016; Markhi et al. 2019). Rill and inter-rill soil erosion under water action play a dominant role on erosion susceptibility in the upland topography caused by terminal velocity of raindrop and surface runoff (Fernandez et al. 2003; Bhattarai and Dutta 2006). Contrastingly, the amount of sediment exported in various ephemeral gully channels helps generate huge sedimentation and also leads to sediment yield generation over a period of time (Fernandez et al. 2003; Samad et al. 2016). PMSE is calculated by the Revised Universal Soil Loss Equation (RUSLE) initially developed by Renard (1997), whereas sediment delivery distributed model (SEDD) is applied to determine the sediment delivery ratio in each morphological cell in a sub-basin (Ferro and Porto 2000). SDR can be estimated through stream order, drainage density, soil type, and watershed size (Corbitt 1990). Sediment yield (SY) is computed as multiply of PMSE and SDR in a sub-basin (Fernandez et al. 2003; Jobin et al. 2018). After detachment and transportation, soil particles are deposited through rill and inter-rill channel driven by raindrop impact and overland flow energy (Mitasova et al. 2013; Fernandez-Raga et al. 2017; Jobin et al. 2018). In this context, USLE and RUSLE do not predict actual estimation of PMSE ignorance of computation about detaches or transported soil particles at large-scale level. Therefore, estimation of SY at sub-basin level faces more complexity following only RUSLE model.

R. K. Bhattacharya (✉) · N. D. Chatterjee · K. Das
Department of Geography and Environment Management, Vidyasagar University, Midnapore, West Bengal, India

© Springer Nature Switzerland AG 2020
P. K. Shit et al. (eds.), *Gully Erosion Studies from India and Surrounding Regions*, Advances in Science, Technology & Innovation, https://doi.org/10.1007/978-3-030-23243-6_10

According to Jobin et al. (2018), the magnitude of PMSE greater than SY which does not consider detachment sediments falls under transportation through rill or gullies. However, SEDD is used for the prediction of effective PMSE and SY before the estimated sediment transport efficiency is incorporated with RUSLE model (Mhangara et al. 2012; Magesh and Chandrasekar 2016). GIS and RUSLE both have been used to estimate the magnitude and distribution of soil erosion (Ganasri and Ramesh 2016; Dissanayake et al. 2018). Sediment transport of all stream is measured by Sediment Delivery Distribution model (SEDD) (Lim et al. 2005; Lu et al. 2006). SDR means the ratio of the SY at a given stream cross-section to the gross erosion from the upstream measuring points of the watershed (Julien 2010). Eroded sediment transferred by delivery outlets is known as ephemeral gully and falls into the river in a drainage basin. SY is an order of lower magnitude than soil erosion rates from steep slope (Jobin et al. 2018). The higher rate of sediment delivery that takes shortest distance along small stream bed slope leads to huge sedimentation, whereas the lower rate of sediment delivery cannot be enough deposited because it takes long distance travel along the stream (Fernandez et al. 2003; Fernandez-Raga et al. 2017; Jobin et al. 2018). Supply of sediment is inversely related to stream distance (Fernandez et al. 2003). The rate of soil erosion accelerates the delivery ratio, but there are several factors which influenced SDR (Magesh and Chandrasekar 2016). Hydrological inputs like rainfall intensity, duration, and landscape parameters such as vegetation, topography, and soil properties with their complex interaction on the land surface have effective impact on sediment delivery (Richards 1993). Catchment area variation is also a dominant factor for the delivery ratio. So, it is more difficult to select the proper factors of hydraulics and landscape parameters. Several variables like basin area, slope steepness, length, and land use/land cover are taken as input parameters to estimate delivery ratio by assigning their significance (Kothyari and Jain 1997).

A similar situation has been found in plateau fringe alluvial non-perennial basin named as Kangsabati basin. There was some literature about soil erosion susceptibility (Shit et al. 2015; Mahala 2018) but lack of literature about the sediment delivery as well as SY estimation. During 1958, after the construction of Mukutmanipur Dam, Kangsabati River considerably declined graded slope which has reduced the higher draining capacity of sediment discharge throughout every rainy season (Mittal et al. 2014). Simultaneously, ephemeral gully channels play a vital role to supply huge sediment at outlet points as well as SY deposition near Kangsabati River. Therefore, the present study estimated sediment delivery ratio and SY using paired model of RUSLE and SDR.

The objective of this chapter is to assess and predict the spatial distribution, rate, and volume of SY along with estimated PMSE as well as SDR of each sub-basin along the rill and gully channel. The source of sediment along the river bed entirely depends on SDR than bed sediment transport following the field observation that gradation of sediment particle diameter is not similar throughout the course.

10.2 Study Area

Kangsabati basin as a transitional watershed covers the plateau fringe area in Eastern Chota Nagpur plateau to floodplain area of Rupnarayan and Khelighai basin in lower Gangetic plain. Maximum coverage area falls in West Bengal including four districts, namely, Purulia, Bankura, Paschim Medinipur, and Purba Medinipur, and the rest parts situated in Jharkhand. Total basin boundary extends latitudinally from $21°45'N$ to $23°30'N$ and longitudinally from $85°45'E$ to $88°15'E$ covering 9658 km^2 (Fig. 10.1). Lithological status in this basin varies from oldest Achaeans (Pre-Cambrian) to the younger Tertiary-Quaternary formations, most of the dominant structure being mica schist which is occasionally garnetiferous and oxidized sand, silt, and clay with in situ caliche groups (Mukhopadhyay 1992). Climatic characteristic indicates sub-humid type where maximum rainfall occurs (1077–1804 mm/year) during monsoon season. Lithological structure and seasonal rainfall help to generate immature soil profile with least organic or vegetable covers in the upper basin, but lateritic cover with transitional geological formation between the plateau and plain in middle site faces more erosion by rill gully formation and rapid deforestation (Ghosh 2015; Shit et al. 2015; Mahala 2018). In the lower basin, conversion of forest to agriculture land, multi-cropping system, and huge groundwater exploitation help to enhance erosion susceptibility (Mondal 2012). Sedimentation status in this basin entirely depends on a number of first stream order, initiated through rill and gully channel, and supplied huge eroded sediment particles. As a result, the magnitude of SY corresponded with soil erosion susceptibility and sediment delivery ratio near ephemeral gully channel (Ghosh 2015). Twenty-seven sub-basins were delineated following the geo-environmental status in the entire study basin: 14 sub-basins fall under the upper part and the other 13 sub-basins are in the lower part (Fig. 10.1).

10.3 Methodology and Mapping

Several types of parameters like rainfall intensity, ASTER GDEM (30 m), soil types (ICAR), and Landsat image (30 m pixel cell size) were required to run RUSLE and SDR model in the entire basin (Table 10.1). LULC map prepared from Landsat image is used to obtain the surface roughness, whereas ASTER GDEM helps to generate stream order,

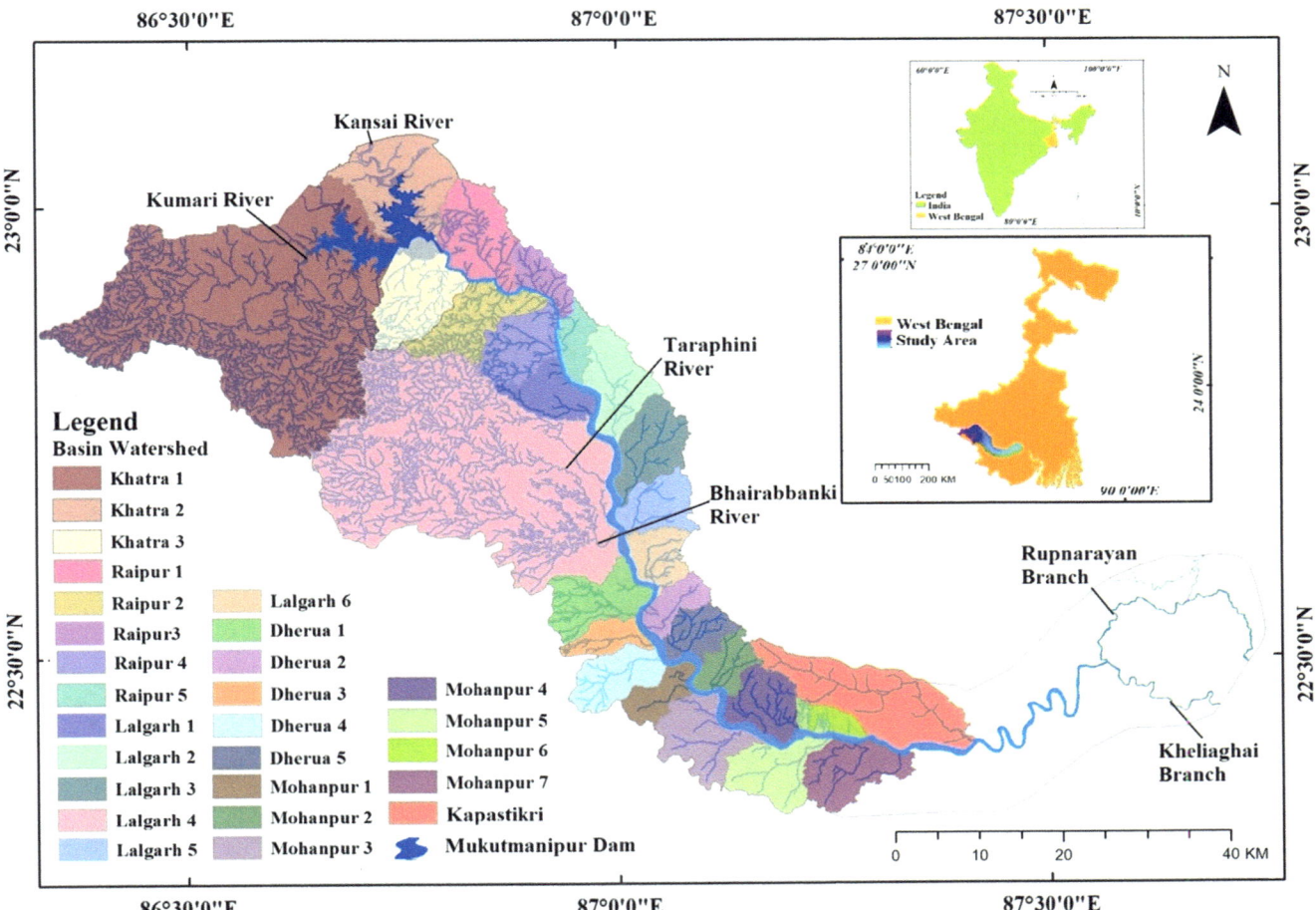

Fig. 10.1 Study area in Kangsabati basin

Table 10.1 Data source and different parameters of RUSLE and SDR model

Data	Date	Specification	Source	
LANDSAT 8	09-12-2016	Row 139; path 45	Global land cover facility (GLCF)	
		Resolution 30 m		
ASTER GDEM	17-10-2011	Pixel 30 m	Advance space bone thermal emission and reflection radiometer	
		Sensor-TIR (10–14)		
Soil sheet	1953	WB soil sheet 1,3,4	National Bureau of soil survey and land use planning (ICAR)	
		Scale 1:500000		
Rainfall	1980–2016	Sixteen rain gauge station	Indian meteorology department (IMD)	
Parameter	Source	File type	Range	References
R	IMD (1980–2016)	Raster	3213–16,345	Renard and Freimund (1994)
K	ICAR (1,3,4)	Raster	0.09–0.44	Wischmeier and Smith (1978)
LS factor	ASTER GDEM	Raster	1.4–9	Wischmeier and Smith (1978)
C factor	LANDSAT TM 8	Decimal	0–0.5	Wischmeier and Smith (1978)
P factor	LANDSAT TM 8	Decimal	0–1	Wischmeier and Smith (1978)
β coefficient	LANDSAT TM 8	Decimal	0.76–6.2 m/s	Haan et al. (1994)
Travel time (t_i)	ASTER GDEM	Decimal	0–23.56 H	Fernandez et al. (2003)

slope steepness, length, flow accumulation, travel time, and stream length following the SOI Toposheet (No 73N-7,8,9,10,12,11; J-15,9,12; I-7,8,10 of 1:50000 scale, 1975). WB soil sheets (1, 3, and 4) and rainfall were used to prepare soil erodibility and rainfall erosivity index throughout the basin.

10.3.1 RUSLE Parameter Estimation

RUSLE model also assesses the significant role of rill and inter-rill erosion on mean PMSE as well as SY deposition (Markose and Jayappa 2016). The model is computed by modifying the different parameters from USLE using the following equation (Renard 1997).

$$A = R \times K \times L \times S \times C \times P \qquad (10.1)$$

where A means the spatial distribution of PMSE in tonnes per hectare per year (t ha^{-1} year^{-1}), R represents rainfall-runoff erosivity in MJ mm ha^{-1} year^{-1}, K means soil erosivity factor in t ha h ha^{-1} MJ^{-1} mm^{-1}, L means slope length factor, S means slope steepness factor, C means cover management factor, and P means support practice factor such as contouring, strip cropping, and terracing. However, L, S, C, and P are considered as dimensionless parameters.

10.3.1.1 Rainfall Erosivity Factor (R)

Daily rainfall intensity helps to detect the spatial distribution of SY by the computation of PMSE (Jain et al. 2001; Dabral et al. 2008). R factor as a most effective index to assigning PMSE followed by Renard and Freimund (1994) formula:

$$R = 0.04830\, P\, 1.610 \ldots \ldots \ldots \ldots P < 850\ \text{mm} \qquad (10.2)$$

$$R = 587.7 - 1.219P + 0.004 \ldots \ldots \ldots \ldots P > 850\ \text{mm} \qquad (10.3)$$

R means annual rainfall erosivity represent in MJ mm ha^{-1} year^{-1} and P means annual precipitation (mm) last 36 year.

R in study area is prepared from Isoerodent map plotted from mean annual rainfall record or isohyet map in 16 rain gauge stations since 1980. Isoerodent map generated on raster surface consisted of 30 m spatial resolution in each grid size adjusted with another thematic map which is required to running up the RUSLE model in this basin. Maximum R value is concentrated in Baka (16,344 mm ha^{-1} year^{-1}) and minimum value found in Mohanpur (4041 mm ha^{-1} year^{-1}) during 1980–2016. Figure 10.2a stated that lower basin has more rainfall intensity than the upper basin.

10.3.1.2 Soil Erodibility Factor (K)

To generate the K factor, values are assigned from respective soil types in soil map; it represents the susceptibility of soil erosion and delivered surface material (Renard 1997; Jobin et al. 2018). Soil textural composition like % silt plus very fine sand, % sand, soil structure, % organic matter, and permeability helps to compute the K value, and then all values are plotted in nomograph to find the structural relationship using Eq. (10.4) (Wischmeier and Smith 1978):

$$K = \left\{ \frac{2.1.10^{-4}(12 - 0M)M^{1.14} + 3.25(S - 2) + 2.5(P - 3)}{100} \right\} \qquad (10.4)$$

M means the product of primary particle size fractions (% of modified silt or the 0.002–0.1 mm size fraction); K means tons acre per erosion index (t ha h ha^{-1} MJ^{-1} mm^{-1}) in SI units; S means soil class for structure; P means the rate of permeability of each soil structure.

K factor denotes wide variation ranging from 0.44 in fine loamy typic ustochreptas-66 to 0.09 in loamy, lethic Haplustaifs-96 in the entire Kangsabati basin. It is a point that low permeability and high organic content concentrated in loamy, lethic Haplustaifs-96, whereas high permeability and low organic content associated in fine loamy typic ustochreptas-66.

10.3.1.3 Slope Length and Slope Steepness Factor (LS)

LS factor in terms of slope length and slope steepness intensifies on topographic role, especially terrain characteristic, and positively influences soil erosion (Jobin et al. 2018). In this study, L factor is computed following the flow accumulation rate and percentage of slope (Wischmeier and Smith 1978; Renard 1997):

$$L = \left(\frac{\lambda}{22.13} \right) m \qquad (10.5.1)$$

where L means slope length; λ means slope length unit in meter; and m value depends on the variation of slope, i.e., 0.5 taking in slopes steeper than 5%, 0.4 for slopes between 3% and 4%, 0.3 for slopes between 1% and 3%, and 0.2 for slopes less than 1%.

Slope steepness and maximum slope direction (9%) signify a higher rate of soil erosion (McCool et al. 1987). S factor is determined from slope steepness algorithm using the following equation (Renard 1997):

$$S = 10.8 \times \sin(M) + 0.03, \text{where slope} < 0.09 \qquad (10.5.2)$$

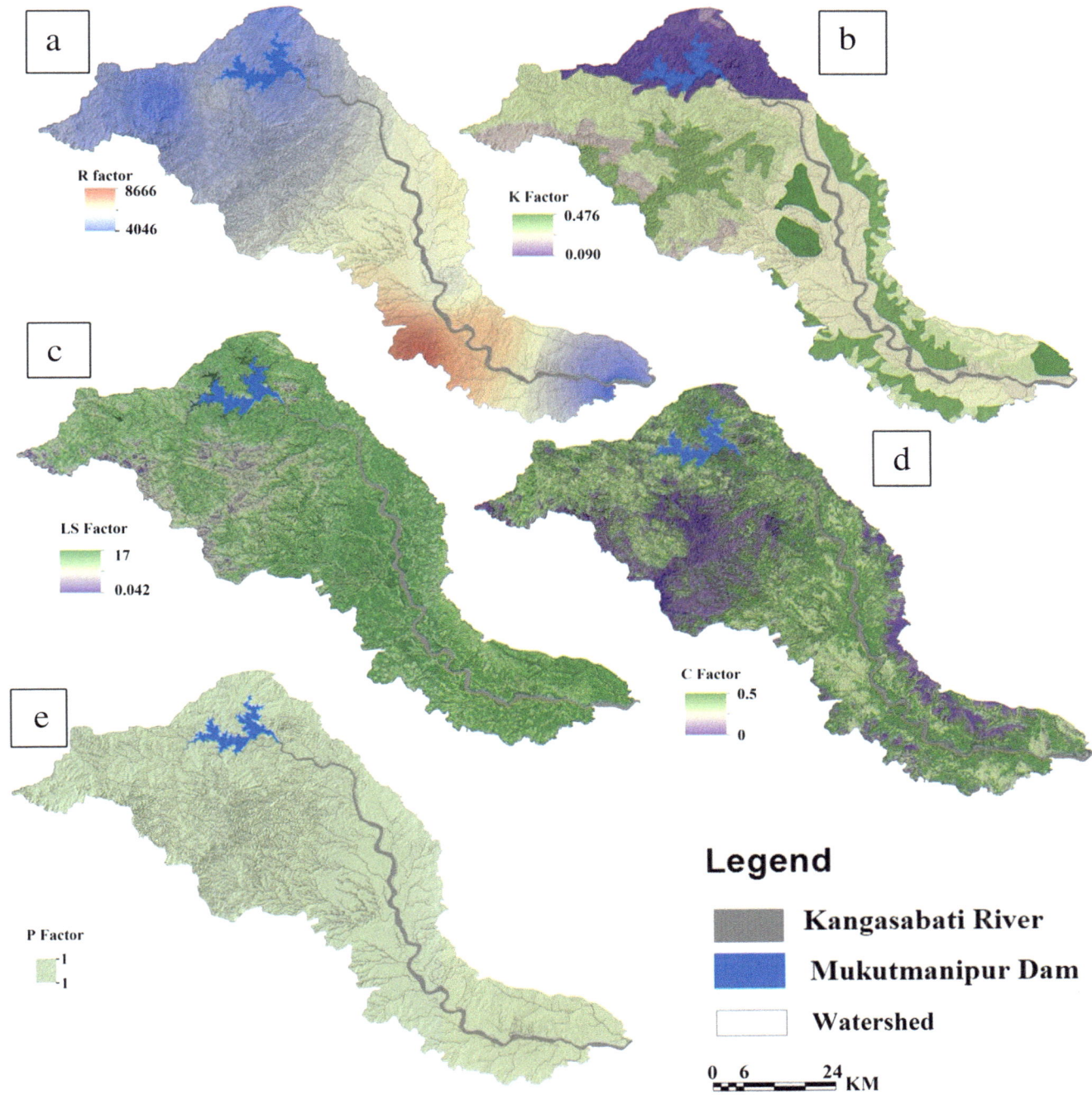

Fig. 10.2 RUSLE estimation parameters: (**a**) *R* factor, (**b**) *K* factor, (**c**) *LS* factor, (**d**) *C* factor, (**e**) *P* factor

$$S = 16.8 \times \sin(M) - 0.5, \text{where slope} \geq 0.09 \quad (10.5.3)$$

where S means slope steepness factor and Θ represents a gradient of slope as in degrees unit.

L and S factor cannot be properly determined by slope variation due to fluctuating slope steepness. Foster and Wischmeier (1974) tried to solve this problem through the division of slope into several numbers of segments. The terrain is divided along a two-dimensional plane following

unit-contributing area. This approach is estimated using Desmet and Govers (1996) equation:

$$L_{ij} = \frac{\left(A_{i,j-in} + D^2\right)ij^{m+1} - A_{i,j-in}^{m+1}}{D^{m+2} \times X_{i,j}^m \times 22.13^m} \quad (10.5.4)$$

$L\_(ij)$ means inlet-wise slope length in grid cell i and j; AAi, $j - in$ means contributing area at the inlet of grid cell (i, j) measured in m^2; D means grid cell size unit in meters; Xi, *j* means sin *aaij* + cos *aai,j*; and *aai,i* means aspect of direction of the grid cell (*i, j*).

To get the *m* and *β* values, Eqs. (10.5.4 and 10.5.5) must be followed and must also consider the *L* and *S* factor to assess the dependency between them. Therefore, Eqs. (10.5.5 and 10.5.6) address the dependency of *m* on the *β* ratio under rill to inter-rill erosion:

$$m = \frac{\beta}{\beta + 1} \qquad (10.5.5)$$

$$\beta = \frac{\frac{\text{Sin}\phi}{0.0896}}{(0.56 + 3) \times (\text{Sin}\phi)^{0.8}} \qquad (10.5.6)$$

where *φ* means slope angle in degrees and *m* value ranges 0–1 where 0 indicates the ratio of rill to inter-rill erosion.

In this study area, *LS* factor ranges from 0.042 to 17; meanwhile higher mean values are assigned from the steep and moderate slope in plateau fringe and undulating topography, whereas lower mean value from gentle slope in low land surface (Fig. 10.2c).

10.3.1.4 Cover Management Factor (*C*)

Cover management factor (*C*) is assigned from land use/land cover patterns (LULC), which reveals the significant role of cropping and management practice in enhancing the soil erosion rate as well as helps to detect comparative role of relative impacts of management options and conservation strategies (Renard 1997; Markose and Jayappa 2016; Jobin et al. 2018). *C* value is interpreted by the following researchers such as settlement considered as 0.2 given by UN-FAO (2001); barren land with laterite as 0.5 following Bakker et al. (2008); degraded forest as 0.05 taken from Bakker et al. (2008) and Jordan et al. (2005); dense forest considered the value of 0.01 taken from UN-FAO (2001), Bakker et al. (2008), and Jordan et al. (2005); value of water body as 0 following Cox and Madramootoo (1998); and single crop and double crop considered the value of 0.2 and 0.31 given by Wischmeier (1960).

In this study, *C* factor is estimated from reclassified land use map including six different classes, i.e., single crop, double crop, fallow land, forest, settlement, barren land, and surface water body, using integrated thematic layers under GIS platform. *C* factor varies from 0.01 to 0.5; prepared from Landsat images (2016) under 94% overall accuracy assessment using supervised classification technique (Fig. 10.2d and Table 10.2). Barren land with laterite outcrop (0.32) and double crop yield (0.28) have maximum *C* value due to lack of plant cover and soil aggregate, whereas dense forest cover (0.06) has minimum *C* value for the presence of maximum organic cover. These results are justified by the following: Wischmeier (1960), Cox and

Table 10.2 *C* factors in different land use/land cover patterns

Land cover/land use	Area (km^2)	Area (%)	*C* factor
Settlement	345.1	11.67	0.2
Barren land with laterite outcrop	487	16.46	0.5
Degraded forest	520.62	17.6	0.05
Dense forest	180	6.08	0.01
Water body	140	4.73	0.0
Single crop	420	14.2	0.2
Double crop	863.92	29.21	0.31

Madramootoo (1998), UN-FAO (2001), Bakker et al. (2008), and Jordan et al. (2005).

10.3.1.5 Support Practice Factor (*P*)

Practice factor (*P*) means support practice factor (contour striping, contour tillage, and terracing system) assigned for land cover management and also modifies soil erosion rate controlled by surface runoff direction, flow pattern, and sediment transport (Wischmeier and Smith 1978; Fernandez-Raga et al. 2017; Jobin et al. 2018; Nasir and Selvakumar 2018). *P* factor is considered as effective soil loss parameter, incorporated with straight-row farming along the up and down slope (Renard 1997). On the other hand, *P* factor is inversely correlated with *C* factor because it takes some conservative management strategies (Mahala 2018). Generally, *P* value is mainly of two types, i.e., good conservation practice (0) and poor conversation practice (1). In this study area, *P* value is considered as 1 for the absence of support practice or management strategies (Fig. 10.2e).

10.3.2 SDR Parameter Estimation

SDR is determined by estimating *β* coefficient and travel time (t_i) following Ferro and Minacapilli (1995) method. They proposed function on the travel time of overland flow within a grid cell in each sub-basin (Fig. 10.3a–d).

$$\text{SDR} = \exp(-\beta t_i) \qquad (10.6)$$

where t_i means travel time (h) for cell I and *β* = basin-specific parameter.

10.3.2.1 Estimation of *β* Coefficient and Travel Time (t_i)

The travel time of runoff water from a grid cell to another grid cell in each watershed depends on flow distance and velocity along the flow path (USDA-SCS 1975; Bao et al. 1997). Cell to cell-based direction of the flow path is done by grid-based GIS analysis using the eight-direction pour point algorithm from segment delivery point of the nearest stream channel (Fernandez et al. 2003). Jain and Kothyari (2000) proposed

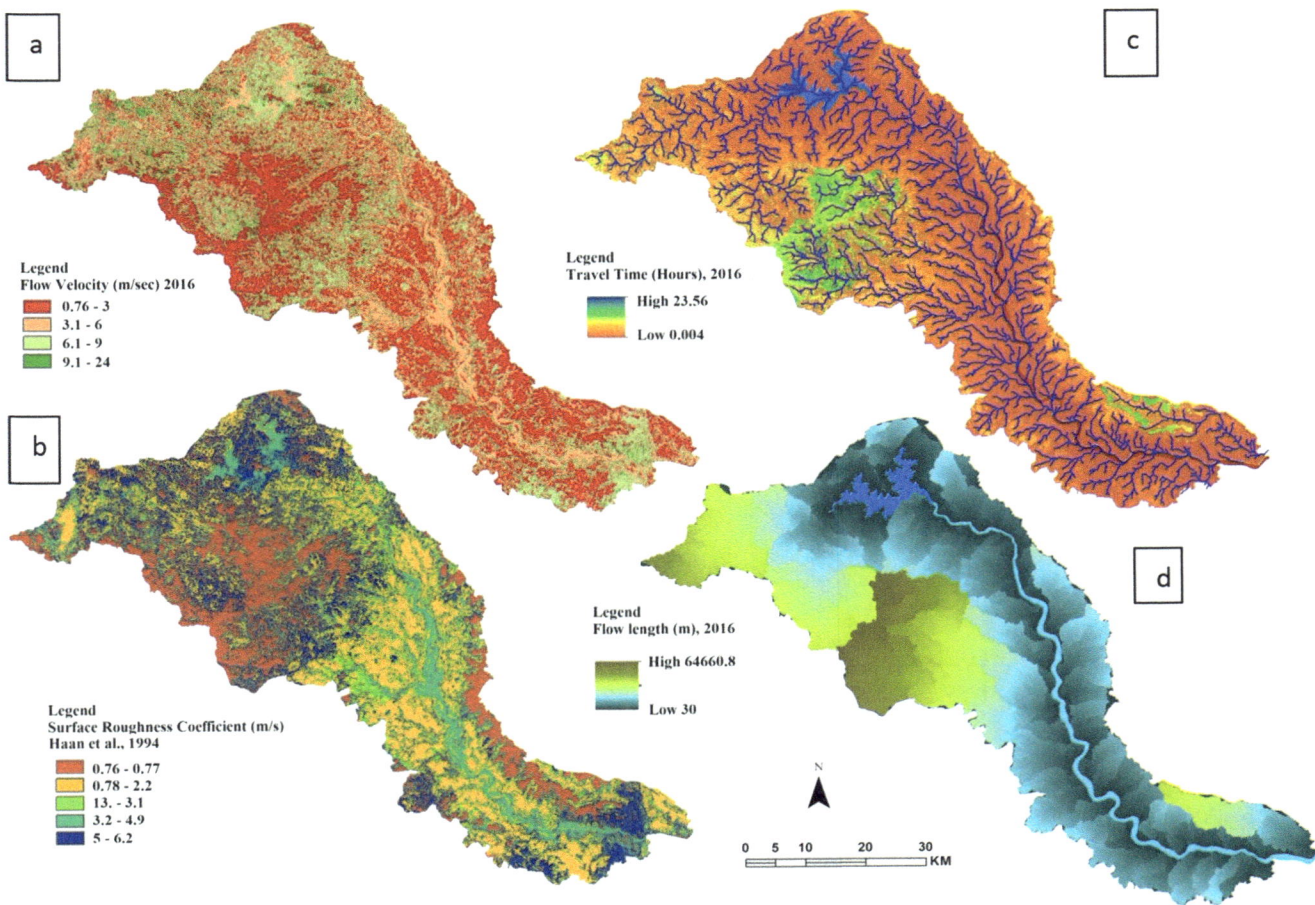

Fig. 10.3 SDR estimation parameters: (**a**) flow velocity; (**b**) surface roughness; (**c**) travel time; (**d**) flow length in the basin

the determination method of the travel time of a channel following Eq. (10.6.1):

$$t_i = \sum_{i=1}^{np} \left(\frac{L_i}{V_i} \right) \qquad (10.6.1)$$

where L_i = segment length denoting I under the flow path of m which is equivalent to the side length or oblique distance of a cell and V_i = cell-wise flow velocity (m/s).

Travel time was measured by taking a specific flow path in I cell time, needed to reach the neighbor channel traverses in Np cell as lies along the flow (Fig. 10.3c). US Soil Conservation Service proposed that flow velocity was determined by land surface slope direction and LULC patterns (SCS 1975).

10.3.2.2 Land Use and Land Cover (à Coefficient)

The value of overland flow and shallow concentrated flow was determined by coefficient number following LULC characteristics. à coefficient was computed from overland flow value following Haan et al. (1994). It includes forest (0.76 m/s), contour (1.56 m/s), strip cropping (1.56 m/s), short grass (2.13 m/s), straight-row cultivation (2.62 m/s),

and paved (6.19 m/s) surface, whereas shallow concentrated flow is associated with alluvial fans (3.08 m/s), grassed waterways (4.91 m/s), and small upland gullies (6.19 m/s) (Fig. 10.3b).

10.3.2.3 Slope Factor (S_i)

Slope plays a crucial role to determine the length of the flow path as well as travel time to reach the nearest channel. This factor can be estimated by the setting of minimum grid cell slope in a small value (Arnold et al. 1995). Slope value in this study considered only 0.3%. Slope factor is another important parameter to determine the à coefficient value.

10.3.2.4 Flow Velocity (V_i)

Flow velocity is classified into two classes, i.e., overland flow and shallow channel flow, according to their LULC patterns (Haan et al. 1994; USDA-SCS-TR-55 1975) (Fig. 10.3a).

$$V_i = d_i s_i^{1/2} \qquad (10.6.2)$$

where s_i = slope of cell i (m/m); d_i = a coefficient for cell i dependent on surface roughness (m/s). d_i is also correlated with $\dot{a}$ coefficient based on overland flow and shallow concentrated flow in different LULC.

10.3.2.5 Length of Segments (L_i)

It is derived from segment length (L_i) along the flow path (m) and is equal to side length or diagonal length of a cell depending on flow direction (Fig. 10.3d).

10.3.2.6 Basin-Specific Parameter (β)

Watershed morphological data can be obtained from basin-specific parameters such as β value using inverse modeling approach (Ferro and Minacapilli 1995). Sediment delivery ratio (SDR) is highly correlated with basin-specific parameters as β value in a watershed. The weighted mean of SDR value is computed following Eq. (10.6.3):

$$\text{SDR}_w = \sum_{i=l}^{N} \exp[\beta t_i] l_i^{0.5} s_i^2 a_i \Bigg/ \int_{i=l}^{N} l_i^{0.5} s_i^2 d_i \qquad (10.6.3)$$

where N = total number of cells over the watershed; l_i = length of cell i along the flow path; s_i = slope of the cell; and a_i = area of the cell.

Field data were collected for stream order, drainage density, soil type, and the size of the watershed and then developed their interrelationship for estimation of SDR_w in each sub-basin (Corbitt 1990). A negative relationship has been found between SDR and drainage area. Therefore, SDR values are validated by USDA (1972), Boyce (1975), and Vanoni's equation (1975) using the following equation:

$$\text{SDR}_w = k(aw)^{-c} \qquad (10.7)$$

where k and c represent dimensionless empirical coefficient values; aw means watershed area (m^2).

After getting SDR_w values, β coefficient can be predicted following Eq. (10.6.3). SY in each grid cell is estimated by an overlay of SDR and PMSE to identify sediment source region in a watershed area (Fernandez et al. 2003; Sahaar 2013).

$$\text{SY} = \text{SDR}_i \times A_i \times \alpha_i \qquad (10.8)$$

where SY means SY; SDR_i means sediment delivery ratio; A_i means mean annual soil loss; α_i means cell area.

Sub-basin-wise sediment source mainly depends on the spatial distribution of SY in the basin. Each sub-basin separately contributes sediment delivery along with a flow path through drainage outlets into the nearest channel segment. All of these steps were adopted to obtain the PMSE, SDR, and SY in each sub-basin in Fig. 10.4.

10.4 Results and Discussions

10.4.1 Estimation of PMSE Using RUSLE

Five dominant factors named as R, K, LS, C, and P were overlaid to produce a spatial distribution of PMSE (30 m cell size) using raster calculator under GIS platform (Fernandez-Raga et al. 2017; Dissanayake et al. 2018). RUSLE estimated PMSE between 0 and 350 t ha^{-1} year^{-1} in the entire basin. In terms of better understanding the real situation and breakdown of the proper interval, PMSE is classified into seven classes (Aiello et al. 2015; Ostovari et al. 2017). The seven assigning erosion classes are very low (0–50 t ha^{-1} year^{-1}), low (51–100 t ha^{-1} year^{-1}), low medium (101–150 t ha^{-1} year^{-1}), medium (151–200 t ha^{-1} year^{-1}), medium high (201–250 t ha^{-1} year^{-1}), high (251–300 t ha^{-1} year^{-1}), and very high (301–350 t ha^{-1} year^{-1}) (Fig. 10.5). According to RUSLE estimation, PMSE in this basin produces 300 t ha^{-1} year^{-1} following all erosion factors in Table 10.3. Most of the erosion susceptibility reaches medium and low medium class (44%), but high erosion class has a limited amount in terms of gross erosion rate (5%).

10.4.1.1 PMSE at Sub-basin Level

Mean annual soil loss indicates that all the requiring layers (R, LS, K, and C) fluctuate more in every sub-basin in accordance with basin area and land use pattern (Ganasri and Ramesh 2016; Mahala 2018). Maximum PMSE occurs in Khatra-1 sub-basin (93 t ha^{-1} year^{-1}) where huge sediments are drained into Mukutmanipur dam. Factors are responsible for large basin area (70,865 ha) with higher stream order and presence of extensive barren land (Tables 10.4 and 10.5). This sub-basin has no contribution to successive deposits along the river bed after completion of the dam in 1958. Another very high class of PMSE is also found in Lalgarh-4 sub-basin (65.58 t ha^{-1} year^{-1}) where maximum stream orders of Taraphini River lead to carrying out of huge sediment into main channel bed within maximum basin area (69,369 ha). On the other hand, lowest PMSE is observed in Dherua 2, Mohanpur 6, and Raipur 3 sub-basins for the presence of dense forest cover and low stream frequency under small basin area. Moderate soil loss ranges from 10.98 to 2.24 t ha^{-1} year^{-1} in the entire basin caused by lower stream order, moderate slope, and single crop practice in the rest of the sub-basins. Contrastingly, maximum PMSE takes place in the upper basin estimated as 75% (226 t ha^{-1} year^{-1}), whereas lower basin produces only 26% (74 t ha^{-1} year^{-1}). Therefore, it can be stated that the upper part faces very severe erosion with the positively dominant RUSLE parameters, but lower part faces very low erosion in terms of negatively dominant estimate parameters.

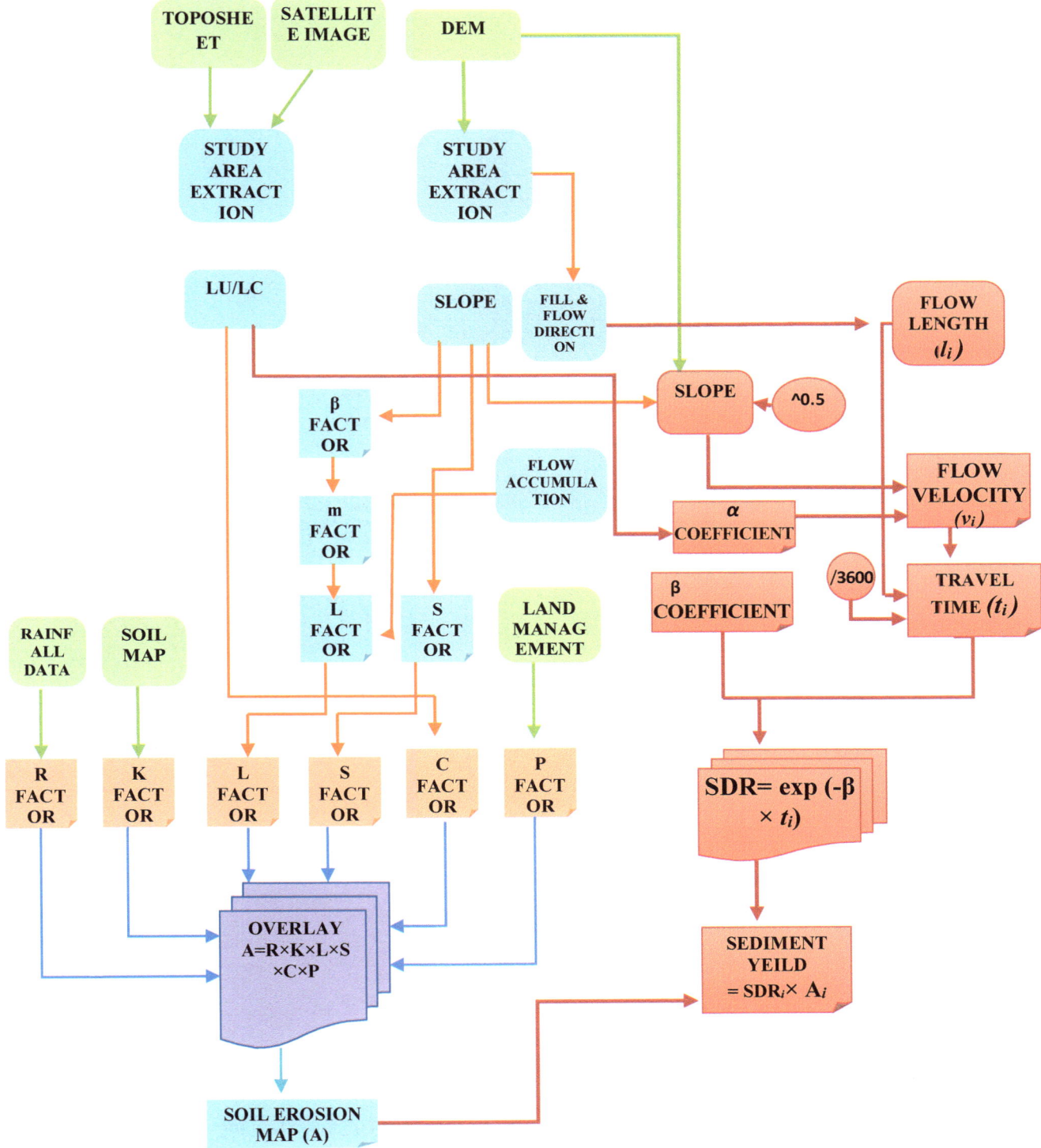

Fig. 10.4 Flowchart of RUSLE and SDR determination of SY

10.4.1.2 Justification of RUSLE Estimation

Due to the lack of previous records data and unavailability of sediment discharge station in this basin, RUSLE model is justified by application of Receiver operating characteristic (ROC) curve. According to Hirzel et al. (2006), the area under the curve (AUC) helps to assign model performance; meanwhile, if the value is 0.5, which means the model is not well-fitted, but the range of value lies between 0.7 and 0.8, it

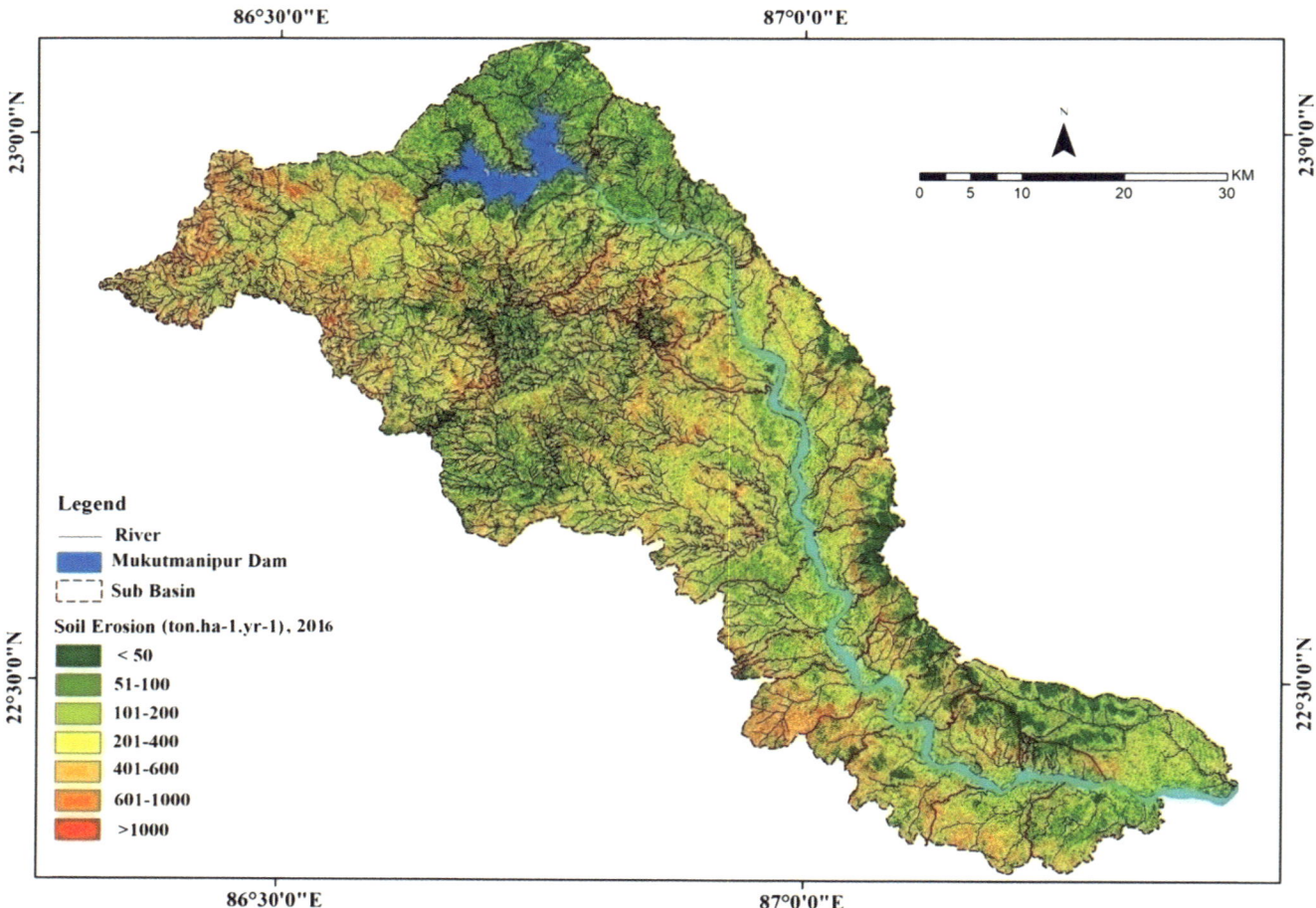

Fig. 10.5 Potential soil loss distribution in the entire Kangsabati basin

Table 10.3 Spatial distribution of PMSE class estimated by RUSLE

Erosion class	Range (t ha^{-1} year^{-1})	Area (km^2)	Percentage area (%)
Very low	<50	570	19.18
Low	51–100	438.35	14.75
Low medium	101–200	635.98	21.39
Medium	201–400	662.62	22.29
Medium high	401–600	297.63	10.01
High	601–1000	231.42	7.78
Very high	>1000	136.63	4.60

indicates successful performance. This curve is plotted from random field observations at 401 points in high erosion zone predicted by RUSLE (Fig. 10.6a). ROC is applied to the determined overall accuracy of the single threshold value in AUC. Figure 10.6b shows that AUC (0.834) is successfully predicted on PMSE susceptibility where true-positive rate (*Y*-axis) is plotted against false-positive rate (*X*-axis or 1-specificity). Moreover, maximum soil erosion prone sites corresponded with high and very high PMSE susceptibility zone.

10.4.2 Delineation of SDR Using SDR Model

Average SDR in all grid cells of the study area is found to be 0.20. Mean SDR is assigned at the outlet points from 27 sub-basins, which is good agreement following Exposed model of USDA (1972), Boyce (1975), and Vanoni's equation (1975). Spatial distribution of SDR ranges from 0.00014 to 1.0. Contrastingly, maximum SDR occurs in the upper part (2.823), but lower part gradually decreases SDR (2.48) as shown in Tables 10.6 and 10.7. Distribution

Table 10.4 Soil loss estimation with RUSLE parameters in upper basin

Sub-basin	Area (km²)	R	K	LS	C	P	Soil loss (t ha⁻¹ year⁻¹)
Khatra-1	708	4955	0.28	1.49	0.25	1	92.87
Khatra-2	113	4795	0.1	0.69	0.27	1	3.36
Khatra-3	96	4941	0.28	1.2	0.23	1	8.737
Raipur 1	90	5183	0.1	0.89	0.29	1	3.816
Raipur 2	75	5283	0.32	1.01	0.25	1	9.726
Raipur 3	32	5772	0.15	0.72	0.3	1	2.105
Raipur 4	63	5691	0.33	1.02	0.24	1	8.388
Raipur 5	29	6068	0.3	0.6	0.26	1	2.796
Lalgarh-1	48	6037	0.32	0.62	0.28	1	5.55
Lalgarh-2	71.39	6318	0.38	0.52	0.23	1	7.008
Lalgarh-3	71.86	6426	0.34	0.56	0.24	1	6.941
Lalgarh-4	693	5923	0.33	1.1	0.21	1	65.58
Lalgarh-5	51.6	6345	0.33	0.66	0.23	1	5.079
Lalgarh-6	43	6103	0.32	0.64	0.21	1	3.7

Table 10.5 Soil loss estimation with RUSLE parameters in lower basin

Sub-basin	Area (km²)	R	K	LS	C	P	Soil loss (t ha⁻¹ year⁻¹)
Dherua 1	69.5	6645	0.28	0.47	0.24	1	5.199
Dherua 2	31.83	6256	0.33	0.54	0.23	1	2.763
Dherua 3	32.55	7239	0.29	0.52	0.27	1	3.477
Dherua 4	60.17	8075	0.37	0.62	0.27	1	10.98
Dherua 5	41.98	6836	0.36	0.57	0.21	1	4.062
Mohanpur 1	34.46	7748	0.35	0.71	0.24	1	4.838
Mohanpur 2	35.6	7022	0.36	0.63	0.2	1	3.523
Mohanpur 3	77	7140	0.34	0.6	0.25	1	9.123
Mohanpur 4	58	6569	0.37	0.59	0.22	1	5.978
Mohanpur 5	67	6144	0.35	0.53	0.29	1	7.444
Mohanpur 6	23	5425	30.0	0.69	0.24	1	2.246
Mohanpur 7	73	5019	0.3	0.5	0.26	1	4.992
Kapastikri-1	159.75	5218	0.33	0.47	0.23	1	9.677

of SDR represents that sediment delivery depends on travel time as a function of flow distance and flow velocity. If the delivery point situated in the maximum distance from the channel bed, the travel time will be increased, and SDR declines (Sahaar 2013). Surface roughness and vegetation type are other important parameters to determine the flow velocity (Randhir et al. 2001; Jobin et al. 2018). Rough surface or dense forest leads to reduction in the flow velocity and extension of the flow path in the upper part, whereas plain surface is comparatively higher due to the presence of impervious and open land surfaces in this basin. The maximum ratio is concentrated in settlement areas (0.34), while its least amount is found in dense forest area (0.049) in the entire basin following C factor of RUSLE model (Fig. 10.7). Ratio result shows integrated potentiality of storing and transportation of eroded soil in 27 sub-basins (Dai and Tan 1996).

10.4.2.1 SDR at the Sub-basin Level

SDR becomes more fluctuating throughout the basin. A higher ratio is seen in Mohanpur 6 sub-basin (0.29) with the presence of least number of stream order (11), while lower values are concentrated in Khatra-1 and Lalgarh-4 sub-basin (0.11) due to a large number of stream order (1409, 1649). Mean annual soil loss relatively reversed relation with SDR in every sub-basin. Most of the SDR concentrated across the channel bed leads to sediment deposition throughout the Kangsabati channel (Fig. 10.8a, b). This situation is found in Mohanpur and Kapastikri delivery points where huge sedimentation occurs on channel bed. Higher drainage frequency in Khatra and Lalgarh sub-basins lead to resist the ratio due to the absence of no specific channel fall into the river bed (Tables 10.6 and 10.7). This result may be explained according to this decision that SDR tends to be more affected by drainage system than land use (Novotny and Chesters 1989; Sahaar 2013).

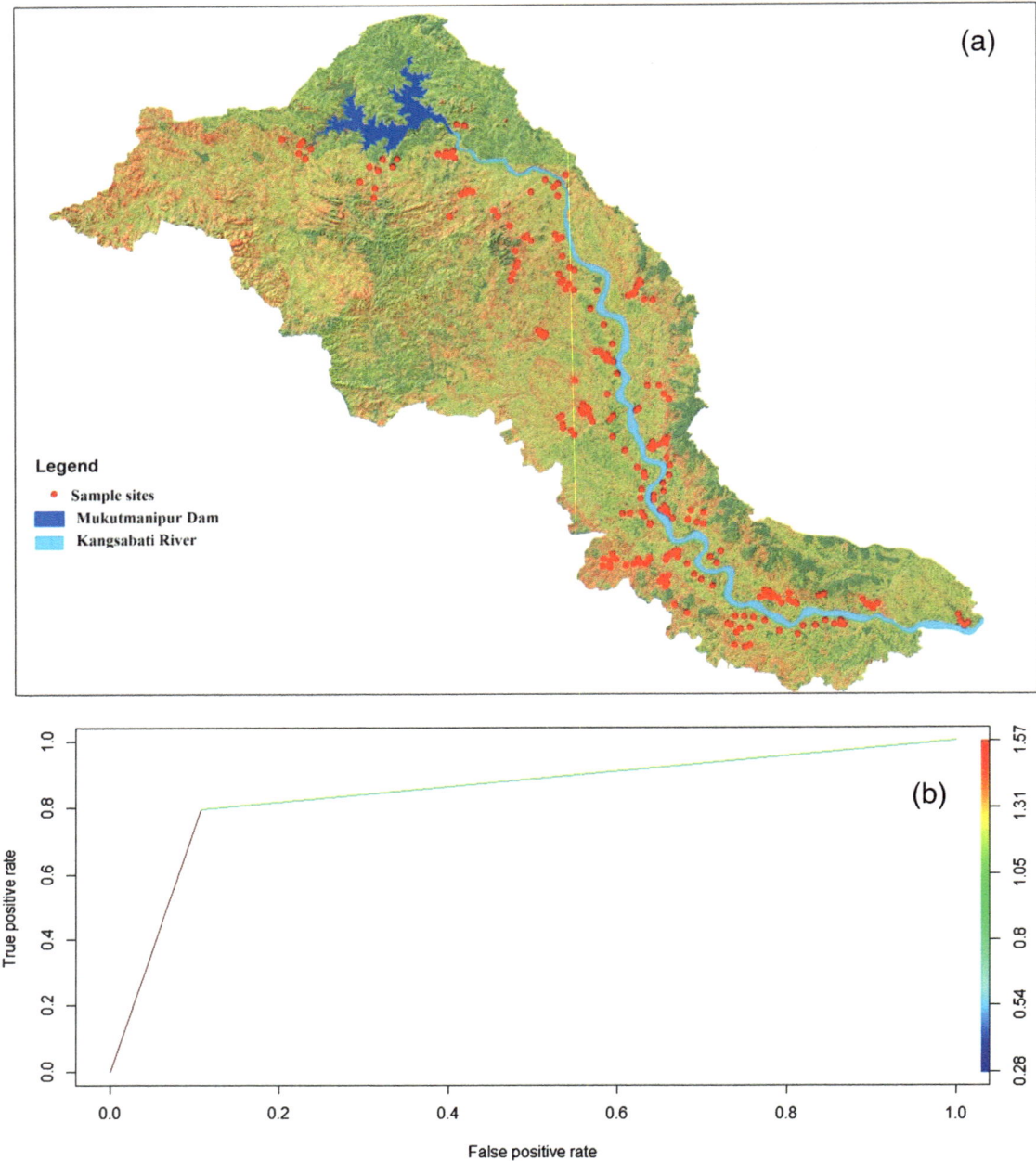

Fig. 10.6 (**a**) Random sample site collected from high susceptible erosion zone, (**b**) ROC curve of RUSLE estimated soil erosion

10.4.2.2 Validation of SDR Estimation

10.4.2.2.1 Drainage Area and SDR

Since the late 1950s, many researchers established the relationship between SDR and drainage area. This relationship depicts a similar trend; watersheds with larger drainage area have lower SDR because large areas have more chance to trap the sediment, whereas chance of sediment reaching along the stream becomes low (Julien 2010). Boyce (1975) established the relationship between SDR and drainage area by compiling and analyzing SY. The following power function is derived from the following equation:

Table 10.6 Estimation of sub-basin-wise SDR and SY in the upper basin

Sub-basin	Gr	L (km^2)	Stream number	First stream	T (M)	SDR	Mean SY (t ha^{-1} year^{-1})	SY (%)	Total SY (t ha^{-1} year^{-1})
Khatra-1	5.8	54.57	1409	1070	2.0	0.11	10.22	22.55	723,962
Khatra-2	9.99	19.32	36	29	0.48	0.21	0.706	1.558	8040
Khatra-3	14.6	17.59	153	115	0.88	0.26	2.272	5.015	21,949
Raipur 1	11.	16.98	99	69	0.54	0.24	0.916	2.022	8315
Raipur 2	9.64	15.29	159	128	0.89	0.24	2.334	5.153	17,615
Raipur 3	15.3	9.54	43	30	0.29	0.24	0.503	1.111	1655
Raipur 4	14.26	13.92	117	89	0.75	0.27	2.265	5.0	14,484
Raipur 5	9.97	8.93	9	8	0.6	0.22	0.61	1.346	1785
Lalgarh-1	7.42	11.95	18	14	0.5	0.22	1.221	2.696	5969
Lalgarh-2	5.76	14.81	13	9	1.08	0.16	1.093	2.413	7804
Lalgarh-3	5.01	14.87	17	12	1.15	0.13	0.902	1.992	6484
Lalgarh-4	4.06	53.91	1649	1227	3.8	0.11	7.214	15.93	500,427
Lalgarh-5	5.91	12.32	8	5	0.98	0.19	0.965	2.13	4980
Lalgarh-6	6.67	11.11	17	12	0.93	0.23	0.851	1.879	3660

Table 10.7 Estimation of sub-basin-wise SDR and SY in the lower basin

Sub-basin	Gr	L (km^2)	Stream number	First stream	T (M)	SDR	Mean SY (t ha^{-1} year^{-1})	SY (%)	Total YS (t ha^{-1} year^{-1})
Dherua 1	6.73	14.59	64	50	0.75	0.2	1.04	2.296	7227
Dherua 2	5.99	9.36	5	4	1.11	0.21	0.58	1.281	1847
Dherua 3	6.15	9.48	31	25	0.81	0.17	0.591	1.305	1924
Dherua 4	4.5	13.44	21	16	0.84	0.18	1.977	4.364	11,896
Dherua 5	7.4	10.96	26	16	1.07	0.25	1.016	2.242	4263
Mohanpur 1	5.8	9.79	2	2	0.82	0.16	0.774	1.709	2667
Mohanpur 2	9.8	9.98	10	7	0.76	0.23	0.81	1.789	2884
Mohanpur 3	6.37	15.47	11	7	1.03	0.17	1.551	3.424	11,953
Mohanpur 4	6.37	13.21	6	5	0.55	0.17	1.016	2.243	5930
Mohanpur 5	4.6	14.32	11	7	0.53	0.15	1.117	2.465	7510
Mohanpur 6	13.9	7.95	38	27	0.61	0.29	0.651	1.438	1554
Mohanpur 7	3.67	15.04	6	5	2.07	0.17	0.849	1.873	6219
Kapastikri-1	1.68	23.41	19	16	0.93	0.13	1.258	2.777	20,096

Gr gradient ratio, *L* basin length in km, *T* travel time (m)

$$\text{SDR} = 0.41 A_t^{-0.3} \qquad (10.9)$$

Following this rule, the correlation value is 0.093 as well as the value of 0.094 of Del Vanoni (1975) (SDR = 1.7935–0.14191 log *A*); 0.094 of Exposed model of USDA (1972) and also SDR observation in this study area predicts that this value as 0.040. Hence, it may be suggested that this model is validated with respect to Boyce, Del Vanoni, and Exposed model proposed by USDA (Fig. 10.9a, b).

10.4.2.2.2   Topographical Factors and SDR

Topographic characteristics of basin area play a vital role to control the SDR. Williams and Berndt (1972) stated that major channel slope has a significant role on SDR assess from this formula:

$$\text{Log(SDR)} = 0.627 \text{SLP}^{0.403} \qquad (10.10)$$

The value of slope factor (S_l) in this basin is justified by a positive linear relationship ($R^2 = 0.674$) with the SDR in Fig. 10.9c. Most of the SDR is identified along the steep channel bed than the gentle bed (Fernandez et al. 2003; Sahaar 2013).

10.4.3   Delineation of SY Zone

RUSLE and SDR estimated that the average annual SY of all cells in this basin is 1.98 t ha^{-1} year^{-1}. The total amount of mean SY is found to be 45 t ha^{-1} year^{-1} in the entire basin. Yield potentiality of the basin is divided into five classes, i.e., very low (0–1 t ha^{-1} year^{-1}), low (1–5 t ha^{-1} year^{-1}),

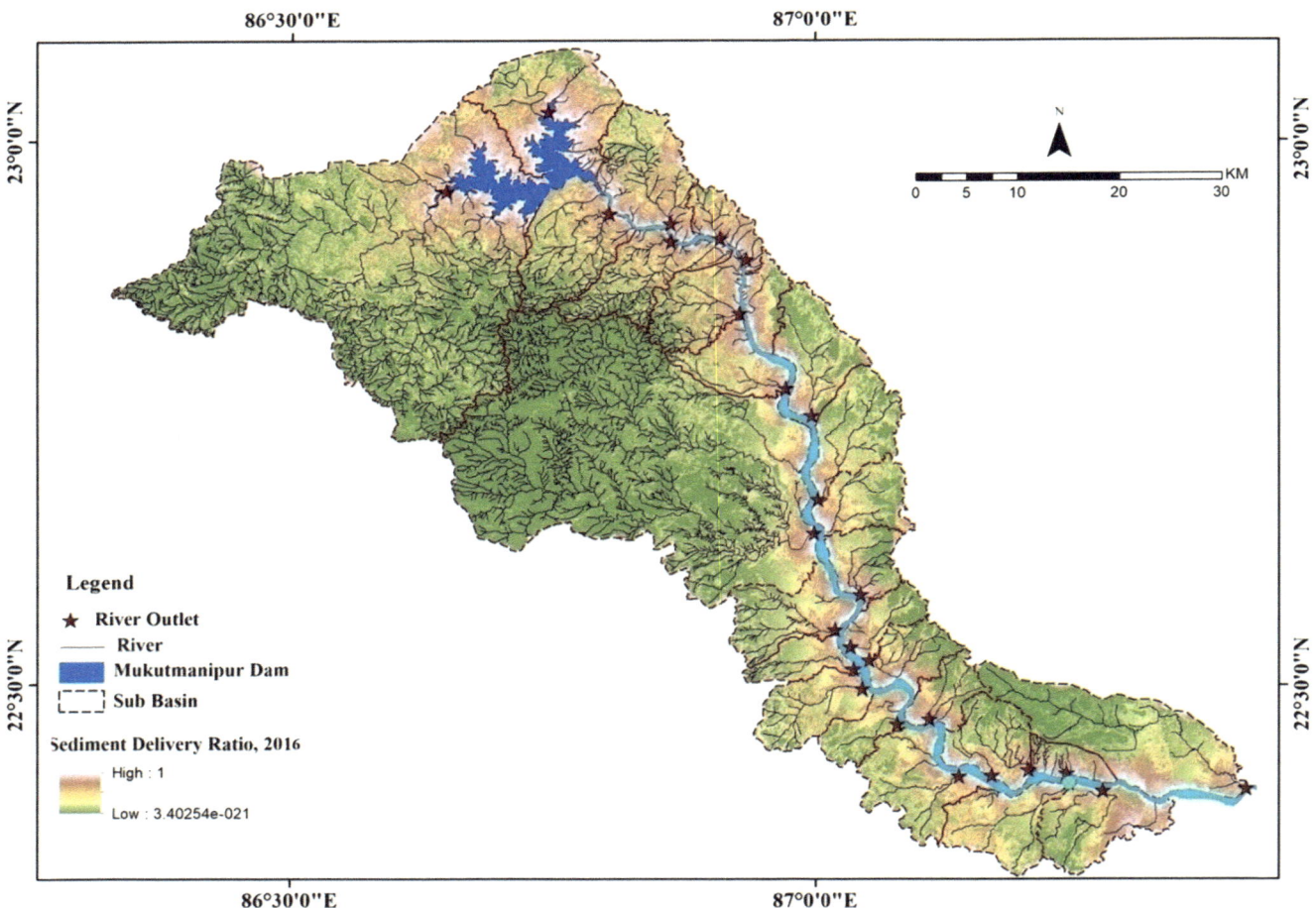

Fig. 10.7 SDR in the entire study area

medium (5–10 t ha^{-1} year^{-1}), high (11–15 t ha^{-1} year^{-1}), and very high (>15 t ha^{-1} year^{-1}). Figure 10.10 indicates that very low and low SY class become most dominant yield zone. Maximum SY deposition is in the upper basin (32 t ha^{-1} year^{-1}) with the presence of maximum number of first stream order (2817) than lower basin (13.23 t ha^{-1} year^{-1}) having low stream number (187). The result indicates that SY and PMSE classes are significantly positively correlated with the PMSE factors in the entire basin (Fernandez et al. 2003; Sahaar 2013).

10.4.3.1 SY at the Sub-basin Level

In terms of gross SY deposition, maximum SY values are high in Lalgarh-4 (500,427 t ha^{-1} year^{-1}) and Khatra-1 sub-basin (723,962 t ha^{-1} year^{-1}), while minimum SY found in Dherua 2 (1847 t ha^{-1} year^{-1}) and Mohanpur 5 sub-basin (1554 t ha^{-1} year^{-1}) in Tables 10.6 and 10.7. Therefore, Khatra-1 and Lalgarh-4 sub-basins (87%) deliver maximum sediment than the rest of the sub-basins (13%). Huge sediment falls down from Khatra-1 sub-basin, but

sediment transport is restricted into the lower segments due to cutoff of sediment supplies after dam construction (1958). Maximum supplies of sediment are derived from Bhairabbanki and Taraphini River and then cumulatively mixed with other drainage outlets (188,722 t ha^{-1} year^{-1}) along the Kangsabati River (Figs. 10.11 and 10.12).

It is pointed that RUSLE estimation parameters positively signify with SY in accordance to basin area and stream number at sub-basin level.

10.4.3.2 Relationship Between Ephemeral Channel (Gully Erosion) and SY

Source of sediment in every segment in Kangsabati River entirely depends on sediment delivery outlets. In Figure 10.12a, b shows that huge sediments (732,003 t ha^{-1} year^{-1}) were silted in Mukutmanipur dam, supplies from delivery outlets of Khatra-1 and 2 but Khatra-3 play dominant role to supplies sediment (21,949 t ha^{-1} year^{-1}) toward downstream. All outlets in Raipur segment deliver sediment near about 43,856 t ha^{-1} year^{-1} toward the

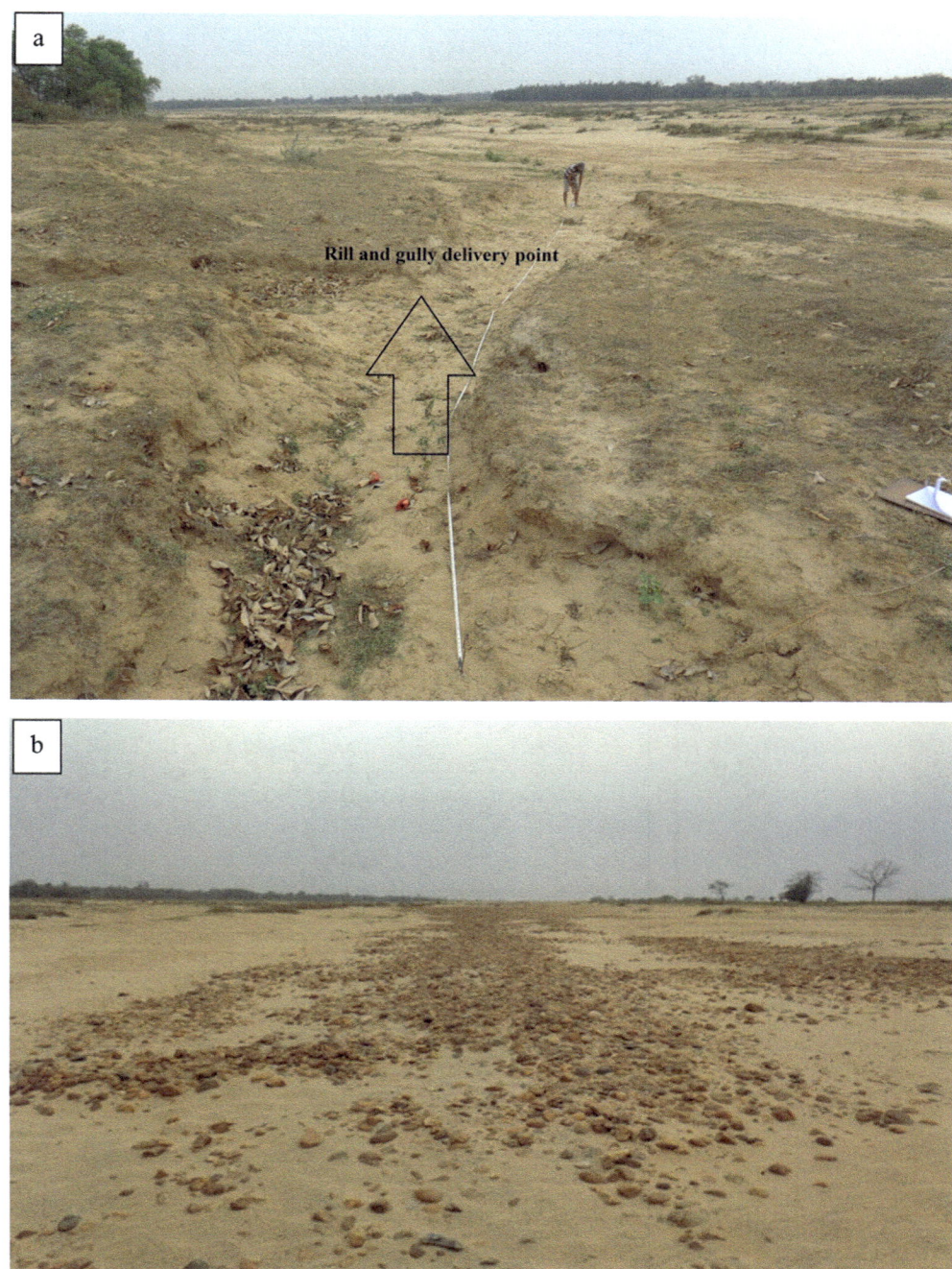

Fig. 10.8 Field photography of SDR: (**a**) ephemeral gully channel; (**b**) delivery outlet across in upper basin

downstream of Kangsabati channel, whereas Taraphini and Bhairabbanki River carried out massive sediment (529,325 t ha^{-1} year^{-1}) from all the sub-basins in Lalgarh segment. This segment reaches highest sediment deposition than others as well as supplies maximum sediment in the entire downstream. Outlet points in Dherua segment lead to sandchar deposition across the channel bank, whereas all the delivery outlets in Mohanpur segment help to generate huge sedimentation from left to right bank of Kangsabati River (38,721 t ha^{-1} year^{-1}) where channel gradient drastically changes due to bridge and dam construction. Massive sedimentation becomes prime source of SY in the entire Mohanpur segment. Channel braiding and sandchar beings are the most common features in this segment. Only single

Fig. 10.9 Validation of SDR following model results from 27 delivery outlets: (**a**) relationship between drainage area and SDR; (**b**) validation against the various models; (**c**) gradient ratio (Gr) and SDR

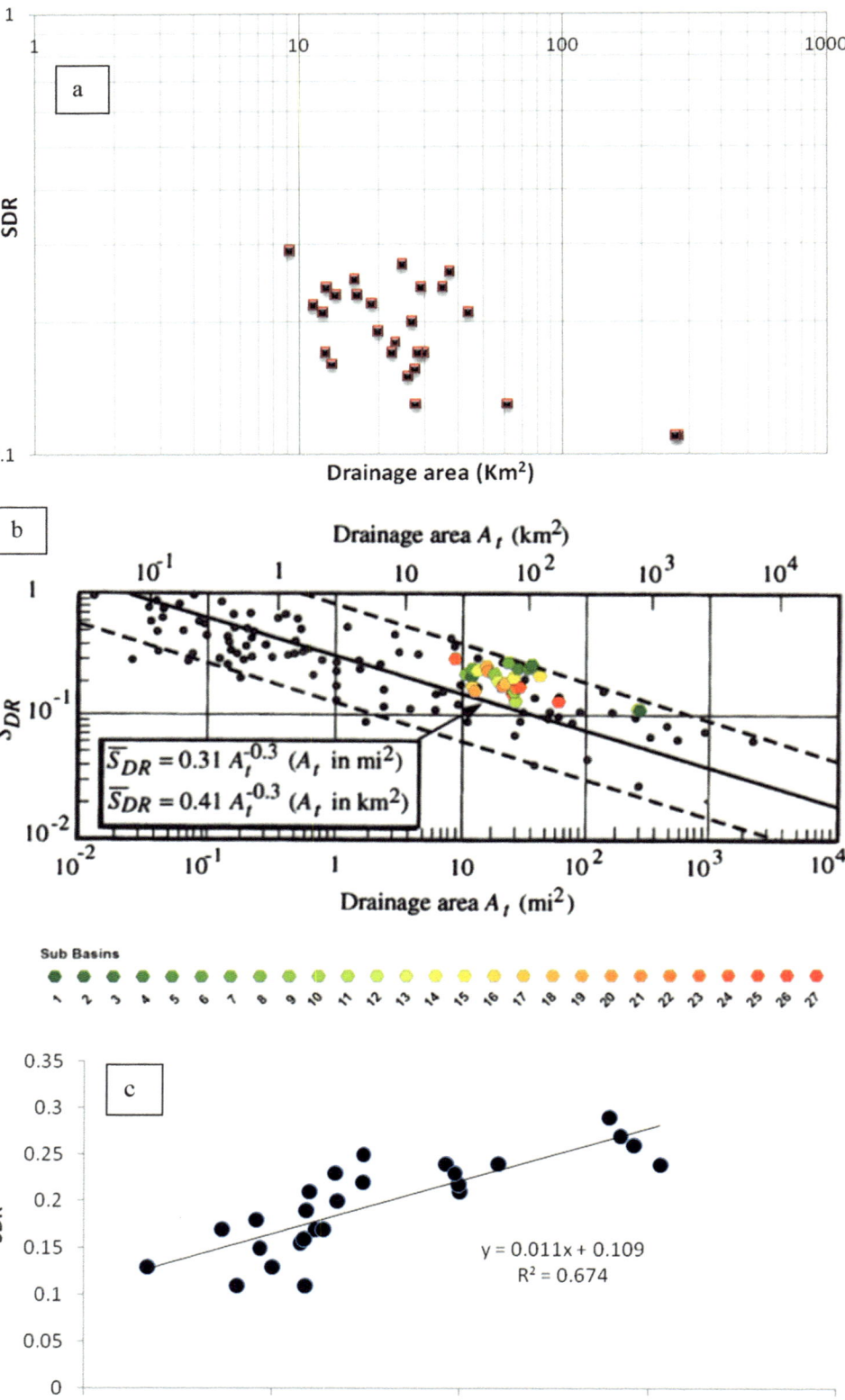

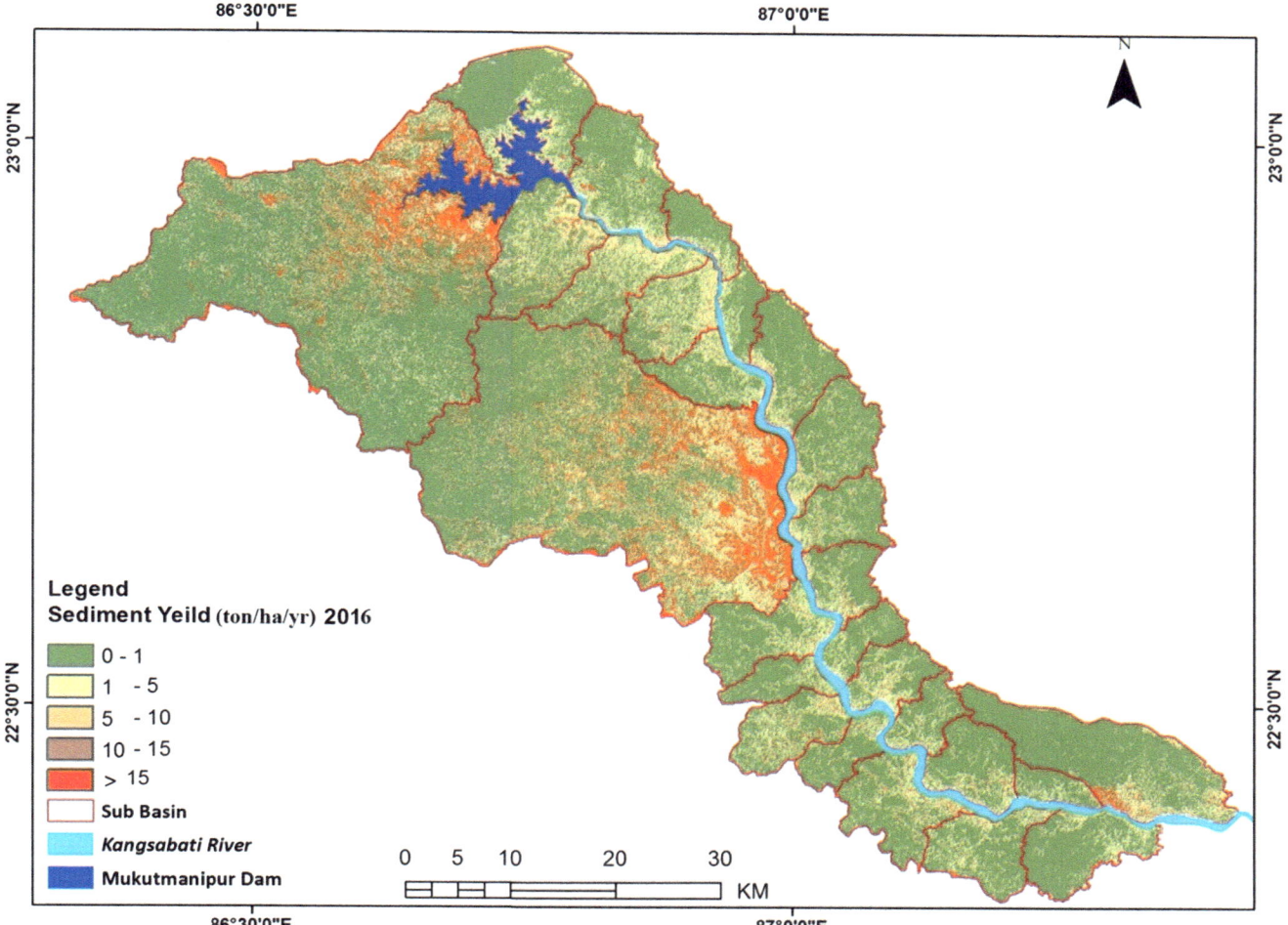

Fig. 10.10 Sub-basin-wise SY distribution in the study area

outlet point supplies sediment in Kapastikri segment (20,096 t ha^{-1} year^{-1}). As a result, channel becomes narrow and deepening in this segment.

In spite of model prediction, some crucial links were assessed like number of first stream order, basin length, and PMSE parameters on SY deposition throughout the basin. Figure 10.13 denotes that first stream number ($R^2 = 0.895$) and basin length ($R^2 = 0.88$) of gully channel positively correlated with sediment deposition enhancing PMSE rate ($R^2 = 0.975$). Possible loss factors affecting SY deposition, mainly R and P factors, have relatively uniform values across the whole sub-basin, and their impacts are similar in the entire basin (Mahala 2018). Three dominant factors, namely, LS, K, and C, are considerable on PMSE as well as SY that means LS, K, and C factors play a crucial role on sediment deposition in accordance with first stream number and basin length (Sahaar 2013). Contrastingly, SDR denotes that most

of the sediment delivery occurs in the upper basin where gradient ratio positively signifies with travel time in each gully channel.

Therefore, SDR gradually reduces by declining gradient ratio in the lower basin. As a result, SY in upper basin deposited 94%, while SY in the lower basin deposited only 6%. In this context, high yield zone is concentrated in Khatra-1 and Lalgarh-4 sub-basin, while low SY is deposited in all sub-basins of Mohanpur and Kapastikri (Fig. 10.12a, b). In this case, specific study denotes that Mukutmanipur dam plays a vital role to cut off the sediment transport from the Kansai and Kumari River, whereas bed sediment transport in Kangsabati River entirely depends on delivery outlets in each sub-basin. Therefore, it is a point that basin length, first stream number, and gradient ratio of gully channel play a dominant role on PMSE, SDR, and SY throughout the basin.

Fig. 10.11 (**a**) SDR; (**b**) SY deposition along river outlet bed

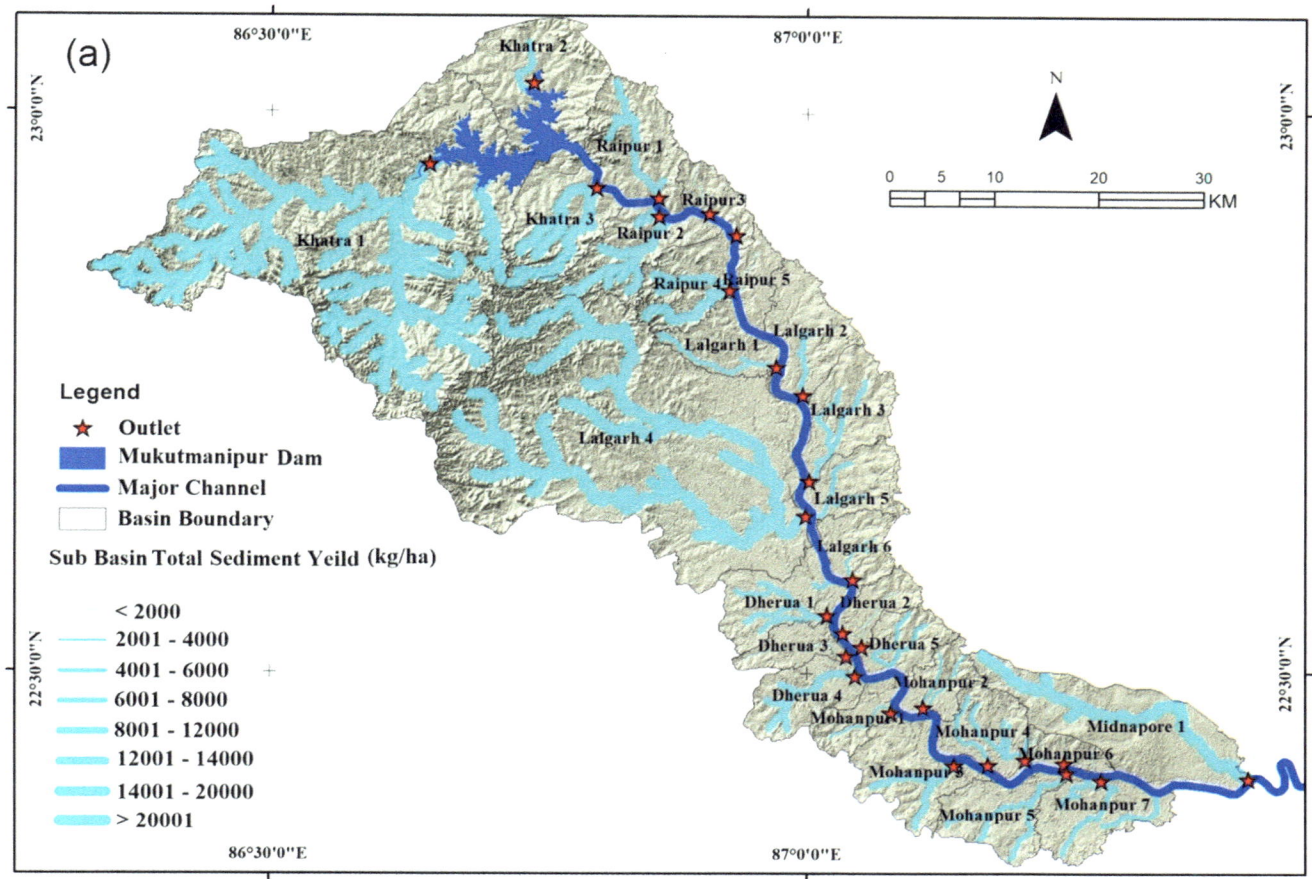

Fig. 10.12 Relation between stream number and SY: (**a**) SY deposition in respect to stream number in 27 sub-basins. (**b**) Source of sediment supplied from outlet points of Mukutmanipur, Khatra, Raipur, Lalgarh, Dherua, Mohanpur, and Kapastikri segments

10.5 Conclusion

RUSLE and SDR are effective tools to detect the spatial distribution of annual soil loss and delivery ratio in 27 sub-basin using the integrated GIS technique. Couple models predict sediment transports along the channel bed depending on rill and inter-rill erosion as well as number of delivery outlets near Kangsabati River. Several factors like R, K, LS, C, and P estimated seven different potential PMSE classes in this basin, namely, very low, low, medium low, medium, medium high, high, and very high erosion class, ranging from 50 to 1000 t ha^{-1} year^{-1} validated by AUC (0.83) in ROC, whereas travel time (t_i) and basin-specific parameter (β coefficient) determined SDR range from 0.00014 to 1 following model result like Boyce, Del Vanoni, and Exposed model (USDA). This integrated approach helps to estimate SY throughout the basin ranging from 0 to 15 t ha^{-1} year^{-1} with five different classes. In terms of gross sediment deposition, SY positively significance with

PMSE enhancing of first stream number, basin length and effected RUSLE parameters like LS, C whereas SDR determined the SY deposition near delivery outlets in accordance with travel time and basin parameter which controlled by gradient ratio (Gr). Therefore, massive gross SY (1,327,136 t ha^{-1} year^{-1}) occurs of in upper basin following maximum amount of significance parameters like stream number (2817), basin length (275 km), LS (11.72), C (3.49) and gradient ratio (0.13) whereas lowest SY deposition (85,976 t ha^{-1} year^{-1}) occurs in lower basin with the following of minimum amount of significance parameters like stream number (187), basin length (167 km), LS (7.44), C (3.15) and gradient ratio (0.08) respectively. On the other hand, negative significant parameters like K gradually reduce the erosion rate or SY that means maximum K factor (4.33) reduces SY in lower basin and minimum K factor (3.88) increases SY in the upper basin. However, in spite of the contribution of RUSLE and SDR parameters, gradient ratio, first stream number, basin length governs major role to determine SY, which controlled by PMSE and SDR. This study

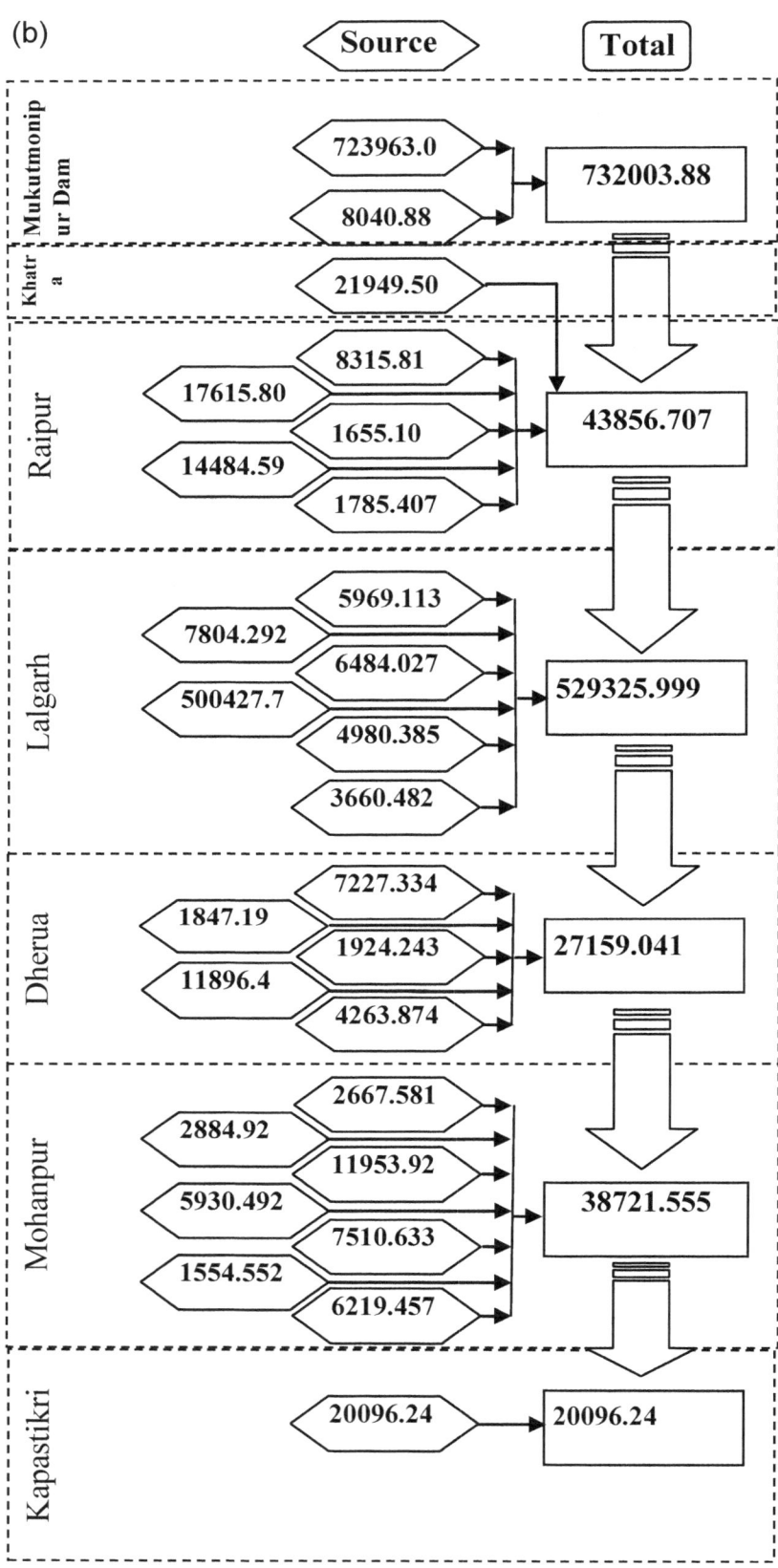

Fig. 10.12 (continued)

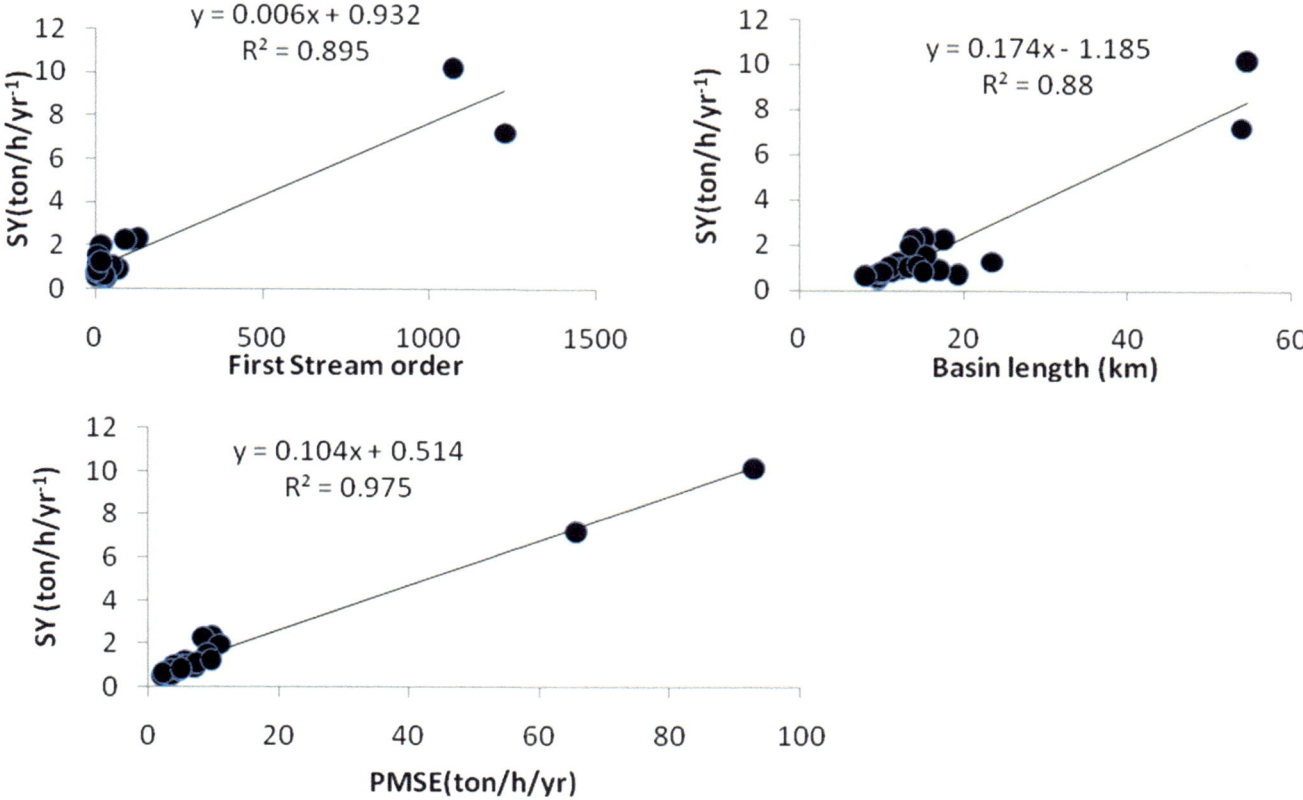

Fig. 10.13 Dominant factors on SY: (**a**) first stream order, (**b**) basin length, (**c**) PMSE

also demonstrated several considerable geo-environmental parameters regarding the sediment deposition in plateau fringe basin under tropical monsoon climate.

Acknowledgment This manuscript is the part of my (Raj Kumar Bhattacharya) Ph.D. thesis (unpublished). We are thankful to Survey of India (SOI) and Irrigation Office of Paschim Medinipur and Bankura for providing the required data. Authors are also grateful to the anonymous reviewers for their valuable comments and suggestions to improve the quality of this chapter.

References

Aiello A, Adamo M, Canora F (2015) Remote sensing and GIS to assess soil erosion with RUSLE3D and USPED at river basin scale in southern Italy. Catena, 131, 174–185

Arnold J, G Williams J R & Maidment D R (1995) Continuous-time water and sediment-routing model for large basins. Journal of Hydraulic engineering, 121(2), 171–183

Bagherzadeh A, Daneshvar M R M (2013) Evaluation of sediment yield and soil loss by the MPSIAC model using GIS at Golestan watershed, northeast of Iran. Arabian Journal of Geosciences, 6(9), 3349–3362

Bakker M M, Govers G, van Doorn A, Quetier F, Chouvardas D, Rounsevell M (2008) The response of soil erosion and sediment export to land-use change in four areas of Europe: the importance of landscape pattern. Geomorphology, 98(3–4), 213–226

Bao J, Maidment D, Olivera F (1997) Using GIS for hydrologic data-processing and modeling in Texas. Center for Research in Water Resources, Texas

Bhattarai R, Dutta D (2006) Estimation of soil erosion and sediment yield using GIS at catchment scale. Water Resources Management, 21(10), 1635–1647

Boyce R C (1975) Sediment routing with sediment delivery ratios. Present and prospective technology for predicting sediment yields and sources, 61–65

Corbitt R A (1990) Standard handbook of environmental engineering

Cox C, Madramootoo C (1998) Application of geographic information systems in watershed management planning in St. Lucia. Computers and Electronics in Agriculture, 20(3), 229–250

Dabral P P, Baithuri N, Pandey A (2008) Soil erosion assessment in a hilly catchment of North Eastern India using USLE, GIS and remote sensing. Water Resources Management, 22(12), 1783–1798

Dai D and Y Tan (1996) Soil erosion and sediment yield in the Upper Yangtze River Basin. Pp. 191–203. *111*: Proceedings of the Exeter Symposium Erosion and Sediment Yield Global and Regional Perspectives, International Association of Hydrological Sciences Publication 236

Desmet PJ1 and G Govers (1996) A GIS procedure for automatically calculating the USLE LS factor on topographically complex landscape units. Journal of Soil and Water Conservation 51 (5):427–433

Dissanayake D M S L B, Morimoto T, Ranagalage M (2018) Accessing the soil erosion rate based on RUSLE model for sustainable land use management: A case study of the Kotmale watershed, Sri Lanka. Modeling Earth Systems and Environment, 1–16

Fernandez C, Wu J Q, McCool D K, Stöckle C O (2003) Estimating water erosion and sediment yield with GIS, RUSLE, and SEDD. Journal of Soil and Water Conservation, 58(3), 128–136

Fernandez-Raga M, Palencia C, Keesstra S, Jordan A, Fraile R, Angulo-Martinez M & Cerda A (2017) Splash erosion: a review with unanswered questions. *Earth-Science Reviews, 171*, 463–477

Ferro V, Minacapilli M (1995) Sediment delivery processes at basin scale. Hydrological Sciences Journal, 40(6), 703–717

Ferro V, Porto P (2000) Sediment delivery distributed (SEDD) model. Journal of hydrologic engineering, 5(4), 411–422

Foster GR and WH Wischmeier (1974) Evaluating irregular slopes for soil loss prediction. Transactions of the American Society of Agricultural Engineers 17(2):305–309

Ganasri B P, Ramesh H (2016) Assessment of soil erosion by RUSLE model using remote sensing and GIS-A case study of Nethravathi Basin. Geoscience Frontiers, 7(6), 953–961

Ghosh D (2015) Mapping and monitoring of the impact of gully erosion in the district of Medinipur (West), West Bengal, India. International Journal of Novel Research in Humanity and. Soc Sci 2(4):73–89

Haan C T, Barfield B J, Hayes J C (1994) Design hydrology and sedimentology for small catchments. Elsevier

Hirzel A H, Le Lay G, Helfer V, Randin C, Guisan A (2006) Evaluating the ability of habitat suitability models to predict species presences; Ecol Model, 199(2), 142–152

Jain M K, Kothyari U C (2000) Estimation of soil erosion and sediment yield using GIS. Hydrological Sciences Journal, 45(5), 771–786

Jain SK, Kumar S, Varghese J (2001) Estimation of soil erosion for a Himalayan watershed using GIS technique. Water Resources Management, 15(1), pp. 41–54

Jordan G, Van Rompaey A, Szilassi P, Csillag G, Mannaerts C, Woldai T (2005) Historical land use changes and their impact on sediment fluxes in the Balaton basin (Hungary). Agriculture, ecosystems & environment, 108(2), 119–133

Julien PY (2010) Erosion and sedimentation. Cambridge University Press

Kothyari U C, Jain S K (1997) Sediment yield estimation using GIS. *Hydrol. Sci. J.* 42(6), 833–843

Lim K J, Sagong M, Engel B A, Tang Z, Choi J, Kim K S (2005) GIS-based sediment assessment tool. Catena, 64(1), 61–80

Lu H, Moran C J, Prosser I P (2006) Modelling sediment delivery ratio over the Murray Darling Basin. Environmental Modelling & Software, 21(9), 1297–1308

Magesh N S, Chandrasekar N (2016) Assessment of soil erosion and sediment yield in the Tamiraparani sub-basin, South India, using an automated RUSLE-SY model. Environmental Earth Sciences, 75 (16), 1208

Mahala A (2018) Soil erosion estimation using RUSLE and GIS techniques—a study of a plateau fringe region of tropical environment; Arab J Geosci, 11(13), 335

Markhi A, Laftouhi N, Grusson Y, Soulaimani A (2019) Assessment of potential soil erosion and sediment yield in the semi-arid N' fis basin (High Atlas, Morocco) using the SWAT model. *Acta Geophysica, 67* (1), 263–272

Markose V J, Jayappa K S (2016) Soil loss estimation and prioritization of sub-watersheds of Kali River basin, Karnataka, India, using RUSLE and GIS. Environmental monitoring and assessment, 188 (4), 225.

McCool DK, LC Brown, GR Foster, CK Mutchler and LD Meyer (1987) Revised slope steepness factor for the Universal Soil Loss Equation. Transactions of the American Society of Agricultural Engineers 30(5): 1387–1396

Mhangara P, Kakembo V, Lim K J (2012) Soil erosion risk assessment of the Keiskamma catchment, South Africa using GIS and remote sensing. Environmental Earth Sciences, 65(7), 2087–2102

Mitasova H, Barton M, Ullah I, Hofierka J, Harmon R S (2013) GIS-based soil erosion modeling. In Treatise on Geomorphology. Elsevier Inc

Mittal N, Mishra A, Singh R, Bhave A G, van der Valk M (2014) Flow regime alteration due to anthropogenic and climatic changes in the Kangsabati River, India. Ecohydrology & Hydrobiology, 14(3), 182–191

Mondal S (2012) Remote sensing and GIS based ground water potential mapping of Kangshabati irrigation command area, West Bengal. Geography & Natural Disasters 1(1):1–8

Mukhopadhyay S (1992) Terrain analysis of river basin, Vora publication, New Delhi

Nasir N, Selvakumar R (2018) Influence of land use changes on spatial erosion pattern, a time series analysis using RUSLE and GIS: the cases of Ambuliyar sub-basin, India. Acta Geophys, 66(5), 1121–1130

Novotny V, Chesters G (1989) Delivery of sediment and pollutants from nonpoint sources: a water quality perspective. Journal of Soil and Water Conservation, 44(6), 568–576

Ostovari Y, Ghorbani-Dashtaki S, Bahrami H A, Naderi M, Dematte J A M (2017) Soil loss prediction by an integrated system using RUSLE, GIS and remote sensing in semi-arid region. Geoderma Regional, 11, 28–36

Randhir T O, O'Connor R, Penner P R, Goodwin D W (2001) A watershed-based land prioritization model for water supply protection. Forest ecology and management, 143(1–3), 47–56

Renard KG (1997) Predicting soil erosion by water: a guide to conservation planning with the revised universal soil loss equation (RUSLE)

Renard KG, Freimund JR (1994) Using monthly precipitation data to estimate the R-factor in the revised USLE. Journal of hydrology, 157 (1–4), pp. 287–306

Richards KS (1993) Sediment delivery and drainage network. Channel network hydrology, pp. 221–254

Sahaar AS (2013) Erosion mapping and sediment yield of the Kabul river basin, Afghanistan (Doctoral dissertation, Colorado State University)

Samad N, Chauhdry M H, Ashraf M, Saleem M, Hamid Q, Babar U, Farid M S (2016) Sediment yield assessment and identification of check dam sites for Rawal Dam catchment. Arabian Journal of Geosciences, 9(6), 466

Shit P K, Nandi A S, Bhunia G S (2015) Soil erosion risk mapping using RUSLE model on Jhargram sub-division at West Bengal in India. Modeling Earth Systems and Environment, 1(3), 28

Thomas J, Joseph S, Thrivikramji K P (2018) Assessment of soil erosion in a monsoon-dominated mountain river basin in India using RUSLE-SDR and AHP. Hydrological Sciences Journal, 63(4), 542–560

UN-FAO (2001) Strategic Environmental Assessment: An Assessment of the Impact of Cassava Production and Processing on the Environment and Biodiversity. Volume 5.Rome. Italy

USDA (1972) National Engineering Handbook. Soil Conservation Service, USDepartment Agriculture, Washington, DC, Section 3

US Department of Agriculture-Soil Conservation Service (USDA-SCS) (1975) Urban hydrology for small watersheds. Technical Release No. 55. U.S. Department of Agriculture Soil Conservation Service, Washington, D.C

Vanoni VA (1975) Sedimentation engineering, ASCE manuals and reports on engineering practice—No. 54. American Society of Civil Engineers, New York

Williams JR and Berndt HD (1972) Sediment yield computed with universal equation. Journal of the Hydraulics Division, 98(Hy 12)

Wischmeier WH (1960) Cropping-management factor evaluations for a universal soil-loss equation. Soil Science Society of America Journal, 24(4), pp. 322–326

Wischmeier WH, Smith DD (1978) Predicting rainfall erosion losses-a guide to conservation planning

Raj Kumar Bhattacharya is currently working as a government approved part time teacher at Sukumar Sengupta Mahavidyalaya, Paschim Mednipur in West Bengal. He obtained his MSc and PhD from Department of Geography and Environment Management in Vidyasagar University (India). His major research interests are Environmental aspects of resource and hazard related Issues, Fluvial Geomorphology, Hydroecology, and Sedimentlogy. His research findings have been published in leading national and international journals, book chapters, and conference proceedings.

Nilanjana Das Chatterjee is an Associate Professor at the Department of Geography and Environment Management, Vidyasagar University in India and an expert in the field of environmental aspects of resource and hazard related Issues, Animal migration and its consequences in forest fringe areas of Bankura District, folk, and tribal culture, and geographical attributes of handloom and cottage industries. She received her PhD from Burdwan University in 2009. She has nearly 20 years of experience in research and education and has authorized more than 30 publications along with honoured many prestigious awards.

Kousik Das is currently working as full time research scholar at the Department of Geography and Environment Management, Vidyasagar University in India, where he completed his MSc in 2015. His research interests lie in the area of Urban Dynamics and its impact on public health and remote sensing and GIS integrated modelling approaches. He has contributed more than five publications and books.

Assessment of Potential Land Degradation in Akarsa Watershed, West Bengal, Using GIS and Multi-influencing Factor Technique

Ujjal Senapati and Tapan Kumar Das

Abstract

Land degradation and gully erosion are the very common and acute geo-environmental problems at the western part of West Bengal. The Akarsa watershed, which is a part of the Dwarakeswar river basin and also a part of Chotanagpur plateau, is highly vulnerable to land degradation. Here, rill and gully erosions are key functions of the land degradation process. In this chapter, delineation of potential land degradation zone (PLDZ) has been mapped by using remote sensing data and geographical information system (GIS) based on multi-influencing factor (MIF) technique for Akarsa watershed in West Bengal. It has been accomplished by integrating and analyzing different thematic maps. The degraded areas were delineated using visual interpretation techniques. Rankings and weights were assigned to each influencing factor for calculating statistically by the multi-influencing factor (MIF) technique. Finally, delineation of the potential land degradation zone (PLDZ) map is executed and classified into five degradation zones, viz., very low 13.74% (47.54 km^2), low 27.52% (95.21 km^2), moderate 38.15% (132.55 km^2), high 16.18% (56.25 km^2), and very high 4.41% (15.24 km^2). Then, the receiver operating characteristic (ROC) curve is applied for validation of the methodology used in this work. The result of AUC (area under the curve) is very good indicating an accuracy of (0.828) 82%. The outcome of this PLDZ can be helpful in land conservation planning and strategy formulation for management in the Akarsa watershed.

Keywords

Potential land degradation zone (PLDZ) · Gully erosion · GIS · ROC

11.1 Introduction

Land bears soil which is a ubiquitous valuable natural resource for human civilization and environment as well as agricultural production. So, land resources are catering to the subsistence of mankind (Minami 2009; Keesstra et al. 2016; Molla and Sisheber 2017). Land degradation is a great universal threat to food security and environment, leading to various problems, mainly, sediment deposition in reservoirs, water logging, soil fertility reduction, salinization, deforestation, and water pollution (Zhao et al. 2013; Cerda et al. 2016; Keesstra et al. 2016; Mahala 2017; Arabameri et al. 2018a). It is a procedure of converting suitable land to unsuitable and unfertile land for humankind, as well as destroying the soil ecosystem, decreasing productivity, and deteriorating the characteristic of usable land (Hill et al. 2005; Naseer and Pandey 2018). Land use/land cover (LULC) change analysis is a major driving factor for the study of land degradation done by several researchers (De Souza et al. 2013; Biro et al. 2013; Pallavicini et al. 2015; Mohawesh et al. 2015), and it has other dimensions of the physical environment, including rocks, soil, ecology, biodiversity, soil moisture, vegetation, geology, and man-made infrastructure (Chatterjee et al. 2014; Arabameri et al. 2018b). LULC analysis is necessary for the development of planning strategies, land management practice, and biodiversity conservation (Ohta and Nakagoshi 2011, Leh et al. 2013). Land degradation is related to soil and gully erosion depending on various geo-environmental factors. Several researchers had used different geo-environmental factors in their study. Physical factors like rainfall, geology, geomorphology, and soil properties in addition to socioeconomic factors like poor standard of

U. Senapati
Department of Geography, Cooch Behar Panchanan Barma University, Cooch Behar, West Bengal, India

T. K. Das (✉)
Department of Geography, Cooch Behar College, Cooch Behar, West Bengal, India

© Springer Nature Switzerland AG 2020
P. K. Shit et al. (eds.), *Gully Erosion Studies from India and Surrounding Regions*, Advances in Science, Technology & Innovation,
https://doi.org/10.1007/978-3-030-23243-6_11

living, poverty, poor health, land fragmentation, low education level, and population density are also studied (Sheng 1989; Bahrawi et al. 2016; Bera 2017). Intense plowing, excessive grazing, lack of vegetation, shifting cultivation, and deforestation are the causes of land degradation in the underdeveloped countries (Ligonja and Shrestha 2015; Arabameri et al. 2018a), while in the other countries, which are developing, the physical factors of land degradation are stimulated by various socioeconomic factors (Feoli et al. 2002; Cerda et al. 2009). Soil erosion including rill and gully formation is the most related form of land degradation (Rahmati et al. 2016). Due to intense land degradation, about 205 million people and 1.9 billion ha of land are vulnerable worldwide (Low 2013), while another source shows 1.5 billion population are affected (Nachtergaele et al. 2010). According to Global Assessment of Soil Degradation (GLASOD), developing countries are facing degradation of around 2 billion ha of land (Scherr and Yadav 1997). According to Barrett-Lennard and Hollington (2006), approximately 10–20 million populations, existent on land, are influenced by salts with very low productivity and alarming threats of environmental degradation. Approximately six million hectares of agricultural land become unproductive due to various processes of land degradation in every year (Asio et al. 2009). As per report of the Indian Council of Agricultural Research (ICAR 2013), about 120.40 million hectares (out of 328.73 million hectares) of land in India are threatened by land degradation. The status of land degradation in India shows that the maximum percentage of area is degraded due to water erosion (10.21%) compared to water degradation (9.63%), wind/aeolian degradation (5.34%), frost shattering (3.1%), salinity/alkalinity (1.6%), mass movement (1.35%), water logging (0.3%), rocky/barren areas (0.5%), and others (0.04%) (Ministry of Environment and Forestry, Govt. of India 2011). Total estimated cost of land degradation in India is about 1.40% of the country's GDP, 3.95% of AGDP, which amounts up to Rs. 25,944 million or US$401,610,007 (Reddy 2003; Mahala 2017). In West Bengal, about 30.10% area has been affected by land degradation (Mahala 2017). According to NBSS & LUP, approximately 10,552.9 km^2 (11.89%) area is moderately and 7366.5 km^2 (8.3%) of total geographical area is highly amenable to soil erosion in West Bengal (Sarkar et al. 2005). According to many researchers, the eastern Chotanagpur plateau region of West Bengal, mainly the Paschim Medinipur, Bankura, Birbhum, and Purulia Districts, is highly affected by land degradation and gully erosion due to water erosion, land use changes, rill and gully erosion, deforestations, rapid growth of agriculture as well as irrigation, livestock's farming, cattle density, population density, laterite soil, the pre-cambrian geology (granite gneiss), undulating topography, high slope, relief, drainage density, and declining groundwater (Shit and Maiti 2012; Gour et al. 2014; Shit et al. 2015; Samanta et al. 2016; Mahala 2017).

As the present study area, Akarsa watershed of West Bengal is a small part of unit of plateau fringe region; gully erosion and land degradation are common characteristic here. So, assessment of gully erosion and delineation of potential land degradation zone (PLDZ) map is essential for it. The specific purpose of this study is to identify the degradation prone area and to assess relationship of geo-environmental factors which are responsible for gully erosion and land degradation process. So, it can be helpful for the planners to minimize or prevent this geo-environment problem by using this potential land degradation zone map.

11.2 Materials and Method

11.2.1 Description of Study Area

The Akarsa watershed, which is a part of the Dwarakeswar river basin, is the present study area (Fig. 11.1). The stream Akarsa originates from the highland of Puruliya District. Geomorphologically, this small Akarsa watershed is a part of the Chotanagpur plateau; for that reason, this area is extremely dissected, with rugged topography and undulating lands (Dey et al. 2009).

The most portion of the research area is covered with pink granite/biotite granite gneiss and migmatite, and some parts are characterized by presence of amphibolites and hornblende schist formations (CGWB & GSI) with western to eastern flowing river system, poor ferruginous, lateritic soil, and hard rock land surface with tropical dry deciduous natural vegetation mainly scrub jungles and sal woods. It has an area of approximately 346 km^2 as calculated by ARC GIS software and the area is characterized by semiarid and dry tropical monsoon climatic condition in which most of the rainfall occurs during July to September (Sen et al. 2004). The average precipitation ranges from low to medium (123–132 cm) and average annual temperature is high (37–42 °C). The study area is situated between 23°09′50″ to 23°20′25″ north latitudes and between 86°37′48″ to 86°55′07″ east longitudes. The Akarsha Watershed is located at the eastern part of Puruliya covering the Hura, Puncha, Kashipur block and the western part of Bankura covering Chhatna, Indpur, Bankura-I block.

11.2.2 Mapping of Gully and Land Degraded Areas

Degraded land area of Akarsa watershed (part of the Dwarakeswar river basin) is mapped using Google Earth Pro of 2018 and by using the visual image interpretation techniques. The created layers are saved as kmz file format from Google Earth imagery, and this kmz file format is converted into shape files format used in Arc GIS. Then

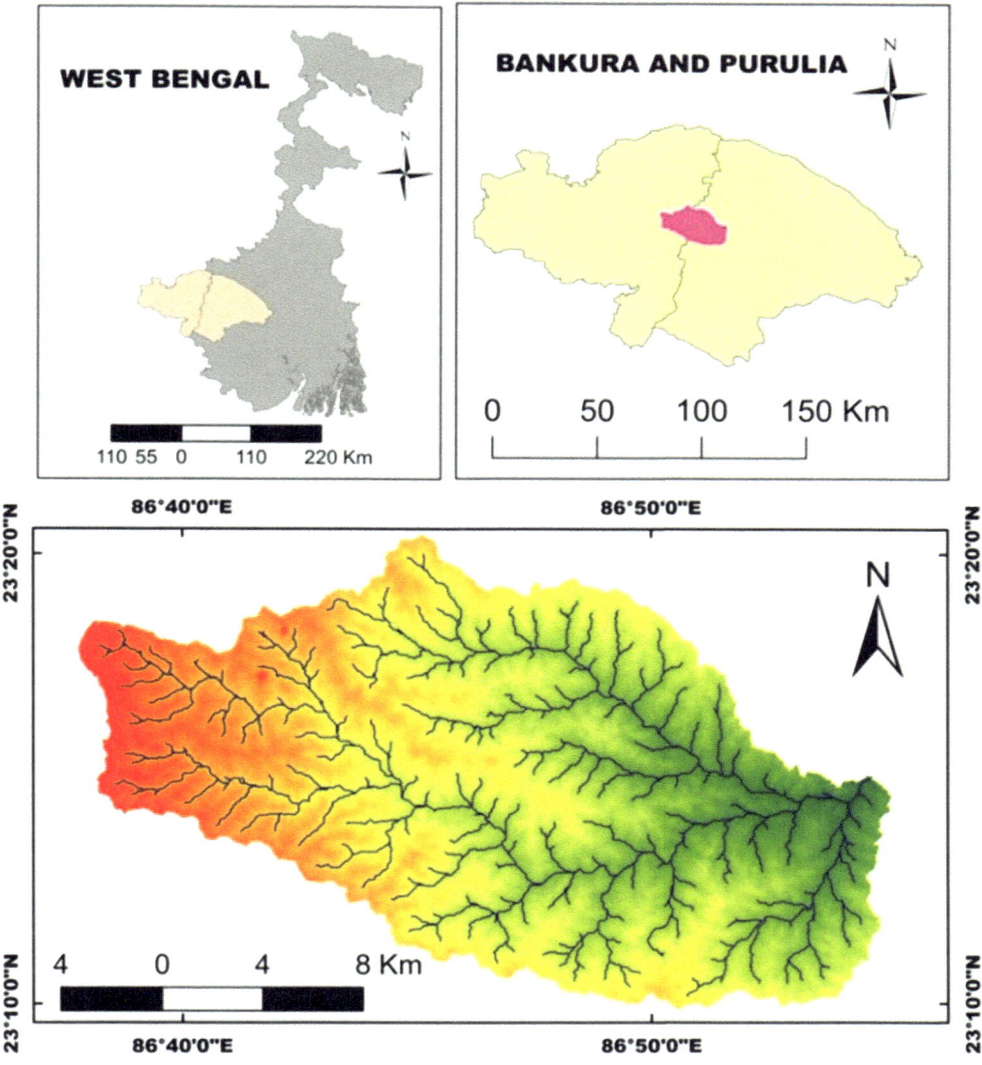

Fig. 11.1 Positional map of The Akarsa watershed

the map is prepared and accuracy assessment is done by field survey during September to November, 2018 (Fig. 11.2).

11.2.3 Data Base

To carry out the total work, remote sensing data, meteorological data, other ancillary data, GIS software (ERDAS IMAGINE 2015, ArcGIS 10.5), Microsoft Office Excel 2007, IBM SPSS Statistics 20, and GPS are used. The specification and data source for delineation of potential land degradation and gully erosion are given in detail in Table 11.1.

11.2.4 Methodology

The method which is used for delineation of potential land degradation zone (PLDZ) of Akarsa watershed is given away in Fig. 11.3. The drainage basin and drainage network are

generated from Digital Elevation Model (DEM) 30 m Spatial resolution [Dem, -fill sink, -flow direction, -flow accumulation, -stream definition (500), -input point data, -delineation basin area] using Arc hydro tool form Arc GIS software.

The aspect, length-slope map, relief, relative relief, slope, and dissection index are prepared form SRTM DEM data. The rainfall and groundwater map are raised using the Inverse Distance Weighted (IDW) method. Landsat 8 OLI satellite images from USGS have been used for preparation of land use/land cover and Normalized Difference Vegetation Index (NDVI) map. Then, different ancillary data, namely, geology, erodibility, practice factor, soil depth, and soil texture map, are geo-referenced (UTM projection, spheroid, and datum WGS 84, Zone 45 North) and digitized. Various thematic layers, viz., aspect, dissection index, erodibility, geology, groundwater, LULC, length-slope (LS) map, practice factor, rainfall, relative relief, relief, slope, soil depth, texture, NDVI, and drainage density factor, are prepared. Then, these are reclassified accepting all thematic maps on

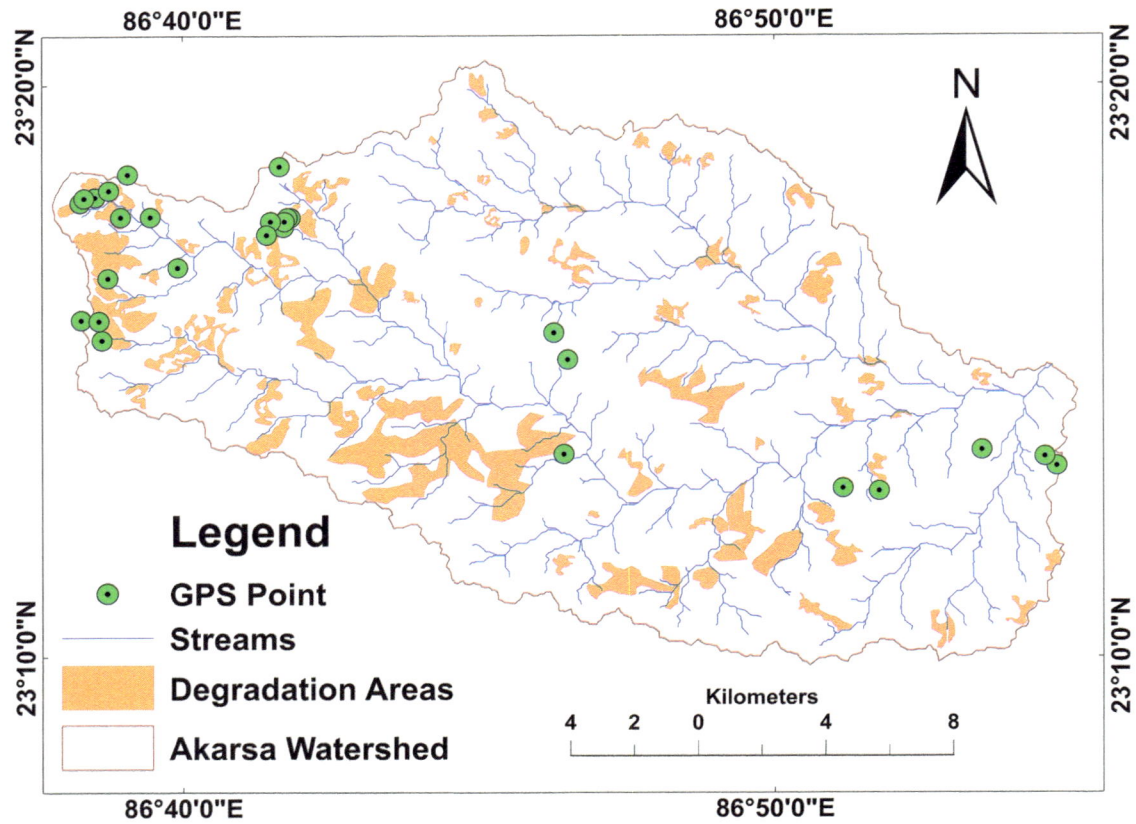

Fig. 11.2 Land degraded areas with ground verification

Table 11.1 Data sources

Data	Source
LANDSAT 8 OLI (spatial resolution 30 m) (LC08_L1TP_139044_20180424_20180502_01_T1)	USGS earth explorer
SRTM DEM (spatial resolution 30 m) (n23_e086_1arc_v3)	USGS earth explorer
Geology map (1:500000)	Central Ground Water Board (CGWB) and Geological Survey of India (GIS)
Soil depth and texture map (1:500000)	National Bureau of Soil Survey and Land use Planning (NBSS and LUP)
Soil erodibility and practice factor	National Bureau of Soil Survey and Land use Planning (NBSS and LUP)
Groundwater data	Mbgl data of Central Ground Water Board (CGWB), India
Rainfall data	Indian Meteorological Department (IMD)

the terms of weighted overlay methods using the MIF Technique.

11.2.4.1 Interrelationship Among Factor Classes and Weighted Calculation

Seventeen influencing factors have been identified to delineate the potential land degradation zone (PLDZ). Interrelationships among these factors are calculated statistically using MIF technique. Each relationship is weighted according to this direct and indirect strength and assigning introduced subclasses based on literature review (relative relation) and field experience (Table 11.2).

Factors having each major and minor effect are assigned a weightage of 1 and 0.5, respectively, and factors having insignificant or no effect are assigned 0 weightage (Magesh et al. 2012; Thapa et al. 2017) (Table 11.2). Higher weight

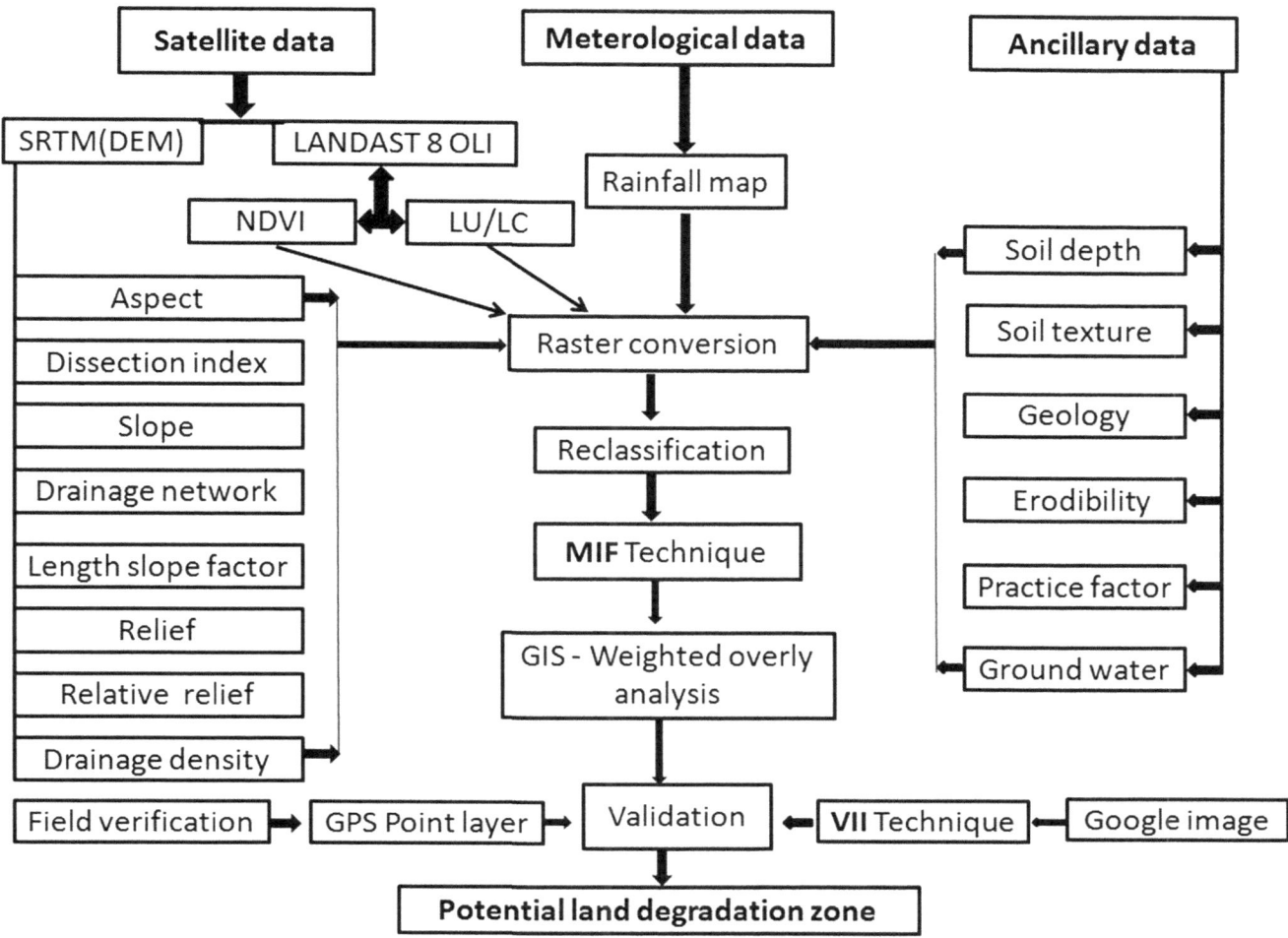

Fig. 11.3 The methodology framework

Table 11.2 Impact of influencing factor, relative rates, and score for each potential factor (Magesh et al. 2012)

Factor	Major effect (X)	Minor effect (Y)	Proposed relative rates ($X + Y$)	Proposed score of each influencing factor
Vegetation cover	11	2.5	13.5	9
Denudated area	11	3	14	9
Soil erodibility	11	1	12	8
Land use and land cover	8	4	12	8
Dissection index	9	2	11	7
Relative relief	6	3	9	6
Drainage density	8	2	10	6
Rainfall	8	2	10	6
Soil depth	7	2.5	9.5	6
Soil texture	3	2	5	3
Geology	6	2.5	8.5	5
Aspect	3	1.5	4.5	3
Groundwater	5	3.5	8.5	5
Length-slope factor	6	2	8	5
Practice factor	6	2	8	5
Slope	5	2.5	7.5	5
Relief	4	2	6	4
			Σ157	Σ100

value means it has a larger impact, and the lowest value shows smaller impact on delineation of PLDZ. The proposed weighted for each influencing factor is calculated by the following formula:

$$\text{Proposed score} = \left[\frac{(X + Y)}{\Sigma(X + Y)}\right] \times 100,$$

where X is direct interrelationship between two factor and Y is indirect interrelationship between two factor.

11.2.4.2 Weighted Overlay Analysis Method

After giving weights and score to factors and their subdivision classes, the potential land degradation zone is delineated through weighted overlay method using the formula as follows:

$$\begin{aligned}
\text{PLDZ} = \sum_{i}^{n} &\text{Vc}^k\text{Vc}^n + \text{Da}^k\text{Da}^n + \text{Er}^k\text{Er}^n + \text{Lu}^k\text{Lu}^n \\
&+ \text{Di}^k\text{Di}^n + \text{Rr}^k\text{Rr}^n + \text{Dd}^k\text{Dd}^n + \text{Ra}^k\text{Ra}^n \\
&+ \text{Sd}^k\text{Sd}^n + \text{St}^k\text{St}^n + \text{Ge}^k\text{Ge}^n + \text{As}^k\text{As}^n \\
&+ \text{Gw}^k\text{Gw}^n + \text{Ls}^k\text{Ls}^n + \text{Pf}^k\text{Pf}^n + \text{Sl}^k\text{Sl}^n \\
&+ \text{Re}^k\,\text{Re}^n,
\end{aligned}$$

where PLDZ represents the potential land degradation zone; "n" and "k" denote "factor classes" and "factor subclasses," respectively; Vc illustrates vegetation cover; Da is denudated area; Er is erodibility; Lu is land use/land cover (LULC); Di is dissection index; Rr is relative relief; Dd is drainage density; Ra is rainfall; Sd is soil depth; St is soil texture; Ge is geology; As is aspect; Gw is groundwater; Ls is length-slope factor; Pf is practice factor; Sl is slope; and Re is relief (Table 11.3).

11.3   Results and Discussion

11.3.1 Vegetation Cover

In this potential land degradation zone (PLDZ), NDVI was found to be a highly influencing factor to land degradation. Vegetation cover is assessed by NDVI formula using Landsat 8 OLI satellite image. NDVI is a susceptible indicator of vegetation coverage, density, and health as well as crop phenology of wide area using spectral reflectance. Following the formula:

$$\text{NDVI} = (\text{NIR} - \text{R})/(\text{NIR} + \text{R})$$

where NIR = near-infrared band and R = red band (Tucker and Sellers 1986).

Here, the relation between vegetation cover and land degradation is negative; this indicates that the area with dense vegetation cover experiences low land degradation and vice versa (Kakembo 2001). The multiple reasons of deforestation are subsistence agriculture, shifting cultivation, overgrazing, hunting, increasing large population, and their immediate need of food (Kelly et al. 2015). Many places of Akarsa watershed experience high deforestation due to the semiarid drought prone climate, agriculture expansion, and settlement. This study area (Fig. 11.4a) shows no vegetation zone (−0.13 to −0), i.e., 1%; very low (0–0.18), i.e., 27 km^2 area (8%); low to moderate (0.18–0.22), i.e., 109 km^2 area (31%); high (0.22–0.26), i.e., 135 km^2 area (39%); and very high (0.26–0.41), i.e,. 73 km^2 area (21%). As the region possesses tropical dry deciduous forest and shrub or grassland, the maximum NDVI value being only 0.41 represents low density forest, whereas areas under NDVI value ranging from 0.22 to 0.26 represent the shrub and grassland.

11.3.2 Denudated Area

Land denudation results in long-term soil erosion which has an effect on environment, ecosystem, and anthropogenic activity. Akarsa watershed, where soil erosion due to rill and gully erosion is a common factor, has experienced different amounts of denudation in different places. The denuded area mapping is prepared using visual image interpretation (VII) techniques from Google earth imagery digitization and field survey with a hand GPS (Global Positioning System) device. The entire 82 identified areas under high denuded zone covering sum total of 42 km^2 area accounts 12.50% of total area. The rest of the areas are considered as moderate to low land denudated area (Fig. 11.4b).

11.3.3 Land Use and Land Cover

The land degradation is very much dependent on the LULC and changing nature of any region (Karamesouti et al. 2015). In Africa, land degradation is found due to increasing population, and in Central America, mainly the forest land converted to agricultural land (Stéphenne and Lambin 2001; Scullion et al. 2014). The LULC play a direct driving and most crucial factor of land degradation. In the Akarsa watershed (Fig. 11.4c), the major LULC types have been occupied by the agricultural land covering 129.84 km^2 (37.52%) area, followed by fallow and pasture land measuring 97.57 km^2 (28.20%), vegetated area 61.3 km^2 (17.72%), settlement 40.50 km^2 (11.70%), and water bodies 16.80 km^2 (4.86%) area. Here, fallow and pasture land and agricultural lands are highly degraded area having high rates of soil erosion. Land

Table 11.3 Classification of weighted factors influencing the PLDZ (potential land degradation zone)

Factor	Subclass	Weightage	Factor	Subclass	Weightage
Vegetation	No vegetation	9	Soil depth	Very deep	1
	Very low	7		Deep-very deep	2
	Low to moderate	5		Moderately shallow-deep	3
	High	2		Shallow-moderately deep	3
	Very high	1		Shallow-moderately shallow	4
Denudated area	High denudated area	9		Very shallow-shallow	5
	Moderate to low denudated	3		Shallow	6
Erodibility	<0.20	1			
	0.20–0.25	2	Soil texture	Loamy sand-sandy clay loam	3
	0.25–0.30	3		Sandy-loamy sand to sandy loam	3
	0.30–0.40	6		Sandy clay loam-sandy loam to sandy clay loam	2
	>0.40	8		Sandy clay loam-clay loam	2
Land use and land cover	Water bodies	1		Gravelly sandy loam-gravely sandy loam	1
	Vegetation	2		Loamy sand to sandy loam-clay loam to clay	1
	Settlement	3		Sandy loam-sandy clay loam	1
	Agricultural land	6		Clay-clay loam	1
	Fallow and pasture land	8		Gravelly sandy loam	1
Dissection index	0.038–0.087	1		Sandy loam-clay loam	1
	0.087–0.102	2		Sandy loam-loam	1
	0.102–0.117	3	Geology	Amphibolite and hornblende schist	1
	0.117–0.137	5		Pink granite/biotite granite gneiss	3
	0.137–0.240	7		Granite gneiss migmatite	5
Relative relief (m)	11–18	1	Aspect	Flat	1
	18–21	2		North	2
	21–24	3		Northeast	2
	24–31	5		East	3
	31–55	6		Southeast	3
Drainage density (km/km^2)	<0.45	2		South	3
	0.45–0.85	3		Southwest	2
	0.85–1.25	4		West	2
	1.25–1.65	5		Northwest	2
	>1.65	6		North	2
Rainfall (mm)	1.230–1.250	2	Length slope factor	<0.05	1
	1.250–1.265	3		0.05–0.81	2
	1.265–1.281	4		0.81–1.50	3
	1.281–1.297	5		1.50–2.00	4
	1.297–1.315	6		>2.00	5
Groundwater (mbgl)	5.28–6.19	1	Practice factor	0.30–0.40	2
	6.19–6.76	2		0.40–0.70	4
	6.76–7.51	3		0.70–1.00	5
	7.51–8.72	4	Slope (degree)	<1	1
	8.72–10.69	5		1–3	2
Relief (m)	<129	1		3–5	3
	129–150	2		5–7	4
	150–170	2		>7	5
	170–193	3			
	>193	4			

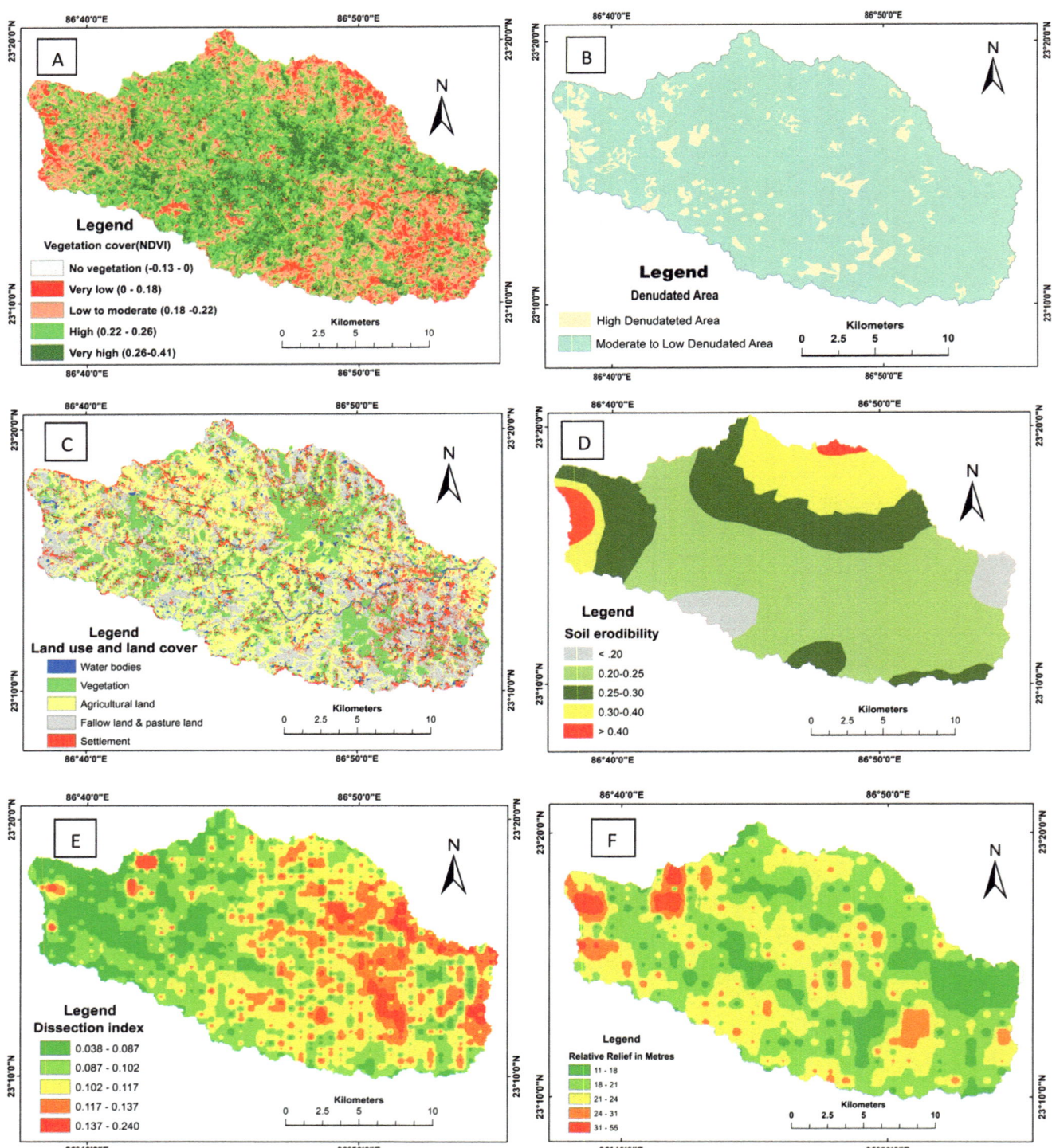

Fig. 11.4 Land degraded factor maps. (**a**) Vegetation cover, (**b**) denudated Area, (**c**) soil erodibility, (**d**) land use and land cover, (**e**) dissection index, (**f**) relative relief, (**g**) drainage density, (**h**) rainfall, (**i**) soil depth, (**j**) soil texture, (**k**) geology, (**l**) aspect, (**m**) groundwater, (**n**) length-slope factor, (**o**) practice factor, (**p**) slope, (**q**) relief map

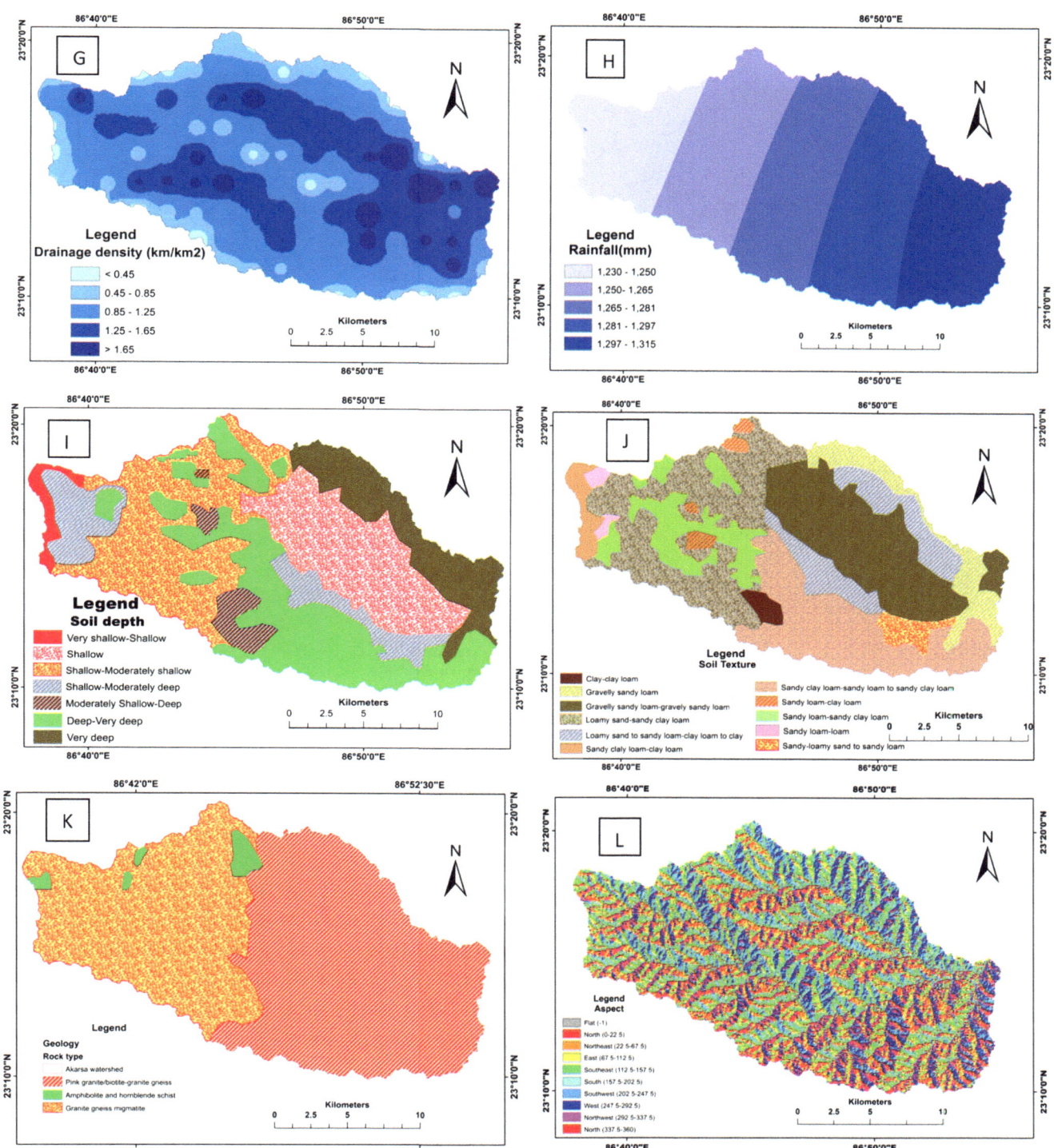

Fig. 11.4 (continued)

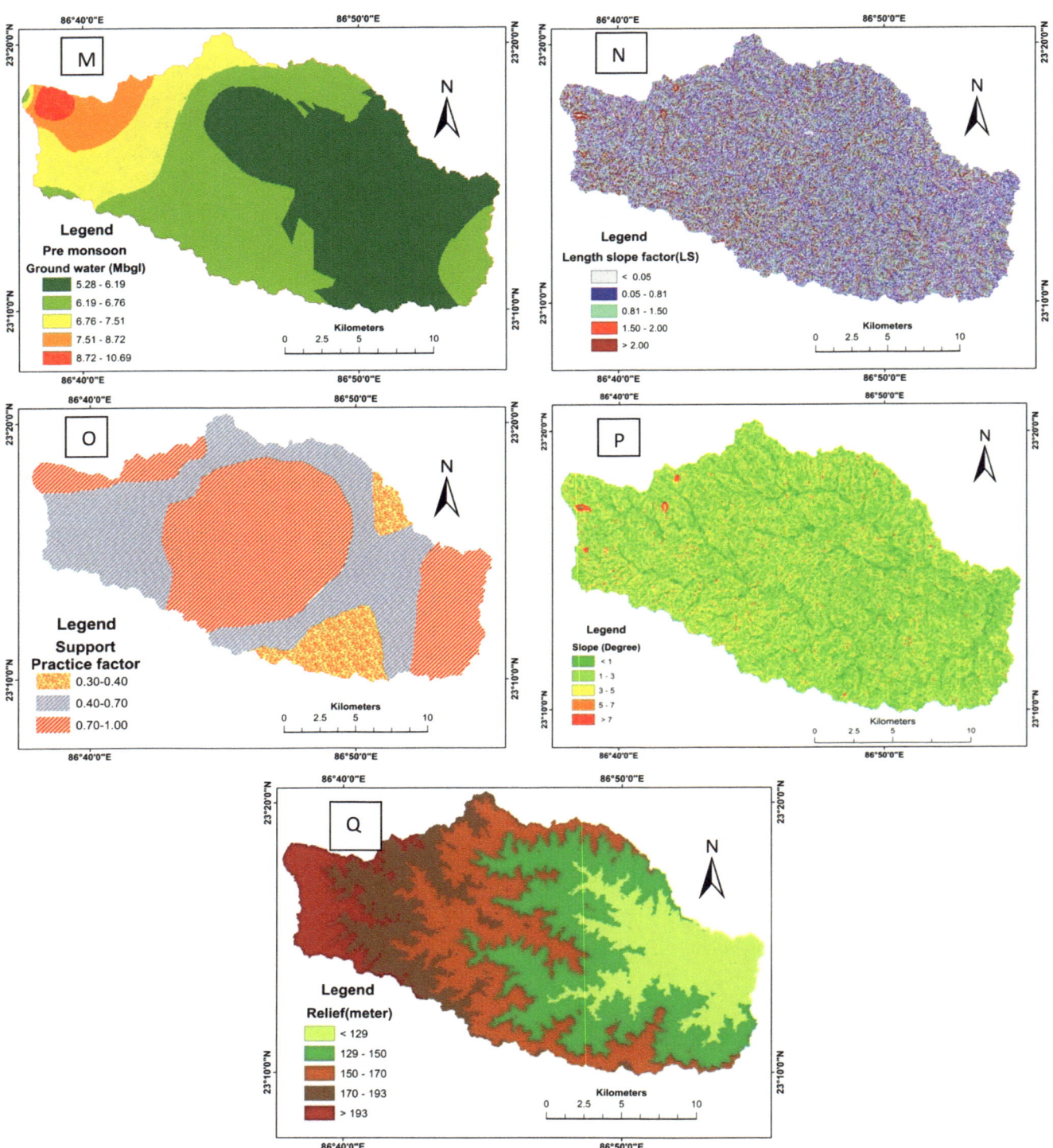

Fig. 11.4 (continued)

degradation occurs moderately in areas of settlement, vegetation, and water bodies where the rate of degradation is comparatively low.

11.3.4 Soil Erodibility

Soil erodibility is another vital factor of land degradation or soil erosion. Soil erodibility is influenced by natural soil properties which include stability of soil structure, soil permeability, percentages of sand, silt, clay, infiltration, organic matter, and soil mineralogy. In this study, erodibility factor "K" is taken after National Bureau of Soil Survey & Land Use Planning (NBSS & LUP), and this "K" is measured by Wischmeir and Smith (1978) with the following equation (Sarkar et al. 2005):

$$K = 1.2917 \left[2.1 \times 10^{-4} M^{1.14} (12 - a) \right. \\ \left. + 3.25(b - 2) + 2.5(c - 3) \right] / 100$$

where

$M = \%$ silt $\times (100 - \%$ clay$)$, $b =$ the soil structure code, $A = \%$ organic matter, $c =$ the profile permeability code

Here, five K values are found (Fig. 11.4d), such as <0.20, 0.20–0.25, 0.25–0.30, 0.30–0.40, and >0.40. The area and percentages are, respectively, 21.50 (6.21%), 185.38 (53.38%), 80.32 (23.21%), 43.25 (12.50%), and 15.45 km^2 (4.47%). High erodibility means high rate of soil erosion and land degradation. So, relation between soil erodibility and land degradation is positive.

11.3.5 Dissection Index

Dissection index is a major influencing factor of land degradation. Dissection index is the measure of amount of dissection done by the river in the area. The ratio between relative relief and maximum altitude is called dissection index. Dissected area is highly vulnerable to land degradation as well as soil erosion. So, relation between dissection index and land degradation is positive. Here, five dissection index classes are developed, i.e., (0.038–0.087), (0.087–0.102), (0.102–0.117), (0.117–0.137), and (0.137–0.240) (Fig. 11.4e). High dissection index means high erosion and vice versa.

This study shows the upper reach of river basin having higher relative relief accounts for lower dissection index values and vice versa. This anomaly occurs due to lack of flow in the tributary in the dry upper reach, but more valley deepening is found due to higher rainfall, drainage density, and concentration of water flow.

11.3.6 Relative Relief

Relative relief is the most important morphometric measurement which shows regional distribution of relief characteristics. The difference between the absolute and lowest altitude in a particular area is called relative relief. In this study, relation between relative relief and land degradation is positive which means higher relative relief value results in higher probability to soil erosion as well as land degradation and vice versa, The relative relief of this study area is divided into five categories, which are "very low" (11–18), "low"(18–21), "medium"(21–24), "high"(24–31), and "very high" (31–55) meter, covering an area of 58, 146, 114, 20, and 7.5 km^2, accounting for 16.76, 42.34, 32.95, 5.78, and 2.17%, of the total area, respectively (Fig. 11.4f).

11.3.7 Drainage Density

According to Horton (1932, 1945), drainage density is the average length of stream channel per unit area expressed as:

$$\text{DD} = \frac{\Sigma \text{I}}{A}$$

where "DD" is the drainage density per unit area

"ΣI" is the total length of drainage
"A" is area

On the basis of drainage density, this Akarsa watershed is divided into five class zones, namely, very low (<0.45), low (0.45–0.85), medium (0.85–1.25), high (1.25–1.65), and very high (<1.65), covering an area of 4.34, 47.15, 161.87, 117.57, 21.44, and 6.78 km^2, respectively. Relation between drainage density and land degradation is positive which means higher drainage density value focuses on the higher rates of land degradation. The morphometric characteristic of drainage ascertains the erosional feature of any region (Frankl et al. 2013) (Fig. 11.4g).

11.3.8 Rainfall

Rainfall is another most important factor of land degradation and gully erosion; the rainfall map is prepared by the interpolation and inverse distance weighted (IDW) method using rainfall data collected from the Indian Meteorological Department (IMD). The study area is situated in tropical hot, dry-subhumid monsoonal climate. The tropical and monsoonal areas of the world have faced intense rill and gully erosion related to soil degradation (Frankl et al. 2013). Akarsa watershed area is divided into five grades on the basis of average rainfall. These are very low

(1230–1250 mm), low (1250–1265 mm), moderate (1265–1281 mm), high (1281–1297 mm), and very high (1297–1315 mm), covering the area of about 51.99 km^2 (14.74%), 75.11 km^2 (21.71%), 82.18 km^2 (23.75%), 76.64 (22.15%), and 60.09 km^2 (17.37%), respectively (Fig. 11.4h).

11.3.9 Soil Depth

Soil depth is a very essential physical factor of land degradation. In this study area, eight soil depth zones are identified using the "National Bureau of Soil Survey & Land Use Planning (NBSS & LUP)" data. Shallow soil zone is a highly erosion-prone area, and deep soil is comparatively a low erosion-prone area. Here, seven soil depth zones are identified. These are very shallow-shallow, shallow, shallow-moderately shallow, shallow-moderately deep, moderately shallow-deep, deep-very deep, and very deep (Fig. 11.4i). Relation between soil depth and land degradation is negative which means lower soil depth causes the higher rates of land degradation and vice versa. In this region, very shallow-shallow, shallow, shallow-moderately shallow, shallow-moderately deep, and moderately shallow-deep classes are highly vulnerable to land degradation and soil erosion.

11.3.10 Soil Texture

Soil texture is a very significant physical parameter for delineating the land degradation and gully erosion vulnerable zones. The assessment of the soil texture discloses that there are 11 categories of soil texture found in the Akarsa watershed, viz., "loam sand-sandy clay loam," "sandy-loamy sand to sandy loam," "sandy clay loam-sandy loam to sandy clay loam," "sandy clay loam-clay loam," "gravelly sandy loam-gravely sandy loam," "loamy-sand to sandy loam-clay loam to clay," "sandy loam-sandy clay loam," "clay-clay loam," "gravelly sandy loam," "sandy loam-clay loam," and "sandy loam-loam" (Fig. 11.4j). As the sandy and loamy texture of soil is highly erosion-prone soil texture than others, loam sand-sandy clay loam, sandy-loamy sand to sandy loam, sandy clay loam-sandy loam to sandy clay loam, and sandy clay loam-clay loam classes have influenced the land degradation in this region.

11.3.11 Geology

Geology plays an essential role in the land degradation and potential gully erosion. Geological formations largely influence the quality of land and land surface environment, i.e., soil, relief, slope, groundwater storage, geomorphology, drainage condition, etc., which are indirectly controlled by the entire land degradation processes (Mahala 2017). The porosity of sand, silt, clay bed, rocks, and alluvium/sediment cover govern the percolation and infiltration of water flow (Shaban et al. 2006). In this Akarsa watershed, three types of lithology are found; they are pink granite/biotite gneiss, amphibolites and hornblende schist, and granite gneiss migmatite. Granite gneiss migmatite is highly erosion-prone rock formation, and in some parts of the study area, lateritic formations having the ability to form rill, gully, and soil erosion result in land degradation (Fig. 11.4k).

11.3.12 Aspect Map

In geology, the compass direction which is faced by a slope is called aspect. Aspect also strongly influences temperature. Vegetation depends on slope aspect because it is related with sunlight and insolation that indirectly influence land degradation process. Aspect is also regarded as a fundamental issue in vulnerability of denudation process (Nagarajan et al. 2000). Here, aspect classes of east, south-east, and south are dominant among the ten classified aspects of the Akarsa watershed. The role of these three aspect classes is crucial for insolation-related land degradation and gully erosion (Fig. 11.4l).

11.3.13 Groundwater

Groundwater means the water which is present or found underground in the cracks and pore spaces in soil, sand, and rock. Groundwater exploitation is related to land degradation (Mahmoud and Alazba 2016). Developing countries are suffering accelerated groundwater reduction (Dedewanou et al. 2015). The groundwater tables are declined due to low depth aquifer as well as low groundwater recharge for existence of hard rock in upper basin area (Mahala 2017). In this study area, pre-monsoon groundwater zones are divided into five classes; these are: very high (5.28–6.19), high (6.19–6.76), moderate (6.76–7.51), low (7.51–8.72), and very low (8.72–10.69) mbgl covering an area of 4.47, 18.31, 48.31, 123.99, and 153.50 km^2, respectively (Fig. 11.4m). Lower groundwater table classes account for the higher rates of land degradation and vice versa, indicating negative relationship between groundwater and land degradation.

11.3.14 Length-Slope Factor

The length-slope factor (LS) is mainly applied to ascertain the impact of topography on erosion and the topographical susceptibility causes on the length of the slope (L) and the steepness of the slope (S) (Shit et al. 2015). The topographic factors are a single parameter created through gradient and length of slope at every position of the grid inspection (NBSS publ.117). The LS has been enumerated based on the following formula explained by Moore and Burch (1986). In this study, SRTM DEM is used to delineate the length-slope factor. Flow direction and flow accumulation are generated using the Arc hydro tool, slope (degree) is calculated using the 3D analyst tools (raster surface), then spatial analyst tools to use raster calculator of Arc GIS software.

$$LS = (fa \times cellsize/22.13)^{0.4} \times (\sin \sigma/0.0896)^{1.3}$$

where fa is flow accumulation derived from the DEM using a GIS accumulation algorithm and r is slope in degrees.

According to length-slope factor, the Akarsa watershed area is divided into five categories which are <0.05, 0.05–0.81, 0.81–1.50, 1.50–2.00, and >2.00 degree covering an area of 156, 84, 57, 19, and 30 km^2, respectively. The land degradation and gully erosion normally occur on high LS values with high slopes (Fig. 11.4n).

11.3.15 Practice Factor (P)

Soil, whenever used for cultivating on a sloping land, requires a support through different practices to slow down the runoff for less soil loss or soil erosion. The different supporting practices in India are terrace cultivation, strip cropping, contour cultivation, field bunding, etc. The practice factor in USLE (universal soil loss equation) is the ratio of soil with a specific supporting practice to the corresponding loss with up and down cultivation. Here, the map is taken from soil erosion of West Bengal (NBSS & LUP) and is also geo-referenced (UTM projection, spheroid and datum WGS 84, Zone 45 North) and finally digitized. In this study area, three practice factor zones are found, which are 0.30–0.40, 0.40–0.70, and 0.70–1.00, covering the areas 31.46, 138.40, and 176.15 km^2, respectively. Higher value of p factor has a high probability to degradation of land or soil erosion. Therefore, the relation between practice factor and land degradation is positive (Fig. 11.4o).

11.3.16 Slope Map

Slope is a significant factor in delineating land degradation. The stability of the weathering rock, debris, and soil materials is directly related to slope. The slopes with higher values in Akarsa watershed have faced higher potentiality of land degradation processes. Here, the slope is classified into five zones, which are $<1°$, 1–3°, 3–5°, 5–7°, and $>7°$ covering an area of 61, 200, 74, 9, and 2 km^2, respectively. High slope has faced high potentiality of erosion and low slope has low potentiality of erosion which means the relation is positive (Fig. 11.4p).

11.3.17 Relief Map

The relief is a factor of physiography which determines the land degradation process of any region. Relief controls directly the surface shape including steepness, height, and landform of any region. Based on the relief feature, the entire geographical area of Akarsa watershed has been separated into five categories; these are very low (<129), low (129–150), moderate (150–170), high (170–193), and very high (>193) meter covering an area 60, 108, 96, 55, and 27 km^2, respectively. The highest altitude is observed in the western part of the study area. Higher relief bears higher potentiality of land degradation as well as gully erosion and lower relief has lower potentiality indicating the positive relationship between these two variables (Fig. 11.4q).

11.3.18 Delineation of Potential Land Degradation Zone

The potential land degradation zone for the Akarsa watershed is generated through the various thematic maps integration, viz., vegetation cover, denudated area, soil erodibility, land use and land cover, dissection index, relative relief, drainage density, rainfall, soil depth, soil texture, geology, aspect, groundwater, length-slope factor, practice factor, slope, relief using remote sensing, GIS, and field survey. On the basis of weighted multi-influencing factor using MIF technique, the potential land degradation zone is assigned (Table 11.3). The final potential land degradation zone map of Akarsa watershed has been classified into five zones, namely, very high, high, moderate, low, and very low in ArcGIS software (Fig. 11.5).

In Akarsa watershed, very high potential land degradation zone (58.36–80.18) is seen across the western plateau. It covers 15.24 km^2 or 4.4% of the entire area of the Akarsa watershed. Small vegetation cover, undulating hard rock geology, and illegal mining of rock or lateritic (Duricust) soil are the causes of very high land degradation. The high potential land degradation zone (52.34–58.35) areas are concentrated in southwestern and various parts of southeastern areas of the watershed about 16.18% of the total area covering an area 56.25 km^2 (Fig. 11.6; Table 11.4). Intensive agriculture and incising high population are the causes of much land degradation here (Shit et al. 2015). 38.15% (132.55 km^2) area falls under moderate potential land

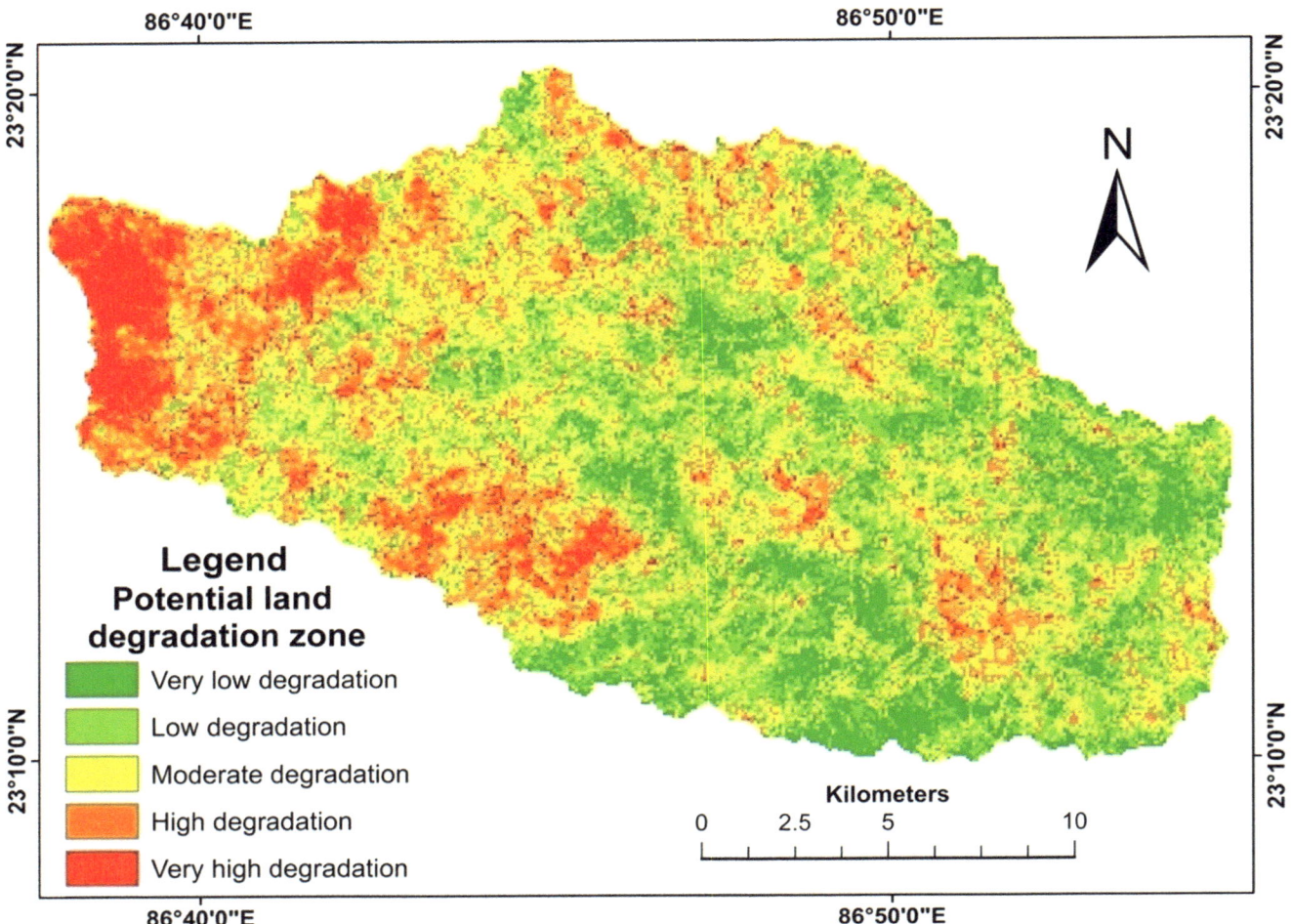

Fig. 11.5 The potential land degradation zone of the Akarsa watershed

degradation zone found in the mid-western and lower middle eastern plain regions of the Akarsa watershed. The land use/land cover change is the most vital reason for land degradation (Halder 2013). About 27.52% of the Akarsa watershed area covering an area of 95.21 km^2 falls under low land degradation zone and 47.54 km^2 (13.74%) of the total study area is delineated as the very low land degradation zone.

11.3.19 Validation of the Results

Visual image interpretation (VII) and Multi-influencing factor (MIF) techniques are used to compare area statistics of google image and final deliniation map of potential land degradation zone respectively. The distribution of positive error in degraded area is [71.49(56.25 + 15.24) − 42] = 29.49 km^2 (approximately), so this 30 km^2 area is highly vulnerable to land degradation.

The final output map of potential land degradation zone is further validated with the data related to field survey (27 GPS point) of different erosion situation and LULC in the Akarsa watershed during September to November 2018 (Fig. 11.7). This validation was performed by the receiver operating characteristic (ROC) curve.

The pictorial representation of prediction rate is shown by blue line for potential land degradation zone (Fig. 11.7). The prediction rate has been evaluated on the basis of the Area under the Curve (AUC) from 0.5 to 1.0. The AUC values can be classified as follows: "0.9–1" excellent, "0.8–0.9" very good, "0.7–0.8" good, "0.6–0.7"average, and "0.5–0.6" poor (Arabameri et al. 2018a; Rahmati et al. 2016).

The area under the curve (AUC) measures the prediction rates for validation of potential land degradation zone (PLDZ). The result of AUC (0.828) and std.error (0.110) under non-parametric assumption denotes that it is a good predictor model having accuracy of 82.8%. Finally, the result

Fig. 11.6 Degraded land areas of Akarsa watershed. (**a**) Measurement of gully erosion (23°17′09″N, 86°38′38″E). (**b**) Excavation of lateritic bare soil (murom) due to anthropogenic activities (23°15′56″N, 86°39′04″E). (**c**) Lithology (23°17′14″N, 86°39′06″E). (**d**) Gully erosion (23°16′59″N, 86°38′45″E). (**e**) Open crust mining of murom and bolder (23°16′59″N, 86°38′32″E). (**f**) Scouring at the base of root system of plants (23°16′59″N, 86°41′43″E)

Table 11.4 Area of different potential land degradation zone of Akarsa watershed

Potential land degradation zone	Value	Area covered (km^2)	Percentage of area (%)
Very low land degradation zone	32.20–43.31	47.54	13.74
Low land degradation zone	43.31–47.26	95.21	27.52
Moderate land degradation zone	47.26–52.34	132.55	38.15
High land degradation zone	52.34–58.36	56.25	16.18
Very high land degradation zone	58.36–80.18	15.24	4.41

Fig. 11.7 Prediction rate curve (ROC)

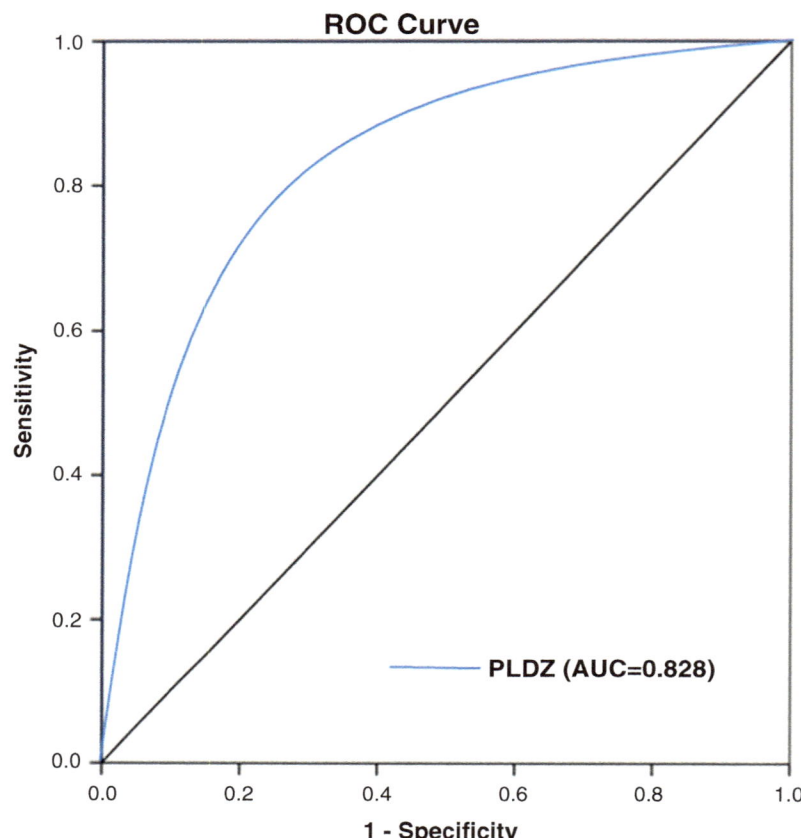

indicates that the total work has done very well to delineate the model of potential land degradation and gully erosion susceptibility zone maps.

11.4   Conclusions

In this study, joint application of GIS (geographic information system), MIF (multi-influencing factor), and RS (remote sensing) techniques is used to develop an effective model in the delineation of potential land degradation zone in Akarsa watershed, West Bengal. Land degradation and gully erosion are vital geo-environmental and semi-hazard problems in the tropical plateau region of India and other developing countries of Asia, Africa, and Latin America. It is found that food scarcity is increasing due to rapid increase of population at drought-prone semiarid Akarsa watershed of West Bengal. So, major area of the region has been transformed from the natural vegetation into agricultural land or built-up area. In the vast area of this watershed, agricultural practice depends on monsoonal rainfall, and during the rest of the year, the land remains a fallow land. This is one of the main reasons of fertile soil loss and land degradation. Thus, land use and land cover (LULC) change due to high population pressure, which is the most influential factor of land degradation, is related to gully erosion. Land degradation and gully erosion in the upper part of this watershed are mainly dominant for physical factors such as geology, vegetation, lithology, relief and relative relief, and length-slope factor; in the middle part, both physical and anthropogenic factors are responsible, and in the lower part LULC changes are mainly working.

The potential land degradation zone (PLDZ) in Akarsa watershed has been prepared based on 17 geo-environmental factors selected by previous literature review, field survey, and accessible data. The various thematic maps have been developed on remotely sensed multi-spectral and secondary data in a GIS platform using Arc GIS software through multiple-criteria decision making (MCDM) giving weightage in MIF techniques and the result was cross-checked with the (ROC) receiver operating characteristic curve for proper validation. The model definitely minimizes the use of time and manpower, making itself cost-efficient and helpful to enable quick decision for sustainable development and land-water resources management. Therefore, the potential land degradation zone (PLDZ) will be an essential tool for planners, engineers, decision-makers, government bodies, and local authorities for suitable management purpose. This method will be broadly applicable to a vast area with plateau and rugged topography for the land degradation, land capability classification, sustainable land use planning, and suitable management of soil erosion and land degradation.

However, in this study many important socioeconomic, anthropogenic factors and the soil physiochemical parameters (pH, EC, organic carbon, alkalinity, etc.) are not included in the methodology. So, next researchers may try to use all factors for conclusion through a holistic approach.

Acknowledgments This work is supported by the UGC (University Grants Commission), and thanks to Amit Bera and Nabin Adhikari for their valuable concepts. The authors are also thankful to USGS, CGWB, GSI, NBSS & LUP, and IMD.

References

Arabameri A, Pradhan B, Pourghasemi HR, Rezaei K (2018a) Identification of erosion-prone areas using different multi-criteria decision-making techniques and GIS. Geomatics, Natural Hazards and Risk 9 (1):1129–1155. doi: https://doi.org/10.1080/19475705.2018.1513084

Arabameri A, Rezaei K, Pourghasemi HR, Lee S, Yamani M (2018b) GIS-based gully erosion susceptibility mapping: a comparison among three data-driven models and AHP knowledge-based technique. Environmental Earth Sciences 77(17):628. doi: https://doi.org/10.1007/s12665-018-7808-5

Asio VB, Jahn R, Perez FO, Navarrete IA, Abit Jr SM (2009) A review of soil degradation in the Philippines. Annals of Tropical Research 31:69–94

Bahrawi JA, Elhag M, Aldhebiani AY, Galal HK, Hegazy AK, Alghailani E (2016) Soil erosion estimation using remote sensing techniques in Wadi Yalamlam Basin, Saudi Arabia. Advances in Materials Science and Engineering 2016:8. doi:https://doi.org/10.1155/2016/9585962

Barrett-Lennard EG, Hollington PA (2006) Development of a national program on saline agriculture for Pakistan. Available at: http://www.cazs.bangor.ac.uk/salinity/reports/National program.htm

Bera A (2017) Assessment of soil loss by universal soil loss equation (USLE) model using GIS techniques: a case study of Gumti River Basin, Tripura, India. Modeling Earth Systems and Environment 3 (1):2

Biro K, Pradhan B, Buchroithner M, Makeschin F (2013) Land use/Land cover change analysis and its impact on soil properties in the northern part of Gadarif region, Sudan, Land Degrad. Dev 24: 90–102. doi:https://doi.org/10.1002/ldr.1116

Cerda A, Gimenez-Morera A, Bodi MB (2009) Soil and water losses from new citrus orchards growing on sloped soils in the western Mediterranean basin. Earth Surface Processes and Landforms 34 (13):1822–1830

Cerda A, Gonzalez-Pelayo O, Gimenez-Morera A, Jordan A, Pereira P, Novara A, Brevik EC, Prosdocimi M, Mahmoodabadi M, Keesstra S, Orenes FG, Ritsema CJ (2016) Use of barley straw residues to avoid high erosion and runoff rates on persimmon plantations in Eastern Spain under low frequency-high magnitude simulated rainfall events. Soil Res 54(2):154–165

Chatterjee S, Krishna AP, Sharma AP (2014) Geospatial assessment of soil erosion vulnerability at watershed level in some sections of the Upper Subarnarekha river basin, Jharkhand, India. Environmental earth sciences 71(1): 357–374. doi:https://doi.org/10.1007/s12665-013-2439-3

De Souza RG, Da Silva DKA, De Mello CMA, Goto BT, Da Silva FSB, Sampaio EVSB, Maia LC (2013) Arbuscular mycorrhizal fungi in revegetated mined dunes, Land Degrad. Dev. 24: 147–155. doi: https://doi.org/10.1002/ldr.1113

Dedewanou M, Binet S, Rouet JL (2015) Groundwater vulnerability and risk mapping based on residence time distributions: spatial analysis for the estimation of lumped parametres. Water Resour Manag 29: 5489–5504. doi:https://doi.org/10.1007/s11269-015-1130-8

Dey S, Ghosh S, Debbarma C and Sarker P (2009) Some observation of regional evidences of Tertiary–Quaternary geo-dynamics in a paleo-coastal of Bengal basin, India. Russian Geol Geophys 50(11)

Feoli E, Vuerich LG, Woldu Z (2002) Processes of environmental degradation and opportunities for rehabilitation in Adwa, Northern Ethiopia. Landsc Ecol 17:315–325

Frankl A, Poesen J, Haile M, Deckers J, Nyssen J (2013) Quantifying long-term changes in gully networks and volumes in dryland environments: the case of Northern Ethiopia. Geomorphology 201:254–263. doi:https://doi.org/10.1016/j.geomorph.2013.06.025

Gour D, Soumendu C, Nilanjana DC (2014) Weathering and mineralogical alteration of granitic rocks in Southern Purulia District, West Bengal, India. Int Res J Earth Sci 2(4):1–12

Halder JC (2013) Land use/land cover and change detection mapping in Binpur-II Block, Paschim Medinipur District, West Bengal: A remote sensing and GIS perspective. IOSR J Hum Soc Sci 8 (5):20–31

Hill MJ, Braaten R, Veitch SM, Lees BG, Sharma S (2005) Multi-criteria decision analysis in spatial decision support: the ASSESS analytic hierarchy process and the role of quantitative methods and spatially explicit analysis. Environ. Modell. Softw 20: 955–976. doi: https://doi.org/10.1016/j.envsoft.2004.04.014

Horton RE (1932) Drainage-basin characteristics. Eos, transactions american geophysical union 13(1):350–361

Horton RE (1945) Erosional development of streams and their drainage basins: hydrophysical approach to quantitative morphology. Geological Society of America Bulletin 50:275–370

ICAR (Indian Council of Agricultural Research): State of Indian Agriculture, 2012–2013, A report of Department of Agriculture and Cooperation, New Delhi, 9, 2013.

Kakembo V (2001) Trends in vegetation degradation in relation to land tenure, rainfall, and population changes in Peddie district, Eastern Cape, South Africa. Environ Manag 28(1):39–46. doi:https://doi.org/10.1007/s002670010205

Karamesouti M, Detsis V, Kounalaki A, Vasiliou P, Salvati L, Kosmas C (2015) Land-use and land degradation processes affecting soil resources: evidence from a traditional Mediterranean cropland (Greece). Catena 132: 45–55. doi:https://doi.org/10.1016/j.catena.2015.04.010

Keesstra S, Pereira P, Novara A, Brevik EC, Azorin-Molina C, Parras-Alcántara, L, Jordan A, Cerda A (2016) Effects of soil management techniques on soil water erosion in apricot orchards. Sci. Total Environ 551:357–366. doi:https://doi.org/10.1016/j.scitotenv.2016.01.182

Kelly C, Ferrara A, Wilson GA, Ripullone F, Harmer N, Salvati L (2015) Community resilience and land degradation in forest and shrubland socio-ecological systems: evidence from Gorgoglione, Basilicata, Italy. Land Use Policy 46:11–20. doi:https://doi.org/10.1016/j.landusepol.2015.01.026

Leh M, Bajwa S, Chaubey I (2013) Impact of land use change on erosion risk: an integrated remote sensing, geographic information system and modelling methodology. Land Degrad. Dev 24:409–421. doi:https://doi.org/10.1002/ldr.1137

Ligonja PJ, Shrestha RP (2015) Soil erosion assessment in kondoa eroded area in Tanzania using universal soil loss equation, geographic information systems and socioeconomic approach. Land Degrad Dev 26(4):367–379

Low PS (2013) Economic and social impacts of desertification, land degradation and drought. White Paper I. UNCCD 2nd Scientific Conference, prepared with the contributions of an international group of scientists. Available at: https://profiles.uonbi.ac.ke/jmariara/files/unccd_white_paper_1.pdf

Magesh NS, Chandrasekar N, Soundranayagam JP (2012) Delineation of groundwater potential zones in Theni district, Tamil Nadu, using remote sensing, GIS and MIF techniques. Geos. Frontiers 3 (2):189–196

Mahala, A (2017) Processes and Status of Land Degradation in a Plateau Fringe Region of Tropical Environment. Environmental Processes 4 (3):663–682.doi:https://doi.org/10.1007/s40710-017-0255-6

Mahmoud SH, Alazba AA (2016) Integrated remote sensing and GIS-based approach for deciphering groundwater potential zones in the central region of Saudi Arabia. Environ Earth Sci 75 (4):1–28. doi:https://doi.org/10.1007/s12665-015-5156-2

Minami K (2009) Soil and humanity: Culture, civilization, livelihood and health. Soil Sci. Plant Nutr 55: 603–615. doi:https://doi.org/10.1111/j.1747-0765.2009.00401.x

Ministry of Environment and Forestry (2011) Elucidation of the 4 National Report submitted to UNCCD Secretariat. Ministry of Environment and Forest. GOI 1–121. http://envfor.nic.in/sites/default/files/unccdreport_0.pdf

Mohawesh Y, Taimeh A, Ziadat F (2015) Effects of land use changes and soil conservation intervention on soil properties as indicators for land degradation under a Mediterranean climate. Solid Earth 6:857–868. doi:https://doi.org/10.5194/se-6-857-2015

Molla T, Sisheber B (2017) Estimating soil erosion risk and evaluating erosion control measures for soil conservation planning at Koga watershed in the highlands of Ethiopia. Solid Earth 8(1):13–13.

Moore ID, Burch GJ (1986) Physical basis of the length-slope factor in the universal soil loss equation. Soil Sci Soc Am J 50:1294–1298

Nachtergaele F, Petri M, Biancalani R, Van Lynden G, Van Velthuizen H (2010) Global Land Degradation Information System (GLADIS), Beta Version, An information database for land degradation assessment at global level, Land Degradation Assessment in Dry lands Technical Report, no. 17, FAO, Rome, Italy, 2010.

Nagarajan R, Roy A, Vinod Kumar R, Mukherjee A, Khire MV (2000) Landslide hazard suspectibility mapping based on terrain and climatic factors for tropical monsoon regions. Bull Engl Geol Env 58:275–287

Naseer A, Pandey P (2018) Assessment and monitoring of land degradation using geospatial technology in Bathinda district, Punjab, India. Solid Earth 9(1):75–90 doi:https://doi.org/10.5194/se-9-75-2018

Ohta Y, Nakagoshi N (2011) Analysis of Factors Affecting the Landscape Dynamics of Islands in Western Japan. In: Hong SK (ed) Ecol. Res. Monogr. Springer, Tokyo, pp 169–185. doi:https://doi.org/10.1007/978-4-431-87799-8_12

Pallavicini Y, Alday JG, Martinez-Ruiz CF (2015) Factors affecting herbaceous richness and biomass accumulation patterns of reclaimed coal mines. Land Degrad. Dev 26:211–217. doi:https://doi.org/10.1002/ldr.2198

Rahmati O, Haghizadeh A, Pourghasemi HR, Noormohamadi F (2016) Gully erosion susceptibility mapping: the role of GIS based bivariate statistical models and their comparison. Nat Hazards 82:1231–1258

Reddy VR (2003) Land degradation in India: extent, costs and determinants. Econ Polit Wkly 38(44): 4700–4713

Samanta RK, Bhunia GS, Shit PK (2016) Spatial modelling of soil erosion susceptibility mapping in lower basin of Subarnarekha river (India) based on geospatial techniques. Modeling Earth Systems and Environment. 2(2):99.doi:https://doi.org/10.1007/s40808-016-0170-2

Sarkar D, Hayak DC, Dutta D, Dhyani BL (2005) Soil Erosion of West Bengal. NBSS publ. No. 117,NBSS & LUP (ICAR), Nagpur, p 1–59

Scherr SJ, Yadav S (1997) Land degradation in the developing world: issues and policy options for 2020. Int Food Policy Res Inst 2020: 44

Scullion JJ, Vogt KA, Sienkiewicz AA, Gmur SJ, Trujillo C (2014) Assessing the influence of land-cover change and conflicting land-use authorizations on ecosystem conversion on the forest frontier of Madre de Dios, Peru. Biol Conserv 171:247–258. doi:https://doi.org/10.1016/j.biocon.2014.01.036

Sen J, Sen S, Bandyopadhyay S (2004) Geomorphological investigation of badlands- a case study at Garbheta, West Medinipur District, West Bengal, India. In: Singh S, Sharma HS, De SK (eds) Geomorphology and environment. ACB publication, Kolkata, p 204–234

Shaban A, Khawlie M, Abdallah C (2006) Use of remote sensing and GIS to determine recharge potential zone: the case of Occidental Lebanon. Hydrogeal J 14(4): 433–443. doi:https://doi.org/10.1007/s10040-005-0437-6

Sheng TC (1989) soil conservation on small farmers in the humid tropics. food and agriculture organization of the United Nations – FAO: Rome.

Shit PK, Maiti R (2012) Rill Hydraulics - An experimental study on gully basin in lateritic upland of Paschim Medinipur, West Bengal, India. J Geogr Geol 4(4):1–11. doi:https://doi.org/10.5539/jgg.v4n4p1

Shit PK, Nandi AS, Bhunia GS (2015) Soil erosion risk mapping using RUSLE model on Jhargram sub-division at West Bengal in India. Model Earth Syst Environ 1(3):28. doi:https://doi.org/10.1007/s40808-015-0032-3

Stéphenne N, Lambin EF (2001) A dynamic simulation model of land-use changes in Sudano sahelian countries of Africa (SALU). Agric Ecosyst Environ 85(1–3):145–161. doi:https://doi.org/10.1016/S0167-8809(01)00181-5

Thapa R, Gupta S, Guin S, Kaur H (2017) Assessment of groundwater potential zones using multi-influencing factor (MIF) and GIS: a case study from Birbhum district, West Bengal. Applied Water Science 7 (7): 4117–4131.doi: https://doi.org/10.1007/s13201-017-0571-z

Tucker CJ, Sellers PJ (1986) Satellite remote sensing of primary production. International journal of remote sensing 7(11): 1395–1416

Wischmeir WH, Smith DD (1978) Predicting rainfall erosion losses-a guide to conservation planning. Predicting rainfall erosion losses-a guide to conservation planning. Agr. Handb. No. 527,USDA, Washington, D.C

Zhao X, Dai J, Wang J (2013) GIS-based evaluation and spatial distribution characteristics of land degradation in Bijiang watershed. International Conference on Combating Land Degradation in Agricultural Areas (ICCLD'10) Zian City, PR China, 11–15 October 2010, Springer Plus, 2, S8, doi:https://doi.org/10.1186/2193-1801-2-S1-S8

Ujjal Senapati is a research scholar of Ph.D. program in the Department of Geography under Cooch Behar Panchanan Barma University, Cooch Behar, West Bengal, and he has been working in the field of agricultural drought in the river basin of West Bengal. He has completed his master's degree (M.A. in Geography) from Rabindra Bharati University (RBU), Kolkata. He has worked on the problems and prospects of floriculture at Paskura Block in Purba Midnapore district. Later, he has worked as a Guest Lecturer in Midnapore College (autonomous) for two years before joining the Ph.D. program.

Tapan Kumar Das is currently working as an Assistant Professor of Geography in Cooch Behar College, Cooch Behar, West Bengal, India.

He obtained his M.Sc. in geography from the University of Calcutta in the year 1997 and M.A. in education from Netaji Subhas Open University, West Bengal, in 2011. He received his Ph.D. in Geography from Vidyasagar University, West Bengal, in 2012, and he has completed his M.Ed. degree from Indira Gandhi National Open University, India, in 2017. His major research area includes river embankment breaching in Sundarbans, sand mining in Dooars, empowerment of female domestic workers in Cooch Behar, drought in Purulia, and some other geomorphological and environmental issues in West Bengal.

Priyank Pravin Patel, Rajarshi Dasgupta, and Sayoni Mondal

Abstract

While high-resolution Digital Elevation Model (DEM) datasets have been captured through space-borne imagery to map badland landscapes at fine scales, these remain essentially top-down views of the surface and cannot adequately capture the micro-morphological features (soil pipes, hollows, earth pillars and rills) which form along the gully flanks and walls. For imaging these, thus, a ground-based, horizontal looking sensor is required, with further post-processing abilities of identifying individual aspects, stitching the images together using common points, generation of sparse and dense point clouds and final creation of meshes with overlain textures to generate viable three-dimensional representations of such surface forms. This paper uses ground-based close-range digital photography along with the structure-from-motion technique to generate high-resolution DEMs and three-dimension (3D) representations of gullies and badland landscapes, which cannot otherwise be captured even by very high-resolution imagery or available satellite-imaged elevation datasets, since these systems lack the required view angle. The structure-from-motion technique has been a recent advancement in this domain, and its mechanism and workflow are explained in detail in this paper. Examples of its implementation in gully mapping are then showcased from the Gangani badland tract in Paschim Medinipur district of West Bengal. Repeated photographic surveys reveal the nature of morphological changes in the micro gully features, documenting erosional activity at fine scales. The imaging and rendering
of these micro features also allow information like the relative lithological hardness to be superimposed on them for a better comparative analysis.

Keywords

Gully · SfM · DEM · Photogrammetry · Badland, Structure-from-motion · Garbeta, Gangani

12.1 Introduction

Gullies are near-ubiquitous features on the Earth's surface, and as such, gully erosion constitutes one of the most prevalent causes of land degradation worldwide. In fact, it accounts for 10–94% of all sediment production and soil loss rates in the world (Poesen et al. 2003). Gully erosion was reported for the first time by Sir Charles Lyell during his visit to the United States in 1846 and the first quantitative assessment was provided by Ireland (1939). However, despite more than a century-long attention, gully erosion remains a complex process and there are still many areas of uncertainties and knowledge gaps that merit immediate attention (see Castillo and Gómez 2016 for a succinct review).

Considerable attention has been given to morphological mapping of gullies using both field-based and remote sensing technologies. While field mapping remains the preferred choice in most investigations, the advent of multi-temporal and multi-resolution satellite imageries, unmanned aerial vehicles (UAVs) and ground-based photogrammetry like LiDAR have enabled better and, often, more detailed assessment of gully dynamics through the use of Digital Elevation Models (DEMs) (Bennett and Wells 2019). The quality, spatial resolution and accuracy of DEMs used in fluvial geomorphology have improved considerably since the advent of aerial and terrestrial laser scanning and digital photogrammetry (Lane 2000; Notebaert et al. 2009), and the use of these datasets to generate topographic information has been

Electronic supplementary material: The online version of this chapter (https://doi.org/10.1007/978-3-030-23243-6_12) contains supplementary material, which is available to authorized users.

P. P. Patel (✉) · S. Mondal
Department of Geography, Presidency University, Kolkata, India
e-mail: priyank.geog@presiuniv.ac.in

R. Dasgupta
Department of Geography, East Calcutta Girls' College, Kolkata, India

investigated in great detail, along with the importance of the resolution of such datasets in eliciting ground information (e.g. Das et al. 2016; Patel and Sarkar 2009, 2010; Patel 2013), and consequently, just in the last decade alone, a large number of studies have investigated gullies using these technologies (e.g. Marzolff and Poesen 2009; Perroy et al. 2010; Shruthi et al. 2015; Liu et al. 2016; Goodwin et al. 2017).

While the use of these techniques has proliferated in the recent years, several problems stand in the way of their effective implementation in all settings. For example, the use of terrestrial laser scanners is expensive, the efficacies of UAVs depend on favourable weather conditions, and more often than not, the resolutions of satellite imageries cannot match with those acquired by ground-based aerial devices (Westoby et al. 2012; Wells et al. 2016). In such instances, close-range digital photogrammetry offers a cost-effective means of monitoring morphological changes. Of the various close-range photogrammetry techniques, Structure-from-motion together with Multi-view stereo (hereafter referred to as Sfm-MVS) provides rapid, high quality, dense three-dimensional point clouds for any surface or object at minimum cost. This technique has enabled the use of commercially sold, handheld digital cameras, smart phones and consumer-grade UAVs to generate DEMs at centimetre- to millimetre-scale resolution with minimal errors for landscape monitoring (James and Robson 2012, 2014; Micheletti et al. 2015; Carrivick and Smith 2019). Thus, Sfm-MVS has found a wide range of use in the geosciences and physical geography, with varied applications in geomorphology (Westoby et al. 2012; Smith et al. 2016). It is however beyond the scope of this paper to discuss the full details, including the working principles, of Sfm-MVS and the interested readers are directed to Carrivick et al. (2016) for a complete overview.

The Sfm-MVS technique has also recently been used in gully mapping using both ground-based and aerial platforms. Gómez-Gutiérrez et al. (2014) have mapped five gully head cuts in southwestern Spain, using a combination of Sfm-MVS and terrestrial laser scanning, and found that they produced comparable results with very low error. Additionally, using the DEM of Difference approach, they also calculated the volume of soil loss at the gully heads. Ground-based Sfm-MVS was also used by Frankl et al. (2015) to assess gully head morphology in Ethiopia and Belgium. Using digital photographs, they constructed 3D models of gully morphology, including finer features like soil pipes. Christian and Davis (2016), on the other hand, used UAV-based Sfm-MVS to study a small gully in California and found that gully morphology can be mapped in more detail using this technique than through LiDAR. In contrast to these studies are those which use both ground- and aerial-based Sfm-MVS to compare the efficacies of the two (Stöcker et al. 2015; Lannoeye et al. 2016; Koci et al. 2017; Glendell et al.

2017). Their results suggest that aerial- and ground-based Sfm-MVS analyses are often complementary to each other and can provide insights that a single platform cannot. For example, during their study in NE Australia, Koci et al. (2017) found that UAV-based 3D modelling is better suited for examining gullies at the hillslope scale, but digital photographs can provide better estimates of the finer aspects of gully morphology. This was also corroborated by Stöcker et al. (2015), who found that UAV-based 3D modelling becomes restrictive in sites with steep walls, and here, digital photography might provide additional information. Lannoeye et al. (2016) used digital photographs and photographs taken from a camera mounted on a pole (as a proxy for UAVs) to assess gully morphology and erosion in northern Ethiopia. Their study showed that Sfm-MVS can be effectively used to study gully erosion at time scales of individual rainfall events. A study comparing 3D models derived from terrestrial laser scanning and both UAV and ground-based Sfm-MVS found that although both the photographic techniques were comparable to the terrestrial laser scanning, the post-processing time for UAVs was shorter than the other two methods used (Glendell et al. 2017). When combined with user-friendly software like FreeXSapp, Sfm-MVS data can also provide useful information on gully cross-sections in remarkably short periods of time (~0.5 h, Castillo et al. 2018). The results of these studies suggest that the Sfm-technique is a cost- and time-effective way to map gullies at very high resolution. Besides, this approach can provide estimates of both vertical erosion and volume loss from gullies, which have important implications for biogeochemical cycling in gullied landscapes (Glendell et al. 2017).

In the present study, we have used the digital photography-based SfM technique to investigate fine-scale rill and gully morphology of a lateritic badland terrain located in the southwestern part of the state of West Bengal in eastern India. Gullied landscapes and the mapping of their morphological aspects in this region have been previously investigated by a number of workers (e.g. Bandhopadhyay 1987; Jha and Kapat 2009; Ghosh and Bhattacharya 2012). The area in focus here, the Gangani Tract (approximately 20 km^2 in area), is situated near the small town of Garbeta (22° 51′ 47″ N, 87° 21′ 13″ E) in Paschim Medinipur District, West Bengal (Fig. 12.1). This site is part of the lateritic uplands situated in the northern and western portions of the district and, as such, is primarily an area of red soils with agriculture being mainly dependent upon intermittently available irrigation. The outstanding characteristic of the area is the series of badlands (intensely gullied surfaces) that have formed, primarily along the right bank of the River Silabati. While this Gangani Tract has been the subject of a few previous works focusing on the gully morphology, most of these have used medium resolution satellite images to study the overall area or ground-based GPS surveys for mapping of

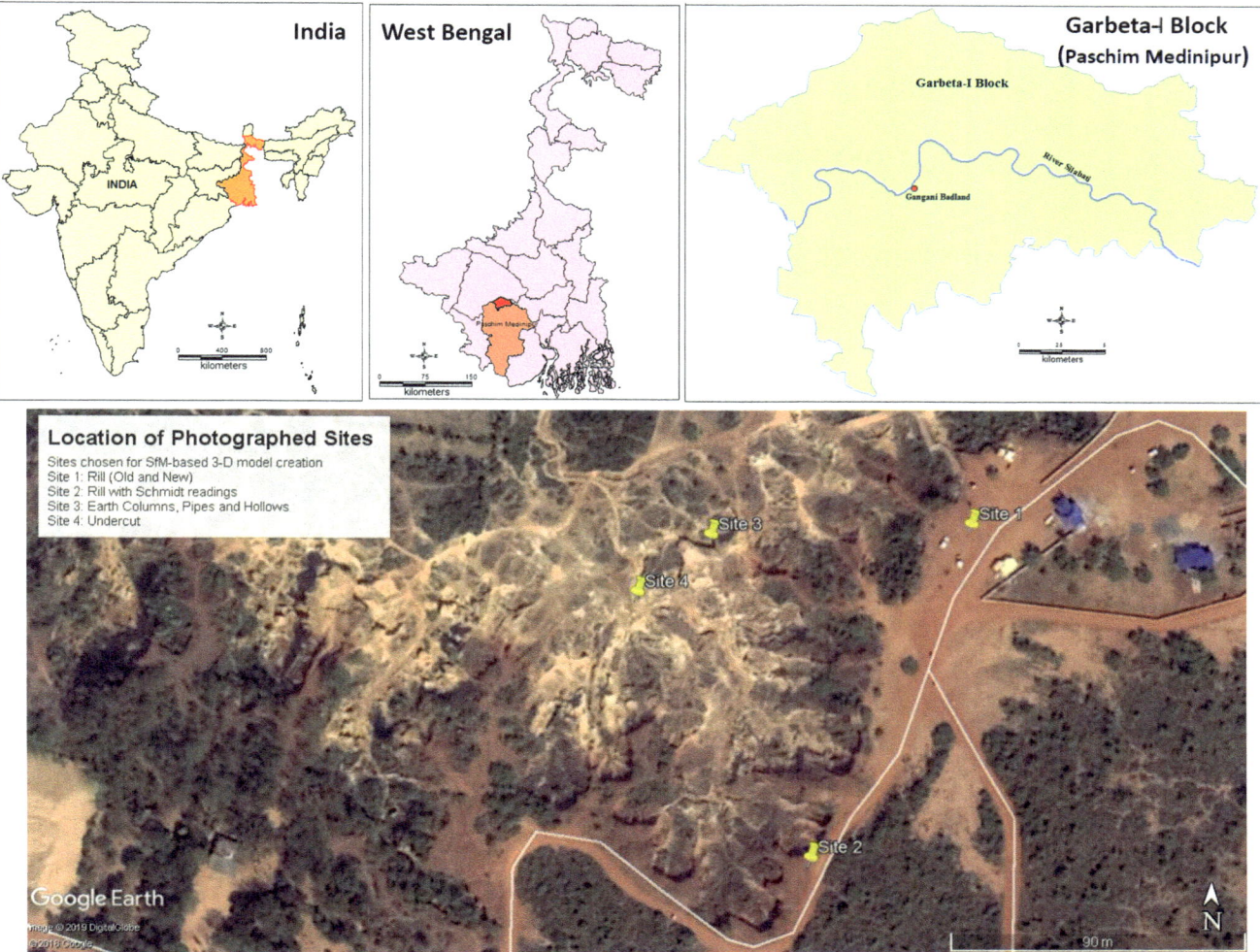

Fig. 12.1 Location of the study area and the photographed sites

small sections (Bandhopadhyay 1988; Patel and Mondal 2019; Shit and Maiti 2012; Shit et al. 2013, 2015). Fine-scaled detailed records of the minor rills and erosional features formed here, particularly along the gully walls and sides, have been rare and no previous studies using UAV or ground-based SfM measurements have been done in this area.

12.2 Materials and Methods

Surveys were conducted across the Garbeta lateritic gully terrain with a digital camera (a Nikon CoolPix series model S5300) and a GPS. Numerous photographs were taken of the different target features, with the aim being to capture the reflected light from the target feature in every possible angle. This was paramount in order to obtain every facet of the feature and not incur data loss in one segment, which would have led to that portion being imaged as a hollow component

with no data points to fill in. The photographs were thus ideally captured during the early morning, to negate the shadow problem from either obscuring the target feature or the shade cast getting photographed and being misrepresented itself. However, this was not possible for all the sites, especially where there were overhanging slopes that cast their shade across the target features and elements of this artefact thus did get incorporated into the final product. A number of Ground Control Point (GCP) tags were stationed along side each target feature, and the positional coordinates of these tags were obtained using a GPS. These were the inputs into the point cloud model generated subsequently for geo-referencing of the targeted features in order to elicit their dimensions. For processing of the images and the generation of their models, Agisoft PhotoScan software was utilised.

The SfM workflow (Fig. 12.2) involved creating a folder in which all the photos of a target feature were kept. Care was taken that the target feature filled up the majority of the portion of each photograph, to avoid computational wastage

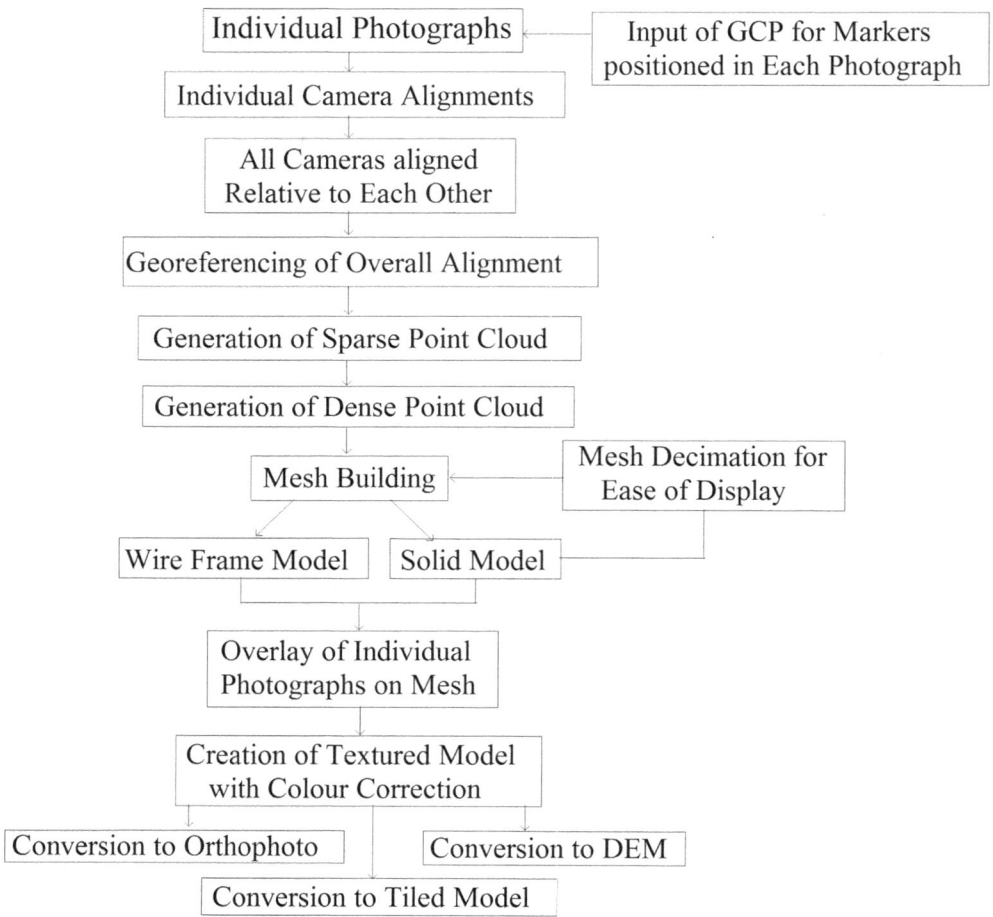

Fig. 12.2 Schematic representation of the workflow involved in generating 3D models via the SfM method

and to reduce the processing time. This also helped to ensure that objects that were not required did not become a part of the generated model, which would otherwise have taken much time and effort to clean. Imaging of plants was avoided as much as possible; however, this was not always feasible. The point cloud features representing leaves and grasses then required careful cleansing from the generated model. The camera specifics were recorded within each image and care was taken that they were not automatically geo-tagged while clicking, in order to remove any inaccuracies that may have crept in from the camera's GPS. Instead, a Total Station and a high-end handheld GPS unit were utilised to obtain the locational information.

Once the photographs were filtered and arranged, the software was used to align them using the highest accuracy mode. This entailed object-based feature recognition between two images and the tie point limit was set at 4000, meaning that up to these many points were sought to be matched between corresponding images. This was a computationally arduous task and its completion elicited plotting of the camera locations with each photograph being positioned relative

to each other in a 3D space. The geo-referencing of this camera alignment was then done using the GCPs obtained during the GPS survey. This required identification of the GCP-tags in each photograph and inputting of its locational coordinates. Subsequent to this, a sparse point cloud was generated from the photographs which gave forth a skeletal representation of the target feature. Computationally, intensive interpolation of this sparse point cloud dataset resulted in the generation of a more complete representation of the target feature with the creation of a dense point cloud. The cleaning of the dense point cloud, to remove extra or not required target features and especially vegetation overhangs, was done next in an intensive manual examination process of the entire structure created so far. The next stage involved creation of a mesh with high face count from this dense point cloud, in order to further fill in the gaps and create a more solid-looking output. The generated mesh could be displayed as either a wireframe model or a solid entity with grayscale hues to represent the variations in the surface form. Since zooming in and out of this model and working with it further were required, the high face count mesh generated for each target feature was decimated to create a more computationally viable model. The individual photographs were then mosaiced together and overlain on top of the created 3D representation, with attendant colour matching, to produce a textured rendering of the target feature, as close to the real world object as feasible. This output was then converted to tiled models and x–y–z representations for exporting to any GIS software for further analysis. They were also exported as 3D Adobe PDF files for ease of display and maneuvering for readers (see the Supplementary Information files).

A number of target features were imaged and created in this manner. They were mostly selected along the gully walls to demonstrate the utility of this method. One target feature was a small rill, which was imaged in October, 2016, and subsequently in September, 2017. Change detection (visual and computational) was performed on this rill feature, which would not be possible otherwise even from very high-resolution images and would require time-consuming and arduous repetitive Total Station or dGPS surveys to otherwise accomplish. Along with the photographic surveys undertaken, we have also used an Type N RockSchmidt Hammer to record and gauge the relative hardness values of the individual strata over which such rills or incipient gullies have developed. Once a target spot was chosen, it was smoothened with a carborundum wheel in order to eliminate any surface irregularities that may skew the rebound value. The plunger of the RockSchmidt was pressed onto the chosen spot and around it from about 7 to 11 times in each case and the rebound values recorded every time. These were averaged to obtain the mean Rebound Value (Q) for each location and then written onto the generated 3D models. The other target features included objects that would be specifically difficult to image from space or airborne vertically downward looking platforms and even from conventional, traditional ground-based surveying implements. These were hollows and cave-like features that had developed into the lateritic scarp, soil erosion pipes and earth pillars and overhang slopes formed by the erosive movement of monsoonal flow along the basal toe sections in these otherwise dry gullies. Each of these otherwise difficult to document and measure objects was photographed from various angles and rendered subsequently for generating their respective 3D models, to demonstrate the SfM technique's utility.

12.3 Results and Discussion

In this section, we present the details of each of the target features that were imaged and recorded in the manner described above. Salient technical details of the photos taken and models generated are also provided (Table 12.1). Each of these features was chosen from within the gully field to demonstrate the quite unique imaging capabilities of the SfM method. These features were as follows:

12.3.1 Upper Surface Rills

A number of small rills have colonised the upper lateritic surface of the Gangani area. These usually begin from the fringes of the lateritic caprock cover in this area and proceed to the edge of the scarp, over which they plunge into a significantly bigger gully head. One such rill, located at 22° 51′ 28″ N, 87° 20′ 25″ E, was selected (Fig. 12.3). The overall dimensions of this rill were in the scale of centimetres (about 3 cm near its head to about 87 cm near its mouth) in terms of width and only a few metres in terms of length (about 20 m), and as such, this feature would have been too minute to image from conventional satellite datasets, even of very high resolution. GCP tags were placed around it and the photographs captured, with shadows being an issue at times in covering up part of the scene. The over 500 photographs captured in October, 2016, of this small target area in this instance were then processed using the matching points in each photo with other photos capturing the same part of the feature (Fig. 12.4a) to obtain the camera alignments (Fig. 12.4b), eventually leading to the generation of the sparse and dense point clouds (Fig. 12.4c, d). The created dense point cloud had over eight million points, which combined to provide a surface perspective of the imaged feature

Table 12.1 Technical details of the generated 3D landform models via SfM

Sl. No.	Landform(s)	No. of photos	No. of points in sparse point cloud	No. of points in sparse point cloud	No. of faces (triangular facets) in 3D mesh	No. of vertices in 3D mesh
1	Rill (old)	563	156,752	8,713,728	193,637	97,967
2	Rill (new)	280	55,861	2,853,877	190,256	95,899
3	Rill (Schmidt)	175	178,811	15,934,539	3,186,496	1,593,453
4	Cave-like form, pipes and pillars	238	129,722	5,976,792	398,451	201,014
5	Overhang and undercut	225	114,540	2,631,289	175,299	88,665

Source: *Compiled by the authors from the photosets imaged and processed by them*

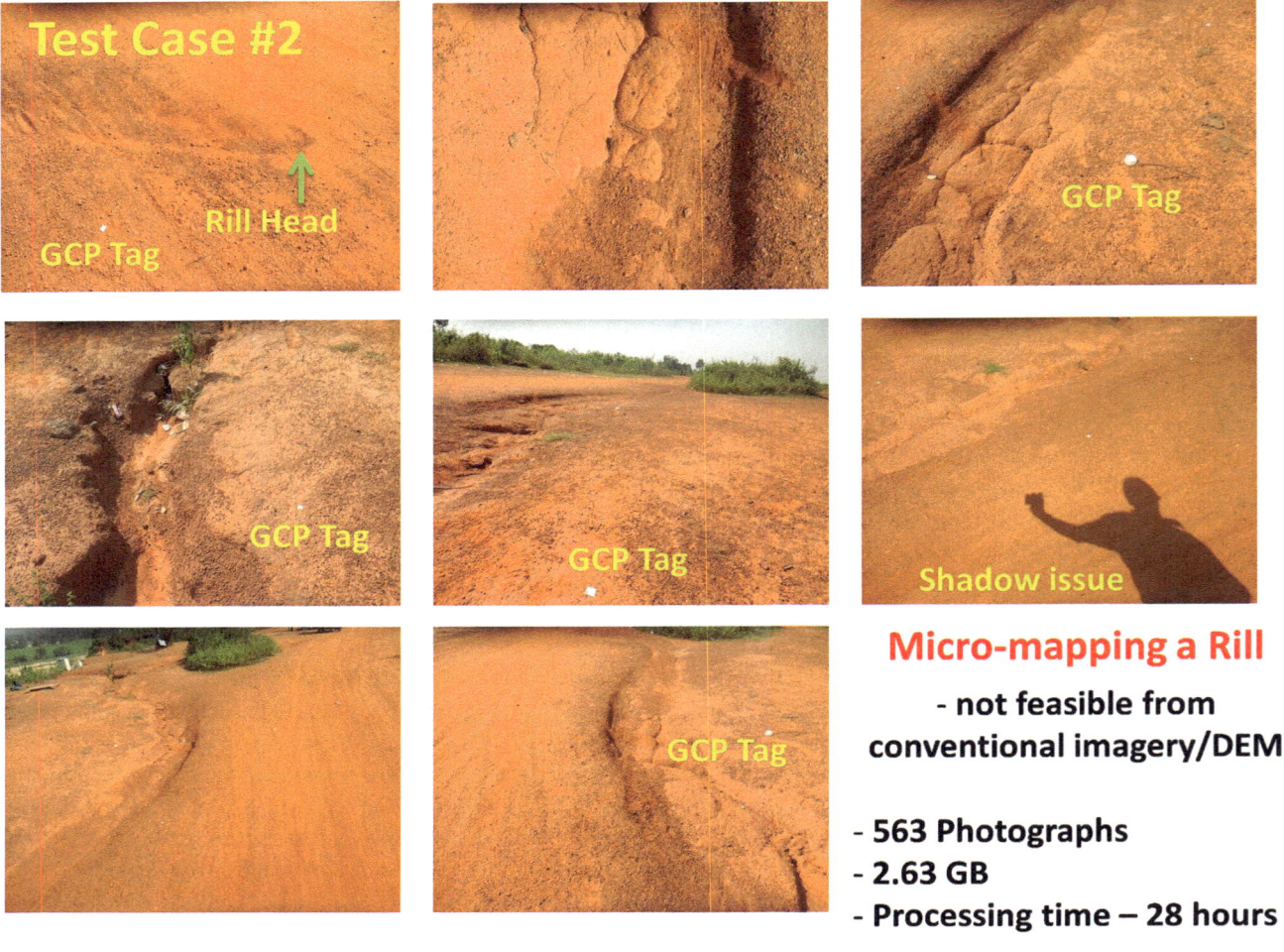

Fig. 12.3 Examples of captured photographs for generating the 3D model of a rill

that could be rotated and viewed from multiple angles (Fig. 12.4e). From the above, the mesh and textured models were rendered, to solidify the point cloud and provide real world colour and lighting to the structure (for interactively accessing, zooming in and rotating this stage of the rill model, see Supplementary Information File SI1). The point clouds and the mesh were then geo-referenced and therefore could be used to extract dimensions as shown (Fig. 12.4f) and also create the DEM for this target rill feature (Fig. 12.4g), which was interpolated from the mesh and provided an extremely high resolution of cell size 0.13 m. This allowed the micro nature of the target feature to be recorded properly for further conventional DEM-based analysis. The same rill was photographed and rendered again in September, 2017 (Fig. 12.5b). Within this time period, from the earlier capture (Fig. 12.5a), substantial land modification had been undertaken at the Gangani badlands to develop it into a potential tourist site. As part of the construction and landscape modification activities undertaken, the aforementioned rill had been covered over

by earth and fragments of *morrum* (hard lateritic caprock), into which some incision had still taken place in the intervening monsoon. Thus, the September, 2017, dataset gave an entirely different representation and perspective of this rill (Fig. 12.5c), with the piled up *morrum* fragments and the small excavated hollows (along the older course of the original rill) being displayed prominently in the rendered model. This comparison of the two different time period photosets elicited information on how changes in the micro-topography had occurred from such activities (for interactively accessing, zooming in and rotating this stage of the rill model, see Supplementary Information File SI2). Measurements elicited and taken at the site during these two time periods (Fig. 12.5d) also show how the nature of the rill surface was modified by the earth-filling process and the small localised erosion that had restarted at some spots along the rill. The overall drop along the rill profile, which had earlier been about 1 m, had been reduced mostly, along the course, except where renewed excavation

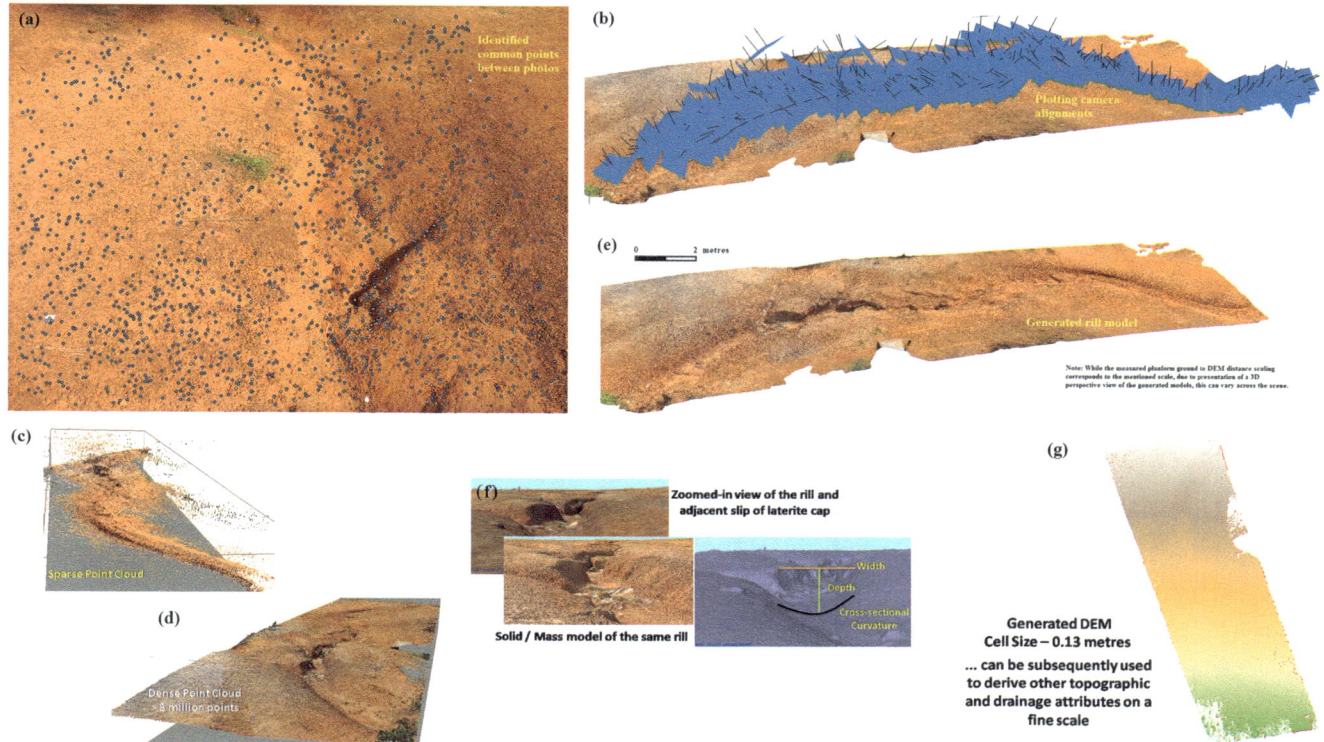

Fig. 12.4 Generating the SfM output for a rill—(**a**) Examples of the point-matching done (similar points in blue) to stitch the photographs together; (**b**) camera alignments performed with camera view axes above the target feature; (**c**) generated sparse point cloud of the rill; (**d**) generated dense point cloud of the rill; (**e**) rotated side-on view of the rill mesh created and (**f**) mesh with zoomed-in insets of the micro-erosional aspects and dimensions; (**g**) generated DEM of the target rill

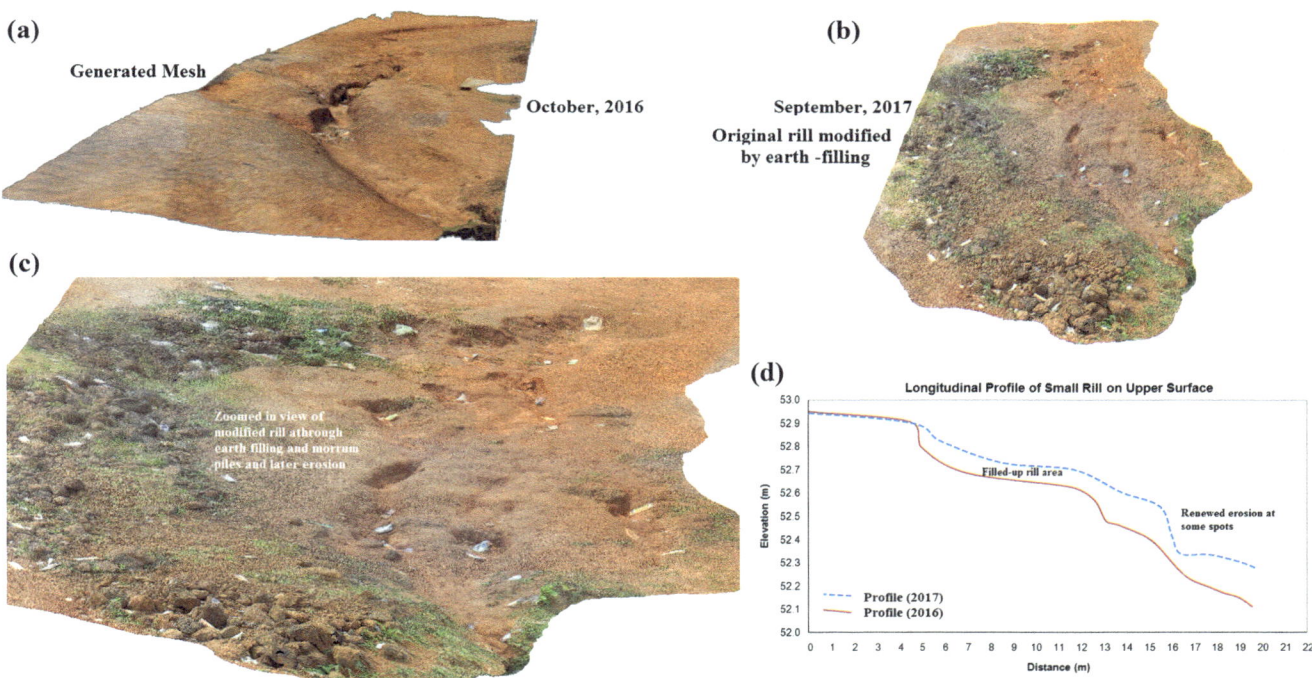

Fig. 12.5 Changes of the rill surface over two time periods—(**a**) The generated 3D model of the rill surface in October, 2016; (**b**) re-photographed model of the same rill in September, 2017, post-earth-filling works; (**c**) zoomed-in view of the piled up earth and micro-erosional aspects in the altered rill surface; and (**d**) changes in the measured rill surface profile

Imaging and constructing a small rill-head through SfM

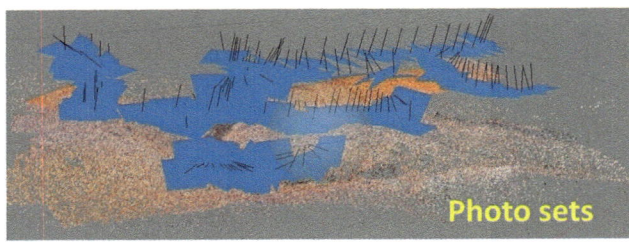

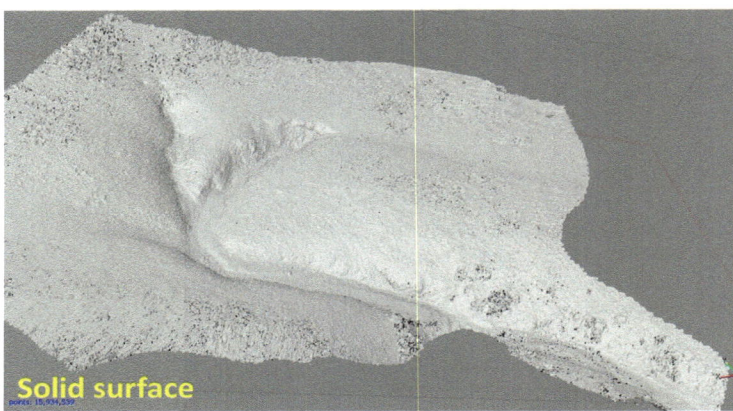

Fig. 12.6 Photographic workflow of the 3D model construction of a small rill-head through SfM

had deepened this small channel again, towards its mouth, where it plunged over the badland scrap into a major gully.

A similar such rill, located at 22° 51′ 23″ N, 87° 20′ 24″ E, was photographed in a slightly different part of the Gangani tract and its 3D model was generated in the same manner (Fig. 12.6). This rill cuts across the upper surface of the lateritic hardpan that had formed here, till its end at a large gully head wall, over which it plunged downwards. A RockSchmidt Hammer was used to obtain the rebound values (Q values) from the different surface lithotypes present in the area (Fig. 12.7). A higher Q value (obtained from the mean of all the observations recorded at each spot for each lithotype) denoted a greater hardness and thereby higher resistance of that stratum. With the area being composed of quite variable lithologies, from very compact and hard caprock to friable and vesicular laterite layers and softer sandstone and compacted clay beds, a considerably wide range of Q values was found across the examined spots. Those that were obtained along the aforementioned rill were then plotted at the tested spots along the rill's 3D mesh and texture model

(Fig. 12.8). It was discernable that as the rill carved its path through and across the initial softer surface veneer and into the lateritic hardpan caprock, the Q values increased. The micro-imaging of such landform allowed the superimposition and visual correlation of these rebound values with the lithology.

12.3.2 Cave-Like Formations and Earth Pillars

The importance of the SfM method could be best gauged from its ability to discern and reproduce features present along the gully walls, which could otherwise have not been imaged from normal sensor positions. Side-looking photographs to generate the sparse point cloud (Fig. 12.9a) were taken of a section of the high scarp at Gangani, located at 22° 51′ 28″ N, 87° 20′ 24″ E, along the right bank of the Silabati River, where a number of cave-like hollows, soil pipes and earth pillars had developed due to basal sapping, percolation and weathering processes and the gradual

Fig. 12.7 Instances of how the Rebound RockSchmidt Hammer was used to extract hardness 'Q' values

washing away of the fines by rains and seepage and subsequent collapse of the remaining structures. The generated mesh model (Fig. 12.9b) and final textured surface (Fig. 12.9c) demonstrated how vividly these features could be captured and represented. The perception of depth was conveyed, and the columnar form of the earth pillars and soil pipes was captured very satisfactorily. Colour variations were also apparent with the shadow effect causing parts of the generated model to be in light or shade (for interactively accessing, zooming in and rotating this model, see Supplementary Information File SI3).

12.3.3 Undercuts and Overhangs

An even more difficult to image and capture aspect of landforms from conventional imagery are the undersides of overhanging slope segments, since views are obscured. The SfM method again provided a viable alternative to this, by the use of side-looking photographs and photographs which were taken with the camera looking upwards towards the slope component while being positioned at the ground level. This was been done for a portion of the gully channel, located at 22° 51′ 27″ N, 87° 21′ 23″ E, where an overhanging slope component had developed due to the basal erosion and undercutting done by ephemeral monsoonal flows. The 3D mesh of this section (Fig. 12.10a) and its wireframe surface (Fig. 12.10b) clearly displayed the overhanging portions and how the channel had cut sideways into the rock at its base. The wireframe surface demonstrated how the edges and vertices were structured and aligned to form the triangular faces that merge with each other to represent the overall landform. The textured model (Fig. 12.10c) displayed the vertical gully wall, the overhanging slope segment and basal channel clearly, while a zoomed in, upward looking view of the same (Fig. 12.10d) showed the hollows that had formed on the underside of the overhang segment from solutional, seepage and weathering processes. It also

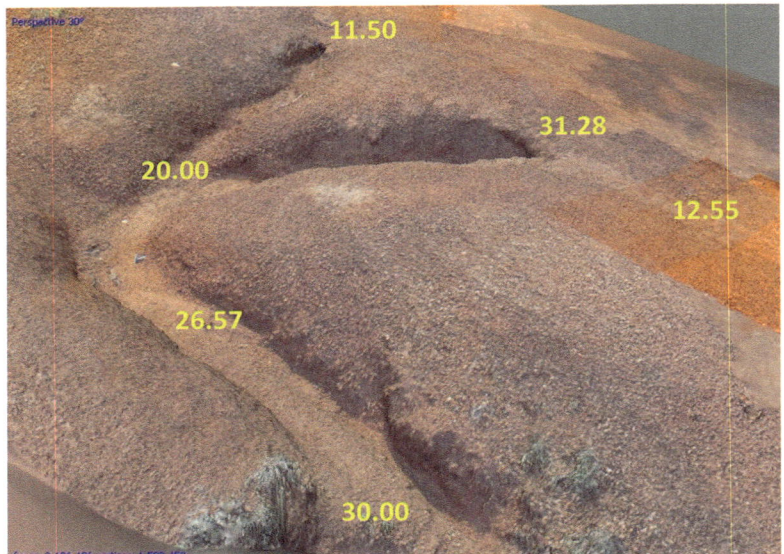

Final Textured Rill head – two views

The "Q"-value [=rebound V divided by inbound V] represents the physical rebound coefficient.

Comparable Q values for other materials obtained during calibration are as follows:
Sandstone block – 32.00
Haematite block – 58.57
Marble block – 72.32

Q values generally increase downstream along the rill from the friable upper surface towards the hardpan caprock over which the rill then plunges

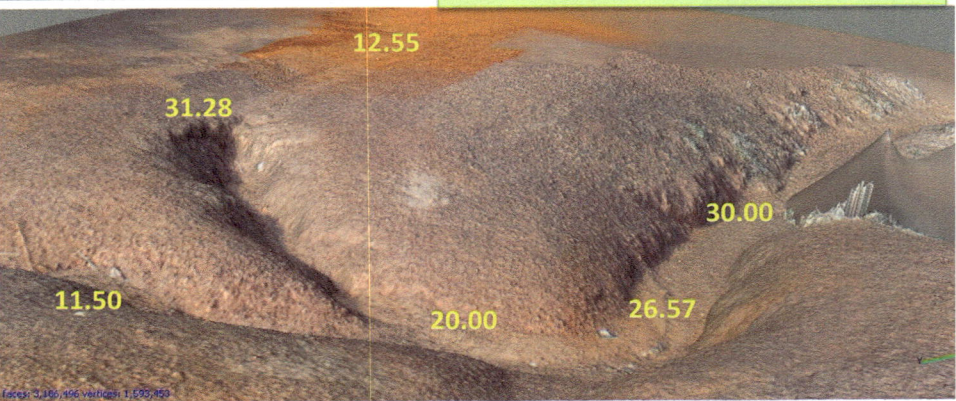

Fig. 12.8 Textured rill model with the 'Q' rebound values showing the variation in hardness along the rill

discerned the light-coloured thin veneer of sand deposited along the thalweg of the gully channel where it had flowed and how this thalweg had abut against and almost intruded within the underside of the bank wall. The rest of the gully channel bed portion, towards the side opposite to this bank wall, was covered by more coarser gains (for interactively accessing, zooming in and rotating this model, see Supplementary Information File SI4).

12.4 Conclusion

The SfM method was able to produce DEMs at much higher resolution than even the highest resolution commercially available DEM. It could generate realistic, true-colour representations of such hard to image and measure features, which would otherwise be overlooked in the usual DEMs produced from space-borne or airborne systems. Use of side-looking photographs enabled generation of features otherwise hidden from the view aspect of usual images. This

method can be thus suitably employed for imaging and mapping gullies and the minor landforms present within larger badland tracts, with this becoming easier if remote-controlled UAV systems are used. The products from both UAV and ground-based SfM methods have been shown to create DEMs at scales comparable to those from Terrestrial Laser Scanner and LiDAR surveys, at far less cost (Westoby et al. 2012; Lucieer et al. 2014), and therefore, this is a very viable method, particularly in difficult to access landscapes, like badlands.

However, the workflow is very computationally intensive and required considerable computer hardware resources to execute satisfactorily. Lesser hardware resources can force the computations and modelling to be done at lower resolutions, thereby increasing the inaccuracies and lowering the sharpness and DEM resolution of the finished product. The generation of higher end models however also requires a considerable investment of time along with the relevant hardware, to allow full processing. In this regard, the use of higher megapixel DSLR cameras, which would elicit larger

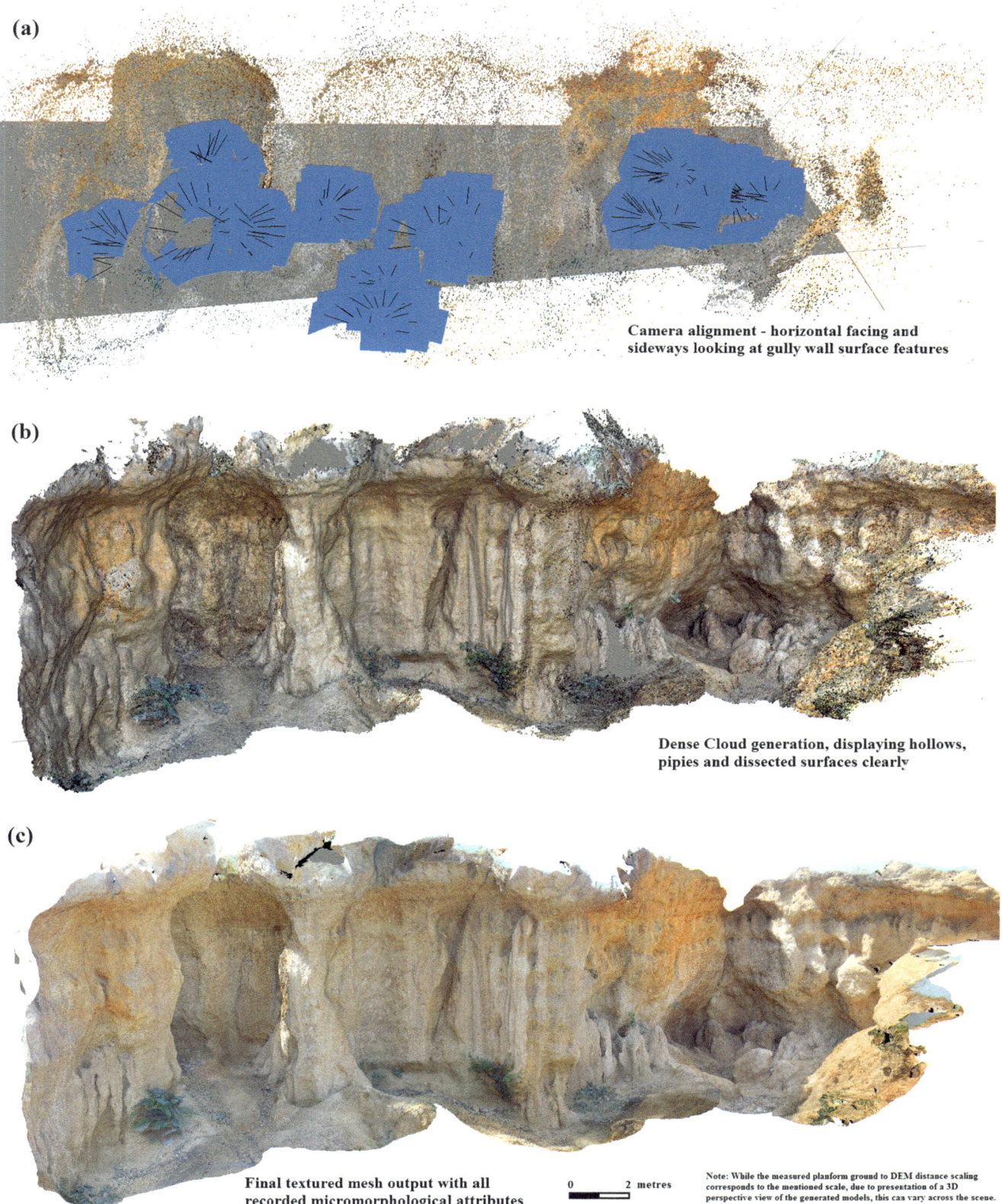

Fig. 12.9 Generation of the SfM model for a gully wall segment—(**a**) Horizontal looking photo capture of landform artefacts on the gully walls and the sparse cloud generated; (**b**) mesh of the cave-like forms, hollows, erosional pipes and earth pillars clearly imaged; and (**c**) textured and colour-corrected model of the features

Fig. 12.10 Generation of the SfM model for an overhanging slope with undercut base—(**a**) 3D mesh of overhang slope and basal undercut by ephemeral flows in a gully channel; (**b**) wireframe representation showing the nodes and vertices that construct the 3D model of the target; (**c**) textured and colour-corrected representation of the target; and (**d**) zoomed-in view of the basal undercut and small hollows formed by weathering and seepage processes

photograph file sizes, can also exert much influence on the amount and type of hardware required for processing and storage of the raw data and final outputs, even though they would obviously provide more fine-scale, detailed models. Photo capture and processing problems can be compounded by shadow issues and overhanging or obscuring vegetation cover that needs to be separated out and cleaned from the point clouds. A further issue encountered at the study site was the presence of garbage and plastic wastes, strewn across the Gangani tract from tourist activities, particularly at the studied rill sites. These needed to be brushed away before photographs could be taken, and their removal entailed some slight modification of the landform artefact itself. Nevertheless, the use of such close-range photogrammetric methods for rendering 3D models of gullies and their minor landforms is the most effective method for their mapping and measurement.

Acknowledgements This study has been funded by a University Grants Commission Start-Up Grant under its Faculty Research Promotion Scheme for early career researchers (Letter No. F.30-78/2014(BSR) dated 22nd January 2015) and by the FRPDF allotment by Presidency University to Priyank Pravin Patel.

References

Bandhopadhyay S. 1987. Man-initiated Gullying and Slope Formation in a Laterite Terrain at Santiniketan, West Bengal, Geographical Review of India, 49(4): 21–26.

Bandhopadhyay S. 1988. Drainage evolution in a badland terrain at Gangani in Medinipur District, West Bengal, Geographical Review of India, 50(3): 10–20.

Bennett SJ, Wells RR. 2019. Gully erosion processes, disciplinary fragmentation, and technological innovation. Earth Surface Processes and Landforms 44: 46–53.

Carrivick JL, Smith MW, Quincey DJ. 2016. Structure from motion in the geosciences. John Wiley & Sons, Chichester.

Carrivick JL, Smith MW. 2019. Fluvial and aquatic applications of structure from motion photogrammetry and unmanned aerial vehicle/drone technology. WIREs Water 6: e1328. DOI: https://doi.org/10.1002/wat2.1328.

Castillo C, Gómez JA. 2016. A century of gully erosion research: urgency, complexity and study approaches. Earth Science Reviews 160: 300–319.

Castillo C, Marín-Moreno VJ, Pérez R, Muñoz-Salinas, Taguas EV. 2018. Accurate automated assessments of gully cross-section geometry using the photographic interface FreeXSapp. Earth Surface Processes and Landforms 43: 1726–1736.

Christian P, Davis J. 2016. Hillslope gully photogeomorphology using structure-from-motion. Zeitschriftfür Geomorphologie Supplementary Issues 60: 59–78.

Das S, Patel PP, Sengupta S. 2016. Evaluation of different digital elevation models for analyzing drainage morphometric parameters in a mountainous terrain: a case study of the Supin - Upper Tons Basin, Indian Himalayas. SpringerPlus, 5: (1544).

Frankl A, Stal C, Abraha A, Nyssen J, Rieke-Zapp D, De Wulf A, Poesen J., 2015. Detailed recording of gully morphology in 3D through image-based modelling. Catena 127: 92–101.

Ghosh S, Bhattacharya K 2012. Multivariate Erosion Risk Assessment of Lateritic Badlands of Birbhum (West Bengal, India): A Case Study, Journal of Earth Systems Science, 121(6): 1441–1454.

Glendell M, McShane G, Farrow L, James MR, Quinton J, Anderson K, Evans M, Benaud P, Rawlins B, Morgan D, Jones L, Kirkham M, DeBell L, Quine TA, Lark M, Rickson J, Brazier RE. 2017. Testing the utility of structure-from-motion photogrammetry reconstructions using small unmanned aerial vehicles and ground photography to estimate the extent of upland soil erosion. Earth Surface Processes and Landforms 42: 1860–1871.

Gómez-Gutiérrez A, Schnabel S, Berenguer-Sempere F, Lavado-Contador F, Rubio-Delgado J. 2014. Using 3D photo-reconstruction methods to estimate gully headcut erosion. Catena 120: 91–101.

Goodwin NR, Armston JD, Muir J, Stiller I. 2017. Monitoring gully change: a comparison of airborne and terrestrial laser scanning using a case study from Aratula, Queensland. Geomorphology 282: 195–208.

James MJ, Robson S. 2012. Straightforward reconstruction of 3D surfaces and topography with a camera: accuracy and geoscience application. Journal of Geophysical Research- Earth Surface 117: F03017. DOI: https://doi.org/10.1029/2011JF002289.

James MJ, Robson S. 2014. Mitigating systematic error in topographic models derived from UAV and ground-based image networks. Earth Surface Processes and Landforms 39: 1413–1420.

Jha VC, Kapat S. 2009. Rill and Gully Erosion Risk of Lateritic Terrain in South-western Birbhum District, West Bengal, India, Sociedade & Natureza, 21(2): 141–158.

Koci J, Jarihani B, Leon JX, Sidle RC, Wilkinson SN, Bartley R. 2017. Assessment of UAV and ground-based structure from motion with multi-view stereo photogrammetry in a gullied savanna catchment. ISPRS International Journal of Geo-Information 6: 328. DOI: https://doi.org/10.3390/ijgi6110328.

Lane SN. 2000. The measurement of river channel morphology using digital photogrammetry. The Photogrammetric Record 16: 937–961.

Lannoeye W, Stal C, Guyassa E, Zenebe A, Nyssen J, Frankl A. 2016. The use of Sfm-photogrammetry to quantify and understand gully degradation at the temporal scale of rainfall events: an example from the Ethiopian drylands. Physical Geography 37: 430–451.

Liu K, Ding H, Tang G, Na J, Huang X, Xue Z, Yang X, Li F. 2016. Detection of catchment-scale gully-affected areas using unmanned aerial vehicles (UAV) on the Chinese loess plateau. ISPRS International Journal of Geo-Information 5: 238. DOI: https://doi.org/10.3390/ijgi5120238.

Lucieer A, De Jong SM, Turner D. 2014. Mapping landslide displacements using Structure from Motion (SfM) and image correlation of multi-temporal UAV photography. Progress in Physical Geography 38: 97–116.

Marzolff I, Poesen J. 2009. The potential of 3D gully monitoring with GIS using high-resolution aerial photography and digital photogrammetry system. Geomorphology 111: 48–60.

Micheletti N, Chandler JH, Lane SN. 2015. Investigating the geomorphological potential of freely available and accessible structure-from-motion photogrammetry using a smartphone. Earth Surface Processes and Landforms 40: 473–486.

Notebaert B, Verstraeten G, Govers G, Poesen J. 2009. Qualitative and quantitative applications of LiDAR imagery in fluvial geomorphology. Earth Surface Processes and Landforms 34: 217–231.

Patel PP, Sarkar A. 2009. Application of SRTM data in evaluating the morphometric attributes: a case study of the Dulung River Basin. Practicing Geographer, 13(2): 249–265.

Patel PP. 2013. GIS techniques for landscape analysis - case study of the Chel River Basin, West Bengal. Proceedings of State Level Seminar on Geographical Methods in the Appraisal of Landscape, held at Dept. of Geography, Dum Dum Motijheel Mahavidyalaya, Kolkata, on 20th March, 2012, 1–14.

Patel PP, Mondal S. 2019. Terrain - land use relations in Garbeta-I Block, Paschim Medinipur District, West Bengal. In: Mukherjee, S (ed) Importance and Utilities of GIS. Burdwan: Avenel Press, 82–101.

Patel PP, Sarkar A. 2010. Terrain characterization using SRTM data. Journal of the Indian Society of Remote Sensing, 38(1): 11–24.

Perroy RL, Bookhagen B, Asner GP, Chadwick OA. 2010. Comparison of gully erosion estimates using air-borne and ground based LiDAR on Santa Cruz Island, California. Geomorphology 118: 288–300.

Poesen J, Nachtergaele J, Verstraeten G, Valentin C. 2003. Gully erosion and environmental change: importance and research needs. Catena 50: 91–133.

Shit PK, Bhunia, GS, Maiti, RK. 2013. Assessment of factors affecting ephemeral gully development in badland topography: a case study at Garbheta badland (Pashchim Medinipur, West Bengal, India). International Journal of Geosciences 4: 461–470.

Shit PK, Maiti RK. 2012. Mechanism of gully-head retreat- a study at Ganganir Danga, Paschim Medinipur, West Bengal. Ethiopian Journal of Environmental Studies and Management, 5(4): 332–342.

Shit PK, Paira R, Bhunia GS, Maiti RK. 2015. Modeling of potential gully erosion hazard using geo-spatial technology at Garbheta block, West Bengal in India. Modeling Earth Systems and Environment 1: (2).

Shruthi RBV, Kerle N, Jetten V, Abdellah L, Machmach I. 2015. Quantifying temporal changes in gully erosion areas with object oriented analysis. Catena 128: 262–277.

Smith MJ, Carrivick JL, Quincey DJ. 2016. Structure from motion photogrammetry in physical geography. Progress in Physical Geography 40: 247–275.

Stöcker C, Eltner A, Karrasch P. 2015. Measuring gullies by synergetic applications of UAV and close range photogrammetry: a case study from Andalusia, Spain. Catena 132: 1–11.

Wells RR, Momm HG, Bennett SJ, Gesch KR, Dabney SM, Cruse R, Wilson GV. 2016. A measurement method for rill and ephemeral gully erosion assessments. Soil Science Society of America Journal 80: 203–214.

Westoby MJ, Brasington J, Glasser NF, Hambrey MJ, Reynolds JM. 2012. 'Structure-from-Motion' photogrammetry: a low-cost, effective tool for geoscience applications. Geomorphology 179: 300–314.

Priyank Pravin Patel is an Assistant Professor of Geography at Presidency University, Kolkata, India. His PhD was on the terrain analysis of river basins for sustainable development planning. His interests within Geography reside primarily in fluvial geomorphology and geotectonics, GIS and remote sensing applications, analysis of cultural landscapes and artefacts and the mapping of urban landscapes. Currently he is engaged in mapping the badlands developed on lateritic terrain in the western part of West Bengal, India for proper land use management.

Rajarshi Dasgupta presently teaches in East Calcutta Girls' College, Kolkata, India. His research interests include application of low temperature geochemistry to geomorphic and environmental problems as well as examining the theoretical and pedagogic aspects of the geosciences.

Sayoni Mondal is a Senior Research Fellow (UGC-NET-JRF) at the Department of Geography, Presidency University, Kolkata. Her ongoing PhD is on the use of ecogeomorphic methods for riverbank protection and flood mitigation in the western part of West Bengal. She has received multiple awards for facets of her research at different Conferences.

Effects of Grass on Runoff and Gully Bed Erosion: Concentrated Flow Experiment

13

Pravat Kumar Shit, Hamid Reza Pourghasemi, and Gouri Sankar Bhunia

Abstract

Soil detachment by overland flow has mainly been studied using small samples with smooth surfaces. However, the inner mechanism for this process remains unclear. In the present chapter, an experimental study was conducted to evaluate the role of vegetation buffer strips on concentrated flow hydraulics and gully bed erosion downstream on gully headcuts. Six gully beds were selected for simulation runoff experiments, with an average vegetation cover of buffer strips of approximately 67%. The flow scouring tests on each gully bed lasted 60 min with a flow discharge of 78.5 l min^{-1}. Hydraulic parameters were calculated, including Reynolds number (Re), Froude number (Fr), hydraulic share tress (τ), Manning roughness coefficient (N), and Darcy-Weisbach friction factor (f). Runoff samples were collected at the bottom part of each gully bed using a 0.5-l plastic bottle during the simulated runoff at 5-min intervals. The average parameter values for the vegetated buffer strips varied between 35% and 50%, which is lower than those for bare gully beds with a minimum value (74.53%) of τ. The vegetated buffer strips in gully beds indicate that total gross deposit of sediment after experiments ranged between 0.02 and 0.1 m^3 with an average of 0.037 m^3, which was 1.6 times higher than that of bare soil gully beds. The net erosion volume shows a negative relationship with time for all the experimental gully beds. Moreover, the area covered with dense vegetation cover had reduced surface runoff, sediment trapped with stems, and reinforcement of soil erosion and stabilization of the gully. Hence, encouraging vegetation in gully beds has been confirmed as an active measure in monitoring the expansion of gullies.

Keywords

Vegetation belts · In situ scouring experiment · Hydraulic properties · Gully bed · Runoff · Land management

13.1 Introduction

Soil or land degradation under rills and gullies is developed by concentrated flow erosion, resulting in the detachment and displacement of topsoil particles (Govers 1991; Govers et al. 2007). The soil loss includes (1) loosening and detachment of soil particles from the soil mass through the process of rilling and gullying, (2) removal and transport of eroded soils downslope and downstream by overland flow, and (3) soil slumping under the impact of the increased volume of detached soils (Singh and Dubey 2002). Gully erosion is one of the forms of accelerated soil erosion. Its occurrence often indicates an extreme form of land degradation warranting special attention (Singh and Phadke 2004).

In India, an area of approximately 4 million ha is affected by rill and gully erosion (Sharda et al. 2007). The Ministry of Agriculture, Government of India (2000) reported the 12 - Indian states that were most seriously affected by rill, gully, and ravine erosion. Gully erosion in India has mainly focused on the morphological characteristics (Bandyopadhyay 1987, 1988; Das and Bandyopadhyay 1996; Sen et al. 2004; Ghosh and Guchhait 2015; Shit et al. 2012), hydrological processes (Shit and Maiti 2013), and controlling measures of hillslope gully and on the development process (Singh and Phadke 2004), features, controlling factors, and field monitoring methods of ephemeral gully channels (Singh and Dubey 2002; Shit et al. 2011, 2012), and the delineation and

P. K. Shit (✉)
Department of Geography, Raja N. L. Khan Women's College (Autonomous), Medinipur, West Bengal, India

H. R. Pourghasemi
Department of Natural Resources and Environmental Engineering, College of Agriculture, Shiraz University, Shiraz, Iran

G. S. Bhunia
Aarvee Associates Architects, Engineers & Consultants Pvt. Ltd, Hyderabad, India

© Springer Nature Switzerland AG 2020
P. K. Shit et al. (eds.), *Gully Erosion Studies from India and Surrounding Regions*, Advances in Science, Technology & Innovation.
https://doi.org/10.1007/978-3-030-23243-6_13

monitoring of gully erosion lands (Pani and Mohapatra 2001; Ghosh 2015; Shit et al. 2015).

Moreover, relatively few studies have examined differences in the causes and influences of gully erosion in a subhumid lateritic belt in West Bengal, India (Shit et al. 2013). In this region, geomorphological features, controlling factors, and monitoring methods for ephemeral gullies have been studied (Sen et al. 2004; Shit et al. 2014; Ghosh 2015). Additionally, Shit et al. (2011) examined headward erosion of gully heads under different discharge condition, and Shit et al. (2013) studied the distribution and influencing factor of vegetation (above biomass and ground biomass) on gully erosion in this region.

Several studies have been carried out on rill and gully erosions, but only a few studies have concerned utilizing natural vegetation to control gully erosion as well as headward erosion. Vegetation has an obvious effect in reducing concentrated flow and controlling soil loss, affecting sediment yield of the overland flow in gully basins (Rey 2003; Spaan et al. 2005; Li et al. 2006; Su et al. 2014; Shit et al. 2012, 2013). However, most of the previous studies of vegetation in soil erosion were mainly conducted on hill slopes with few examining the effects on vegetation in gully erosion. Numerous researchers reported that the vegetation cover has a significant influence on surface runoff volume and sediment (Braud et al. 2001). Spaan et al. (2005) and Rey (2003) reported that vegetation could effectively slow the velocity of the flow, trap sediment, build up backwater, and promote sedimentation to reduce gully erosion. However, few studies focus on rill-gully erosion mechanisms and particularly the impact of vegetation cover on hydraulic properties of surface runoff in gully channel and beds, because of the complex development processes and the many influential factors compared with sheet erosion.

In this chapter we investigated the role of a vegetation belt on gully erosion. For this aim, six ephemeral gully channels were selected to (1) examine the dynamic changes of gully hydraulic properties and sediment yield in response to the grass cover in the gully beds and (2) analyze the relationship between runoff properties and sediment content.

13.2 Materials and Methods

13.2.1 Study Area

All experiments were conducted at the Rangamati badland area ($22°23'$ to $22°24'$ N latitude and $87°17'$ to $87°18'$ E longitude), which is a typical dissection of the landscape by dense and deep rills and gully erosion in the western part of West Bengal (Fig. 13.1). The average slope of this area is between 25% and 35%. The most frequent landforms are complex slopes and gullies. The average annual precipitation and temperature are 1500 mm and 23.5 °C, respectively. The dominant soils at the surface of the study area are dry red soil and vertisols (USDA soil taxonomy). Dry red soils often have high sand content by iron and silica; vertisols often have high clay content with a mean bulk density of 1.34 to 1.86 cm^{-3}. According to a previous study (Shit and Maiti 2014), the region is dominantly herb species with *Eragrostis cynosuroides*, *Lantana camara*, and *Andropogon aciculatus* vegetation species in the gully bed (Shit and Maiti 2014). *Eragrostis cynosuroides* is a local species that can grow rapidly on gully beds in a subhumid tropical environment (Shit et al. 2013).

13.2.2 Field Monitoring and Measurement

Six gully beds were selected for simulation runoff experiments. The gully beds were 18 m long with width varying from 0.25 to 0.50 m, with an average slope of 18–20° on the gully bed. The six gully beds have similar topographical conditions, and headcut height was 0.3 m. The grain size distribution of soil is mainly gravel (30%), sand (40%), and silt and clay (30%). Dry soil bulk density ranges from 1.34 to 1.78 g/cm^{-3}.

Eragrostis cynosuroides was growing naturally in four gully beds with similar density and an average height of vegetation of 20–50 cm during the experiments, and the average vegetation cover of buffer strips was 67%. However, vegetation buffer strips were selected at the bottom end of the gully beds extended up to headward direction. The only bare gully bed was selected for control measurement and also compared with the buffer gully beds. Six strip widths were established on gully beds: bare (0), 3, 6, 9, 12, 15, and 18 m, which were named Gully-A, Gully-B, Gully-C, Gully-D, and Gully-E, respectively (Fig. 13.2).

The experiment on each gully bed consisted of five simulation runoff tests in the dry season (November 2018), during which no natural rainfall occurred. Clean water pumped through a 2-HP diesel pump from the nearby pond and a nylon water pipe was used to connect the adjustable opening to a flowmeter to measure the flow discharge during the experiment (Su et al. 2014). The flow scouring tests on each gully bed at lasted 60 min with flow discharge of 78.5 l min^{-1}. The discharge was adjustable according to average precipitation in the region. During these experiments, the hydraulic parameters and sediment yield were measured. The hydraulic parameters included the flow velocity (v in ms^{-1}), flow width (w, in meters, m), flow depth (d, in meters, m), and temperature (in °C). The flow velocity was measured using the color tracer method (potassium permanganate solution); flow width and depth were measured with a ruler and measuring tapes (Shit and Maiti 2014). Water temperature was measured with a centigrade thermometer. All parameters were measured three times for each gully bed

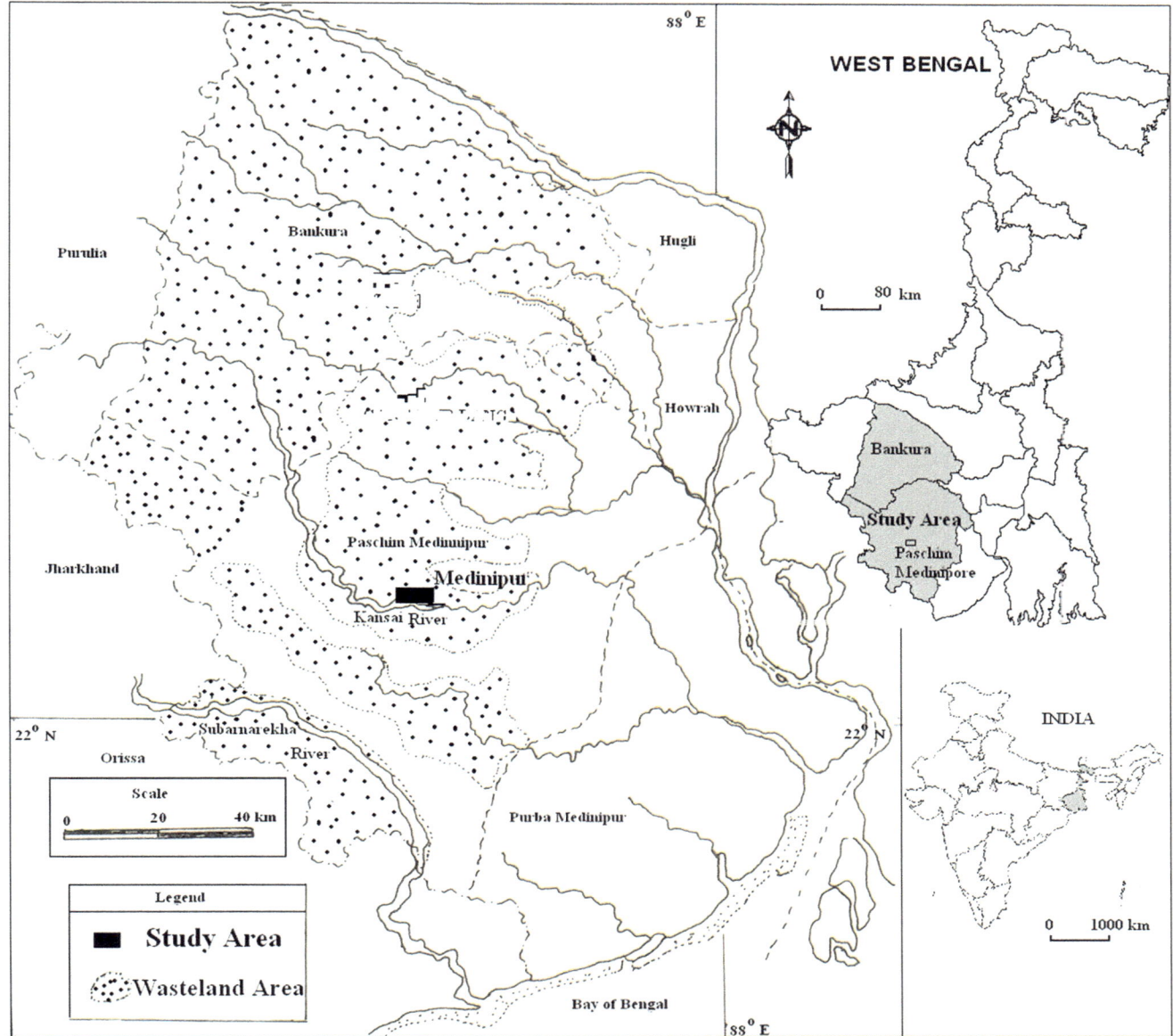

Fig. 13.1 Location of lateritic belt area (i.e., rill-gully-prone area) and experimental site (Shit and Maiti 2014)

section at 10-min intervals during the experiments (Fig. 13.3). Additionally, hydraulics parameters, including Reynolds number (Re), Froude number (Fr), hydraulic share stress (τ), Manning roughness coefficient (N), and Darcy-Weisbach friction factor (f) were calculated according to Eqs. (13.1)–(13.4):

$$Re = \frac{vr}{\mu} \tag{13.1}$$

$$Fr = \frac{v}{\sqrt{gh}} \tag{13.2}$$

$$\tau = \rho grs \tag{13.3}$$

$$f = \frac{8grs}{v^2} \tag{13.4}$$

Where v is the flow velocity (m s^{-1}), r is the hydraulic radius (m), s is the surface slope angle (m m^{-1}), g is the gravitational acceleration (m s^{-2}), h is the flow depth (m), ρ is

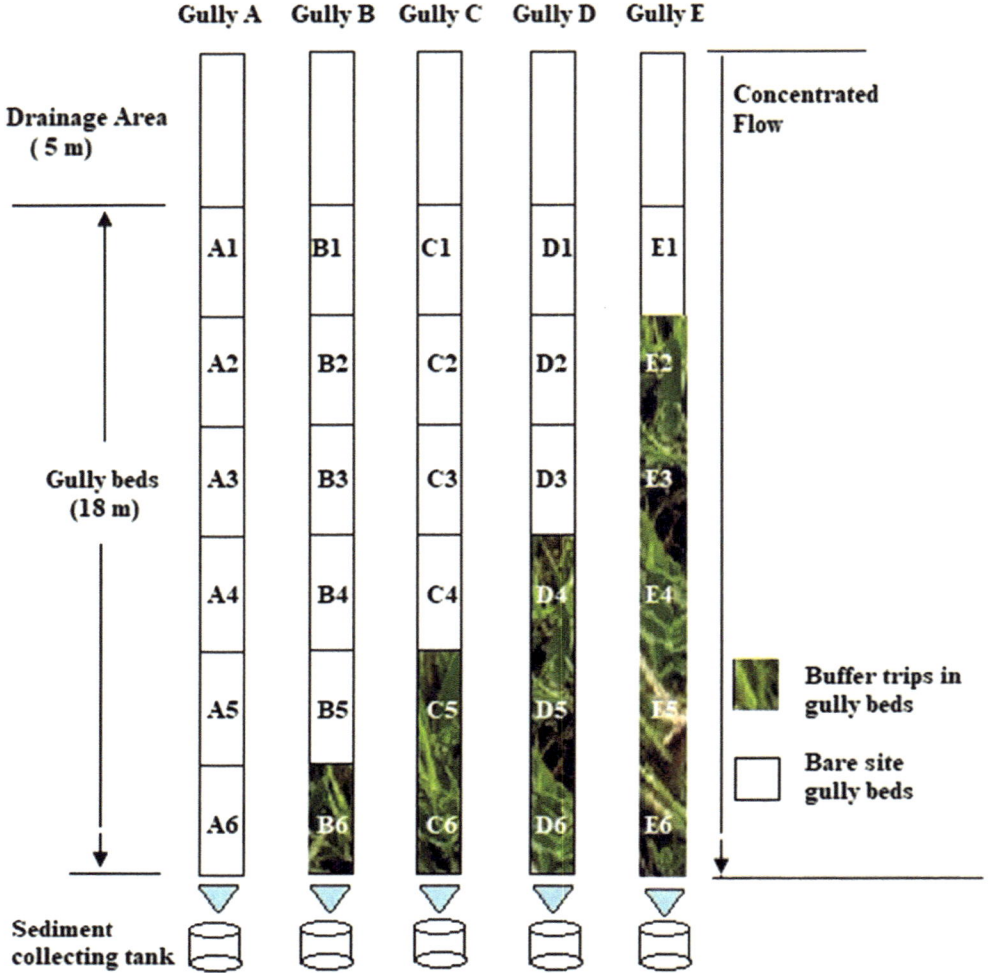

Fig. 13.2 Sketch of experimental simulation concentrated flow and vegetation distribution of gully beds. The length of each section is 3 m, gully width varies from 0.2 to 0.25 m, and average slope of the gully bed is 18–20°

the water density (kg m^{-3}), and μ is the water kinematic viscosity coefficient (m^2 s^{-1}).

Runoff samples were collected at the bottom part of each gully bed using a 0.5-l plastic bottle during the simulation runoff at 5-min intervals. These samples were deposited for 24 h and the clear water decanted; then, the sediment was oven dried at 105 °C for 12 h and weighed to determine the sediment loss (Kodamatani et al. 2017). Gully bed topography was monitored by paired photographs with field measurements based on terrain height (points) per square meter. Gully bed topography changes, including width, depth, volume, and elevation, were analyzed by subtracting the volumes of the digital elevation model (DEM) at 10 m spatial resolution before and after scouring using Raster math tools and spatial analyst tools of ArcGIS software. All statistical analysis was carried out using MS Excel and Origin 6.1 software.

13.3 Results and Discussion

13.3.1 Vegetation Effects on Concentrated Flow Hydraulics in Gully Beds

Summaries of the statistics of all experimental results are presented in Table 13.1. Hydraulics parameters (i.e., v, Re, Fr, F, t, and N) for vegetation buffer strips were clearly shown to be smaller than those for bare gully beds. The average parameter values for the vegetated buffer strips varied between 35% and 50%, which is lower than those for bare gully beds with a minimum value (74.53%) of τ (Table 13.1). Concentrated flow resistance values were presented by Darcy-Weisbach friction factor (f) and Manning roughness coefficient (N). The results showed that these values for

Fig. 13.3 Layout of experimental plots: (**a**) simulation of water supplied by 2-hp diesel pump; (**b**) Gully-A; (**c**) Gully-C; (**d**) Gully-B; (**e**) Gully-E, (**f**) Gully-D

Table 13.1 Summary of the statistics of hydraulic parameters on bare sites and gully beds with vegetation buffer strips

Hydraulic parameters	Descriptive statistics	Bare gully beds	Buffer strips in gully beds
Number of the section	–	19	11
v (m s^{-1})	Mean	0.66	0.47
	Range	0.31–1.04	0.29–0.61
	SD	0.24	0.10
h (m)	Mean	0.05	0.04
	Range	0.03–0.11	0.03–0.06
	SD	0.02	0.01
w (m)	Mean	0.24	0.54
	Range	0.12–0.52	0.28–0.67
	SD	0.24	0.21
Re	Mean	13,822	8817
	Range	6478–19,321	4841–12,873
	SD	4153	2724
Fr	Mean	0.63	0.51
	Range	0.29–1.27	0.39–0.63
	SD	0.27	0.09
τ (N m^{-2})	Mean	207.14	74.53
	Range	54.35–354.19	53.87–97.80
	SD	98.05	14.69
f	Mean	7.17	9.95
	Range	0.78–17.24	3.45–16.84
	SD	4.59	4.38
n	Mean	0.15	0.18
	Range	0.04–0.27	0.12–0.26
	SD	0.10	0.12

vegetated buffer sites were 1.2 and 1.4 times higher than those for bare gully sites, respectively (Table 13.1).

13.3.2 Relationship of Different Hydraulic Factors Between Vegetated and Bare Sites

Figure 13.4 represents a relationship between Reynolds number (Re) and several hydraulic factors in vegetated and bare gully beds. The increasing regression lines of Re-v were illustrated for both bare soil and vegetated gully beds, but the Re-v trends line for bare soil has a steeper gradient than that of vegetated buffer gully beds (Fig. 13.4a). The equal value was also calculated for Re-Fr (Fig. 13.4b). The increasing trends lines with values of flow shear stress for bare sites of gully were significantly higher than that for vegetated buffer strips (Fig. 13.4c). Figure 13.4d shows a negative relationship between Re-f and indicated that bare sites have a steeper slope than vegetated sites. However, these results revealed that vegetated buffer strips usually have higher

values of hydraulic flow resistance and lower values of flow hydraulics parameters (i.e., Re, Fr, t) than bare gully beds.

13.3.3 Role of Vegetation in Soil Erosion of Gully Beds

Table 13.2 illustrates the erosion processes of gully beds during concentrated flow experiments. All gully sections studied were clearly eroded after six experiments: average gully erosion volume was about 0.33 m^3 and 0.27 m^3 for bare soil gully beds and vegetated gully beds, respectively. These results indicated that sedimentation was 1.24 times higher than that in vegetated buffer strips in respect to bare soil gully beds. The vegetated buffer strips in gully beds indicate that total gross deposit of sediment after experiments ranged between 0.02 and 0.1 m^3 with an average of 0.037 m^3, which was 1.6 times higher than that of bare soil gully beds. The net erosion volume ranged from 0.22 m^3 to 0.55 m^3 via an average of 0.31 m^3 in bare soil gully beds, whereas vegetated buffer strip beds varied from 0.05 m^3 to 0.15 m^3 by

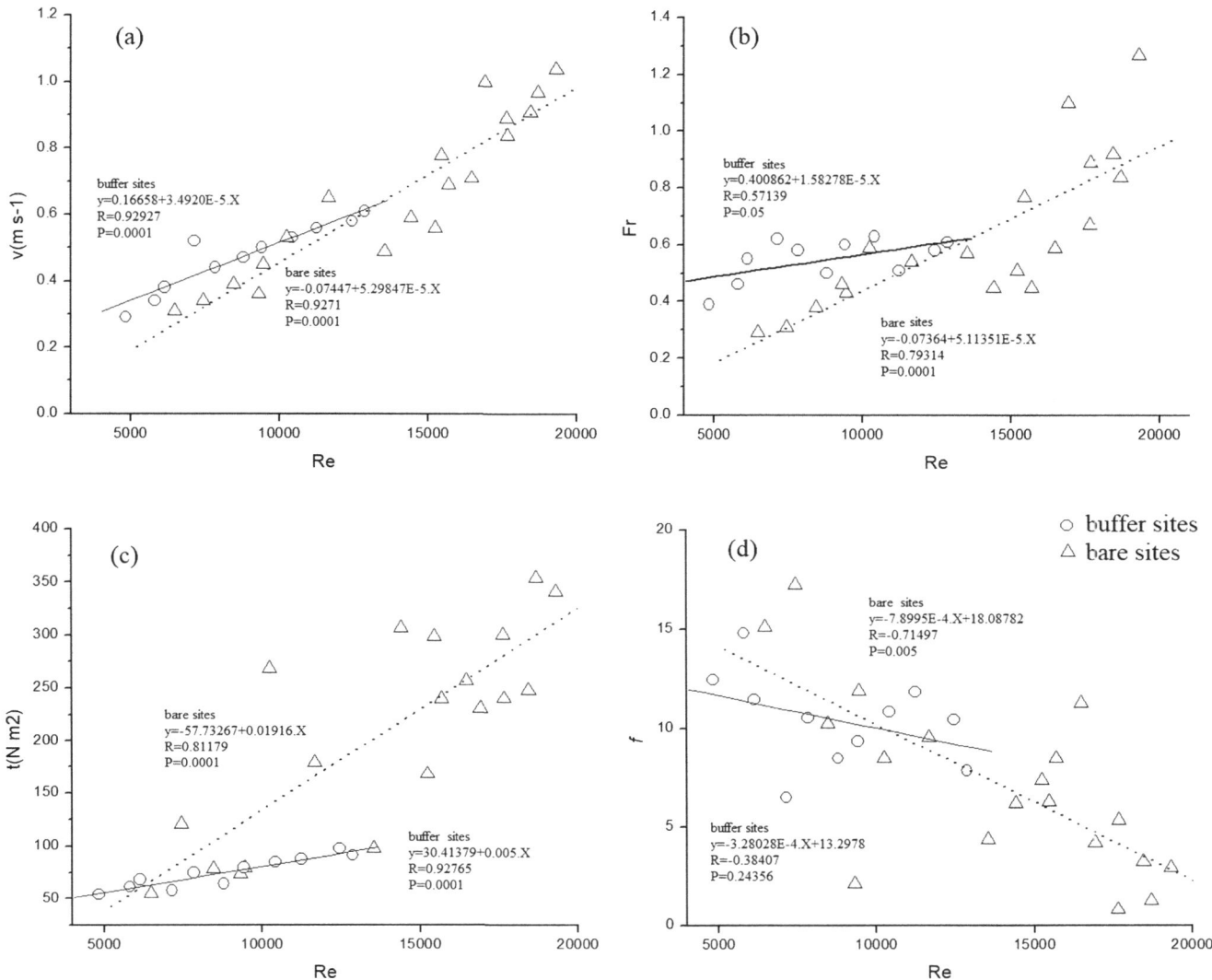

Fig. 13.4 Relationship between vegetated sites and bare sites for different hydraulic factors: (**a**) Reynolds number (Re), (**b**) Froude number (Fr), (**c**) flow shear stress (t), and Darcy-Weisbach friction factor (f) (**d**)

an average of 0.23 m^3 (Table 13.2). These results showed that vegetated buffer strips significantly reduce flow velocity, control gully incision, and trap sediment in gully beds.

13.3.4 Relationship Between Experimental Time (Minutes) and Soil Erosion

Figure 13.5 illustrates the experimental gully beds, both vegetated and bare soil, sediment concentration, and net erosion volumes against concentrated flow times. The net erosion volume shows a negative relationship with time for all the experimental gully beds. Higher erosion at the beginning time was observed in Gully-A. All gully beds were

relatively stable during the experimental time, except Gully-A (Fig. 13.5a). The presence of the vegetation buffer did not significantly reduce soil erosion during the first 30 min. Similarly, a negative relationship could be observed on sediment concentration against time (Fig. 13.5b). The maximum total erosion volume was observed at Gully-A, whereas vegetation cover had zero percentage (i.e., bare gully beds), and minimum total erosion volume was observed at Gully-E, whereas vegetation cover was 66.67% (Table 13.3). The vegetation buffer strips showed that erosion rate is 3.1 times less than in bare soil gully beds (i.e., Gully-A). These results indicated that the percentage of vegetation buffer significantly controls erosion rate and sediment deposit in gully beds (Table 13.3).

Table 13.2 Landscape changes at different sites of gully beds during experiments

Gully sites	Gully section ID	Gross erosion volume (m³)	Gross deposition volume (m³)	Net erosion volume (m³)	Upstream net erosion volume (m³)
Bare gully beds	A1	0.58	0.03	0.55	0.12
	A2	0.34	0.05	0.29	0.10
	A3	0.54	0.02	0.52	0.21
	A4	0.46	0.00	0.46	0.13
	A5	0.31	0.02	0.29	0.14
	A6	0.28	0.03	0.25	0.10
	B1	0.45	0.01	0.44	0.08
	B2	0.4	0.04	0.36	0.12
	B3	0.58	0.00	0.58	0.07
	B4	0.35	0.02	0.33	0.20
	B5	0.28	0.03	0.25	0.34
	C1	0.24	0.00	0.24	0.08
	C2	0.31	0.01	0.3	0.14
	C3	0.26	0.03	0.23	0.16
	C4	0.23	0.01	0.22	0.17
	D1	0.29	0.00	0.29	0.11
	D2	0.28	0.00	0.28	0.10
	D3	0.23	0.01	0.22	0.09
	E1	0.26	0.00	0.26	0.11
Vegetation buffer strips/ gully beds	B6	0.18	0.10	0.08	1.02
	C5	0.16	0.08	0.08	1.00
	C6	0.15	0.10	0.05	1.23
	D4	0.22	0.07	0.15	0.85
	D5	0.15	0.08	0.07	0.75
	D6	0.12	0.06	0.06	1.10
	E2	0.23	0.03	0.20	0.23
	E3	0.18	0.04	0.14	0.82
	E4	0.12	0.03	0.09	0.76
	E5	0.1	0.05	0.05	0.14
	E6	0.11	0.02	0.09	1.00
Drainage area	A	0.67	0.01	0.66	–
	B	0.54	0	0.54	–
	C	0.49	0.02	0.47	–
	D	0.37	0.01	0.36	–
	E	0.32	0	0.32	–
Average bare gully beds		0.33381	0.023333	0.310476	–
Average vegetated gully beds		0.269286	0.037143	0.232143	–

13.3.5 Effects of Vegetation on Hydraulic Processes and Soil Erosion of Gully Beds

Figure 13.6 represents the relationship between hydraulic parameters and net erosion volume on gully beds. Runoff velocity (v), Reynolds number (Rn), and Froude number (Fr) showed negative trends with net erosion volume. A similar relationship between hydraulic parameters and soil erosion resulted from vegetation buffer strips, but the relationships were not statistically significant (Fig. 13.6). A similar poor relationship was found for friction factor (f) and erosion loss (Fig. 13.6d). Our result was in line with a previous study by Gimenez and Govers (2008), who showed that a poor correlation existed between flow hydraulic parameters and rill detachment rate in buffer strips. Figure 13.7a clearly represents that during the rainy season abrupt gully incision occurs without vegetation cover, but Fig. 13.7b shows that vegetation is growing along gully beds. These photographs indicate that increasing vegetation cover also reflects the increasing root system (i.e., length, density, mass) in gully beds. Moreover, the area covered with dense vegetation had reduced surface runoff, sediment traps by stems, and reinforcement of soil erosion and stabilization of the gully (Fig. 13.7c).

Fig. 13.5 Relationship between experimental time (min) and soil erosion: (**a**) net erosion volume; (**b**) sediment concentration (g l^{-1})

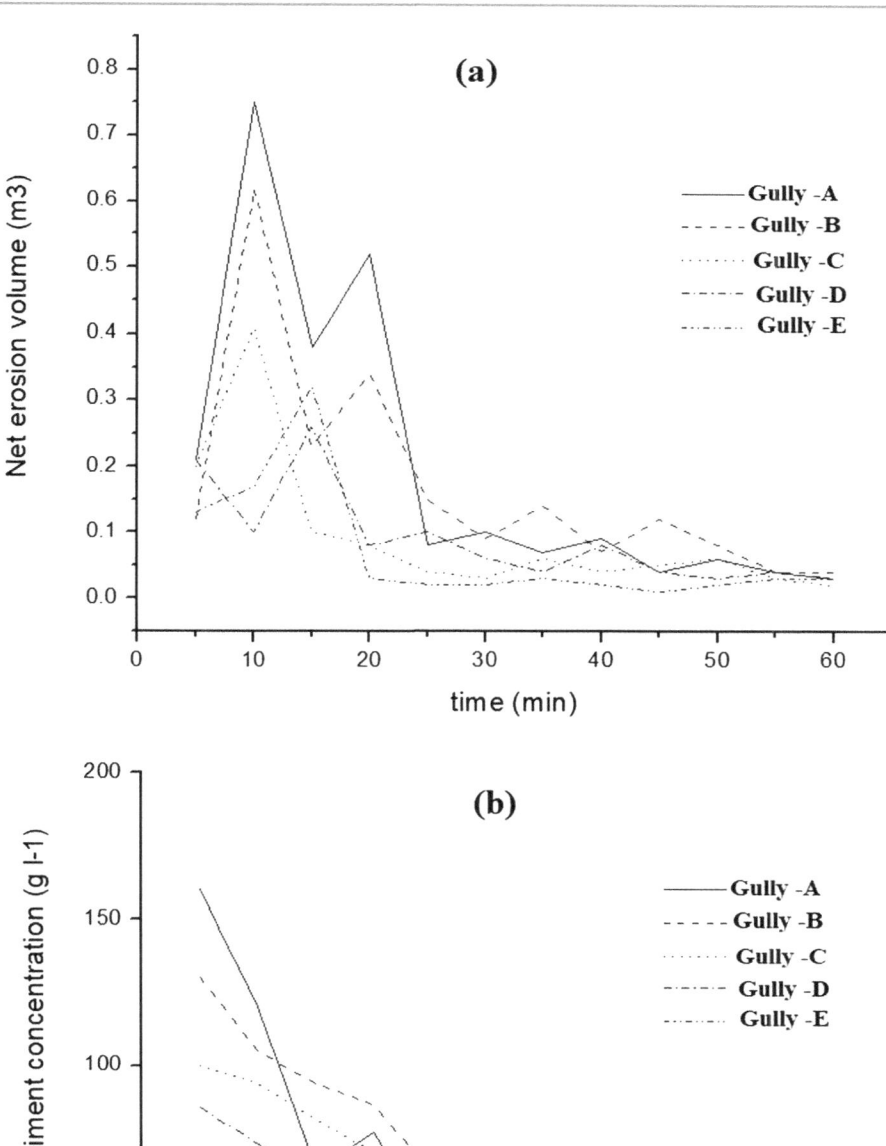

13.4   Conclusion

Our field experimental results showed that a vegetation buffer on the gully floor significantly influences hydraulics parameters, and it could increase resistance to concentrated flow. The consequences of sediment entrainment rates are reduced, and sediment is trapped by vegetation roots in gully beds. Results of the hydraulics parameters (velocity, Reynolds number, Froude number, and Darcy-Weisbach friction factor) of bare soil gully beds showed the vegetation buffer strips in gully beds are 35–50% smaller and have 1.2 to

Table 13.3 Net erosion volumes and erosion rate of each gully beds under flow discharge (78.5 l min^{-1}) during 1 h

Gully sites	Percentage of vegetation buffer strips in gully beds	Total erosion volume (m^3)	Total deposition volume (m^3)	Net erosion volume (m^3)	Erosion rate (m^3 h^{-1})
Gully-A	0.00	2.51	0.15	2.36	0.31
Gully-B	16.67	2.24	0.2	2.04	0.27
Gully-C	33.33	1.35	0.23	1.12	0.21
Gully-D	50.00	1.29	0.22	1.07	0.14
Gully-E	66.67	1.00	0.17	0.83	0.10

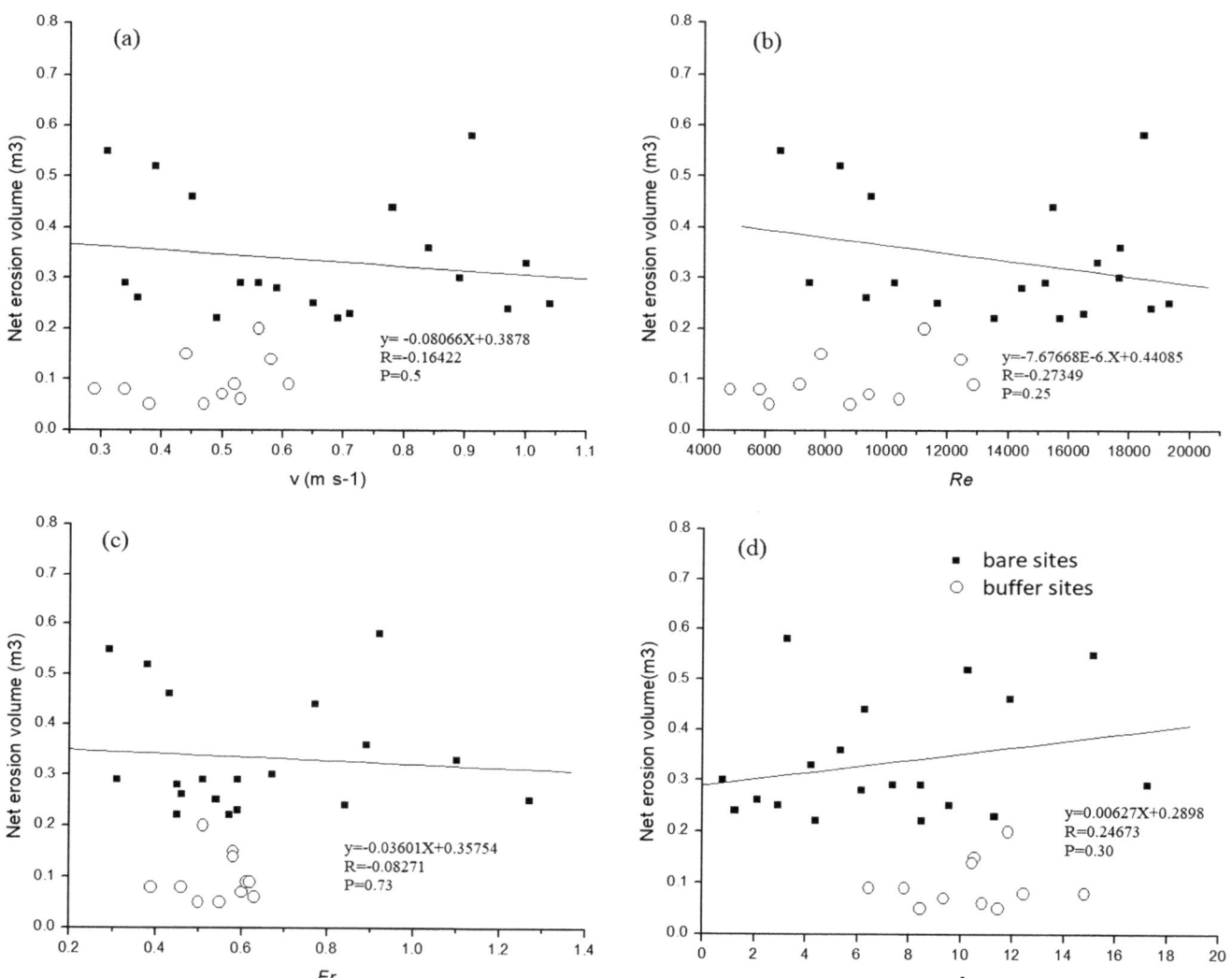

Fig. 13.6 Relationship between hydraulic parameters and net erosion volumes: runoff velocity (v) (**a**), Reynolds number (Re) (**b**), Froude number (Fr) (**c**), and Darcy-Weisbach friction factor (f) (**d**)

1.4 times higher concentrated flow resistance in Darcy-Weisbach friction factor (f) and Manning roughness coefficient (N), respectively. The total sediment deposit and net erosion volume on five gully beds produced a significant negative exponential, decreasing with increasing experimental time and vegetation cover. The vegetation cover shows that the erosion rate is 3.1 times less than in bare soil gully beds. Thus, this study showed that vegetated buffer strips significantly reduce flow velocity, control gully incision, and sediment loss in gully beds.

Fig. 13.7 Influence of vegetation cover at different location of Rangamati badland area: (**a**) gully incision without vegetative cover, (**b**) gully stabilized by gully beds vegetation, and (**c**) reduced gully erosion by growing the vegetation along the gully channel and beds

Acknowledgments We are grateful to the Department of Geography, Raja N.L. Khan Women's College (Autonomous), affiliated to Vidyasagar University, Midnapore, West Bengal, India for supporting this research. The author (P.K. Shit) gratefully acknowledges University Grant Commission (UGC-ERO, Kolkata), Govt. of India for financial support through Minor Research Project [No. F.PHW-171/15-16 (ERO)]. Many people helped with the originally field data collection, including Sumanta Dinda, Sumana Jana, and Bhanishikha Bhattacharya.

References

Bandyopadhyay, S. (1987) Man-Initiated Gullying and Slope Formation in a Lateritic Terrain at Santiniketan, Geographical Review of India. 49 (4), 21–26.

Bandyopadhyay, S. (1988) Drainage Evolution in Badland Terrain at Gangani in Medinipur District, West Benagl. Geographical Review of India. 50 (3), 10-20.

Braud, I., Vich, A. I., Zuluaga, J., Fornero, L., & Pedrani, A. (2001) Vegetation influence onrunoff and sediment yield in the Andes region: Observation and modelling. Journal ofHydrology, 254, 124–144.

Das K., and Bandyopadhyay, S. (1996) Badland Development in a Lateritic Terrain: Santiniketan, West Benagl. National Geographer. XXXI (1&2). 87–103.

Ghosh D (2015) Mapping and Monitoring of the Impact of Gully Erosion in the District of Medinipore (west), West Bengal, India. International Journal of Novel Research in Humanity and Social Sciences, 2(4):73-89.

Ghosh S, Guchhait SK (2015) Characterization and Evolution of Laterites in West Bengal: Implication on the Geology of Northwest Bengal Basin. Trans 37(1):93-119.

Gimenez, R., Govers G. (2008) Effects of freshly incorporated straw residue on rill erosion and hydraulics, Catena 72(2):214-223. DOI: https://doi.org/10.1016/j.catena.2007.05.004.

Govers G. (1991) Rill erosion on arable land in central Belgium: rates, controls and predictability. Catena, 18:133–155.

Govers G., Giménez R. and Oost, K. Van. (2007) Rill erosion: exploring the relationship betweenexperiments, modelling and field observations. Earth Science Reviews, 84 (3–4): 87–102.

Kodamatani H, Maeda C, Balogh SJ, Nollet YH, Kanzaki R, Tomiyasu T. (2017) The influence of sample drying and storage conditions on methylmercury determination in soils and sediments. Chemosphere. 173:380-386

Li, P., Cai, W. B., & Zheng, L. Y. (2006) Effects of vegetative cover on runoff hydrauliccharacteristics and erosion. Science of Soil and Water Conservation, 4, 55–59.

Pani P, Mohapatra SN (2001) Delineation and monitoring of gullied and ravenous lands in a part of lower Chambal valley, India using remote sensing and GIS. In: Proc. ACRS. Singapore

Rey, F. (2003) Influence of vegetation distribution on sediment yield in forested marly gullies. Catena, 50, 549–562.

Sen, J., Sen, S., Bandyopadhyay, S. (2004) Geomorphological investigation of badlands- a case study at Garbheta, West Medinipur District, West Bengal, India. In Singh, S., Sharma, H.S., and De, S.K (eds). Geomorphology and Environment. acb publication, Kolkata. 204-234.

Sharda, V.N., Juyal, G.P., Prakash, C., and Joshi, B.P. (2007) Training Manual Soil conservation and watershed management, Vol-II, (soil water conservation engineering).Central Soil and Water Conservation Research and Training Institute (CSWCRTI), Dehradun, Uttaranchal, pp. 1–410.

Shit, P.K, Maiti R., Pati C.P. (2011) Methodological Framework to Select Grass Species for Controlling Rill and Gully Erosion. Science

Cruiser, Institute of Science, Education and Culture. Vol-25(2), 32-37.

Shit, P.K., Bhunia G.S., Maiti R. (2012) Effect of Vegetation Cover on Sediment Yield: An Empirical Study through Plots Experiment, Journal of Environment and Earth Science, Vol 2, No 5 , 32-40.

Shit P.K., Bhunia G.S., and Maiti R. (2013) Assessment of factors affecting ephemeral gully development in badland topography: A case study at Garbheta badland (Pashchim Medinipur, West Bengal, India), International Journal of Geosciences. 4 (2) 461–470. doi: https://doi.org/10.4236/ijg.2013.42043.

Shit P.K., Bhunia G. S, and Maiti R. (2014) Morphology and Development of selected Badlands in South Bengal (India), Indian Journal of Geography and Environment, 13, 161-171.

Shit P.K., and Maiti R. (2013) Management Techniques of Rill-Gully Erosion in Badland Topography: Experimental Research at Laboratory, Plot and Catchments Scale. LAP Lambert Academic Publishing, Germany. (ISBN: 978-3-659-32409-3), pp 136.

Shit PK, Nandi AS, Bhunia GS (2015) Soil erosion risk mapping using RUSLE model onjhargram sub-division at West Bengal in India. Model Earth Syst Environ 1(28):1-12

Shit P.K. and Maiti R. (2014) Gully Erosion Control: Lateritic Soil Region of West Bengal, Indian Science Cruiser, 28(3), 54-61.

Singh S., Dubey A. (2002) Gully erosion and management methods and applications (A FieldManual). New Academic Publishers, Delhi, 1 - 248 pp.

Singh, R, Phadke, V.S. (2004) Assessing soil loss by water of the erosion in Jamin river basin, Bundelkhand region, India. Geographical Review of India, Vol. 66, No-2, pp. 143-152.

Spaan, W. P., Sikking, A. F. S., & Hoogmoed, W. B. (2005) Vegetation barrier and tillageeffectson runoff and sediment in an alley crop system on a Luvisol in Burkina Faso. Soil & Tillage Research, 83, 194–203.

Su, Z. A., Xiong, D. H., Dong, Y. F., Li, J. J., Yang, D., Zhang, J. H., & He, G. X. (2014) Simulated headward erosion of bank gullies in the Dry-hot valley region of Southwest China. Geomorphology, 204, 532–541.

Pravat Kumar Shit received his PhD in Geography (Applied Geomorphology) from Vidyasagar University (India) in 2013, MSc in Geography and Environment Management from Vidyasagar University in 2005, and PG Diploma in Remote Sensing and GIS from Sambalpur University in 2015. He is Assistant Professor in the Department of Geography, Raja N. L. Khan Women's College (Autonomous), Gope Palace, Midnapore, West Bengal, India. His main fields of research are soil erosion spatial modeling, badland geomorphology, gully morphology, water resources and natural resources mapping, and modeling. He has published more than 45 international and national research articles in various renowned journals; also, he has published three books. His research work has been funded by the University Grants Commission (UGC), India and Higher Education Science and Technology and Biotechnology, Government of West Bengal. He is Associate Editor and on the editorial boards of three international journals in geography and earth environmental sciences.

Hamid Reza Pourghasemi is an Associate Professor of Watershed Management Engineering in the College of Agriculture, Shiraz University, Iran. He has a BSc in Watershed Management Engineering of the University of Gorgan (2004), Iran, an MSc in Watershed Management Engineering, from Tarbiat Modares University, Iran (2008), and a PhD in Watershed Management Engineering from the same University (Feb 2014). His main research interests are GIS-based spatial modelling using machine learning/data mining techniques in different fields such landslide, flood, gully erosion, forest fire, land subsidence, species distribution modelling, and groundwater/hydrology. Also, Hamid Reza works on Multi-Criteria Decision Making methods in Natural Resources and Environment. He has published more of 90 peer reviewed papers in high-quality journals, with three chapters in Springer. Also, he published two books in Springer (https://www.springer.com/gp/book/9783319733821) and Elsevier (https://www.elsevier.com/books/spatial-modeling-in-gis-and-r-for-earth-and-environmental-science/pourghasemi/978-0-12-815226-3).

Gouri Sankar Bhunia received his PhD from the University of Calcutta, India, in 2015. His PhD dissertation work focused on environmental control measures of infectious disease (visceral leishmaniasis or kala-azar) using geospatial technology. His research interests include kala-azar disease transmission modeling, environmental modeling, risk assessment, data mining, and information retrieval using geospatial technology. He is Associate Editor and on the editorial boards of three international journal in Health GIS and Geosciences. He worked as a 'Resource Scientist' in Bihar Remote Sensing Application Centre, Patna (Bihar, India). He is the recipient of the Senior Research Fellow (SRF) from Rajendra Memorial Research Institute of Medical Sciences (ICMR, India) and has contributed to multiple research programs kala-azar disease transmission modeling, development of customized GIS software for kala-azar 'risk' and 'non-risk' area, and entomological study.

Anand Verdhen

Abstract

Gullies facilitate drainage system, but also accelerate erosion, generate silt, and damage valuable land. On one hand it leads to land degradation and plant devastation, structural damage, and activities disruption, while on the other hand it provides a clearway to flowing water as natural drainage channel. Spatial survey of the entire area concerned helps in distinguishing different zones and processes involved in the formation and features of the gully. It is a catastrophic and surface flow (melt/rain water)-induced erosion. At present awareness toward the development, impact, and management of gullies is important to formulate the plan and policy in order to protect sustainable nature and mitigate the harmful impact covering diverse geographical and man-made conditions. Studies are limited and it is difficult to assess the contribution of sediment from gully erosion being area-specific. The objective of the study is to assess the nature of gully formation and sediment/runoff generation in Himalayan watershed, plateau, and alluvial plains.

Typical zones of ice and snow dominating mountains, plateau watershed, alluvial plains, roads, and agricultural land have been assessed. It has been observed that process of gully formation differs in study zones and at outfall. Sediment production by gullies on the minimum global scale in North Bihar Rivers at foothill is four to six times higher than the average value (33 t/km^2/year) of India but matching at downstream, resulting in rise in riverbed and building-up of alluvial plains. Pernigaon nala in snowy catchment of 2.81 km^2 above Jammu-Srinagar road (NH1A) at 218.293 m (old chainage) has been considered for hydrological and hydraulic analysis which reveals that Lacey's linear waterway is not in agreement with the actual waterway.

Keywords

Snow and ice melt · Mountainous region · Alluvial plains · Gully erosion · Sediment yield · Linear waterway · Highways drainage

14.1 Introduction

Gully erosion is a natural process accelerated and decelerated as per the nature of the surface topography, outcrop, climate, unstable/narrow channel, flowing concentrated surface water, land use, and any other living and nonliving activities. Cut above 30 cm deep into the ground is defined as 'gully formation'. Montgomery and Dietrich (1988) stated that the process of gully erosion in erodible hillslope is driven by an excessive critical-flow shear stress on the soil surface. Overland sheet flow through hill slopes produces sheet erosion carrying fertile soil with nutrients; further down rill flow produces rill erosion, which together accelerates the gully erosion. The formation of rills and small channels may be due to increase in runoff velocity of overland flow. The region between rills is known as inter-rill areas. Gregory and Walling (1973) described it as dynamic system that transfer or store sediments along their channel. Ephemeral gullies appear on the same location each year when the concentrated flow is large enough to create large channels and normally difficult to be filled in by tillage. Vashisth (1962) defined it as an obsequent stream proceeding to subsequent stream, if remains connected. Subsequent gullies flow join together further down the slope to generate stream flow or river flow.

Desmet et al. (1999) indicated that the slope gradient is a controlling factor of gullies formation. Slope steepness is highly related to the existence of gullies (Meyer and

A. Verdhen (✉)
SASE (DRDO), Manali, Himachal Pradesh, India

NIH Roorkee, Roorkee, Uttar Pradesh, India

Centre for Water Resources Studies, Patna University, Patna, Bihar, India

ICT & CES Pvt. Ltd., Delhi, India

DCE, Gurgaon, Haryana, India

Independent Researcher at Sadarpur, Bihar Sharif, Bihar, India

Martinez-Casasnovas 1999). At some depressed location, weak soils tunnels develop as piping beneath the surface. Brooks et al. (2007) revealed that the extent of land impacted by the gullies is increasing and even land management practices and development of roads augment the formation. Animals in way of movement and grazing disturb the soil directly, resulting in decreased infiltration and increased run-off as well as sediment yield (Dunne et al. 2011), while the vegetation helps in stabilizing gully and aggradation to it (Sandercock and Hooke 2011). Herzig et al. (2011) and Marden et al. (2011) estimated through model studies that afforestation efforts have reduced the sediment yield up to 38%.

Renard et al. (1997) and Yu (1998) expressed the rainfall erosivity as the product of daily sum of 6 hourly kinetic energy times the daily maximum 30-min rainfall intensity to relate with the measured daily/periodical soil erosion. Contemporary hydrological regimes are required to be established (Ward et al. 2011). Meandering due to obstacle or alluvial plains of moderate slope may increase erosion of the banks, usually on the outside bends.

Finding an erosion contribution, as whether it is dominated by the watershed and its nature, gullies, or by the rivers, is uncertain due to various factors and the problem being is area-specific. The objective of the study is to assess the nature of formation of gullies in Himalayan watershed, plateau, and alluvial plains and to identify its proneness to yield sediment in relation to melt and rainfall-runoff.

14.2 Material and Methods

The basis of the study has been assessment of available literature and filling the gaps to explore further. The materials are available generally for the hillslopes and cultivable lands, while the mountains and temporary and permanent snow-ice-covered areas are dynamic in the formation of permanent and ephemeral gullies with enormous sediment yield. Study-based results of sediment yield from the Himalayan typical watersheds have been presented.

Garde and Kothyari (1987) presented regression analysis of Indian streams and catchments, which include the followings:

$$SY_{va} \text{ (Garde and Kothyari)}$$
$$= 1.067 \times 10^{-6} \times P^{1.38} \times F_c^{2.51} \times D_d^{0.4}$$
$$\times CA^{1.29} \times S^{0.13} \qquad (14.1)$$

where SY_{va} = annual sediment yield (Mm^3); CA = catchment area (km^2); P = annual mean precipitation (cm); F_c = Forest

cover factor; D_d = drainage density (km^{-1}); S = catchment slope.

Equation (14.1) is modified to suit the Himalayan and global conditions to get power coefficient of CA ($a = 1.03$ to 1.29) and of S ($b = 0.08$ to 0.13) by simulation.

$$SY_{va} \text{ (Verdhen et al. 2012)}$$
$$= 1.1 \times 10^{-6} \times P^{1.33} \times F_c^{2.46} \times D_d^{0.4}$$
$$\times \left(CA^a \times S^b \right) \qquad (14.2)$$

Langbein and Schumm (1958) equation in FPS has been converted into SI unit. SY_{wa} (Langbein et al.) = R (annual runoff)/V (mass density of vegetation). This relationship was developed by trial-error and graphical methods:

$$Sy_{wa} \text{ (Langbein et al.)} = \left(10 \times 8.53 \times P^{2.3} \right) /$$
$$\left(1 + 0.0007 \times 8.53 \times P^{2.3} \right) \quad (14.3)$$

where SY_{wa} is in 0.3861 t/km^2 (assuming a density of 60 lb./ft^3 or 961 kg/m^3) and precipitation P is in cm, R in km^3, and V in t/km^3. The first factor of Eq. (14.3) describes the erosive action of rainfall in the absence of vegetation. The second factor represents the protective action of vegetation.

Deforestation rate has imbalanced the ecological, environmental, and climatic conditions and further forestation may not serve the purpose, as a new sapling cannot compete the grown-up vegetation and old aged trees. Detailed studies are needed to compare the past and present conditions and its impact for any future course of action.

Decrease in area of ditch, ponds, and lake-like water bodies has enhanced the possibilities of gullies formation in plains including alluvial tracks. Even the natural drainage has been blocked due to unauthorized encroachment and authorized restricted waterway in cross-drainage structures. High embankment and roads are prone to have gully formation and failure during the monsoon and flood to flash flood-like situation. Warming has increased the uncertainty of rainfall intensity and storm events. A fresh look on formation, control, and management of gullies is required in near future.

To address the above deficiency and problems, the status of sediment yield and main factors, time of concentration, rainfall-runoff, HFL, regime depth, and scour depth methodology have been highlighted. A stream/gully, Pernigaon in snowy catchment has been considered to assess and demonstrate the design discharge and hydrological and hydraulic features at crossing of Jammu-Srinagar road (NH1A) at 218.293 m. The 24 hourly rainfalls in the region extracted from Flood Estimation Reports (CWC 1984) pertaining to hydrological zone/subzone of CWC, published jointly by IMD, MoRST, and IMD for 50 years' return period, is 180 mm (18 cm).

14.2.1 Methodology for Discharge Computation

Design discharge to be computed by using various methods (i.e., Empirical Formula, Rational method, and Slope-Area method, as recommended in IRC: SP-13 (2004) and IRC: SP-5 (1998)) is briefed below.

14.2.1.1 Dicken's Formula

Dicken's empirical formula is commonly used for flood discharge based on catchment area

$$Q = C_1 A^{0.75} \qquad (14.4)$$

where Q = runoff (m^3/s); A = Catchment area (km^2); and C_1 = Runoff coefficient which depends on the topography, type of soil, vegetation, ground slope, climate of the region, etc.

14.2.1.2 Rational Formula

Rational empirical formula is commonly used having catchment area of less than 100 ha for flood discharge due to rainfall intensity of desired return period

$$Q = 0.028 \, PfAI_C \qquad (14.5)$$

where Q = maximum runoff (m^3/s); A = catchment area (ha); Ic = critical intensity of rainfall (cm/h) at desired return period (25/50 years in case of cross-drainage works); P = coefficient of runoff for the given catchment characteristics; and F = spread factor for converting point rainfall into area mean rainfall.

$$I_c = (F/T) * (T + 1)/(T_c + 1) \qquad (14.6)$$

where F = total rainfall of T hours (24 h) duration corresponding to 50 years return period (cm) and T_c = time of concentration (h).

$T_c = (L_c/V_c) + (B/(2*V_B))$ and/or $T_c = [0.87(L^3/H)]^{0.385}$ where L_c = main channel length (km); V_c = mean velocity of flow along the main channel [1.75 m/s]; B = average width of catchment contributing flow (km); V_B = mean velocity of the lateral flow (1.0 m/s); L = length of catchment (km); and H = elevation difference (m). It is better to have an average T_c out of both. Total rainfall in 24 h is adjusted corresponding to T_c for finding critical rainfall intensity, I_c.

14.2.1.3 Slope-Area Method

This method is based on conveyance factor (K) and the slope (S) of stream. For calculation of the conveyance factor, two or more cross-sections at u/s (up stream) and d/s (down stream) of

the structure are used. Slope of the channel will be determined using survey data of lowest bed level. The discharge is calculated by the Manning's formula given below:

$$Q = Ke * S^{1/2} \qquad (14.7)$$

$$Ke = \text{Equivalent conveyance rate} = (K_1, K_2 \ldots K_n)^{1/n}$$

$$K_n = (1/N_n) \, A_n R_n^{2/3} \ (n = 1, 2, 3 \ldots n) \qquad (14.8)$$

where Q = discharge in (m^3/s); A = cross-sectional area of flow (m^2); R = hydraulic mean depth = A/P(m); P = wetted perimeter (m); S = mean longitudinal slope of the channel [fraction]; K = conveyance factor and n refers to no. of cross-section; N = rugosity coefficient as per IRC: SP-13 (2004); and n = nos. of cross-sections.

14.2.1.4 Scour Depth

Lacey's equation is adopted for estimating regime water way, regime depth, and normal scour depth as per IRC: SP-5 (1998).

$$\text{Wr} = 4.5 * Q^{0.5} \text{(a)}; R_1 = 0.473 \, (q/f)^{1/3} \text{ (b)}; R_2$$
$$= 1.34 \, (q^2/f)^{1/3} \text{ (c)}; f = 1.76 \, (d_{50})^{1/2} \text{ (d)} \, (14.9a\text{–}d)$$

where Wr = Lacey's regime waterway (m); R_1 = Lacey's regime depth (m); R_2 = scour depth below HFL (m); q = discharge intensity (m^2/s); f = silt factor; and d_{50} = sediment size (mm).

Normal scour depth and observed depth (HFL–LBL)/1.27 are compared to adopt higher value.

14.2.1.5 Velocity (V)

Velocity, V, is calculated using the relation of free flow:

$$V = Q/(D_u * W') = Q/A_u \qquad (14.9)$$

where W' = channel width at u/s and A_u = cross-section at u/s (m^2). Afflux (h) given by;

$$h = (D_u - D_d) = (\text{U/S H.F.L.} - \text{D/S HFL}) \qquad (14.10)$$

where, D_u and D_d = depths of flow u/s and d/s of the structure measured above a given datum.

14.2.1.6 Froude Number (F)

Froude no., F, has been computed for checking the flow (subcritical, critical, or super critical), using the channel flow in normal condition:

$$F = V/(gR_1)^{0.5} \quad (F = 1\,\text{critical}, < 1\,\text{subcritical}, > 1\,\text{supercritical})$$
$$(14.11)$$

Having consideration of snowmelt flood enhancing probability (Verdhen 2018), the discharge and other parameters have been computed for a typical mountainous gully having small catchment area, nearly 3 km^2 above Jammu and Srinagar highway.

14.2.2 Gully Development

Top soil gets eroded if it is not resistant to erosion due to the concentrated runoff, traversing the available slope. Eroded portion generates waterfall eroding gully head which advances toward the upper edge of the watershed. Consequently, upstream movement of the gully head and enlargement of the gully in width and depth occur with removal of material rapidly along with the flow of water. The gully having dendrites and joining perpendicularly is called trellis-type gully (Bertrand and Woodburn 1964). Gully gains its regime attaining stable slope and walls depending on the flow conditions. Vegetation cover may develop over the gully surface to anchor the soil with time.

14.2.3 Gully Classification

The gullies can be classified based on shape (U, V, or Trapezoidal shaped) which generally depends upon the resistance against erosion of top and subsoils. Where top and subsoils have same resistance U-shape, relatively more resistant subsoil V-shape, and getting bottom of the gully harder and relatively less in slope, a Trapezoidal shape of gully may occur (Bennett 1939; USDA 1954). Gullies may be continuous with a main channel and many mature or immature branches or discontinuous type and independent, which may generally develop on hillsides after landslides. It has been also defined as small, medium, or large with reference to gullies depth up to 1 m, in between 1 and 5 m or above 5 m, respectively, corresponding to the contributing drainage area up to 2 ha, between 2 and 20 ha, or above 20 ha (USDA 1954). However, gully erosion categories of low, medium, high, and very high, based on gully density per square km (Sargeant 1984), are up to 1 km for low and above 1 to 5 km for high. While in foothills of the Shiwalik in Punjab (Kukal et al. 1991), these are low up to 4 km and very high above 15 km, and in between defined as medium or moderate. Hauge (1997) distinguished gullies from rills by a critical channel cross-sectional area per square foot. Rasool et al. (2011) estimated the density between 22 and 35 km/km^2 in Kandy area of Punjab. Gully may be linear, branching, or dendritic of narrow and uniform width and active only during rains (Michael and Ojha 2012). A ravine is a deep narrow gorge larger than a gully.

14.2.4 Gully Erosion

Gully erosion occurs when rainfall, topography, pedology, land use, and hydraulic threshold are attained or exceeded (Moghaddam and Saghafi 2008). The basin shape and its exposure to sun are important to consider for hydrological analysis (Horton 1932; Jarvis 1976; Gardiner 1981) as fragile catchment for shorter time of concentration is more dynamic to erosion. Poesen et al. (2003) reported soil loss and sediment production due to gully erosion in the range of 10 to 94% at global scale. It is estimated that about 4 mha of land in India are affected by gully erosion. In India rate of soil erosion from gullies is 33 t/ha/year in ravine regions (Shekinah and Saraswathy 2005). Arc-View GIS-based hydrological modeling in delineating the contributing area and flow generated are in practice using HEC-HMC, SWAT, and SRM watershed models for simulation and mapping, but too much has to be done for using it as a prediction and application model (Torkashvand 2008; Verdhen et al. 2013).

14.2.5 Control Measure

Prevention of gully formation is always easier than controlling after its formation. Basically the melt- and rainfall-generated flows are needed to be diverted to acceptable drainage channel and streams. Good crop management, forestation, and soil conservation practices by strip cropping, contouring, terracing, bunding, diversion ditches, etc. provide added value. If it is not possible to either retain or divert the flow, then gully control structures are inevitable. It includes stabilizing the gullies and construction of series of designed check dams/weirs such that it may reduce the erosion and sedimentation and harvest water with suitable openings, flumes, stepped chutes, and spillways to handle the peak flows and spaced in such fashion that the crest level of one will be the same as the bottom elevation of the adjacent structure upstream with sufficient water cushion preventing any probability of scour.

14.2.6 Slope Stabilization

The shoulder and embankment of the road face gully erosion, like distresses in the form of cut and erosion marks, resulting in slope failure and drops. It may be significant to moderate and very frequent. For such affected sections, investigations reveal the problem of poor drainage over road. Slope

Fig. 14.1 Fluvial gullies and channel observed in August 1988: (**a**) on snow-free surface of Chhota Shigri glacier at 4700 m asl and (**b**) at the tongue of glacier ahead Zoji La/Pass of Jammu and Kashmir

The snow and ice melt and rainfall generating concentrated flow over glacier surface create gully similar to sloping soil layer with an additional most sensitive factor, temperature of water/weather/dust particle. It may be covered by fresh snow and melt-freeze phenomena of permeable overloading snow mass blockage.

(a)

Gully formation by snow and glacier melt flow carrying high silt load and deposition of debris and moraine left with retreat in glacier tongue.

(b)

stabilization measures to increase anchorage including safe and adequate drainage are required to avoid failure of slope and landslide occurrences.

14.3 Results and Discussion

Gully formation (Fig. 14.1a) on Chhota Shigri glacier (North-Westerly) is due to runoff of snow- and glacier-melt water including rain, where melt-freeze phenomenon follows by day-night cycle during July–September. The tongue of another glacier in Jammu and Kashmir (Fig. 14.1b) has retreated a little toward u/s, leaving moraine in the gully with almost nil melt flow in the morning. While 1500 m downhill on the other side and with the same range of the Himalayas, Solang Valley, oriented southerly (Fig. 14.2) with snow-covered gullies, melts away by the month of April. On the other hand, non-snowy hills and plateau fallow (Fig. 14.3a, b) have different patterns of gullies.

A typical sketch (Fig. 14.4; Verdhen 2010) represents status of weekly melt and discharge from mid-February with mean air temperature predominantly from the south facing the catchment area of 67 km^2 (Kothi) and 194 km^2 (Solang Nala) above Manali within an altitude of 2250 to 5000 m in the Pirpanjal range of the Himalaya. During melt, soil surface under snow cover, after 2 °C of weekly mean air temperature, remains wet and erodible. The melt water flows through the interface of snow and soil and getting sloping

Fig. 14.2 Portion of the Beas
subbasin in Pirpanjal Range of the
Himalayan mountains: (**a**)
drainage network and catchment
of Solang and Kothi Nalas jointly
at Kulang, Himachal Pradesh,
using Arc-View GIS and SWAT
model with catchment delineation
environment (Verdhen et al.
2013), and (**b**) upper right Valley
Solang Nala at Dhundi Thatch
showing snow-covered gullies
and melt-water flow through
channel of rocky bed

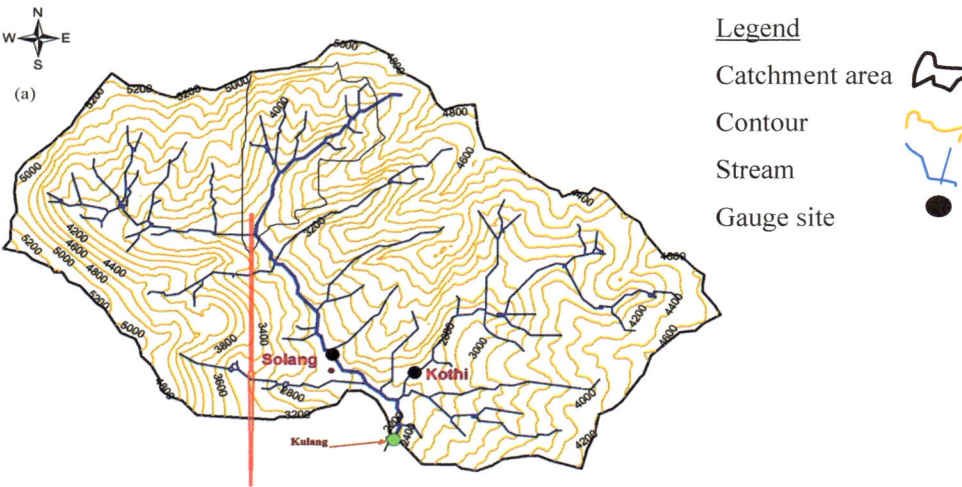

paths to erode the underneath soil forming a rill in upstream
to gullies and nala downstream, which are more prone to
erosion having rainfall-runoff.

Sediment in glacial melt streams ranges from few to 1.8
gm/l of water with increased discharge during sunny days
which carries several hundred times larger sediment than in
winter. The following factors mainly affect the rate of sediment yield:

Terrain slope and drainage density
Land use and soil cover type, heave, snowmelt, and rainfall
Cloud burst and lake outburst, glacier melt, and glacier
    movement
Weathering and wind/gust, landslides, and avalanches
Deforestation and tillage and civil works/construction
    activities

To reduce the bed load in avalanche- and landslide-prone
zones (Verdhen 2015), suitable anchorage, reinforcement,
and control structures like rakes, snow bridges, diversion
dam, etc. are required to be considered.

Tywoniuk and Fowler (1972) and Verdhen (1991) have
shown that there was no sediment production in winter but
high sediment load concentration in spring. Karaushev et al.
(1974) found soil wash in spring snowmelt flood period.
Walling and Kleo (1979) developed relations between sediment yield, annual precipitation, and runoff. Hewitt (1982)
advocated role of catastrophic floods as a result of river
damming by landslide and glacier ice within the highly
glacierized Karakoram Himalaya.

Gupta (1977) and Verdhen (1991) revealed that sediment
yields in the Himalayan Rivers are 7 to 70×10^2 t/km^2 and
lesser than 7×10^2 t/km^2 in non-Himalayan rivers. The

The banks of gully are not defined in the rocky area having gentle slope. The computation of discharge based on conveyance section using slope-area method may not yield

Fig. 14.3 (**a**, **b**) Depict gully features on hillslopes and foothill of Western plateau plains. Gullies are broken or on slopes but continuous on plains and tough to erode or carry sediment load having limited velocity and remain wider with shallow depth. The Lacey's theory of regime (IRC: SP-132004) is not valid for such region and yet to be developed but being used modifying the coefficients. It also shows the deforestation of hills and bushes afforestation in plains. Further, an unplanned development may also do injustice with the existing natural drainage pattern

design annual sediment yield rate for the Trans and Greater Himalayan regions is estimated between 1600 and 3200 t/km², respectively (u/s region may be considered double the regional value of d/s). The water and sediment yield annually during the monsoon (June–September) are 73.7% and 92.10%, respectively.

Locations of representative Figs. 14.1, 14.2, 14.3, 14.4, 14.5, 14.6, and Fig.14.7 are shown on the downloaded Index

Fig. 14.4 NH48 highways portion toward Mumbai, preceding India's first Mumbai-Vadodara Expressway: (**a**) erosion of high embankment and gully formation, (**b**) protection of slope by stepped chute, and (**c**) trace of typical gully formation on the rocky river bed during lean and dry flow

map (Fig. 14.8). Figures 14.1 and 14.2 represent the mountainous region covered with seasonal or permanent snow covers. The Western Ghats have plateau hills, flat terrain, and sea coast which are less venerable to deep erosion. The third area is the alluvial plain of Bihar, where gullies formation is common as such of that in the hill sloping contour farming plots.

Prasad and Verdhen (1992) computed monsoon sediment based on year 1970 to 1979 flood flow data of North Bihar which resulted in sediment load of 11×10^2 to 20×10^2 t/km^2 of Indo-Nepal rivers. Average monsoon sediment concentration ranged from 0.94 gm/l (Bagmati at Hayaghat) to 1.9 gm/l (Gandak at Triveni) producing nearly 2000 t/km^2, whereas, discovered relatively less, 1400 t/km^2 by the Mahananda and 1200 t/km^2 by the Kosi at Brahkshetra. Further it gets reduced as flow traverses downstream, Bagmati to 300 t/km^2 at Hayaghat and Kosi to 400 t/km^2 after 300 km d/s at Baltara, due to flatter slope, apparently responsible for aggradations of

rivers and building up of the alluvial tracts. Overall, monthly sediment yield distribution computed as 6% in June, 24% in July, 32% in August, 22% in September, and 12% in October.

Poesen et al. (2003) reported soil loss and sediment production due to gully erosion in the range of 10 to 94% at global scale and reported rate of soil erosion in India is 33 t/ha/year from gullies in ravine regions (Shekinah and Saraswathy 2005). Consequently, considering even at the lowest global range of 10%, the sediment yield through gullies erosion of North Bihar Rivers at foothill of the Himalayas would be more, approximately 4 to 6 times higher than that of the average value of India. However, it almost matches at the downstream. It reveals that the alluvial plains of North Bihar are building up and rivers are in the process of aggradation.

Gullies on the glacier and under snow cover (Figs. 14.1 and 14.2) during snow and ice melt are basically due to temperature and radiation, augmented by the rain, if any. These temporary gullies may disappear with change in

Fig. 14.5 Development of drainage pattern like gully formation and conditions of unmaintained roads of alluvial tracks exhibit:(**a, b**) poorly paved road prone to erosion due to rain water flow network formation, and (**c, d**) unpaved roads facing tragic and unnatural longitudinal and transverse gully formation affecting traffic and pedestrians

season or experiencing heavy snowfall. A stanza of Magahi poem (Verdhen 2017) has been found relevant to present with translated version on such phenomenon:

Kiya pura jatan, laga lage lagan,
(Committed with best efforts, felt emotionally involved),

Bin baris bahe, bin jaldhi bahe,
(flows without precipitation or any water body),

Lahrate, ghargharate, balkhate, ithlate,
(sinuously meanders, cynically, wavy-falls, plausibly),

Yaisan sarpat chale, shristi hath male,
(move so much rapidly that creator fails to control),

Jaise parwat sikhar ke pighalahe taan,
(as body is melting away from mountain's peaks),

Dekh sundar saman, jag niyara lage,
(observing pleasant landscape, planet turns gorgeous),

Khulal failal asman sitara lage,
(open and wide spread sky looks like galaxy of stars),

Magahi tara lage ek dhruvatara lage,
(mother tongue or land brains star, a polar star),

Appan kasaba ke bhasaba pyara lage.
(language/nature of own region wits indeed lovely).

Similarly, for the non-snowy plateau and alluvial plains (Figs. 14.3 and 14.7), a portion of poem, 'Garmi ke Sukhard Dharati' (heated and dry land) by Damodar Prasad (Verdhen 2017) is relevant in order to have dynamics of interaction between rain spell and dry land:

Fig. 14.6 Features of gullies in plain barren and agricultural land and its outfall in consequent/subsequent stream, (**a**) by formation of conjunctive fan-out and plummet due to rise and fall of stage-discharge in the river, and (**b**) through cultivable land erosion and encroached seasonal water bodies and channel

Garmi ke sukhal dharati par,	(The dry and deserted land due to hotness),
pawas ke giral pahala bund,	(received first drop of rain in rainy season).
Maan mast kar deva hey,	(which delights the mind fully),
sugandh se masti bhar deva hey.	(rainy-soil-scent spreads happiness all around),
Dharti ke ang-ang angra hey,	(and the every part of the land starts yawning),
premi premi ke rang mein ranga hey,	(the lovers get colored together by the color of each other), like rainwater and soil get intermixed.

The resultant eroded soil mass gets carried away under the carrying capacity of the flowing water. The erosivity of fresh and sediment-saturated water is limited due to lack in its abrasion or transportation capacity. However, the gully/channel on snow and glacier mass is due to temperature (Fig. 14.9)/heat energy available in flowing water to further

Fig. 14.7 Various other examples of erosion, cutting, and sedimentation: like (**a**) top left to right gully affects agricultural land and practices in Sekhpura, (**b**) river bank cutting and flood plains, (**c**) uncared urban sewerage on the bank of Ganges, and (**d**) sea coast of Andaman and Nicobar

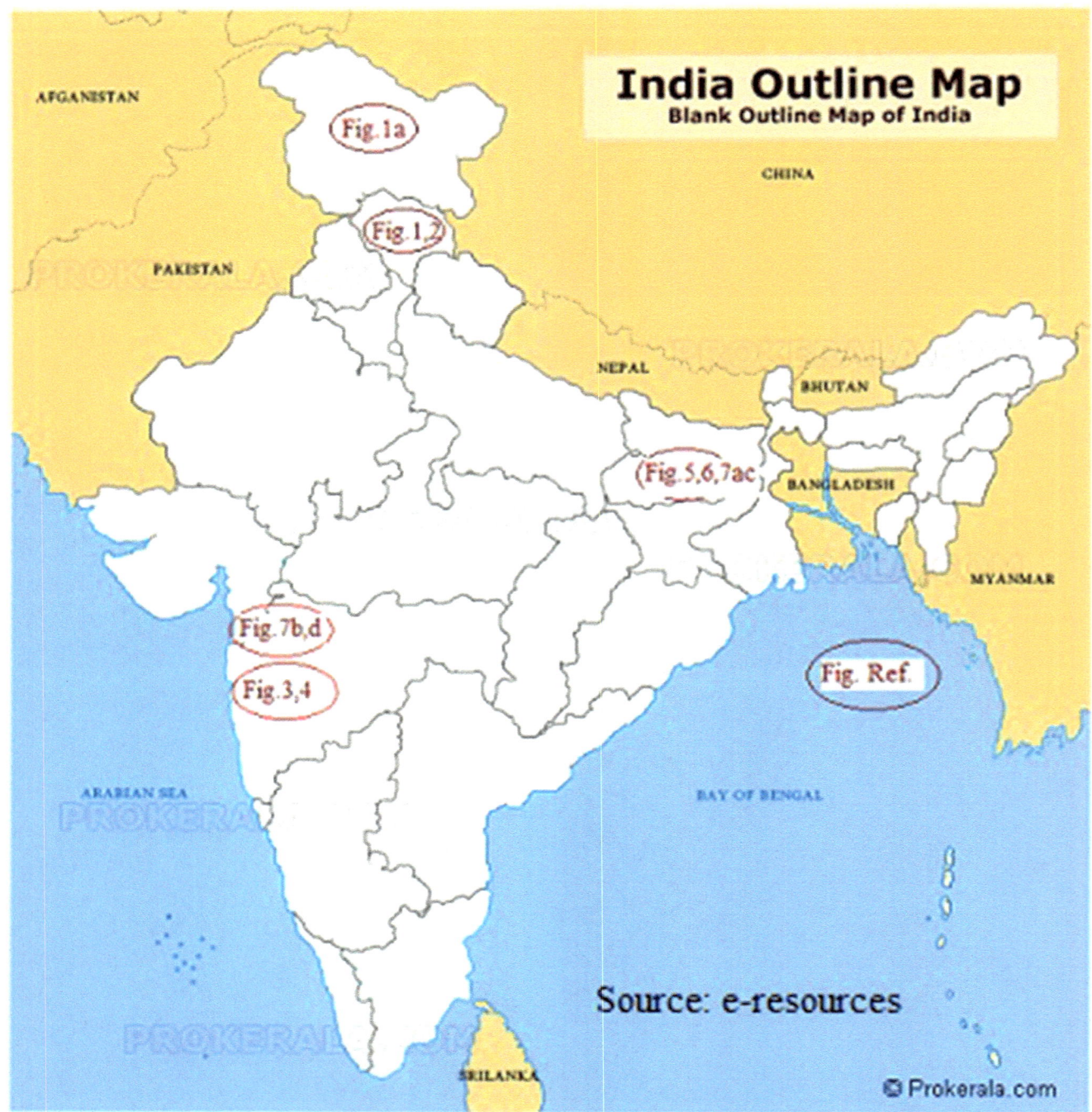

Fig. 14.8 Index map (downloaded from e-resources) to indicate the locations of Figs. 14.1 and 14.2 (Himachal and Jammu and Kashmir); Figs. 14.3, 14.4, 14.6, and 14.7 (between Surat and Mumbai); and Fig. 14.4, 14.5, 14.6, and 14.7 (Patna, Nalanda and Sekhpura District)

Fig. 14.9 Winter ending weekly discharge of Kothi and Solang nala of the Beas basin with mean air temperature at Solang (After Verdhen 2010) exhibits melt after the sixth week (mid-February)

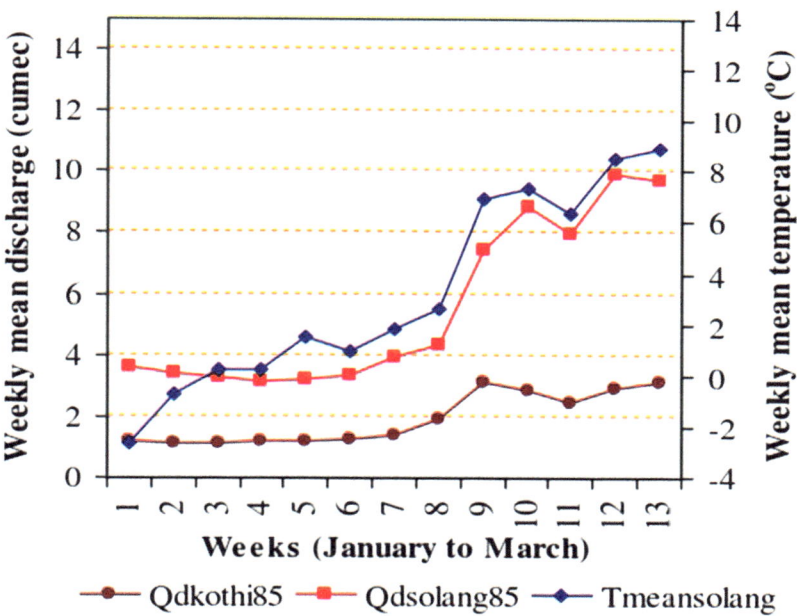

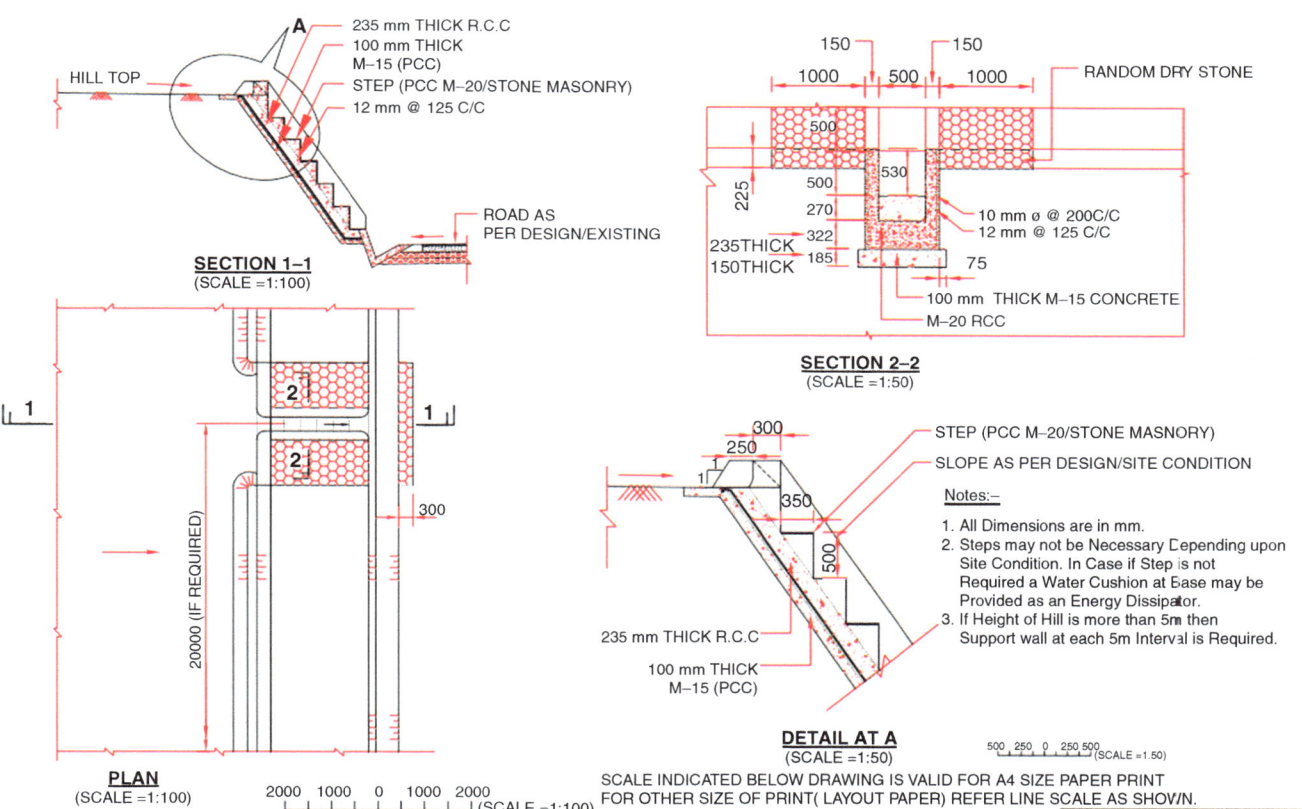

Fig. 14.10 Stepped chute with hillside-road or highways embankment drainage system (After Verdhen 2015)

melt away/erode (Figs. 14.1 and 14.2) in contrast to Figs. 14.3 and 14.4 and Figs. 14.5, 14.6, and 14.7. Figure 14.10 is a typical sketch of stepped chute and its components to protect the hillslope from land sliding and erosion of high embankment/road (Fig. 14.4b) by restricting seepage and dissipating flowing water energy generation.

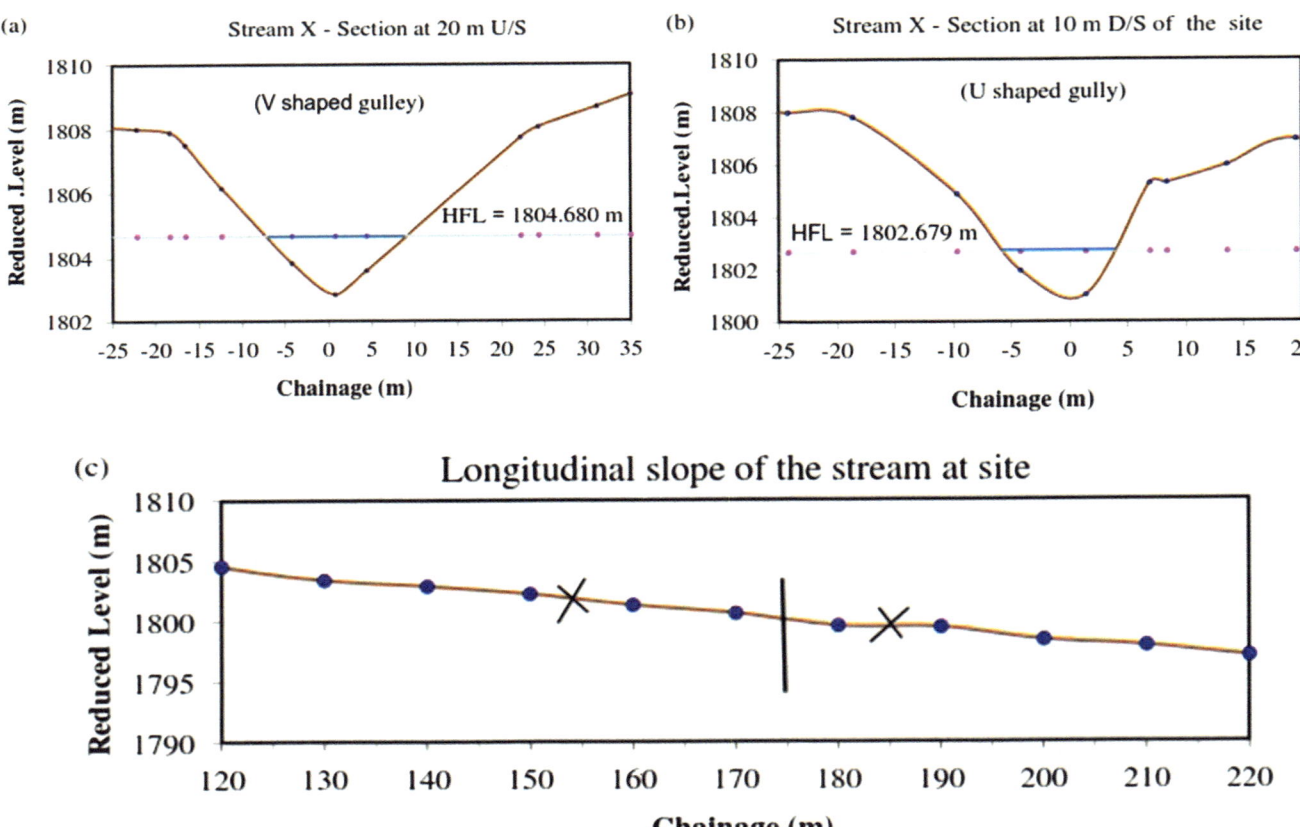

Fig. 14.11 Prenigaon stream's gully cross-sections at (**a**) 20 m u/s and (**b**) 10 m d/s from the NH1A highways to Srinagar at CD 219/1 including its (**c**) long section having slope of 1 in 15

Occurrences of temporary gullies are not uncommon if proper and adequate longitudinal or cross-drainage structures are not provided along and across the paved or unpaved road (Fig. 14.5). The suitable drainage structure keeps the health and life of the road preserved providing an efficient and effective utility scenario.

Hydrology and hydraulic studies are essential for proposing measures to control gullies in agricultural field, diverting the surplus flows, protecting the banks and unstable slopes, and even in providing check dams or regulating the safe drainage under diverse natural or manmade conditions (Figs. 14.6 and 14.7). Typical hydrological and hydraulic analysis for Pernigaon stream of 2.81 km² catchment area above NH1A at CD 291/1 has been carried out using Eqs. (14.4–14.11).

The parameters C, B, Lc, L, H, and F used in the concerned equations are 18, 0.84 km, 6.25 km, 3.35 km, 1200 m, and 18 cm, respectively. The cross-sectional area at 20 m upstream and 10 m downstream of the site along with long section (Fig. 14.11) reveals an interesting result. The u/s section is of V shape, while the d/s section is of nearly U shape.

Relevant features (Table 14.1), discharge computation result at 50 years return period including regime depth,

scour depth, HFL, area of flow, velocity, waterway, Froude no., etc., reveal that the slope is gentle, the designed peak flow remains subcritical, and the linear waterway at HFL is less than half the waterway computed by Lacey's Equation (14.9a-d). It indicates that the formation of section in stable channel is not in agreement with the alluvial channel, defined by the Lacey's theory of aggradations and degradation in open channel flow.

14.4 Conclusion

Hydrology and hydraulic studies are essential for assessing the reasons behind the gullies formations or proposing any suitable measures to control gullies in the agricultural field, divert flow, protect banks and unstable slopes, or provide check dams for regulating the flow under diverse natural or manmade conditions. Arc-View GIS-based hydrological modeling in delineating the contributing area and simulating rainfall and runoff are in practice. Sediment yield by empirical relation (Garde and Kothyari 1987) and its modified form (Verdhen et al. 2012) are found within acceptable error of estimation. The erosivity of fresh and sediment-saturated

Table 14.1 Features of the channel, its lowest bed level, highest flood level, silt factor, design discharge, flow area, velocity, Froude no., linear waterways, regime, and scour depth of the stream

Name of the Gully/Stream	Restricted water way (m)	D_{50} (mm)	Normal HFL (by SA method)	LBL (m)	Depth (HFL-LBL)	Catchment area (km^2)	Computed discharge (m^3/s)					
							Empirical method	Rational method	Slope-area method	Weir/Orifice formula	Unit hydrograph method	Design flow (m^3/s)
(1)	(2)	(3)	(4)	(5)	(6)	(7)	(8)	(9)	(10)	(11)	(12)	(13)
Prenigaon	12.00	1.00	1803.346	1801.734	1.612	2.810	39.0	71.0	78.87[a]	66.00	–	71

Silt factor, f	Design HFL (m) U/S	Design HFL (m) D/S	Slope of the stream	Design discharge (m^3/s)	Scour depth (m)	Regime depth (m)	Area of flow (m^2)	Velocity at the NH1A site	Regime waterway (m), Wr	Actual waterway at HFL (m), W	Froude's No.	Fluming ratio
(14)	(15)	(16)	(17)	(18)	(19)	(20)	(21)	(22)	(23)	(24)	(25)	(26)
1.760	1804.68	1802.679	0.067	71.000	3.631	1.622	30.0	2.367	37.918	15	0.59	0.80

[a]Discharge by slope-area method is ignored as revised HFL is 1804 m; NHFL = 183.346 m

water is limited due to lack in its abrasion or transportation capacity. However, the gully/channel on snow and glacier mass is due to temperature/heat energy potential in flowing water to further melt or erode.

Sediment yield is found to be up to 700 t/km^2 in non-Himalayan Rivers and up to 7000 t/km^2 in the Himalayan Rivers. The estimated annual sediment yield rates for the Trans and Greater Himalayan regions are 1600 and 3200 t/km^2, respectively (more yield in the higher region). Considering minimum global range of 10% contribution of sediment from gullies, the sediment shared by the gullies of North Bihar Rivers at foothill of the Himalayas would be of 4 to 6 times higher than that of the Indian average value, but it matches at downstream. These indicate that alluvial plains of North Bihar are building up and rivers are in the process of aggradation.

Stepped chute helps to protect the seepage-induced landslides or high embankment/road from erosion by dissipating water energy to erode. Adequate longitudinal and cross-drainage structures are essential along and across the paved or unpaved roads in alluvial plains. The suitable drainage structures keep the health and life of the road, providing an efficient utility scenario. Study indicates that the formation of gully section in stable channel is not in agreement with the alluvial channel, defined by the Lacey's theory of aggradations and degradation in open channel flow.

Acknowledgments The author acknowledges the organizations that gave the opportunity to work on massive field and design-based research to contribute them to science and the society. The author also wishes to thank Mr. Aarsh Verdhan who has taken interest to proofread this manuscript. Further, the author wishes to thank the editor, publisher, and reviewers for the opportunity, sample papers, and valuable suggestions.

References

Bennett HH (1939) Soil Conservation. McGraw-Hill Book Company Inc., New York, p 993

Bertrand AR, Woodburn R (1964) A fresh look at gully erosion in the South. J. Soil Water Conserv 19:173–175

Brooks A, Spencer J, Knight J (2007) Alluvial gully erosion in Australia's tropical rivers: a conceptual model as a basis for remote sensing mapping procedure. In: Proc. of the 5th Australian stream management conference on Australian Rivers making difference (ed: Wison et al.), Charls Sturt University, Thurgoona, New South Wales

CWC (1984) Flood Estimation Report of J&K (Zone-7). Jointly Prepared with RDSO, IMD and MoRST by Directorate of Hydrology, Central Water Commission, New Delhi, India

Desmet PJ, Poesen J, Govers G, Vandaele K (1999) Importance of slope gradient and contributing area for optimal prediction of the initiation and trajectory of ephemeral gullies. CATENA 37:377–392

Dunne T, Western D, Dietrich WE (2011) Effects of cattle tramping on vegetation, infiltration, and erosion in tropical rangeland. Journal of Arid Environment 75(1):58–69

Garde RJ, Kothyari UC (1987) Sediment yield estimation. J. Irrigation and Power (India), 44(3):97–123

Gardiner V (1981) Drainage basin morphometry. In: Geomorphological Techniques, Goudie, A. (Ed.). Allen and Unwin, London, UK, pp 47–55

Gregory KJ, Walling DE (1973) Drainage basin form and processes. Edward Arnold Scrap Processors Inc., London

Gupta GP (1977) An appraisal of the sedimentation problem in the country and measures to combat it. In: Proc. of Symp. on silting of reservoirs with special reference to estimating the life of Reservoirs and Measures to Arrest the Rate of Sedimentation, ISI, Manak Bhawan, New Delhi, CBIP Pub. No. 126, pp 1–9

Hauge C (1997) Soil erosion definitions. California Geol. 30:202–203

Herzig A, Dymond JR, Marden M (2011) A gully-complex model for assessing gully stabilization strategies. Geomorphology. doi:https://doi.org/10.1016/j.geomorph.2011.06.012

Hewitt K (1982) Pakistan case study, catastrophic floods. pp 131–135

Horton, R.E., 1932, Drainage-basin characteristics. Trans. Am. Geophys. Union 13:350–361

IRC: SP-13 (2004) Guidelines for the Design of Small Bridges and Culverts (First Edition), Indian Roads Congress, Sahjahan Road, Delhi, India

IRC: SP-5 (1998) Standard Specification and Code of Practice for Road Bridges, Section I- General Features of Design (7th Edition), Indian Roads Congress, Sahjahan Road, Delhi, India

Jarvis RS (1976) Classification of nested tributary basins in analysis of drainage basin shape. Water Resour. Res. 12:1151–1164

Karaushev AV, Bogoliubova IV, Bobrovitslsaya NN (1974) Water erosion and sediment discharge. IAHS Publ. No. 113:46–52

Kukal SS, Sur HS, Gill SS (1991) Factors responsible for soil erosion hazard in sub-mountain Punjab, India Soil Use Manage. 7:38–44

Langbein WB, Schumm SA (1958) Yield of sediment in relation to mean annual precipitation. Trans. Amer. Geophys. Union 39 (6):1076–1084

Marden M, Herzig A, Arnold G (2011) Gully degradation, stabilization and effectiveness of reforestation in reducing gully-derived sediment, East Coast region, North Island, New Zealand. Journal of Hydrology (New Zealand) 50(1):19–36

Meyer A, Martinez-Casasnovas JA (1999) Prediction of existing gully erosion in Vineyard Parcels of the NE Spain: A logistic modeling approach, Soil. Soil. Till. Res. 50:319–331

Michael AM, Ojha TP (2012) Principles of Agricultural Engineering. Vol 2. Jain Bros., p 655

Moghaddam MHR,Saghafi M (2008) Gully erosion monitoring on shakhen drainage basin, Southern Khorasam Province, Iran, J. Applied Sci. 8:946–955

Montgomery DR, Dietrich WE (1988) Where do channels begin. Nature 336:232–234

Poesen J, Nachtergaele J, Verstraeten G, Valentin C (2003) Gully erosion and environmental change: Importance and research needs. CATENA 50:91–133

Prasad G, Verdhen A (1992) Sediment transport measurement in North Bihar Rivers indicating erosion from Indo- Nepal basins. In: IAHS Proc. of Oslow Int. Symp. on erosion and sediment transport monitoring programme in River basins, Norway, 1992, pp 117–125

Rasool IU, Khera KL,Gul F (2011) Proliferation of gully erosion in the submontane Punjab, India. Asian Journal of Scientific Research 4:287–301. doi:https://doi.org/10.3923/ajsr.2011.287.301

Renard KG, Foster GR, Weesies GA, McCool DK, Yoder DC (1997) Predicting soil erosion by water: A guide to conservation planning with the RevisedUniversal Soil Loss Equation. US Department of agriculture, Agricultural handbook 703, Washington, DC, p 404

Sandercock PJ, Hooke JM (2011) Vegetation effect on sediment connectivity and processes in an ephemeral channel in SE Spain. Journal of Arid Environment 75:239–254

Sargeant IJ (1984) Gully erosion in Victoria. In: Proceedings of the natural soils conference, May 13–18, Brisbane, Australia

Shekinah DM, Saraswathy R (2005) Impacts of soil erosion by water- A review. Agric. Rev. 26(3):195–202

Torkashvand AM (2008) Investigation of some methodologies for gully erosion mapping. J. Applied Sci. 8:2435–2441

Tywoniuk N, Fowler JL (1972) Winter measurement of suspended sediment. In: Proc. of Symp. on the role of snow and ice in hydrology, Banoff symposia Vol. 1, pp 814–827

USDA (1954) A Manual of Conservation of Soil and Water, Handbook of Agricultural Workers. USDA, SCS, Agricultural handbook no. 61, Washington, D.C., 1954, p 208

Vashisth RS (1962) Bhugol ke Bhautik Adhar (H). Atmaram and Sons, Kashmiri Gate, Delhi

Verdhen A (2010) Hydrological investigation challenges of transboundary watershed aquifer in the Himalayan region. In: International Conference on Trans-boundary Aquifers, Challenges and new directions, ISARM-2010, UNESCO, Paris, 6–8 December, P 92 (1–10)

VerdhenA (1991) Hydrological study in a glaciated mountain stream. In: Proceedings of the national meet on Himalayan glaciology, Meet held in June 1989 at DST, New Delhi

Verdhen A (2015) Rain induced landslide mitigation through technological intercession. In: Proc. of National Seminar on Landslide: management & mitigation strategies (LAMAMIS-2015), DTRL (DRDO), Metcalf House, New Delhi, 5–6 Feb. 2015

Verdhen A (2017) Bhasba anmol ratana (Lingua gem).Magahi Bayar (Magahi flow, poems), Nalanda District Hindi Sahitya Sammelan and Sudama-Parmeshwar Sahitya Sansthan, Bihar Sharif, Nalanda, India

Verdhen A (2018) Rain and snowmelt augmented design flood for highways bridges in snowy mountains. J of Hydrogeology and Hydrologic Engineering 7:2. doi:https://doi.org/10.4172/2325-9647.1000171

Verdhen A, Chahar B, Sharma OP (2012) Suspended load yield in glaciated streams in the Himalayas. In: Proc. of BIS seminar on suspended sediments for dams in Himalayan regions, Bureau of Indian Standards, New Delhi, India, 30 November 2012

Verdhen A, Chahar B, Sharma OP (2013) Snowmelt runoff simulation using HEC-HMS in a Himalayan watershed. In: Proc. of world environmental & water resources congress, EWRI of ASCE, Ohio, May 2013

Walling DE, Kleo AH (1979) Sediment yields of rivers in areas of low precipitation, a global view. In: Proc. of Symp. on the hydrology of areas of low precipitation, pp 479–493

Ward DJ, Berlin MM, Anderson RS (2011) Sediment dynamics below retreating cliffs. Earth Surface Processes and Landforms 36 (8):1023–1043

Yu B (1998) Rainfall erosivity and its estimation for Australia's tropics. Australian Journal of Soil Research 36(1):143–165

Anand Verdhen is civil engineering graduate from RIT JSR (Ranchi University) and M.Tech (81CEM45F) and Ph.D. from IIT Delhi. He served in Snow and Avalanche Study Establishment (DRDO, Govt. of India) as Scientist "B" and Team Leader (1983–1987); NIH, Roorkee (Scientist C: 1987–1989); CWRS, PU (WRE/Assoc. Prof., 1989–2000), sent on LWP, Since 2004 NITP damaged services by not taking back); ICT, CES(I), and Egis (I) Pvt. Ltd at Delhi as Hydrologist/GM/TL (2000–2010) covering the field of snow, meteorology, avalanche, glacier, hydrological and climate modeling, flood, conjunctive irrigation, wetland, highways, and drainage. He headed research and was Professor (CE Dept.) at DCE, Gurgaon (Dec. 2013–18 Apr. 2015). To his credit he has more than 50 research reports and above 50 research papers. Recieved best paper award on Flood Flow Behavior of the Rivers in North Bihar (CBIP, 1993).

Influence of Road-Stream Crossing on the Initiation of Gully: Case Study from the Terai Region of Eastern India

15

Suvendu Roy

Abstract

The undersized road-stream crossing (RSC) creates the severe problem of soil loss over the landscape of Terai Region, especially in and around the tea estate due to the unconsolidated and fragile condition of the underlying geology. Artificial channel routing for the irrigation of tea gardens has increased the drainage density and decreased the concentration time of runoff at the inlet of RSC. Due to the underestimation of typical hydrology of upstream tea gardens at Site C, the downstream of studied RSC faces unexpected changes in channel geometry between upstream and downstream. The typical hydraulics of Site C has induced an increase in the channel width (480%), depth (560%), and cross-section area (3728%) in the downstream in comparison with its upstream and formed a 981 m-long enormous gully with an area of 5.49 ha, average depth of 12 m, and width of 14–49 m. The comparative study between three RSCs in the same region shows that the process of gully erosion varies with the size of crossing structure and the nature of upstream drainage networks and condition of the riparian zone.

Keywords

Road-stream crossing · Gully · Terai region · Soil loss · Drainage network

15.1 Introduction

Gully erosion is the most effective and severe problem of land degradation in any country (Poesen et al. 2003). Gully erosion works as a major non-point source of sediments to the streams and severely affect the river health physically as well as biologically (Valentin et al. 2005), which seems to be accelerated in near future by changing the climate, extreme events, changing land use pattern, and other anthropogenic activities. Globally about 75 billion metric tons of soil is removed from land annually, of which the average rate of soil erosion is 40 tons/ha/year in Asia and about 17 tons/ha/year in the United States and Europe (Pimentel et al. 1995). However, Borrelli et al. (2017) have challenged the previous results through a high-resolution (250 × 250 m) global potential soil erosion model, using a combination of remote sensing, GIS modelling, and census data. Borrelli et al. (2017) have estimated that globally the annual soil loss amount is 35.9 Pg/year in 2012, which is at least two times lower than the previous estimation. The study has also claimed that changing the land-use pattern is the principal cause for the 2.5% increase in the global soil loss since 2001. In addition, the study refers to India as a hot spot (higher than 20 Mg/ha/year) of soil erosion along with China, Brazil, and African territories located along the equator and others. In general, the study has predicted that the less-developed economies show the highest prediction of annual total soil erosion (20.7 Pg/year), equal to 59.2% of the global soil erosion.

India faces the problem of land degradation for its 29.32% of geographical area during 2011–2013, of which the principal share is due to water-mediated erosion (10.98%) followed by vegetation degradation (8.91%) and wind action (5.55%), as per the report published by the Space Application Centre, Indian Space Research Organization (SAC-ISRO 2018). According to the Central Soil Water Conservation Research and Training Institute (CSWCRTI), Dehradun, India, loses 5334 million tones of soil every year, especially due to the excess use of fertilizer, pesticides, and wrong practice of irrigation in the agricultural sector (The Hindu 2010). Among the agricultural practices in India, the land under plantation farming, especially in the tea estate (about 566 thousand ha, as Tea Board of India in 2015–2016), recently faces a serious problem of land degradation due to huge soil loss. The growing number of tea estate on rain-fed

S. Roy (✉)
Department of Geography, Kalipada Ghosh Tarai Mahavidyalaya, Darjeeling, West Bengal, India

P. K. Shit et al. (eds.), *Gully Erosion Studies from India and Surrounding Regions*, Advances in Science, Technology & Innovation,
https://doi.org/10.1007/978-3-030-23243-6_15

(1150–6000 mm) sloping land without any soil and water conservation is the major cause of such degradation (Sahoo et al. 2016). The estimated rate of soil loss in the Nilgiris (South India) is about 28–40 tonnes/ha/year during the initial phase of tea plantation (Chinnamani 1977; Madhu and Tripathi 1997). In southern Jiangsu Province of China, Yan et al. (2003) have estimated that the amount of net soil loss is 1946 tonnes/km^2 at the bottom of the tea gardens, and the amount has increased toward upland at the Yixing Tea Planation region using ^{137}Cs tracer method. Due to the soil loss in a tea estate, Ananda (2014) estimated about 12,331 kg/ha/year (at 95% confidence level) of yield loss from the high-grow region of tea plantation in Sri Lanka. Intensive tea plantation also reduces the soil nutrients, organism and organic matter, soil elasticity, chemical composition, and water-holding capacity (Zichenga et al. 2012; Sahoo et al. 2016), which have collectively affected the soil stability and increased the soil loss.

The effects of undersized road-stream crossing on the channel geomorphology, hydrology, and fluvial connectivity are well documented in literature (Gregory and Brookes 1983; Wellman et al. 2000; Hancock 2002; Resh 2005; Wheeler et al. 2005; Merril and Gregory 2007; Blanton and Marcus 2009, 2014; Roy and Sahu 2017, 2018). Some studies have also demonstrated about the road (paved and unpaved) surface drainage and channel modification, especially through sediment and water dynamics (Montgomery 1994; Wemple et al.1996; Croke and Mockler 2001; Poesen et al. 2003; Thomaz and Peretto 2016). Development of rills and gullies around the transport infrastructures is an emerging field of research worldwide (Croke and Mockler 2001; Nyssen et al. 2002; Jungerius et al. 2002; Pathak et al. 2005; Takken et al. 2008; Raiter et al. 2018). Formation of gullies and associated soil loss around the road network is now a big concern for every road engineer (Jungerious et al. 2002), whereas the problem becomes severe for the developing countries due to lack of maintenance and provision to save outlets from excessive runoff (Adams and Watson 2002). Mati (1984) has estimated that about 50% of gullies in the Kiambu District, Kenya, were formed by road drainage. According to Roy and Sahu (2018), road-stream crossing works as an artificial knick-point and accelerates the flow velocity to the downstream of the crossing structure.

However, the specified field still demands more investigation in different land use practices and lithology across the world. Artificial channel routing for the tea plantation leads to a higher concentration of runoff on the sloppy ground and induces to cross the threshold limit between soil resistivity and critical shear stress (Montgomery 1994). The transport network is expanding rapidly over North Bengal, which also influences to change the land use practices and increase the interaction with the local landscape. In this consideration, the present study focuses on the problem of gully initiation due to road-stream crossing in-and-around tea plantation region of eastern India, especially in the Terai region of West Bengal. The Terai region is also a crucial belt to study for its unconsolidated lithology being a part of sub-Himalayan alluvial upper fan region (Bandyopadhyay et al. 2014) and one of the major tea production regions of India. The primary objective of the present study is to identify and explain the role of road-stream crossing on the initiation of gullies in the Terai region and the effect of artificial drainage system in the tea estate on this process of soil loss.

15.2 Materials and Method

15.2.1 Study Area and Selected Sample Sites

Extension of the Terai region for the present study has been defined based on the topography at the sub-Himalayan fan (upper) region of Darjeeling District, West Bengal, covering an area of ~1464 km^2. The region has been delineated by the 100 m contour in the southern side and 300 m contour in the northern side as the boundary between hilly and piedmont regions. The Tista and Mechi rivers are used as the boundaries of the eastern and western sides of the study area, respectively (Fig. 15.1). From the physiographical perspective, the region is the result of partly coalesced alluvial fans dominated by the Tista megafan with an average slope of 1.25° (Bandyopadhyay et al. 2014). The older alluvium of middle to upper Pleistocene dominates the geology of the region, and soil texture varies from gravelly in the north to sandy in the south, whereas the lithological characteristics suggest the nature of the deposition is unconsolidated and the product of periglacial, fluviatile, and product of sub-areal erosion (Bhattacharya 1993; Bandyopadhyay et al. 2014) (Fig. 15.2).

The characteristics of the annual rainfall and temperature of the Terai region significantly differ from the hilly region of Darjeeling District (Bhattacharya 1993; Sarkar and De 2017). The Terai region receives almost 2735 mm of rainfall annually with a mean temperature of 25 °C, which is about 3382 mm and 17 °C in the hilly region, respectively (Sarkar and De 2017). The vegetation of the study region is mainly composed of tropical moist deciduous type, dominated by the tree species of Khair (*Accacia catechu*), Sissu (*Dalbargia sissoo*), Simul (*Bombam malabaricum*), Siris (*Albizza* spp.), Sal (*Shorea robusta*), etc. Central Ground Water Board, India (2014), has recognized the presence of alluvium aquifer in the study area, where the depth of the water table is about 5–10 m bgl (below ground level) during pre-monsoon period and <5 m bgl during post-monsoon season. Surface hydrology of the region shows within the interfluves of Tista and Mechi Rivers a number of minor streams are flowing from north to south with their different tributaries.

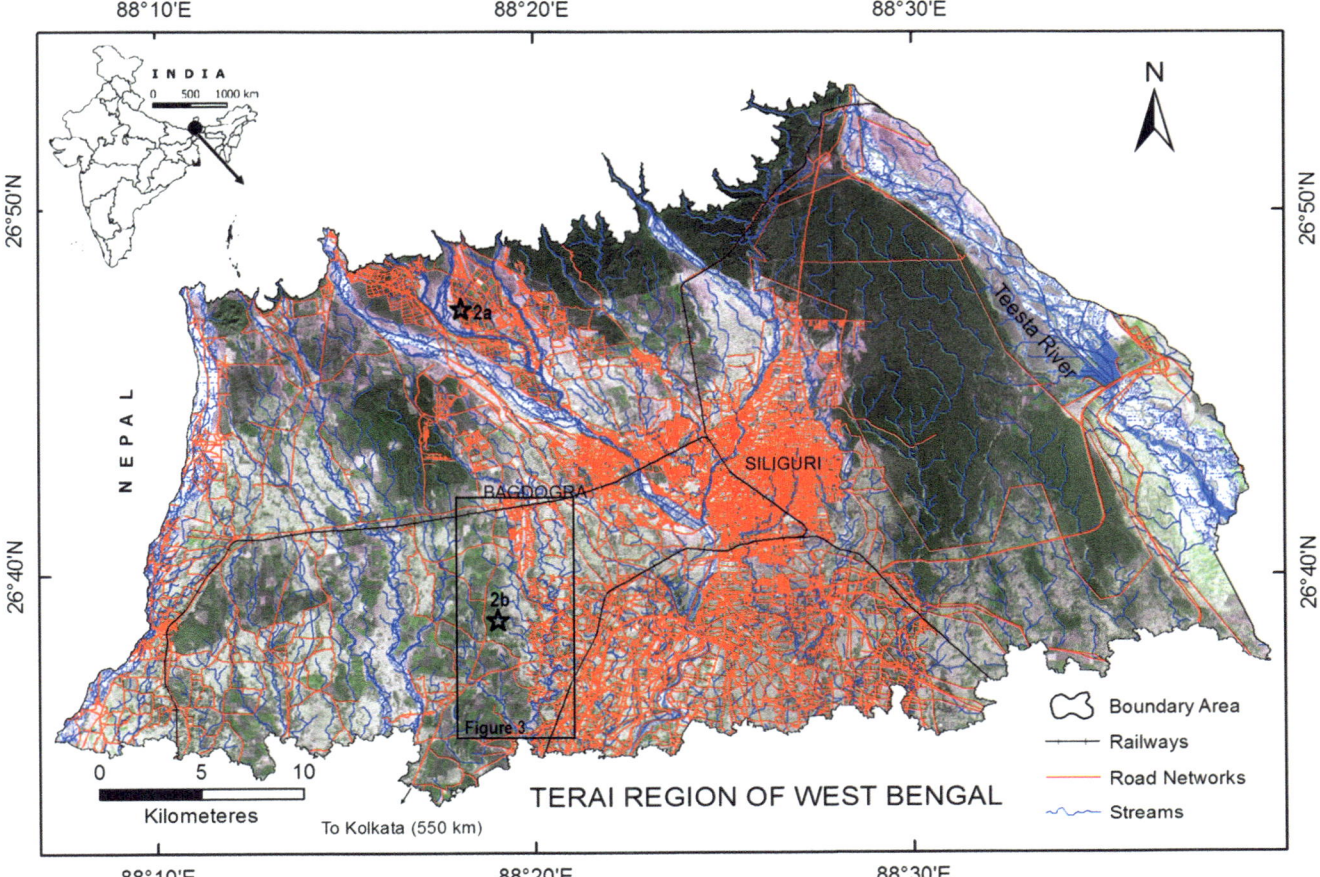

Fig. 15.1 Location map of the study area (2a and 2b indicate the plates provided in figures 15.2a & 15.2b)

The region is also called the Dooars-Terai Tea region, where the tea estates are situated from 90 to 1150 m above the sea level. As per the Tea Board of India (2017), the spatial extent of this region is about 140.44 thousand ha and the production of tea from this region (24.57 M kg) shares almost 35% of total tea production in India (71.39 M kg) during 2017–2018. In addition, with the best tea production belt of India, the region is also famous for its natural beauty and climate. Therefore, the region becomes an important economic province for the production of the cash crop and tourism industry. As a result, the transport network and urbanization of the region are also expanding day by day. Siliguri Metropolitan is the major urban center of this region in addition to a number of sub-urban centers of development, e.g., Uttar Bagdogra. The region is also characterized by extended railway tracks and the Asian Highway (AH) 02, National Highway (NH) 31, 31C, and State Highway (SH) 12, 12A with a number of minor paved and unpaved roads across the tea estates, forests, and defense area. Therefore, there are many intersection points between stream and road networks over the study area.

Three intersection sites have been selected for the present study considering the objectives with a special focus on the structure of the road-stream crossing, their upstream land use, and drainage network. All three sites have been selected on the NH 31, the road which connects Kolkata from Uttar Bagdogra. The names of the sites are given as A, B, and C, from north to southward (Fig. 15.3). The upstream land use of all three sites is dominated by tea plantations; however, differences have been observed in the drainage system, crossing structure, irrigation system, and controlling measures (Table 15.1). The drainage system of all three sites is part of Buribalason River system, which is also a major tributary of Mahananda River.

15.2.2 Procedures of Mapping, Field Estimation, Applied Equations

Special attention has been given to digitize the drainage lines and to delimit their catchment area within the tea garden using high-resolution Google Earth Image using WebGIS

Fig. 15.2 Lithological sections show the variation of underlying texture from north to south of the Terai region (position highlight by star marks in Fig.15.1); (**a**) presence of granular structure in the north and (**b**) presence of coarse sandy texture in the south; 2b also proves the unconsolidated nature of underlying soil profile at study sites. Length of helmet (30 cm) in left photo and hammer (33 cm) can be used as scale of reference (Source: Field Survey 2019)

mode in ArcGIS 10.4. Other streamlines were also digitized from Google Erath Image and Sentinel 2A satellite data. All the major and minor transport networks have been extracted from Open Street Map (OSM) using QGIS 2.14.

The geometry of all three crossing structures and corresponding channels has been estimated by intensive field investigation using tape and leveling instrument and staff. The measurements regarding the morphometry of gully were also calculated by field investigation using leveling instrument and GPS. Two cross-sections from each site (one in the upstream and another in the downstream) have been surveyed at 20 m upstream and 20 downstream of the crossing structure. One longitudinal profile for each site has also been surveyed from 20 m upstream to 20 m downstream including 8 m-long crossing structure. All the cross-profiles and long-profiles have been drawn using HEC-RAS 5.0.1 software. The shape of the catchments was also analyzed using the Form Ratio Index (Horton 1932) and Circularity Ratio Index (Miller 1953).

The structure efficiency of studied box culverts has been estimated using the standard protocol proposed by FHWA (2014) and CDF (2004). FHWA (2014) predicted the minimum bed width in any culvert by (Eq. (15.1):

$$W_{\text{culvert bed}} = 1.2 W_{\text{ch}} + K \qquad (15.1)$$

where K = a constant value, i.e., 0.6 m (2 ft), and W_{ch} = the width of the bankfull channel, m (ft).

CDF (2004) has also calculated the area of the required culvert opening (A_c) as follows:

$$A_c = 3 A_{\text{bf}} \qquad (15.2)$$

where A_{bf} is an area of the channel at its bankfull stage.

Fig. 15.3 Selected study sites (A, B, C) and their drainage pattern, land use, catchment area, and transport networks superimposed on the false color composition of Sentinel 2A data

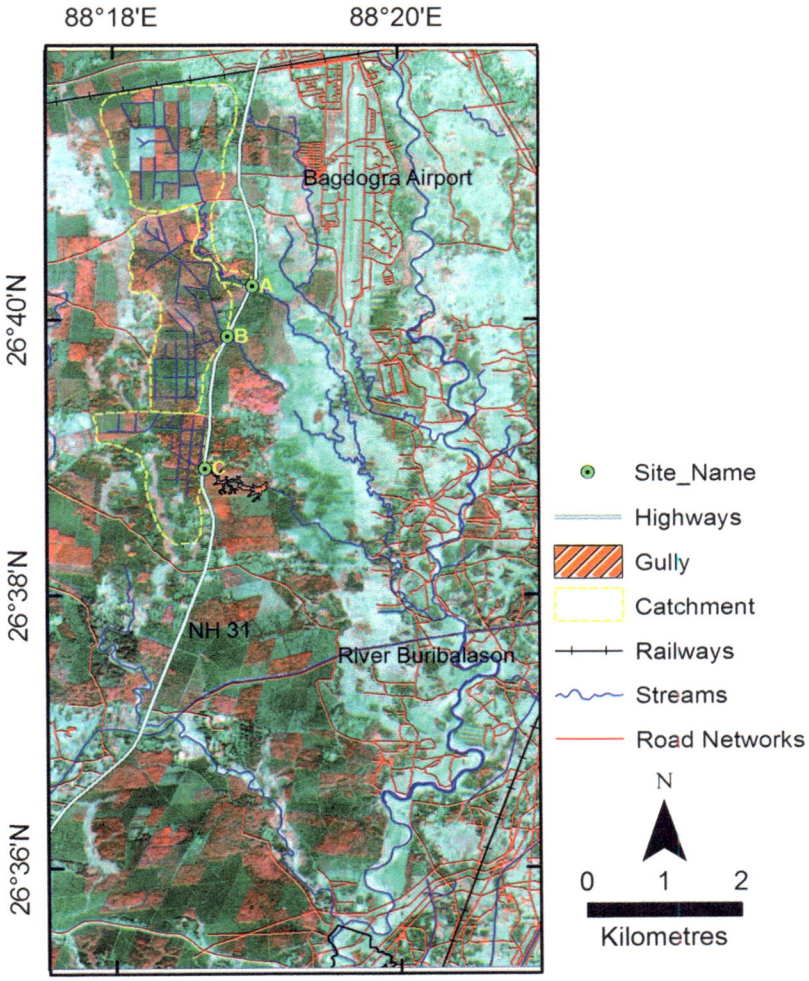

Table 15.1 Typical features of the study sites

Site code	Location	Crossing type	Catchment area[a] (km^2)	Drainage system	Drainage density (km/km^2)	Upstream land use	Basin shape analysis Form ratio (Fr) (Horton 1932)	Circularity ratio (Cr) (Miller 1953)	Problem of gully initiation
A	26°40'15.27"N; 88°18'59.07"E	Box culvert	2.33	Natural + artificial	6.27	Tea garden with limited natural forest	0.25	0.41	No gully has been observed still now
B	26°39'52.70"N; 88°18'48.04"E	Box culvert	2.07	Artificial	8.50	Tea garden	0.66	0.49	The initial phase of gully development
C	26°38'55.46"N; 88°18'37.81"E	Box culvert	1.36	Artificial	9.71	Tea garden	0.65	0.55	A large extended gully has been developed and still expanding through its tributaries

[a]Catchment area measures up to the crossing sites

15.3 Results

15.3.1 Characteristics of Upstream Drainage System

To retain a good drainage practice within the tea estates, the upstream of all three sites has been characterized by artificial drainage pattern with a number of straight drains from a different direction as per the suitability of irrigation to the different parts of tea gardens. However, in the upstream of Site A, a 2.4 km-long natural stream in the meandering pattern (Sinuosity Index = 1.93) has been observed up to the tea garden from the crossing site (Fig.15.3). Significant differences in drainage density have been estimated among the upstream catchments of three sample sites. The estimated values of drainage density are 6.27, 8.50, and 9.71 km/km^2 for the catchments A, B, and C, respectively. Hence, the density values are negatively correlated ($r = -0.91$; $p = 0.01$) with their respective catchment area (Table 15.1). Since belonging to a similar litho-topic unit, higher drainage density might be induced to form extended gully in the downstream of Site C and initiation of gully in the downstream of Site B. However, no sign of gully formation has been observed in the downstream of Site A with relatively low drainage density and part of natural meandering stream than Sites B and C. The shape analysis of all catchment shows that the catchment of B (Fr—0.66; Cr—0.49) and C (Fr—0.65; Cr—0.55) sites are nearly circular with higher form ratio and circularity ratio, whereas catchment area of A (Fr—0.25; Cr—0.41) site is nearly elongated with lower values of form and circularity ratios (Table 15.1).

15.3.2 Efficiency Level of Road-Stream Crossing Structure

All three culverts are box-type culverts; however, their structural dimensions are significantly reduced from Site A to C as mentioned in Table 15.2. Figure 15.4 and Table 15.2 also reveal that all the crossing structures are undersized in reference to the international protocols for culvert construction based on the upstream channel geometry. However, the difference between required and actual culvert width (w) is

that cross-section areas (a) of crossing structures markedly vary among the crossing sites. The culvert exit widths of Sites A, B, and C are undersized about 22%, 31%, and 33%, respectively, than required widths. The present culvert's cross-section areas at Sites B and C are also undersized about 8% and 69%, respectively, than required. Nevertheless, the existing cross-section area of crossing structure at Site A is about 110% surplus than the required amount (Table 15.2).

15.3.3 Comparison of Channel Geomorphology Around the Crossing Structure

Site-wise channel cross-section data and their longitudinal profiles show a significant difference in channel geometry between upstream and downstream of the crossing structures (Figs. 15.5 and 15.6). At Site A, both the upstream and downstream sections are part of the natural stream with higher (>10) width-depth ratio (Table 15.3). However, due to the presence of culvert, the downstream section is slightly incised with a drop height of 1.62 m at the outlet of the culvert. The changes in channel width (14%), mean depth (25%), and cross-section area (42%) between upstream and downstream section at Site A are very low in comparison with the other two sites (Table 15.3). The asymmetry in cross-profile defines the presence of a natural character in the stream of Site A, whereas symmetric nature of channel both in the upstream and downstream of Sites B and C reveals the anthropogenic modification in channel form (Knighton 1981). At site C, an unexceptional change in channel depth (560%), width (480%), and cross-section area (3728%) has been observed from its upstream to downstream, which clearly expose the severe condition of land degradation (Table 15.3 and Fig. 15.5).The changes in channel depth (59%) and cross-section area (86%) of Site B are also matter of concern because the values are indicating the process of gully initiation like Site C (Fig. 15.5). In addition to typical changes of channel geometry at Sites B and C, very low (<2) width-depth ratio also reveals the deep incision of channels with high vertical erosion (Table 15.3).

The constructed longitudinal profiles of all three sample sites have visualized the changes in channel bed height from

Table 15.2 Upstream channel geometry and differences between existing and required culvert width and cross-section area at sample sites in reference to the international protocols

Site ID	Upstream channel geometry			Dimensions of box culverts			Required width of culvert inlet (m) (FHWA 2014)	Required area of culvert inlet (m^2) (CDF 2004)
	w (m)	d (m)	a (m^2)	w (m)	h (m)	a (m^2)		
A	7.1	0.65	4.62	7.3	4.0	29.2	9.12	13.46
B	2.4	1.20	2.88	2.0	4.0	8.00	3.48	8.64
C	2.0	1.20	2.40	1.5	1.5	2.25	3.00	7.20

w width of channel and culvert, d mean depth of channel, a cross-section area of channel and culvert, h height of the culvert (Source: Filed Survey and Author's Calculation 2019)

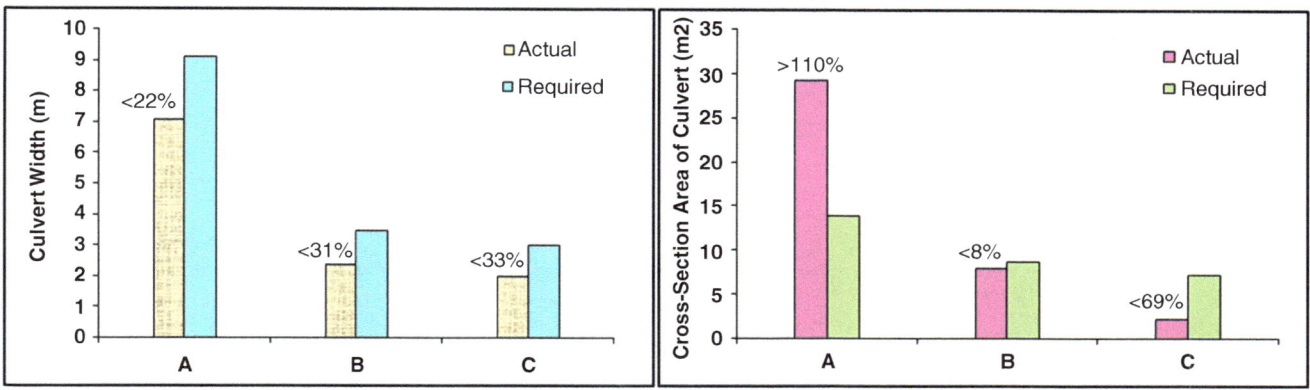

Fig. 15.4 Level of the efficiency of crossing structures based on culvert width (left) and cross-section area (right). Values at the top of actual's bar indicate the level of undersized (<) or surplus (>) than required dimensions

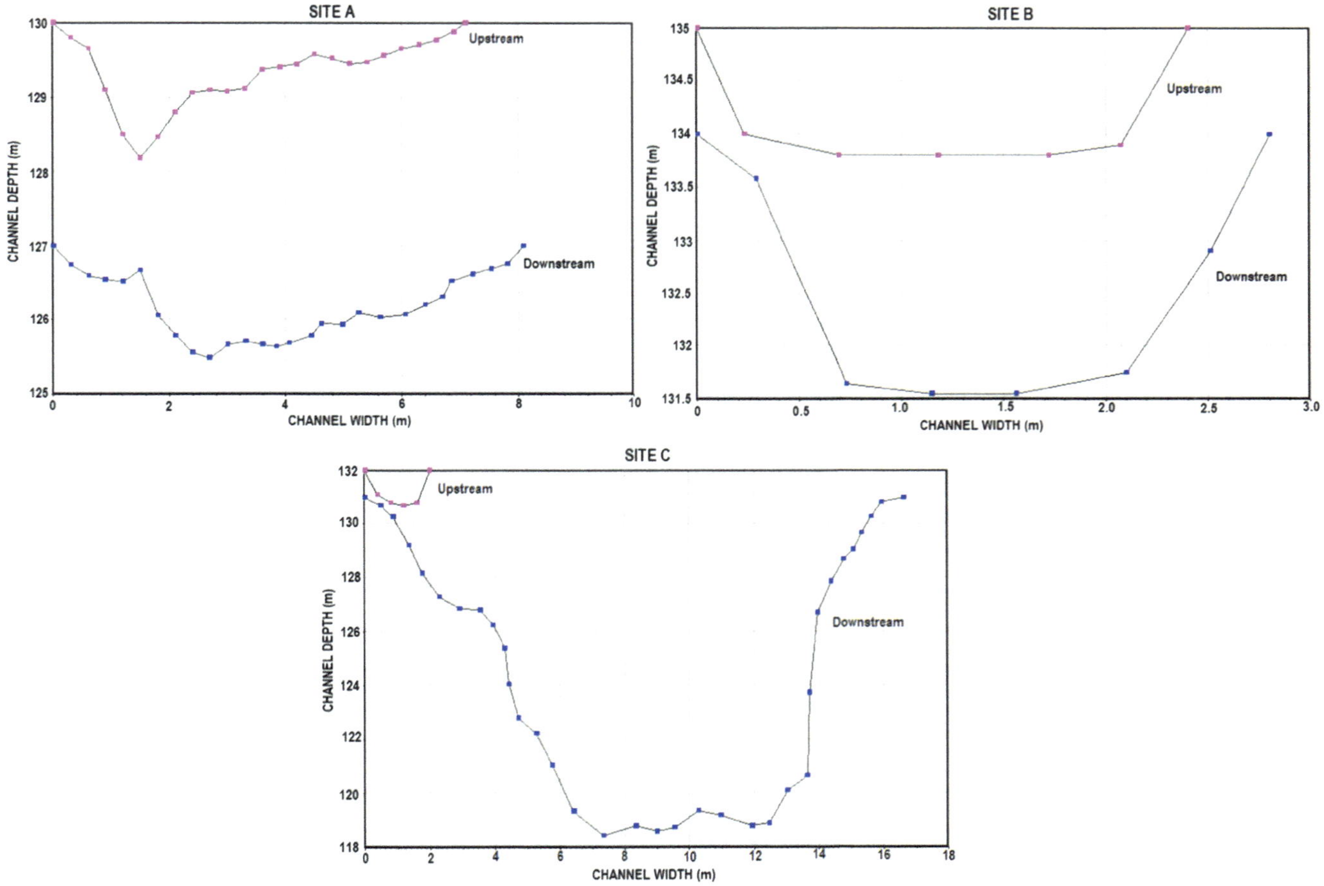

Fig. 15.5 Diagrammatic comparison of channel form of three sites from their upstream to downstream (Source: Filed Survey 2019)

20 m upstream to 30 m downstream of streams from the crossing sites (Fig. 15.6). The long profile of Site A shows an undulating nature of channel bed with the value of channel gradient being 4.2%. At the immediate downstream of the crossing structure (8 m), a rapid vertical fall in bed height has been observed for all three sites, which is actually the drop

height of respective crossing structure. Immediate after the drop height, the concave shape of the channel bed indicates the magnitude of scouring at the downstream of crossings. Due to the artificial channel in the upstream of Sites B and C, a flat channel bed has been observed (Fig. 15.6). However, significant fall has been estimated in the downstream with

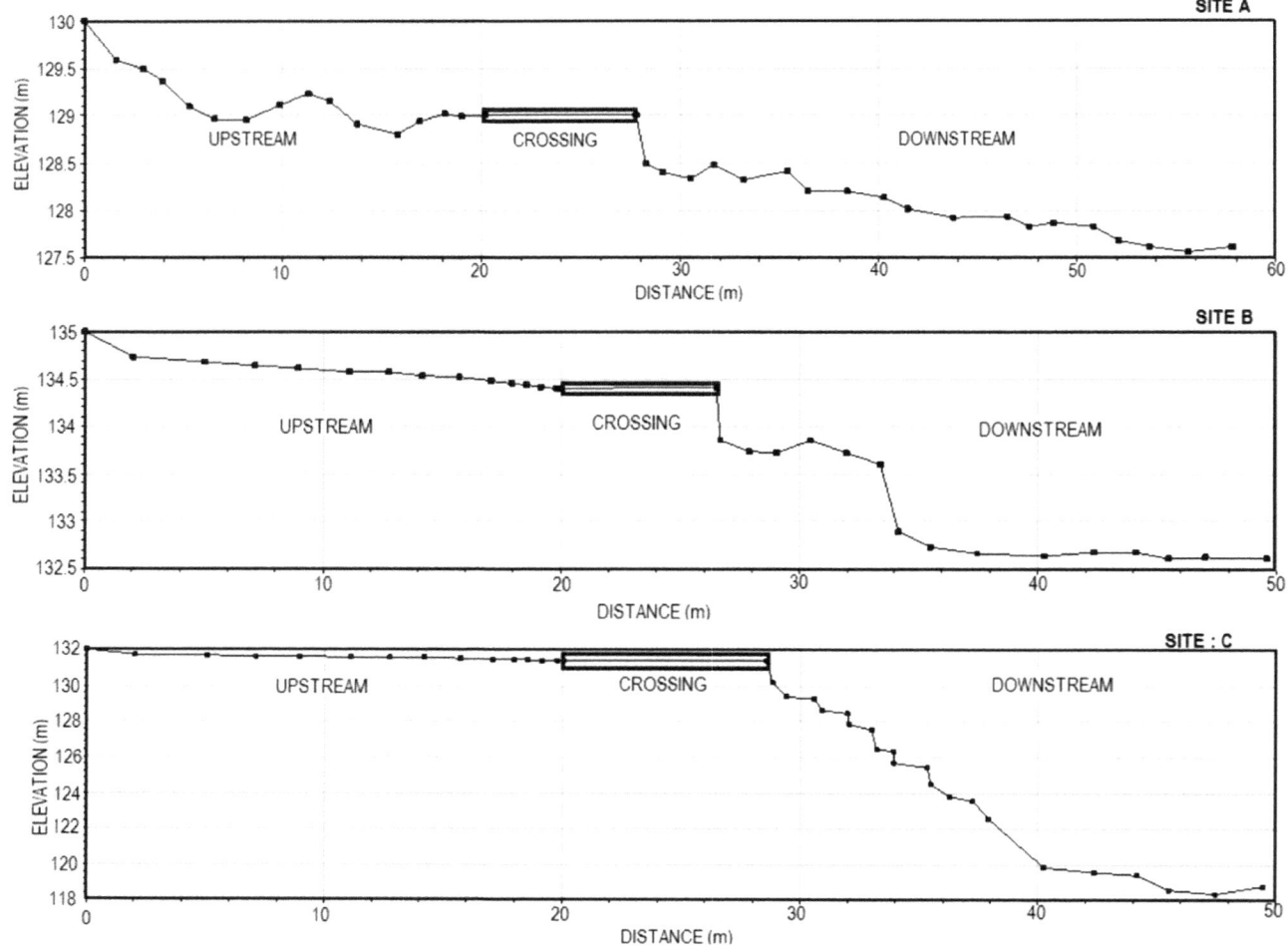

Fig. 15.6 Longitudinal profiles of the study sites from upstream to downstream of the crossing sites (Source: Filed Survey 2019)

Table 15.3 Variation in channel geometry from upstream to downstream of crossing sites

Site ID	Channel width (m)			Channel mean depth (m)			Channel cross-section area (m²)			Width-depth ratio (w/d)	
	Upstream	Downstream	Change (%)	Upstream	Downstream	Change (%)	Upstream	Downstream	Change (%)	Upstream	Downstream
A	7.1	8.1	14	0.65	0.81	25	4.62	6.56	42	10.92	10.00
B	2.4	2.8	17	1.20	1.91	59	2.88	5.35	86	2.00	1.46
C	2.0	11.6	480	1.20	7.92	560	2.40	91.87	3728	1.67	1.46

Source: Field Survey (2019)

higher channel gradients, i.e., 5% for Site B and 28% for Site C. A step-like form on channel bed in the downstream of Site C indicates the artificial concrete stairs prepared to reduce bank erosion and headward progress of gully head (Fig. 15.7).

15.3.4 Dimension of Existing Gully

In the downstream of site C, about 981 m-long gully has been developed, which is covering an area of about 5.49 ha with the perimeter of 4.75 km (Fig. 15.8). The width of the studied gully varies from 14 m in the headward region to about 49 m in its middle section with an average depth of 12 m. The shape of the gully is trapezoidal to U-shaped as the bank side material is sandier and non-cohesive soils from its top to bottom. To control the head erosion, a 10 m-long stairway-like channelization has been constructed from the outlet of road-stream crossing (Fig. 15.7). Unfortunately, such construction leads to increase of the flow velocity with 28% of gradient and influence on deep downcutting in the downstream section.

Fig. 15.7   Ground level view of all three box culverts and condition of channel form in their upstream and downstream. Man (1.62 m) standing in front of the culvert and channel can be used as scale of reference to understand the magnitude of gully (Source: Field Survey 2019)

15.4   Discussion

Characteristics of upstream catchment, especially channel network, land use, and slope, are a very sensitive aspect to understand the transformation of water input into discharge (Wharton 1994). Length, number, and platform of the channel network are effective factors to control the time of concentration and peak flow magnitude (Gregory and Gradiner 1975, 1979; Pallard et al. 2009). Therefore, higher drainage density in a catchment indicates short travel time of channelized flow and higher magnitude of the flood (Pallard et al. 2009). In addition, the shape of catchment also plays a crucial role in travel time and hydrograph (Sassolas-Serrayet et al. 2018). The circular shape of basin reduces the time of concentration and time to reach peak flow condition and vice versa for the elongated shape of the basin (Chow 1964). In these circumstances, the process of gully development in the downstream of Sites B and C is mainly due to the typical features of upstream drainage network, their density, and shape of the catchment. The circular shapes with higher drainage density in the upstream of B and C sites have been accelerated by the rapid flow concentration and higher magnitude of discharge during the rainy season. Although all sites belong to the similar lithological condition, the process of gully formation has been accelerated with a higher value of drainage density and higher circularity; for example, no sign of gully formation has been observed in Site A due to lower drainage density and elongated catchment. The dimension of gully also varies between Site B and C for the same reason. In the upstream of Site A (Fig. 15.9a, b), the 2.4 km-long meandering path with natural riparian vegetation might be responsible for the gully not developing in its downstream by reducing the flow magnitude with higher lateral connectivity between floodplain and channel (Roy and Sahu 2016).

The structural inefficiency of installed culverts is the main cause for the problem of bed degradation and

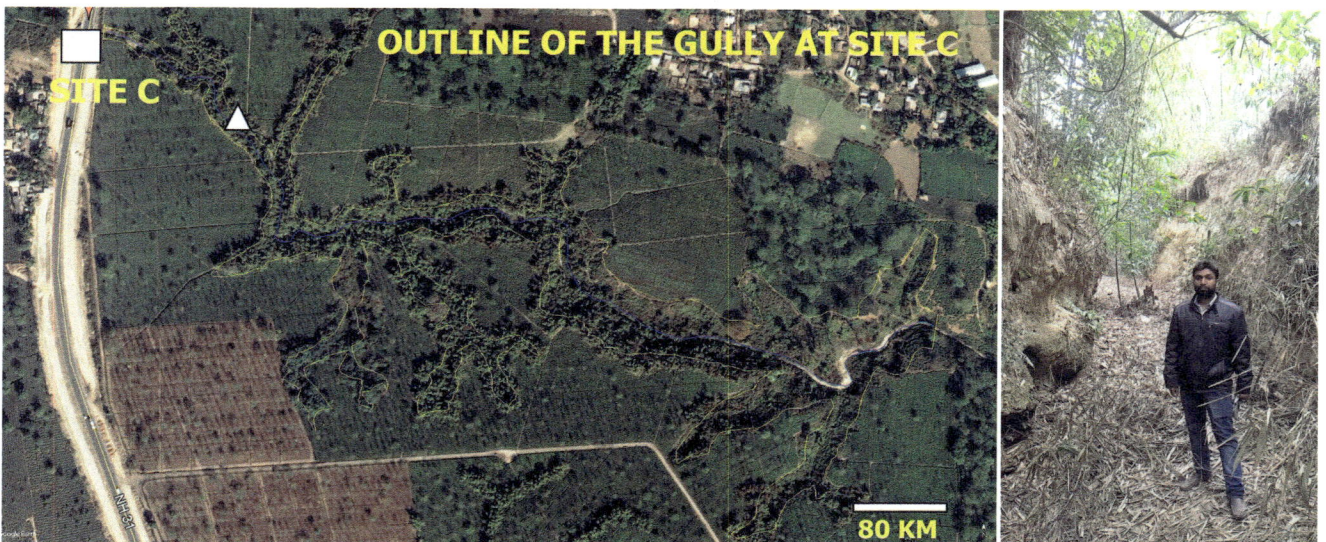

Fig. 15.8 Areal view of the gully area in the downstream of Site C and the real picture of gully inside at the triangle located in satellite image; height of the man (1.62 m) can be used to access the inside

configuration of the formed gully (Source: Google Earth Pro and Field Survey 2019)

associated gully initiation in the downstream of studied sites, which is clearly revealed through the cross-sections and longitudinal profiles of sample sites. Higher flow pressure from the upstream through undersized culverts has increased the velocity in the downstream, increased the shear stress, and associated bank erosion (McKenney et al. 1995; Robert 2003; Wang et al. 2010). Nevertheless, the notable drop height of culvert at immediate downstream and scouring also have influenced to increase the ability of bed degradation and formation of gully for such unconsolidated basement (Roy and Sahu 2018). According to Simon and Johnson (1999), scours have a long-term impact on bed degradation and can affect the entire channel reach and the exact problem has been demonstrated in the study area. The underestimation of runoff capacity from the smaller tea estate might be the cause to build such undersized culverts for Sites B and C. Conversely, with the surplus capacity to pass peak flow at Site A crossing, gully initiation has not taken place in its downstream. Worthwhile, smaller catchment with higher drainage density is more effective to generate a large volume of discharge with quick peak flow, especially when the channels are artificial in nature (Gregory and Gardiner 1979; Pallard et al. 2009).

The mechanism of erosion is the main factor to control the rate gully formation (Milton 1971). Due to the unconsolidated nature of soil profile in the study area, sidewall collapse mechanism and slumping are the most effective processes of gully formation followed by the water-

mediated transfer of such collapsed materials and downcutting (Fig. 15.9c). The symmetric nature of channel form at Sites B and C has been also indicated by the presence of steep bank slope and a higher probability of bank failure.

15.5 Conclusion

Artificial channel routing for the irrigation facility in the tea estate has increased the drainage density and reduced the travel time of rainwater and unusually increased the peak discharge. Construction of existing culverts was undersized due to the underestimation of upstream hydrological behaviors. The negative effect of inefficient crossing structure on the landscape has been admitted through the formation of the large gully in the downstream and resulted in land degradation. The preventive role of channel side forest cover and meandering platform in reducing flow velocity and related bed degradation have been observed at Site A. The severity of soil erosion comes from the appearance of unconsolidated lithology after removing the top soil by increased flow velocity in the downstream of the road-stream crossing. Alternative irrigation system for tea garden can reduce the artificial channel routing and the problem of unusual discharge generation. Road engineers should give special attention to the construction of road-stream crossing for passing water from the tea estates. Preventive measures to control

Fig. 15.9 (**a** and **b**) Presence of dense forest and natural stream characteristics with pool, riffles, and lateral connectivity between channel and floodplain in the upstream area of Site A. (**c**) The hanging section of gully's bank is awaiting for collapse during the next rainy season, a common feature along the studied gully at the downstream of Site C (Source: Filed Survey 2019)

gully enhancement should be inaugurated immediately with special attention on fragile lithology.

Acknowledgement Infrastructural facilities were provided by the Department of Geography, Kalipada Ghosh Tarai Mahavidyalaya (KGTM), Bagdogra, Siliguri, West Bengal. Assistance from Mr. Chinmoy Sarkar and Dr. Shayam Charan Barma, Assistant Professors of the Department of Sociology and Economics, respectively, KGTM, during fieldwork was gratefully acknowledged.

References

Adams WM, Watson EE (2002) Soil Erosion. Indigenous Irrigation and Environmental Sustainability, Marakwet, Kenya.

Ananda J (2014) Soil Erosion Damage Function for Smallholder Tea in Sri Lanka: An Empirical Estimation. Retrieved from https://www.researchgate.net/publication/228431373 on 01-02-2019.

Bandyopadhyay S, Kar NS, Das S, Sen J (2014) River Systems and Water Resources of West Bengal: A Review. Geological Society of India Special Publication 3:63–84.

Bhattacharya SK (1993) A Comprehensive Study of the Problems of Management of the Rakti Basin in the Darjeeling Himalaya. Ph.D. Thesis, Department of Geography, North Bengal University.

Blanton P, Marcus WA (2009) Railroads, roads and lateral disconnection in the river landscapes of the continental United State. Geomorphology 112: 212–227.

Blanton P, Marcus WA (2014) Roads, railroads, and floodplain fragmentation due to transportation infrastructure along rivers. Annals of the Association of American Geographers 104(3): 413–431. https://doi.org/10.1080/00045608.2014.892319.

Borrelli P, Robinson DA, Fleischer LR, Lugato E, Ballabio C, Alewell C, Meusburger K, Modugno S, Schütt B, Ferro V, Bagarello V, Oost KV, Montanarella L, Panagos P (2017) An assessment of the global impact of 21st century land use change on soil erosion. Nature Communications 8(2013): 1–13. https://doi.org/10.1038/s41467-017-02142-7.

CDF (California Department of Forestry) (2004) Designing watercourse crossings for passage of 100-year flood flows, wood, and sediment. Report No. 1, Sacramento, CA.

CGWB (Central Ground Water Board) (2014) Aquifer Systems of West Bengal. Ministry of Water Resources, Govt. of India, Eastern Region, Kolkata.

Chinnamani S (1977) Soil and water conservation in the hills of Western Ghats. Soil Conservation Digest 5(1): 25–33.

Chow VT (1964) Handbook of applied hydrology. New York: McGraw Hill.

Croke J, Mockler S (2001) Gully Initiation and Road-To-Stream Linkage in a Forested Catchment, Southeastern Australia. Earth Surf. Process. Landforms 26: 205–217

FHWA (Federal Highway Administration) (2014) Design for fish passage at roadway – stream crossings: synthesis report. Available from: http://www.fhwa.dot.gov/engineering/hydraulics/pubs/07033/7.cfm.

Gregory KJ, Gardiner V (1979) Comment on Drainage density and stream flow: a close look by S.L. Dingman. Water Resource Research 15: 1662–64.

Gregory KJ, Gradiner V (1975) Drainage density and climate. Zeitschurift fur Geomorphology 19: 287–98.

Gregory KJ, Brookes A (1983) Hydrogeomorphology downstream from bridges. Applied Geography 3: 145–159.

Hancock PJ (2002) Human impacts on the stream-groundwater exchange zone. Environmental Management 29 (6): 763–781.

Horton RE (1932) Drainage basin characteristics. Transactions of American Geophysics Union 13: 350–361.

Jungerious PD, Matundura J, JAM van de Ancker (2002) Road construction and Gully Erosion in West Pokot, Kenya, John Wiley and Sons Ltd.

Jungerius PD, Matundura J, De Ancker JAMVan (2002) Road construction And Gully Erosion In West Pokot, Kenya. Earth Surf. Process. Landforms 27: 1237–1247.

Knighton AD (1981) Asymmetry of river channel cross-section: part 1, Quantitative indices. Earth Surface Processes and Landforms 6: 581–588.

Madhu M, Tripathi KP (1997) Soil and water conservation practices in tea at Nilgiris: some facts. Indian Farming 47(1): 33–36.

Mati BM (1984) A technical evaluation of soil conversation activities in Kiambu District. Soil and Water Conservation Brach, Ministry of Agriculture, Nairobi.

McKenney R, Jacobson RB, Wetheimer RC (1995) Woody vegetation and channel morphogenesis in low gradient gravel-bed streams in the Ozark plateaus, Missouri and Arkansas. Geomorphology 13: 175–198.

Merril MA, Gregory J (2007) The effects of culverts and bridges on stream geomorphology. In: J.F. Levine, et al. eds. A comparison of the impacts of culverts versus bridges on stream habitat and aquatic fauna, Technical Report (FHWA/NC/2006–15), North Carolina State University, Department of Forestry and Environmental Resources, Raleigh, 15–45.

Miller V C (1953). A quantitative geomorphic study of drainage basin characteristics in the Clinch mountain area. New York: Department of Geology, ONR, Columbia University, Virginia and Tennessee, Project NR389–402, Technical Report 3

Milton LE (1971) A Review of Gully Erosion and Its Control. Soil Conservation Authority, Victoria.

Montgomery DR (1994). Road surface drainage, channel initiation, and slope instability. Water Resources Research 30(6): 1925–1932.

Nyssen J, Poesen J, Moeyersons J, Luyten E, Veyret-Picot M, Deckersn J, Haile M, Govers G (2002) Impact of Road Building on Gully Erosion Risk: A Case Study From The Northern Ethiopian Highlands. Earth Surf. Process. Landforms 27: 1267–1283.

Pallard B, Castellarin A, Montanari A (2009) A look at the links between drainage density and flood statistics. Hydrology and Earth System 13: 1019–1029.

Pathak P, Wani SP, Sudi R (2005) Gully control in SAT watersheds. Global Theme on Agroecosystems Report no. 15. Patancheru 502 324, Andhra Pradesh, India: International Crops Research Institute for the Semi-Arid Tropics. 28 pp.

Pimentel DC, Harvey K, Resosudamo K, Sinclair K, Kurz M, McNair S, Crist L, Shpritz L, Friton R, Staffouri R, Blaiv R (1995) Environmental and Economic Costs of Soil Erosion a conservation benefits. Science 276:1117–1123.

Poesen J, Nachetergaele J, Verstraeten J, Valentin C (2003) Gully erosion and environmental change: importance and research needs. CATENA 50(2–4):91–134.

Raiter KG, Prober SM, Possingham HP, Westcott F, Hobbs RJ (2018) Linear infrastructure impacts on landscape hydrology. Journal of Environmental Management 206: 446–457.

Resh VH (2005) Stream crossings and the conservation of diadromous invertebrates in South Pacific island streams. Aquatic Conservation: Marine and Freshwater Ecosystems 15: 313–317.

Robert A (2003) River processes. London, Arnold.

Roy S, Sahu AS (2018) Road-stream crossing an instream intervention to alter channel morphology of headwater streams: case study, International Journal of River Basin Management 16(1): 1–19. https://doi.org/10.1080/15715124.2017.1365721

Roy S, Sahu AS (2016) Effect of land cover on channel form adjustment of headwater streams in a lateritic belt of West Bengal (India). International Soil and Water Conservation Research 4: 267–277. https://doi.org/10.1016/j.iswcr.2016.09.002

Roy S, Sahu AS (2017) Potential interaction between transport and stream networks over the lowland rivers in Eastern India. Journal of Environmental Management 197: 316–330. https://doi.org/10.1016/j.jenvman.2017.04.012

SAC-ISRO (2018) Desertification and Land Degradation Atlas of Selected Districts of India (Based on IRS LISS III data of 2011-13 and 2003- 05), Volume-1, Space Applications Centre, ISRO, Ahmedabad, India, 148 pages.

Sahoo DC, Madhu MG, Bosu SS, Khola OPS (2016) Farming methods impact on soil and water conservation efficiency under tea [*Camellia sinensis* (L.)] plantation in Nilgiris of South India. International Soil and Water Conservation Research 4:195–198. https://doi.org/10.1016/j.iswcr.2016.07.002.

Sarkar S, De, SK (2017) B3: Geomorphological Field Guide Book on Darjeeling Himalaya. 9th International Conference on Geomorphology by Indian Institute of Geomorphoogist (IGI), New Delhi.

Sassolas-Serrayet T, Cattin R, Ferry M (2018) The shape of watershed. Nature Communication 9(3791): 1–8.

Simon A, Johnson PA (1999) Relative roles of long term channel adjustments processes and scour on the reliability of bridge foundations. In: E.V. Richardson and P.F. Lagasse, eds. Stream stability and scour at highway bridges. Fort Collins: American Society of Civil Engineers, 151–165.

Takken I, Croke J, Lane P (2008) Thresholds for channel initiation at road drain outlets. Catena 75: 257–267.

Tea Board of India (2017) http://www.teaboard.gov.in/; http://www.teaboard.gov.in/pdf/Estimated_production_for_December_2018_pdf8735.pdf. Retrieved on 31/01/2019.

The Hindu (2010) Agriculture: India losing 5334 million tonnes of soil annually due to erosion: Govt., 26th November. https://www.thehindu.com/sci-tech/agriculture/India-losing-5334-million-tonnes-of-soil-annually-due-to-erosion-Govt/article15717073.ece. Retrieved on 31/01/2019.

Thomaz EL, Peretto GT (2016) Hydrogeomorphic connectivity on roads crossing in rural headwaters and its effect on stream dynamics. Science of the Total Environment 550: 547–555. https://doi.org/10.1016/j.scitotenv.2016.01.100

Valentin C, Poesen J, Yong L (2005) Gully erosion: Impacts, factors and control. CATENA 63:132–153. https://doi.org/10.1016/j.catena.2005.06.001

Wang J, Edwards PJ, Goff WA (2010) Assessing changes to instream turbidity following construction of a forest road in West Virginia. TMDL 2010 Watershed Management to Improve Water Quality CD-ROM Proceedings, ASABE Publication Number 711P0710cd.

Wellman JC, Combs DL, Bradford CS (2000) Long-term impacts of bridge and culvert construction or replacement on fish communities and sediment characteristics of streams. Journal of Freshwater Ecology 15 (3): 317–328. https://doi.org/10.1080/02705060.2000.9663750

Wemple BC, Jones J.A, Grant GE (1996). Channel network extension by logging roads in two basins, Western Cascades, Oregon. Water Resources Bulletin 32(6): 1195–1207.

Wharton G (1994) Progress in the use of drainage network indices for rainfall-runoff modelling and runoff prediction, Prog. Phys. Geog. 18: 539–557.

Wheeler PA, Angermeier PL, Rosenberger AE (2005) Impact of new highways and subsequent landscape urbanization on stream habitat and biota. Reviews in Fisheries Science 13(3): 141–164.

Yan Z, Hong Z, Bu-zhuo P, Hao Y (2003) Soil erosion and its impact on environment in Yixing Tea Plantation of Jiangsu Province. Chinese Geographical Science 13(2): 142–148.

Zichenga Z, Xiaolingb HE, Tingxuanc LI (2012) Status and Evaluation of the Soil Nutrients in Tea Plantation, 2011 International Conference on Environmental Science and Engineering. Procedia Environmental Sciences 12: 45–51.

Suvendu Roy is an Assistant Professor, Department of Geography, Kalipada Ghosh Tarai Mahavidyalaya, Bagdogra, Siliguri, West Bengal, India. His research interest is in the interface between anthropogenic activities and changing channel geomorphology, which includes forest river geomorphology, anthropogeomorphology, and archaeogeomorphology. He did his B.A. degree in geography, from Burdwan Raj College, the University of Burdwan, and M.A. in geography (specialized in advanced geomorphology), from the University of Burdwan (India). He has been awarded Ph.D. in geography (anthropogeomorphology), by the University of Kalyani (India). He has more than 20 research publications in various journals of international and national repute.

Land Degradation Processes of Silabati River Basin, West Bengal, India: A Physical Perspective

16

Avijit Mahala

Abstract

Degradation of land indicates a definite decline in productive capacity and environmental function of land. Almost half of the total terrestrial land faces land degradation due to different processes. The tropical, temperate and arid environment lead to different processes of land degradation. Water-related soil erosion is deemed to be one of the dominant processes for the degradation of lands in the tropical and subtropical areas of the world. Arid and semi-arid regions, on the other hand, face vulnerability owing to wind or vegetal degradation related land degradation processes. Tropical plateaus of the world face a distinct type of land degradation which is generally created by water erosion processes. The present study involves understanding the physical processes of land degradation in tropical plateau environment. Chotanagpur plateau is one of the most degraded region in tropical eastern India; mostly eroded by water. Granite–gneiss geological formation, low-to-medium developed soil cover, undulating lateritic uplands, high drainage density, semi-humid climate (35–40 °C of average monthly temperature; 100–140 cm of annual rainfall) and dry tropical deciduous forest areas make the River Silabati basin a true representative of plateau region of tropical environment. Erosion by water, degradation of vegetation and declining soil quality are the major processes of land degradation in the Silabati basin. Different physical parameters causing the degradation of the land include topographic features, edaphic maturity, hydrological features and vegetation cover span. Remote sensing (multispectral information), elevation data (DEM), meteorological data, field observation and thematic maps are being used to unveil the possible mechanisms of land degradation in tropical plateaus. Granite–gneiss geological formation is the foundation for developing an undulating topography in the basin. Less developed soil profile, low organic matter and poor structure of soil cause high soil erosion. The dissected highland and crests of undulating plateau cause topographic hindrance in productivity of land. High drainage density and frequency in rugged upland cause high soil erosion. Decreasing rainfall and increasing aridity (P/Potential evapotranspiration (PET)) cause threats of water stress condition in the region. Green biomass cover area is also continuously declining. Through overlaying the different physical factors (geological formation, soil characteristics, geomorphological characteristics, etc.) of considerable importance in geographic information system (GIS) environment, the variability in the level of land degradation is mapped. It is found that with intense eroded laterite soil cover, middle reaches of Silabati basin are more susceptible to soil erosion within the whole basin.

Keywords

Land degradation · Soil erosion · Tropical plateau · Chotanagpur plateau · Lateritic upland · Silabati basin · Remote sensing · GIS

16.1 Introduction

A temporary or a prolonged decrease in the functioning of the ecosystem and its productive capacity is termed land degradation. In general, it indicates reduction in the benefits to humanity. It negatively impacts the livelihood of billions of people (Nkonya et al. 2015). The loss of productivity of soil characterized by reduced soil fertility, derogation of natural resources and biodiversity is termed as degradation of land (Thiombiano and Tourino-Soto 2007). Expanding population enhances the demand for fertile agricultural land for food and fuel leading to land degradation, a topic which is the centre of much attention in contemporary times. The reduction of productivity in terms of forest resource and pasture land is also termed as degradation of land. One-fifth of total

A. Mahala (✉)
Center for Study of Regional Development (CSRD), Jawaharlal Nehru University (JNU), New Delhi, India

© Springer Nature Switzerland AG 2020
P. K. Shit et al. (eds.), *Gully Erosion Studies from India and Surrounding Regions*, Advances in Science, Technology & Innovation,
https://doi.org/10.1007/978-3-030-23243-6_16

cultivated area, one-tenth of total grassland and one-third of total forest area across the world have faced the brunt of land degradation. Degradation encompasses dry land desertification as well as deforestation. Around 43% of total terrestrial vegetated surface faces declining capacity of production due to land degradation. Only 17% of the earth's vegetated land faces vulnerability to soil degradation owing to anthropogenic activities. Around 73% of global rangeland which covers 88% of dryland, faces moderate-to-high desertification-related land degradation problem (Daily 1995). The coverage of degraded land is extensive. Approximately 23% of the total geographical area is categorized as degraded land. It is estimated that nearly about 75% of the arid lands in the world are facing the problem of acute desertification (Gibbs and Salmon 2015). The given estimate of degraded land for different continents is Asia (912 m ha), Africa (660 m ha), North America (469 m ha), South America (398 m ha), Australia and Pacific (236 m ha), Europe (65 m ha) and an overall figure for the world comes to 2740 m ha as given by GLADA (Global Assessment for Land Degradation) (Gibbs and Salmon 2015).

The Chotanagpur plateau of eastern tropical India is characterized by severe water erosion (Mahala 2017; NRSC 2011). More than 30% geographical area of the states of Odisha, West Bengal and Jharkhand is degraded (ISRO 2016). Gondwana geological formation dominates Chotanagpur plateau. Land productivity gets hindered due to the granite–gneiss in the geological makeup of the land and the undulating terrain of the plateau (Dolui et al. 2014). The soil of the region is characterized by less mature development (Ghosh and Guchhait 2015). Badland topography develops on the lateritic cap uplands due to gully expansion and head retreat (Shit and Maity 2012). Further, water stress conditions also develop due to high drainage density and increasing aridity. Cumulatively, these factors make Chotanagpur plateau a distinct and true representative of tropical plateau with much of its land degraded. This is a question which needs to be addressed.

16.2 Literature Review and Statement of Problem

Many studies have been conducted in the past to understand the land degradation processes in this region. Most of these studies focus on water erosion and vegetation degradation. Their focus is to understand the degradation mechanism in a specific morpho-climatic set-up, but fails to depict the patterns found in degraded land in the whole world. There is a lack of research on the processes of land degradation on tropical plateau environment. Most studies have laid emphasis on the land degradation processes in temperate climatic region, Europe particularly. There is, however, a growing inclination towards understanding the processes of

desertification in the tropics particularly in the middle and west Asia (Babaev et al. 2015). Most of the land degradation studies within Asia have focused on Middle East (Faour 2014). Some have also looked into the temperate regions of China and Mongolia (Gong et al. 2014). Thus, it can be comprehended that tropical plateau lands within the tropics have remained relatively understudied in terms of research on land degradation.

In India, a lot of work has been done to give insight into the mechanisms of land degradation. As per ISRO (2007, 2016) studies, there is a marked 2% hike in different land degradation processes over a span of less than a decade. But in India, majority of the studies on land degradation are in the light of the economic aspect (Reddy 2003; Mythili and Goedecke 2016). Few have focused the state of degraded land status (Ajai et al. 2009; NRSC 2011). In western dry and semi-dry areas of India, substantial studies have been done on wind erosion and the desertification processes (Singh et al. 1994; Chauhan 2003; Kundu et al. 2015). With the change in environmental settings, intensity of the aeolian and fluvial processes also change. Ranga et al. (2016) tried to comprehend the dynamism of water erosional process as well as badland topography of semi-dry areas in the lower Chambal river plain, which is unique and markedly different from the tropical areas of Chotanagpur plateau. Most of this work focuses on the key role of human activities over environmental factors contributing to degradation of land (Pani and Carling 2013). There has been only little work aimed at unfolding frost erosion and degradation of lands in the higher alpine regions of the mountain (Rashid et al. 2011). To conclude, it can be stated that there is acute dearth of detailed studies to unravel the land degradation processes in tropical Chotanagpur plateau of India. Water erosion processes and their impact on the laterite soil cap of Chotanagpur plateau have been analysed in some of the studies (Sarkar et al. 2014). Thus, there is substantial shortage of research works done on land degradation eroded by water in Chotanagpur plateau of tropical eastern India.

Different data sources in large volume are a great requirement in order to study land degradation processes. GIS and remote sensing are of great importance for this. Greater spatial–temporal access of remote sensing technology is required to provide larger sets of information. To evaluate soil structure and its properties, reflectance spectroscopy has been used widely (Santra et al. 2015). Multispectral optical images taken from various satellites have been used to map degraded areas in dry and semi-dry regions (Baroudy 2011). Algorithm of Normalized Difference Vegetation Index (NDVI) has been used to map degraded lands. Remote sensing has been widely recommended to validate laboratory results in several studies (Vågen et al. 2013). MODIS data availability at regular intervals is of great use to monitor land degradation (Eckert et al. 2014). There are different

Algorithms used with multiple spectral bands coupled in such a way that it enhances the value of the study, specifically to project soil salinity in dry and semi-dry areas of the planet (Mandal et al. 2009; Allbed and Kumar 2013). Still, soil that lead to the degradation of the land has been rarely studied because of data sets shortage and deficit in knowledge of modelling tools and techniques.

Remote sensing tool with varied data sources, namely thematic maps, meteorological data and vegetation kinetics put together play a pivotal role in understanding the scenario of degraded lands in tropical plateau. Thus, the present study aims to understand the processes of land degradation in a tropical plateau area. In tropical Chotanagpur plateau, enriched geology of granite–gneiss, laterite soil and undulating terrain of the plateau are the major geographical features. Hydrologically, the region has high drainage density and increasing aridity. Silabati river basin eastern Chotanagpur plateau, chosen in the present study truly represents all of the above-mentioned characteristics to be depicted as the microtropical Chotanagpur plateau.

16.3 Location and Description of the Study Area

River Silabati basin has been picked to study the features of tropical Chotanagpur plateau as it is the representative unit of the area to be studied. The upper part of the basin falls under the proper plateau region while the lower part falls under the peripheral part of the Chotanagpur plateau. The basin is part of the Ganga drainage system. River Silabati originates in the middle highlands of the Chotanagpur plateau region and flows in the east direction going through the undulating uplands of the plateau fringe (Fig. 16.1). The area under study is part of the granite–gneiss geological division in the upper part and lateritic in the middle reaches (Dolui et al. 2014). Gondwana coal deposits are also found in the upper part of the Silabati basin. It constitutes a sloping landscape with undulating surface, intersected with hills here and there in the upper part of the basin while lateritic soil in the lower reach is a characteristic feature of the region (Shit and Maity 2012). A poorly developed soil profile with abundance of

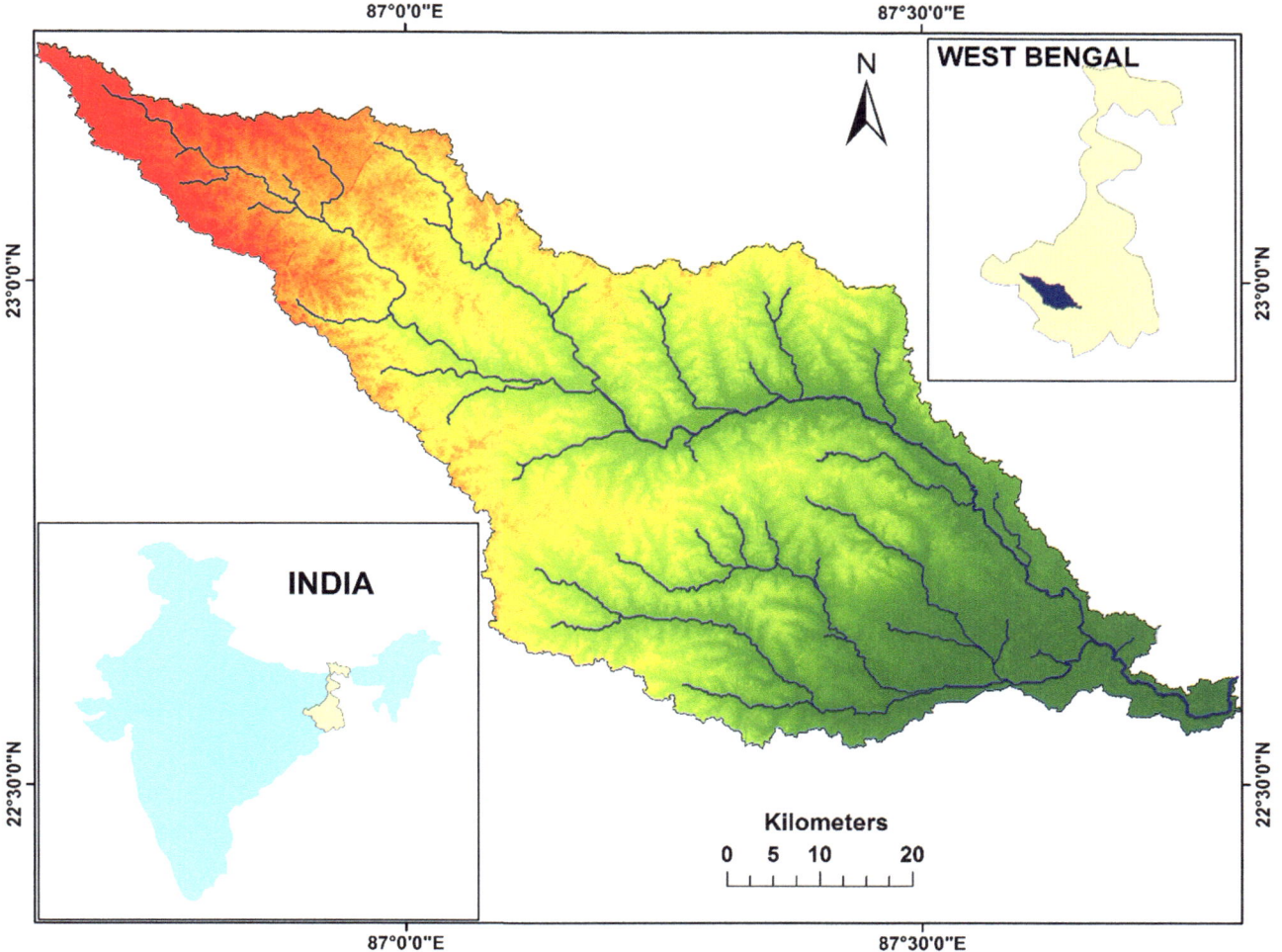

Fig. 16.1 Location map of Silabati river basin

silica content, large textured soil grains with low organic matter is a major characteristic feature of the plateau areas. The lower-middle part shows primary as well as secondary lateritic formations (Ghosh and Guchhait 2015). In the upper-middle reaches, gully development amidst lateritic soil cover is a unique characteristic (Shit and Maity 2012). Overall, the basin falls within the subhumid tropical climatic zones (with average annual rainfall of 100–150 cm and average temperature 32–37 °C).

In the flood plain areas, extensive agriculture is practised where there is abundance of population in the middle and lower parts of the basin. This has led to widespread deforestation in these areas of the basin even though a thick chunk of dry deciduous type of vegetation in the middle-upper part of the Silabati basin remains. All these factors cumulatively make the region a true replica of the whole of the tropical Chotanagpur plateau area.

16.4 Materials and Methods

Different physical factors can be used to influence land degradation processes. The geological formations of any region characterize its geomorphological setting as well as the pedological characteristics of the region. Different geomorphological formations like undulations and slopes accelerate or decelerate the processes of land degradation (Bhan 1988; Ghose et al. 1977). The processes of land degradation and its relation to rainfall intensity and aridity (P/PET) are also well-known (Boschetto et al. 2010; Alves and Azevedo 2015). Gully land development is also known to accelerate the land degradation processes in both arid and tropical areas (Frankl et al. 2013). Also, increasing drainage density enhances the soil erosion process (Girmay et al. 2009). Finally, the loss of green biomass cover has accelerated the land degradation exponentially (Higginbottom and Symeonakis 2014). There is, therefore, a need for an integrative approach to address all these factors that degrade the lands in tropical plateau areas.

To understand the land degradation in tropical areas, several indicators have been taken into consideration. In the work of Reza et al. (2017) emphasis was on the degradation of varied alluvial soil types in the tropical areas. Most studies use remote sensing to identify the processes of land degradation (Hereher and Ismael 2016). But it is not possible to take into account all the factors that eventually lead to degradation of the land. Identification of land degradation using spectral indices has also been a trend but its sceptics remain (Kannan et al. 2015).

Desertification and associated land degradation processes in the tropical environment have also been the subject of study by several scholars (Allbed and Kumar 2013). However, this method cannot be entirely applied in the tropical sub-humid environment. Climate is a major factor to assess land degradation (Abdelrahman et al. 2015). In the tropical plateaus, degradation of land is a result of complex processes, namely physiographical, pedological, hydrological and climatic factors. A single factor cannot be overemphasized to the disadvantage of others. Also, remote sensing and GIS techniques continue to be crucial to any study of the processes of land degradation (Diodato and Ceccarelli 2004).

Pedology, groundwater table, drainage conditions and mineral availability are dependent on geological characteristics which subsequently control the processes of land degradation. Hydrological parameters include fluctuations in groundwater, drainage density and drainage frequency deciding the mechanism of land degradation which ultimately decide the course of overland flow. Variability in climatic characteristics like aridity (P/PET) has also been taken into account along with soil type and vegetation cover. These factors need to be studied individually and scaled for the whole basin to decipher their role in land degradation processes in the region. Without this detailed insight, this theme will not be covered properly. For the present study, remote sensing tools and techniques along with thematic maps and point data sets of information have been taken into account (Table 16.1).

The aforementioned parameters have a key role in determining the capacity and probability of land degradation in the

Table 16.1 Data sources and techniques for different physical factors of land degradation

Physical factors	Data types and year	Techniques
Geological formations	Map of 'Geological Survey of India' 2011	Vector mapping for different geological formations
Soil characteristics	Map of 'National Bureau of Soil Science' 2013	Vector mapping for different soil groups
Relief characteristics	ASTER DEM 30 m 'USGS Earth Explorer' 2011	Relief raster creation for different relief classes
Slope formations	ASTER DEM 30 m 'USGS Earth Explorer' 2011	Slope (θ) raster creation for different slope formations
Geomorphological features	ASTER DEM 30 m 'USGS Earthexplorer' 2011	Vector mapping for different geomorphological features
Groundwater fluctuation	MBGL data of 'Central Ground Water Board' 2011	MBGL raster creation for different groundwater availability classes
Drainage density	ASTER DEM 30m 'USGS Earth Explorer'. 2011	Drainage density raster creation by (drainage length/Sq km)
Aridity conditions	Daily temperature and rainfall data of 'Indian Meteorological Dept'. 1960–2015	Aridity (P/PET) raster creation for different aridity classes
Green biomass cover	Landsat8 OLI image 'USGS Earth Explorer' 2016	NDVI raster creation for different vegetation cover areas

Table 16.2 Weightage for different physical factors of land degradation

Land degradation factors	Subcriteria	Score	Land degradation factors	Subcriteria	Score
Geological formations	Younger alluvium	10		Gently sloping upland	40
	Older alluvium	20		Dissected highland	50
	Secondary laterite	40		Crest of undulating plateau	60
	Primary laterite	50		Residual hillocks and mountains	70
	Gneiss	70		Dissected plateau	80
	Gneiss and schist	90		Badland	90
Soil characteristics	Ustifluvents	10	Groundwater fluctuations	<5.5	20
	Haplaquepts	20		5.5–5.7	30
	Ustorthents	30		5.7–5.9	40
	Ochraqualfs	30		5.9–6.1	50
	Ustrothents	40		6.1–6.3	60
	Haplaquepts	50		6.3–6.5	70
	Haplaqualfs	60		>6.5	80
	Paleuustalfs	70	Drainage density	<1.5	10
	Ustochrepts	80		1.5–1.7	20
	Haplustalfs	90		1.7–1.9	30
Relief characteristics	<50 m	10		1.9–2.1	50
	50–100 m	20		2.1–2.3	70
	100–150 m	40		>2.3	90
	150–200 m	60	Aridity Index (P/PET)	>0.75	10
	>200 m	80		0.7–0.75	30
Slope formations	<2	10		0.66–0.7	50
	2–4	20		0.6–0.65	70
	4–6	30		<0.6	90
	6–8	40	Green Biomass cover (NDVI)	>0.24	10
	8–10	50		0.2–0.24	120
	10–20	60		0.16–0.20	140
	20–30	70		0.12–0.16	160
	>30	80		0.08–0.12	180
Geomorphological features	Flood plain and low land	10		<0.08	190
	Lower alluvial plain	20			
	Upper undulating alluvial plain	30			

tropical plateau areas (Table 16.1). All the above-mentioned factors are equally important to comprehend the overall picture of land degradation or its susceptibility in a specific morpho-climatic setup (Hassan et al. 2015). Every individual factor, namely geological formation, soil types and aridity spatially, has its own unique role in enabling land degradation. But, we can classify the above-mentioned factors, assigning them weightage according to their land degradation potential (Prakash et al. 2016; Kosmas et al. 2015; Liu et al. 2003). Presently, a multifactor vulnerability model has been developed to know the present situation of degraded land (Eq. 16.1) (Rashid et al. 2011; Hassan et al. 2015). Using the above equation, these factors have been classified into different classes according to their land degradation potential (Table 16.2) (Kosmas et al. 2015; Liu et al. 2003). These factors are measured and recorded in their units and are subsequently, standardized into a common value in order to execute Weightage Overlay method. Alternatively, they are converted in standardized units for better understanding (Pramanik 2016; Prakash et al. 2016). Factors with '0' indicate the minimal land degradation potential with '100' being the highest land degradation potential (Table 16.2) (Liu et al. 2003; Kosmas et al. 2015). All weightages are listed in multiples of '10' viz: 10, 20, 30, etc. The same Weightage rasters (ranging from '0' to '100'; '0' = minimum degradation, '100' = maximum degradation) have been listed for the different physical factors of the basin individually. Further, weightage of each of the physical factors has been reclassified to be used as an input data for weightage overlay that was subsequently multiplied using raster calculator tool in ArcGIS (Eq. 16.1). Output raster represents land degradation susceptibility of the basin because of different geophysical factors (Table 16.3).

Table 16.3 Area under different Physical Vulnerability Index of land degradation for Silabati river basin

Physical Vulnerability Index of land degradation	Score	Area (km^2)	Percentage of area
Very low	<120	3.258583	0.090768
Low	121–130	174.43	4.858775
Moderate	130–140	910.6781	25.36708
High	141–150	1730.691	48.20866
Very High	>150	770.9424	21.47472

Physical factors induced Land Degradation Vulnerability Index
$$= (\text{Geological formations} * \text{Soil characteristics}$$
$$* \text{Relief characteristics} * \text{Slope formations}$$
$$* \text{Geomorphological features}$$
$$* \text{Groundwater (MBGL) fluctuation}$$
$$* \text{Drainage density} * \text{Aridity conditions (P/PET)}$$
$$* \text{Green biomass covers})^{1/9} \qquad (16.1)$$

Weightage of each factor after multiplication gives a final raster using Eq. (16.1). Final raster values range from '0' to '100' indicating low to very high potentiality, respectively (Table 16.3) (Prakash et al. 2016; Hassan et al. 2015) with '0' being the least vulnerability of land degradation while '100' being very high.

16.5 Result and Discussions

16.5.1 Mechanism of Physical Land Degradation of River Silabati Basin

Land degradation due to the physical factors in Silabati is discussed below.

16.5.1.1 Geological Formations
The geological formation and related mineralogy control land fertility and production capacity. The land surface conditions, that is relief characteristics, slope formations and geomorphological features are also controlled by geological formations. Pedology, groundwater table, drainage conditions and mineral availability is dependent on geological characteristics which subsequently control the processes of land degradation. Granite- and gneiss-dominated archean rock system forms a region of the world where the soil has not developed properly; therefore, it is known as immature soil (Bocco et al. 2001). The calcareous geological formations aid water erosion-related degradation.

The upper part of the Silabati basin is covered with granite, gneiss and schist geology (Fig. 16.2a) (Dolui et al. 2014).

Low weathering characteristics in granite–gneiss–schist geology in the upper region cause major hindrance in soil development. This causes less developed soil profile, undulating terrain as well as high water-related erosion in the studied basin. Laterite formation covers the middle and upper-lower part of the basin. This formation is characterized by a thick laterite cover which is the dominant cause for rill and gully erosion (Shit and Maity 2012). The lower part of the basin covered by alluvium is also affected by sheet erosion and water-accumulation-related problems.

16.5.1.2 Soil Characteristics
Soil characteristics of any region impact land degradation processes of a region greatly. The soil and land degradation occur simultaneously. Soil degradation promotes land degradation, while land degradation causes the soil to degrade as well (Lal 2001). Physical parameters of soil such as compactness, hardness, infiltrative capacity and density determine land characteristics (Bready and Well 2005). The high and low bulk densities of soil cause specific types of soil degradation (Keller and Hakansson 2010). Soil texture determines other characteristics of soil (Regelink et al. 2015). Soil structure ultimately determines land characteristics (Askari et al. 2013). The soil chemical characteristics like cation exchange capacity (CEC), pH, total dissolved salt (TDS), respectively, determine the available C, N, P and K in the soil (Behera and Shukla 2015). The nutritional and biological health of soil ultimately characterizes the land quality (Heshmati et al. 2012). Organic content of soil reflects the land characteristics and water-holding capacity of any region (Riezebos and Loerts 1998; Rajan et al. 2010). Chotanagpur plateau is an area vulnerable to intense soil erosion (Chatterjee et al. 2013).

The coarse-grained soil (mainly sandy loam) in the upper Silabati basin is characterized by low humus content, high silica content and low water-holding capacity (Mahala 2017) (Fig. 16.2b). All these characteristics cause hindrance in land productivity. The laterite soil in the middle undulating plateau, that is the upland areas of the basin have high iron (Fe) content, high leaching capacity, low pH and low water-holding capacity, causes land to degrade (Ghosh and Guchhait 2015). Important nutrient materials are leached out of these soils, assisted by rill and gully formation (Shit et al. 2015). High clay content in the soil of the lower reaches of the basin results in high water-holding capacity of the soil, which is indirectly responsible for flooding. The soils in most parts of the basin are vulnerable owing to their immature characteristics.

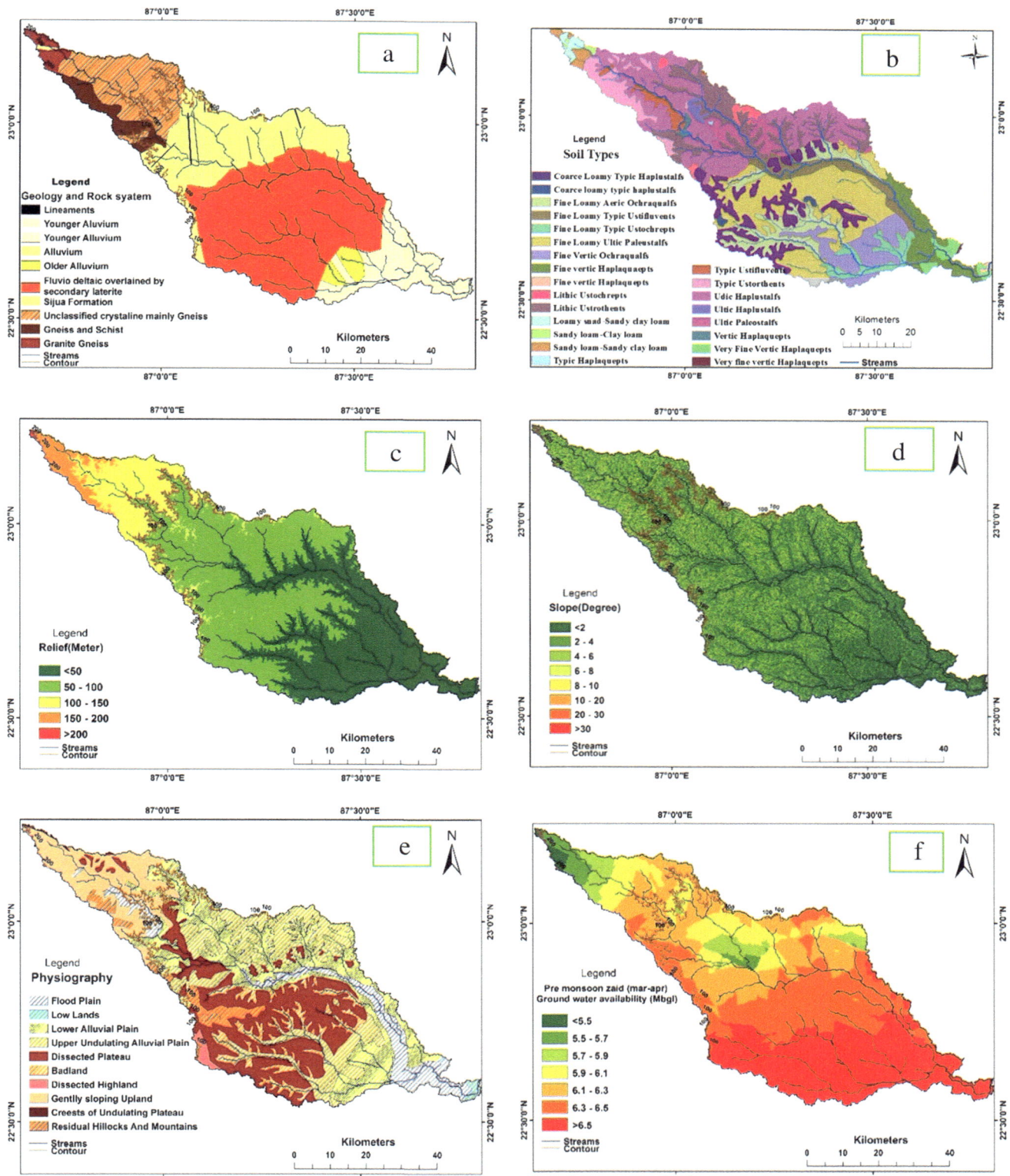

Fig. 16.2 Physical factors and vulnerability of land degradation of River Silabati basin: (**a**) Geological formations, (**b**) Soil characteristics, (**c**) Relief characteristics, (**d**) Slope formations, (**e**) Geomorphological features, (**f**) Groundwater fluctuation, (**g**) Drainage density, (**h**) Aridity conditions (P/PET), (**i**) Green biomass cover, (**j**) Physical Vulnerability Index

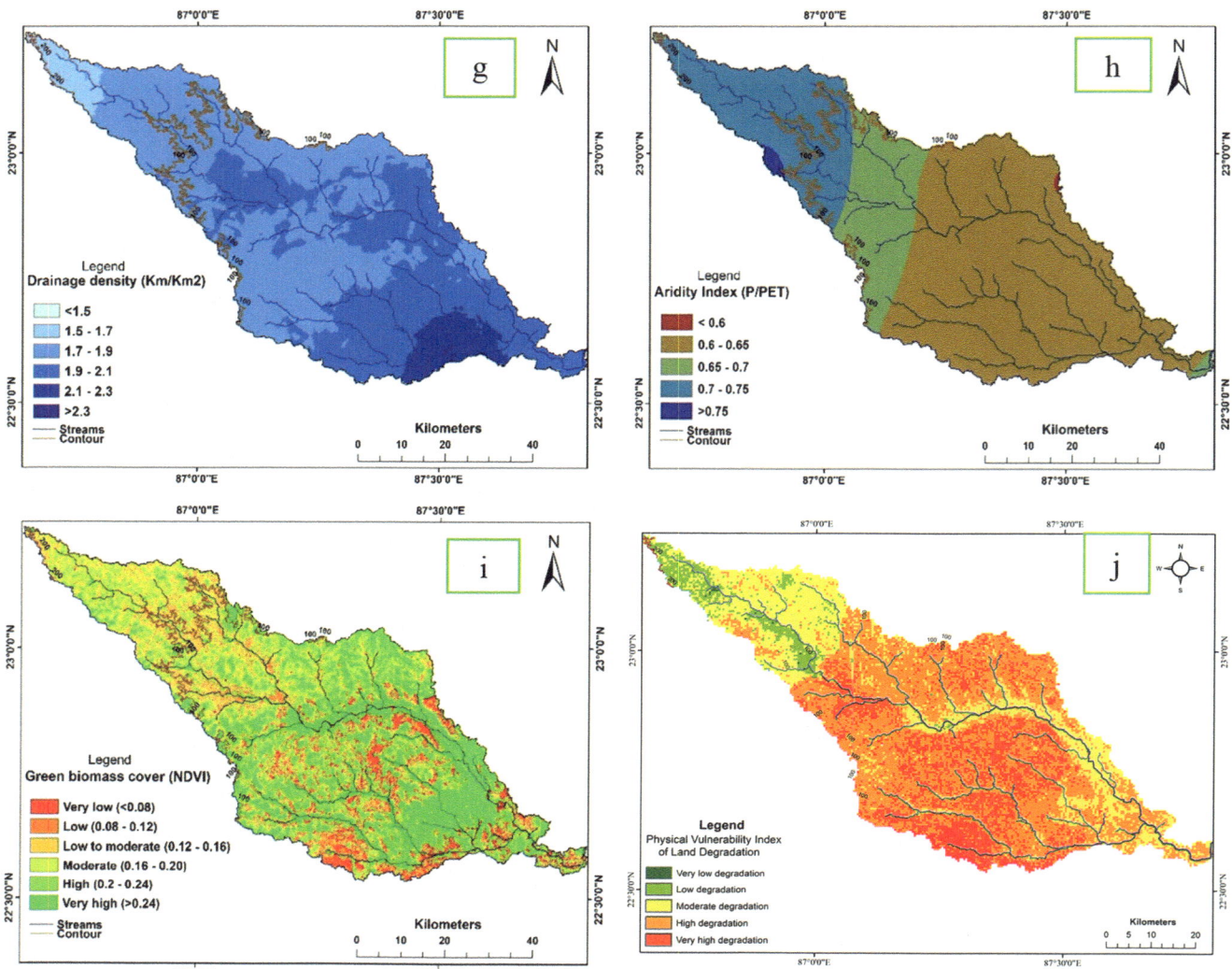

Fig. 16.2 (continued)

16.5.1.3 Relief Characteristics

The surface relief characteristics of any region determine the nature of land degradation. High relief (>600 m) characterized by high slope surface can give rise to land slides, avalanches, glacial hazards and debris flow-related problems in mountain regions of world (Myint and Thinley 2006). The high relief regions are often characterized by steep slopes where soil profiles are poorly developed. These relief characteristics also hinder construction activity on a large scale (Norbu et al. 2003).

In Silabati basin, high relief area (>200 m) largely covers the upper reaches of the basin (Fig. 16.2c). This area is characterized by undulating topography, gently sloping upland and pediment formation in foothill zones. All these characteristics hinder land productivity. The middle reaches of the basin are characterized by medium relief (100–200 m) which also displays undulating topography. The general

relief is characterized by dissected plateau fringe topography in this part of the basin. Soil development gets hindered over this dissected plateau fringe region (Ghosh 2015; Shit and Maity 2012). The areas characterized by low relief (<50 m) in the lower reaches of the basin have the problem of water-accumulation-related land degradation.

16.5.1.4 Slope Formations

Land degradation characteristics are determined by slope characteristics of any region. Sloping surfaces exceeding 30° (>30°) give rise to a lowered potentiality of the soil in stabilizing materials. Immature soil on unstable sloping surfaces in the absence of vegetation faces high vulnerability of land degradation. Areas characterized by high slope values, mostly occurring in the mountainous regions face slope failure, failure of shear strength, landslides and other related problems (Rashid et al. 2011). Land degradation processes of the Himalayan regions are mostly due to slope failure (Naz and Romshoo 2012).

The upper reaches of Silabati basin have a high slope (>8°) which aims at high soil erosion rates as well as low soil moisture retentive capacity (Fig. 16.2d). Sloping areas around isolated hills and pediments are not suitable for deep soil profile development. High slope in the right bank of basin poses a greater vulnerability to soil erosion throughout the expanse of the basin. Formation of rills and gullies is prominent in high slope areas of the right bank of Silabati (Ghosh 2015). Very low slope (<2°) in the lower reaches of the basin causes water to accumulate and gives rise to related problems.

16.5.1.5 Geomorphological Features

Land characteristics of any region are determined by geomorphological features. In the mountain environment, presence of hill and valley topography causes hindrance to land quality development. Landslides, glacial avalanches and flash floods are common in the Himalayas or any other mountainous area (Rashid et al. 2011). The surfaces characterized by high slope and varied geomorphic features at higher altitudes are less conducive for soil profile development. Due to the high slopes, the eroding capacity of rivers is also enhanced which results in high amounts of soil loss. The foothills of mountainous regions also suffer from debris flow-related land degradation and landslides (Norbu et al. 2003). In the plateau regions, different dissected surface features cause hindrance in land quality development (Sarkar et al. 2014). The crests of the undulating plateau, undulating hills and dissected highlands give rise to badland characteristics in any region.

The mountains, residual hillocks, gently sloping uplands and crests in the upper reaches of the Silabati basin represent an undulating topography (Fig.16.2e). Dissected highlands and occasional badlands in the middle reaches of the basin reflect low land quality. Lateritic soil caps have developed in the interfluvial areas of the basin (Gulati and Rai 2014). This area is characterized by high rates of water erosion mostly characterized by rill and gully formation (Lenka et al. 2014). Thick laterite cover has developed in the upper undulating alluvial plain (Mahala 2017). The floodplain and lowlands of the basin are also prone to flood-related land degradation.

16.5.1.6 Groundwater Fluctuation

Groundwater availability with high water table is one of the requisite features for enhancing land productivity and soil moisture. Imbalances between recharge and extraction of groundwater and shifting of groundwater recharge source result in groundwater decline. All these factors contribute to declining land productivity (Nag and Ghosh 2013). Groundwater is an important source of soil moisture. The decline in water table causes soil dryness, which in turn leads to soil

degradation (Cosby et al. 1984). In arid and semiarid areas of the world, agriculture systems depend on groundwater. Continuous extraction of groundwater increases soil salinity in these parts of the world (Mahmoud and Alazba 2016; Besser et al. 2017). With increasing pressure on land, the agriculture density has increased dramatically, therefore necessitating the greater dependence of population on land productivity. The water table has decreased dramatically due to extraction of water for irrigation in double and triple cropping areas. On the other hand, the decreasing rainfall rates in the recent past do not permit groundwater to recharge sufficiently. So the water table goes down annually. Arid and semiarid areas of the world face intense problems related to groundwater decline (Mahmoud and Alazba 2016; Kallioras et al. 2011; Nag and Ghosh 2013).

Groundwater conditions in the Silabati basin vary. Very deep or very high values of metres below ground level (>6.5 mbgl) is characteristic of the groundwater level in the lower reaches of the Silabati basin (Fig. 16.2f). Continuously expanding demands and decline of groundwater table are the resultant features of intensive agricultural practices in the lower reaches of the basin. The continuous decrease of groundwater table in areas of intensive agriculture (lower reaches of the basin) poses serious possibilities of land degradation. The middle reaches of basin show medium availability of groundwater (5.7–6.1 mbgl). It is due to medium intensive agricultural practices and comparatively lower groundwater extraction rate. Upper reaches of the basin show good groundwater availability conditions (<5.7 mbgl). Lower intensity of agricultural practices and higher forest cover are deemed to be the contributing factor for this phenomenon.

16.5.1.7 Drainage Density

The drainage network characteristics of any region control the land degradation processes both directly and indirectly. Broadly, drainage channels determine the slope of the region, accumulation and erosion characteristics. Drainage network is guided by geomorphology and geological characteristics of the region. Drainage frequency, density, profile and length are the major parameters which are generally used to determine the erosional stages along the course of a stream (Frankl et al. 2013). The work of water as an erosional agent also depends on basin characteristics like basin slope, form and size. High frequency of primary order streams in sloping surfaces causes high rates of soil erosion-related land degradation (Girmay et al. 2009). Continuous headward erosion of first-order streams in tropical plateau fringe lateritic region aids rill formation. Over some time, these rills mature into gullies and ravines which are characteristic of badland topography (Shit and Maity 2012).

In River Silabati basin, high drainage density (>2.3 km/km²) is a feature of the lower-middle interfluvial areas

(Fig. 16.2g). High slopes on the right bank of the River Silabati, thick laterite cover and continuous leaching activity are the major causal factor for high drainage density. This area is mostly dominated by rill and gully erosion-related badland area formation. Most of the middle-upper reaches of basin show medium drainage density (1.5–2 km/km^2) due to less porous soil with medium vegetative cover. Most parts of the upper reaches of the basin show very low drainage density (<1.5 km/km^2) due to the presence of hard rock formations underneath the undulating plateau landscape.

16.5.1.8 Aridity Conditions

The climatic characteristics of any area (temperature, rainfall, sunshine, duration of the day, wind direction, moisture) determine the land quality. The surface of the land has a very important role to play in the climatic system. The land surface and climatic interactions involve multiple processes and feedbacks. Rainfall plays a prime role in vegetal distribution. The variability and extremity of the events of rainfall-induced soil erosion and land degradation. Soil erosion characteristics in tropical plateau region are greatly determined by rainfall amount, intensity and frequency (Wessels et al. 2007). Different temperature characteristics (extremity, range, annual average) determine land quality in desert and semidesert type environment of the tropical regions (Boschetto et al. 2010). Rainfall and temperature also determine the water balance conditions of any region. The aridity of any region is determined by the potential evapotranspiration characteristics (Xu and Singh 2002). Climatologists estimate water stress condition by comparing the ratio of precipitation to that of potential evapotranspiration (Thornthwaite 1948).

The humid Silabati basin reflects characteristic low aridity and high P/PET ratio (>0.7) in upper reaches of the basin (Fig. 16.2h). The high forest cover in the higher reaches of the basin exhibits positive correlation with the precipitation rate. The aridity condition is low in this region. Recent deforestation and related agricultural land expansion pose a threat of increasing aridity. The middle and lower reaches of Silabati basin show medium aridity or low P/PET ratio (<0.65). In spite of proximity to the Bay of Bengal which is a continuous source of moisture as well as tropical cyclones, threats to widespread aridity remain. Intensive agricultural practice and continuous decline of forest land through agricultural expansion are the major causes of declining precipitation. All these issues ultimately pose a real possibility of land degradation.

16.5.1.9 Green Biomass Cover

Different external factors influencing land degradation are less effective over densely vegetated land. So it is predicted that the land covered by vegetation are less prone to degradation (Higginbottom and Symeonakis 2014). Open lands are affected by direct rainfall as it causes soil detachment. Open land is also affected by low soil moisture due to direct sun rays which results in becoming frail and vulnerable to degradation. The humus content is greater on land covered by green biomass. Green biomass also acts as an agglomerating agent for the soil, increasing compactness and in turn indirectly increasing fertility (Kakembo 2001). The soils of open land undergo degradation due to the lack of the agglomerating effect of vegetation roots. Land covered by green biomass is less prone to soil loss (Balpande et al. 1996, Higginbottom and Symeonakis 2014). All these factor increase soil fertility. The continuous decline of green cover in arid and semiarid environments increases the danger of land degradation (Kairis et al. 2014; Kiage 2013). Very less biomass productivity in desert regions increases land degradation vulnerability. Tropical areas of the world are commonly affected by deforestation-related land degradation (Cornelio 2010).

Silabati basin shows very low green biomass cover in middle reaches of the basin (NDVI < 0.08) (Fig. 16.2h). Continuous deforestation and agriculture expansion are deemed to be the causative factor. Due to low vegetation cover, this region is highly affected by rill and gully erosion. Upper parts of basin show low vegetation cover (NDVI 0.12–0.24). Rocky, barren, immature surfaces do not present a suitable ground for lush vegetation growth in the upper part of the river basin. Lower part basin shows high vegetation cover (NDVI > 0.24) due to intense agriculture. The vulnerability of soil erosion is greater here in the cropping gap periods.

16.5.2 Physical Vulnerability Index of Land Degradation

Through the multiplication of different physical factor weightage rasters, the environmental vulnerability index of land degradation for Silabati basin has been established (Table 16.3, Fig. 16.2j). It shows very high land degradation vulnerability in the interfluvial lateritic areas. Deep lateritic cover with prominent rill and gully erosion is the major cause behind this. The river flood plain area in the middle and lower reaches of the basin reflects medium levels of land degradation. Deforestation and related agricultural expansion play major roles in determining land vulnerability trends. The upper reaches of basin show very low levels of land degradation due to the presence of lush vegetative cover.

16.6 Conclusion

The land degradation processes in tropical plateau areas are not similar to those found in other parts of the world. Old geological formations, immature soil profile development or presence of highly eroded soil cover, presence of dissected

Fig. 16.3 Land degradation through gully erosion in upper course (**a**, **b**) and Lower course (**c**, **d**) of Silabati river basin

geomorphological landforms are the major physiographic characteristics of land degradation. We have tried to see the processes of land degradation and its vulnerability through an integrative study of all individual factors and multiplication of all of these factors simultaneously. Remote sensing and GIS play an effective role in accessing this information. There is the presence of granite–gneiss–schist geological formation in the upper reaches of Silabati basin which lies underneath the undulating terrain characterized by high relative relief and shallow soil profile development. On the other hand, the lateritic geology in lower reaches causes high intensity of erosional risks. High relief and sloping areas cause shallow soil cover, high rates of erosion and high drainage density. The dissected high land, crests of undulating plateau and denudational hills of the Silabati basin are natural topographic deterrent factors in productivity (Fig. 16.3). The very low groundwater table in hilly upland areas, continuous decline of the same in agriculture-rich areas and enhanced groundwater exploitation lead to water stress conditions in soil. High drainage density and frequency in undulating upland areas result in high rates of erosion. Decreasing rainfall and increasing aridity (low P/PET) cause water stress in the ecological productivity of the area. The green biomass cover is also observed to be continuously declining in this area. The overall Environmental Vulnerability Index is greater in lower-middle reaches of the basin due to high water-related rill and gully erosion in lateritic soil cover.

Finally, it can be concluded that soil erosion is the major process causing land degradation in the tropical plateau region. The result can be used for sustainable management of land degradation. Different soil management processes like mulching, terrace farming and contour bunding in hilly upper reaches of the basin should be focused upon. Some conservational issues like conservative tillage, use of bands, protect overland flow and avoidance of construction works in sloping areas should be implemented. Increased biomass cover in the open and barren land, minimum water use for irrigation, water conservation through the pond and deep area construction should be considered corrective measures for combating with the risks of land degradation.

Acknowledgement The paper is a part of work done by the author under M.Phil. Program of Jawaharlal Nehru University. The work has been conducted with the financial support of University Grants Commission. The author is also thankful to his supervisor Dr. Padmini Pani (Associate Professor, JNU) for continuous guidance and support.

References

Abdelrahman MA, Natarajan A, Hegde R (2015) Climate and its impact on soil biological degradation using GIS and remote sensing (II): biological degradation as a result of removing organic matter by water erosion. J Soil Biol Ecol 208-216

Ajai, Arya AS, Dhinwa PS, Pathan SK, Raj KG (2009) Desertification/land degradation status mapping of India. Current Science 97 (10):1478-1483

Allbed A, Kumar L (2013) Soil Salinity Mapping and Monitoring in Arid and Semi-Arid Regions Using Remote Sensing Technology: A Review. Advances in Remote Sensing 2:373-385

Alves TLB, Azevedo PVd (2015) Influence of climate variability on land degradation (desertification) in the watershed of the upper Paraíba River. Theor Appl Climatol 127(3-4):741–751

Askari MS, Cui J, Holden NM (2013) The visual evaluation of soil structure under arable management. Soil and Tillage Research 134:1-10

Babaev MP, Gurbanov EA, Ramazanova FM (2015) Main Types of Soil Degradation in the Kura–Aras Lowland of Azerbaijan. Eurasian Soil Science 48(4):501-512

Balpande SS, Deshpande SB, Pal DK (1996) Factors and processes of soil degradation in vertisols of the Purna Valley, Maharashtra, India. Land Degrad Dev 7:313-324

Behera SK, Shukla AK (2015 Spatial Distribution of Surface Soil Acidity, Electrical Conductivity, Soil Organic Carbon Content. Land Degrad Dev 26(1):71–79

Besser H, Mokadem N, Redhouania B et al (2017) GIS-based evaluation of groundwater quality and estimation of soil salinization and land degradation risks in an arid Mediterranean site (SW Tunisia). Arab J Geosci 10:350-370

Bhan C (1988) Spatial Analysis of Potential Soil Erosion Risks in Welo Region Ethiopia: A Geomorphological Evaluation. Mountain Research and Development 8(2/3):139-144

Baroudy AA (2011) Monitoring land degradation using remote sensing and GIS techniques in an area of the middle Nile Delta, Egypt. Catena 87:201–208

Ranga V et al. (2016) Detection and analysis of badlands dynamics in the Chambal River Valley (India), during the last 40 (1971–2010) years. Environ Earth Sci 75(183):1-12

Bocco G, Mendoza M, Velazquez A (2001) Remote sensing and GIS-based regional geomorphological mapping—a tool for land use planning in developing countries. Geomorphology 39 (3-4):211–219

Boschetto RG, Mohamed RM, Arrigotti J (2010) Vulnerability to Desertification in a Sub-Saharan Region: A First Local Assessment in Five Villages of Southern Region of Malawi. Ital J Agron 5 (2S):91-101

Bready NC, Well RR (2005) The Nature and Properties of Soil. Pearson Prentice Hall, Singapore

Chatterjee S, Krishna AP, Sharma AP (2013) Geospatial assessment of soil erosion vulnerability at watershed level in some sections of the Upper Subarnarekha river basin, Jharkhand, India. Environ Earth Sci 71(1):357–374

Chauhan SS (2003) Desertification Control and Management of Land Degradation in the Thar Desert of India. The Environmentalist 23 (3):219–227

Cornelio DL (2010) Land Use Conversion and Agricultural Intensification in Tropical Hill Slopes: A Geographical Approach. In: Gökçekus H, Türker U, LaMoreaux J W (ed) Survival and Sustainability. Springer, Berlin, pp 401-421

Cosby BJ, Hornberger GM, Clapp RB, Ginn TR (1984) A Statistical Exploration of the Relationships of Soil Moisture Characteristics to the Physical Properties of Soils. Water Resour Res 20(6):682–690

Daily GC (1995) Restoring value to the world's degraded lands. science 269(5222):350-354

Diodato N, Ceccarelli M (2004) Multivariate indicator Kriging approach using a GIS to classify soil degradation for Mediterranean agricultural lands. Ecol Indic 4(3):177-187

Dolui G, Chatterjee S, Das Chatterjee N (2014) Weathering and Mineralogical Alteration of Granitic Rocks in Southern Purulia District,

West Bengal, India. International Research Journal of Earth Sciences 2(4):1-12

Eckert S, Hüsler F, Liniger H, Hodel E (2014) Trend analysis of MODIS NDVI time series for detecting land degradation and regeneration in Mongolia. J Arid Environ 13:16-28

Faour G (2014) Detection and Mapping of Long-Term Land Degradation and Desertification in Arab Region Using MODERSAT. Lebanese Science Journal 15(2):119-131

Frankl A, Poesen J, Haile M, Deckers J, Nyssen J (2013) Quantifying long-term changes in gully networks and volumes in dryland environments: The case of Northern Ethiopia. Geomorphology 201:254–263

Ghose B, Singh S, Kar A (1977) Desertification around the Thar - A geomorphological interpretation. Ann Arid Zone 16(3):290-301

Ghosh D (2015) Mapping and Monitoring of the Impact of Gully Erosion in the District of Medinipore (west), West Bengal, India. International Journal of Novel Research in Humanity and Social Sciences, 2(4):73-89

Ghosh S, Guchhait SK (2015) Characterization and Evolution of Laterites in West Bengal: Implication on the Geology of Northwest Bengal Basin. Trans 37(1):93-119

Gibbs HK, Salmon JM (2015) Mapping the world's degraded lands. Appl Geogr 57:12-21

Girmay G, Singh BR, Nyssen J, Borrosen T (2009) Runoff and sediment-associated nutrient losses under different land uses in Tigray, Northern Ethiopia. J Hydrol 376:70–80

Gong JR, Wang Y, Liu M, et al. (2014) Effects of land use on soil respiration in the temperate steppe of Inner Mongolia, China. Soil Till Res 144:20-31

Gulati A, Rai SC (2014) Cost estimation of soil erosion and nutrient loss from a watershed of the Chotanagpur Plateau, India. Res Commun 107(4):670-674

Hassan M, Mahmud-Ul-Islam S, Rahman MT (2015) Integration of Remote Sensing and GIS to Assess Vulnerability of Environmental Degradation in North-Western Bangladesh. Journal of Geographic Information System 7:494-505

Hereher ME, Ismael H (2016) The application of remote sensing data to diagnose soil degradation in the Dakhla depression – Western Desert, Egypt. Geocarto Int 31(5):527-543

Heshmati M, Arifin A, Shamsuddin J, Majid NM (2012) Predicting N, P, K and organic carbon depletion in soils using MPSIAC model at the Merek catchment, Iran. Geoderma 175-176:64-77

Higginbottom TP, Symeonakis E (2014) Assessing Land Degradation and Desertification Using Vegetation Index Data: Current Frameworks and Future Directions. Remote Sens 6:9552-9575

ISRO (2007) Dessertification and Land Degradation Atlas of India. Goverment of India, Space Application Center. Ahmedabad. https://www.isro.gov.in/desertification-and-land-degradation-atlas-released

ISRO (2016) Dessertification and Land Degradation Atlas of India. Government of India, Space Application Center, Ahmedabad. https://www.isro.gov.in/desertification-and-land-degradation-atlas-released

Kairis O, Kosmos C, Karavitis C, Salvati L (2014) Evaluation and Selection of Indicators for Land Degradation and Desertification Monitoring: Types of Degradation, Causes, and Implications for Management. Environ Manage 54:971–982

Kakembo V (2001) Trends in vegetation degradation in relation to land tenure, rainfall and population changes in Peddie district, Eastern Cape, South Africa. Environ Manage 28(1): 39–46

Kallioras A, Pliakas F, Skias S, Gkiougkis I (2011) Groundwater vulnerability assessment at SW Rhodope aquifer system in NE Greece. In: Lambrakis N, Stournaras G, Katsanou K (Ed) Advances in the Research of Aquatic Environment. Springer, Berlin, pp 351-358

Kannan B, Krishnan R, Jagadeeswaran R, Ragunath KP, Kumaraperumal R (2015) Development of Spectral Index for Discriminating Degraded Lands. Madras Agric J 102(1-3):89-91

Keller T, Hakansson I (2010) Estimation of reference bulk density from soil particle size distribution and soil organic matter content. Geoderma 154(3-4):398-406

Kiage LM (2013). Perspectives on the assumed causes of land degradation in the rangelands of Sub-Saharan Africa. Prog Phys Geog 37(5), 664-684

Kosmas C, Kairis O, Karavitis C, Acikalin S, Alcalá M, Alfama P (2015) An exploratory analysis of land abandonment drivers in areas prone to desertification. Catena 128:252–261

Kundu A, Patel NR, Saha SK, Dutta D (2015) Monitoring the extent of desertification processes in western Rajasthan (India) using geo-information science. Arab J Geosci, 8(8):5727 – 5737

Lal R (2001) Soil degradation by erosion. Land Degrad Dev 12(6):519–539

Lenka NK, Mandal D, Sudhishri S (2014) Permissible soil loss limits for different physiographic regions of West Bengal. Res Commun 107(4):665-670

Liu Y, Gao J, Yang Y (2003) A Holistic Approach Towards Assessment of Severity of Land Degradation Along the Great Wall in Northern Shaanxi Province, China. Environ Monit Assess 82:187–202

Mahala A (2017) Processes and Status of Land Degradation in a Plateau Fringe Region of Tropical Environment. Environ Process 4:663–682

Mahmoud SH, Alazba AA (2016) Integrated remote sensing and GIS-based approach for deciphering groundwater potential zones in the central region of Saudi Arabia. Environ Earth Sci 75:344-372

Mandal AK, Sharma RC, Singh G (2009) Assessment of salt affected soils in India using GIS. Geocarto Int 24(6):437-456

Myint M, Thinley P (2006) Mapping Potential Land Degradation in Bhutan. ASPRS 2006 Annual Conference, Reno

Mythili G, Goedecke J (2016) Economics of Land Degradation in India. In: Nkonya E, Mirzabaev A, Braun JV (ed) Economics of Land Degradation and Improvement – A Global Assessment for Sustainable Development, Springer, pp 431-469

Nag SK, Ghosh P (2013) Delineation of groundwater potential zone in Chhatna Block, Bankura District, West Bengal, India using remote sensing and GIS techniques. Environ Earth Sci 70:2115–2127

Naz SN, Romshoo SA (2012) Assessing the geoindicators of land degradation in the Kashmir Himalayan region, India. Nat Hazards 64:1219–1245

Nkonya E et al. (2015) Global Cost of Land Degradation. In: Economics of Land Degradation and Improvement – A Global Assessment for Sustainable Development. Springer, Cham, pp. 117-165

Norbu C, Baillie I, Dema K et al. (2003) Types of Land Degradation in Bhutan. Journal of Bhutan studies 8:88-114

NRSC (2011) Wastelands Atlas of India. Govt. of India, Ministry of Rural development, Hyderabad.

Pani P, Carling P (2013) Land degradation and spatial vulnerabilities: a study of inter-village differences in Chambal Valley, India. Asian Geographer 30(1):65-79

Prakash S, Sharma MC, Kumar R, Dhinwa PS, Sastry KL, Rajawat AS (2016) Mapping and assessing land degradation vulnerability in Kangra district using physical and socio-economic indicators. Spat Inf Res 24:733–744

Pramanik MK (2016) Site suitability analysis for agricultural land use of Darjeeling district using AHP and GIS techniques. Model Earth Syst Environ 2(56):1-22

Rajan K, Natarajan A, Kumar KS, Badrinath MS, Gowda RC (2010) Soil organic carbon - the most reliable indicator for monitoring land degradation by soil erosion. Curr Sci India 99(6):823-827

Rashid M, Lone MA, Ramshoo SA (2011) Geospatial tools for assessing land degradation in Budgam district, Kashmir Himalaya, India. J Earth Syst Sci 120(3):423-433

Reddy VR (2003) Land Degradation in India: Extent, Costs and Determinants. Econ Polit Weekly 38(44):4700-4713

Regelink IC, Stoof RC, Rousseva S, Weng L et al (2015) Linkages between aggregate formation, porosity and soil chemical properties. Geoderma 247-248:24-37

Reza SK, Nayak DC, Mukhopadhyay S, Chattopadhyay T, Singh SK (2017) Characterizing spatial variability of soil properties in alluvial soils of India using geostatistics and geographical information system. Aarch Agron Soil Sci 63(11):1489-1498

Riezebos HT, Loerts AC (1998) Influence of land use change and tillage practice on soil organic matter in southern Brazil and eastern Paraguay. Soil Till Res 49(3):271–275

Santra P, Singh R, Sarathjith MC (2015) Reflectance spectroscopic approach for estimation of soil properties in hot arid western Rajasthan, India. Environ Earth Sci 74(5):4233–4245

Sarkar D, Mandal D, Halder A (2014) Soil maturity assessment along a topo sequence in Chotanagpur Plateau, West Bengal using inorganic soil phosphorus based weathering index, soil taxonomy and other chemical indices: A comparative study. Agropedology 24(1):82-94

Shit PK, Nandi AS, Bhunia GS (2015) Soil erosion risk mapping using RUSLE model on jhargram sub-division at West Bengal in India. Model Earth Syst Environ 1(28):1-12

Shit PK., Maity R (2012) Rill Hydraulics - An Experimental Study on Gully Basin in Lateritic Upland of Paschim Medinipur, West Bengal, India. Journal of Geography and Geology 4(4):1-11

Singh S, Kar A, Joshi DC, Kumar S, Sharma KD (1994) Desertification problem in Western Rajasthan. Ann Arid Zone 33(3):191-202

Thiombiano L, Tourino-Soto I (2007) Status and Trends in Land Degradation in Africa. In: Sivakumar MV, Ndiang'ui N (ed) Climate and Land Degradation. Springer, Berlin, pp 39-53

Thornthwaite CW (1948) An Approach toward a Rational Classification of Climate. Geogr Rev 38(1): 55-94

Vågen TG, Winowiecki LA, Abegaz A, Hadgu KM (2013) Landsat-based approaches for mapping of land degradation prevalence and soil functional properties in Ethiopia. Remote Sens Environ 134:266–275

Wessels KJ, Prince SD, Malherbe J, Small J, Frost PE, VanZyl D (2007) Can human-induced land degradation be distinguished from the effects of rainfall variability? A case study in South Africa. J Arid Environ 68(2):271-297

Xu C-Y, Singh VP (2002) Cross Comparison of Empirical Equations for Calculating Potential Evapotranspiration with Data from Switzerland. Water Resour Manage 16(3): 197-219

Avijit Mahala is currently working as a Ph.D. scholar in the Center for Study of Regional Development (CSRD) at Jawaharlal Nehru University (JNU), New Delhi, India. He is also a Senior Research Fellow (SRF) under the UGC NET-JRF program. He obtained his M.Phil. from JNU in 2016 and continuing Ph.D. from 2016. He has been awarded by many organizations like CSIR-JRF and UGC-JRF and is M.A. first class first. His major research interests are land degradation, soil erosion, soil geochemistry, soil fertility, and fluvial geomorphology.

Assessment of Gully Erosion and Estimation of Sediment Yield in Siddheswari River Basin, Eastern India, Using SWAT Model

17

Amit Bera, Bhabani Prasad Mukhopadhyay, and Swagata Biswas

Abstract

Gullies are widened rills and are a manifestation of soil erosion. Adverse impact of natural agents and anthropogenic exploitation brings about significant changes in surface soil and degrades its quality subsequently leading to erosion. Siddheswari basin is a soil erosion-prone region heavily cross-cut by gullies. For this study area, SWAT model has been selected. SWAT is a physically based model used for sediment yield analysis and hydrology modelling. SWAT model uses information derived from remotely sensed data like climate, soil, land use/land cover and Digital Elevation Model (DEM). For SWAT modelling, the entire basin was subdivided into five sub-basins. The outcome results show that maximum sediment yield took place in the year 2011 which was 851.521 t/ha and the following year simultaneously showed a high rate of precipitation and runoff (7207.619 mm). Predicted average annual soil loss and gully erosion susceptibility map of Siddheswari river basin has been classified into three categories according to the intensity of soil loss. Under limited availability of input data, SWAT paired with GIS proves to be an effective tool for simulation and quantitative analysis. The obtained results will be useful for planning of mitigation measures and soil and water conservation and management.

Keywords

SWAT · Simulation · Sediment yield · Surface runoff · Siddheswari basin

17.1 Introduction

Gully erosion is the most destructive form of soil-water erosion that results in displacement of huge masses of soil and also is a potential threat for habitats and croplands. Gully erosion is a global issue and attracts an increasing number of researchers worldwide. Ephemeral gullies are the ones with more width than depth. Gullies are formed as a combined erosional effect of soil and water erosion and are one of the primitive and major aspects of soil degradation. Increasing deforestation and improper farming techniques applied in agriculture increases the rate of overland flow, which further encroaches towards the marginal lands. Overland flow highly influences gully erosion, soil piping and subsurface water movement adds in. Gully erosions are fairly unpredictable; they have a characteristic trait of retreating in a region where high-intensity rainfall and rainstorms prevail. Gully formation takes place through subsequent stages of initiation, advancement and stabilization (Bastola et al. 2018). Each stage is impacted by a multitude of factors—lithology, slope, catchment area, land use and cover. Brice (1966) described gully as a drainage channel recently extended with steep sidewalls and head scarp, which is of depth > 2 ft. and width > 1 ft. A change in the pattern of land use causes an initiation of gullies and alters the overland flow component of hydrological factors and infiltration. Downward incision and channel widening cause further progress in the formation of gullies. This is the advancement stage. Destabilizing forces such as formation of tension cracks, seepage undercutting of slope and erosion of plunge pools further pours into the advancement stage. Soil creeps of steepened sidewalls and slab failure aid in the growth of vegetation in the closure stage of stabilization of gullies.

Various soil erosion models include Universal Soil Loss Equation (USLE), Areal Nonpoint Source Watershed Environmental Response Simulation (ANSWERS), Agricultural Non-Point Source Pollution Model (AGNPS), Erosion Productivity Impact Calculator (EPIC), Water Erosion

A. Bera (✉) · B. P. Mukhopadhyay · S. Biswas
Department of Earth Sciences, Indian Institute of Engineering Science and Technology, Shibpur, West Bengal, India
e-mail: amit.rs2017@geology.iiests.ac.in

© Springer Nature Switzerland AG 2020

P. K. Shit et al. (eds.), *Gully Erosion Studies from India and Surrounding Regions*, Advances in Science, Technology & Innovation,

https://doi.org/10.1007/978-3-030-23243-6_17

Prediction Project (WEPP), Hydrologic Simulation Program-Fortran (HSPF) and Soil and Water Assessment Tool (SWAT). In this present research, SWAT model has been used for effectively evaluating the sediment loss and delineating the gully erosion-prone regions. A physically distributed, continuous, long-term model which aids in predicting the effects of land management practices on sediment yield, hydrological modelling in agricultural watershed and hydrology referred to as SWAT (Arnold et al. 1998; Duru et al. 2018). Previously researchers across the world have used SWAT model for sediment yield during calibration, and few researchers have worked on demarking the gully erosion-prone regions for predicting the sediment loss. Shen et al. (2009) compared WEPP and SWAT model for estimating the soil erosion in Zhangjiachong Watershed, China. Singh et al. (2012) did a comparative study of the MLP and SWAT models for evaluating the sediment yield in Nagwa watershed, Jharkhand, India. Mosbahi et al. (2013) used SWAT model to predict soil erosion-prone zones and to assess the surface runoff along with sediment loss of the region. Yang et al. (2013) used SWAT model for assessing agricultural BMPs in the Watershed of Gully Creek. Li and Gao (2015) applied SWAT model for studying the case of Xichuan Watershed, China, and researched on sediment yield and runoff variation corresponding with precipitation changes. Allen et al. (2018) used SWAT model for predicting head cut advancement of gullies in the continental United States. Very few research works were conducted in Indian subcontinent for delineating gully erosion-prone areas and assessing sediment yield using SWAT model.

Gully erosion of the area seriously affects the biodiversity of the region, degrades soil fertility, lowers the productivity of soil, reduces groundwater recharge, and degrades water supply quality. Furthermore, channels and reservoirs get overloaded with silt, and this causes an increased tendency of flooding. The present study area, Siddheswari river basin, is located along the eastern fringe of Chota Nagpur Plateau. The region is an active soil erosion zone and highly prone to formation of rills and gullies. Gully head bundhs (reservoirs) are a result of excessive soil and gully erosion. These gully head bundhs were constructed under Massanjore Dam projects for checking soil erosion and reducing sedimentation rate. Previous researches regarding gully and soil erosion made on this area are by Jha and Kapat (2003, 2011), Shit et al. (2014, 2016), Ghosh and Guchhait (2015), Ghosh and Saha (2015), Debanshi and Pal (2018), Pal and Debanshi (2018) and Sutradhar (2018). SWAT model applied in this area uses daily and annual sediment yield data for generating simulated output results. The main objective of the study conducted is to assess the gully erosion-prone regions and estimate the sediment yield of that region using SWAT model.

17.2 Study Area

The present study was conducted on Siddheswari river basin, a distributary of Mayurakshi River which is further a tributary of River Bhagirathi. Geographically the region is located between the latitudes $23°56'32''$ N to $24°16'18''$ N longitudes $86°53'40''$ E to $87°21'58''$ E with an area of 809.9 km^2 (Fig. 17.1). River Siddheswari joins with its master stream at 98.4 km, the Mayurakshi River. The river covers parts of Dumka, Jamtara and Deoghar districts of Jharkhand and also a petty portion of Birbhum district of West Bengal. Hence, the region is characterized by varying lithologies and soil types. The region is highly affected by gullies and prone to debris flow and shallow level erosion. About 62% of the river basin is covered by laterites, 28% is covered by granite gneiss and meta-sediments, and the rest 10% is covered by sandstones, silts and augen gneiss. Summers are dry and hot, and climate is tropical with rainfall occurring in the months of July–September. Heavy rainfall washes off the top soil cover, and this causes extensive formation of gullies.

17.3 Materials and Methods

For preparing the base map of the study area, the topographic maps sheet no. 72 P/1, 72 P/4, 72 P/8, 73 I/13, 73 M/1 and 73 M/5 (of scale 1:50,000) have been referred. Some factors influencing gullying of a region are geology, land use, land cover, rainfall, slope, temperature, elevation and solar radiation. ArcGIS 10.1 was used for analysing geospatial data. The drainage map of the catchment area was prepared from the SOI topo-sheets and validated with the ASTER DEM data. Geological map of the study area was collected from the Geological Survey of India (GSI) to delineate the lithological variations. Land-use/land-cover map was prepared from LANDSAT 8 data. Siddheswari basin has very limited number of rain gauge stations; hence, to prepare rainfall distribution map, grid-based rainfall data (www.worldclim.org) were selected. To understand the monthly rainfall characteristics, rainfall data of Deoghar, Jamtara and Dumka districts were collected from the Indian Meteorological Department (IMD).

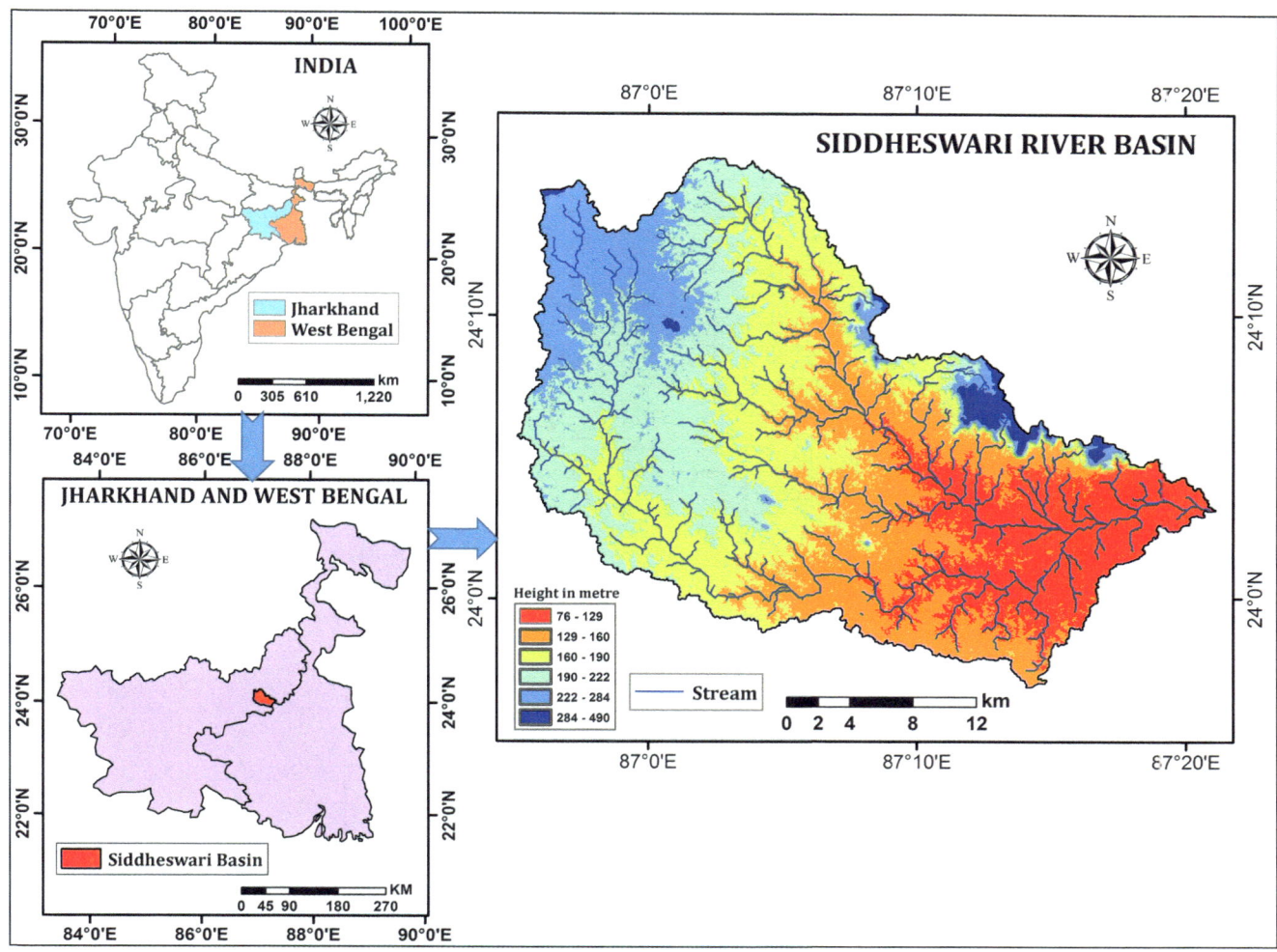

Fig. 17.1 Location of the study area

17.3.1 SWAT Model Development

The SWAT model was originally developed by the US Department of Agriculture-Agricultural Research Service (USDA-ARS). The SWAT is a model designed to simulate water, chemical and sediment fluxes in large catchment areas with varying geomorphological characters and land-use management (Arnold et al. 1998; Srinivasan et al. 1998). The sub-basin consists of hydrological response units (HRU) which are land areas consisting of unique soil management and land cover combinations regardless of their spatial position. Each HRU separately calculates the sediment yield and surface runoff rate. ArcSWAT is an extensional tool of ArcGIS-ArcView software. The SWAT model has been integrated with ArcGIS10.1 software. Modified Universal Soil Loss Equation (MUSLE) aids in estimation of sediment yield (Fig. 17.2). The model uses inputs of DEM, soil, LULC and weather information such as temperature and daily

precipitation. Sediment budget, runoff determined by the hydrology component of the model. The MUSLE model makes the sediment estimation a nonlinear function of HRU area (Phuong et al. 2014; Prabhanjan et al. 2015). At HRU level, simulations are performed and summarized at each sub-watershed (Fig. 17.3). The MUSLE equation is given by:

$$\text{sed} = 11.8 \times \left(Q_{\text{surf}} \cdot q_{\text{peak}} \cdot \text{Area}_{\text{hru}}\right)^{0.56} \times K_{\text{USLE}}$$
$$\times C_{\text{USLE}} \times P_{\text{USLE}} \times \text{LS}_{\text{USLE}} \times \text{CFRG} \qquad (17.1)$$

where sed is sediment yield for a given day (t/d), Q_{surf} is surface runoff volume (mm/ha), q_{peak} is peak runoff rate (m³/s), Area$_{\text{hru}}$ is area of HRU (ha), K_{USLE} is Universal Soil Loss Equation (USLE) soil erodibility factor, C_{USLE} is the USLE cover and management factor, P_{USLE} is the USLE support practice factor, LS$_{\text{USLE}}$ is the USLE topographic factor and CFRG is the coarse fragment factor.

Fig. 17.2 Methodological flowchart for SWAT model

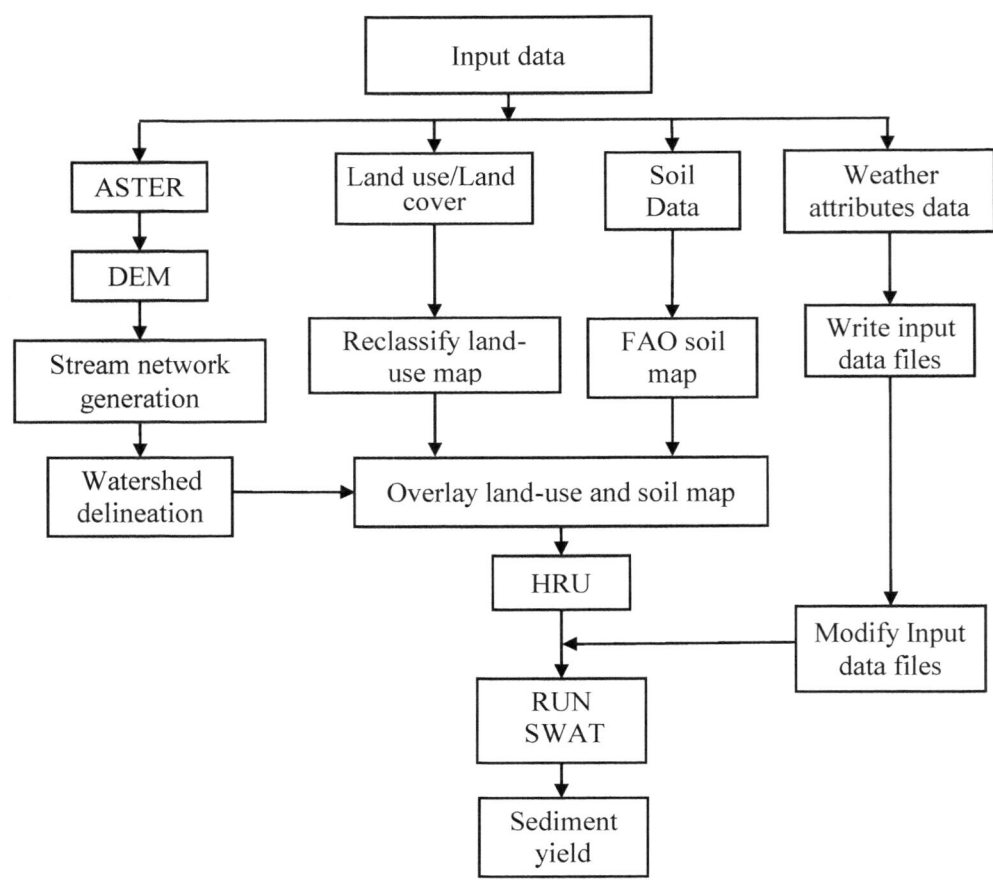

17.3.2 Model Input

This model is capable of simulating effects of change in agricultural and land-use management with available inputs. It requires information on land use, drainage, climate, soil pattern, topography and sediment yield on sub-watershed basin (Fig. 17.4).

Digital Elevation Model (DEM)

Based on Advanced Spaceborne Thermal Emission and Reflection radiometer (ASTER) data, a Digital Elevation Model (DEM) was prepared of the study area. DEM is a raster data. It consists of an array of pixels or cells containing elevation values. For this work, 30 m × 30 m resolution was downloaded from ASTER. It is projected to co-ordinate system (WGS 1984 UTM Zone 45N) and processed in an ArcGIS environment using a watershed delineator. Slope map has been prepared form DEM (Fig. 17.5).

Weather Data

SWAT model requires data from a weather generator model. The various weather variables used for driving the sediment yield are maximum/minimum temperature, daily precipitation, relative humidity, solar radiation and wind speed for the

years 1979–2014. The data were obtained from the Climate Forecast System Reanalysis (CFSR), National Centers for Environmental Prediction (NCEP) (https://globalweather.tamu.edu).

Land Use/Land Cover

The land use/land cover was classified by supervised classification into five classes, namely, water bodies (WATR), natural vegetation (FRST), build-up area (URML), agricultural land (AGRL) and fallow land/pasture land (PAST) (Fig. 17.5 and Table 17.1). The LULC map through the processing of satellite Landsat 8 image of 2016 having a spatial resolution of 30 m is extracted.

Soil Data

The soil data was obtained mainly from the World Soil Database developed by the Food and Agriculture Organization of the United Nations (FAO-UN) (Nachtergaele et al. 2009). Three different soil classes were identified in the study area, namely, sandy loam (Lf32-1b-3788), clay loam (Ne56-2b-3827) and sandy clay loam (Ne59-2b-3834) (Fig. 17.6). Among these, sandy clay loam dominates and covers the maximum area of watershed.

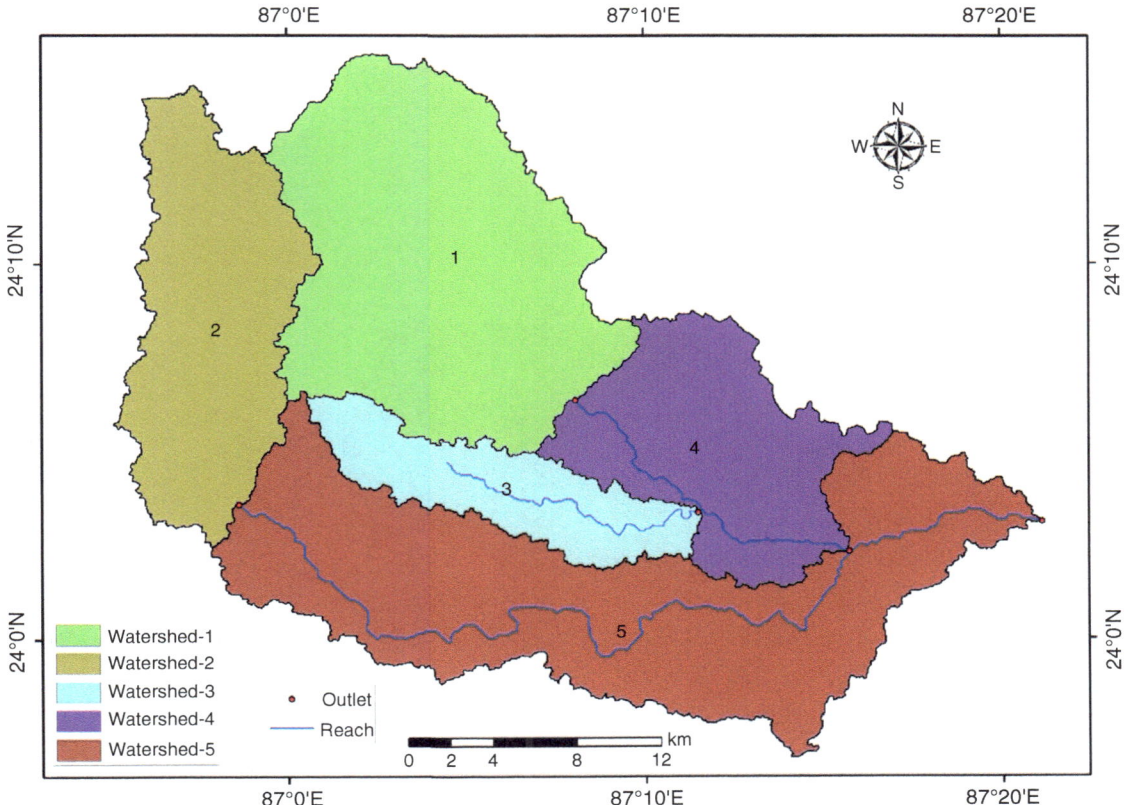

Fig. 17.3 Sub-watershed of the Siddheswari basin

17.3.3 Data Analysis and SWAT Model Setup

ERDAS 2014, ArcGIS 10.1, ArcSWAT2012 model and Microsoft Excel are few software and applications mainly required for executing, processing and analysing collected data. Cloud-free digital data from Landsat 8 with 30 m spatial resolution downloaded from USGS website (https:// earthexplorer.usgs.gov/) for the year 2016 were used to generate the land-use/land-cover map of Siddheswari basin. In the present LULC map, overall classification accuracy and kappa coefficient were 78.02% and 0.74, respectively.

The SWAT model was setup using ArcSWAT 2012.10.1.18. Watershed delineation and parametering of stream reaches and sub-catchment geomorphology were automatically done using these interfaces. For deriving flow direction and accumulation, DEM-based stream definition was used. In SWAT model, individual natural homogeneous areas are referred to as hydrologic response unit (HRU). The objective of HRU definition was to decrease the heterogeneities due to soil types, climate, geology and topography influencing hydrologic response (Sisay et al. 2017). The next step after importing the climatic data was to build few additional inputs for running the model. Finally, the

SWAT model was run for simulating the various hydrological components (Fig. 17.2).

17.4 Result and Discussion

17.4.1 Runoff and Sediment Yield

Five different watersheds were identified in the study area. From the output diagram (Fig. 17.7), it can be concluded that the study area has mean surface runoff 903.45 mm/year. The average upland sediment yield is 114.99 Mg/ha. The inlet/point sources sediment is nil, and in-stream sediment change is −101.75 Mg/ha. The maximum upland sediment yield is 395.8 Mg/ha.

Texture of soil strongly influences the soil erodibility of the region concerned. Coarse-textured soil is more prone to erosion. Low runoff is indicated by coarse texture, and high surface runoff is indicated by fine soil texture. Rainfall is a prime factor that needs to be considered in case of soil erosion and gullying. The intensity and distribution of rainfall determine soil erosivity and runoff generation. The higher the rainfall, the higher the rate of runoff and vice versa. Even

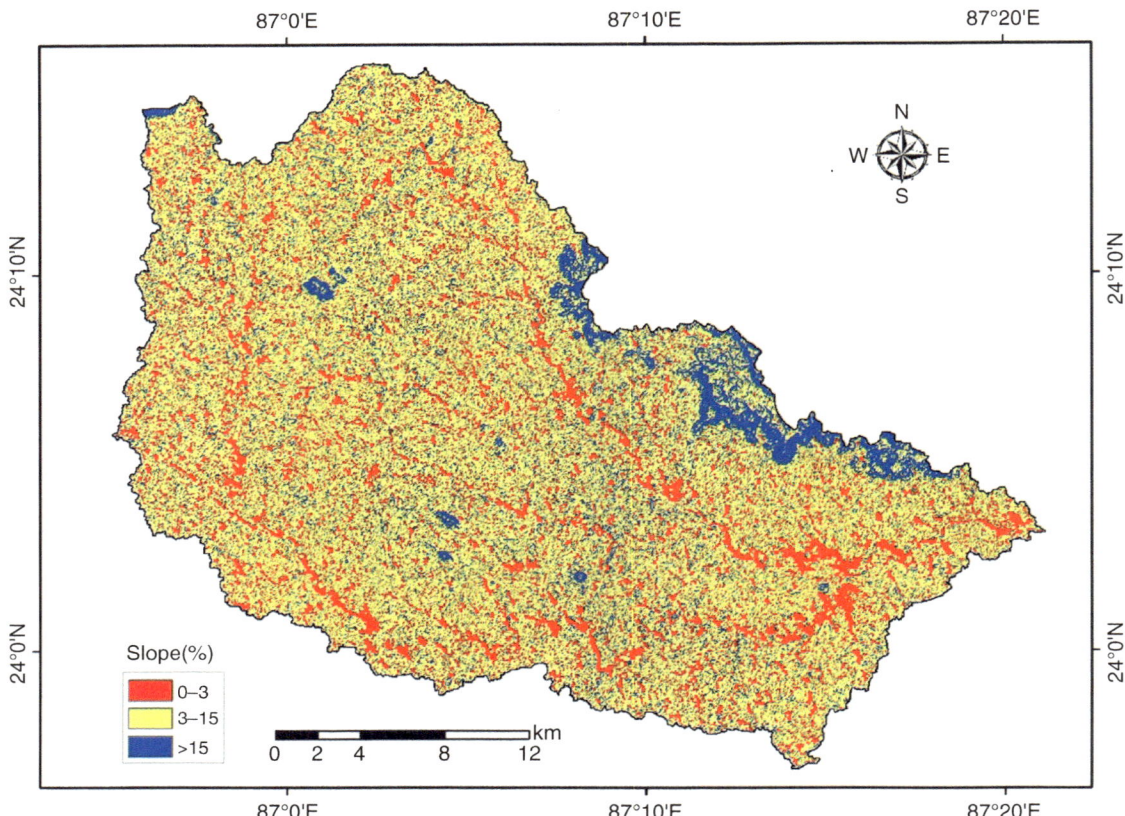

Fig. 17.4 Slope map of the study area

distribution paired with constant rate of rainfall makes the soil less prone to erosive agents. Soil surface with stagnant water level is not a favourable site for formation of rills or gullies. From the daily surface runoff graph (Fig. 17.8), it can be inferred the runoff rate is highest in October and January to June experience very low rate of runoff. Similarly, the mid of October experiences the highest sediment yield (Fig. 17.9), and in the monsoon, the runoff and rainfall being high, sediment yield is also high.

The graph (Fig. 17.10) illustrates sub-watershed-wise annual rainfall variation. In the year 2011, maximum amount of rainfall (2645.8 mm) took place at watersheds 4, 3 and 1. Lowest rainfall (787.8 mm) recorded was on 2014 at watershed 5. The north-eastern part of the basin area receives highest rainfall; the eastern and south eastern part receives moderate rainfall. The catchment area receives an average of 1100–1206 mm rainfall during the monsoon seasons (June–September), and during non-monsoons (October–May), the region receives showers ranging from 236.89 to 292.95 mm. Surface runoff graph (Fig. 17.11) shows that in 2011 and 2013, the runoff rate was high, that is, 7207.619 mm and 6111.788 mm, respectively. Figure 17.12 shows the annual sediment yield graph. In a similar way to that of the runoff, it can be seen that high rate of sediment yield took place in the years 2011 and 2013. Hence, it can be inferred that the

sediment yield, rainfall and runoff are positively correlated. The details of watershed-wise yearly rainfall, surface runoff and sediment yield rate are given in Tables 17.2, 17.3 and 17.4.

17.4.2 Soil Loss Mapping and Identification of Gully Erosion Susceptibility Zones

Database is aptly sorted and formatted after erosion. Soil loss map is simulated with the aid of ArcGIS 10.1 software. Average annual soil loss of Siddheswari river basin has been classified into three erosional intensity classes to assess gully erosion susceptibility. Spatial distribution of the gully head bundhs in the basin are displayed in Fig. 17.13. The high rate (>750 t/ha/year) of soil erosion was found along the major stream and steep slope region of the basin. Fragile lithological setup, coarse-textured soil with rich ferrous mineral and bare terrain with sparse vegetation are responsible for such high rate of soil loss and gully erodibility. From Fig. 17.14, it can be seen that majority of gully head bundhs are located in the high soil erosion-prone zones. The lower catchment zone is gently sloped and less prone to surface runoff, soil erosion and developing gullies (Fig. 17.15).

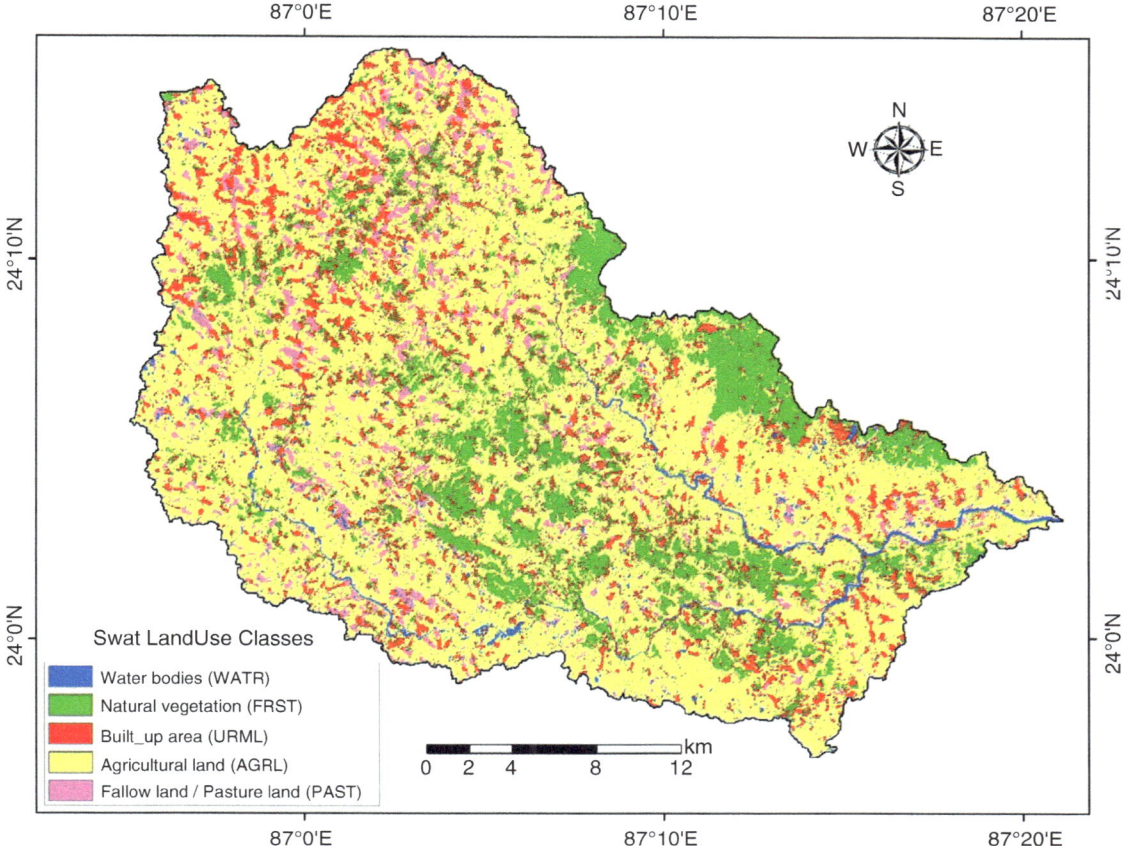

Fig. 17.5 Land-use/land-cover map of the study area

Table 17.1 Detailed land use/soil/slope distribution of the catchment

| Watershed | Land use | | Soil type | | Slope | |
	LULC type	Area coverage (%)	Soil class	Area coverage (%)	Slope class limit (%)	Area coverage (%)
Watershed 1	WATR	0.86	Clay loam	23.59	3	21.53
	FRST	16.44				
	URML	13.12			3–15	10.96
	AGRL	59.80	Sandy clay loam	76.41	>10	67.51
	PAST	9.78				
Watershed 2	WATR	1.92	Clay loam	22.82	3	23.85
	FRST	8.10				
	URML	13.77	Sandy clay loam	77.18	3–15	7.74
	AGRL	67.76			>10	68.40
	PAST	8.44				
Watershed 3	WATR	1.10	Sandy loam	26.33	3	22.10
	FRST	25.58				
	URML	10.93	Clay loam	35.49	3–15	10.81
	AGRL	56.89	Sandy clay loam	38.18	>10	67.09
	PAST	5.49				
Watershed 4	WATR	3.13	Sandy loam	41.54	3	20.86
	FRST	24.75				
	URML	11.18			3–15	21.40
	AGRL	57.83	Sandy clay loam	58.46	>10	57.74
	PAST	3.11				
Watershed 5	WATR	2.27	Sandy loam	23.87	3	23.44
	FRST	15.88				
	URML	12.03	Clay loam	28.16	3–15	11.15
	AGRL	63.02	Sandy clay loam	47.97	>10	65.42
	PAST	6.80				

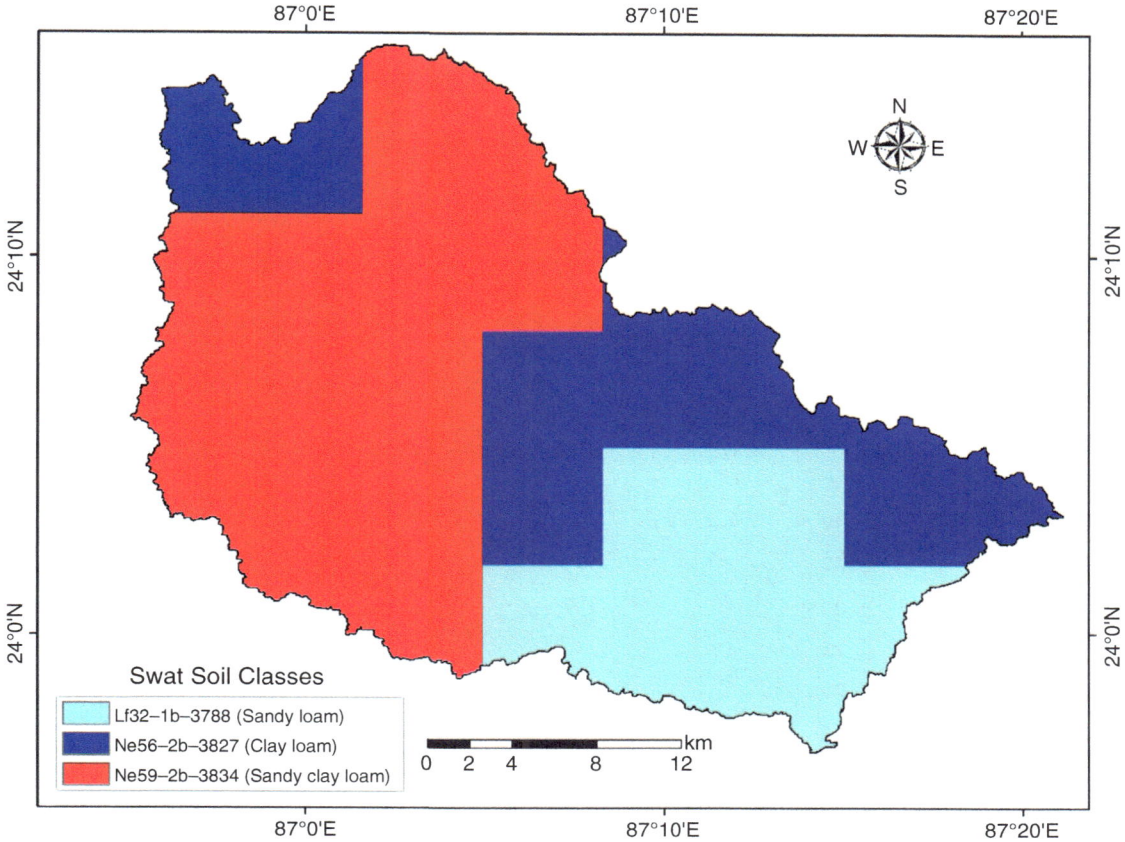

Fig. 17.6 Soil map of the study area

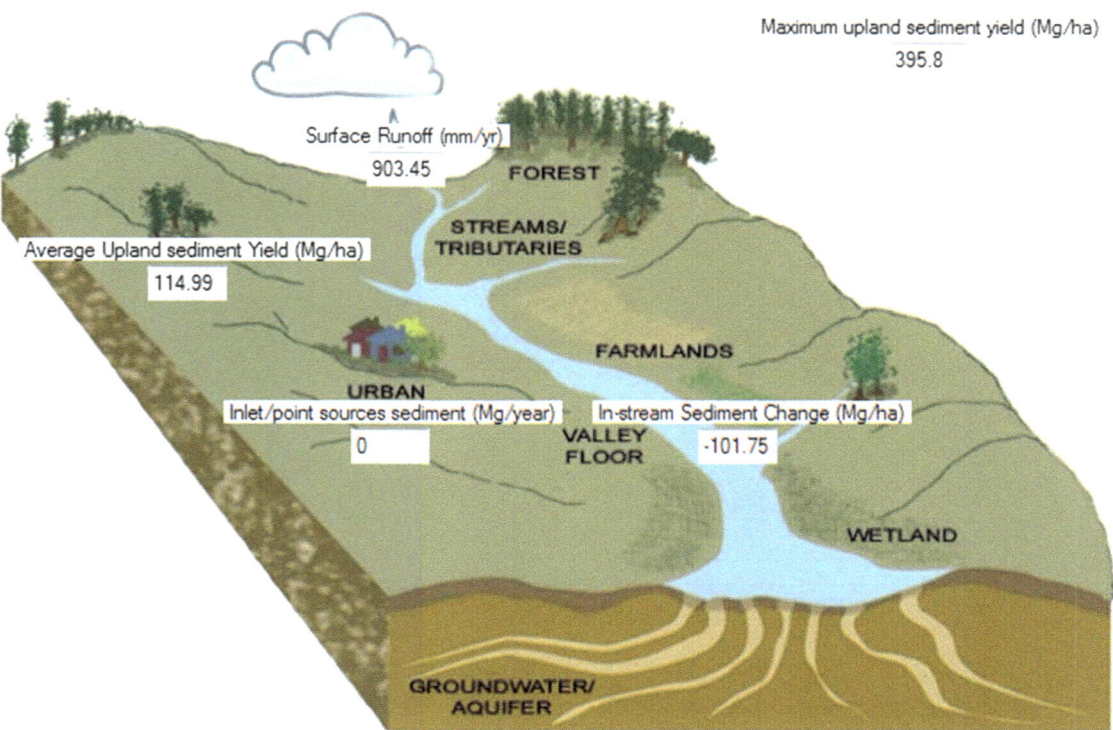

Fig. 17.7 A schematic diagram illustrating the sediment inflow, outflow and surface runoff of the study area

Fig. 17.8 Daily surface runoff of year 2013 of the catchment area

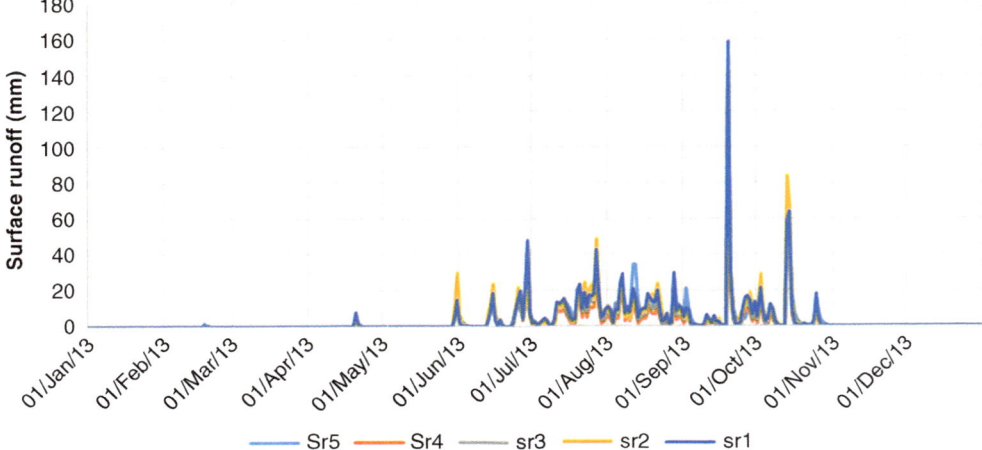

Fig. 17.9 Daily sediment yield of year 2013 of the catchment area

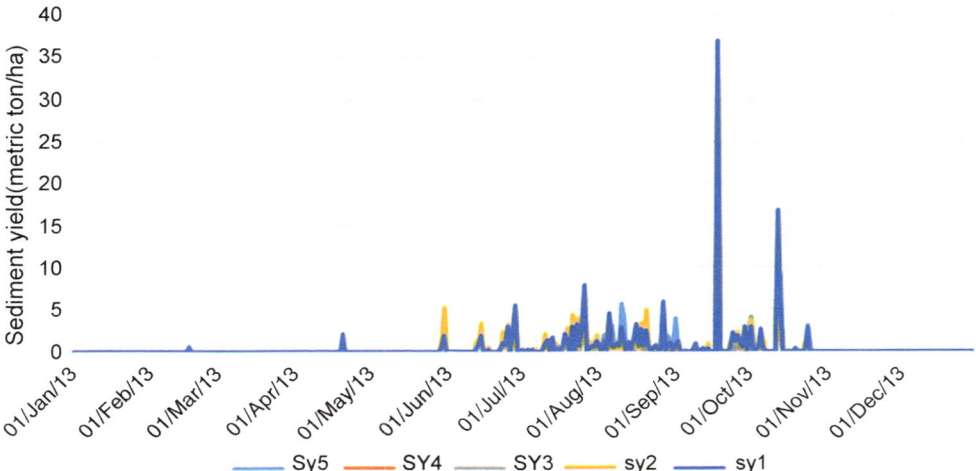

Fig. 17.10 Temporal variation of annual rainfall

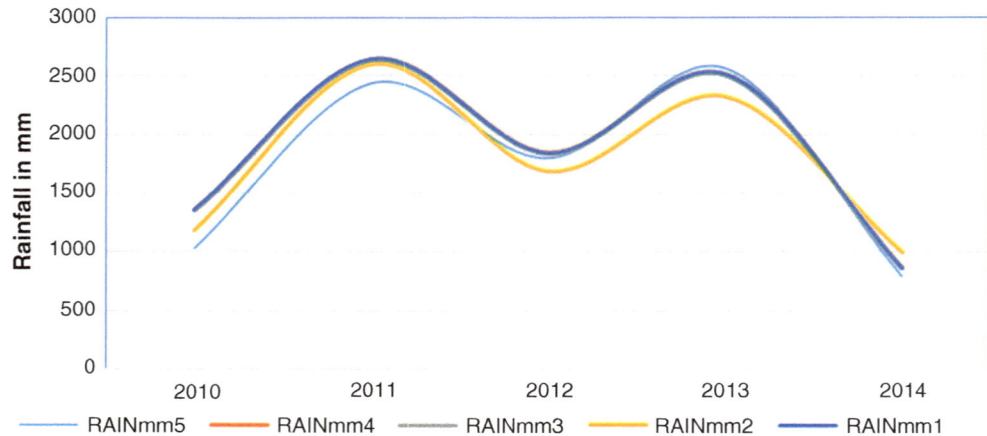

Fig. 17.11 Temporal variation of annual surface runoff rate

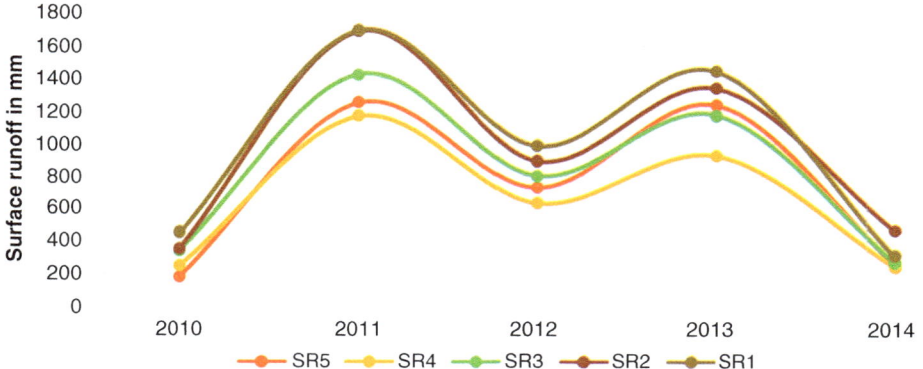

Fig. 17.12 Temporal variation of annual sediment yield rate

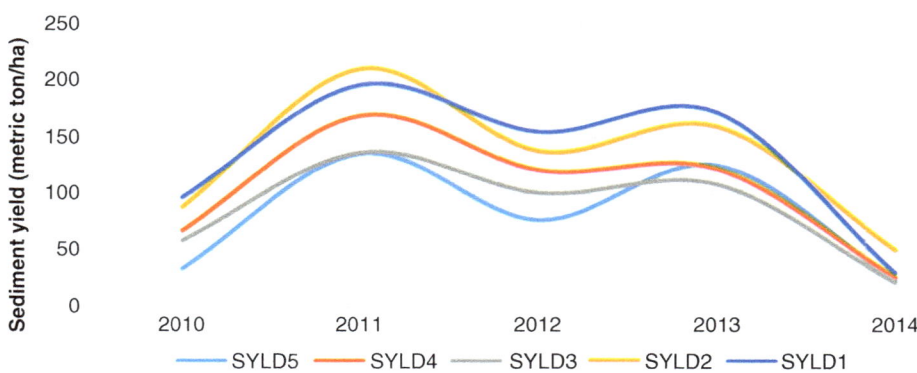

Table 17.2 Watershed-wise rainfall distribution from 2010 to 2014

Year	Rainfall (mm)				
	Watershed 5	Watershed 4	Watershed 3	Watershed 2	Watershed 1
2010	1025.7	1354	1354	1173.1	1354
2011	2443.101	2645.8	2645.8	2598.9	2645.8
2012	1803.801	1848.6	1848.6	1681.9	1848.6
2013	2570.101	2514.6	2514.6	2319.4	2514.6
2014	787.8	861.6	861.6	982.9	861.6

Table 17.3 Watershed-wise surface runoff distribution from 2010 to 2014

Year	Surface runoff (mm)				
	Watershed 5	Watershed 4	Watershed 3	Watershed 2	Watershed 1
2010	182.077	248.884	340.851	352.546	449.339
2011	1248.406	1163.873	1417.908	1684.361	1693.071
2012	728.289	629.327	797.415	889.112	984.977
2013	1236.77	924.026	1171.086	1336.649	1443.257
2014	266.622	243.726	276.161	468.171	316.035

Table 17.4 Watershed-wise sediment yield rate from 2010 to 2014

Year	Sediment yield (t/ha)				
	Watershed 5	Watershed 4	Watershed 3	Watershed 2	Watershed 1
2010	33.575	67.647	58.5	88.346	96.89
2011	136.077	169.769	137.165	211.193	197.317
2012	76.888	121.317	101.043	138.047	155.691
2013	126.052	122.713	108.92	160.177	173.531
2014	23.952	26.599	21.523	50.562	30.402

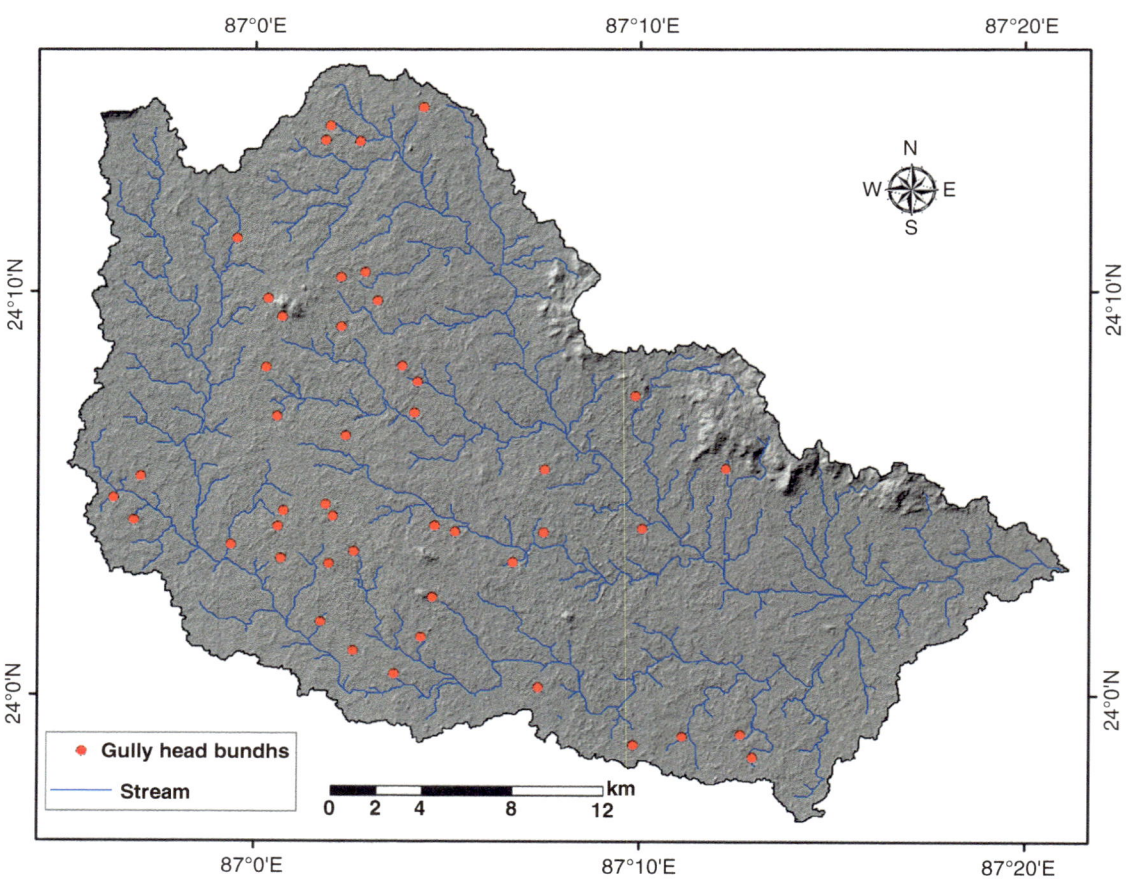

Fig. 17.13 Location of the gully head bundhs in the basin

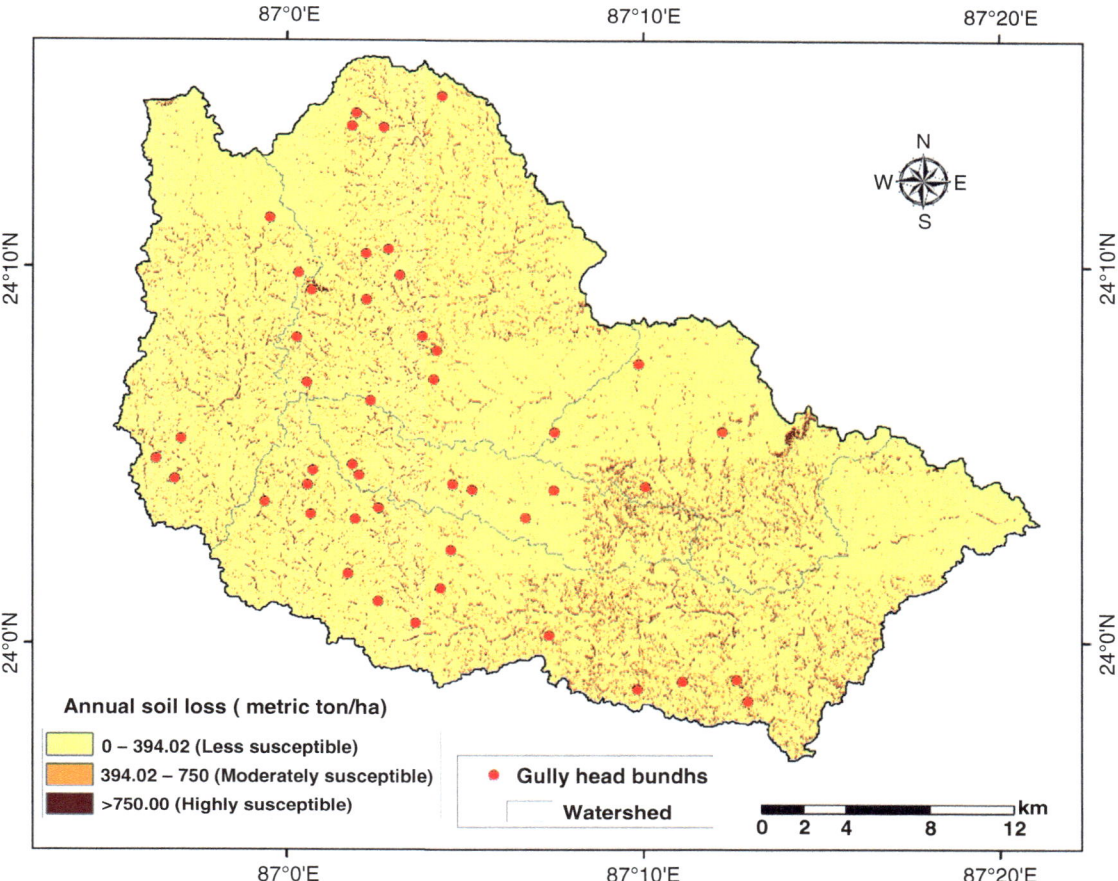

Fig. 17.14 Average annual soil loss and gully erosion susceptibility map of Siddheswari river basin

Fig. 17.15 Photograph illustrating gully erosion in different locations of Siddheswari basin. (**a**) A typical rill on a lateritic land, (**b**) gully-prone bare lateritic topography and (**c**) man-made check dams for preventing soil loss on a gully-affected region

17.5   Conclusion

A daily as well as annual simulation of data was done for analysis and SWAT modelling of the study area. SWAT modelling was done to simulate the water balance components understand processes operating there and assessing the sediment yield of the catchment area. In this research work, SWAT 2012 along with GIS was effectively applied as a time saver approach. The distribution of sediment yield and runoff in the Siddheswari basin shows the major sediment inflow occurs in the summer season and the monsoons. Since the river is dominantly dependent on rain showers, the winters experience low runoff rate, and therefore, the sediment yield is also insignificant. During the years 2010–2014, more amount of sediment yield (653.83 t/ha)

took place in the upper catchment region (watershed 1) and the lowest (396.55 t/ha) took place in lower catchment region (watershed 5). The Siddheswari river has numerous gully bundhs located on its upper catchment region, and the region also experiences high rate of sediment yield. Hence, proper maintenance of the bundhs according to area-wise susceptibility of gullying would prove to be an effective measure for checking gully erosion. The outcome results were pretty reliable and can be of great help for policy makers, resource managers, and stake holders for building scenarios.

Acknowledgements The authors are thankful to the Indian Institute of Remote Sensing (IIRS), Indian Space Research Organization (ISRO) and Indian Meteorological Department (IMD) for continuous support during the work. We are thankful to Dr. Pravat Kumar Shit (Editor, Gully erosion studies from India and surrounding regions) for suggesting modifications, which improved our manuscript. The authors also extend their thanks to anonymous reviewers for the valuable comments and suggestions.

References

Allen PM, Arnold JG, Auguste L, White J, Dunbar J (2018) Application of a simple headcut advance model for gullies. Earth Surface Processes and Landforms, 43(1), 202-217. https://doi.org/10.1002/esp.4233

Arnold JG, Srinivasan R, Muttiah RS, Williams JR (1998) Large area hydrologic modeling and assessment part I: model development. Journal of the American Water Resources Association 34(1):73-89

Bastola S, Dialynas YG, Bras RL, Noto LV, Istanbulluoglu E (2018) The role of vegetation on gully erosion stabilization at a severely degraded landscape: A case study from Calhoun Experimental Critical Zone Observatory. Geomorphology 308:25-39. https://doi.org/10.1016/j.geomorph.2017.12.032

Brice JC (1966) Erosion and deposition in the loess-mantled Great Plains, Medicine Creek drainage basin, Nebraska. U.S. Geol. Survey Prof. Paper 352-H:255-339

Debanshi S, Pal S (2018) Assessing gully erosion susceptibility in Mayurakshi river basin of eastern India. Environment, Development and Sustainability 1-32. https://doi.org/10.1007/s10668-018-0224-x

Duru U, Arabi M, Wohl EE (2018) Modeling stream flow and sediment yield using the SWAT model: a case study of Ankara River basin, Turkey. Physical Geography 39(3):264-289. https://doi.org/10.1080/02723646.2017.1342199

Ghosh, K. G., & Saha, S. (2015). Identification of soil erosion susceptible areas in Hinglo river basin, eastern India based on geo-statistics. Universal Journal of Environmental Research and Technology, 5(3), 152–164.

Ghosh, S., & Guchhait, S. K. (2015). Characterization and evolution of laterites in West Bengal: Implication on the geology of northwest Bengal Basin. Transactions, 37(1), 93-119.

Jha, V. C., & Kapat, S. (2003). Gully erosion and its implications on land use: A case study of Dumka block, Dumka district, Jharkhand. In V. C. Jha (Ed.), Land degradation and desertification (pp. 156–178). Jaipur: Rawat Publications

Jha VC, Kapat S (2011) Degraded lateritic soils cape and land uses in Birbhum district, West Bengal, India. Revista Sociedade and Natureza 23(3):545–556

Li T, Gao Y (2015) Runoff and sediment yield variations in response to precipitation changes: A case study of Xichuan watershed in the loess plateau, China. Water 7(10):5638-5656. https://doi.org/10.3390/w7105638

Mosbahi M, Benabdallah S, Boussema MR (2013) Assessment of soil erosion risk using SWAT model. Arabian Journal of Geosciences 6 (10): 4011-4019. https://doi.org/10.1007/s12517-012-0658-7

Nachtergaele F, van Velthuizen H, Verelst L, Batjes N, Dijkshoorn K, Van Engelen V, . . . & Prieler S (2009) Harmonized world soil database (version 1.1). FAO, Rome, Italy & IIASA, Laxenburg, Austria.

Pal S, Debanshi S (2018) Influences of soil erosion susceptibility toward overloading vulnerability of the gully head bundhs in Mayurakshi River basin of eastern Chottanagpur Plateau. Environment, Development and Sustainability 1-37. https://doi.org/10.1007/s10668-017-9963-3

Phuong TT, Thong CVT, Ngoc NB, Chuong HV (2014) Modeling Soil Erosion within Small Moutainous Watershed in Central Vietnam Using GIS and SWAT. Resources and Environment 4(3): 139-147. https://doi.org/10.5923/j.re.20140403.02

Prabhanjan A, Rao EP, Eldho TI (2015) Application of SWAT model and geospatial techniques for sediment-yield modeling in ungauged watersheds. Journal of Hydrologic Engineering 20(6):C6014005

Shen ZY, Gong YW, Li YH, Hong Q, Xu L, Liu RM (2009) A comparison of WEPP and SWAT for modeling soil erosion of the Zhangjiachong Watershed in the Three Gorges Reservoir Area. Agricultural Water Management 96(10):1435-1442. https://doi.org/10.1016/j.agwat.2009.04.017

Shit PK, Bhunia GS, Maiti R (2014) Morphology and development of selected Badlands in South Bengal (India). Indian Journal of Geography and Environment 13:161-171

Shit PK, Bhunia GS, Maiti R (2016) An experimental investigation of rill erosion processes in lateritic upland region: A pilot study. Eurasian Journal of Soil Science 5(2):121-131

Singh A, Imtiyaz M, Isaac RK, Denis DM (2012) Comparison of soil and water assessment tool (SWAT) and multilayer perceptron (MLP) artificial neural network for predicting sediment yield in the Nagwa agricultural watershed in Jharkhand, India. Agricultural Water Management 104:113-120. https://doi.org/10.1016/j.agwat.2011.12.005

Sisay E, Halefom A, Khare D, Singh L, Worku T (2017) Hydrological modelling of ungauged urban watershed using SWAT model. Modeling Earth Systems and Environment 3(2):693-702. https://doi.org/10.1007/s40808-017-0328-6

Srinivasan R, Arnold JG, Jones CA (1998) Hydrologic modelling of the United States with the soil and water assessment tool. International Journal of Water Resources Development 14(3):315-325

Sutradhar H (2018) Surface Runoff Estimation Using SCS-CN Method in Siddheswari River Basin, Eastern India. Journal of Geography, Environment and Earth Science International 1-9. https://doi.org/10.9734/JGEESI/2018/44076

Yang W, Liu Y, Simmons J, Oginskyy A, McKague K (2013) SWAT Modelling of Agricultural BMPs and Analysis of BMP Cost Effectiveness in the Gully Creek Watershed. University of Guelph, Guelph, Ontario. xi:1-161

Amit Bera graduated in Geography from Calcutta University. He completed his masters in Geography and Disaster Management from Tripura University and has specialized in Fluvial Geomorphology. Currently he is a research scholar of Earth Sciences Department in Indian Institute of Engineering Science and Technology, Shibpur. He has several publications in well renowned SCI journals. His research areas of interest are hydrogeology, environmental hazards, water resource management, GIS and remote sensing applications, geomorphology and soil sciences.

Bhabani Prasad Mukhopadhyay is currently Professor in the Department of Earth Sciences, Indian Institute of Engineering Science and Technology, Shibpur. His research areas are sedimentology and hydrogeology. At present he is working on the quantitative and qualitative aspects of subsurface water in different districts of West Bengal. He has published a good number of research papers in journals of national and international reputation. He has completed more than five projects funded by DST, UGC, ONGC, AICTE and Government of West Bengal. He has participated as Indian Collaborator in International Project funded by National Science Foundation, USA. He has published a good number of books on popular science in vernacular medium and has been awarded 'Satyendra Puroskar' by the Department of Science and Technology, Government of West Bengal in 2005.

Swagata Biswas has graduated in Geological Sciences from Durgapur Government College. She received her master's degree from Indian Institute of Engineering Science and Technology, Shibpur in the year 2018. Currently, she is pursuing her PhD in Hydrogeology from the same university. Her interests are in Hydrogeology, GIS and Remote Sensing, Geomorphology and Environmental Geology.

Pravat Kumar Shit, Hamid Reza Pourghasemi, and Gouri Sankar Bhunia

Abstract

The roots of plant can be very potent in alleviating the soil against concerted flow erosion. However, disposed to incisive erosion practices, very limited number of research subsists on the effectiveness of plant roots in plummeting strenuous flow erosion rates in lateritic soils. Therefore, presented study represents the effect of fine-branched root systems of native plants on the resistance of topsoil to flow volume is quantified through the relationship between relative soil detachments rate (RSD) and roots density (RD) by flume experiment. To carry out the experiment, the undisturbed 21 topsoil samples were collected from 3 different places considering bare soil samples, grass samples with low sowing density, and grass samples with high sowing density. Laboratory experiments on simulated concentrated flow were conducted with a flume similar to standard method. Results of the analysis showed soil erosion is decreasing with increasing grass roots density. Nonlinear regression analysis represented the decreasing trend of RSD ($R^2 = 0.73067$) with increasing RD. Our results also showed roots of less than 1 mm in diameter varied greatly for different vegetation and have the unequal impact on soil anti-scouribility. The present information provides an important and useful result for understanding the effects of the plant roots parameter of vegetation on erosion mechanisms that may create a generalizable soil erosion model in lateritic region.

P. K. Shit (✉)
Department of Geography, Raja N.L. Khan Women's College, Medinipur, India

H. R. Pourghasemi
Department of Natural Resources and Environmental Engineering, College of Agriculture, Shiraz University, Shiraz, Iran

G. S. Bhunia
Aarvee Associates Architects, Engineers & Consultants Pvt. Ltd, Hyderabad, India

Keywords

Soil erosion · Soil detachment rate · Vegetation root properties · Lateritic region

18.1 Introduction

Rills and gullies are developed due to concentration of erosive flow, resulting from the detachment and displacement of soil particles (Govers et al. 1990; Shit et al. 2013, 2014, 2015). Recent studies indicated that rill-gully erosion represents an important sediment source in any environment (Poesen et al. 2003). The use of traditional techniques to prevent and control concentration of erosive flow, above the ground biomass (such as shrub hedges), has been given enough importance, whereas the role of the below-ground biomass (i.e., roots system) is not studied till date (Shit et al. 2013). However, vegetation cover can be very limited in lateritic undulating surface and is often the above-ground biomass which can temporally disappear because of overgrazing or shoots due to high temperature. On the contrary, roots can play a crucial role in controlling soil erosion rates during concentrated flow.

Below-ground biomass that is root systems act form anchors that can stabilize loose soil (Gyssels and Poesen 2003; Knapen et al. 2007; Zuazo and Pleguezuelo 2008). Roots played an important role in controlling of soil properties including aggregate stability, infiltration capacity, soil bulk density, soil texture, organic and chemical content, and shear strength (Miller and Jasirow 1990; Reubens et al. 2007). Many studies have been conducted on investigation of the effects of plant roots on soil aggregate stability (Monroe and Kladivko 1987), soil shear strength (Waldron 1977; Tengbeh 1993), soil penetrability (Wu et al. 2000), soil erodibility (Mamo and Bubenzer 2001b), and soil anti-scouribility (Li and Xu 1992; Wu et al. 2000; Zhou and Shangguan 2005). Li and Xu (1992) and Wu et al. (2000)

© Springer Nature Switzerland AG 2020
P. K. Shit et al. (eds.), *Gully Erosion Studies from India and Surrounding Regions*, Advances in Science, Technology & Innovation, https://doi.org/10.1007/978-3-030-23243-6_18

suggested that the density of fine roots, defined as the number of roots <1 mm in diameter per unit soil volume, has a considerable effect on soil anti-scouribility. Mamo and Bubenzer (2001a, b) reported that soil erodibility decreased sharply with increasing root length density (cm of root length per cm^3 of soil).

Roots reduce soil erosion by binding the soil particles at the ground surface and so will reduce surface runoff velocity (Greenway 1987). The soil adjacent to the roots is affected by both hydrologically and mechanically in terms of aggregate stability (Wang et al. 2011), infiltration capacity, soil bulk density, soil texture, organic and chemical content, and shear strength (Amezketa 1999; Morgan 1995; De Baets et al. 2007a, b). The mechanics of how the plant roots reinforce the soil are twofold; at first, roots and root remnants physically bind soil particles that are formed mechanical barriers for soil and water movement (Tengbeh 1993). Major controlling parameters of the mechanical influence of roots are diameter, degree of bifurcation, appearance of root hairs, friction between root and soil, and, obviously, root network distribution (Abe and Ziemer 1991). Shallow interlocking root networks can substantially contribute to mechanical reinforcement of soils, acting as an anchored net of densely interwoven roots (Sidle et al. 1985; Preston and Crozier 1999). Dense root mats aid to carpet the ground and provide the substantial soil cohesion, which ultimately limits erosion by overland flow (Prosser et al. 1995; Sidorchuk and Grigorev 1998). Moreover, living and dead root systems can provide subsurface water flow pathways by creating biopores and thus will reduce the amount of erosive overland flow. Secondly, roots and root remnants excrete binding agents and form a food source for microorganisms that in turn produce other organic bindings (Reid and Goss 1987). These bindings increase the amount of stable soil aggregates in the long term and thus reduce soil erodibility (Hartman and De Boodt 1974).

In recent years, research on root-soil interaction has increased considerably. Although quantitative data on the effects of roots mainly result from research on slope stability, some experimental studies investigate the effects of plant roots on water erosion processes. These studies often concentrated on roots from tops, whereas root properties and root effects of native vegetation covers are much less studied. Gyssels et al. (2005a, b) indicated that investigated the impact of roots on the resistance of the topsoil to concentrated flow generally used root density (RD, kg m^{-3}) or root length density (RLD, km m^{-3}) as a root parameter to predict the water erosion reducing effect. Zhou and Shangguan (2005) and Li et al. (1991) reported that RLD is a good parameter for expressing the root-soil contact area, but their results showed that the root effect cannot be precisely predicted if only the number of roots <1 mm is used. Zhou

and Shangguan (2005) stated that root diameter distribution has to be taken into account and prediction of erosion intensity by roots during concentrated flow and therefore proposed to use root surface area density ($RSAD$, cm^2 cm^3) for predicting the soil anti-scouribility. De Baets et al. (2007a, b) reported that root diameter is an important variable in explaining relative soil detachment rates as the effect of RD decreases with increasing root diameter. There are numerous reports on the profound effects that living plant roots have on the formation of aggregate structure and maintaining soil stability (Reid and Goss 1980, 1981; Habib et al. 1990; Dexter 1991). Mamo and Bubenzer (2001a) indicated that concentrated flow of erodibility decreased over time with increasing root length density. Gyssels et al. (2005a, b) and Gyssels and Poesen (2003) observed an exponential decline in relative soil loss with increasing root density. However, root properties (RD, RLD) must be considered as an important parameter of root on soil erodibility. So, in this study, the effect of fine-branched root systems of native plants on the resistance of topsoil to flow volume is quantified through the relationship between relative soil detachments rate (RSD) and roots density (RD) by flume experiment.

18.2 Material and Methods

18.2.1 Sample Collection

Twenty-one topsoil samples were taken by the monolith method of Bohm in lateritic environment from three different places, i.e., bare soil, scatter grass roots, and densely grass roots in the rill- and gully-affected areas of Rangamati, Paschim Medinipur (22°24.697′ N, 87°17.798′ E to 22°24.798′ N, 87°17. 895′ E) (Fig. 18.1), located about 1 km southwest of the town of Medinipur in West Bengal, nearer to Vidyasagar University. The climate is semiarid, with a high interannual variability of rainfall. Mean annual rainfall is about 1850 mm, and potential evaporation reaches to 1692.3 mm. Mean annual temperature is about 28.4 °C, and the average summer (May) and winter (December) temperatures are 40.9 °C and 7.5 °C, respectively. The region has a typical undulating lateritic landforms and topography, with an altitude ranging from 20 to 85 m m.s.l. The soil is sandy loam with sand, silt, and clay contents of 56%, 24%, and 20%, respectively, and bulk density ranges from 0.82 to 1.43 (g/cm^3) (Table 18.1).

18.2.2 Soil Sample Collection and Preparation

Undisturbed 21 topsoil samples were collected from 3 different places: a set consisted of 7 bare soil samples, 7 grass

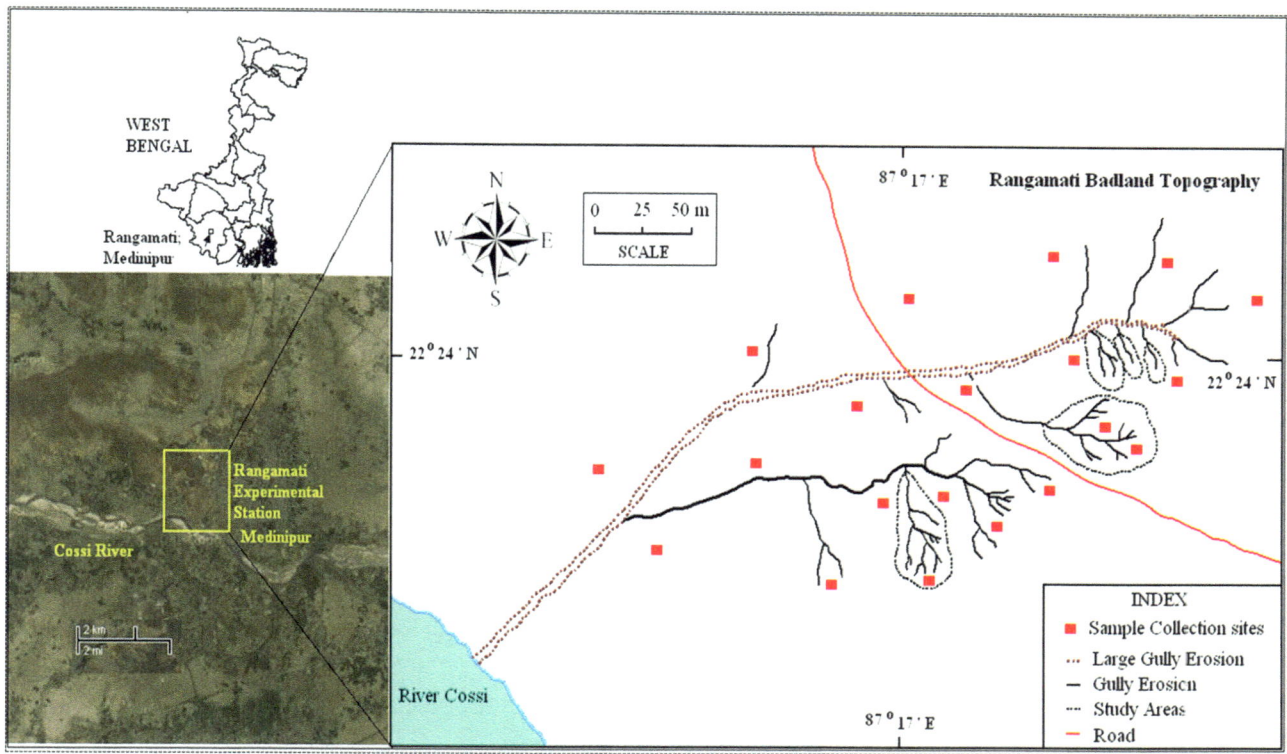

Fig. 18.1   Location of the sample collection sites for flume experiment

Table 18.1   Calculate the soil texture (%) and bulk density of 21 topsoil samples

Soil properties	Sand (0.06–0.002) mm	Silt loam and clay (<0.06 mm)	Bulk density (g/cm^3)
Micro-aggraded sandy loam soil	67.12	32.88	0.82–1.43

samples with low sowing density, and 7 grass samples with high sowing density for the experiments. The mentioned grasses were a mixture of the following species:

23% of *Andropogon aciculatus* (Poaceae), 40% of *Eragrostis cynosuroides* (Poaceae), 17% of *Panicum maximum* (Poaceae), and 20% of *Saccharum munja* (Poaceae)

The sampling of rectangular soil monoliths was done using boxes. The metal boxes (0.35 m long, 0.09 m wide, and 0.08 m deep) were driven into the soil using a hammer. To protect the top of the metal box, a wooden plank was placed on the top during the hammering. After sampling, the above-ground biomass (i.e., stem) was clipped level at the soil surface, and the samples were placed in a container with a constant water level 3.5 cm below the soil surface to allow for slow capillary rise for 12 h, in order to obtain similar soil moisture contents for all samples (Fig. 18.2). In this process, all samples were stored in a normal room temperature (20–25 °C). Twelve hours before the experiments, the soil samples were taken out of the water to drain and to get

prepared for testing their susceptibility to concentrated flow erosion.

18.2.3 Experimental Design

Laboratory experiments on simulated concentrated flow were conducted with a flume similar to standard method of Wang et al. (2011) (Fig. 18.3, length = 1.82 m, width = 0.094 m). The flume is contained an opening at its base, equaling the size of the metal sample box (length = 35 cm, depth = 9 cm, width = 8 cm), so that the soil surface of the sample was at the same level of the flume surface. When inserted into the flume, the surface of the soil sample in the steel box forms a continuum with the bed of the flume. To reduce the edge effects, water losses were prevented at the contact between the soil sample box and the flume by sealing with painter's mastic. Soil surface slope (S), flow discharge (Q), mean bottom flow velocity, water temperature, and sediment concentration were measured. Seven slope percentages were set between 0–5%, 5–10%, 10–15%, 15–20%, 20–25%,

Fig. 18.2 Preparation of hydraulic flume experiment. (**a**) Sample collection and ground biomass (i.e., stem) were clipped on soil surface, (**b**) the samples placed in a container with a constant water level 3.5 cm, and (**c**) samples stored in a normal room temperature (20–25 °C), and (**d**) hydraulic flume was setup

25–30%, and 35%. Flow discharge (Q) was measured 25 times before and after the experiment by collecting the volume of water during 15 s time. Simulated flow discharge ranged between 0.000492 and 0.00064 m^3 s^{-1}. Values of total flow shear stress (τ, Pa) were calculated (Sidorchuk and Sidorchuk 1998) using the following equation:

$$\tau = \rho_w gRS \qquad (18.1)$$

where τ is mean bottom flow shear stress (Pa); ρ_w is water density (kg m^{-3}) with temperature being taken into account; g is acceleration due to gravity (m s^{-2}); S is sin ($\alpha°$), in which α is slope angle of soil surface (°); and R is hydraulic radius (m). Equation 18.2 is used to calculate R value:

$$R = \frac{w \times d}{w + 2d} \qquad (18.2)$$

where w is flume width (i.e., 0.094 m) and d is depth of the water flow in experimental flume (m) according to Eq. (18.3):

$$d = \frac{q}{u} \qquad (18.3)$$

where q is unit flow discharge (m^3 s^{-1}) and u is average flow velocity (m s^{-1}). Flow velocities were measured during the experiment by recording the travel time ($n = 20$) of a dye tracer technique (potassium permanganate solution) over a distance of 1.82 m. Water depths ranged from 0.003 to 0.004 m. For sediment concentration measurement, runoff

Fig. 18.3 (**a**) Illustration of the topsoil area occupied by fine roots (indicated by arrows) for plants with fine-branched root systems, (**b**) setup of hydraulic flume, and (**c**) hydraulic flume used to measure detachment rates from root-permeated topsoil samples

water and detached sediments are collected 10 times during the experiment run (each experimental run lasts 1'50") in 210 buckets at outlet of the flume. The mean soil detachment (MSD) rate (kg s^{-1} m^{-2}) is calculated by multiplying the mean gravimetric sediment concentration of the 10 buckets (kg l^{-1}) with average runoff discharge (l s^{-1}) divided by soil sample surface area (m^2). Since small soil samples were exposed to clear water in this study, the measured mean detachment rate is equal to detachment capacity.

18.2.4 Measurements After Experiment

18.2.4.1 Soil Erosion

The collected sediment was separated from the water by settling during at least 12 h and then by decanting the water. The sediments were oven dried at 105 °C and dry sediment was weighted.

The absolute soil detachment rate (ASD; kg m^{-2} s^{-1}) was calculated (Shit and Maiti 2012) as following Eq. (18.4):

$$\text{ASD} = (\text{SC} \times Q)/A \qquad (18.4)$$

where SC is sediment concentration (kg l^{-1}), Q is flow discharge (l s^{-1}), and A is area of soil sample surface (m^2).

The ASD values were calculated for each runoff sample taken every 15 s during 150 s. Since the ASD varies over time, the mean and standard deviation of ASD were calculated for each sample and used as an indicator of soil erosion susceptibility during concentrated flow (Tables 18.2 and 18.3).

Soil anti-scouribility index (ASI; L/g) was expressed by soil erodibility according to Eq. (18.5):

$$\text{ASI} = \frac{f \times t}{W} \qquad (18.5)$$

where f is the flow rate (in L/min), t is the washing time (min), and W is the weight of the oven-dried sediment (g). The higher soil of ASI has the lower soil erodibility.

18.2.4.2 Measurement of Root Properties

After each experiment, roots were separated from the soil by wet handwashing method (Schuurman and Goedewaagen 1965), and the root samples were washed and sieved using a 0.5 mm sieve, by sprinkling water at low water pressure (De Baets et al. 2006) as seen in Fig. 18.4. Root density properties/indices of each sample were measured manually using flowing criteria:

(a) *Root density (RD) is ratio of root mass and volume of soil sample.*

$$\text{RD} = \frac{M_\text{D}}{V} \qquad (18.6)$$

where M_D is dry living root mass (g) and V is volume of the sample box (cm^3).

(b) *Root length density (RLD)* is the total length of the roots divided by the volume of the root-permeated soil sample (Smit et al. 2000):

$$\text{RLD} = \frac{L_\text{R}}{V} \qquad (18.7)$$

where L_R is length of the living roots (cm).

(c) *Root surface area density (RSAD)* is the total surface of contact of roots with sample soil, which is calculated by the following equation (De Baets et al. 2007a, b):

$$\text{RSAD} = N\frac{D \times \text{RL}}{2} \qquad (18.8)$$

where N is number of roots, D is mean root diameter, and RL is mean root length.

In general, all these analyses were performed using Microsoft Excel (ver. 7.0) and Origin-8 program. Significant differences among treatments for soil detachment rate and aggregate stability were determined using the LSD (least significant difference) procedure for a multiple range test at the 0.05 significance level. The relationships between relative soil detachment rate (*RSD*) and root density (*RD*) were analyzed by a nonlinear regression method. And other significant differences among treatments for anti-scouribility and plant root properties were determined using the LSD (least significant difference) procedure for a multiple range test at the 0.01 significance level. The relationships between anti-scouribility and *RD*, *RLD*, and *RSD* were analyzed by a nonlinear regression method. The regression results were evaluated by the coefficient of determination.

Table 18.2 Roots traits (root density, RD; root length density, RLD; root area ratio, RAR); absolute soil detachment rate (ASD) and relative soil detachment rate (RSD) for topsoil samples with experiment at 25° slope gradient

Resistance against concentration flow erosion (roots indicators)	Bare soils		Scatter grass roots topsoils		Densely grass roots topsoils	
	Mean	SD	Mean	SD	Mean	SD
RD (kg kg/m^{-3})	–	–	2.093	0. 2978	4.36	0.585
ASD (kg m^{-2} s^{-1})	0.00592	0.00312	0.00528	0.00302	0.00363	0.00077
RSD	1	–	0.85636	0.05997	0.61739	0.1021

Table 18.3 Roots traits (root density, *RD*; root length density, *RLD*; root area ratio, *RAR*); absolute soil detachment rate (*ASD*) and relative soil detachment rate (*RSD*) for topsoil samples with experiment at 35° slope gradient

Resistance against concentration flow erosion (roots indicators)	Bare soils		Scatter grass roots topsoils		Densely grass roots topsoils	
	Mean	SD	Mean	SD	Mean	SD
RD (kg kg/m^{-3})	–	–	2.124	0.3183	4.87	0.674
ASD (kg m^{-2} s^{-1})	0.0169	0.0151	0.00823	0.0039	0.0041	0.0017
RSD	1	–	0.57980	0.13088	0.13088	0.1073

Fig. 18.4 Soil and root properties measurement after flume experiment: (**a**) sediment collection after decanting of water; (**b**) sediment collection after experiment; (**c**) after each experiment, roots were separated from the soil by wet handwashing method (Schuurman and Goedewaagen 1965), and the root samples were washed and sieved using a 0.5 mm sieve, by sprinkling water at low water pressure (De Baets et al. 2006); (**d**) soil samples after the flume experiments

18.3   Results and Discussion

18.3.1 Absolute Soil Detachment Rate

Figure 18.5a, b shows the comparison in the Absolute Soil Detachment Rate (*ASD*) during the experiment with scatter grass roots and densely grass roots and bare topsoils. Mean *ASD* decreased exponentially with time due to soil consolidation, which is increasing in soil stability caused by effective stresses in soils (Nearing et al. 2005). Cruse and Larson

(1977) indicated that these stresses increase the soil resistance to splash detachment by raindrops. According to Nearing et al. (2000), soil erodibility also decreases due to increase soil shear strength that resulted from stresses induced by draying. Soil consolidation also involves a decreasing in the void ratio by squeezing of water from pores. The present results showed at 25° of slope, mean absolute soil detachment rate (kg m^{-2} s^{-1}) for bare soil was 0.005915; for scatter grass roots, it was 0.00527125; and for densely grass roots, the value was 0.003627083, respectively (Table 18.2). In another experiment, results with 35° slope

A. Compare the absolute soil detachment rate (ASD, Slope 25 degree)

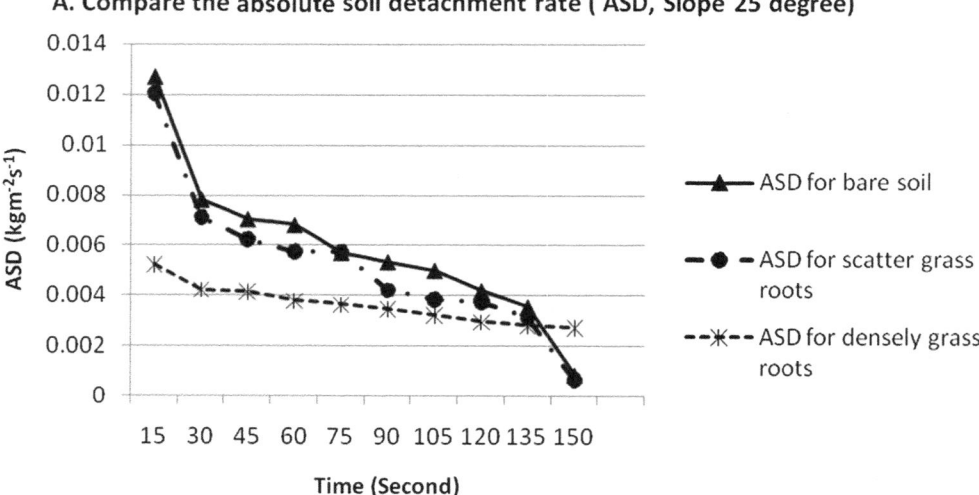

B. Compare the absolute soil detachment rate (ASD, slope 35 degree)

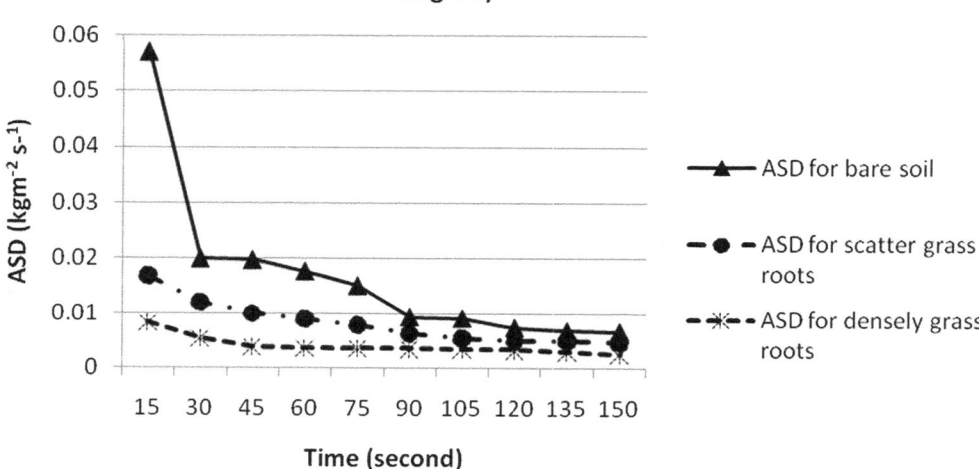

Fig. 18.5 (**a**) Variation in absolute soil detachment rate (ASD, kg m^{-2} s^{-1}) during three experiments with 25° slope angle. (**b**) Variation in absolute soil detachment rate (ASD, kg m^{-2} s^{-1}) during three experiments with 35° slope angle

angle showed that the mean absolute soil detachment rate (kg m^{-2} s^{-1}) for bare soil was 0.016963333, whereas its value was 0.008227917 and 0.004047083 for scatter grass roots and densely grass roots, respectively (Table 18.3). It is clear that soil erosion is decreasing with increasing grass roots density. This variation in ASD of the bare topsoil justifies the use of RSD to compare the effects of different root densities on the soil detachment rates.

18.3.2 Relation Between Root Density (*RD*) and Relative Soil Detachment Rate

The Relative Soil Detachment Rate (RSD) values were calculated in Tables 18.2 and 18.3. A nonlinear regression analysis

showed the decreasing trend of RSD ($r^2 = 0.73067$) with increasing RD (Fig. 18.6). This was probably due to living plant roots that improved soil physical properties and increased aggregate stability and infiltration (Wu et al. 2000; Mamo and Bubenzer 2001a, b), resulting in reductions of runoff and sediment yields. The increase in aggregate stability and infiltration, according to Monroe and Kladivko (1987), could be attributed in relation to the physical reinforcement of aggregates and adhesion of soil particles by roots, which were closely related to the root area in contact with soil particles. Thus, RD is an important indicator of these behaviors. The greater is the mean value of RD, the larger the area of soil in contact with the roots. Concentrated flow erosion is reduced by the presence of roots in the topsoil. This result is also line to Knapen et al. (2007). However, the

Fig. 18.6 Relationship between root density (*RD*) and relative soil detachments (*RSD*)

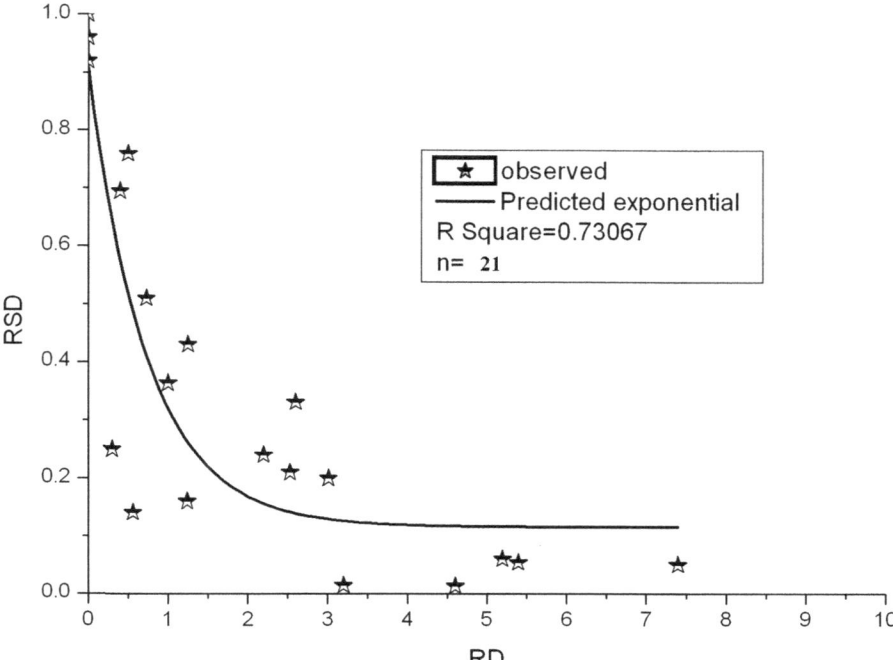

effect of roots was considered in Eq. (18.9). It seemed desirable to take into account the root density in the concentrated flow erosion prediction model. In this study, no crop residue was present in the soil samples, and a significant negative exponential relationship was observed between relative soil detachment rate (RSD) and root density (RD). The value was obtained at the 5% significance level ($R^2 = 0.73067$, $P = <0.05$) (Fig. 18.6). A new equation for relative soil detachment rate (*RSD*) in concentrated flow is developed as follows:

$$Y = a \times e^{(-x/t)} + Y_o$$

$$\mathrm{RSD} = 0.79038 e^{-1.3537\mathrm{RD}} + 0.06458$$
$$\left(R^2 = 0.73067, P < 0.05\right) \tag{18.9}$$

The results of the current study are indicated that soil detachment rate was closely related to root density. Equation (18.9) was convenient and effective for predicting soil detachment rate in lateritic topsoil from Western part of West Bengal.

18.3.3 Effect of Root Properties on Soil Anti-scouribility

The value of the intensified soil anti-scouribility (*AS*; L/g) for describing the effect of roots on soil anti-scouribility was calculated by subtracting the anti-scouribility value (*AS*) for root density (*RD*), root length density (*RLD*), and root surface area density (*RSAD*). Figure 18.6 shows that *AS* increased

with an increasing in *RD*, *RLD*, and *RSAD*, respectively. The relationship among *AS* versus *R*, *RLD*, and RSAD for different parameter is given in Table 18.4 and according to Eq. (18.10). Before the present study, some other similar studies investigated the influence of plant roots on soil anti-scouribility. According to Li et al. and Wu et al. (2000), soil anti-scouribility is a function of the number of roots of less than 1 mm in diameter per unit of soil volume. Meanwhile, Mamo and Bubenzer (2001a, b) established a relationship between soil anti-erodibility and root length density. The calculations in these studies (either root number or root length) have some advantages for understanding the effects of roots on soil anti-scouribility or anti-erodibility. However, there are some disadvantages. For example, root diameters, even those of roots of less than 1 mm in diameter, varied greatly for different vegetations and have the different impact on soil anti-scouribility. So, it will not be precise if the intensified soil anti-scouribility effect of the roots is

Table 18.4 Equation expressing the correlations among *RD*, *RLD*, and *RSDA* with *AS*

Parameter	RD	RLD	RSAD
Y_o	10.07548	10.25888	12.65134
A_1	−3.37055	−8.11837	−2.06437
t_1	−2.48883	−1.93043	−0.15205
A_2	−3.37055	−8.11837	−4.02216
t_2	−2.48912	−1.93043	−3.62438
A_3	−3.37055	−8.11837	−3.5681
t_3	−2.4890	−1.93043	−0.15183
R^2	0.901	0.902	0.981

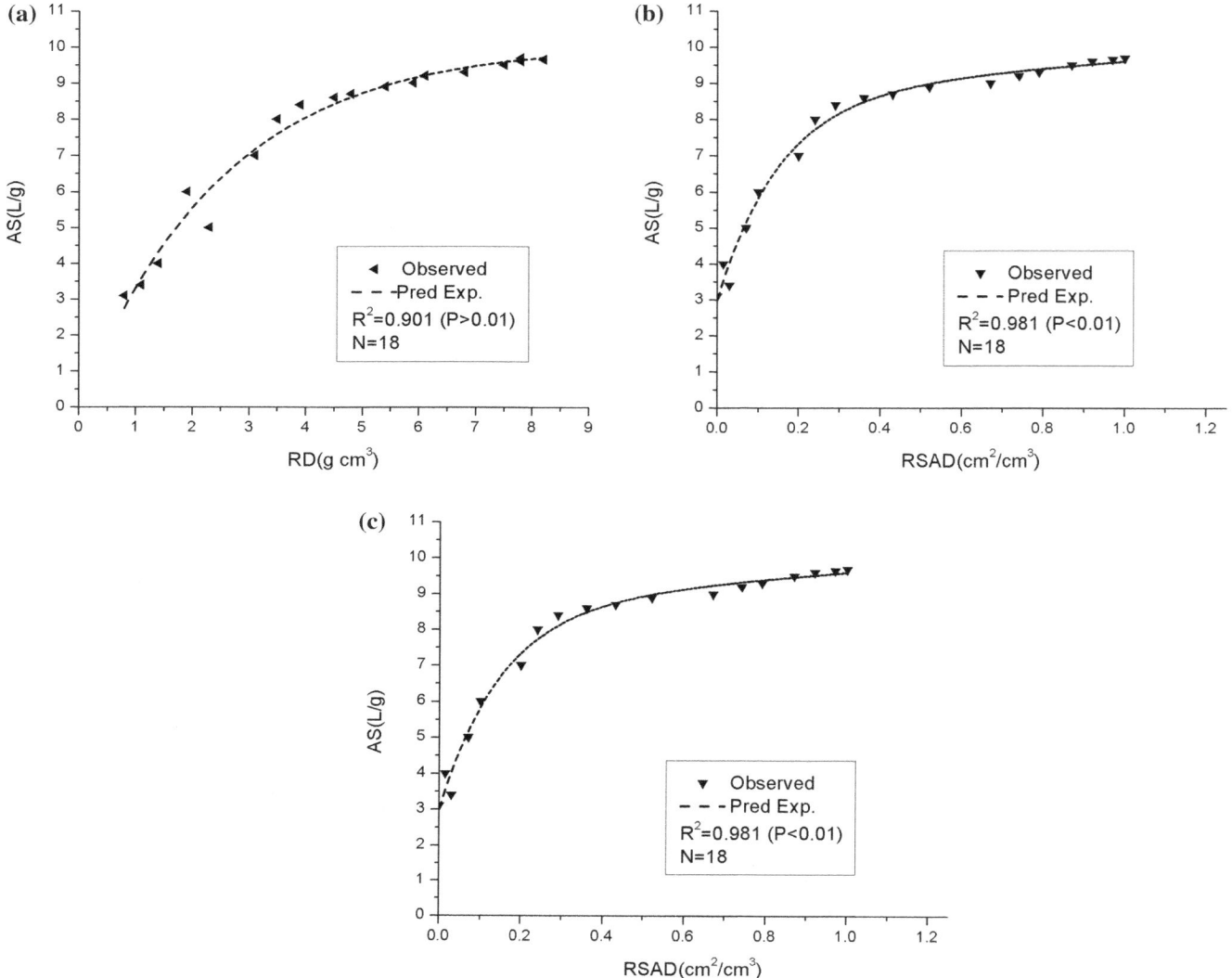

Fig. 18.7 (**a**) Relationships between intensified soil anti-scouribility (AS) and root density (RD), (**b**) relationships between intensified soil anti-scouribility (AS) and root length density (RLD), and (**c**) relationships between intensified soil anti-scouribility (AS) and root surface area density (RSAD)

measured only with the number of roots of less than 1 mm in diameter per unit of soil volume. Meanwhile, some researchers (Ghidey and Alberts 1997) had pointed out that the increase of soil anti-scouribility is due to root anchoring and the adherence of root exudates to soil particles. The roles of both anchoring and adhering are related to the soil-contacting area of the roots. The concept of *RSAD* can correctly describe the soil-contacting area of the roots. A higher RSAD indicates a larger area of contact of between the roots and the soil and more rhizospheres. According to Reid et al. and Habib et al. (1990), soil aggregate stability increases because soil particles are anchoring and cohered by roots in rhizospheres. Therefore, the soil anti-scouribility is increasing. Thus, it is reasonable to use the RSAD to calculate the value of AS through Eq. (18.10) established in the present study. After analysis of the results in Origin 0.8

version platform, the following exponential equation (Eq. 18.10; Fig. 18.7a–c) is set with different values of parameters (Table 18.4) for *RD*, *RLD*, and *RSAD*.

$$\Delta \mathrm{AS} = Y_o + A_1 \exp(x/t_1) + A_2 \exp(x/t_2) + A_3 \exp(x/t_3)$$

$$(18.10)$$

18.4 Conclusion

Roots play an important role in increasing the topsoil resistance against erosion by concentrated flow. Laboratory experiments were conducted to establish a significant relation between relative soil detachment and plant root density by concentrated runoff of undisturbed lateritic topsoils. In this study, a significant negative exponential relationship is obtained between relative soil detachment (RSD) and plant root density (RD) at the 5% level ($R^2 = 0.73067$, $P = 0.05$). The soil anti-scouribility (AS) is increased as RD, RLD, and RSAD have been increased. The effect of plant roots on intensifying soil anti-scouribility could be expressed by an exponential growth function. The AS is increased exponentially with the amplification of RSAD value as well as RD and RLD. The results of the present study provide important and useful information for understanding the effects of the plant roots parameter of vegetation on erosion mechanisms. The relationship among RSAD, RD, and RLD and AS can be used in the erosion models for the prediction of soil erosion.

Acknowledgments The authors thank Mr. Kartic Rishi, Raj Kumar Bhattacharya, Sonom Lama, and Mr. Nitaynanda Sar for their help in the field data collection and experiment. The author (Pravat Kumar Shit) is also thankful to his supervisor Prof. Ramkrishna Maiti (Professor, Vidyasagar University) for continuous guidance and support during the preparation of this manuscript.

References

Abe K, Ziemer RR. 1991. Effect of tree roots on shallow-seated landslides. USDA Forest Service General Technical Report 22 (2):95–108.

Amezketa E. 1999. Soil aggregate stability: a review. *J. Sustain. Agr.* 14: 83–151.

Cruse, R.M., W.E. Larson. 1977. Effect of soil shear strength on soil detachment due to raindrop impact. *Soil Sci. Soc. Am. J.* 41, 777–781

De Baets S, Poesen J, Knapen A, Galindo P. 2007a. Impact of root architecture on the erosion-reducing potential of roots during concentrated flow. *Earth Surf Process Landform* 32:000–000

De Baets S; Poesen J; Kanpen A. 2006. Effects of grass roots the erodibility of topsoil during concentrated flow. *Geomorphology.* 76, 54–67.

De Baets, S.; Poesen, J.; Kanpen A.; Barbera G.G.; Navarro J.A. 2007b. Root characteristics of representative Mediterranean plant species and their erosion-reducing potential during concentrated runoff. *Plant Soil.* 294: 169–183.

Dexter AR. 1991. Amelioration of soil by natural processes. *Soil Tillage Research* 20: 87–100.

Ghidey F., Alberts E.E. 1997. Plant root effects on soil erodibility, splash detachment, Soil strength and aggregate stability. *Trans. Am. Soc. Agr. Eng.* 40: 129–135.

Govers G., Everaert W., Poesen J., Rauws G., De ploey J., Latridou J.P. 1990. A long flume study of the dynamic factors affecting the resistance of a loamy soil to concentrated flow erosion. *Earth Surface Processes and Landforms.* 11: 515–524.

Greenway DR. 1987. Vegetation and slope stability. In Slope Stability, Anderson MG, Richards KS (eds). John Wiley & Sons: Chichester; 187–230.

Gyssels, G. and Poesen, J. 2003. The importance of plant root characteristics in controlling concentrated flow erosion rates. *Earth Surf. Process. Landforms.* 28: 371–384.

Gyssels, G., Poesen, J., Bochet, E. and Li, Y. 2005a. Impact of plant roots on the resistance of soils to erosion by water: A review. *Progress in Physical Geography* 29: 189–217.

Gyssels, G., Poesen, J., Liu, G., Van Dessel, W., Knapen, A., De Baets, S. 2005b. Effects of cereal roots on detachment rates of single- and double-drilled topsoils during concentrated flow. *European Journal of Soil Science* 57(3): 381–391.

Habib L, Morel JL, Guckert A, Plantureux S, Chenu C. 1990. Influence of root exudates on soil aggregation. *Symbiosis* 9: 87–91.

Hartman R, De Boodt M. 1974. The influence of the moisture content, texture and organic-matter on the aggregation of sandy and loamy soils. *Geoderma* 11(1): 53–62.

Knapen A, Poesen J, Govers G, Gyssels G, Nachtergaele J. 2007. Resistance of soils to concentrated flow erosion: a review. *Earth-Science Reviews* 80: 75–109.

Li Y, Zhu X, Tian J. 1991. Effectiveness of plant roots to increase the anti-scourability of soil on the Loess Plateau. *Chinese Science Bulletin* 36: 2077–2082.

Li, Y. and Xu, X. Q. 1992. The intensifying effect of plant roots on the soil anti-scouribility on the Loess Plateau. *Science in China (Series B) (in Chinese).* 3: 254–259.

Mamo M., Bubenzer G.D. 2001a. Detachment rate, soil erodibility and soil strength as influenced by living plant roots: Part I, Laboratory study, *Am. Soc. Agr. Eng.* 44: 1167–1174.

Mamo M., Bubenzer G.D. 2001b. Detachment rate, soil erodibility and soil strength as influenced by living plant roots: Part II, Field study, *Am. Soc. Agr. Eng.* 44: 1175–1181.

Miller, R.M., and Jasirow, J.D. (1990) Hierarchy of root and mycorrhizal fungal interactions with soil aggregation. *Soil Biol. Biochem.* 22: 579–584.

Monroe, C. D. and Kladivko, E. J. 1987. Aggregate stability of a silt loam soil as affected by roots of corn, soybeans, and wheat. Comm. Soil Sci. *Plant Anal.* 18: 1 077–1 087.

Morgan RPC. 1995. Soil Erosion and Conservation (second edition). Longman Group: Harlow.

Nearing M.A., Jettenb V., Baffautc, C., Cerdand, O., Couturierd, A., Hernandeza, M., Bissonnaise, Y. Le., Nicholsa, M.H., Nunes, J.P. Renschlerg, C.S., Souche'reh, V., van Oosti K. (2005). Modeling response of soil erosion and runoff to changes in precipitation and cover. *Catena* 61.131–154

Nearing, M.A., Romkens, M.J.M., Norton, L.D., Stott, D.E., Rhoton, F. E., Laflen, J.M., Flanagan, D.C., Alonso, C.V., Binger, R.L., Dabney, S.M., Doering, O.C., Huang, C.H., McGregor, K.C., Simon, A. 2000. Measurements and models of soil loss rates. *Science* 290, 1300–1301.

Poesen J, Nachtergaele J, Verstraeten G. 2003. Gully erosion and environmental change: Importance and research needs. *Catena,* 50 (2–4): 91–133.

Preston NJ, Crozier MJ. 1999. Resistance to shallow landslide failure through root-derived cohesion in East Coast Hill Country soils, North Island, New Zealand. *Earth Surface Processes and Landforms* 24: 665–675.

Prosser IP, Dietrich WE, Stevenson J. 1995. Flow resistance and sediment transport by concentrated overland flow in a grassland valley. *Geomorphology* 13: 71–86.

Reid BJ, Goss MJ. 1987. Effect of living roots of different plant species on the aggregate stability of two arable soils. *Journal of Soil Science* 32: 521–541.

Reid JB, Goss MJ. 1980. Changes in the aggregate stability of sandy loam effected by growing roots of perennial ryegrass (Lolium perenne). *Journal of the Science of Food and Agriculture* 31:325–328.

Reid JB, Goss MJ. 1981. Effect of living roots of different plant species on the aggregate stability of two arable soils. *European Journal of Soil Science* 32: 521–541.

Reubens B.; Poesen J.; Danjon F., Geudens G.; Muys B. (2007). The role of fine and coarse roots in shallow slope stability and soil erosion control with a focus on root system architecture: a review. Trees. 21:385–402. DOI https://doi.org/10.1007/s00468-007-0132-4

Schuurman, J. J. & Goedewaagen, M. A. J. 1965. Methods for the Examination of Root Systems and Roots. *Wageningen.*

Shit PK, Bhunia G, Maiti R (2013) Assessment of Factors Affecting Ephemeral Gully Development in Badland Topography: a Case Study at Garbheta Badland (Pashchim Medinipur). Int J Geosci 4 (2):461–470. doi:https://doi.org/10.4236/ijg.2013.42043

Shit PK, Bhunia G, Maiti R (2014) Morphology and development of selected Badlands in South Bengal (India). Indian J Geogr Environ 13:161–171

Shit PK, Maiti R. (2012). Effects of plant root density on the erodibility of lateritic topsoil by simulated flume experiment. Int. J. Forest, Soil and Erosion, 2 (3): 137–142

Shit PK, Paira R, Bhunia GS, Maiti R (2015) Modeling of potential gully erosion hazard using geo-spatial technology at Garbheta block, West Bengal in India. Model. Earth Syst. Environ. 1:2, DOI https://doi.org/10.1007/s40808-015-0001-x

Sidle RC, Pearce AJ, O'Loughlin CL. 1985. Hillslope Stability and Land Use. American Geophysical Union: Washington, DC.

Sidorchuk A, Grigorev V. 1998. Soil erosion on the Yamal Peninsula (Russian Arctic) due to gas field exploitation. *Advances in Geo-Ecology* 31: 305–811.

Sidorchuk A, Sidorchuk A. 1998. Model for estimating gully morphology. Modelling Soil Erosion, Sediment Transport and Closely Related Hydrological Processes (Proceedings of a symposium held at Vienna, July 1998). IAHS Publ. no. 249, 1998.

Smit, A.L.; Bengough, A.G., Engels, C., van Noordwijk, M., Pellerin, S., van de Deijin, S.C. 2000. Root Methods, a Handbook. Springer-Verlag, Heidelberg

Tengbeh GT. 1993. The effect of grass roots on shear strength variations with moisture content. *Soil Technology* 6: 387–295.

Waldron, L. J. 1977. The shear resistance of root-permeated homogeneous and stratified soil. *Soil Sci. Soc. Am. J.* 41: 843–849.

Wang J. G., Li Z, Cai C, Yang W, Ma R. and Zhang G. 2011. Predicting physical equations of soil detachment by simulated concentrated flow in Ultisols (subtropical China). *Earth Surf. Process. Landforms.* (wileyonlinelibrary.com) DOI: https://doi.org/10.1002/esp.3195.

Wu, W. D., Zheng, S. Z. and Lu, Z. H. 2000. Effect of plant roots on penetrability and anti-scouribility of red soil derived from granite. *Pedosphere.* 10(2): 183–188.

Zhou, Z. C. and Shangguan, Z. P. 2005. Soil anti-scouribility enhanced by plant roots. *J. Integrative Plant Biol.* 47(6): 676–682.

Zuazo, V. H. D., Pleguezuelo C. R.R. 2008. Soil-erosion and runoff prevention by plant covers - A review. *Agronomy for Sustainable Development* 28(1): 65–86

Pravat Kumar Shit received his PhD in Geography (Applied Geomorphology) from Vidyasagar University (India) in 2013, MSc in Geography and Environment Management from Vidyasagar University in 2005, and PG Diploma in Remote Sensing and GIS from Sambalpur University in 2015. He is Assistant Professor in the Department of Geography, Raja N. L. Khan Women's College (Autonomous), Gope Palace, Midnapore, West Bengal, India. His main fields of research are soil erosion spatial modeling, badland geomorphology, gully morphology, water resources and natural resources mapping, and modeling. He has published more than 45 international and national research articles in various renowned journals; also, he has published three books. His research work has been funded by the University Grants Commission (UGC), India and Higher Education Science and Technology and Biotechnology, Government of West Bengal. He is Associate Editor and on the editorial boards of three international journals in geography and earth environmental sciences.

Hamid Reza Pourghasemi is an Associate Professor of Watershed Management Engineering in the College of Agriculture, Shiraz University, Iran. He has a BSc in Watershed Management Engineering of the University of Gorgan (2004), Iran, an MSc in Watershed Management Engineering, from Tarbiat Modares University, Iran (2008), and a PhD in Watershed Management Engineering from the same University (Feb 2014). His main research interests are GIS-based spatial modelling using machine learning/data mining techniques in different fields such landslide, flood, gully erosion, forest fire, land subsidence, species distribution modelling, and groundwater/hydrology. Also, Hamid Reza works on Multi-Criteria Decision Making methods in Natural Resources and Environment.He has published more of 90 peer reviewed papers in high-quality journals, with three chapters in Springer. Also, he published two books in Springer (https://www.springer.com/gp/book/9783319733821) and Elsevier (https://www.elsevier.com/books/spatial-modeling-in-gis-and-r-for-earth-and-environmental-science/pourghasemi/978-0-12-815226-3).

Gouri Sankar Bhunia received his PhD from the University of Calcutta, India, in 2015. His PhD dissertation work focused on environmental control measures of infectious disease (visceral leishmaniasis or kala-azar) using geospatial technology. His research interests include kala-azar disease transmission modeling, environmental modeling, risk assessment, data mining, and information retrieval using geospatial technology. He is Associate Editor and on the editorial boards of three international journal in Health GIS and Geosciences. He worked as a 'Resource Scientist' in Bihar Remote Sensing Application Centre, Patna (Bihar, India). He is the recipient of the Senior Research Fellow (SRF) from Rajendra Memorial Research Institute of Medical Sciences (ICMR, India) and has contributed to multiple research programs kala-azar disease transmission modeling, development of customized GIS software for kala-azar 'risk' and 'non-risk' area, and entomological study.

Bamboo-Based Technology for Resource Conservation and Management of Gullied Lands in Central India

19

S. Kala, A. K. Singh, B. K. Rao, H. R. Meena, I. Rashmi, and R. K. Singh

Abstract

This vast tract of existing gully lands is posing potential threat to nearby productive lands because of overexploitation and poor management. Situation-specific cost-effective viable technologies for reclamation and productive utilization of gullied lands are highly essential. The study designed and evaluated bamboo-based technology for resource conservation and protective utilization of gullied lands for the analysis of hydrological behaviour, growth and economic analysis. Among three treatments, the growth performance of bamboo plants showed maximum average culm height and culm collar diameter of 11.76 m and 42.11 mm. The average crown size and number of culms per clump were recorded to be 7.27 m and 29.60 numbers, respectively, at Manikpura village watershed. Technology of planting bamboo (*D. strictus*) with suitable moisture conservation practice proved as a viable alternative on ravines for gully beds stabilization, control sloping land erosion through good soil-binding effect and fast-growing vegetative cover. Hydrological results revealed that runoff was reduced from 9.6 to 1.8% and soil loss from 4.2 to 0.6 t/ha/year in the last 4 years. The economic analysis suggested a cash outflow of Rs. 48,000 ha^{-1} from seventh year onwards to the stakeholders in the region. The study revealed that cultivation of bamboo in gullied lands of ravine area has the potential for good earning to the resource-poor farmers and improving livelihood.

Keywords

Bamboo · Clumps · Gully beds · Livelihood improvement · Resource conservation · Soil erosion · Runoff · Vegetative cover

S. Kala (✉) · A. K. Singh · H. R. Meena · I. Rashmi · R. K. Singh
ICAR—Indian Institute of Soil and Water Conservation, Research Centre, Kota, Rajasthan, India

B. K. Rao
Water and Land Management Training and Research Institute, Hyderabad, Telangana, India

19.1 Introduction

Degraded non-arable lands cover vast tract in the country and are a major source for supplying fuel, fodder and timber and other minor products to the population. Gully and ravine lands are highly degraded lands and it has several constraints for vegetation growth due to vicarious climate, poor soil fertility, low soil moisture, extreme variation in temperature and heavy biotic pressure. Gully erosion is an advance stage of water-induced sheet and rill erosion which eventually leads to extreme physical landscape deformation. A ravine system demonstrates all forms of water-induced erosion. There are surface and sub-surface erosion processes which eventually lead to gully formation.

Ravine or gully erosion-infested landscapes are also known as *lavaba* (French), *vossoroca* or *bocoroea* (Brazilian) and *arroyo* (Central Russia and Southwestern America). In Indian context, the word ravine refers to a network of gullies which are generally spread along any river system. The gully is an erosion channel developed by ephemeral streams with steep banks and a nearly vertical deep gully head enough to create hindrance to the normal tillage operations. Usually, these are deeper than 0.3 m. The rate of gully erosion depends on the runoff producing characteristics of the watershed which are governed by the size and shape of drainage area, soil characteristics, the alignment, size, shape and the gradient of the gully channel. Severe terrain deformation and land degradation have occurred along some of the major river systems of India. The Yamuna-Chambal Zone along the interstate boundaries of Rajasthan, Uttar Pradesh and Madhya Pradesh is a spectacular and extreme example of water-induced soil erosion. The gullies are widening and ravines are extending at the rate of 8–9 m per annum with average soil loss of more than 17 t/ha/year (Singh et al. 2014). The hot, semi-arid ravine tracts of the Indo-Gangetic region pose very harsh climatic conditions like low and erratic rainfall and high summer temperature. This vast tract of existing ravine lands in northwestern part of

© Springer Nature Switzerland AG 2020
P. K. Shit et al. (eds.), *Gully Erosion Studies from India and Surrounding Regions*, Advances in Science, Technology & Innovation,
https://doi.org/10.1007/978-3-030-23243-6_19

India is posing potential threat to nearby productive lands because of overexploitation and poor management. Therefore, there is an urgent need to detain these problems and protect both the arable and non-arable land from further degradation. Situation-specific cost-effective viable technologies and dissemination of suitable technologies for reclamation and productive utilization of ravine lands in India are highly essential for arresting gully extension, reclamation of gullied and ravine lands and improving production potential.

The ravine regions are characterized with very scarce natural vegetation largely due to much eroded soils having low fertility status as well as very unsuitable climatic and topographic conditions for most of the economical tree/grass species. Champion and Seth classified the natural vegetation of ravines under type 6 BC2: Northern tropical ravine thorn forests. The climate of the ravine region in Rajasthan, Madhya Pradesh, Uttar Pradesh and Gujarat is semi-arid to sub-humid (agro-climatic zone IV) with annual average rainfall of 600–800 mm, mostly received from July to October in intense storms. The temperatures may vary from 3 to 47 °C during coldest to hottest months. Humidity may be as low as 7% and evaporation very high during May and June (Bhushan and Saxena 1984; Singh et al. 1972, 1976; Prajapati et al. 1977; Prakash and Rao 1986). Rao et al. (2013) reported that stream bank erosion is a major cause of land degradation, leading to deteriorated drainage systems and causing loss of topsoil and/or terrain deformation. In Uttar Pradesh, ravine lands are more prominent along the banks of Yamuna river and its tributaries. In the ravine area along the banks of river Yamuna and its tributaries, soils are excessively drained, coarse-loamy/fine loamy (calcareous/non-calcareous), sandy loam, poor in water-holding capacity and rich in kanker granules with hard sub-soil. The bulk density of sub-soil is high ranging from 1.60 to 1.72 (Anon 2011). Ravine lands of India have fluvial systems which are more sensitive to rainfall and important indicator of climate change. The yield and water use efficiency of crops are severely affected by low moisture content in soil profile. At many places, these ravines are the abode of many unsocial elements. The ravine control and reclamation of this area is one of the major problems of the state.

Bamboo has evergreen leaves, dense canopy and numeral culms, which can help to intercept considerable amount of rainfall. Falling raindrops change their direction and ways, and reduce velocity, and therefore decrease the direct soil erosion after multiple interceptions by tens of shoot layers and larger amount of culms. They respond to the variability in precipitation by alternating their plan form, channel geometry and sedimentation pattern which in turn influence floods and associated landscape evolutions (Rao et al. 2012a, b). Several scholars (Simon and Collison 2002; Nath and Krishnamurthy

2008; Kurothe et al. 2012) have clearly established the importance of bamboo plant as effective means resource conservation. Impacts of bamboo planting include raised groundwater level, increased land productivity, improved micro-climate and improved socio-economic conditions. In comparison to deciduous and coniferous plantations, bamboo plants are hydrologically best-suited plantation in degraded ravine lands (Sharda et al. 1982). Bahadur et al. (1980) reported that improved farm practices and land treatment by bamboo, grasses and legumes on bunds, sloppy lands and terraced lands have been found to be effective means of soil and water conservation. Singh et al. (2015a, b) reported that plantation of *D. strictus* was quite successful in increasing the economic productivity of gullied beds. Pande et al. (2012) worked out the economics of bamboo plantation in ravines and suggested policy measures to development and finance institutions for large greening of ravines in the country. A lot of studies showed that most of the rhizomes and roots, around 80%, were present in the upper 0–30 cm soil layer that is the area where roots and rhizomes serve best in controlling soil erosion. The extensive underground root and rhizome system has a significant capacity to bind the topsoil. A study estimated that a single bamboo plant can bind up to 6 m^3 of soil (Anonymous 1997). Because of this, it is perfect for arresting the ravages of water erosion in areas prone to it (such as slopes and lowlands). Bamboo planted along stream and river banks grows particularly well because of a more even and abundant supply of moisture. The fibrous mass of roots binds the soft banks, and the thick culms arrest strong currents during flood periods. The soil behind the revetment is reinforced further by plantings of bamboo, thus building a solid wall of living plant material on the banks of the river, and solved the problem forever.

19.2 Materials and Methods

Bamboo is known to be one of the fastest-growing plants in the world, with a growth rate ranging from 30 to 100 cm per day in growing season. It can grow to a height of 36 m with a diameter of 1–30 cm. Considering the above characteristics, it is easy to understand that bamboo is the fastest-growing and highest-yielding plant in the world. National Bamboo Mission, Ministry of Agriculture, GOI, has clearly emphasized the conservation value of bamboo plantations. Ravines are the network of gullies running parallel and discharging into river. Very extensive degradation of land has occurred along some of the major river systems of the country. For identifying relative land suitability for bamboo, ravine lands can be classified into the following two major groups such as (1) gully beds (2) gully banks and gully heads. The Research Project on Hydrologic and Economic

Evaluation of Bamboo Plantations in Gullied Lands under Major Ravine Systems of India started at three research centres in 2008–2009 and completed in 2013–2014.

Therefore, a study was initiated in 2008–2009 to evaluate the effectiveness of bamboo in conserving soil and water resources and its supportive role in increasing the life and efficiency of small earthen structures and peripheral bunds. As part of the programme, implemented at the Yamuna ravines at Agra (UP), various treatment combinations were imposed to generate relevant information on conservation value of bamboo. Four small catchments of 3–5 ha size were selected, and four treatments such as plantation of rows of bamboo in the upstream and downstream side of small earthen gully plugs (T_1), plantation of bamboo in staggered manner in the gully bed (T_2), two rows of bamboo planted in staggered manner as vegetative barrier (T_3) and a control catchment having no intervention (C) were imposed. One ravenous watershed of 9.8 ha at Manikpura village (26° 49′–26° 51′ N, 77° 32′ 30–77° 35′ 30″ E and 168 m msl), block Pinhat, Tehsil Bah, district Agra, UP, India, by ICAR—Indian Institute of Soil and Water Conservation (IISWC), Research Centre, Agra (Table 19.1). This watershed is a part of Utangan river, which is a tributary of Yamuna river system (Fig. 19.1). Plantation in all the three treatments has already been completed in 2009–2010. Three treatments were imposed for studying the effectiveness of bamboo in stabilizing gully head extension (Table 19.2).

Before planting the bamboo in gullied and ravine lands, the bushes should be removed. In all the interventions, 45 × 45 × 45 cm size of pits has been made uniformly. One-year-old bamboo seedlings having a height range of 50–60 cm were selected from nursery or should be planted with the ball of the earth, 4–6 cm below the ground level at the centre of the pit or trench after filling the excavated soil with 1 or 3 kg of FYM and then topsoil firmly compacted to prevent evaporation of soil moisture. Planting should be done in the early part of the rainy season, but after the soil has become sufficiently moist. Particularly for rhizome planting, special care will be taken for life-saving watering after planting. During dry periods, incidence of termite attack increases resulting in higher mortality and reduced growth of the planted seedlings and young plants. Use of insecticides powers like endosulfan or Folidol or solution of termicide (chlorpyrifos) may be applied as in the soil mixture during filling of pits shall help in reducing termite incidence. Periodic drenching with the solution of these insecticides helps in reducing the recurrence of termite attack. Where termite infestation is high, 2.5 L of solution may be applied to the soil near the planted saplings (Fig. 19.2).

The growth period of bamboo is 2–3 months after rains, during which time they attain their full height and diameter. The development of lateral braches takes place during the second season of growth. After rainy season, 8–12 life-saving irrigation of 10–12 L per plant is provided for survival of bamboo in the initial 2 years (Figs. 19.3 and 19.4). During the first year, frequent weeding should be done for improving survival and growth of plants. Special care needs to be taken to ensure that plants do not suffer prolonged drought during first year of planting. Manuring is essential for increased productivity. Apart from fertilizer application at the time of planting, one more dose after one month is needed. The doses need to be repeated 5–6 months after end of monsoon or during shoot bud initiation period. Soil working before emergence of new culms helps in improving the emergence of new culms. Organic fertilizers should be applied during winter or dry period. The bamboo rhizomes need loose and well-aerated soil for good growth and production of new culms. Since rain is a major source of water in the region, every drop of it must be conserved in the soil through in situ rainwater conservation. All treatment implemented and control microravenous watersheds having a gauging structure with triangular weir were constructed at the outlet of watershed to collect data on runoff and soil loss of every runoff producing rainfall event. Hence, an experiment was undertaken to evaluate bamboo-based resource conservation techniques in the Yamuna ravines by IISWC, Research Centre. The study is based on both primary and secondary data. The secondary data were supported with the primary data being collected from fields in a research project being carried out at the Research Centre, Agra (UP). Four years (2010–2013) data of rainfall, runoff, soil loss and bamboo plant growth have been collected from the gauged ravenous micro-watershed

Table 19.1 Data of the catchments selected for the study

Location of experiment	Treatments	Latitude and Longitude	pH	EC (milli mhos/cm)	Infiltration rate (cm/ha)	Organic carbon (%)
Research Farm, Chhalesar, Agra (UP)	T_1	27:12:55 N 78:05:11 E	8.38	0.152	6.81	1.034
	T_2	do	8.38	0.182	4.28	1.067
	T_3	do	8.47	0.140	7.16	0.606
	Control	do	8.64	0.160	1.86	0.606
Manikpura, Pinhat, Agra (UP)	Gullied land at Manikpura	26:56:16 N 78:24:15 E	7.86	0.222	2.24	0.806

Soil properties represent soils of gully bed

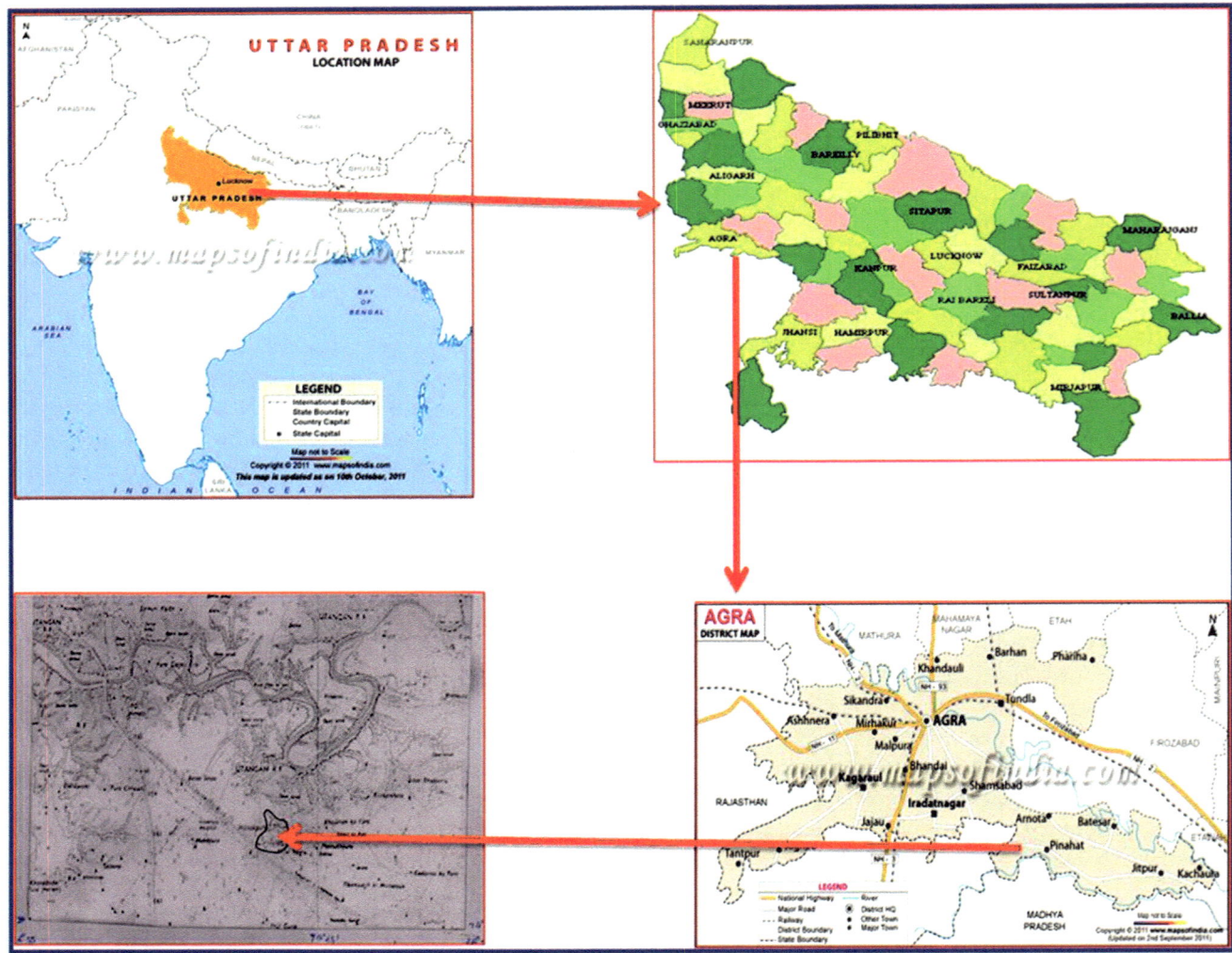

Fig. 19.1 Location map of the project site at Manikpura–Agra (UP) India

Table 19.2 Details about the treatments selected for the study at project site

Particulars	T_1	T_2	T_3	Control	Manikpura (UP) India
Area (ha.)	6.8	4.4	2.8	3.3	9.8
Seedlings planted	315	650	1045	0.00	1050
Distance (m)	4×4	4×4	2×2	0.00	4×4
Weeding (no.)	3	8	7	0.00	7
Urea (gms/plants)	20	20	20	0.0	20
Irrigation (No.)	2	7	7	0.00	7
No. of plants selected for data collection	100	100	100	0.00	100

and analysed. Daily rainfall charts from recording type rain gauge have been collected from the meteorology observatory, Agra (UP). Runoff charts of stage level recorders from runoff producing storm were analysed for runoff calculation. Runoff water samples (500 mL) were collected manually for calculation of soil loss. Basic plant growth (average culm height, culm diameter, crown length and no. of culm/clump) parameters were also recorded in all treatments.

19.3 Results and Discussion

Survival of bamboo plants from various treatments in ravines varies from 50 to 80%. Based on field data, it is observed that the highest survival (up to 80%) was recorded in gully beds treated with bamboo plantation with supportive staggered trenches followed by bamboo plantation with earthen check

Fig. 19.2 View of degraded
Yamuna Ravine before planting

Fig. 19.3 Transportation of
bamboo seedling

dams. The higher survival in these treatments is mainly due to increased moisture availability. The months of June–July–August are the main period of rainy season which helps continuous vegetative activity which indicates higher growth of bamboo plants in the field. The results of three of experimentation to study the growth performance and efficiency of bamboo in conserving runoff and soil loss and improvement in soil health in comparison to control plots have been observed and explained below.

19.3.1 Growth Performance of Bamboo Plantation

Table 19.3 represented the bamboo growth performance under different treatments, and the maximum value of Av. culm height and Av. culm collar diameter (mm) has been observed in T_2 treatment (3.80 m, 22.50 mm) followed by T_1 (2.05 m, 13.36 mm) at our research farm. The value of Av. crown size and no. of culms per clump were recorded maximum at T_2 with value of 3.93 m and 18.02, respectively.

Fig. 19.4 Initial stage life-saving irrigation to seedlings

In Manikpura village (Table 19.3), culm height increased twofold from 5.12 to 11.76 m in 4 years (Table 19.3; Figs. 19.5, 19.6 and 19.7). Similarly, average culm collar diameter and average crown size were 42.11 mm and 7.27 m, respectively, which showed more than 200% growth in 4 years. The average numbers of culms/clump were 29.60. The excellent growth of bamboo plants, viz. Av. culm height, Av. culm collar diameter, average crown size and maximum no. of culms per clump, was observed in the study area (Figs. 19.8 and 19.9). The support of moisture conservation techniques and high in situ moisture availability directly augment the overall growth performance of bamboo plants under ravines region over the years. Better growth of bamboos may be attributed to decreased soil loss, which might have brought changes in moisture and silt retention owing to the growth of a dense vegetative cover in the gully beds. Apart from that, bamboo acts as a good soil binder owing to its dry and hardy nature, peculiar dense clump formation and extensive interlocking fibrous root systems and natural capacity to regenerate through its rhizomes which play an important role in preventing erosion and increasing water-holding capacity and nutrient cycling under gully beds. Similar results on use of bamboo plantation for reclaiming Mahi and Yamuna ravine area were also reported by Rao et al. (2012b). Pande et al. (2012), Singh et al. (2015a, b) and Yadav also reported that higher level of moisture content and soil accretion remain under bamboo (*D. strictus*) plantation than other species.

19.3.2 Hydrological Performance of Bamboo Plantation

The soil physical and chemical properties were considerably influenced by bamboo plants under treatment imposed ravines lands. The soil studies revealed that low soil pH (8.63) and high soil organic carbon content of (0.86%) with respect to initial values (0.62%), reducing soil loss pattern, were observed from Manikpura ravine watershed. Four-year (2010–2013) runoff and soil loss data from the Manikpura ravine watershed revealed that average seasonal rainfall in the study site was 427 mm (Table 19.4). Runoff reduced from 9.65% in the first year to 1.81% in the fourth year due to bamboo plantation in ravine bed. Soil loss over the 4 years also comes down from 4.27 to 0.60 t/ha/year. The results are in conformity with findings of Rao et al. (2012b) who reported that runoff and sediment behaviour under the bamboo plantation-based interventions absorbs more than 80% of rainfall. Due to influence of vegetation on soil, permeability of the soil was increased resulting in reduced surface runoff, soil loss, evaporation and better water penetration into soil as well as increased drainage capacity of soil.

19.3.3 Yield and Benefits

The analysis carried out using data from the Yamuna ravine system suggests a cash outflow ranging from Rs. 33,550 to

Table 19.3 Mean value of growth parameters of bamboo plants at research farm and Manikpura watershed in Agra district (U.P)—India

Treatments	Av. Culm height (m)	Av. Culm Col. Dia. (mm)	Av. Crown size (m)	No. of Culms/clump
T_1: Earthen gully plugs	2.05	13.36	2.31	16.01
T_2: Trenches	3.80	22.50	3.93	18.02
T_3: Live check dams	2.35	9.80	2.73	10.95
Manikpura watershed	11.76	42.11	7.27	29.60

Fig. 19.5 Bamboo plant after 1-year-old plantation

Fig. 19.6 Close planting of bamboo as a vegetative barrier

48,000 ha^{-1} from seventh year onwards to individual stakeholders in the region, in addition to the benefits accrued to society at large in terms of enhanced soil health (Table 19.5). Harvesting commences from seventh year onwards. The sale price per piece of bamboo is considered at Rs. 35–40. Hopefully, bamboo planted in ravines for purpose to meet twin benefits like soil conservation and economic remunerations. So considering the above objective,

Fig. 19.7 Growth of bamboo plant after 3–4 years

Fig. 19.8 Complete gully stabilization by bamboo

it could be possible if we adopt 10% harvest matured culms/clump (i.e. two matured culms/clump) from seventh year onwards to protect the ravine ecosystem on sustainable way (Figs. 19.10 and 19.11).

In this kind of degraded ravine lands, bamboo acts as a good soil binder owing to its dry hardy nature, peculiar dense clump formation and extensive interlocking fibrous root system and natural capacity to regenerate through its rhizomes which play an important role in preventing erosion and increasing water-holding capacity and nutrient cycling under gulley beds. This result proved that bamboo is an ideal tropical hardy bamboo species for massive rehabilitation of degraded ravine lands. This moisture conservation technique is substantially quite effective in reducing runoff volume and soil loss. The better in situ rain water harvesting and moisture conservation techniques helps bamboo plants in sustaining their growth during hot and drought period. The highest average culm height and culm diameter and no. of culms/clump were recorded in T_1 which may be attributed to high in situ moisture availability in sub-soil (Fig. 19.12). Using bamboo-based conservation technology is quite promising in biomass productivity and enhances the in situ moisture use efficiency. Through this finding, we could reclaim major ravine land ecosystem using bamboo

Fig. 19.9 Measurement of growth parameters

Table 19.4 Runoff and soil loss under bamboo plantation

Year	Seasonal rainfall (mm)	Runoff from bamboo-planted area (mm)	% runoff	Soil loss (t/ha/year)
2010	456	44.0	9.65	4.27
2011	226	6.04	2.67	0.66
2012	531	14.50	2.73	0.78
2013	494	8.96	1.81	0.60

Table 19.5 Expected Yield and Income of bamboo plantations in different ravine systems

Year	Yield (Poles No ha^{-1}) in Yamuna	Net income (Rs. ha^{-1}) in Yamuna
VII	1200	39,550
VIII to X	1200	34,000
XI year on wards	1600	48,000

Source: Singh et al. (2015a, b)

Fig. 19.10 Reclaimed Yamuna Ravine through bamboo plants

Fig. 19.11 View of Yamuna
Ravine after the project

Fig. 19.12 Growth and development of bamboo plantation under different stages

along with recommended moisture conservation measures. It has been identified as potential species for efficient resource utilization in non-arable lands, vulnerability analysis of livelihood support systems, etc. This measure is highly applicable to light-textured well-drained soils including reclaimed ravine land where moisture and nutrient resources are scarce.

Fig. 19.13 Popularization of bamboo technology through various HRD programmes During 2012–2015 at Manikpura (UP), India

The technology can suitably be adopted on the non-cultivable wastelands distributed in the northwestern part of India and in other areas having similar agro-climatic and soil conditions.

Bamboo harvest cycle continues for a long time in ravines if a recommended practice of harvesting one-third culms per clump is followed. The soil carbon build-up would enhance with the age of plantation due to litter fall. Based on the evidence in ravines, this conservative value of enhanced soil carbon adds up to a net present value of benefits to the economy. Cost of bamboo plantation can be a constraint to small and medium local stakeholders located in the vicinity of the ravine lands. Initial high cost of establishment can be met through subsidies and banks' financial inclusion programme for such stakeholders (Fig. 19.13). Government's

waste land development programme can be an appropriate mechanism to address this problem. The substantial amount of public funding can also be routed through appropriate budgetary provisions in the development plans of corporate entities involved in the rural development in the country.

The intangible benefits or ecosystem benefits like carbon sequestration and prevention of soil erosion are the added benefits. Protection from biotic interference during the first 2 years especially from wild pig and porcupine, which damage the roots and browsing animals, is important. Measures for soil moisture management may improve the survival and growth of bamboo in ravines. The gully head and gully stabilization measures with bamboo interventions have been found useful for the ravine areas. A good plantation can be established with the help of local untrained manpower who can be taught the necessary skills. Due to the size of the plantation, it may be established as part of a coordinated regional or local bamboo development venture including bamboo processing units that the plantation could supply. In this case, it may be preferable to establish it with the assistance of NGOs or state agencies to ensure the proper infrastructural facilities and linkages are in place. Apart from this, very good awareness creation activities like meetings and exposure visit to the plantation site were also arranged for various stakeholders from different sectors during project period.

19.4   Conclusion

The study has revealed that planting of bamboo with suitable moisture conservation practice proved as a viable alternative on ravine lands for gully beds stabilization, control sloping land erosion through good soil-binding effect and fast-growing vegetative cover. The technology reduces soil erosion, runoff, soil pH and nutrient losses considerably by imposition of trenches, basins and vegetative barriers due to improvement in infiltration rates that results in higher water use efficiency. The technology requires fairly less initial investment. The returns from bamboo culms are also accrued after a gestation period of 7 years. Therefore, this technology would be more suitable to medium and large farmers who have fairly large holdings, greater capacity of capital investment and better access to credit facilities. This technology is applicable to light-textured well-drained soils including reclaimed ravine land where moisture and nutrient resources are scarce. The technology can suitably be adopted on the non-arable wastelands and gullies of Uttar Pradesh, Madhya Pradesh, Rajasthan and Gujarat states. The techniques are very much useful to various stake holders and officials of forest and agriculture department and other user agencies for improving livelihood through reclamation and productive utilization of ravine lands. Therefore, the use of bamboo in watershed management, soil and water conservation and rehabilitation of degraded land could be possible, and bamboo farming and material processing are well suited to this region due to twin concerns of livelihood enhancement and environmental protection—the key components for developing these resource-poor lands. Hence, an effective organized massive cultivation of this species through community participation could be helpful to increase the green cover and establish industries that provide sustainable livelihood development to the local community.

References

A.K. Singh, S. Kala, S.K. Dubey, B. Krishna Rao, M.L. Gaur, K.P. Mahopatra and Prasad B (2014): Evaluation of bamboo based conservation measures for rehabilitation of degraded Yamuna ravines. Indian Journal Soil Conservation, Vol. 42, No. 1, Pp. 80–84.

Anon. (2011). Annual Report. Central Soil and Water Conservation Research and Training Institute, Dehradun.

Anonymous. 1997. Healing degraded land [J]. INBAR Magazine, 5(3): 40–45.

Bahadur Prakash, Satish Chandra and Gupta, DK (1980). Hydrological studies on experimental basins in the Himalayan region, The influence of man on the hydrological regime with special reference to representative and experimental basin. Symposium-IAHS-AISH Publ. No. 130.

Bhushan LS, Saxena SC (1984). Rainfall erosion index for Agra, Indian Journal of Soil Conservation, 12 (2 & 3): 24–29.

Kurothe, RS, Gaur, ML, Rao, BK, Parandiyal, AK, and Singh, AK (2012). Conservation and Production Potentials of Bamboo in Ravine Lands, CSWCRTI, Dehradun, ISBN 978-81-924172-1-9: 160p

Nath, S and Krishnamurthy, R (2008). Nutrient cycling in plantation stands under laterite soils of South West Bengal, India, Proc. International Conference on Improvement of Bamboo Productivity & Marketing for Sustainable livelihood, 15th -17th April, 2008, New Delhi.

Pande VC, Kurothe, RS., Rao, BK, Kumar, Gopal, Parandiyal, AK, Singh, A. and Ashok Kumar (2012). Economic Analysis of Bamboo Plantation in Three Major Ravine Systems of India. Agricultural Economics Research Review, 25 (1): 63–73.

Prajapati MC, Agarwal MC, Bhaskar KS (1977). Rainfall features and agricultural droughts at Agra. Annals of Arid Zone Research, 16(2): 176–184.

Prakash C, Rao DH (1986) Frequency analysis of rainfall data for crop planning – Kota. Indian Journal of Soil Conservation, 14(2): 23–26.

Rao BK, Gaur ML, Kumar G, Kurothe RS and Tiwari SP (2013). Morphological characterization and alterations in cross section of different order streams of Mahi Ravines, Indian J. Soil Conservation, 41(1): 20–24.

Rao, BK, Kurothe, RS, Pande, VC and Kumar, Gopal (2012a). Throughfall and stem flow measurement in bamboo (Dendrocalmus strictus) plantation, Indian Journal of Soil Conservation. 40 (1): 60–64.

Rao BK, Kurothe, RS, Singh, AK, Parandiyal, AK, Pande, VC, Kumar, Gopal. (2012b). Bamboo plantation based technological interventions for reclamation and productive utilization of ravine lands, CSWCRTI, T-62/V-4, 30p.

Sharda, VN, Bhushan, LS and Singh, Raghuvir. (1982). Hydrological behaviour of ravinous watersheds under different land uses. Proc. International symposium on Hydrological aspects of mountainous watershed: 14–18.

Simon A, Collison AJC. (2002). Quantifying the mechanical and hydrologic effects of riparian vegetation on stream-bank stability. Earth Surface Processes and Landforms. 27: 527–546.

Singh A, Shah CM, Dayal R (1972). Point rainfall analysis of Soil Conservation Research Centre, Vasad (Gujarat). Indian Forester 98 (9): 514–519.

Singh A, Shah CM, Kamannavar HK (1976). Rainfall erosivity analysis of soil Conservation Research Centre, Vasad (Gujarat). Indian Forester 102 (2): 126–132.

Singh AK, Kala S., Dubey SK, Rao BK and Mishra PK (2015a). Bamboo based resource conservation – A viable technology for reclamation of Yamuna ravine. Technical bulletin No. T-67/A-01. IISWC, Research Centre, Agra -6 (U.P.).

Singh AK, Kala, S Dubey SK, Pande VC, Rao BK, Sharma KK and Mahapatra, KP (2015b). Technology for rehabilitation of Yamuna ravines cost-effective practices to conserve natural resources through bamboo plantation. Current Science, 108 (8):1526–1533.

S. Kala is presently working as a Scientist (Forestry) at ICAR—Indian Institute of Soil and Water Conservation, Research centre, Kota (Rajasthan). She was born in 1982. She completed PhD (Forestry) at Forest College and Research Institute (TNAU), Mettupalayam in Tamil Nadu with brilliant academic records. She has joined in ICAR in November, 2009. She has more than 8 years of experience in the field of natural resource management (NRM). She has been actively engaged in research and development work covers under NRM programmes. She has published more than 25 research papers in reputed journals. She has published more than 15 book chapters, one book and a bulletin exclusively from her research findings and experiences in the field of forestry and agroforestry sciences.

A. K. Singh is presently working as Principal Scientist (SWCE) at ICAR—Indian Institute of Soil and Water Conservation, Research Centre, Kota (Rajasthan). He did his PhD in Soil and Water Conservation Engineering from Maharana Pratap University of Agriculture and Technology, Udaipur, Rajasthan in 2009. He has given research focus mainly on soil loss, watershed management in red and black soils of South India as well as ravine region of central India. He has written more than 60 research papers, 5 technical bulletins, 7 review papers, 5 popular articles, contributed 25 chapters in different books and written 5 books which are National and International references.

B. K. Rao completed his B.Tech (Ag. Engineering) from ANGRAU, Hyderabad, and master's and PhD in Soil and Water Conservation Engineering from Water Technology Centre, Indian Agricultural Research Institute, New Delhi. He is currently working as Director (Agriculture & Research) in WALAMTARI. He has developed decision support system (CWREDSS) for increasing water use efficiency in canal command areas, cost-effective and innovative plastic check dams and groundwater recharge filters for rainwater harvesting, bamboo-based bio-engineering interventions for reclamation and productive utilization of degraded lands and in situ moisture conservation measures for enhancing the productivity of farm lands. He has 100 publications to his credit including research papers, bulletins, books, book chapters, popular articles, etc. He received NK Roy Gold Medal of IARI, New Delhi; National Academy of Agricultural Sciences (NAAS)—Young Scientist Award; ICAR—Outstanding Interdisciplinary Team Research Award; Indian Association of Soil & Water Conservationists (IASWC) —Young Scientist Award; IASWC—Gold Medal; and best paper awards for outstanding contributions in land and water management in command as well as catchment areas.

H. R. Meena (PhD) joined as a Scientist during November, 1999 at ICAR—Central Institute of Postharvest Engineering and Technology (CIPHET), Regional Station, Abohar (Punjab) and presently working on the capacity of Senior Scientist (Horticulture) at ICAR—Indian Institute of Soil and Water Conservation (IISWC), Research Centre Kota, Rajasthan (India). He has 20 years of experience in the field of postharvest management (PHM) and natural resource management (NRM). Ha has been actively engaged in research, development and extension works under institute mandates. He has completed 13 research projects at both the institutes. He has published more than 35 research papers in reputed national and international journals.

I. Rashmi is working as Scientist (Soil Science and Agricultural Chemistry) at the ICAR—Indian Institute of Soil and Water Conservation, Research Centre, Kota, Rajasthan, India. Dr Rashmi's research areas include soil phosphorus dynamics, chemical analysis and testing, soil and water conservation, biochar nutrient dynamics, zeolite soil interaction, etc. She has credited a number of research articles in various reputed national and international journals. She has also participated in many national and international conferences. She is a member of various scientific bodies including Indian Society of Soil Science, Indian Association of Soil and Water Conservation, Indian Science Congress, etc.

R. K. Singh is presently working as a Principal Scientist and Head at ICAR—Indian Institute of Soil and Water Conservation, Research centre, Kota (Rajasthan). He has 38 years of experience in research, training, and extension in the field of natural resource management. Dr. Singh has been actively engaged in research, extension and development activities involved in Natural Resource Management programmes offered by various funding programmes for effective utilization of ravine lands in India.

R. Gobinath, G. P. Ganapathy, Isaac I. Akinwumi, E. Prasath, G. Raja, T. Prakash, and G. Shyamala

Abstract

Soil erosion is a major challenge, especially in hilly areas and countries. This study presents the results of erosion control in Nilgiris area, one of the hilly regions in India located at Western Ghats of Tamilnadu. This area experiences gully erosion that causes slope instability. In this study, an attempt was made to investigate the effect of rainfall in causing gully soil erosion in this landslide-prone area, while considering factors such as varying soil density, slope angle and rainfall intensity. The study also investigated the remediation of gully erosion using vetiver, a soil bioengineering approach. A soil simulator filled with the landslide-prone soil with and without the plant roots was used at different slopes. The rate of soil erosion was studied. Soil samples were collected from the simulator and tested for its engineering properties. Results obtained showed that the slope angle has an effect on soil erosion and erosion varies inversely with density. Pore water pressure generation during rainfall creates instability among the topsoil surface, which becomes susceptible to soil erosion. HydroCAD was used in modelling the runoff over the soil surface. Soil erosion was found to occur with an increase in velocity of rainwater, which erodes the top fertile soil and triggers sheet erosion. Generally, the vetiver root was found to be efficient for the minimization of gully erosion in the Nilgiris district.

Keywords

Bioengineering · Environment · Gully erosion · Landslide · Slope instability

20.1 Introduction

Whenever studies about soil degradation are carried out, soil erosion induced by water has been the major process of focus in many parts of the world (Poesen et al. 2003; Valentin et al. 2005; Vanmaercke et al. 2016). Soil erosion creates several impacts such as soil structure loss and sedimentation in rivers. Researchers have been interested in investigating how gully erosion causes slope failure and sedimentation (Haregeweyn et al. 2017; Ionita et al. 2015; Poesen et al. 1996). Some study results marked that the soil loss rate due to gully erosion varies regionally under different environmental conditions (i.e. topography, lithology, gully type, soil properties, land use and climate), and the contribution of gully erosion to the total sediment yield varies considerably. A study in America found that gully erosion contributes two-thirds of the annual sediment in channels (Zhang et al. 2017). Gully erosion in Europe ranged from 10% to 94% of total sediment caused by water erosion (Poesen et al. 2003). The relatively high soil-loss variability is due to the wide contrast in environmental variables of the catchments, such as terrain, lithology, surface cover and how prone they are to gully erosion. Ethiopia estimated tolerable erosion loss of about 2–18 tonne per hectare per year (Hurni 1985). In India, steep or unstable slopes formed by landslides have been widely used for cultivation which causes a decline in productivity as a result of gully surface erosion (Partap and Watson 1994). Erosion in slope surface as a result of rainfall is controlled by slope gradient and length and soil vegetation.

R. Gobinath (✉) · G. Shyamala
Department of Civil Engineering, SR Engineering College, Warangal, Telangana, India

G. P. Ganapathy
Center for Disaster Mitigation and Management, Vellore Institute of technology, Vellore, India

I. I. Akinwumi
Department of Civil Engineering, Covenant University, Ota, Ogun State, Nigeria
e-mail: isaac.akinwumi@covenantuniversity.edu.ng

E. Prasath · G. Raja
Department of Soil Mechanics & Foundation Engineering, Karur College of Engineering, Karur, Tamilnadu, India

T. Prakash
Department of Civil Engineering, Jayshriram Group of Institutions, Tirupur, Tamilnadu, India

P. K. Shit et al. (eds.), *Gully Erosion Studies from India and Surrounding Regions*, Advances in Science, Technology & Innovation,
https://doi.org/10.1007/978-3-030-23243-6_20

In this study, a laboratory-scale simulation of erosion of slope as a result of rainfall was carried out. This study investigated erosion control using vetiver plant root (a soil bioengineering approach) by varying the slope angles (30°, 40°, 50°, 60° and 65°) and determination of the engineering properties of the soil after the erosion studies.

20.2 Soil Bioengineering

Soil bioengineering is the concept of using living plants to perform technical engineering functions. It has been an effective tool for stabilizing unstable and eroding slope sites. Treatments using soil bioengineering systems can range from simple live staking to complex systems designed to stabilize steep, eroding or unstable slopes (Schiechtl and Stern 1997). The plant materials, generally torpid cuttings, used in soil bioengineering systems germinate and take root, establishing early ecological sere perfect for the treatment of damaged slope sites (Polster 1989). By using early seral species, the restored site was re-aligned with the natural successional trajectory for the area. A discussion of plant materials that can be used in soil bioengineering projects is presented initially followed by information on the various techniques that can be used and the situations in which they are appropriate.

In hilly regions, slope instability and soil erosion by water and wind are major environmental hazards that have been occurring every year. They are the result of natural geomorphological processes, such as human activities and natural activities (Polster n.d.). In nature, vegetation is acting a role of maintaining equilibrium in the landscape between destructive forces and regenerative forces of stability (Fattet et al. 2011). The dangers of slope failure and erosion are enhanced by the removal of vegetative cover. This paper aims at tackling these important issues by studying the mechanisms by which vegetation plays its role. Vegetative cover functions as soil reinforcement and drainage medium to reduce instability and erosion. In bioengineering, the slope can be stabilized without making any changes in the soil mass by introducing vegetation (Howell n.d.). Through the hydrological cycle, bioengineering influences the transfer of water from the atmosphere to the earth's surface, soil and underlying rock. It, therefore, influences the volume of water contained in rivers, lakes, soil and groundwater reserves. The aboveground elements of the vegetation, such as leaves and stems, partially absorb the energy of the erosive agents of water and wind, so that less is directed at the soil, while the below-ground elements, comprising the root system, contribute to the mechanical strength of the soil.

Evapotranspiration is the combined process of the removal of moisture from the earth's surface by evaporation and transpiration from the vegetation cover.

Evapotranspiration from plant surfaces is compared to the equivalent evaporation from an open water body (Norris and Greenwood 2006). The two rates are not equivalent as a result of the energy balances of the surfaces that are markedly different.

For erosion to take place, some amount of runoff must occur. The amount of runoff generated is closely related to the infiltration rates (unsaturated and saturated hydraulic conductivity) of the soil, the antecedent moisture content and, indirectly, to the direction of water flow within the soil. When rainwater reaches the ground underneath vegetation, it stands a better chance of infiltrating the soil than on unvegetated soil. Organic matter, root growth, decaying roots, earthworms, termites and a high level of biological activity in the soil help to maintain a continuous pore system and thereby a higher hydraulic conductivity. Through an increase in the infiltration rate, and perhaps also in the moisture storage capacity of the soil, vegetation may decrease the amount of runoff generated during a storm; it will most likely additionally increase the time taken for runoff to occur.

20.3 Gully Erosion

Deep watercourses within soils, which develop from the erosion of a soil, after heavy rainfall or flooding are called gullies, and their consequence may be onsite and offsite damage. It causes rapid soil transportation from top to low-lying places. Gully erosion is the removal of soil on drain lines by surface water runoff. Once started, gullies will continue to move by headward erosion or by slumping of the side-walls unless steps are taken to stabilize the disturbance. Gully erosion happens once water is channelled across unprotected land and washes away the soil on the drain lines. Under natural conditions, a runoff is tempered by vegetation that usually holds the soil along, protecting it from excessive runoff and direct precipitation. In India, gully erosion control methods include creating contour bunding, constructing a check dam, through aerial seeding and land-levelling. The micro-level observation of erosion spots is important to evaluate and compare various land reclamation measures. Government agencies take various strategies to control and prevent such degradation in erosion-prone zones (Pani 2016).

20.4 Description of Study Site

20.4.1 Study Area

This research was conducted in the Nilgiris district situated in Tamilnadu state, India. Nilgiris literally means blue hills, or in Tamil, it's called as "Neelamalai", the name originated owing to the blue-coloured haze of considerable quantum.

This hill is surrounded by Kerala and Karnataka states on the west and northern sides and by Coimbatore district on the eastern side, and it lies between latitude $11°30'00''$ and $19°30'00''$ North and between longitude $76°29'52''$ and $76°36'00''$ East. One of the peaks, Doddabetta, has an altitude of 2595 m and is the highest point in this area. Some major towns like Kotagiri, Coonoor and Gudalur also lie at a height of 1983, 1858 and 1117 m above mean sea level, respectively. The district has numerous waterfalls and almost between every pair of mountain runs a river or a stream. The Nilgiris district covers an area of 2543 km^2, in that gross area under cultivation is nearly 77,520 ha (Fig. 20.1). Rainfall in the district is bountiful and affected majorly by monsoons from southwest and northeast; data available show that more rainfall occurs during southwest monsoon (up to 50%) than northeast (40%).

20.4.2   Field Investigation

Site inspections were conducted for two consecutive years during monsoon and in the dry season to ascertain the field density and soil moisture content. Sampling was done at 49 places within the study area. The density in field condition was found to vary between 1200 kg/m^3 and 1450 kg/m^3. The water content of the soil during the rainy season varied from 12% to 43%. Mostly in the areas where small-to-medium landslides occurred, the water content ranged from 18% to 31%. The soil sizes were found to be predominantly fine-grained.

20.4.3   Climate Condition

The district receives rainfall from both southwest and northeast monsoons. The southwest monsoon is additional active contributory nearly fifty percent (50%) within the west and forty percent (40%) within the east. The northeast monsoon is reported as moderate, contributory nearly forty percent (40%). The precipitation of rainfall gradually decreases from west to east. The rains throughout the winter and summer periods were measured. Rainfall data from four stations over the period of 1991–2001 were utilized and the analysis shows that the mean annual rainfall of the district is 1200 mm. It is minimum around Ootacamundalam (1376.20 mm) in the eastern part of the district. It gradually increases towards the west and attains a maximum around Gudalur (2269.00 mm). The high elevation of this district results in low temperature, which is further lowered by the excessive moisture content of the atmosphere resulting from the exhalation by the vegetation in this area. The day temperature in the district ranges from 22.1 °C in summer to 5.1 °C in winter. The night temperature sometimes reaches 0 °C.

The summer begins early in March with the highest temperature reached in April and May. The weather cools down increasingly from the middle of June, and by January, the mean daily temperature drops to 5.1 °C.

20.4.4   Rainfall Intensity and Runoff

Both rain and runoff factors should be thought-about in assessing water erosion drawback. Raindrops breakdown soil particle aggregation on the soil surface and disperse the particles. Lighter mixture materials like terribly fine sand, silt, clay and organic matter can be easily removed by the raindrop splash and runoff water; greater raindrop energy or runoff amounts can be needed to manoeuvre the larger sand and gravel particles. Soil movement by rain (raindrop splash) is typically greatest and most noticeable throughout the short length, high-intensity thunderstorms. Although the erosion caused by durable and fewer intense storms is not as spectacular or noticeable as that made throughout thunderstorms, the amount of soil loss can be significant, especially when compounded over time. The runoff will occur whenever there's excess water on a slope that cannot be absorbed into the soil or cornered on the surface. The amount of runoff will be multiplied if infiltration is reduced because of soil compaction, crusting or freezing. Runoff from the agricultural land is also greatest throughout spring months once the soils' square measure is sometimes saturated, snow is melting and vegetative cover is stripped. Table 20.1 presents the data of rainfall experienced in the study area.

From the field visit all over the erosion-prone areas and its neighbouring residential zones, it was found that in 80% of the locality, the slopes have mostly small trees and shrubs. From previous erosion in the area brought about by rainfall during the rainy seasons, more than 120 small-to-medium slips were noticed along transportation paths. At least 30% of this slip obstructed transportation paths. Soil samples, both disturbed and undisturbed, were collected from all the areas visited to ascertain their geotechnical characteristics.

20.5   Materials and Methods

20.5.1   Simulator

To study the soil erosion characteristics of the Nilgiris district soil, a low-cost modified soil erosion and rainwater simulator (Fig. 20.2), which can hold up to 150 kg of soil with density varying from 1250 to 1500 kg/m^3, was used in this research work. The slope of the soil bed in the simulator can be varied from 30° to 85° with the attached circular angle changing setup (Fig. 20.2). Rainfall intensity on the soil in the simulator can be varied with the pumping and

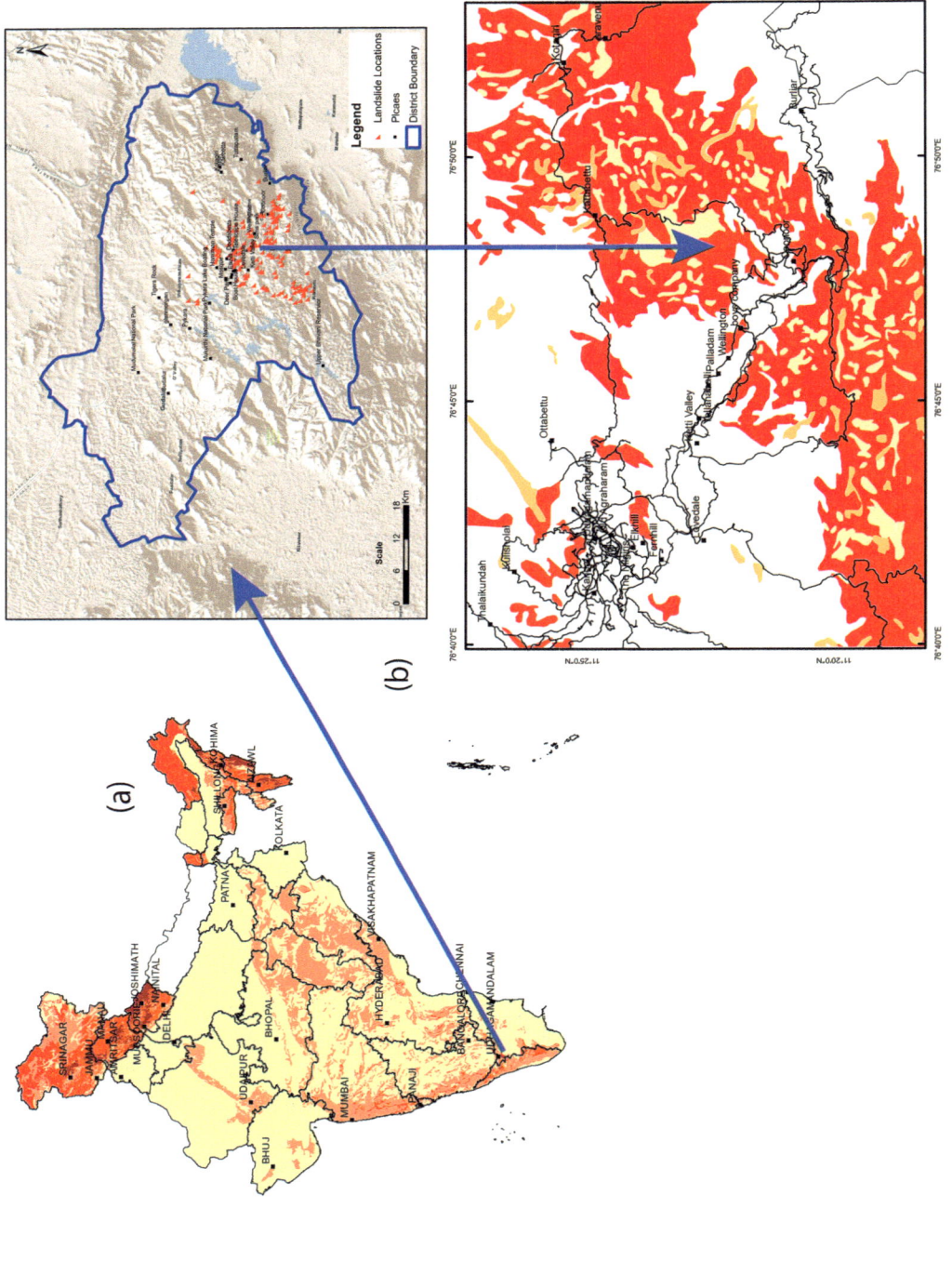

Fig. 20.1 (a) Study area—Nilgiris district and map showing the Nilgiris district. (b) Spatial distribution of past decade landslide-affected areas

Table 20.1 Time series data of rainfall by seasons in the Nilgiris area

Sl. No.	Year	Southwest monsoon		Northeast monsoon		Winter season		Hot weather season	
		Normal	Actual	Normal	Actual	Normal	Actual	Normal	Actual
1	1995	1022.2	950.6	494.1	338.2	62.5	38.6	278.1	186.9
2	1996	1022.2	882.9	494.1	523.0	62.5	73.2	278.1	223.2
3	1997	1022.2	866.7	494.1	530.2	62.5	31.1	278.1	134.9
4	1998	1022.2	1042.6	494.1	679.4	62.5	32.6	278.1	146.5
5	1999	1022.2	610.0	494.1	645.9	62.5	33.9	278.1	199.9
6	2000	1022.2	928.5	494.1	386.4	62.5	46.2	278.1	207.0
7	2001	1022.2	799.1	494.1	385.3	62.5	14.0	278.1	287.9
8	2002	1022.2	602.9	494.1	236.7	62.5	16.2	278.1	192.6
9	2003	1060.0	577.0	367.7	471.3	62.5	20.4	278.1	203.0
10	2004	1060.0	943.7	367.7	564.8	30.8	49.2	237.2	427.9
11	2005	1060.0	1032.5	367.7	557.2	30.8	23.2	237.2	307.7
12	2006	1060.0	653.9	367.7	656.1	30.8	23.8	237.2	324.7
13	2007	1060.0	1142.6	367.7	515.5	30.8	35.9	237.2	178.8
14	2008	1060.0	1067.9	367.7	516.3	30.8	193.6	237.2	438.8
15	2009	1060.0		367.7		30.8	1.7	237.2	286.5

Source: Assistant Director of Statistics, the Nilgiris

delivery attachments, which will be measured by using the attached rain gauge. The Nilgiris soil and soil mixed with roots and having different densities were placed in the simulator. The slope angle was varied according to the slope angle obtained in the field during field visits. The rainfall intensity was fixed for a particular duration (mostly 10 min of flow is allowed to obtain required rain intensity) and the amount of soil that is eroded is calculated by collecting the eroded soil.

After rainfall simulation on soil, various strength parameters were determined for the top, middle and the bottom portions of the soil in the simulator. To get an easy flow and full-sprinkling on the simulator bed, a gardening sprinkler was used and the rate of flow was measured using a half-inch pipe as 917 ml/s. To cover simulator bed of area 1.08 m², pipes were positioned at a height of 2.18 m from the ground surface. Before starting the experiment in the simulator, the artificial rainfall setup to simulate the Nilgiris district field condition had been finalized after many trials of rainfall setup.

20.5.2 Roots

Using available literature, previous studies and from a field visit (Fig. 20.3), vetiver was found to be the best root plant having good morphology characteristic against soil slope failure and easily available in the study area. So, vetiver roots collected from their original places were collected and preserved.

20.6 Results and Discussion

20.6.1 Engineering Properties of Soil

The results of various geotechnical tests as per Indian Standard (IS) carried on the collected Nilgiris district soil are tabulated (Table 20.2). Samples were collected, sealed in air-tight bags and then transported to the laboratory. We used a rapid moisture tester to determine the field moisture content and obtained values ranging from 12.3% to 19%. In the laboratory, this soil was tested to determine its physical and engineering properties, such as specific gravity, liquid limit (LL), plastic limit (PL), shrinkage limit and shear tests using standard procedures specified in IS2720. Table 20.2 shows that the soil can be classified as clay with low plasticity (CL).

20.6.2 Soil Erosion Study

Using the rainfall data of the previous decade, a lab-scale rainfall simulator was designed to study the erosion of soil as a result of rainfall in the Nilgiris District. The study was carried out on the collected soil compacted in the simulator and the soil reinforced with vetiver grass roots. With the help of rain gauge, the amount of artificially created rain was measured for each slope angle. Table 20.3 presents the summary of the results of the height of rain gauge from the ground, the amount of rainfall and the rate of soil erosion for each of the soil and vetiver root-reinforced soil.

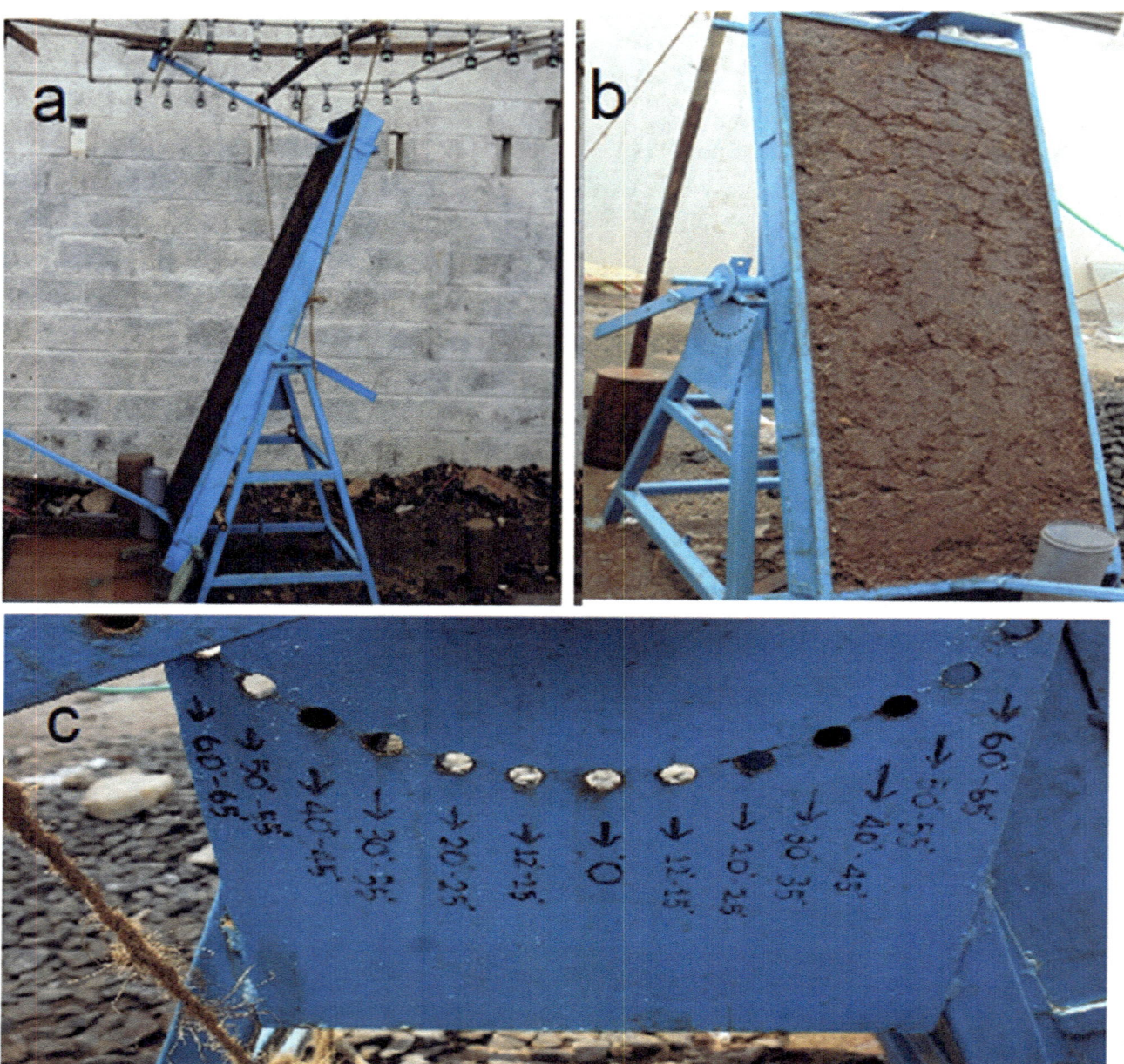

Fig. 20.2 Erosion study simulator bed with artificial rainfall setup with different slope angles

Table 20.3 shows that the erosion of topsoil (kg/m^2) by the simulated runoff decreased for the soil having the vetiver root and for each of the slope angles. The presence of gully is a great indicator that erosion is out of control and the slope is getting into a critical phase that leads to slope failure. These collapsing erosion and gullies in downslope are mostly associated with a high intensity of runoff and sheet/rill erosion processes within the catchments. Sheet erosion, rill and gully formation in simulator bed are shown in Fig. 20.4.

However, the eroded soil from the simulator bed was collected in the chamber for easy estimation of the rate of erosion that occurred. Experimental results showed gully erosion occurring for the steep angles of 40°, 50°, 60° and 65°. This study, which was repeated for soil with 0.5% vetiver root engineered soil and the result as presented in Table 20.3, showed a decrease in the rate of soil erosion.

Figure 20.5 presents a graphical illustration of the rate of soil erosion. It shows that the rate of soil erosion is lower for the soil with the vetiver root than those of the undraped soils. The incorporation of the root network and the presence of soil fauna opened up the pore system and contributes to the stability of the soil aggregates by increasing the density and

Fig. 20.3 Field visit and vetiver root collection

Table 20.2 Engineering properties of Nilgiris district soil

Properties	Nilgiris district soil	IS 2720
Specific gravity	2.55	Part 3
Liquid limit	34%	Part 5
Plastic limit	22.72%	Part 5
Plasticity index	11.28%	Part 5
Maximum dry density	1.575 g/cm^3	Part 7
Bulk density	1.796 g/cm^3	Part 7
Optimum moisture content	10.2%	Part 7
Cohesion	5.140 kN/m^2	Part 11
The angle of internal friction	29°	Part 11
Total stress	8.117 N/cm^2	Part 11
Shear strength	13.53 N/cm^2	Part 11
Unconfined compressive strength (UCS)	1.227 N/cm^2	Part 10
Co-efficient of permeability	9.5×10^{-3} mm/s	Part 36
California bearing ratio (Unsoaked)	7.10%	Part 16
North Dacota test (Bearing capacity)	20.0 kg/cm^2	–
Vane shear test	16.10 kg/cm^2	Part 30
Consolidation test	1.090 cm^2/s	Part 15

lowering the permeability which also acts as an anchorage in shearing zone. Results clearly show that reduction of gully erosion was due to soil reinforcement by plant roots that prevents crack formation at the gully head crown. Figure 20.6 shows the change in density of simulator bed after rainfall.

The variation of the density of the soil samples collected from the top, middle and bottom of the simulation bed did not follow any defined pattern.

Figure 20.7 presents the results of the unconfined compressive strength (UCS) tests on the soil samples collected from the simulator. The sample for the determination of the UCS of simulator bed soil after rainfall was collected using a core-cutter (Fig. 20.8) and directly subjected to the test in the soil compression machine. Figure 20.7 shows distinct variation in the UCS when compared with the results of the Nilgiris soil sample without the vetiver plant. The UCS results were generally low. Comparing the UCS of samples taken from the top, middle and bottom of the simulator show that samples collected from the middle had the highest UCS for slope angles less than 50°. This may be attributed to the consolidation of the soil in the middle of the simulator.

Figure 20.9 shows how the shear parameters of the soil vary. Since shear strength affects the resistance of the soil to detachment by raindrop impact (Cruse and Larson 1977;

Table 20.3 Summarized report of rain intensity and erosion

Angle of simulator	Location in simulator	Height of rain gauge from the ground (cm)	Rainfall (cm) Avg	Rate of soil erosion (kg/m^2) Soil	Soil with vetiver root
30	Top	164	29.67	6.26	3.89
	Middle	160			
	Bottom	155			
40	Top	174.5	29.13	5.41	2.987
	Middle	131			
	Bottom	85			
50	Top	183.5	29.9	5.083	2.96
	Middle	133.4			
	Bottom	71.2			
60	Top	193	30.5	4.65	2.82
	Middle	132.7			
	Bottom	69			
65	Top	194	30	4.537	2.645
	Middle	132.2			
	Bottom	52			

Fig. 20.4 (**a**) Sheet erosion, (**b**) rill erosion and (**c**) gully erosion

Al-Durrah and Bradford 1982), the susceptibility of the soil to rill erosion (Laflen et al. 1987; Rauws and Covers 1988) and the likelihood of mass soil failure, root systems can have a considerable influence on all these processes. The maximum effect on resistance to soil failure occurs when the tensile strength of the roots is fully mobilized and that, under strain, the behaviour of the roots and the soil is compatible. This requires roots of high stiffness or tensile modulus to mobilize sufficient strength and the 8–10% failure strains of most soils. The tensile effect is limited with shallow-rooted vegetation where the roots fail by pullout, slipping due to loss of bonding between the root and the soil before peak tensile strength is reached (Waldron and Dakessian 1981). The tensile effect is most marked with trees where the roots penetrate several metres into the soil and their tortuous paths around stones and other roots provide good anchorage. Root failure may still occur, however, by rupture, breaking of the roots when their tensile strength is exceeded. The strengthening effect of the roots will also be minimized in situations where the soil is held in compression instead of tension, at the bottom of hillslopes. Root failure here occurs by buckling. From the result presented in Fig. 20.9, the soil samples with plant roots generally had shear strengths greater than the shear strength (13.53 N/cm^2) of the soil without plant roots. Consequently, vegetation on an erosion-prone soil contributes to increasing the stability of the soil. Bioengineering the soil shows wide variation in shear strength property even after rainfall.

Root systems lead to an increase in soil strength through an increase in cohesion brought about by their binding action

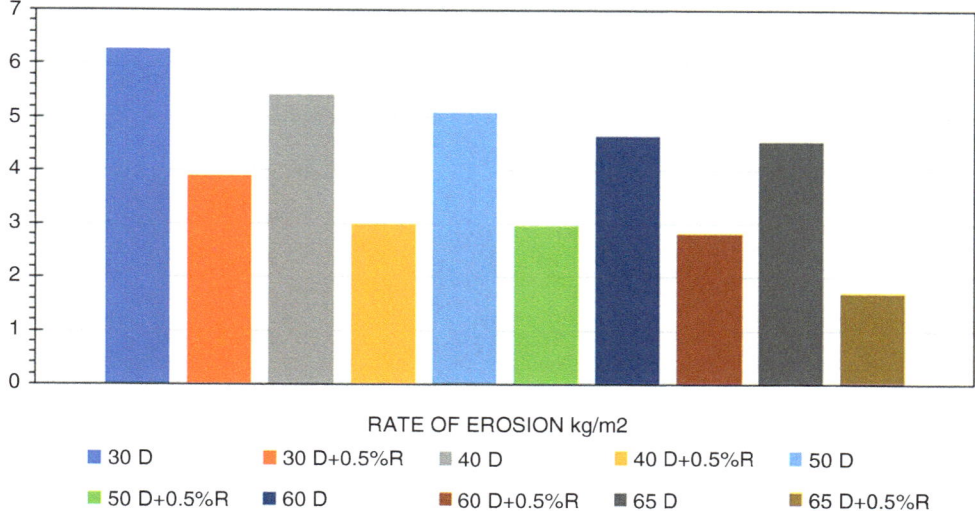

Fig. 20.5 Rate of erosion

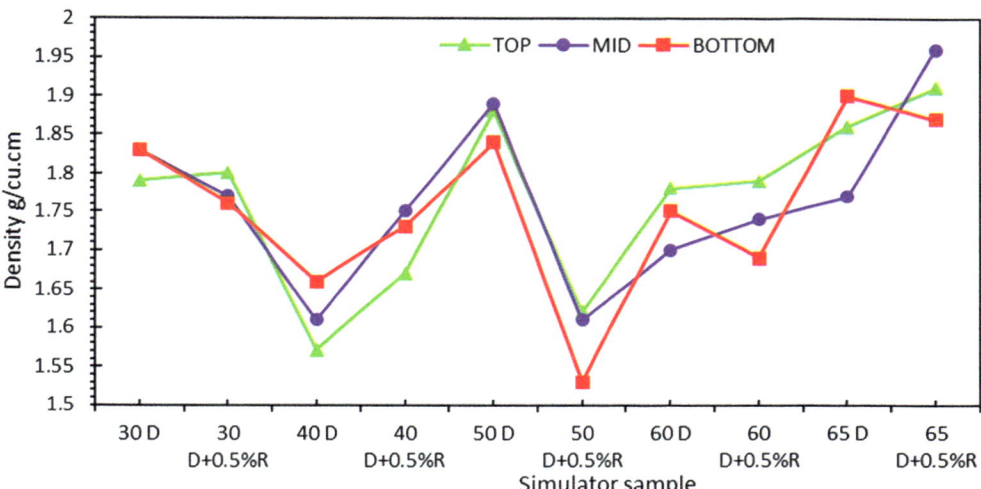

Fig. 20.6 Density variation in simulator after rainfall

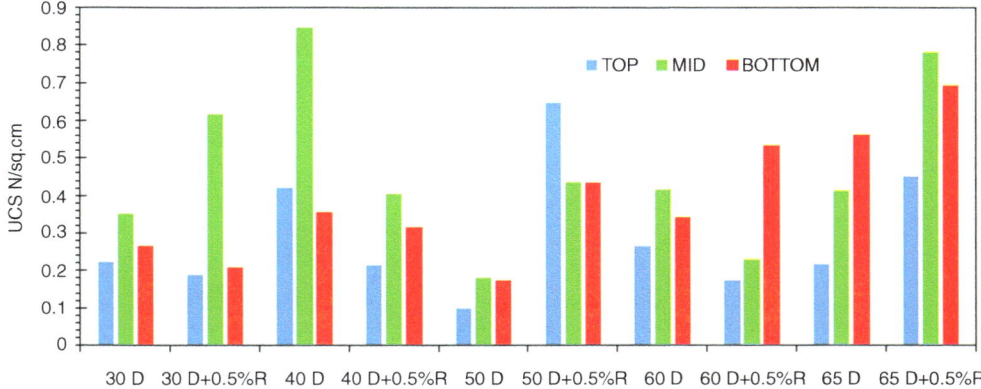

Fig. 20.7 Unconfined compression strength of simulator bed soil

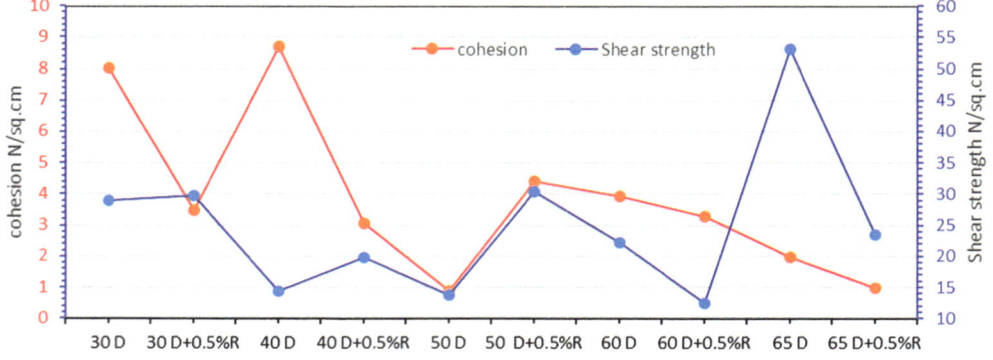

Fig. 20.8 (**a**) Undisturbed sampling from simulator bed. (**b**) Various samples for simulator laboratory testing

Fig. 20.9 Comparison of simulator soil shear parameter

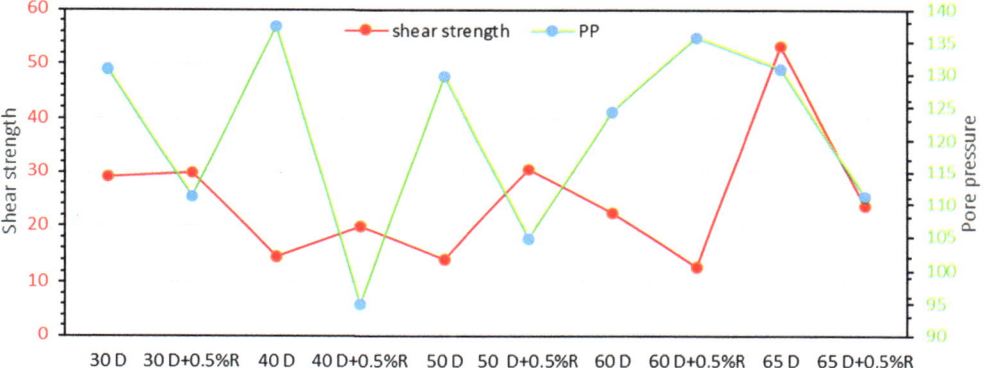

Fig. 20.10 Comparison of shear strength vs. pore pressure

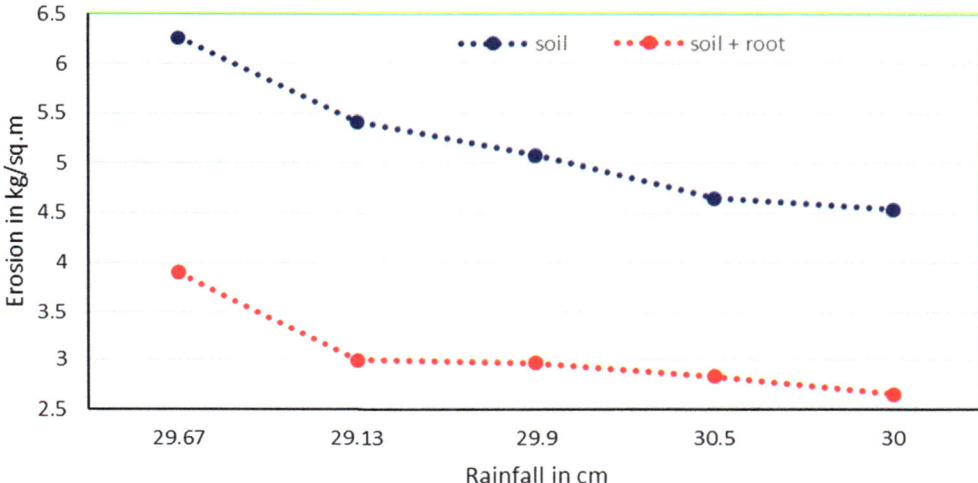

Fig. 20.11 Rainfall intensity versus erosion plot

in the fibre/soil composite and adhesion of the soil particles to the roots is tested and the results are shown in Fig. 20.9. It is generally held that roots have no effect on soil friction angle but that grassroots increased the angle of internal friction of a sandy soil but had no such effect on a sandy clay loam (Tengbeh 1989).

From Fig. 20.9, it can be seen that the cohesion of the soil generally decreased for the soil having vetiver roots while there was no well-defined pattern for the effect of the vetiver root on the shear strength of the soil.

The surcharge becomes critical when rainfall of several hundreds of millimetres occurs in a wet spell for a few days. Sometimes interception and evapotranspiration are reduced virtually to zero and the soil is unable to either dry out or drain so pore pressure change is tested using triaxial setup and the result is shown in Fig. 20.10. The critical factors here may well be the low angle of internal friction of the soil material, which, when close to waterlogging, is reduced to less than 20°, and the steepness of the slopes, which are over 20°. In contrast, the surcharge is beneficial when cohesion is low and groundwater levels are high provided that the angle

of internal friction of the soil is also high and the slope angles are low.

The amount of soil erosion in relation to the amount of rainfall was compared in Fig. 20.11 for both the soil and soil with vetiver root. For the soil with vetiver root, the amount of soil erosion decreased remarkably to 3.89, 2.987, 2.96, 2.82 and 2.645 kg/m^2 (which corresponds to 37.85, 44.78, 41.76, 39.35 and 41.76% decrease). Hence, from Fig. 20.11, it can be concluded that the use of vetiver root to reinforce the soil was effective at reducing the rate of gully erosion head.

20.6.3 Variation in the Flow of Stormwater at Different Slope Angles Obtained in Nilgiris District

HydroCAD is a commonly used software to generate runoff graph from hydrological and site date to verify the adequacy of the drainage system and to predict whether flooding or erosion problems will occur during heavy rainfall. The

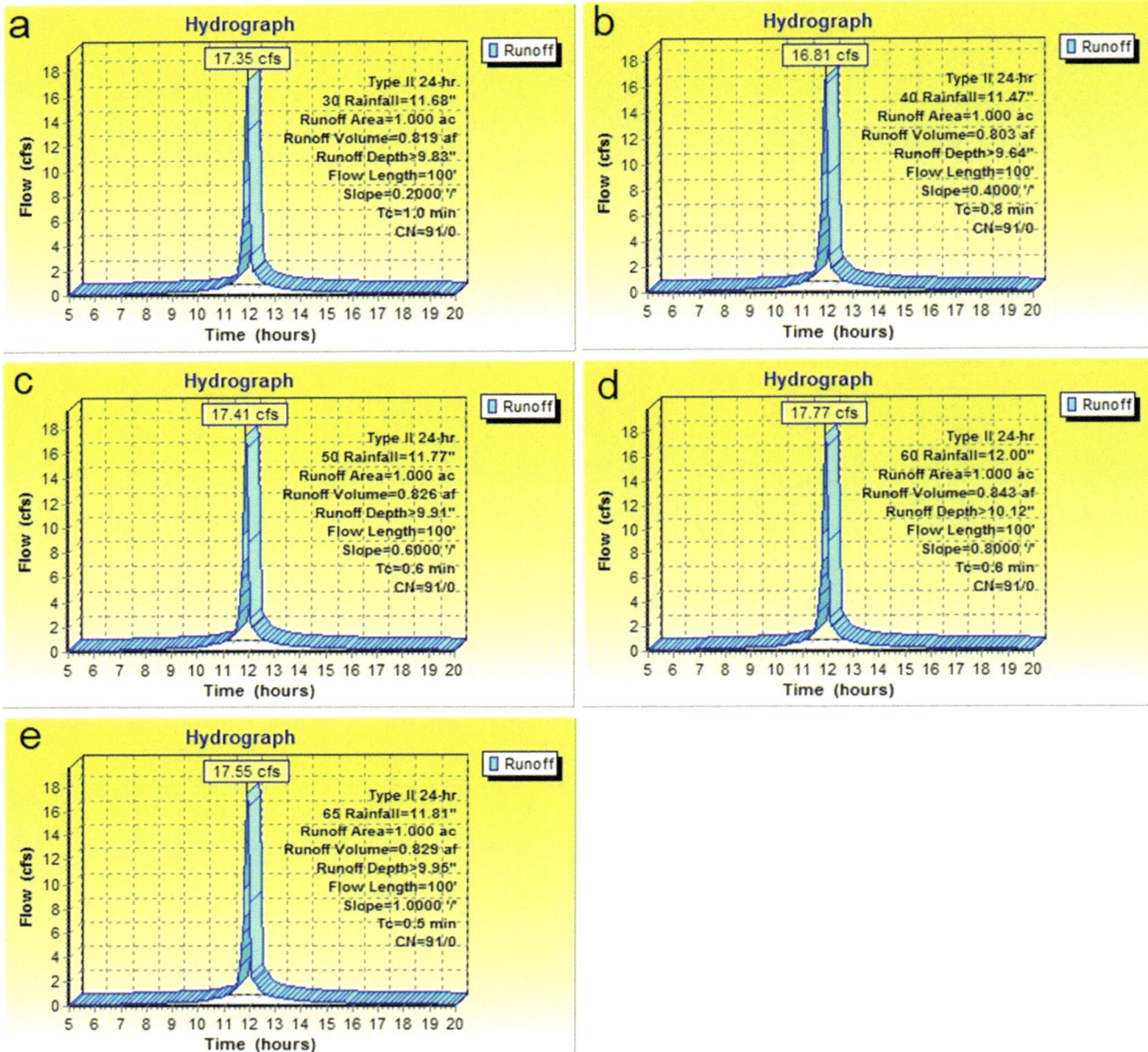

Fig. 20.12 Output of hydrograph for rain water flow in uncovered soil surface at various slope angles: (**a**) 30°, (**b**) 40°, (**c**) 50°, (**d**) 60°, (**e**) 65° slope

investigation of runoff to the catchment area was studied in the form of precipitation sums and flow rates. The data of rainfall intensity for every slope angle with and without the vetiver root was used in modelling the runoff from the slope surface, using the HydroCAD. The input quantity for the HydroCAD model simulation of this study used a period of 48 h. Thus, the observation of intensity at different slope angles obtained from real-time study in the rainfall simulator served as its input. HydroCAD is not only for runoff management, it is also used to study the erosion loss along

commercial and residential zones. It uses the rational method to calculate the maximum rainfall intensity and its flow with intensity-duration-frequency plot. For the slope models water collected in the collection chamber is arrived by calculating the detention period together with tail water, with this multiple rainfall models is generated.

Input to model the simulator bed sample are variation in intensity, mean values of runoff curve numbers, the values change with the change in vegetation/root cover in the soil (Figs. 20.12 and 20.13). Accelerated storm duration of 24 h

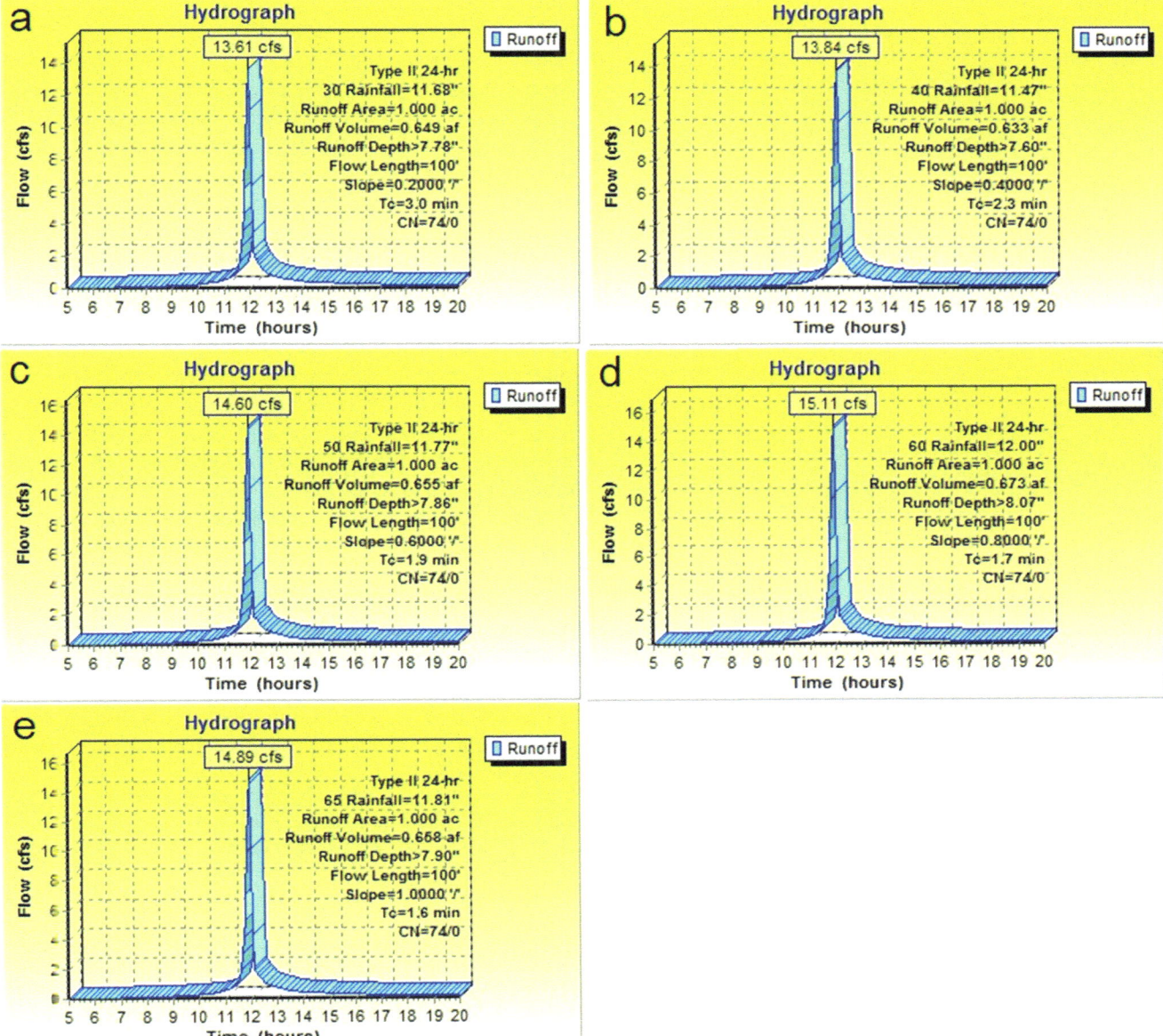

Fig. 20.13 Output of hydrograph for rain water flow in bioengineered soil surface at various slope angles: (**a**) 30°, (**b**) 40°, (**c**) 50°, (**d**) 60°, (**e**) 65° slope

was used, and within the duration, complete saturation state was reached. Exfiltration for this study was assumed to be the minimum value of 1 cm/h. For a different flow gradient, the data off low of water in soil for the soil and the soil having vetiver roots were calculated and the infiltration of water into the soil was studied. In Table 20.4, time-dependent runoff and stormwater flow from HydroCAD were tabulated.

Using the HydoCAD, the rate of infiltration after heavy rainfall and causes of erosion after percolation were studied. Figure 20.14 shows the rate of percolation based on slope

Table 20.4 CFS values from HydroCAD

Angle of slope	Flow of storm water (CFS)	
	Nilgiris soil	Nilgiris soil with dense grass surface
30	17.35	13.61
40	16.81	13.84
50	17.41	14.6
60	17.77	15.11
65	17.55	14.89
Runoff volume (of)		
30	0.819	0.649
40	0.803	0.633
50	0.826	0.655
60	0.843	0.673
65	0.829	0.668
Time of concentration (min)		
30	1	3
40	0.8	2.3
50	0.6	1.9
60	0.6	1.7
65	0.5	1.6

angle variation. Figure 20.14 demonstrates that an increase in the inclination of slope gives rise to a reduction in percolation rate for the soil and lesser amount of infiltration rate for the soil with vetiver root.

20.7 Conclusion

In this work, an attempt is made to elucidate the usage of plant for soil erosion protection; since not many studies are available which are conducted in lab scale, consequently this study is taken up. Through this study, it is proved that rate of erosion decreases with the root content and also it is affected by the slope angle, more slope angle promotes more erosion, but it can be controlled by planting properly selected plants. This chapter focused on the simulation of gully erosion produced by artificially generated rainfall on the Nilgiris district soil and how the reinforcement of the soil with vetiver root may minimize the erosion. Yet this study has its own limitations since the natural rainfall has varying characteristics than the simulated rainfall. The study used a lab-scale simulator to model the soil properties, slope and rainfall characteristics to investigate the erosion. Results obtained showed that the slope angle has an effect on soil erosion and erosion varies inversely with density. Pore water pressure generation during rainfall creates instability among the topsoil surface, which becomes susceptible to soil erosion. HydroCAD was used in modelling the runoff over the soil surface. Soil erosion was found to occur with an increase in velocity of rain water, which erodes the top fertile soil and triggers sheet erosion. Generally, the vetiver root was found to be effective for the minimization of gully erosion in the Nilgiris district. Also, several indigenous plants that are available in the area where

Fig. 20.14 Rate of infiltration vs. slope angle

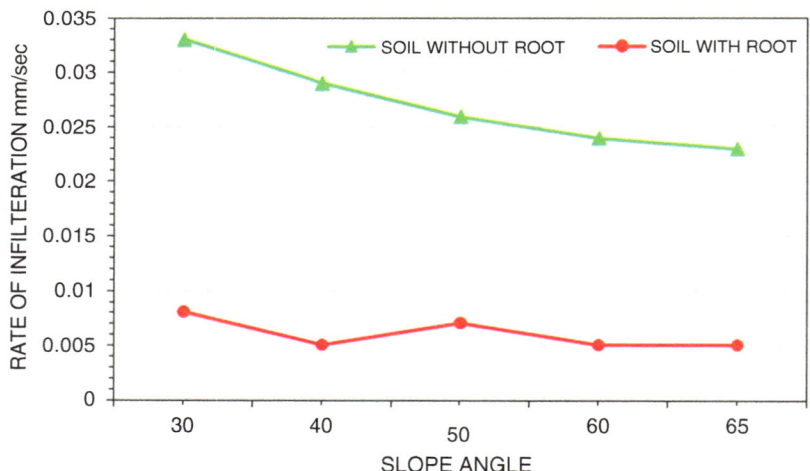

the measures are to be taken can also be utilized; researchers propose further location-specific research work with locally available soil and locally available plants which will be more suitable.

References

Al-Durrah, M.M. and Bradford, J.M., 1982. The Mechanism of Raindrop Splash on Soil Surfaces 1. *Soil Sc. Soc. of America Jr. 46*(5): pp. 1086-1090.

Cruse, R.M. and Larson, W.E. (1977) Effect of soil shear strength on soil detachment due to raindrop impact. Soil Sci. Soc. Am. J., 41, 777–81.

Fattet, M. et al. 2011. "Effects of Vegetation Type on Soil Resistance to Erosion: Relationship between Aggregate Stability and Shear Strength." Catena. DOI: https://doi.org/10.1016/j.catena.2011.05.006

Haregeweyn, N., Tsunekawa, A., Poesen, J., Tsubo, M., Meshesha, D. T., Fenta, A. A., ... Adgo, E. (2017). Comprehensive assessment of soil erosion risk for better land use planning in river basins: Case study of the Upper Blue Nile River. Science of the Total Environment, 574, 95–108

Howell, J H. "Introducing Bio-Engineering to the Road Network of Himachal Pradesh."

Hurni, H. (1985) Erosion-productivity-conservation systems in Ethiopia. In: Proceedings of paper presented at the 4th international conference on soil conservation, Maracay, Venezuela

Ionita, I., Fullen, M. A., Zgłobicki, W. and Poesen, J. 2015. Gully erosion as a natural and human-induced hazard. Natural Hazards, 79(S1), 1–5. doi: https://doi.org/10.1007/s11069-015-1935-z

Laflen, J.M., Thomas, A.W. and Welch, R., 1987. Cropland experiments for the WEPP project. American Society of Agricultural Engineers (Microfiche collection)(USA). no. fiche no. 87-2544.

Norris, Joanne E, and John R Greenwood. "Assessing the Role of Vegetation on Soil Slopes in Urban Areas."In Proceedings of the 10th IAEG International Congress, IAEG2006 (2006), 744 by Joanne E. Norris, John R. Greenwood

Pani, Padmini. 2016. "Development in Practice Controlling Gully Erosion : An Analysis of Land Reclamation Processes in Chambal Valley, India." 4524 (November). doi:10.1080/09614524.2016.1228831.

Partap, T. and Watson, H.R., 1994. Sloping agricultural land technology (SALT): A regenerative option for sustainable mountain farming. available at http://agris.fao.org/agris-search/search.do?recordID=QZ1998000050 accessed during February 2019

Poesen, J., Nachtergaele, J., Verstraeten, G., & Valentin, C. (2003). Gully erosion and environmental change: Importance and research needs. CATENA, 50, 91–133. doi:https://doi.org/10.1016/S0341-8162(02)00143-1

Poesen, J., Vandaele, K. and van Wesemael, B., 1996. Contribution of gully erosion to sediment production in cultivated lands and rangelands. IAHS Publ., 236, 251–266.

Polster, D.F. 1989. Successional reclamation in Western Canada: New light on an old subject. Paper presented at the Canadian Land Reclamation Association and American Society for Surface

Polster, David F. "Soil Bioengineering for Slope Stabilization and Site Restoration 1."

Rauws, G. and Covers, G., 1988. Hydraulic and soil mechanical aspects of rill generation on agricultural soils. Jr. Of Soil Sci, 39(1), pp. 111-124.

Tengbeh, G.T., 1989. The effect of grass cover on bank erosion, Ph.D thesis, Silsoe College, Cranfield Institute of Technology

Schiechtl, H.M. and R. Stern. 1997. Water Bioengineering Techniques for Watercourse, Bank and Shoreline Protection. Trans. By L. Jaklitsch. Blackwell Scientific. Oxford, U.K. 185 pp.

Valentin, C., Poesen, J. and Li, Y. 2005. Gully erosion: Impacts, factors and control. Catena, 63(2–3), 132–153. doi: https://doi.org/10.1016/j.catena.2005.06.001

Vanmaercke, M., Poesen, J., Van Mele, B., Demuzere, M. and Bruynseels, A., 2016. How fast do gully headcuts retreat?. Earth-Science Reviews, 154, pp. 336-355.

Waldron, L.J. and Dakessian, S., 1981. Soil reinforcement by roots: calculation of increased soil shear resistance from root properties. Soil science, 132(6), pp. 427-435.

Zhang, Xiyu, Jianrong Fan, Qing Liu, and Donghong Xiong. 2017. "The Contribution of Gully Erosion to Total Sediment Production in a Small Watershed in Southwest China." Physical Geography 3646 (July): 0. doi:https://doi.org/10.1080/02723646.2017.1356114.

Ravindran Gobinath is Professor in civil engineering department of S R Engineering College; he is involved actively in soil bioengineering research. His area of research includes soil stabilization, root reinforcement, soil bioengineering and environmental geotechnology. To his credit he had published more than 100 research papers in peer-reviewed journals, and also he serves as editorial board member and reviewer of journals. He is registered life member of several professional bodies including Indian Society of Technical Education (ISTE), Indian Society of Remote Sensing (ISRS) and Indian Society of Geomatics (ISG).

G. P. Ganapathy is presently working as Professor and Director, Center for Disaster Mitigation and Management, VIT, Vellore. He is an expert in the field of geohazards and micro-zonation studies. He is an active researcher collaborating with researchers across the globe; to his credit he has more than 50 publications in peer-reviewed journals and had authored more than 5 books.

Isaac Akinwumi is a Lecturer in Civil Engineering at Covenant University, Ota, Nigeria. He is involved in research works on soil stabilization and the protection of the environment. He is an author or coauthor of over 50 peer-reviewed scientific papers and a reviewer for some reputable international journals. He is a Registered Engineer with the Council for the Regulation of Engineering in Nigeria (COREN), a Corporate Member of the Nigeria Society of Engineers (NSE), and a Member of the International Association for Engineers (IAENG), Nigerian Young Academy (NYA) and the American Society of Civil Engineers (ASCE).

E. Prasath is an active geotechnical researcher who works in the field of soil stabilization using novel materials; he is presently pursuing master's in soil mechanics engineering in Karur College of Engineering. His area of interest includes soil stabilization, soil bioengineering and highway planning.

G. Raja is a geotechnical engineering students who is pursuing master's in the field of soil mechanics from Karur College of engineering. His area of interest includes seismic analysis of soils, soil stabilization, and soil bioengineering. He is presently working in the field of unsaturated soil mechanics.

T. Prakash is a civil engineering graduate presently working as quality officer in a leading highway materials testing company. His area of interest includes soil erosion studies along slopes, and he has worked in several such projects analyzing soil samples from Tamil Nadu region.

G. Shyamala is an environmental engineering professional who holds a doctorate degree with specialization in environmental engineering. Her study involves environmental geotechnology, and mostly she works in the field of fluid flow in saturated and unsaturated soils. She is an expert in modeling of fluid flow in the soil; her present work involves soil erosion studies in the hilly regions of Tamil Nadu and Telangana.

Planning, Designing and Construction of Series of Check Dams for Soil and Water Conservation in a Micro-watershed of Gujarat, India

Deodas Meshram, S. D. Gorantiwar, Saurabh Samadhan Wadne, and K. C. Arun Kumar

Abstract

Management of soil and water resources for enhancing horticulture production is of growing concern worldwide and this is especially true for developing countries like India. Efficient management and utilization of these resources are very important to increase the horticulture/agriculture production and productivity per unit area. One of the principal reasons for the low productivity in horticulture/agriculture is the progressive deterioration of soil due to erosion. The factors for soil erosion in this area are excessive deforestation, overgrazing and faulty practices. Consequently, valuable top soil is lost and its fertility gets depleted resulting in poor horticulture/agriculture yield. Soil erosion mainly occurs due to high velocity of runoff flowing over the land surface. It is dependent on land slope, crop cover and rainfall characteristics in a micro-watershed. The results show that the various soil and water conservation structures have good potential for conserving the soil and water.

Keywords

Micro-watershed · Series of water-harvesting structures · Soil and water conservation techniques · Soil erosion

21.1 Introduction

The ministry of agriculture, Government of India (GOI), has identified watershed management as the most rational unit for planning, design and implementation of the programs, dealing with the agriculture as well as horticulture production. Check dams are built using ancient techniques. India's first check dam, the Grand Anicut ("Kallanai" in the Tamil language), was built by an ancient Chola king named Karikalan in the Cauvery River delta in Tamil Nadu. It is the world's oldest water diversion structure still in use (Govindasamy 2007). It was constructed across the ephemeral streams to intercept runoff from catchments and store it for optimum utilization (Khan 1992). These structures are suitable where ephemeral streams are available in catchments with good runoff-producing characteristic and are widely adopted in Gujarat (Anon 1988). Only a portion of the rainwater gets stored in the soil profile and excess runoff water needs to be harvested in structures to meet the irrigation requirement of crops and other water needs in the area (Pillai 1987; Natividad and Woodrige 1997). In designing spillways and outlets or waterways, peak rate of runoff value is required; for assessing storage value in earthen dam, tanks, ponds, etc., estimated runoff volume is required. Singh (1987), in their study on conservation structures in catchments of the river valley projects and flood-prone river, urged the necessity of standardizing the planning and designing of soil and water conservation in India in order to make it cost-effective. Narayana (1962) observed that check dams constructed by local farmers are sometimes either big or smaller than the requirements and are not located at a proper place and not designed appropriately.

Different water conservation practices apply to reduce surface runoff and divert it to recharge zones in order to increase moisture as well as crop production (Lesschen et al. 2007; Krois and Schulte 2014). The National Water Policy of 2002 emphasized on water management through extensive soil conservation, catchment area treatment, preservation of forests and increase of the forest cover and the construction of check dams for conservation of the water in the micro-watershed (Ibid 2002).

Check dams are of greater importance as it addresses water conservation as well as soil erosion. In general, check

D. Meshram (✉) · S. S. Wadne
Land and Water Management Engineering, ICAR-NRC on Pomegranate, Kegaon, Maharashtra, India

S. D. Gorantiwar
Dr. A.S. Shinde College of Engineering and Technology, Rahuri, Ahmednagar, Maharashtra, India

K. C. Arun Kumar
YP-II, GIS, ICAR-NBSS & LUP, Nagpur, Maharashtra, India

© Springer Nature Switzerland AG 2020
P. K. Shit et al. (eds.), *Gully Erosion Studies from India and Surrounding Regions*, Advances in Science, Technology & Innovation,
https://doi.org/10.1007/978-3-030-23243-6_21

dams are usually suggested for lower-order streams (up to third order), and the slope of the terrain should be between flat and gentle slope in order to retain a maximum possible quantity of water which comes under the L-section. These structures are usually proposed where water table fluctuations are very high and the stream is influent and or internally effluent. For an economical design, the catchment area should be more than 25 ha. This structure is suitable for soil type of less to medium permeable, to allow infiltration to the downstream side of the dam if necessary. It should be located in the area which has a higher potential for crop production and settlement areas to allocate the harvested water (Rao et al. 2005).

In horticulture, to increase the fruit production and to meet the flood, fodder and fibre requirements of ever-growing population, proper planning, designing and construction of check dams for soil and water conservation resources on the watershed basis is desirable.

21.2 Materials and Methods

21.2.1 Study Area

The study area is bounded by latitude 22°41′38″N and longitude 73°33′22″E and is located in Central highland of Panchmahal district, Gujarat, with a total area of 0.667 km² (Fig. 21.1). The micro-watershed is mostly plane with altitude ranging from 110 m at S/H No. 5 to 115 m amsl at Kharashalia Village. It is a part of Ruparail tributaries of Goma River, Mahi basin of Gujarat state having conspicuous physiographic variations comprising undulating plan and basalt tracts. Drainage pattern is mostly dendrite and jointly controlled by parallel drainage pattern at some places. The area has sandy clay loam soil; highly truncated (Fig. 21.2) and gently sloping terrain on both sides of river result in quick buildup of runoff into the Mahi river (Fig. 21.3). This

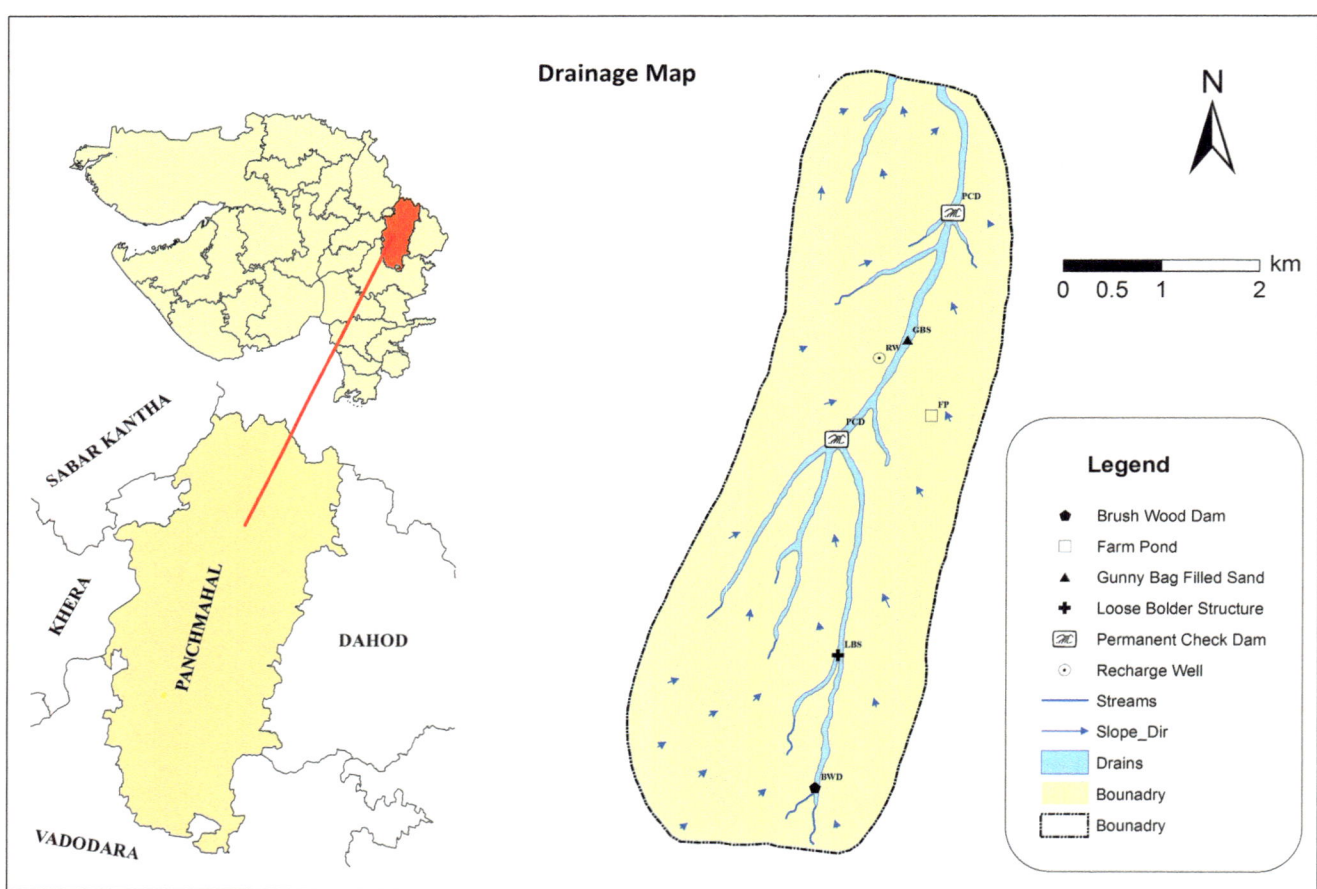

Fig. 21.1 Location map and depicted series of water-harvesting structures of the study area

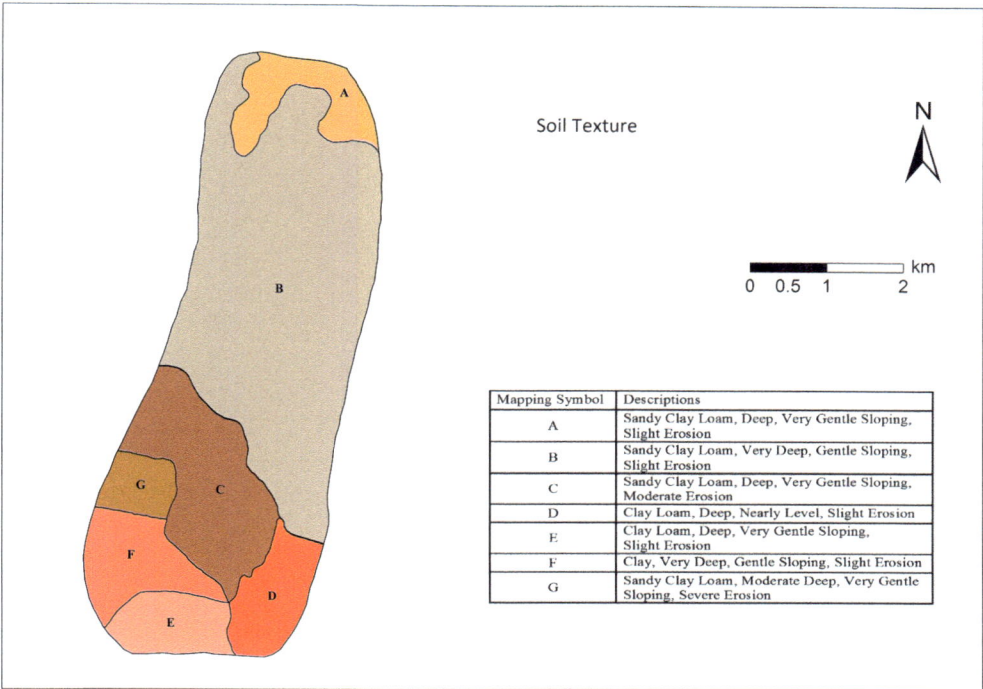

Fig. 21.2 Soil texture of micro-watershed

Mapping Symbol	Descriptions
A	Sandy Clay Loam, Deep, Very Gentle Sloping, Slight Erosion
B	Sandy Clay Loam, Very Deep, Gentle Sloping, Slight Erosion
C	Sandy Clay Loam, Deep, Very Gentle Sloping, Moderate Erosion
D	Clay Loam, Deep, Nearly Level, Slight Erosion
E	Clay Loam, Deep, Very Gentle Sloping, Slight Erosion
F	Clay, Very Deep, Gentle Sloping, Slight Erosion
G	Sandy Clay Loam, Moderate Deep, Very Gentle Sloping, Severe Erosion

Fig. 21.3 Topography of the actual side area of micro-watershed

Table 21.1 Monthly mean of the weather parameters

Months	Weather parameters						
	T_x (°C)	T_n (°C)	R (mm)	Eva (mm day^{-1})	RH (%)	W_s (kmph)	SS (h)
January	28.2	12.5	2.0	4.1	63.3	3.87	9.41
February	32.2	14.3	0.0	5.6	56.7	4.5	9.39
March	35.7	17.9	0.2	8.0	55.9	4.47	9.29
April	38.6	22.0	1.7	9.2	62.6	5.56	10.10
May	39.6	25.5	7.0	10.5	69.5	10.5	10.20
June	37.1	27.8	142.3	7.8	79.5	10.9	7.78
July	32.0	25.2	307.3	3.8	86.2	8.33	3.84
August	30.7	24.5	265.8	3.7	85.9	6.45	3.73
September	32.4	24.0	108.9	5.7	86.5	5.26	5.75
October	35.3	20.3	22.4	8.1	71.7	3.15	8.10
November	32.6	17.3	9.20	8.8	66.4	3.34	8.83
December	30.0	14.8	6.20	7.6	63.0	3.51	7.64
SD	3.5	5.08	11.27	6.1	11.2	5.8	7.84
CV (%)	10.6	24.9	152.7	38.2	15.9	46.7	28.8

Source: Gujarat Agricultural University, Agrometeorology Station, Derol and Collector office of Panchamahal district

region is characterized by semi-arid climate with little or no water surplus (Bhattacharjee et al. 1982).

The climate of the study area is semi-arid eco-region with well-defined summer (March–June), rainy season (July–October) and mid winter season (November–February). The mean annual precipitation is 873 mm of which nearly 98% is received during south-west monsoon period (July–October). The mean annual air temperature is 27.5 °C. The mean air summer temperature is 29 °C, while mean winter air temperature is 24 °C. Table 21.1 indicates that the area falls under hypothermic soil temperature regime.

21.2.2 Criteria for Site Selection

- The place where plugging is proposed must produce some direct or indirect benefits to the villagers or farmers.
- Slope of the nalla bed should be less than 30°.
- Plug should not be proposed at the curve of the nalla.
- Width of the nalla should be minimum possible where the structure is to be proposed.

21.2.3 Methodology

The general topography of the study area is plane and rolling with gully along drainage channels of the watershed. Land slope varies from 0 to 5% with low drainage density and high drainage frequency. The micro-watershed land-use and land-cover pattern along with geo-morphological parameters is shown in Table 21.2.

Detailed studies were made to determine land use/land cover, soil type, drainage pattern, current erosion condition and slope of the micro-watershed. Based on these data,

Table 21.2 Geomorphologic parameters of the micro-watershed

S. No.	Aspects	Parameters	Value
1	Number of first-order stream	N_1	12
2	Number of second-order stream	N_2	3
3	Number of third-order stream	N_3	1
4	Length of first-order stream	L_1	1.645 km
5	Length of second-order stream	L_2	0.55 km
6	Length of third-order stream	L_3	0.66 km
7	Length of main stream	L_{mg}	0.8855 km
8	Perimeter of the basin	L_p	0.6 km
9	Length of centroid	L_{ca}	0.775 km
10	Length of overland flow	L_o	0.153 km
11	Elongation ratio	R_e	0.952
12	Circulatory ratio	R_c	0.67
13	Form factor	R_f	0.265
14	Drainage density	D_d	3.298 km km^{-2}
15	Drainage frequency	D_f	28.23
16	Relief ratio	R_f	0.012
17	Ruggedness number	R_N	0.085
18	Average slope	S_1	0.013
19	Mean elevation of watershed	E	113.5 m (above BM)

decisions are made including excessive meandering of the drainage and wetness in the bed or expected pounding area and hard strata of site for the construction of check dams (temporary/permanent) and recharging well. A number of check dams were planned to arrest erosion and to plug the small gully at appropriate places.

Soil erosion can be measured by the small-size runoff plot test experiments in actual field condition and is shown in Fig. 21.4. The small-size runoff plot test uses a plot of 1 × 1 m,

Fig. 21.4 Small-size runoff plot test at actual field site condition

for example. It does not measure the actual amount of soil erosion but the effect of experimental treatments. Common experimental treatments are the percent of soil surface cover both by residue and vegetation, slope, rainfall intensity, the type of soil, the depth of impermeable layer or groundwater level, and the degree of compaction by tractor.

Any intervention such as construction of a sequence of check dams along a river or stream in complex ecosystem will affect the ecological stability and balance of its biotic communities. To minimize the ecological disturbance, the construction of engineering structures and other measures should be as compatible as possible with the stream's natural tendency to reach a stable configuration over a long period. Such criteria should be applied with care and be accompanied by traditional interventions.

Thematic map were prepared in 1:5000 scale on land use/land cover, soil and water resource action plan with drainage network and planning, design and construction of various types of series of water-harvesting structures (i.e. ED—Earthen Dam; LBS—Loose Bolder Structure; BWD—Brush Wood Dam; GBSD—Gunny Bag sand dam; PCD-1—Permanent Check Dam-1; and PCD-2—Permanent Check Dam-2) (Figs. 21.1 and 21.5) using survey of India Toposheet No. 46F/10, actual ground truth and analysis of collected data.

21.3 Results and Discussion

For the micro-watershed (length 1.654 km and catchment area 66.74 ha), the hydrogeomorphic features are presented in Table 21.2. The design peak rate of discharge and storage capacity of each structure in Table 21.3 indicates that the storage capacity and peak rate of discharge of permanent check dam are higher as compared to others. It is revealed that the total volume of storage is 4.55 ha at a cost of Rs. 6.75/1000 l and is estimated to protect 1607 m^3 of top productive soil from flowing out of the area. The design dimensions of drop spillway at each water-harvesting structure are presented in Table 21.4.

The analysis of erosion due to rainfall indicated that more than 2.2 cm rainfall storm is erosive. In general, 42% total rain storms cause erosion in this area. The slope and erosion range at each structure indicate that when the slope is more, the rate of erosion is more. In this area, the slope varies from 0 to 5% with soil erosion of 0–80 Mg ha^{-1} year^{-1}. The soil erosion under brush wood dam, gunny bag, filled with sand dam and Permanent Check Dam-2 were found to be 40--80 Mg ha^{-1} year due to high slope and poor land-use/land-cover pattern.

The soil is generally derived from basic rocks which are calcareous in nature. It is shallow to very deep, clay loam to clay having a calcium carbonate layer at different depths depending upon the topography, thus necessitating careful irrigation scheduling and adequate drainage measures (Fig. 21.2). The soil is generally saline sodic which needs amelioration by addition of calcium-bearing amendments. Micro-watershed soil is considerably eroded and occurrence of pebbles and boulders is a common phenomenon. The cost of the total storage per cubic metre (Table 21.5) was calculated as Rs. 4.12, 4.05, 6.72, 3.35, 9.70 and 10.49 for earthen dam, loose bolder structure, brush wood dam, gunny bag filled with sand dam, Permanent Check Dam-1 and Permanent Check Dam-2, respectively. The series of structure is also used to increase ground water potential and recharging water in downstream side of the well. It can be used as barrier at regular distance in gully and small streams. The tributaries of big streams if provided with a series of structure at a

Fig. 21.5 Various structures constructed in micro-watershed

Table 21.3 Peak rate of discharge at structure through rational formula ($Q = $ CL A/360)

Parameters	ED	LBS	BWD	GBSD	PCD-1	PCD-2
C	0.23	0.23	0.20	0.25	0.20	0.30
I (mm/h)	90.00	120.0	120.0	120.0	120.0	120.0
A (ha)	12.00	8.0	6.00	8.0	15.00	17.90
T_c (h)	0.20	0.30	0.20	0.20	0.30	0.40
Q (m³/s)	0.84	0.61	0.40	0.66	1.00	1.80
SA (ha)	0.56	0.48	0.32	0.38	1.25	1.56
D (m)	3.10	3.30	2.50	2.8	3.20	3.60
SC (m³)	6944	6336	3200	4256	18,000	22,464

Notation: C runoff coefficient, I rainfall intensity (mm h^{-1}), A area (ha), T_c time of concentration (h), Q peak discharge (m³/s), SA submerged area, D maximum depth of water, SC storage capacity

Table 21.4 Design dimension of drop spillway at various WHS

Parameters	ED	LBS	BWD	GBSD	PCD-1	PCD-2
Q (m³/s)	0.84	0.61	0.40	0.66	1.00	1.80
L (m)	1.00	1.00	1.00	1.00	1.50	2.50
H (m)	0.70	0.60	0.40	0.60	0.60	0.60
F (m)	0.50	0.50	0.50	0.50	0.75	1.25
C (m)	1.114	1.114	1.114	1.114	1.112	1.007
E (m)	2.70	2.40	1.80	2.40	2.40	2.40
L_B (m)	1.85	1.68	1.172	1.628	1.758	2.018
J (m)	1.40	1.20	0.80	0.20	1.20	1.20
S_l (m)	0.175	0.15	0.10	0.15	0.15	0.15
S_t (m)	0.23	0.20	0.13	0.12	0.20	0.20

Notation: Q peak discharge (m³/s), L length of weir, H depth of weir, F net drop, C correction factors, E minimum length of head wall extension, L_B length of apron, J height of wing wall and side wall junction, S_l height of longitudinal sill, S_t height of transverse sill

regular interval can help in preventing silting of big dams. Hence, a series of water-harvesting structures are useful in conservation of water and preventing loss of productive soil in this area. The excess rainfall storage due to structure is also useful for irrigation during less rainfall periods. It was found effective for conserving soil and water, protecting the productive soil and regulating the flow of water from horticulture land. These structures are helpful for soil and water

Table 21.5 Average cost of construction of various water-harvesting structure

| Parameters | CA (ha) | Slope | | | Erosion | |
		Slope class	Slope range (%)	Soil loss Class	Soil loss (Mg ha^{-1} year^{-1})	
ED	12	Nearly level	0–1	Slight	0–5	
LBSD	8	Very gentle	1–3	Moderate	5–10	
BWD	6	Gentle	3–5	Severe	40–80	
GBSD	8	Gentle	3–5	Severe	40–80	
PCD-1	15	Very gentle	1–3	Moderate	5–10	
PCD-2	17.74	Gentle	3–5	Severe	40–80	

conservation, preventing bed scoring and new gully formation in the horticulture land.

21.4   Conclusion

The series of water-harvesting structures are useful in conservation of water and preventing loss of productive soil in a micro-watershed. The excess rainfall storage due to structure is also useful for irrigation during fewer rainfall periods. It was found effective for conserving soil and water, protecting the productive soil and regulating the flow of water from horticulture land. These structures are helpful for soil and water conservation, preventing bed scoring and new gully formation in the horticulture land.

Acknowledgements   The authors express their appreciation to ICAR-CHES (CIAH), for full support of the project. It is necessary to thank Shri. R.B. Rathava and VV AppRao for their valuable help and support.

References

Anonymous (1988) Seminar on Water Harvesting System and Management. National Drinking Water Mission, Bangalore

Bhattacharjee JC, Roy Choudhary C, Landey RJ, and Pandey S (1982) Bio climatic Analysis of India N.B.S.S. and L.U.P. Bull No.7, Nagpur. Water Balance of Panchmahals

Govindasamy Agoramoorthy (2007) India Second Green Revolution Needs to Transform the Dryland, 92 Current Science. 157, available at http://www.ias.ac.in /currsci/jan 252007.pdf

Ibid (2002) National Water Policy: Government of India, New Delhi

Khan MA (1992) Development of surface water resources In: Rehabilitation of Degraded Arid Ecosystem. Scientific Publishers, Jodhpur, p 136–143

Krois J and Schulte A (2014) GIS-based multi-criteria evaluation to identify potential sites for soil and water conservation techniques in the Ronquillo watershed, northern Peru. Applied Geography, 51, p 131–142.

Lesschen JP, Kok K, Verburg, PH and Cammeraat, LH (2007). Identification of vulnerable areas for gully erosion under different scenarios of land abandonment in Southeast Spain, Catena, 71 (1),110–121

Narayana VV (1962) Mechanical measures in gully control. Indian J. Conser., 10 (3 and 4): p 41–52

Natividad RA, Woodrige R (1997) Planning and management procedures in irrigation system with mixed cropping. Asian regional Symposium on maintaining and operation of Irrigation, China, May24–27, p 147–163

Pillai KM (1987) Water Management and planning, Himalaya Publishing House Bombay, p 148

Rao K.D, Rao V.V., and Roy P.S. (2005). Water resources development-Role of Remote sensing and Geographical information system.

Singh S (1987) Conservation structures in catchment of river valley projects and flood prone rivers Indian J. Soil Conser., Vol 153. p 8–16.

Deodas Tarachand Meshram is working as a Sr. Scientist at Land and Water Management Engineering division in ICAR-NRC on Pomegranate (Solapur, India). He received his M.Tech (1997) and Ph.D. (2010) in irrigation and water management engineering in IIT, Kharagpur, and MPUAT, Udaipur, respectively. He has published many different articles in indexed journals and 6 books.

S. D. Gorantiwar is working as a Professor & Head at Dr. A. S. Shinde College of Agriculture Engineer & Technology, MPKV, Rahuri, India. He received his M.Tech. (1984), IIT, Kharagpur, Ph.D. (1996), Post Doctorate (2002) and Post Doctorate (Advance) (2006) in Water Engineering & Development, Loughboroagh University, UK

Wadne Saurabh Samadhan worked as SRF at Land and Water Management Engineering in ICAR-National Research Centre on Pomegranate, Solapur, Maharashtra (India), from January 2018. He completed his postgraduation degree of M.Sc. Agriculture in the subject of soil science and agriculture chemistry at the College of Agriculture, Latur (VNMKV, Parbhani, Maharashtra) in the year 2017.

Arun Kumar K. C. received his M.Sc. in geography from the Department of Geography, Kannur University, Kerala, India, in 2016 and M. Tech in geoinformatics from Bharathidasan University, Tamil Nadu, India, in 2018. Currently, he is a Young Professional-II (RS&GIS) with ICAR-National Bureau of Soil Survey and Land Use Planning (NBSS&LUP), Maharashtra, India. His research interests include remote sensing and GIS applications to agriculture, agro-disasters monitoring, and regional planning for agricultural development.

Avijit Kar, Deep Sankar Chini, Manojit Bhattacharya, Basanta Kumar Das,
and Bidhan Chandra Patra

Abstract

At present, gully erosion contributes one of the most significant environmental soil degradation problems related to the laterite zones in West Bengal, India. Erosion also causes undesirable sedimentation that directly or indirectly affects the native aquatic environment. Such environmental variability is widely reflected through the anomalous distribution of confined fish species aggregations. In addition to the effect of gully erosion, a trend to decline in water quality controls the availability of plankton in certain areas, as well as being responsible for the habitat loss of fish and other aquatic organisms. Thus, the present study endeavored to depict the affected riverine areas through gully erosion and characterization of such environmental factors to prevent ichthyofaunal loss. Such work also suggested a future strategy development for local people to lead to the impact assessment of gully erosion.

Keywords

Gully erosion · Habitat loss · Ichthyofauna · Water quality

22.1 Introduction

In our diverse, yet increasingly interdependent world, it is observed that environmental problems are plaguing the nations in different ways. These multi-faceted environmental issues include global warming, pollution, escalated desertification, soil and land degradation, massive extermination of floral and faunal diversity, increased flooding, unnatural erosion, and other unwholesome anthropogenic activities that have, over time, significantly mutilated human society as well as the defacing the Earth's surface (Ibáñez et al. 2016). In addition, the diverse effects of gully erosion and the gradual corruption of submerged low land causes wanton obliteration of soils and farmlands as well as local habitat, which directly or indirectly affects marginal lives (Mukai 2017). Gully erosion reflects a clear form of soil degradation processes that are restricted in an open, unstable incised/worn channel that is 25–30 cm deep. Also, the uncontrolled increase of gullies is responsible for poor land topography along with ecological degradation and degradation of the economy of the affected confined area (Cavey 2006). Such natural hazards as gully erosion damage the surrounding agricultural land, gradually decaying water quality by impure contamination, which causes degradation of the environment and the native ecosystem (Chen et al. 2016). Therefore, incorporation of multiple factors by gully erosion alters the physicochemical nature of the lotic and lentic freshwater systems, which is crucial in the life cycle of aquatic and semi-aquatic beings in their natural habitat (Conoscenti et al. 2014). However, river water is essential to conserve and manage a variety of economically significant aquatic animals, particularly fishes, leading to essential nutrition and maintaining the promising income generation source for the people (Mohanty et al. 2010). In addition to aquatic biodiversity, ichthyofauna, which belong to the top trophic level, maintain and stabilize the natural ecosystem along with indicating overall environmental quality for respecting the intrinsic worth of all species (Kolding and Van Zwieten 2006; Bhattacharya et al. 2018). Accordingly, several scientific studies support that soil erosion is related to many individual gullies that are consolidated into a significantly crucial factor of land degradation on a local-scale basis (Poesen et al. 2003). Subsequent to such previous work, enhanced overflow by runoff from sewage, mutilation of laterite soil from several parochial channels, and

A. Kar · D. S. Chini · B. C. Patra (✉)
Centre for Aquaculture Research, Extension & Livelihood, Department of Zoology, Vidyasagar University, Midnapore, West Bengal, India

M. Bhattacharya · B. K. Das
ICAR—Central Inland Fisheries Research Institute, Barrackpore, Kolkata, West Bengal, India

© Springer Nature Switzerland AG 2020
P. K. Shit et al. (eds.), *Gully Erosion Studies from India and Surrounding Regions*, Advances in Science, Technology & Innovation,
https://doi.org/10.1007/978-3-030-23243-6_22

sedimentation from upper slopes are considered the main causative agents for gully erosion, which has become an anxiety and a prime issue at the respective regions of Paschim Medinipur river sites in West Bengal, India (Shit et al. 2015b). Moreover, the Paschim Medinipur district is a rich source of freshwater fisheries, having rivers, perennial water bodies, wetlands, irrigation channels, creeks, and culturing areas (Kar et al. 2017; Patra et al. 2017). Subsequently, the Kangsabati and the Silabati are the two leading rivers that flow through this region, having immense scope for freshwater fisheries development. Furthermore, a wider range of wild and small indigenous freshwater fishes are found in the surroundings of the Kangsabati and Silabati river basin and adjoin regions. Consequently, in the past few decades, gully erosion has threatened river fisheries by highly undesirable discharge that includes negative impacts on the physicochemical nature of the riverine ecosystem (Sovacool and Linnér 2016). As a consequence, increased turbidity, conductivity, salinity, biological oxygen demand (BOD), and fluctuation of pH level in the this ecosystem result in habitat loss of primary producers, algae, and plankton that are considered key feeding resources, as well as an assemblage of ichthyofauna (Kisku et al. 2017). In that scenario, it is urgently required to estimate the subsequent effects of gully erosion that change the stocking density of freshwater fishes associated with habitat loss. Regular increase in sediment delivery rates from gullies has a negative impact on the local riverine ecology and degrades environmental variables where fish utilize their breeding and feeding grounds (Shit et al. 2014). Thus, the present work monitors the drainage network dynamics and successive effects of gully erosion through field study and assessment of the fish diversity fluctuation in relationship to the physicochemical properties of the catchment area of the Kangsabati and the Silabati river basin. Current investigation assesses the interworking of processes leading to the development of sedimentation rates from the confluence of gullies into the rivers, causing degradation of riverine water quality. That degradation surely will have subsequent effects of seasonal abnormality on fish

stocks in the specified area of both the Kangsabati and the Silabati rivers.

22.2 Description of the Study Area

The existing research design has been focused on the lateritic part of the Kangsabati and the Silabati river basins in West Bengal, India. Survey-based analysis of the two gullies in the degraded regions located at the Midnapore zone was confined to the Kangsabati river basin and the Gangani connected area enriched by the Silabati River in Paschim Midnapore district. The Gangani area situated near Garhbeta town within West Midnapore District is remarkably noted as the Grand Canyon of India, almost 70 ft deep, and formed by massive erosion of the Silabati riverbank. Simultaneously, the Kangsabati River passes through from Bankura District, then enters Jhargram and Paschim Medinipur region and continues to flow to the different parts of the district. Three sampling sites in each river were taken for the superior investigation of the present work where the Rangamati study site at the Kangsabati riverbank is located behind the district town of Paschim Medinipur and covered by laterite soil. Dherua and Munibgar include the clay alluvial type of soil of the Kangsabati riverbank that may help in irrigation, agriculture, and also river fisheries. All sampling points were recorded in real time corresponding to GPS location (Table 22.1).

22.3 Physiographic and Surface Soil Characteristics at the Study Area

The River Kangsabati and Silabati (locally named as Silai) is a seasonal freshwater river, originating from the Choto Nagpur plateau in the state of Jharkhand, where both rivers stretch approximately 465.23 km and 91.3 km through the districts of Purulia, Bankura, Jhargram, and Paschim Medinipur (Dey et al. 2009). The climate of the two study regions becomes tropical in type with hot, dry summers

Table 22.1 GPS location of different gully eroded study sites at the Kangsabati and Silabati river basin

Different study sites at two river basin		GPS position of different sampling station	
		Latitude (east)	Longitude (north)
Kangsabati River	Munibgar (SI)	87°22′45″	22°23′58″
	Rangamati (SII)	87°17′51″	22°24′34″
	Dharua (SIII)	87°05′39″	22°29′22″
Silabati River	Lakhiyapal (SI)	87°11′19″	22°53′31″
	Gangani (SII)	87°20′37″	22°51′25″
	Parkhanda (SIII)	87°24′38″	22°52′23″

(pre-monsoon period, March–June), with precipitation in West Bengal state concentrated in the monsoon period (July–October) (Sen et al. 2004). The land surface of the Silabati study site is characterized by hard and rocky uplands, barren lateritic-covered areas, and nonarable lands. whereas the Kangsabati River site has a consistently alluvial pattern of soil, except in the Rangamati area. The erosive nature of both landscapes at the study areas become noticeably influenced by runoff water in the rainy season that alters soil characteristics, pH, deposited organic matter, ionic concentration, cation exchange (Na^+, K^+, Ca^{2+}), and other vital physicochemical properties at lower sections of the gully catchments (Shit et al. 2015a, b).

22.4 Materials and Methods

22.4.1 Field Observation and Water Quality Assessment

Three study sites were selected from the Kangsabati and the Silabati: Site I (SI) for the upper stream, Site II (SII) for the middle stream, and Site III (SIII) for the lower stream from each river. Seasonal investigation also determined the geometric growth and dimension of the gully, initial length of gullies, and observed the soil loss as well as discharge of undesirable gradients of the gullies from each selected study site. Real-time observation was confined to a 51-km stretch of the Kangsabati River and a 28.5-km stretch of the Silabati River in Paschim Medinipur. Riverine water samples from the six respective sampling sites were collected at around 10 cm from 30 cm depth in the three major seasons: pre-monsoon (March–June), monsoon (July–October), and post monsoon (November–February), at regular intervals during 2015–2018. After collection, physicochemical parameters of each sample (pH, conductivity, salinity, dissolved oxygen, BOD, etc.) were analyzed with a water analyser kit (model Systronics 371).

22.4.2 Fish Sample Collection and Laboratory Procedure

The fish specimens were captured and collected with the assistance of a gill net or drag net, or scoop net, as well as with hooks and lines through the active participation of the local people who are experienced in fishing with relative mass size, during our study period from the two river basins. In all cases, sampling was done during early morning hours and the specimens were preserved immediately in 4% formaldehyde for further studies in the laboratory. The fish specimens were identified following keys provided by Jhingran (1991). Talwar and Jhingran (1991), and Jayaram

(1999) in addition to online databases (Froese and Pauly 2019).

22.4.3 Analysis of Field Data

After collection of primary data through by real-time field observation, all the data were statistically analyzed to estimate the interrelationship between soil erosion and such environmental variables as pH, turbidity, conductivity, phosphate, dissolved oxygen (DO), and BOD that directly or indirectly affect the fish assemblage.

22.4.4 Statistical Analysis

The present research also analyzed the ichthyofaunal diversity indices to assess the ecological rarity or commonness of species in a community to understand the community structure of the concerned study areas.

To better assess the diversity profile, we used Shannon's diversity index (H), Simpson's index (Baker et al. 2007), Simpson's index of diversity (1-D), Pielou's evenness index (J'), Chao-1 diversity index, individual rarefaction model, the α-diversity profile, and SHE analysis.

The *Shannon diversity index (H)* was used to characterize the species diversity in a community. Shannon's index accounts for both abundance and evenness of the species present.

$$H = -\sum_{i=1}^{S} p_i \ln p_i$$

where H = Shannon's diversity index, S = total number of species in the community, Pi = proportion of S including the ith species, and ln = natural logarithm.

Simpson's index (Baker et al. 2007) measured the probability that any two individuals drawn at random from an infinitely large community will belong to the same species. There are two versions of the formula for calculating D. Thereafter, *Simpson's index of diversity (1-D)* was calculated to represent the probability that two individuals randomly selected from a community will belong to different species. The value of the (1-D) index ranges varies from 0 to 1.

$$D = \frac{\sum n(n-1)}{n(n-1)}$$

where n = the total number of individuals of each species and N = the total number of organisms of all species.

Pielou's evenness index (J') was calculated to represent the evenness of a community.

$$J' = \frac{H'}{H'_{max}}$$

where H' is derived from the Shannon diversity index and H'_{max} is the maximum possible value of H' (if every species was equally likely) and calculated as

$$H'_{max} = -\sum_{i=1}^{s} \frac{1}{S} \ln \frac{1}{S} = \ln S$$

where S is the total number of species.

SHE analysis examines the relationship between S (species richness), H (information—the Shannon–Wiener diversity index), and E (evenness as measured using the Shannon–Wiener evenness index (Pielou J') in the samples.

Chao-1 analysis was calculated for the estimation of total species richness.

$$Chao1 = S + \frac{F1(F1-1)}{2(F2+1)}$$

where $F1$ = singleton species number, $F2$ = doubleton species number, and S = taxa number.

Individual rarefaction index was analyzed to assess species richness from the results of sampling in each study site.

$$E(S_n) = \sum_{i=1}^{S} \left[1 - \frac{\binom{N-N_i}{n}}{\binom{N}{n}} \right]$$

where $E(S_n)$ = expected species number.

22.5 Results and Discussion

In addition to water quality, analysis at six sampling sites revealed that the Rangamati and Gangani sampling sites represent a higher range of pH, turbidity, conductivity, total dissolved solids (TDS), chloride, and phosphate and a lower value of dissolved oxygen (DO) resulting in a gradual increase of biological oxygen demand (BOD), which directly or indirectly affect the habitat loss of aquatic organisms to the particular study areas (Table 22.2). During the survey, pH value showed a lower range in Gangani (SII) and Rangamati (SII) from the Silabati and the Kangsabati, respectively, where a favorable range was found in Ramnagar (SIII), Dherua, within the Kangsabati river basin. Turbidity, conductivity, and TDS also measured higher during the monsoon season and lower in the post-monsoon period, but at Rangamati and Gangani regions a higher range was seen compared to the other study areas of the two river basins. Therefore, estimated DO was lower at SII regions of both river basins, and BOD remained higher than at other study sites (Table 22.3).

Sediment volumes of clay, silt, sand, and different cations were also measured at six respective study sites in three different seasons (pre-monsoon, monsoon, and post-monsoon) for 3 years. Thus, the results were also compared between the inflow rate of different gullies, and the highest level of sediment volume was measured in the Gangani and Rangamati basin rather than at other study sites. However, heavy sedimentation in the respective two study areas caused a high cation-exchange rate, which increases water turbidity. The confluence of different gullies into both the Kangsabati and the Shilabati riverine water causes a large influx of laterite soil, which is considered as one of the major reasons for geo-environmental degradation in native aquatic habitats (Shit et al. 2015b). The local arrangement of gullies entails an alteration of overland flow and gradual increase of both runoff water and sediment volume, resulting in low densities of planktonic diversity (Valentin et al. 2005). Thus, the growing interest on the relationship of gully erosion and degradation of aquatic habitat reflects the need to develop definite knowledge of the ecological impacts as well as such environmental factors as control riverine fish diversity. Along with the comparison between sedimentation rate and different environmental variables, the sedimentation rate may vary in different seasons (pre-monsoon, monsoon, and post-

Table 22.2 On the basis of modes and conditions of formation and common advance mechanism (Ezechi and Okagbue 1989), gullies can be divided into the following types and effect-prone area

Gully type	Modes and condition of formation	Common advance mechanism	Affected study site Kangsabati River	Silabati River
Base level	Groundwater flow	Slope undermining, sliding and slumping	SI, SII	SI, SII, SIII
Scarp	Runoff and slope change	Slope undermining, sliding/slumping, toppling	SII	SII
Fracture	Runoff and shrinkage fracture	Collapsing, also block failure	SI, SII	SI, SII
Incidental	Runoff concentration and vulnerable soil exposure by humans	Common sliding/slumping	SII	SII

Table 22.3 Physicochemical parameters of six different sampling sites fort Kangsabati and Silabati riverine water

Sampling site	pH			Turbidity (NTU)			Conductivity ($\mu\Omega$-1 cm^{-1})			DO (mg/l)			TDS (mg/l)			Chloride (mg/l)			Phosphate (mg/l)			BOD (mg/l)		
	Pr. M	M	P. M	Pr. M	M	P. M	Pr. M	M	P. M	Pr. M	M	P. M	Pr. M	M	P. M	Pr. M	M	P. M	Pr. M	M	P. M	Pr. M	M	P. M
Kangsabati River																								
SI(U)	8.16	7.87	8.09	3.21	17.84	2.74	65.58	95.47	46.39	10.38	8.56	13.86	97.1	123.04	74.5	225.28	297.74	195.37	0.08	0.37	0.07	1.29	2.03	1.2
SII(M)	7.89	6.92	7.51	10.75	26.87	8.45	120.23	167.5	105.22	7.23	6.37	7.95	126.36	197.55	102.35	265.28	307.74	215.37	0.13	0.98	0.12	2.22	3.07	1.91
SIII(L)	7.96	7.02	8.14	4.27	19.65	3.75	69.25	102.45	51.35	10.08	8.01	14.06	95.37	119.05	76.5	227.85	289.55	192.35	0.09	0.28	0.065	1.35	2.09	1.32
Silabati River																								
SI(U)	8.06	7.17	7.98	2.36	19.05	2.85	70.25	90.15	49.05	09.28	8.75	12.97	92.25	136.01	76.95	220.35	298.05	185.9	0.09	0.22	0.05	1.31	2.73	1.4
SII(M)	7.41	6.57	6.85	14.45	39.85	11.25	110.35	175.55	115.86	8.12	6.07	6.85	132.57	205.45	115.64	265.28	307.74	225.85	0.23	1.07	0.72	2.48	4.12	2.97
SIII(L)	7.81	7.02	7.95	3.01	14.74	4.17	71.56	100.87	62.45	10.08	7.42	12.85	82.46	108.75	65.38	205.35	265.85	165.35	0.03	0.09	0.02	1.08	2.01	0.92

monsoon) at the study regions (Fig. 22.1d). The results also support that the sedimentation rate is at a high level at the Gangani and Rangamati study sites compared to other sites and was very high in the monsoon season. Thus, these two regions showed lower planktonic concentrations because the heavy sedimentation causes aquatic environment pollution through higher levels of BOD and turbidity and a lower level of DO, with a negative role in fish diversity in those two study sites.

In addition to the regular field observations and analysis of the physicochemical properties of riverine water during the whole survey period, in total 42 fish species belonging to 30 genera, 7 orders, and 16 families were recorded within the respective study area of the two river basins. Current investigation also revealed that the greatest number of species present were in the order Cypriniformes (16), with the family Cyprinidae showing dominance (15). On the basis of real-time investigation for the 3 years, in total 6769 fish samples were captured from the six sampling sites of the Kangsabati and Silabati river basin. In addition to the collection of fishes, the Rangamati site represents lower species richness (923) within the Kangsabati River; SI and SIII represent higher species richness (935 and 933, respectively) than in the Gangani sampling station (502). Results also support that a few fish species, including *Labeo calbasu*, *Labeo catla*, *Lepidocephalichthys guntea*, *Chitala chitala*, and *Glossogobius giuris*, were not found in either the Gangani or Rangamati sampling stations, whereas *Osteobrama cotio*, *Xenentodon cancila*, *Gudusia chapra*, *Channa marulius*, *Channa striata*, *Sperata aor*, and *Rita rita* were absent only from the Gangani sampling sites (Table 22.4). Results also supported that most of the collected fish species utilized the herbivore type of feeding habit, and only a few species were omnivorous. Hence, statistical analysis explored fish availability in relationship to the species richness fluctuation of the six different study sites. According to the individual rarefaction index, a minimum of 20 species were found in all sampling station but richness was high at the SI and SIII sites of the Kangsabati River basin and SI of the Silabati River basin (<38 species) (Fig. 22.2a). According to statistical analysis in both study zones of the Kangsabati and Silabati river basin, measured by Shannon–Wiener index, Simpson's diversity index, SHE analysis, and Chao-1 analysis, these symbolize a scientific co-ordination to the habitat distribution of fishes and habitat disturbances caused by erosion. The value range of the Shannon–Wiener index was 3.02 and 3.41 (range, 1–5), whereas Simpson's diversity index was 0.935 and 0.961 (ranges, 0–1) for each respective study site (Fig. 22.2b).

In relationship to environmental variables and biodiversity, aquatic environments are alarming; a serious threat, with a warning, both to aquatic organisms and to the ecosystem sustainability strategies that have been projected to overcome

Fig. 22.1 (**a**) Study area, (**b**) study sites at the Kangsabati River basin, (**c**) study sites at the Silabati River basin, and (**d**) gully erosion bank of River Silabati at Gangani

this crisis throughout the world. Such prolonged environmental stress is caused by ongoing increases in road construction, urbanization, unauthorized barrier creation, unscientific drainage systems for sewage runoff, and other anthropogenic activities (Conoscenti et al. 2014). As a result, the quality of lotic and lentic freshwater ecosystems, and of rivers, ponds, canals, and creeks, becomes gradually degraded, generating negative impacts to all aquatic organisms, especially fish fauna. However, favorable physicochemical factors and nutritional condition regulate the productive efficiency of plankton populations that initiate the elementary nodes of the aquatic food chain and enhance species dominance in the concerned water bodies. Current research clarifies that fish species with the herbivore type of feeding habit would be much more numerous (45%) than carnivorous (31%) and omnivorous (24%) feeders of the fish species recorded from the selected study sites (Fig. 22.3). However, on the basis of the ecological aspects, fish species richness and plankton density have a positive co-relationship that ensures

nutritional demand unless the undesirable influx of some physicochemical factors to the aquatic ecosystem penetrates the productive balance. Usually, the river acquires such hazards from local sewage or increased runoff that causes the planktonic populations to fluctuate, having a negative impact on herbivorous fishes; that problem was observed especially at the SII sampling sites of the Kangsabati and Silabati river basin.

Therefore, a significant question arises: why are fish assemblages lowest at the Rangamati and Gangani sampling sites? Evidence from the survey reports that physicochemical ranges in pH, TDS, BOD, chloride, phosphate, and conductivity, as well as rate of soil sedimentation if also higher, decrease the amount of DO throughout all seasons at the Rangamati and Gangani study sites. Thus, problems associated with herbivores require herbivorous fish to shift to other favorable sites of the river stretch to ensure their food availability. Accordingly, such undesirable variations in this particular aquatic ecosystem cause negative impacts to the

Table 22.4 Organic checklist with scientific name, local name, IUCN status, human use, feeding habit, and catch data during survey period (2015–2018)

Order	Family	Scientific name	Local name	IUCN	Human use	Feeding type	Kangsabati River Site			Silabati River Site		
							Site I (U)	Site II (M)	Site III (L)	Site I (U)	Site II (M)	Site III (L)
Cypriniformes	Cyprinidae	1. *Osteobrama cotio* (Hamilton, 1822)	Keti	LC	Ornamental/commercial	Herbivore	04	01	03	03	00	00
		2. *Devario devario* (Hamilton, 1822)	Techokha	LC	Ornamental	Herbivore	97	58	91	78	47	64
		3. *Danio rerio* (Hamilton, 1822)	Zebrafish	NT	Ornamental	Herbivore	149	97	110	123	81	98
		4. *Pethia ticto* (Hamilton, 1822)	Punti	LC	Ornamental/commercial	Herbivore	41	22	35	58	37	42
		5. *Puntius sophore* (Hamilton, 1822)	Sar punti	LC	Ornamental/commercial	Herbivore	26	20	63	28	21	23
		5. *Puntius phutunio* (Hamilton, 1822)	Punti	LC	Ornamental/commercial	Herbivore	41	31	54	27	22	25
		6. *Pethia conchonius* (Hamilton, 1822)	Kanchan Punti	VU	Ornamental/commercial	Herbivore	33	29	38	16	09	17
		7. *Salmostoma bacaila* (Hamilton, 1822)	Chela	LC	Commercial	Herbivore	47	31	49	44	27	29
		8. *Labeo calbasu* (Hamilton, 1822)	Kalbose	LC	Aquaculture/commercial	Herbivore	03	00	01	03	00	00
		9. *Labeo bata* (Hamilton, 1822)	Bata	LC	Aquaculture/commercial	Herbivore	19	13	28	14	03	11
		10. *Labeo rohita* (Hamilton, 1822)	Rohu	LC	Aquaculture/commercial	Herbivore	37	16	48	09	03	14
		11. *Cirrhinus mrigala* (Hamilton, 1822)	Mrigal	LC	Aquaculture/commercial	Omnivore	28	08	31	08	03	11
		12. *Labeo catla* (Hamilton, 1822)	Catla	LC	Aquaculture/commercial	Herbivore	09	02	21	07	00	11
		13. *Amblypharyngodon mola* (Hamilton, 1822)	Mourola	LC	Commercial	Herbivore	137	89	168	75	31	103
		14. *Esomus danrica* (Hamilton, 1822)	Darkina	LC	Ornamental/commercial	Herbivore	42	21	51	24	8	25
		15. *Salmostoma phulo* (Hamilton, 1822)	Chela	LC	Ornamental/commercial	Herbivore	37	12	43	13	03	09
	Cobitidae	16. *Lepidocephalichthys guntea* (Hamilton, 1822)	Guntey	LC	Ornamental/commercial	Omnivore	05	00	09	02	00	00
Cyprinodontiformes	Aplocheilidae	17. *Aplocheilus panchax* (Hamilton, 1822)	Techoukka	LC	Ornamental/commercial	Herbivore	57	31	68	31	17	26
Beloniformes	Belonidae	18. *Xenentodon cancila* (Hamilton, 1822)	Kakila	LC	Aquaculture	Omnivore	17	03	29	08	00	03
Clupeiformes	Clupeidae	19. *Gudusia chapra* (Hamilton, 1822)	Khaira	LC	Commercial	Herbivore	09	04	15	01	00	01
Osteoglossiformes	Notopteridae	20. *Chitala chitala* (Hamilton, 1822)	Chital	NT	Ornamental/commercial	Omnivore	00	00	02	01	00	00
		21. *Notopterus notopterus* (Pallas, 1769)	Pholui	LC	Ornamental/aquaculture	Carnivore	21	02	33	12	01	09

(continued)

Table 22.4 (continued)

Order	Family	Scientific name	Local name	IUCN	Human use	Feeding type	Species availability (sampling time)					
							Kangsabati River Site			Silabati River Site		
							Site I (U)	Site II (M)	Site III (L)	Site I (U)	Site II (M)	Site III (L)
Perciformes	Channidae	22. *Channa punctata* (Bloch, 1793)	Lata	LC	Ornamental/aquaculture	Carnivore	29	45	43	21	17	23
		23. *Channa marulius* (Hamilton, 1822)	Sal	LC	Ornamental/aquaculture	Carnivore	01	03	01	03	00	01
		24. *Channa gachua* (Hamilton, 1822)	Chang	LC	Ornamental/aquaculture	Carnivore	26	29	30	10	02	21
		25. *Channa striata* (Bloch, 1793)	Shol	LC	Ornamental/commercial	Carnivore	07	12	14	01	00	04
	Gobiidae	26. *Glossogobius giuris* (Hamilton, 1822)	Bele	LC	Ornamental/Commercial	Herbivore	03	00	05	02	00	03
	Nandidae	27. *Nandus nandus* (Hamilton, 1822)	Bheda	LC	Ornamental/commercial	Carnivore	56	28	87	19	07	27
	Ambassidae	28. *Parambassis ranga* (Hamilton, 1822)	Chanda	NE	Ornamental/commercial	Omnivore	91	51	109	54	29	47
		29. *Chanda nama* (Hamilton, 1822)	Chanda	LC	Ornamental/commercial	Omnivore	71	36	106	33	18	53
	Osphronemidae	30. *Trichogaster fasciata* (Bloch & Schneider, 1801)	Khalisa	LC	Ornamental	Omnivore	79	32	94	51	18	38
		31. *Trichogaster lalius* (Hamilton, 1822)	Khalisa	LC	Ornamental	Omnivore	27	09	37	10	05	17
Siluriformes	Bagridae	32. *Mystus cavasius* (Hamilton, 1822)	Tengra	LC	Commercial	Carnivore	118	38	92	43	18	57
		33. *Sperata aor* (Hamilton, 1822)	Aard	VU	Ornamental/commercial	Carnivore	37	21	52	02	00	00
		34. *Sperata seenghala* (Sykes, 1839)	Tangra	LC	Commercial/aquaculture	Carnivore	27	12	33	14	06	28
		35. *Mystus tengara* (Hamilton, 1822)	Tangra	LC	Commercial/aquaculture	Carnivore	61	29	73	31	16	39
		36. *Mystus vittatus* (Bloch, 1794)	Tangra	LC	Commercial/aquaculture	Carnivore	31	11	33	24	15	27
		37. *Rita rita* (Hamilton, 1822)	Reta	LC	Ornamental	Herbivore	17	04	10	03	00	00
	Claridae	38. *Clarias batrachus* (Linnaeus, 1758)	Magur	LC	Ornamental/commercial	Carnivore	14	23	19	08	11	07
	Pangasidae	39. *Pungitius pungitius* (Linnaeus, 1758)	Pangus	LC	Ornamental/commercial	Omnivore	16	11	28	07	05	03
	Siluridae	40. *Wallago attu* (Bloch & Schneider, 1801)	Boal	NT	Commercial	Carnivore	02	01	07	01	00	05
		41. *Heteropneustes fossilis* (Bloch, 1794)	Singi	LC	Commercial	Carnivore	03	14	04	01	07	03
	Mastacembelidae	42. *Macrognathus pancalus* (Hamilton, 1822)	Pankal	LC	Commercial	Omnivore	14	24	20	12	15	09

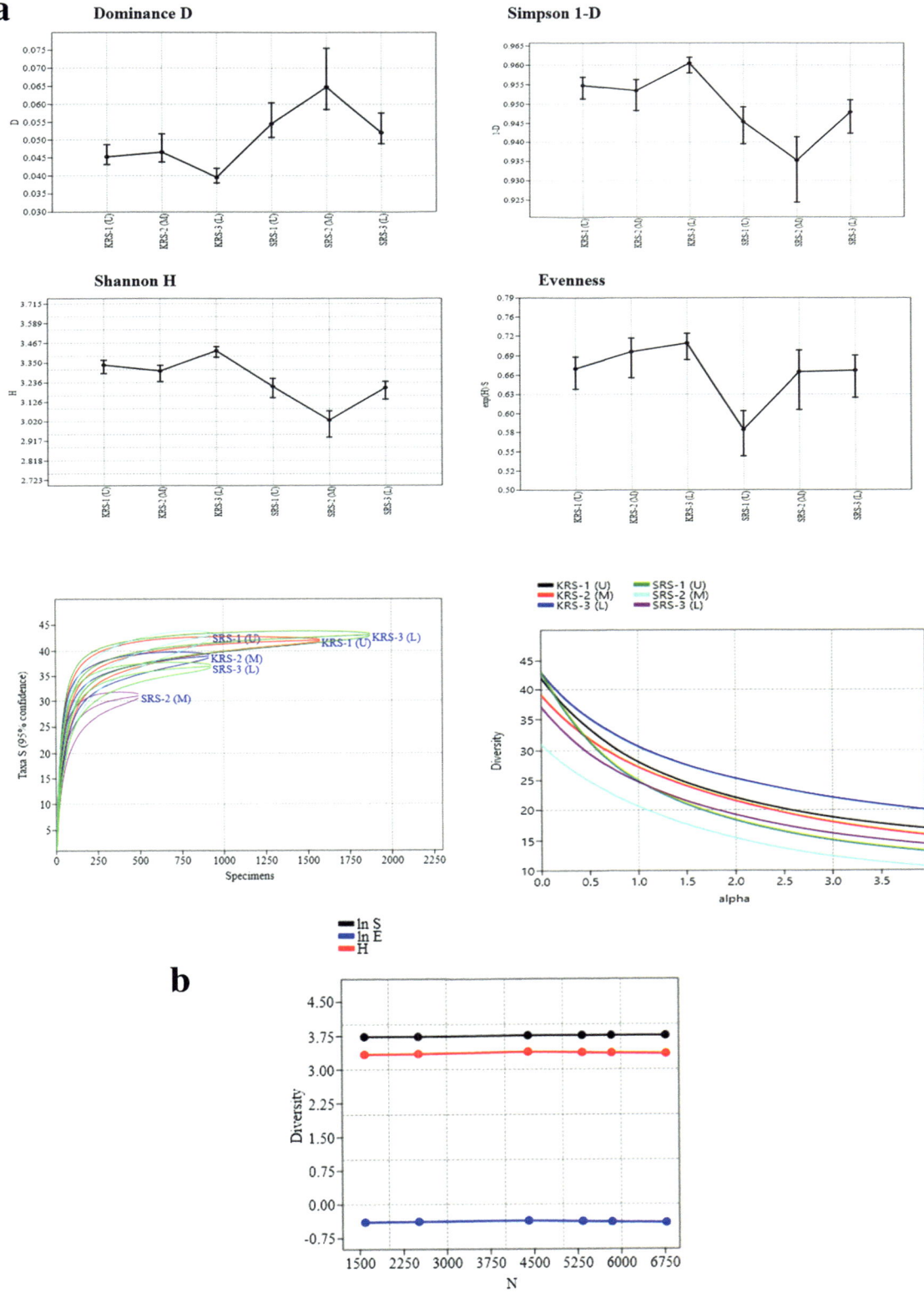

Fig. 22.2 (**a** and **b**) Fish diversity index and profile of fish species available in different study sites at Kangsabati and Silabati river basin

Fig. 22.3 Feeding habits of
available fish species in different
study sites of Kangsabati and
Silabati river basin

Feeding types

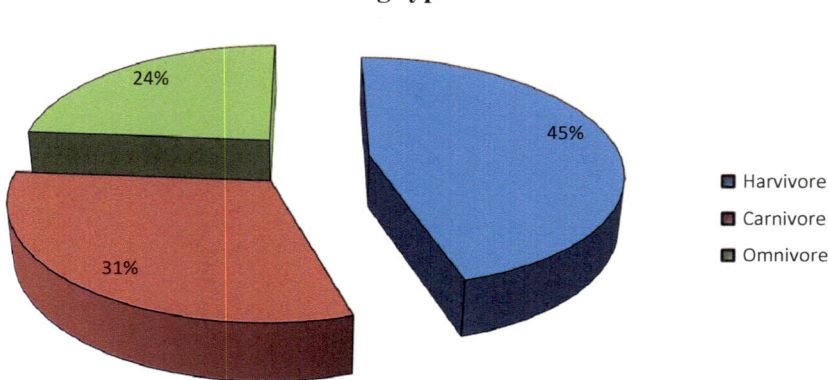

current fish faunal diversity, as well as planktonic density, at
the Rangamati and Gangani sites.

22.6 Conclusion

The present research study clarified a considerable degree of
laterite exposures that are directly related to sewage outfall,
morphometric characters of drainage systems, soil erodibility
with texture loss, surface runoff, and susceptibility to gully
and rill erosion in the respective river basins of concern. In
addition to the positive correlation of soil sedimentation rate
with different environmental variables, a lower range of
planktonic availability results that requires a proper management policy to overcome the present fluctuation of the fish
assemblage in certain areas of the study regions. Significantly, human interference coupled with intrinsic and morphometric characteristics of the lateritic landscape result in
gully erosion that is largely responsible for varying piscine
diversity as well as altering the physicochemical nature of the
concerned aquatic habitat. Consequently, such phenomena
also transfer numerous negative impacts to human society,
resulting in biased conditions. Therefore, proper management
and conservation policies are required to protect the riverine
areas on an urgent basis where biodiversity fluctuates from
certain intrinsic effects of undesirable changes in physicochemical factors. Besides such findings, it is also shown that
gully erosion leads to changes in different physicochemical
characteristics of native aquatic ecosystems that may be
directly or indirectly responsible for the floral and faunal
diversity linked to habitat loss.

References

Baker, C., R. L. Lawrence, et al. (2007). "Change detection of wetland
    ecosystems using Landsat imagery and change vector analysis."
    Wetlands **27**(3): 610-619.

Bhattacharya, M., D. S. Chini, et al. (2018). "Assessment and modeling
    of fish diversity related to water bodies of Bankura district, West
    Bengal, India, for sustainable management of culture practices."
    Environment, Development and Sustainability: 1-14.
Cavey, B. (2006). "Gully erosion, Department of Natural Resources and
    Water." Fact sheets L **81**: 1-4.
Chen, Z., W. Chen, et al. (2016). "Effects of polyacrylamide on soil
    erosion and nutrient losses from substrate material in steep rocky
    slope stabilization projects." Science of the total environment **554**:
    26-33.
Conoscenti, C., S. Angileri, et al. (2014). "Gully erosion susceptibility
    assessment by means of GIS-based logistic regression: a case of
    Sicily (Italy)." Geomorphology **204**: 399-411.
Dey, S., S. Ghosh, et al. (2009). "Some observation of regional
    evidences of Tertiary–Quaternary geo-dymanics in a paleo-coastal
    of Bengal basin, India." Russian Geology and Geophysics.
Froese, R. and D. Pauly. Editors. 2019. FishBase. World Wide Web
    electronic publication. www.fishbase.org, version (02/2019).
Ibáñez, J., J. L. Contador, et al. (2016). "Evaluating the influence of
    physical, economic and managerial factors on sheet erosion in
    rangelands of SW Spain by performing a sensitivity analysis on an
    integrated dynamic model." Science of the total environment **544**:
    439-449.
Jayaram, K. (1999). "The fresh water fishes of the Indian Region,
    Narendra Publisting house." Delhi-551.
Jhingran, V. (1991). "Fish and Fisheries of India 3rd Edition, Hindustan
    Pub." CorporationNew Delhi.
Kar, A., M. Bhattacharya, et al. (2017). Ichthyofaunal Diversity of
    Kangsabati River at Paschim Medinipur District, West Bengal,
    India. Proceedings of the zoological society, Springer.
Kisku, S., D. S. Chini, et al. (2017). "A cross-sectional study on water
    quality in relation to fish diversity of Paschim Medinipur, West
    Bengal, India through geoinformatics approaches." The Egyptian
    Journal of Aquatic Research **43**(4): 283-289.
Kolding, J. and P. Van Zwieten (2006). "Improving productivity in
    tropical lakes and reservoirs." Challenge Program on Water and
    Food–Aquatic Ecosystems and Fisheries Review Series **1**: 139.
Mohanty, B., S. Mohanty, et al. (2010). Climate change: impacts on
    fisheries and aquaculture. Climate change and variability, InTech.
Mukai, S. (2017). "Gully erosion rates and analysis of determining
    factors: a case study from the semi-arid main ethiopian rift valley."
    Land Degradation & Development **28**(2): 602-615.
Patra, B. C., A. Kar, et al. (2017). "Freshwater fish resource mapping
    and conservation strategies of West Bengal, India." Spatial Informa-
    tion Research **25**(5): 635-645.
Poesen, J., J. Nachtergaele, et al. (2003). "Gully erosion and environmen-
    tal change: importance and research needs." Catena **50**(2-4): 91-133.

Sen, J., S. Sen, et al. (2004). "Geomorphological investigation of badlands: A case study at Garhbeta, West Medinipur District, West Bengal, India." Geomorphology and Environment: 204-234.

Shit, P., G. Bhunia, et al. (2014). "Vegetation Influence on runoff and sediment yield in the lateritic region: An experimental study." J Geogr Nat Disast **4**(116): 2167-0587.100011.

Shit, P. K., A. S. Nandi, et al. (2015a). "Soil erosion risk mapping using RUSLE model on Jhargram sub-division at West Bengal in India." Modeling Earth Systems and Environment **1**(3): 28.

Shit, P. K., R. Paira, et al. (2015b). "Modeling of potential gully erosion hazard using geo-spatial technology at Garbheta block, West Bengal in India." Modeling Earth Systems and Environment **1**(1-2): 2.

Sovacool, B. and B.-O. Linnér (2016). The political economy of climate change adaptation, Springer.

Talwar, P. K. and A. G. Jhingran (1991). Inland fishes of India and adjacent countries, CRC Press.

Valentin, C., J. Poesen, et al. (2005). "Gully erosion: impacts, factors and control." Catena **63**(2-3): 132-153.

Avijit Kar is currently pursuing his doctoral degree in the Department of Zoology, Vidyasagar University, West Bengal, India, and working as a Research Fellow under SEED Division, DST, Government of India-sponsored research project. At present, his research activities focus on ecological modeling, diversity, conservation, and management aspects of fishery resources. He has more than five years of research experience in different research projects under DBT, DST-SERB of Government of India.

Deep Sankar Chini currently served as research scholar and is pursuing doctoral program at the Department of Zoology, Vidyasagar University, in fisheries and aquaculture. His field of interest and published work includes fish diversity assessment, conservation biology, and ecohydrological aspects also.

Manojit Bhattacharya, SERB-National Postdoctorate Fellow, is working at ICAR-Central Inland Fisheries Research Institute, Barrackpore, Kolkata, West Bengal, India. For the last seven years, he is working on molecular genetics and computational biology platforms and has published 30 scientific research articles. Dr. Bhattacharya has wide expertise in different aspects of bio-resource mapping and characterization.

Basanta Kumar Das is the Director, ICAR-Central Inland Fisheries Research Institute, Barrackpore, Kolkata, West Bengal, India. He has worked on fish health management, fish genetics, and microbial molecular analysis and published more than 120 international and national research articles in various reputed, high-impact journals. He also has 22 years of research experience and notable contributions in aquaculture sectors. Dr. B. K. Das's research focuses on aquaculture resources mapping, management, and modeling aspects.

Bidhan Chandra Patra is Senior Professor of Zoology, Vidyasagar University, West Bengal, India. He worked on fisheries science and nutrition biology and published more than 120 scientific research articles in several leading journals. Professor Patra is having more than 30 years of active, operational experience in research and development sectors. At present, he is working on fisheries data management, coastal fishery, computational biology, and advanced geoinformatics techniques.

Gully Erosion in I. R. Iran: Characteristics, Processes, Causes, and Land Use

23

Majid Soufi, Reza Bayat, and Amir Hossein Charkhabi

Abstract

Gully erosion due to land destruction and depletion of soil moisture, especially in drought periods, has an important role for decreasing biomass production in Iran. The aim of this study was to determine the characteristics, processes, and the main causes of gully incision. Satellite images, aerial photos, and anecdotal evidence with field measurements were used to obtain information to answer the research questions. The results of this study indicated that Iranian gullies with an area more than 1,420,000 ha were distributed from coastal zones to highlands with an altitude higher than 3000 m above the sea level. They occurred mostly in the areas with precipitation of 100–300 mm and 500–1000 mm. Gullies are dominant more in semi-arid, arid, and Mediterranean climates. Most of the studied gullies are continuous and valley side. The survey of the view plans of the gullies indicated that surface runoff was the dominant hydrological process for gully incision. The results revealed that the rangelands and forests area decreased while rain-fed farms and barren land increased during the last decades. Most Iranian gullies were located in the altitude of 0–500 and 1000–2000 m above the sea level. Overgrazing and land-use change from rangeland and forests to cultivated lands were the main causes of gully erosion. Data showed that Iranian gullies with an average length and depth equal to 570 m and 2.8 m produced 21 m³/m sediment per unit gully length.

Keywords

Gully erosion · Land use · Erosion processes · Iran

23.1 Introduction

Soil erosion is recognized as the most important factor for soil degradation and many environmental problems in the world (Kropacek 2019; Fang et al. 2019; Gutierez et al. 2009). Gully erosion is more important than other types of water erosion because of limited research, its unknown aspects, and higher contribution for sediment production and damage (Poesen et al. 2003, 2017; Zhang et al. 2019). Gully erosion causes on-site and off-site problems, damages such less trafficability due to breaking the roads and bridges (Soufi 2005; Soufi 2009; Soufi et al. 2017; Soufi and Bayat 2015, 2016; Rey et al. 2019) and decreased biomass production of croplands and rangelands (Nyssen 2001; Avni 2005). Due to the lack of a detailed data bank for gully erosion in the world (Poesen et al. 2003) and also Iran about gully characteristics, its environment and impact on decreasing the biomass and water depletion, especially during drought periods, this study was conducted throughout Iran.

Gullies are classified based on different factors. They are classified based on the shape of the cross-section and its location in the landscape (Imeson and Kwaad 1980; Brice 1966; Deng et al. 2015), slope of gully banks (Crouch and Blong 1989; Ahmadi 1999), gully length and depth (Campos et al. 2000; Sun et al. 2014), and gully area and depth (FAO 1982). The gully cross-section was divided into V, U, and trapezoid or intermediate (Deng et al. 2015). Deng et al. (2015) emphasized the importance of gully cross-section not only for computing the volume and rate of erosion, but also for understanding the relationship of gullying processes, landforms, land use, and erosional features. Different classifications were used for the gully depth. For example, a small gully is defined with 0.9 m (Ahmadi 1999), 2 m (Refahi

M. Soufi (✉)
Fars Research and Education Center for Agriculture and Natural Resources, Department of Soil Conservation and Watershed Management, Shiraz, Iran

R. Bayat · A. H. Charkhabi
Department of Soil Conservation Engineering, Institute of Soil Conservation and Watershed Management, Tehran, Iran

© Springer Nature Switzerland AG 2020
P. K. Shit et al. (eds.), *Gully Erosion Studies from India and Surrounding Regions*, Advances in Science, Technology & Innovation,
https://doi.org/10.1007/978-3-030-23243-6_23

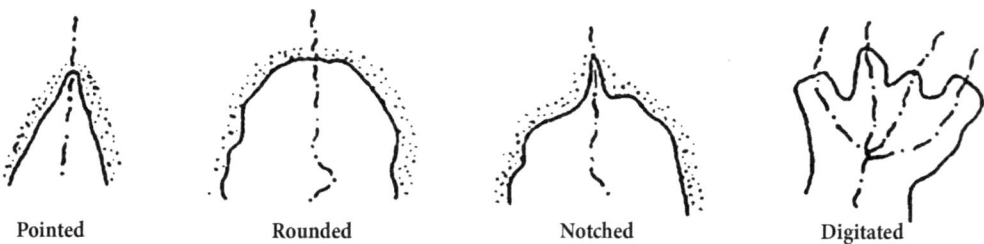

Fig. 23.1 Classification of the view plan of the gully heads (Ireland et al. 1939)

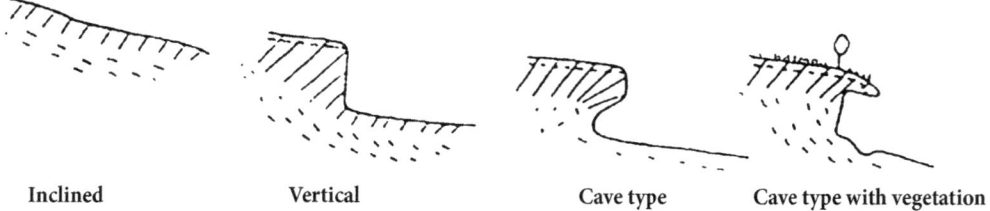

Fig. 23.2 Classification of the longitudinal profile of the gullies (Ireland et al. 1939)

2000), and 1 m by FAO (1982). The medium size gully is defined as 1–5 m depth and the large size was introduced as 5–10 m depth (FAO 1982; Ahmadi 1999; Refahi 2000). Some studies have divided the gullies into continuous (old) and discontinuous (young) in the USA (Leopold and Miller 1965; Heede and Mufich 1974), valley side, and valley floor based on their location (Brice 1966; Imeson and Kwaad 1980). Ireland et al. (1939) classified the view plan of the gullies into six classes including linear, bulbous, dendritic, trellis, parallel, and compound; and the view plan of the gully heads was categorized into four classes including pointed, rounded, notched, and digitated in the USA (Fig. 23.1). They also classified the gully heads based on their longitudinal profile into inclined, vertical, cave type, and cave type with vegetation cover (Fig. 23.2).

Some studies (e.g., Poesen and Govers 1990; Poesen 1989, 1993; Moore et al. 1988) have classified the gullies into ephemeral and permanent based on the possibility of omission, length of life, morphology, location, land use, and dominant hydrologic processes in Europe croplands, Australia and South Africa (Kakembo et al. 2009). The width-to-depth ratio of the gully was used practically by some researchers (Poesen and Govers 1990; Poesen 1993; Deng et al. 2015). Ratios bigger than 1 imply more damages to the cropland and is used for prioritization of its control. Morgan (1995) classified the gullies based on their soil texture and process of gullying. He classified the gullies into three categories including *axial* parallel gullies in coarse soil texture, digitate with vertical head cuts in the loam-clay soil, and frontal with rilled banks and created by tunneling. Some researchers such as Morgan (1995) and Chaplot (2013) introduced categories of gullies based on the mechanism of their formation including those formed by surface runoff, seepage and piping, and mass movement and their

combination. Dietrich and Dunne (1993) classified the gully heads into gradual, step (depth < 1 m), and head cut (depth > 1 m). Ahmadi (1999) classified the gullies of Iran based on their length into three categories, small ($L < 120$ m), medium ($L = 120$–240), and large ($L > 240$ m).

Understanding the causes of gully erosion was a big challenge and controversy to develop strategies for gully control during the last century (Graf 1983; Soufi 1997). Causes of gully erosion were of three types (Soufi 1997), (1) deterioration of ecosystem by human impacts (Nachtergaele 2001; Faulkner 1995; Oostwoud Wijdenes et al. 2000; Bork et al. 2001; Soufi 2004; Harvey 1996; Valentin et al. 2005; Castillo and Gomez 2016; Abi et al. 2018; Yu et al. 2019; Okuh and Osumgborogwu 2019; Rong et al. 2019); (2) climate change (Balling and Wells 1990; Castro et al. 2000; Vanmaercke et al. 2016); and (3) intrinsic or random change in the ecosystem (Starr 1989; Soufi 1997). Gully incision and development are attributed to severe grazing by cattle and climate change in southwest USA (Webb and Hereford 2001), after European settlement in the east of Australia since the last 200 years (Prosser and Winchester 1996), change of vegetation cover by humans in England (Harvey 1996), high pressure on land use, and intense rainfall in fourteenth century in Germany (Bork et al. 2001). Pine plantation after clearance of the *Eucalyptus* forest and soil plowing was introduced as the main factor for massive gully erosion in southeast Australia (Soufi 1997). Monsiers et al. (2015) found that cropland with exclusive drainage ditches was most vulnerable to gully development. Increase in the cultivation of corn in central Belgium (Nachtergaele 2001) or increased almond plantation without terracing after destruction of native Mediterranean vegetation cover or increased plantation areas within the catchments of SW Spain (Gutierez et al. 2009) have been proposed as one of

Fig. 23.3 The provinces studied for gully erosion in Iran

the main factors for gully erosion in South Spain (Faulkner 1995; Oostwoud Wijdenes et al. 2000). Svoray (2009) believed that unpaved roads had a greater influence on the location of gully heads than tillage direction in Northern Israel. A recent review by Vanmaercke et al. (2016) indicated that gully head retreat was significantly correlated to the runoff contributing area and rainy day normal. They found that land use and soil type had no significant correlation with gully head retreat in different parts of the world. They believe that gully erosion will become more intense and widespread in the following decades as a result of climate change.

Nowadays, human impacts are introduced as the most important factor for gully incision. In a large scale, land use plays an important role in gully formation. For example, there is less gully erosion in the forests, but it is more in rangelands and croplands. The reason is decline in the range condition due to overgrazing and reduced resistance of the soil surface in croplands. Therefore, changes in the land use from natural to cultivated land have increased the risk of gully incision and development such as southeast highlands of Vietnam (Valentin et al. 2005). In similar land uses, gully erosion depends on topography. That is, lands with steeper gradient might have more gully erosion. Although human impact or climate change had an important role in gully erosion in some parts of the world, coincidence of these two factors has created a complex situation for determining the dominant factor. Therefore, it was stated that the combination of human impacts such as land-use change with intense rainfall is followed by gully formation and/or development (Valentin et al. 2005; Vanmaercke et al. 2016). The type of gully affects the rate of soil erosion (Deng et al. 2015). The rate of gully erosion in ephemeral gullies was 10 times more than the lateral ones in central Belgium (Poesen et al. 1996). The share of gully erosion also depends on the soil texture and

rock fragments on the soil surface. Evans (1993) found that the share of gully erosion was more in the heavy soil textures in England. Poesen et al. (1998) stated that gully erosion is dominant on the soils with high rock fragments and in homogeneous soils. Field observation indicated that gully erosion had more share in grasslands (Bradford and Piest 1980) and poor rangelands (Soufi 2004). Data collected from the highlands of Ethiopia (Nyssen 2001) indicated that the contribution of gully erosion increased from 33% to 55% after road construction due to runoff concentration.

23.2 Materials and Methods

23.2.1 Study Area

The study area covered Iran with 15° latitude (25°–40°N) and 20° longitude (44°–64°E) (Fig. 23.3). Mountains such as Alborz and Zagrus and Caspian sea, Persian Gulf, and Oman sea play important roles in local variation in precipitation and temperature throughout the country. Therefore, different climate zones are the result of this variation. The mean of annual precipitation, temperature, and slope are 236 mm, 18.4 °C, and 8.7%, respectively. Based on DeMarton's method, there are six arid climates; semiarid climates are dominant and the other including the Mediterranean, humid, semi-humid, and extra-humid cover a limited area in the country.

23.2.2 Methodology

Regions with gully erosion were determined in four steps. At first, we talked to the provincial experts of natural resources office to collect the name and location of the regions with gully erosion that had an area equal or larger than 500 ha in order to show them on a topographic map with a scale of 1:250,000. The second step consisted of using the satellite images and recent aerial photos obtained in 1984–1987 to review the gully sites mentioned by experts. The third step was field surveying and recording the position of gully boundaries with GPS. In the fourth step, the position of gully erosion in each region was determined on available topographic maps with scales 1:250,000, and map of gully erosion for Iran was prepared using GIS. Permanent gullies were mapped by digitizing orthophoto maps in Arc/info 3.5.2 GIS and converting them to shape files using Arcview 3.2 GIS. Then, the type of the climate for each gullied region was determined, using developed DeMarton's climate map prepared by Jamab office (Ministry of energy). At least two gully regions were selected in each climate zone in each province. In the next step, three representative gullies were selected in each gully region of each climate zone in each province. A questionnaire was filled out with field survey and lab activity for each representative gully. Location of the gully system on the valley side or valley floor and evolutionary period of each gully system, continuous (old) or discontinuous (young), were determined by field observation. The altitude of the gully system from the sea level was recorded using GPS. The shape of the gully view plan was classified based on Ireland et al.'s (1939) classification. Morphometric measurement of the gullies including length, depth, top, and bottom width was carried out using tape meter in the field. Measurement was carried out in each uniform reach and also in the gully head, 25, 50, and 75% of the gully length from the head cut. Soil characteristics were analyzed from the soil samples collected from the gully heads and gully banks from the surface layer where dominant plant roots existed. Particle size and bulk density were determined using a hydrometer and steel cores, respectively. The cores were dried up in an oven for 24 h at 105 °C. Bulk density was calculated by dividing the dry weight by the volume of the cores. Organic carbon was measured using the Walkley-Black method. Ec and pH were measured by Ec and pH meter device, using saturated soil extract. The levels of potassium and sodium were measured, using a flame photometer. Calcium and magnesium levels were measured, using titration with EDTA (Handbook no. 467, Soil and Water Research Institute 2008). The longitudinal profile of the gullies was surveyed using theodolite. Slope gradient of areas draining to gully heads and banks was measured using Sento clinometers. The current and previous land uses were determined by field observation and anecdotal evidence (aerial photos and talking with old residents in the region), respectively. The causes of gully incision and development were determined using anecdotal evidence and field survey. The map of gully erosion for Iran was prepared using GIS. Statistical parameters were calculated using Excel 2007 and SPSS 23.

23.3 Results and Discussion

23.3.1 Area of Gully Erosion

Gully erosion was studied in all provinces of Iran, but the results of two provinces Theran and Lorestan were not presented (Fig. 23.4). Khorasan Province is presented as a unified province in data tables because it was not separated during our study, but the location of gully erosion is indicated throughout northern Razavi and southern Khorasan provinces on the digital map of gully erosion (Fig. 23.4). Totally, 141 regions of gully erosion were surveyed in the presented provinces of Iran. The total area of gully erosion in Iran is more than 1,429,954 ha. Kermanshah and Isfahan with an area equal to 409,895 ha and 256.52 ha had the maximum and minimum areas of gully erosion among the provinces in Iran (Fig. 23.5).

23.3.2 General Characteristics of the Gullies

Results of this study indicated that 59% of the gullies are located on the valley sides (Fig. 23.6a). Eighty-four percent of them are continuous (old) and only 16% of them are young (Fig. 23.6b), which implies most of the sediments from the gullies (old one) are transported to downstream and reservoirs. Therefore, alternatives for gully control should be done for young (discontinuous) gullies. Comparison of geologic formation indicated that 51% of gullies are located on Marls or older formations (Fig. 23.6c), which represent rangelands or rain-fed farms and the remainder is located on the quaternary. Results indicated that gully heads had 40% and 33% pointed and digitated view plans, respectively (Fig. 23.6d). This means 73% of gully incision and/or development is caused by uncontrolled surface runoff. The survey of general view plan of the gullies revealed that 60% of them had dendritic view plan (Fig. 23.6e); this indicates the surface runoff as dominant hydrological process for gully incision. About 38% of the gully view plans belong to linear view plan (Fig. 23.6e); which means this sort of gullies were formed due to flood concentration such as culverts or bridges construction without stabilization alternatives at downslopes or breaking of earth dikes due to floods.

About 38% of the gully heads had vertical long profiles (Fig. 23.6f); this means they have a plunge pool and need emergency measures to mitigate their development. More than 68% of the gully cross-sections had trapezoid and V shape (Fig. 23.6g), showing that there is a resistance layer in

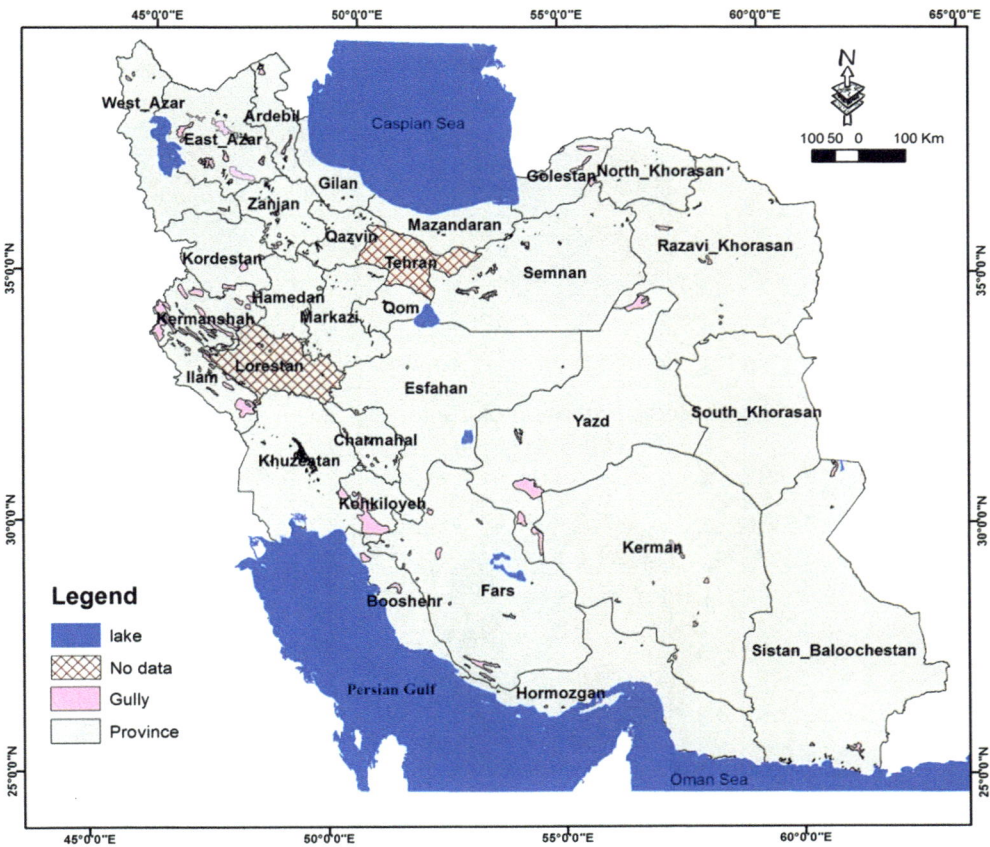

Fig. 23.4 Map of gully erosion in Iran

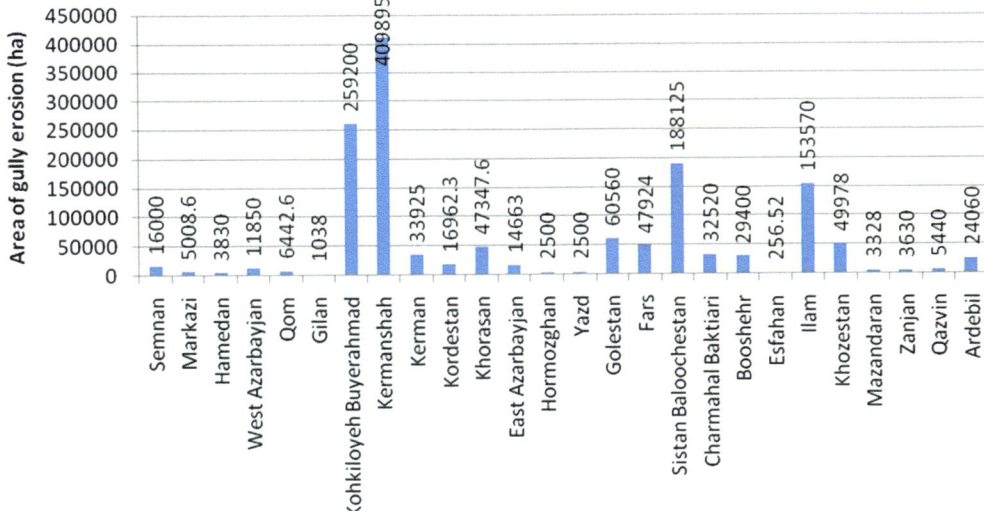

Fig. 23.5 Comparison of the area of gully erosion in different provinces in Iran

Fig. 23.6 Comparison of Iranian gullies regarding their location in the landscape (**a**), state of evolution (**b**), geologic formation (**c**), gully head view plan (**d**), general view plan of gullies (**e**), long profile of gully heads (**f**), and shape of the cross-sections (**g**)

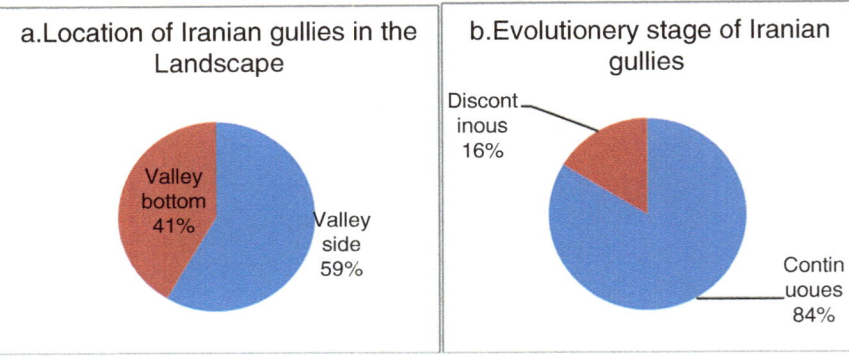

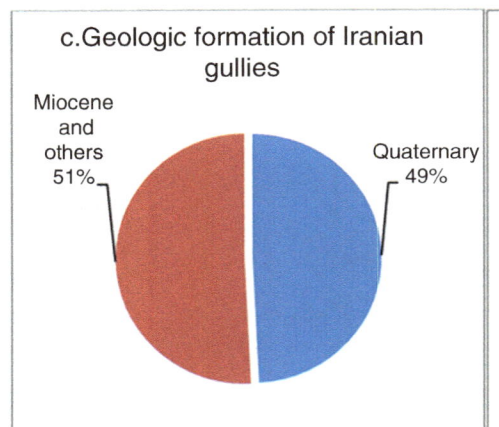

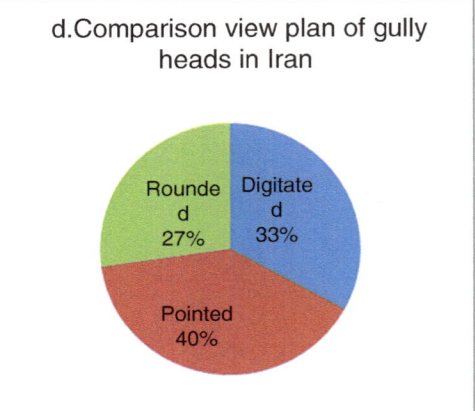

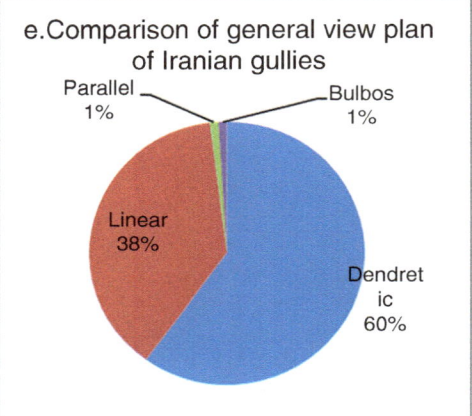

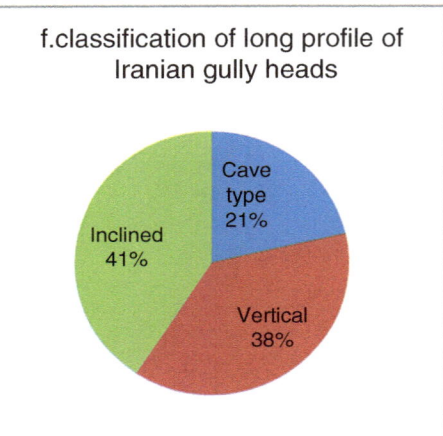

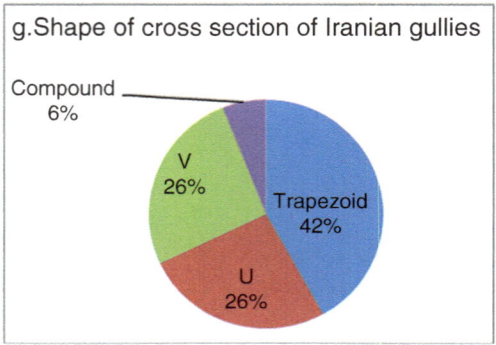

Fig. 23.7 Climate of the regions
with gully erosion

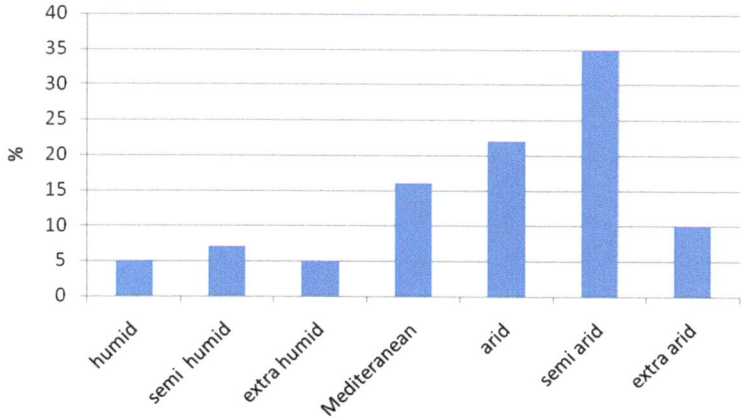

Fig. 23.8 Distribution of Iranian
gullies in different classes of
annual precipitation

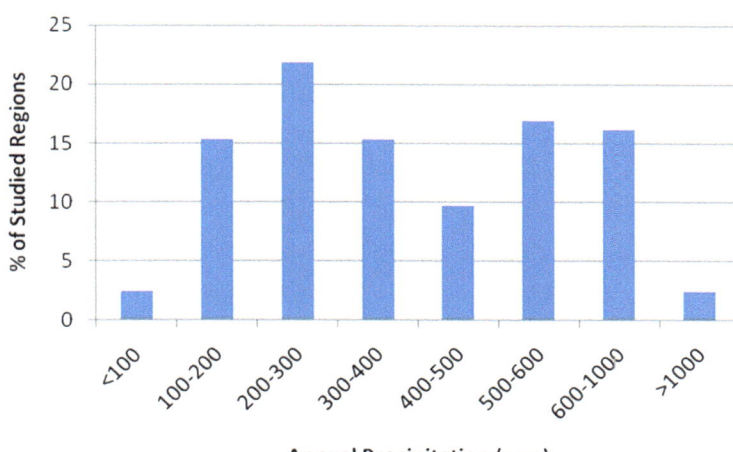

the subsurface, revealing the necessity of an alternative to maintaining the resistance of soil surface to prevent the gully incision, especially in cultivated lands. This result is in the same line with that of Deng et al.'s study (2015). They found that 78% of 456 studied gullies in Yuanmou Dry-Hot Valley, China had intermediate (trapezoid) cross-section. They believed that gully cross-sections were controlled by weathering crusts, soil properties, and vegetation cover.

23.3.3 Climate of Regions with Gully Erosion

Results of this survey indicated that 35% of the Iranian gullies are located in a semi-arid climate. Then, they are distributed in arid and Mediterranean climates with 22% and 16%, respectively (Fig. 23.7). These results with our field observations indicate two facts. The first is related to the land-use change in semi-arid and Mediterranean climates from rangeland to rain-fed farms and gardens and the second belongs to overgrazing of arid rangeland and changes in the range condition to poor and very poor conditions.

23.3.4 Precipitation

Gully erosion occurred in the lands below 100 mm and above 1000 mm annual precipitation. The least portion of gullies (2.5%) belong to the class of precipitation below 100 mm and above 1000 mm both with gully incision occurred mostly in 200–300, 500–600, and 600–1000 mm with 22%, 17%, and 16%, respectively (Fig. 23.8). Class 200–300 mm refers to arid rangelands and class 500–1000 refers to new croplands and gardens due to the land-use change.

23.3.5 Land Use

Results in Fig. 23.9 indicate that the rangelands and forests were reduced and cultivated lands including rain-fed farms and gardens increased. Results revealed that rain-fed farms and gardens increased twofolds. Barren lands increased four times. Rangeland decreased from 54% to 35% and forest decreased from 11% to 1%. Deteriorated rangeland and forest increased by 2.5- and 5-folds, respectively (Fig. 23.9).

Fig. 23.9 Comparison of land-use change in regions with gully incision

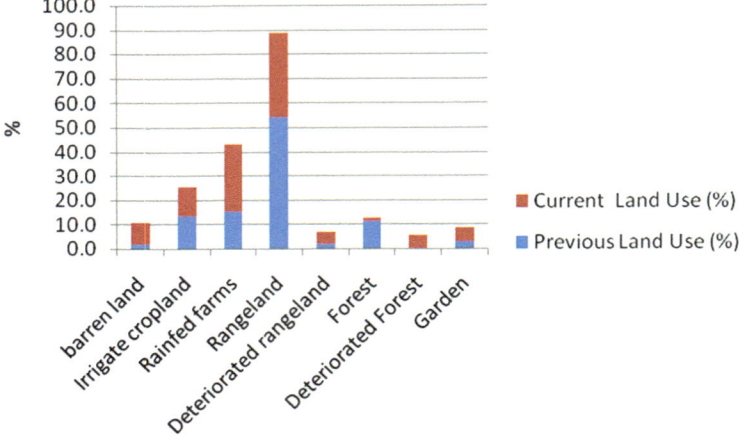

Fig. 23.10 Distribution of Iranian gullies at different altitudes above the sea level

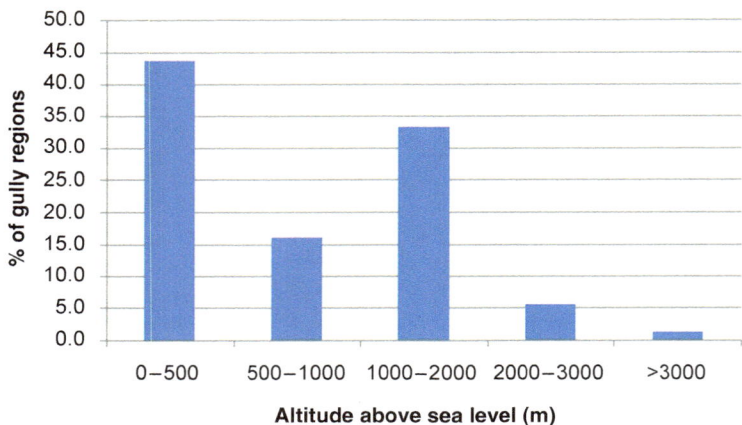

Kakembo et al. (2009) found that 75% of the gullied area occurred on abandoned lands in the Eastern Cape province, South Africa. Gutierez et al. (2009) indicated that gully erosion was closely related to land use, especially with the extent of cultivation areas in the SW Spain. They also stated that there was no clear relationship between the evolution of the gullied area and rainfall amounts.

23.3.6 Altitude Above the Sea Level

Figure 23.10 indicates that gully erosion happened in lands around the Persian Gulf, inland lakes, the Caspian sea and in highlands with over 3000 m altitude. Most of the Iranian gullies were distributed in classes 0–500 m and 1000–2000 m above the sea level with 44% and 33%, respectively (Fig. 23.10). Human interference including overgrazing and increased overland flow due to urban development is the main cause of gully formation in the altitude of 0–500 m above the sea level. The main causes of gully formation in the altitude of 1000–2000 m above the sea

level were overgrazing and change in the land use from rangeland and forest to cultivated lands. Kakembo et al. (2009) stated that gully erosion was dominant in concave bottomlands of the Eastern Cape province, South Africa. Mararakanye and Sumner (2017) found that gully formation was associated with duplex soils on colluviums and alluvial deposits on a lower slope where overland flow converges and accumulates in South Africa.

23.3.7 Soil Properties

Soil texture of Iranian gullies is displayed in Fig. 23.11. Seven hundred thirty-two soil samples were collected from the heads and banks of Iranian gullies. Results indicated that loam, sandy loam, clay loam, and silt loam were the most common soil texture with 28%, 24%, 13%, and 10%, respectively (Fig. 23.11). Loam belongs to the gullies of southwest and southeast provinces, sandy loam to coastal gullies, clay loam to the northwest and northern gullies, and silt loam to northeast gullies.

Fig. 23.11 The soil textures of the samples collected from Iranian gullies

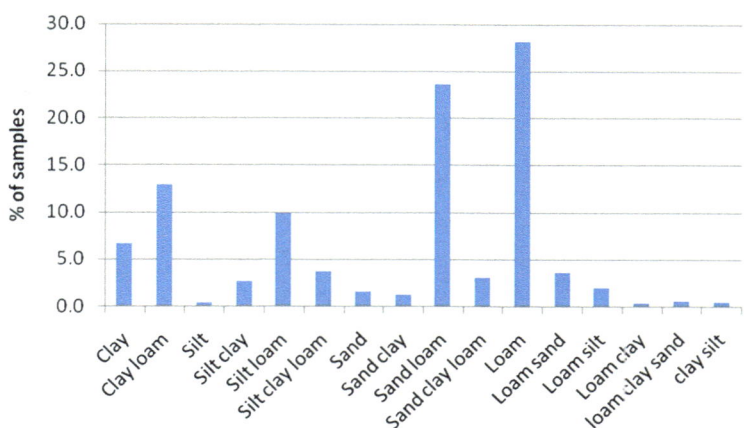

23.3.8 Morphometric Characteristics of the Gullies

Table 23.1 compares the dimensional factors and soil loss from gullies in different provinces of Iran. Results indicated that the average length of gullies was equal to 517.4 m. Maximum and minimum gully lengths belonged to Mazandaran with 2900 m and Booshehr with 52.1 m, respectively (Table 23.1). These results show that the classification of Iranian gully length by Ahmadi (1999) could be changed. Ahmadi (1999) classified gullies with a length of 120 m and 240 m as medium, while in this survey the gullies with a length around 500 m are in the medium class. The average top and bottom widths were 7.2 m and 3.0 m, respectively. The maximum top and bottom widths belonged to Yazd and Sistan and Baloochestan provinces with 15.3 m and 11.7 m, respectively (Table 23.1). The minimum top and bottom widths of the gullies with 2.0 m and 0.7 m, respectively, belonged to Semnan province (Table 23.1). The average depth of the gullies was equal to 2.8 m, which belongs to the medium size gully. The standard deviation for the gully depth presents the medium size gullies with a class of 1.2–4.2 m for this study. Maximum and minimum depth of the gullies belonged to Yazd and Semnan with 8.0 and 0.8 m, respectively (Table 23.1). Results indicated that the average soil erosion by gully was 21.2 m^3/m of gully length (Table 23.1). The maximum and minimum of soil erosion by gullies belonged to Yazd and Semnan with 151.9 m^3/m and 1.4 m^3/m, respectively. Comparison of data showed that gully erosion in some provinces such as Mazandaran, Sistan and Baloochestan, Kohkiloye and Boyerah, Kordestan, Esfahan, Khorasan, Golestan, Yazd, and Hormozghan was higher than the average rate (21.2 m^3/m) of erosion (Table 23.1).

23.3.9 Causes of Gully Erosion

Figure 23.12 shows the causes of gully erosion in Iran. Results indicated that 65% of data belonged to human impacts and 35% to natural factors such as rainfall intensity and soil erodibility. Among different aspects of human impacts, overgrazing with 32% and changes of rangeland and forest with 17% had a high contribution to the causes of gully erosion in different parts of Iran (Fig. 23.12).

23.4 Conclusion

This study indicated that gully erosion distributed throughout different parts of Iran with an area more than 1,420,000 ha from the coastal zones to highlands with an altitude higher than 3000 m above the sea level. Although gullies are observed in different climate zones, most of them are located in semi-arid and arid zones. Eighty-four percent of the gullies are continuous (old). This indicates that most of the sediments from the gullies were transported downslope. Therefore, the priority for gully control belongs to saving lands, infrastructures, and residential areas. The dominant general gully (60% dendritic) and heads (73% digitated and pointed) view plans show the action of the surface runoff as the dominant hydrological process for gully incision and/or development. Forty-one percent of Iranian gullies had loam and clay-loam soil texture. The results of gully view plan and soil texture are in the same line with Morgan's (1995) and Deng et al. (2015) conclusions. The results indicate that gully incision coincides with changing rangeland and forest to cultivated land including croplands and gardens. Overgrazing and changing the rangeland and forest were the dominant causes of gully erosion in Iran.

Table 23.1 Comparison of the average length, top width, bottom width, volume and volume per unit length of Iranian gullies in different provinces

Provinces	Ave. length (m)	Ave. top width (m)	Ave. bottom width (m)	Ave. depth (m)	Ave. volume (m³)	Ave. vol./unit length (m³/m)
Yazd	650.0	15.3	5.5	8.4	98,707.8	151.9
Hormozghan	116.0	9.3	8.4	1.9	3281.4	28.3
Golestan	274.8	9.8	3.6	3.2	7995.3	29.1
Kerman	135.1	5.2	1.7	1.4	1059.9	7.8
Qom	140.3	10.5	1.0	2.4	1970.0	14.0
Fars	127.1	7.3	2.0	2.6	2510.5	19.8
Semnan	79.3	2.0	0.7	0.8	109.1	1.4
Khorasan	217.7	4.5	2.7	3.0	4426.3	20.3
Zanjan	564.3	5.3	1.5	1.7	3853.9	6.8
Charmahal and Baktiari	762.5	6.1	1.9	2.6	9970.5	13.1
Booshehr	52.1	5.2	2.0	1.8	772.8	14.8
Esfahan	244.1	11.2	1.2	2.4	6966.4	28.5
Kermanshah	678.2	7.2	1.4	2.3	6592.1	9.7
Kordestan	2183.3	8.0	4.0	3.7	48,469.3	22.2
Gilan	293.5	4.2	2.6	1.2	1147.6	3.9
Markazi	83.8	6.7	1.6	3.2	1101.3	13.1
Kohkiloyeh and Buyerahmad	300.0	10.3	5.1	4.4	10,190.4	34.0
West Azarbayjan	214.0	6.4	1.4	2.1	1727.1	8.1
East Azarbayjan	265.2	4.6	2.2	2.6	2294.7	8.7
Hamedan	439.0	4.1	1.4	1.5	1809.5	4.1
Sistan and Baloochestan	1690.0	14.3	11.7	3.2	69,258.7	41.0
Qazvin	162.8	3.0	1.3	1.6	532.0	3.3
Ardebil	113.9	7.0	3.1	2.9	1664.5	14.6
Ilam	392.9	10.1	1.7	2.7	6219.9	15.8
Khuzestan	183.3	6.5	3.0	2.4	2085.4	11.4
Mazandaran	2900.0	6.9	3.5	7.2	108,576.0	37.4
Zanjan	707.0	4.3	4.3	2.4	7330.2	10.4
Average	517.4	7.2	3.0	2.8	15,208.2	21.2
Max.	2900.0	15.3	11.7	8.4	108,576.0	151.9
Min.	52.1	2.0	0.7	0.8	109.1	1.4
SD	681.2	3.2	2.4	1.6	29,718.8	28.2

Fig. 23.12 Comparison of the causes of gully erosion in Iran

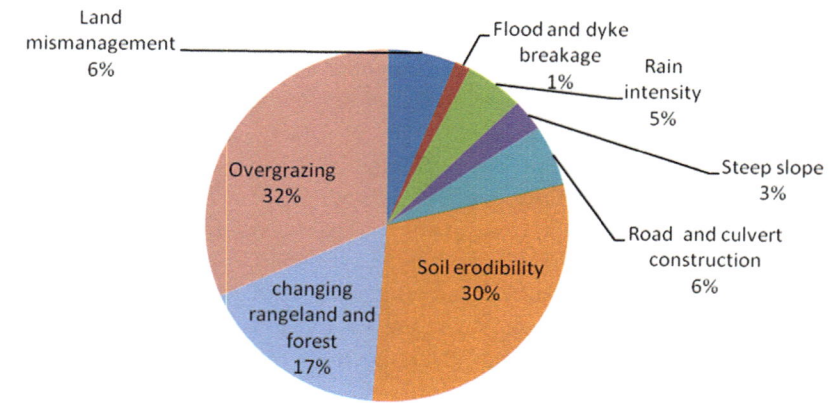

Acknowledgments This project is sponsored by the Institute of Soil Conservation and Watershed Management in Iran. We acknowledge their support and also the Research and Education Centers in Various Provinces in Iran.

References

Abi M, Kessler A, Oosterveer P, Tolossa D (2018) Understanding the spontaneous spreading of stone bunds in Ethiopia: Implications for sustainable land management. Sustainability 10: 2666

Ahmadi H (1999) Applied geomorphology, Volume 1: Water erosion, Tehran University, Tehran, Iran (In Persian)

Avni Y (2005) Gully incision as a key factor in desertification in an arid environment, the Negev highlands, Israel. Catena 63:185–220

Balling R C; Wells S G (1990) Historical rainfall patterns and Arroyo activity within the Zuni River Drainage Basin, New Mexico. Annals of the Assoc. of Am. Geogr. 80 (4): 603–617

Bork H R; Li Y; Zhao Y; Zhang J; Shiquan Y (2001) Land use changes and gully development in the upper Yangtze River Basin, SW-China. Journal of Mountain Science 19(2): 97–103

Bradford J M; Piest R. F (1980) Erosion development of valley-bottom gullies in the Upper Mid Western United States, In Coates D R and Vitek J D (eds), Thresholds in geomorphology, Allen and Unwin, Shubbery

Brice J B (1966) Erosion and deposition in the loess-mantled great plains Medicine Creek drainage basin, Nebraska. U. S. Geological Survey Professional Paper 352H: 235–339

Campos A B; Castro S S; Casseti V; Santos R R; Martins M S; Silva A A (2000) Geological and topographic indicators of the gully erosion at the upper Araguaia river basin, Brazil, International Symposium on Gully Erosion under Global Change, 2000

Castillo C, Gomez J A (2016) A century of gully erosion research: Urgency, complexity and study approaches, Earth Science Reviews 160; 300–319

Castro N; Auzet A V; Chevallier P; Leprun J C (2000) Influence of extreme rainfall events (El Nino) on a gully system typical of the basaltic plateau of Southern Brazil, International Symposium on Gully Erosion under Global Change, 2000

Chaplot V (2013) Impact of terrain attributes, parent material and soil types on gully erosion, Geomorphology 186;1–11

Crouch R J; Blong R J (1989) Gully sidewall classification: Methods and application, Zeitschrift Fur Geomorphologe, N.F. Supplement Band 33(3):291–305

Deng Q, Qin F, Zhang B, Wang H, Luo M, Shu C, Liu H, Liu G (2015) Characterizing the morphology of gully cross-sections based on PCA: A case of Yuanmou Dry-Hot Valley, Geomorpholog 228; 703–713

Dietrich W E; Dunne T (1993) The Channel Head in Keith, B. and Kirkby, M.J. (eds.), Channel Network Hydrology, John Wiley and Sons. Chichester, England

Evans R (1993) Extent, frequency and rates of rilling of arable land in localities in England and Wales. In Wicherek S (ed.). Farm land erosion in temperate plains environment and hills, Elsevier, Amsterdam

Fang N F; Zeng Y; Ni L S; Shi Z H (2019) Estimation of sediment trapping behind check dams using high–density electrical resistivity tomography. Journal of Hydrology 568:1007–1016

FAO (1982) Gully erosion control. Rome

Faulkner H (1995) Gully erosion associated with the expansion of unterraced almond cultivation in the coastal sierra de lujar, S. Spain. Land Degradation and Rehabilation 9:179–200

Graf W L (1983) The arroyo problem-paleohydrology and paleohydraulics in the short term. In: G.K. Gregory, (ed.),

Background to paleohydrology, John Wiley and Sons, London,, pp 262–303

Gutierez A G, Schnabel S, Contador F L (2009) Gully erosion, land use and topographical thresholds during the last 60 years in a small rangeland catchment in SW Spain, Land Degradation and Development 20(5);535–550

Harvey M A (1996) Holocene hillslope gully systems in the Howgill fells, Cumbria. In: M.G. Anderson, Brooks, S. M. (eds), Advances in hillslope processes 2:731–752

Heede B H; Mufich J G (1974) Field and computer procedures for gully control by check dams. Journal of Environmental Management 2;1–49

Imeson A C; Kwaad F J P M (1980) Gully types and gully prediction. KNAG Geografisch Tijdschrift XIV 5:430–441

Ireland H A; Sharpe C F S; Eargle D H (1939) Principles of gully erosion in the Piedmont of South Carolina. USDA Technological Bulletin 633

Kakembo V, Xanga W W, Rowntree K (2009) Topographic thresholds in gully development on the hillslopes of communal areas in Ngqushwa Local Municipality, Eastern Cape, South Africa, Geomorphology 110(3–4); 188–194

Kropacek J (2019) Erosion dynamics in the southern Tibetan Plateau at a century time scale from historical photographs. Journal of Arid Environments 161:47–54

Leopold L B; Miller J P (1965) Ephemeral streams-hydraulic factors and their relation to the drainage net, Physiographic and Hydraulic Studies of Rivers

Mararakanye N, Sumner P D (2017) Gully erosion: A comparison of contributing factors in two catchments in South Africa, Geomorphology 288;99–110

Monsiers E, Poesen J, Dessie M, Adgo E, Verhoest N E C, Deckers J and Nyssen J (2015) Effects of drainage ditches and stone bunds on topographical thresholds for gully head development in North Ethiopia, Geomorphology 234; 193–203

Moore I D; Burch G J; Mackenzie D H (1988) Topographic effects on the distribution of surface soil water and the location of ephemeral gullies, Transaction of ASAE 1098–1107

Morgan R P C (1995) Soil Erosion and conservation, second edition, Longman

Nachtergaele J (2001) A spatial and temporal analysis of the characteristics, importance and prediction of ephemeral gully erosion, Ph.D Thesis, Dept. of Geography, University of Leuven, Belgium

Nyssen J (2001) Erosion processes and soil conservation in a tropical mountain catchment under threat of anthropogenic desertification—a case study from Northern Ethiopia. Unpubl. PhD thesis, Dept. Geography–Geology, K.U. Leuven, Belgium

Okuh D; Osumgborogwu I E (2019) Adjustments to hazards of gully erosion in rural southeast Nigeria: A case of Amucha communities, Applied Ecology and Environmental Sciences 7(1):11–20

Oostwoud Wijdenes D J; Poesen J; Vandekerckhove L; Ghesquiere M (2000) Gully-head activity and sediment contribution to a Mediterranean channel, International Symposium on Gully Erosion under Global Change, 103

Poesen J (1989) Conditions for gully formation in the Belgian Loam Belt and some ways to control them. Soil Technology Series 1:39–52

Poesen J (1993) Gully typology and gully control measures in the European loess belt. In: Wicherek, S.(ed), Farmland erosion in temperate plains environment and hills. Elsevier, Amsterdam

Poesen J (2017) Soil erosion in the Anthropocene: research needs. Earth Surface Processes and Landforms 43:64–84

Poesen J; Govers, G (1990) Gully erosion in the loam belt of Belgium: Typology and control measures. In: J. Boardman, Foster, I.D.L. and Dearing, J. A. (eds), Soil Erosion on Agriculture Land, John Wiley and Sons, Chichester, England

Poesen J; Vandaele K; van Wesemael B (1996) Contribution of gully erosion to sediment production in cultivated lands and rangelands, IAHS publications 236:251–266

Poesen J; Vandaele K; Wesemael B (1998) Gully erosion: importance and model implication. In:Boardman J; Favis-Mortlock D T(eds) Modelling soil erosion by water Springer-Verlag, Berlin NATO-ASI Series I-55: 258–311

Poesen J; Nachtorgale J; Verstrac G (2003) Gully erosion and environmental change: importance and research needs, Catena 50:91–133

Prosser I P; Winchester S J (1996) History and processes of gully initiation and development in Eastern Australia. Aeitschrift fur Geomorphologie, N. f. Supplement Band, 105:91–109

Refahi H G (2000) Waer erosion and it's control. Tehran university, Tehran (In Persian)

Rey F, Bifulco C B, Bischetti G B, Bourrier F, Cesare G De, Florineth F, Grat F, Marden M, Mickovski S B. Philips C, Peklo K, Poesen J, Polster D, Preti F, Rauch H P, Raymond P, Sangalli P, Tardio G, Stokes A (2019) Soil and water bioengineering: Practice and research needs for reconciling natural hazard control and ecological restoration. Science of the total Environment 648: 1210–1218

Rong LI; Duan X; Zhang G; Gu Z; Feng D (2019) Impacts of tillage practices on ephemeral gully erosion in a dry-hot valley region in southwestern China. Soil and tillage Research 187:72–84

Soil and Water Research Institute (2008) Manual for laboratory analysis of soil and water samples, no. 467

Soufi M (1997) Processes and rates of gully development in Pine Plantations, Southeastern new South Wales, Ph.D. Dissertation, Univ. of N.S.W. Sydney

Soufi M (2004) A study of morphoclimatic characteristics of gullies in fars province, final report of research plan, institute of soil conservation and watershed management, SBN 83.1153

Soufi M (2005) Impacts of vegetation cover and urban development on the gully development in south of Fars province, Proceeding of third national conference of erosion and sediment 350–355

Soufi M (2009) A study of influential factors on the initiation and development of gullies in different climates of Fars province, Final research report, Institute of soil conservation and watershed management, SBN 49477

Soufi M, Bayat R (2015) Morphoclimatic classification of gullies in different climates of I.R. Iran (phase 2). Organization of research education and extension for agriculture. Ministry of Jihad –E-Agriculture. Iran, SBN no. 48474 (In Persian)

Soufi M, Bayat R (2016) Morphoclimatic classification of gullies in different climates of I.R. Iran (phase 3). Organization of research education and extension for agriculture. Ministry of Jihad –E-Agriculture. Iran, SBN no. 50689 (In Persian)

Soufi M, Bayat R, Charkhabi A H (2017) Morphoclimatic classification of gullies in different climates of I. R. Iran (phase 1). Organization of research education and extension for agriculture. Ministry of Jihad –E-Agriculture. Iran, SBN no. 51544(In Persian)

Starr B M (1989) Anecdotal and relic evidence of the history of gully erosion and sediment movement in Michelago Creek Catchment Areas NSW

Sun W; Shao Q; Liu J; Zhai J (2014) Assessing the effects of land use and topography on soil erosion on the Loess Plateau in China, Catena 121;151–163

Svoray T (2009) Catchments scale analysis of the effect of topography, tillage direction and unpaved roads on ephemeral gully incision, Earth Surface Processes and Land forms 34(14):1970–1984

Valentin C; Poesen J; Li Y (2005) Gully erosion: Impacts, factors and control, Catena 63:132–153

Vanmaercke M, Poesen J, Van Mele B, Demuzere M, Bruynseels A, Golosov V, Rodrigues Bezerra J F, Bolysov S, Dvinskih A, Frankl A, Fuseina Y, TeixeiraGuerra A J, Haregeweyn N, Ionita I, Imwangana F M, Moeyersons J, Moshe I, Nazari Samani A, Yermolaev O (2016) How fast do gully headcuts retreat? Earth Science Reviews 154;336–355

Webb R H; Hereford R (2001) Floods and geomorphic changes in the southwestern United States and historical perspective, Proc. Seventh federal interagency sedimentation conference, Nevada, USA

Yu Y; Wei W; Cheng L; Feng T; Daryanto S (2019) Quantifying the effects of precipitation, vegetation, and land preparation techniques on runoff and soil erosion in a Loess watershed of China. Science of the total environment 652: 755–764

Zhang B; Zhang G; Yang H; Wang H (2019) Soil resistance to flowing water erosion of seven typical plant communities on steep gully slopes on the Loess Plateau of China, Catena 173:375–383

Majid Soufi is an Associate Professor at the Department of Watershed Management and Soil Conservation, Fars Research and Education Center for Agriculture and Natural Resources, Shiraz, Iran. He got his Ph.D. in 1997 from the University of New South Wales in Sydney, Australia. He did his master's in watershed management in 1990 and B.S. in range and watershed management engineering in 1985 from Tehran University. He has professional experience of more than 30 years in the field of natural resources management and soil conservation, especially in gully erosion research and mitigation. He has presented 250 research papers in the national and international conferences and has published 55 research papers in scientific national and international journals. He was consultant of many Iranian projects in watershed and drought management which cooperated with UNDP, UNESCO, and FAO.

Reza Bayat is an Assistant Professor in the Department of Water and Soil Conservation Engineering at Soil Conservation and Watershed Management Research Institute, where has been a faculty member since 2001. He is Ph.D. candidate at Lorestan University and completed his M.Sc. studies at Tehran University and undergraduate studies at Gilan University. His research interests lie in the area of soil management, especially in soil erosion and conservation, and GIS application in this area. In recent years, he has focused on gully erosion and rainfall simulation, especially works with some mulches. He has collaborated actively with researchers in several other disciplines of environmental science, particularly watershed management and pollution.

Amir Hossein Charkhabi is an Associate Professor. He has received Ph.D. and M.Sc. from Iowa State University, USA, in 1995 and 1990, respectively. His research interest is in soil erosion and environmental science. At present, he is working as free environmental consulting expert.

Narges Kariminejad, Mohsen Hosseinalizadeh, Hamid Reza Pourghasemi,
Majid Ownegh, and Mauro Rossi

Abstract

Gully-head has been observed in a wide range of continuous and categorical conditioning factors in different countries. This study aimed to examine the association of gully-heads with the most effective hydrologic factors via univariate and bivariate analyses in the standard mode. A 2700 ha area in the loess-covered region of Iran was selected and the point map of 287 gully-heads prepared by unmanned aerial vehicle (UAV) images. The pattern of gully-heads was evaluated using univariate tests ($O(r)$ & $g(r)$). The occurrence of gully-heads in relation to the linear features including road networks (RN_S) and stream networks (SN_S) was assessed using bivariate correlation tests ($O_{12}(r)$ $g_{12}(r)$). The analysis mode in mark correlation function ($k_{mm}(r)$) was applied for soil particles categorized into three groups by size including clay, sand, and silt content. The Mont Carlo simulation intervals were also conducted based on fifth highest and lowest values of the summary statistic of 199 simulated null model data sets. According to the results of the univariate spatial statistics, gully-heads had an aggregated distribution. The bivariate O-ring and pair correlation ($g_{12}(r)$) test revealed that gully-heads had positive interactions with RN_S and SN_S. Based on mark correlation function $k_{mm}(r)$, clay content of nearby gully-heads was consistently smaller than the mean value of clay content ($\mu^2 = 22.93\%$) in the study area. However, the silt contents of nearby gully-heads were significantly larger than the mean value of silt content ($\mu^2 = 64.58\%$). The mean sand contents ($\mu^2 = 14.75\%$) do not differ from the mean sand contents taken over all pair gully-heads. Consequently, compared to other interoperation, the suggested approach prepares a proper technique to erosion research community which would be of interest to policy makers and geomorphologists.

Keywords

Gully-head · Spatial modeling · Road networks · Stream networks · Soil texture · Iran

24.1 Introduction

Gully erosion is one of the erosion processes in which runoff water cumulates and mostly revolves in thin channels and, after short periods, takes away the soil from this thin area to remarkable depths (Poesen et al. 2003); so, it is known as surface water erosion process. Gullies are mostly distinguished by activity of headcut retreats. A gully-head is a vertical drop which is often extended in a bed of gully channel elevation (Poesen et al. 2003; Valentin et al. 2005). This is the uppermost part of the gully, where gully starts on the slope. Gully-head is joined to whole gully. Further, there is a surface connection between gully-head and gully.

Comprehensive information of the process of gully-heads is very important to not only scientists, but also land managers and farmers (Shruthi et al. 2011). Considering the gully-head-affected areas, the development and arraignment of gully-heads are controlled exceedingly by the linear features. Hence, stream networks are considered as the ground skeleton in gully properties mapping (Liu et al. 2018). Road networks are also considered as the primary reason for gully-retreating process (Jungerius et al. 2002; Makanzu Imwangana et al. 2018). Thus, in the present study, the stream network and road network derived from the high-resolution digital orthophoto map based on UAV

N. Kariminejad · M. Hosseinalizadeh · M. Ownegh
Department of Watershed and Arid Zone Management, Gorgan
University of Agricultural Sciences and Natural Resources, Gorgan, Iran

H. R. Pourghasemi (✉)
Department of Natural Resources and Environmental Engineering,
College of Agriculture, Shiraz University, Shiraz, Iran
e-mail hr.porghasemi@shirazu.ac.ir

M. Rossi
Department of Research Institute for Geo-Hydrological Protection IRPI,
Perugia, Italy

© Springer Nature Switzerland AG 2020
P. K. Shit et al. (eds.), *Gully Erosion Studies from India and Surrounding Regions*, Advances in Science, Technology & Innovation,
https://doi.org/10.1007/978-3-030-23243-6_24

images were selected as the main factors controlling gully-head activities.

Moreover, soil factors are essential in controlling or satisfactorily modeling the activity of gully-heads (Gan et al. 2018; Choubin et al. 2019). Normally, gully-head activity has been related to different soil properties including soil texture. In other words, the total growth of each gully-head has been linked with the rates of soil erodibility (Ollobarren Del Barrio et al. 2018). The studies which determined the influence of soil erodibility on gully-head activity are rare. This is due to the complex interactions and relationship between soil properties and the erosive process. Thus, the occurrence of gully-heads in relation to the particle size distribution should be considered to identify the soil properties—easy to determine—which best reflect soil erodibility on gully-heads.

Summary functions as a new approach characterize properties of the spatial arrangement applying pairs of gully-heads which are located nearly at different distance *r* of each other (Genet et al. 2014). Because the spatial factors controlling the distribution of gully-heads are related to linear phenomena such as road networks and stream networks, univariate and bivariate correlation test was applied to estimate the distribution correlation of gully-heads. As soil particles were also recognized as the hot spots of gully-head density, mark correlation test was utilized to compare the value of particle size distribution of nearby gully-heads with the mean value of all gully-heads distributed in the study area.

24.2   Case Study

The examination was obtained in one of the greatest erosional areas in the Iranian Loess Plateau (ILP), with an area of approximately 2700 ha in the northeast of Golestan Province, Iran (Fig. 24.1). It is located between $55° 36'$ to $55° 40'$ E longitude and $37° 37'$ to $37° 40'$ N latitude (Fig. 24.2). The selected area, which is so-called Iranian Loess Plateau (ILP), has an annual potential evapotranspiration and average rainfall of 880 mm and 370 mm, respectively. Temperature regime is Termic with a semiarid climate (Jafari Shalamzari and Zhang 2018). Conducted on a hilly outlook of the ILP with maternal materials, the studied region consisted mostly of loess deposits. Commonly, loess is deposited along the Kopeh Dagh and Alborz Mountains of Iran (Keshavarzi 2014). Generally, the thickness of loess deposits in Golestan province increases gradually from southwest to northeast. The study area consists of four soil taxonomies, including, clay loam (0.11%), loam (47.78%), silt loam (40.27%), and silty clay loam (11.85%). Also, this study comprises a region with two different land uses, namely agricultural (58.81%) and rangeland (41.19%). Agriculture is the main occupation of the majority of people in this area. However, 10.49% of gully-heads located in the agricultural land and the remains (89.51%) are in the rangeland.

24.3   Methods

24.3.1   Field Measurements

Gully-head modeling was performed by 1 m DEM-resolution. The methodology flowchart for investigating our concept is reported in Fig. 24.3. In the first part, a UAV system was employed to acquire aerial photographs. The UAV considered over the field of erosion was DJI Phantom-4quadcopter with a size of 4000×3000 pixels. The maximum time for each flight length was around 30 min. The locations of all gully-heads were collected from the UAV images and field survey during the summer in 2018. Then, the data related to the location of all 287 investigated gully-heads were prepared in Excel format. Two linear indicators (road networks and stream networks) were employed independently for spatial modelling of gully-heads. The stream networks and road networks were created from the high-resolution digital orthophoto map based on UAV Images (Fig. 24.4). They were conducted as the ground skeleton that controls gully-head bodies. The investigation of 570 soil samples was made in relation to the outlet–inlet location of gully-heads. During the fieldwork, 2 kg materials were taken from each soil sample and transferred to the laboratory. The taken samples were air-dried in the laboratory and used for further analyses. The hydrometer method was applied for analysis of soil particle size (Beretta et al. 2014) without the elimination of carbonates, secondary oxides, and organic matter. Data layers were prepared using ArcGIS 10.2.2 with the pixel size of 1 m. The *Programita* Software (http://programita.org/), developed by Wiegand and Moloney (2004) and spatstat package in R 3.5.1 software (Baddeley 2018) were utilized to perform these analyses.

24.3.2   Spatial Analysis

The univariate spatial statistic was utilized to investigate the spatial patterns of gully-heads. While a range of different measures are available for analyzing point patterns, Ripley's derivatives including $g(r)$ and $O(r)$ were selected as they are based on distances between all gully-headcut coordinates, rather than just the nearest-neighbor, and use the intensity of locations within the area, allowing tests against different null models (Wiegand and Moloney 2014; Hosseinalizadeh et al. 2018). The g-function, the sensitive version of Ripley's K-function, is calculated to stabilize the variance (Eq. 24.1).

$$g(r) = \frac{\sqrt{K(r)}}{\pi - r} \tag{24.1}$$

where, g is square root transformation of K, which was transformed to ease linear interpretation; K is the expected number of coordinates of a pattern within a given distance r of an accidental point of that pattern, and r is the calculated

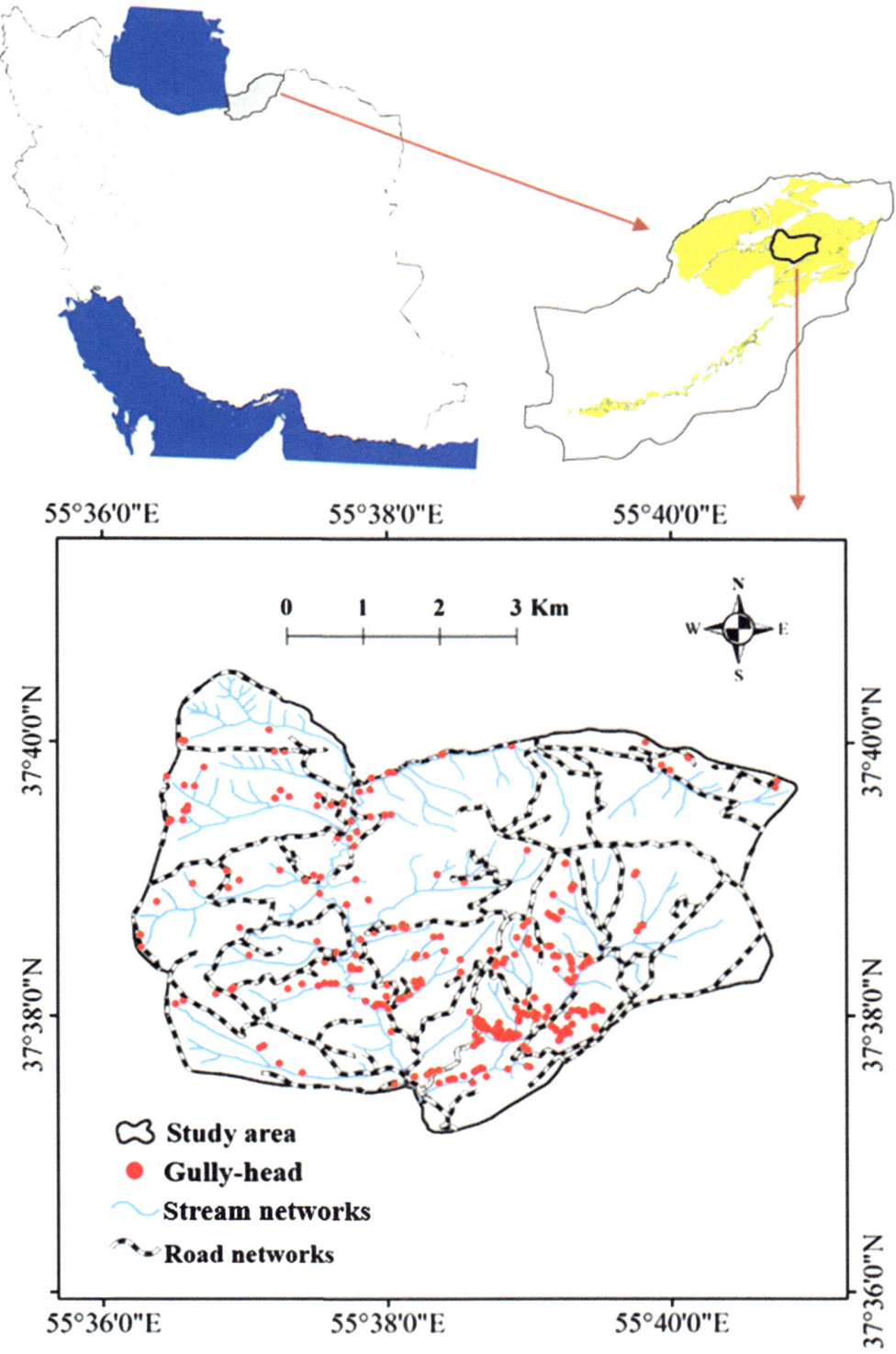

Fig. 24.1 Location map of the study area

spatial scale (meters in this study). Additionally, an O-ring statistic was computed, as this is a complementary measure to Ripley's g-function (Wiegand and Moloney 2014). Whereas the g-function applies all the locations on the certain distance,

the O-ring statistic only applies locations that are found at that special distance, thereby avoiding cumulative effects (Wiegand and Moloney 2004). Moreover, the bivariate distribution test including $g_{12}(r) \; O_{12}(r)$ was utilized to find the

Fig. 24.2 The gully-head forms in the study area

relationship between gully-headcuts with linear phenomena including roads and stream networks, separately.

The mark correlation function (MCF) was utilized for the persistent mark of the circle. It is very similar to the univariate pair correlation test $g(r)$. However, the MCF examines not only the distance correlation, but also the soil correlation at different radius r. If the soil indicates no spatial point correlations, it means $k_{mm}(r) = 1$, and if $k_{mm}(r)$ be less than1, it means marks (soil particle size distribution in this study) of gully-heads which are more nearby together are lower than the average. If $k_{mm}(r)$ be more than1, it means marks of gully-heads which are more nearby together are greater than that of isolated gully-heads (Wiegand and Moloney 2014; Getzin et al. 2015). To examine departures from the considered null model, approximately 95% simulation intervals were used with the fifth highest and lowest values of 199 Monte Carlo simulation intervals (Wiegand and Moloney 2014).

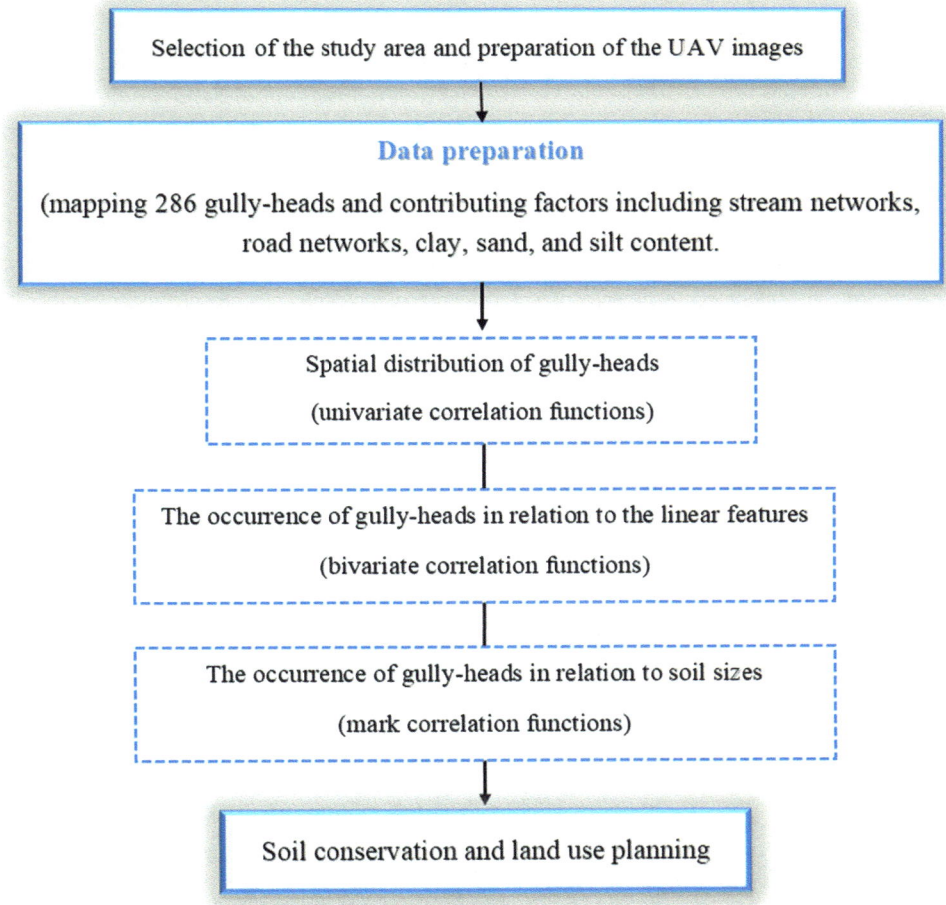

Fig. 24.3 Flowchart of the study area

24.4   Results

In the present study, the spatial distribution of gully-heads was analyzed using univariate correlation functions in the standard mode. Regarding the results of pair coloration test $g(r)$, the accumulation of gully-heads was observed in the study area. Whenever the function line was unfitted to be in the bounds of 95% simulation intervals, the accumulation of gully-heads was a significant departure from random labeling in all distances from 5 m to 355 m (Fig. 24.5A). Completely adjust with the $g(r)$ graph, the results of univariate O-ring test implied the spatially aggregated pattern of gully-heads in all of the distance scales (5 m–355 m). Also, the observed line was out of the Monte Carlo intervals which implied statistically significant accumulation of 199 simulated patterns of gully-heads and the fifth lowest and highest values as calculated intervals (Fig. 24.5B). Further, the establishment of gully-heads was much closer in the specific parts of area comparing to the other parts (Fig. 24.6).

According to the results of the bivariate tests, $g_{12}(r)$ and $O_{12}(r)$, the new aspects of the interaction between gully-heads and stream networks were shown in the study area. Regarding gully-heads and stream networks, the results of the bivariate correlation tests were considerably different from random labelling. As shown in Fig. 24.7, a positive interaction was confirmed for gully-heads and SN_S at different scales. This investigation showed that gully-heads were located next to the stream networks. Similar to the results; the SN_S map confirmed that gully-heads and stream networks had positive relationship at small scales (Fig. 24.8).

The results of bivariate correlation tests $g_{12}(r)$ and $O_{12}(r)$ for gully-heads and road networks exemplify significant departure from random labeling at all scales which present positive associations (Fig. 24.9). Moreover, similar to the results of test, positive spatial interaction was confirmed between gully-head locations and the map of road networks at all distances (Fig. 24.10). The correlation between gully-heads and RN_S was completely significant at level of 5% because the test line was out of the Monte Carlo intervals.

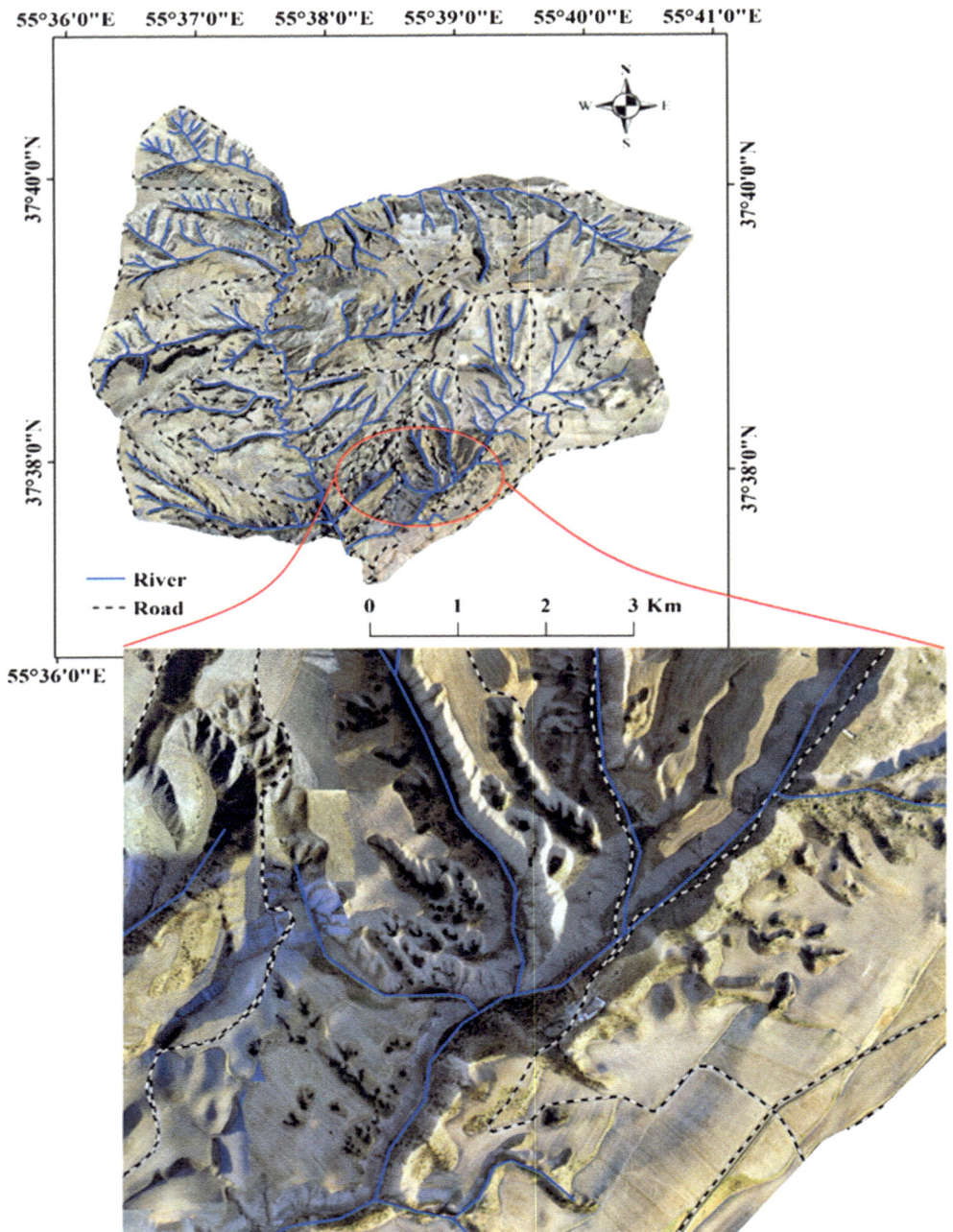

Fig. 24.4 The map of linear features created from the high-resolution digital orthophoto

Thus, based on the results of bivariate tests, gully-heads and RN$_S$ were positively correlated at all scales. Additionally, the spatial association of gully-heads and linear phenomena was presented in Table 24.1. According to Table 24.1, distance to stream networks and road networks was evaluated using restrictive buffer zones. The comparison of the test results and observed gully-heads within study area confirmed that there was a highly significant association between observed and recognized gully-heads. Furthermore, the spatial accumulation of gully-heads from stream networks was supposed to be *roughly* rising at the scales between 5 to 355 m. An enhancement happened in the number of gully-heads from RN$_S$ at the small scales (0–355 m). Similarly, the degree of the spatial associations from RN$_S$ enhanced at the small distances, implying that positive association was shown at the scales of 0–355 m.

Fig. 24.5 Spatial pattern analysis of gully-heads based on the univariate summary statistics (A: pair correlation function (*r*); B: *O*-ring function *O*(*r*))

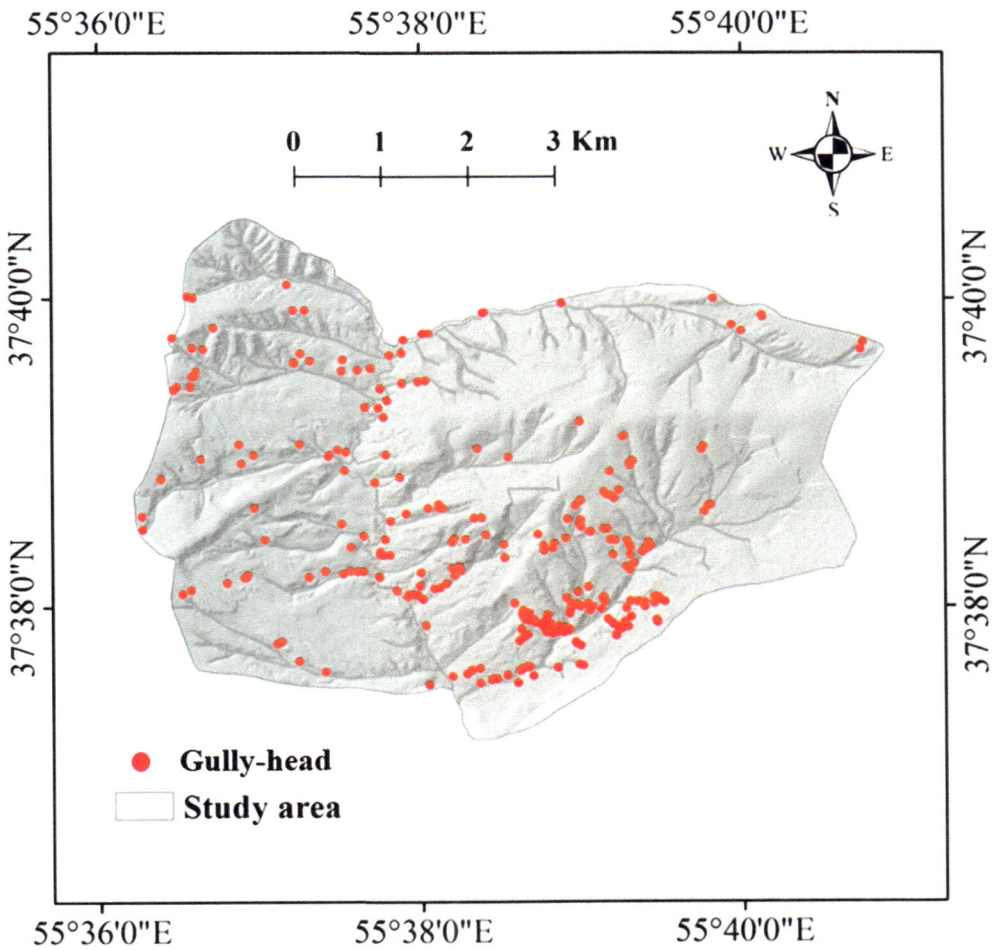

Fig. 24.6 The distribution map of gully-heads in the study area

The different size of soil particles related to gully-head locations in the study region was represented in Table 24.2. Soil texture as one of the spatial arrangements of soil particles may potentially have a significantly impact on gully-head activity. Regarding the results of mark correlation test, clay content of gully-heads was consistently smaller than the mean value of clay content ($\mu^2 = 22.93\%$) in the study area (Fig. 24.11A). The mean sand contents ($\mu^2 = 14.75\%$) do not differ from the mean sand contents taken over all pair gully-heads (Fig. 24.11B). Also, gully-heads and silt content were

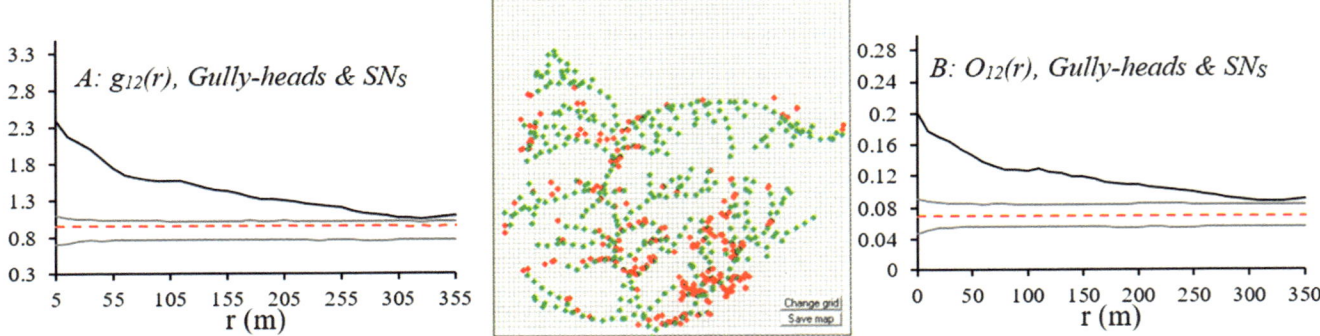

Fig. 24.7 Analyses of the occurrence of gully-heads in relation to the stream networks (SN$_S$) using bivariate summary statistics (A: $g_{12}(r)$ and B: $O_{12}(r)$)

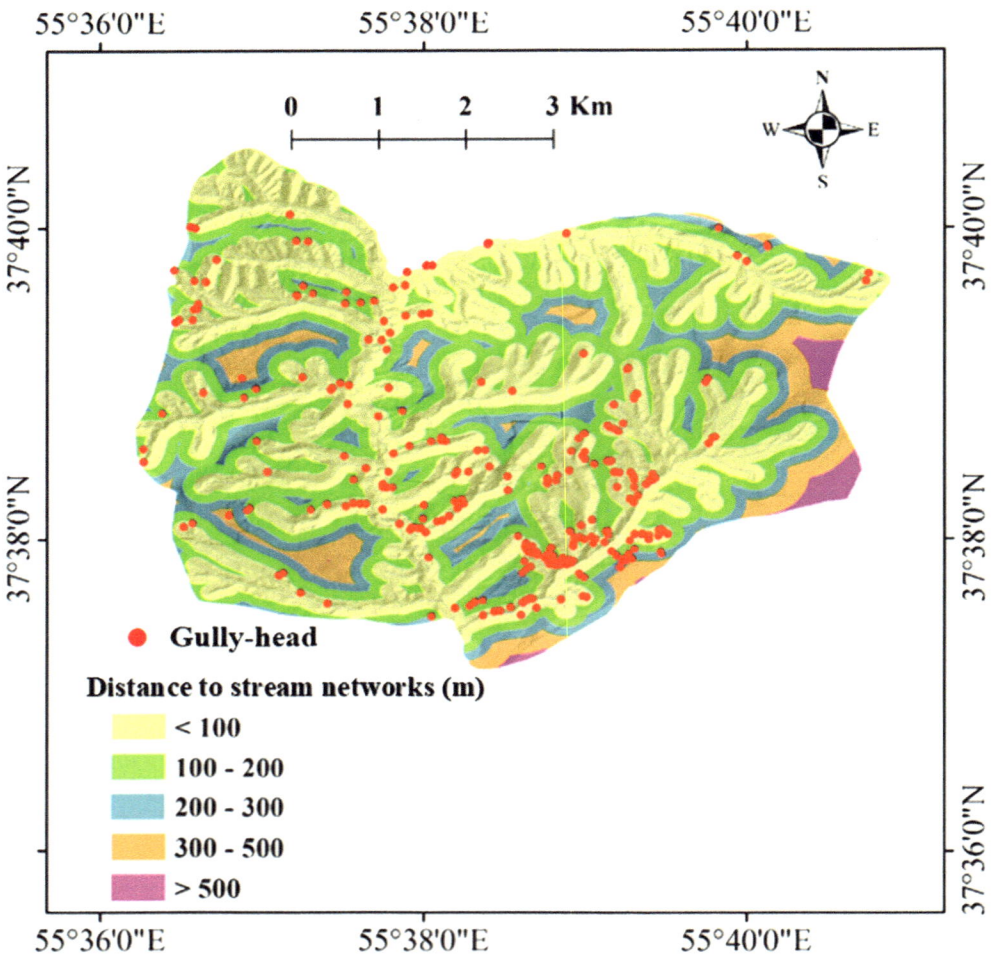

Fig. 24.8 The map of gully-heads occurrence in relation to the stream networks (SN$_S$) in the study area

highly correlated. The silt contents of nearby gully-heads were significantly larger than the mean value of silt content ($\mu^2 = 64.58\%$). This analysis shows that a very strong spatial correlation occurs between the development of gully-head and the more values of silt content (Fig. 24.11C). As the test failed to be in the lines of respective 95% simulation intervals, the spatial aggregation of gully-heads was a significant departure from the random labelling at level of 0.05. Thus, loess and silt content in any topographic site was mostly to be influenced by gully-head.

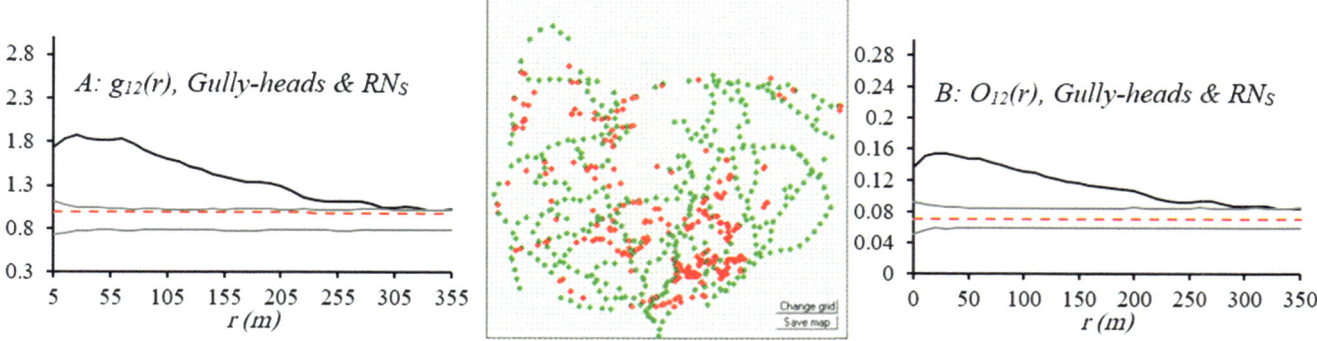

Fig. 24.9 Analyses of the occurrence of gully-heads in relation to the road networks (RN$_S$) using bivariate summary statistics (A: $g_{12}(r)$ and B: $O_{12}(r)$)

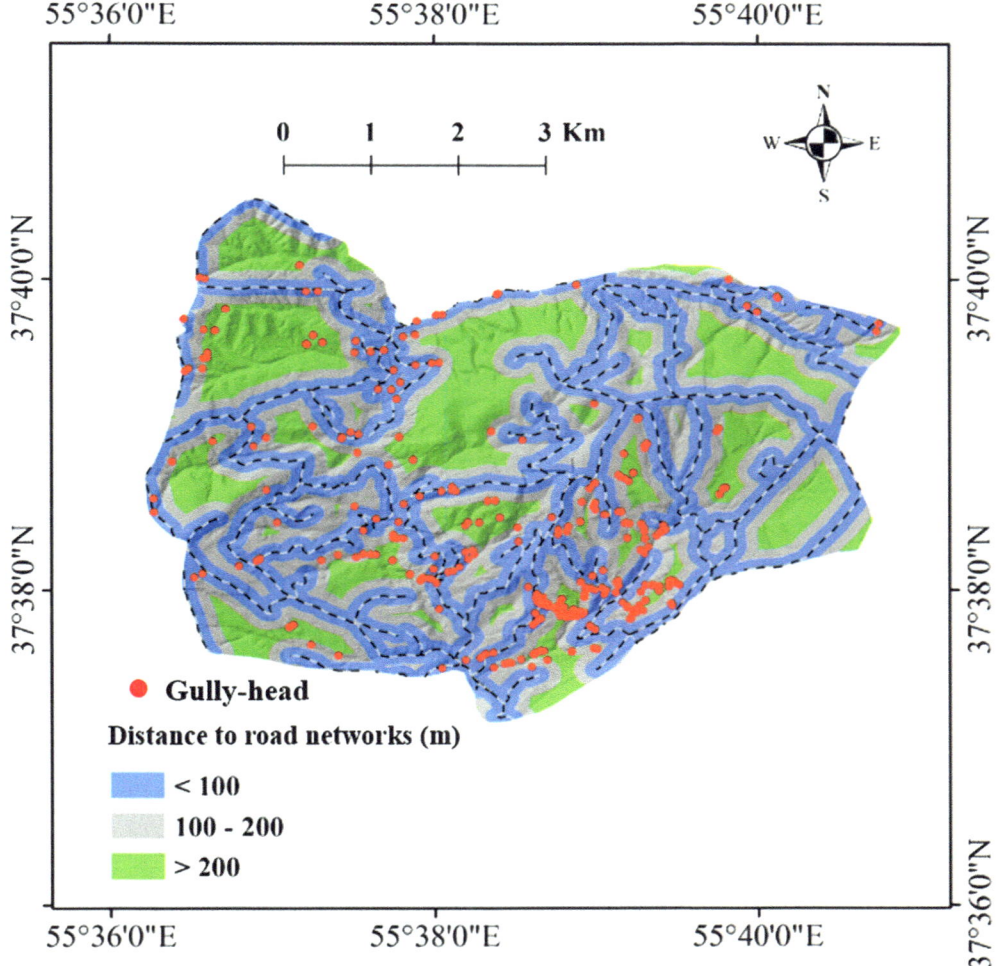

Fig. 24.10 The map of the occurrence of gully-heads in relation to the road networks (RN$_S$) in the study area

24.5 Discussion

Statistical methods applied in the spatial point pattern analysis can be successfully used to study distribution and spatial properties of gully-heads, as it was recently done in piping erosion studies (Hosseinalizadeh et al. 2018). The univariate gully-head tests only provide one-dimensional gully distributions information. In this research, two univariate tests were used to distinguish the spatial point pattern of gully-heads at various scales. An aggregated pattern was confirmed in the spatial pattern of gully-head applying two

Table 24.1 The occurrence of gully-heads in relation to the linear features in the study area

Conditioning factors	Classes	Number of gully-head	Percentage of gully-head	Area (ha)	Percentage of area
Distance from stream networks (*m*)	<100	221	77.27	1401.22	50.6
	100–200	49	17.13	813.07	29.4
	200–300	14	4.90	343.66	12.4
	300–500	2	0.70	172.20	6.2
	>500	0	0.00	39.88	1.4
Distance from road networks (*m*)	<100	128	44.76	1266.12	45.7
	100–200	104	36.36	837.26	30.2
	>200	54	18.88	666.64	24.1

Table 24.2 The occurrence of gully-heads in relation to the three soil textures in the study area

Soil factors	Classes	Number of gully-head	Percentage of gully-head	Area (ha)	Percentage of area
Clay content (%)	<10	2	0.7	80.12	2.9
	10–20	91	31.8	2375.62	87.0
	10–30	156	54.5	259.40	9.5
	>30	37	12.9	14.17	0.5
Sand content (%)	<10	34	11.9	8.27	0.3
	10–15	111	38.8	1496.80	54.8
	15–20	114	39.9	1211.33	44.4
	>20	27	9.4	12.91	0.5
Silt content (%)	<50	7	2.4	2.04	0.1
	50–60	87	30.4	938.45	34.4
	60–70	160	55.9	1785.43	65.4
	>70	32	11.2	3.38	0.1

univariate functions. The results of this study are in line with Tonini et al. (2012), who determined that the landslides analyzed applying mathematical models were not randomly distributed. Recently, Hosseinalizadeh et al. (2018) emphasized that gully-heads are considerably aggregated in their region. The suggested method allowed automatically to discover gully-head gaining in accuracy compared to the visual technique. Further, development of the research is needed to be considered for analyzing multi-dimensional point patterns.

Due to the importance of the occurrence of gully-heads in relation to the linear features, statistically bivariate methods were also applied to find a positive/negative relationship of linear features with gully distribution. Stream networks, which have a powerful interaction with gully-head distribution, were chosen as one of the investigated factors. Based on the different spatial relationships, SN$_S$ should reflect the majority of the gully-head regions to gain suitable performance (Liu et al. 2018). As a matter of fact, the landform and gully-head morphology in each regional unit are mostly consistent. It means that there is a highly positive interaction between gully-heads and SN$_S$. The results of this study are in accordance with Liu et al. (2018), who explained that the extracted drainage networks are derived from DEM areas' ground skeleton because they can control the bodies of gully.

Regarding the results, the correlation between gully-heads and road networks was significantly positive. The results of these tests were also exemplified by the RNs map; as explored by the RNs map, the gully-heads and RNs had remarkable spatial correlations in the finer distances. They are considered as the primary reason for gully-heads. It is because road networks generate by far most runoff. It means that roads are the most important generator of continuous runoff (Makanzu Imwangana et al. 2018). According to Jungerius et al. (2002), the design of the road can control the occurrence of soil erosion at the time of construction. However, the amount of gully-head development along the road is obviously out of control and the outcomes for road network cannot be anticipated.

Additionally, the development of gully-heads is controlled by soil factors; thus, their distribution cannot be random. In this study, soil particles which are categorized into three groups by size are recognized as the hot spots of gully-head density. Based on mark correlation test $k_{mm}(r)$, the values of clay and silt content of nearby gully-heads were statistically smaller and larger than the mean value of clay ($\mu^2 = 22.93\%$) and sand ($\mu^2 = 64.58\%$) content, respectively. The mean sand content ($\mu^2 = 14.75\%$) does not differ from the mean sand contents taken over all pair gully-heads. From Table 24.2, whole study area is covered by soils characterized

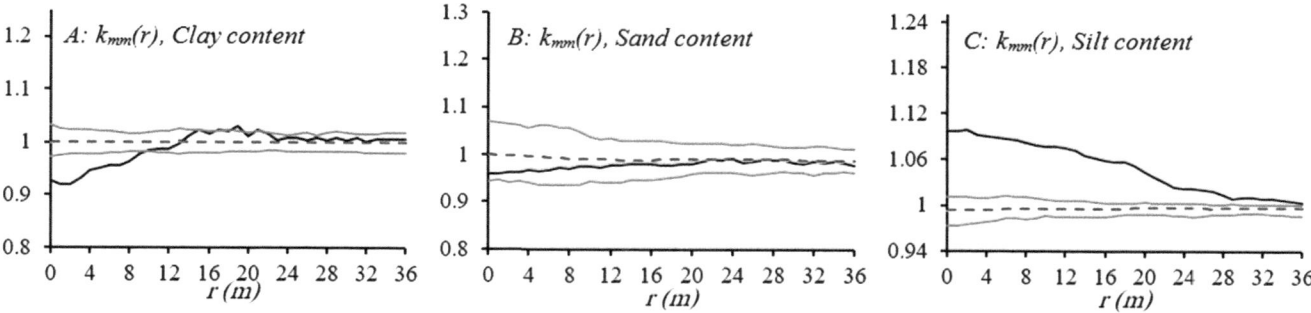

Fig. 24.11 Spatial impact of soil factors on the distribution of gully-heads based on the mark correlation function

by high silt content (more than 50% of silt). It is because the study area is a part of the Iranian Loess Plateau built of silt-rich sediments (Maleki et al. 2017). It also showed that the explanatory value of the silt content for the aggregated spatial location of gully-heads is limited. Silty texture of the analyzed soils generally favors gully-head locations, but it is not the only factor accountable for their formation. Based on soil texture, the most affected by gully-heads' location were soils characterized by silt content. Mostly, silty soils are considered as highly erodible and it is admitted that soils with a silt content more than 40% are highly erodible (Vandekerckhove et al. 2000; Morgan 2005). The other study claimed that soils with a clay content between 9 and 30% are the most susceptible to erosion (Evans 1980). This is in accordance with the results, as the soils in the study area with the silt content above 50% and clay content between 10 and 30% are characterized by the highest number of gully-heads (Table 24.2). However, based on the mark correlation function, the clay content had less value in the area with gully-heads (Fig. 24.11A). It means that the more values of clay content significantly limit the number of gully-heads. It is also confirmed that closer packing and smaller particle sizes generate higher apparent cohesion and greater capillary stress (Stefanovic and Bryan 2007). Further, the clay content affects, but it is not the main factor controlling gully-heads' occurrence in the study area.

24.6 Conclusion

This study applied statistically different spatial models, namely univariate, bivariate, and mark correlation tests. The spatial distribution of gully-heads was evaluated by univariate tests ($g(r)$ and $O(r)$). Bivariate correlation tests ($O_{12}(r)$ $g_{12}(r)$) were applied to assess the occurrence of gully-heads in relation to the linear features including road networks (RN$_S$) and stream networks (SN$_S$). The mark function was applied to find the spatial effect of soil features on the distribution of gully-head occurrence. In this regard, the distribution of gully-heads was significantly aggregated based on the results of univariate tests. A significantly positive correlation was found among gully-heads with road networks and stream networks in the study area. The $k_{mm}(r)$ test showed that the mean sand contents do not differ from the mean sand contents taken over all pair gully-heads. A significantly smaller than overall averages was observed between the gully-heads density and clay content. However, the values of silt content were higher than the mean. Further, the produced models in this study can be used for soil conservation and land use planning of gully-heads and, consequently, for sustainable development in the study area.

References

Baddeley, A., 2018. spatstat. local: Extension to'spatstat' for local composite likelihood. URL https://CRAN.R project.org/package=spatstat.local, r package version 3.5–7

Beretta, A.N., Silbermann, A.V., Paladino, L., Torres, D., Bassahun, D., Musselli, R., García-Lamohte, A., 2014. Soil texture analyses using a hydrometer: modification of the Bouyoucos method. Ciencia e Investigación Agraria, 41: 25–26

Choubin, B., Rahmati, O., Tahmasebipour, N., Feizizadeh, B., Pourghasemi, H.R., 2019. Application of fuzzy analytical network process model for analyzing the gully erosion susceptibility. In Natural hazards gis-based spatial modeling using data mining techniques (pp. 105–125). Springer, Cham

Evans, R. 1980. Mechanics of water erosion and their spatial and temporal controls: an empirical viewpoint. In: Kirkby MJ and Morgan RPC (eds) Soil erosion. Wiley: Chichester, 109–128

Gan, M., Jia, Y., Shao, M. A., Guo, C., Li, T., 2018. Permanent gully increases the heterogeneity of soil water retention capacity across a slope-gully system. Agriculture, Ecosystems & Environment.

Genet, A., Grabarnik, P., Sekretenko, O. Pothier, D., 2014. Incorporating the mechanisms underlying inter-tree competition into a random point process model to improve spatial tree pattern analysis in forestry. Ecological modelling, 288, pp.143–154

Getzin, S., Wiegand, K., Yizhaq, H., Hardenberg, J. Meron, E., 2015. Adopting a spatially explicit perspective to study the mysterious fairy circles of Namibia. Ecography, 38,1–11

Hosseinalizadeh, M., Kariminejad, N., Campetella, G. Jalalifard, A., 2018. Spatial point pattern analysis of piping erosion in loess-derived soils in Golestan Province, Iran. Geoderma, 328: 20–29

Jafari Shalamzari, M., Zhang, W., 2018. Assessing water scarcity using the water poverty index (WPI) in Golestan province of Iran. Water, 10: 1–22

Jungerius, P. D., Matundura, J., Van De Ancker, J.A.M., 2002. Road construction and gully erosion in West Pokot, Kenya. Earth Surface Processes and Landforms, 27, 1237–1247

Keshavarzi B. 2014. A possible link between mineralogy of loess deposits and high incidence rate of esophageal cancer in Golestan province of Iran. Iranian Journal of Science and Technology 38: 281–287

Liu, K., Ding, H., Tang, G., Song, C., Liu, Y., Jiang, L. Zhao, B., Gao, Y., Ma, R., 2018. Large-scale mapping of gully-affected areas: An approach integrating Google Earth images and terrain skeleton information. Geomorphology, 314, 13–26

Makanzu Imwangana, F., Moeyersons, J., Ozer, P., Ntombi, M., & Dewitte, O., 2018. Factors controlling and triggering urban gullies in the high town of Kinshasa (DR Congo). In Geophysical Research Abstracts (Vol. 20, pp. EGU2018–7037). European Geophysical Society

Maleki, S., Khormali, F., Bodaghabadi, M.B., Mohammadi, J., Kehl, M., Hoffmeister, D., 2017. Geological controlling soil organic carbon and nitrogen density in a hillslope landscape, semiarid area of Golestan province, Iran. 2: 221–228

Morgan RPC. 2005. Soil erosion and conservation. Blackwell Publishing: The United Kingdom.

Ollobarren Del Barrio, P., Campo-Bescós, M. A., Giménez, R., Casalí, J., 2018. Assessment of soil factors controlling ephemeral gully erosion on agricultural fields. Earth Surface Processes and Landforms

Poesen, J., Nachtergaele, J., Verstraeten, G., Valentin, C., 2003. Gully erosion and environmental change: importance and research needs. Catena 50: 91–133

Shruthi, R.B., Kerle, N., Jetten, V., 2011. Object-based gully feature extraction using high spatial resolution imagery. Geomorphology, 134: 260–268

Stefanovic, J.R., Bryan, R.B. 2007. Experimental study of rill bank collapse. Earth Surface Processes and Landforms 32: 180–196

Tonini, M., Abellan, A., Pedrazzini, A., 2012. Cluster analysis of geological point processes with R free software. Open Source Geospatial Research and Education Symposium, Switzerland

Valentin, C., Poesen, J., Li, Y., 2005. Gully erosion: Impacts, factors and control. Catena, 63: 132–153

Vandekerckhove, L., Poesen, J., Oostwoud Wijdenes, D., Gyssels, G., Beuselinck, L., De Luna, E., 2000. Characteristics and controlling factors of bank gullies in two semi-arid mediterranean environments. Geomorphology, 33: 37–58

Wiegand, T. and Moloney, K.A., 2004. Rings, circles, and null-models for point pattern analysis in ecology. Oikos 104, 209–229

Wiegand, T. and Moloney, K.A., 2014. Handbook of spatial point-pattern analysis in ecology. CRC Press, New York, 538 p

Narges Kariminejad is a Ph.D. student of Desert Control and Management in Department of Watershed and Arid Zone Management, Gorgan University of Agricultural Sciences and Natural Resources, Gorgan, Iran. She has received B.Sc. degree in Environmental Science from Shiraz University, Iran. She has published some research articles in various international reputed journals.

Mohsen Hosseinalizadeh received Bachelor degree in Range and Watershed Management in 2002 from Yazd University and graduated from Gorgan University of Agricultural Sciences and Natural Resources in Watershed Management. He was awarded Ph.D. degree from University of Tehran. Presently, he is working as assistant professor at Gorgan University of Agricultural Sciences and Natural Resources (GUASNR, IRAN). His research interests include spatial statistics, bioengineering especially in Gully erosion, Geostatistics, and landslide.

Hamid Reza Pourghasemi is an associate professor of Watershed Management Engineering in the College of Agriculture, Shiraz University, Iran. He has a B.Sc. in Watershed Management Engineering of the University of Gorgan (2004), Iran; an M.Sc. in Watershed Management Engineering, from Tarbiat Modares University, Iran (2008); and a Ph.D. in Watershed Management Engineering from the same University (Feb 2014). His main research interests are GIS-based spatial modelling using machine learning/data mining techniques in different fields such landslide, flood, gully erosion, forest fire, land subsidence, species distribution modelling, and groundwater/hydrology. Also, Hamid Reza works on Multi-Criteria Decision Making methods in Natural Resources and Environment.

He has published more than 100 peer reviewed papers in high-quality journals, with three chapters in Springer. Also, he published two books in Springer (https://www.springer.com/gp/book/9783319733821) and Elsevier (https://www.elsevier.com/books/spatial-modeling-in-gis-and-r-for-earth-and-environmental-science/pourghasemi/978-0-12-815226-3).

Majid Ownegh (Ph.D.) born in Palmah Peykar in 1956 and completed his higher education at Ferdowsi University of Mashhad, Shahid Beheshti, and a full-time faculty member in Watershed Management and management of desert areas of the University of Agricultural Sciences and Natural Resources of Gerakan. After completing a course on natural hazards management at the University of Newcastle, Australia, in March 1983, he was promoted to an Associate Degree, and with the development of the "Hazard-Theory", in September 2002, he was promoted to Geomorphology and Land Management, and the owner of the model and new theory in the field of science of geomorphologic development of Gorgan plain, integrated management of natural disasters and land degradation and sustainable development of Golestan province.

Mauro Rossi is a research scientist from the "Consiglio Nazionale delle Ricerche" (CNR) in Roma, Italy. He is pursuing his research at the "Istituto di Ricerca per la Protezione Idrogeologica" (IRPI) in Perugia, Italy. Though he has diversified research interests, he mainly focuses on mapping, modelling, and forecasting of landslides, floods and erosion processes in different geo-environmental and anthropic contexts. Mauro Rossi has developed (1) new methodologies for statistical and deterministic analysis of the susceptibility and hazard posed by different geo-hydrological phenomena and for the estimation of their impacts, (2) new approaches to the definition of rainfall thresholds for triggering landslides, (3) early warning systems, and (4) approaches to the design optimal models for estimating landslide susceptibility and for the assessment of social risk posed by landslides and floods. He has also developed specific software for the landslide susceptibility modelling, for the landslide magnitude modelling and for the joint modelling of landslides and erosion processes in relation to different scenarios of geomorphological, climatic, vegetation, and anthropic changes, in order to adequately characterize the hill slopes and the hydrological basins dynamics. He has published more than 150 research papers in many international journals.

Majid Soufi, Reza Bayat, Aliakbar Davudirad, Majid Zanjanijam,
and Hossein Esaei

Abstract

The relationship between the slope gradient (S) and drainage area (A) upslope the gully head or point of gully incision represents the topographic conditions for development and positions of gully erosion in different environments. In this study, 300 gullies from 4 provinces including Fars, Markazi, Zanjan and Golestan were selected, and their physical characteristics such as watershed, soil, ground surface and their dimensions were measured. A digital elevation model (DEM) was produced with ArcGIS 9.3, using topographical maps with 1:25,000 scale. The location of gullies was recorded using GPS in the field and transferred on the DEM; also, the boundary of the gullies was depicted using Av. SWAT. Upslope drainage area and slope of the soil surface of gullies were measured using DEM and field survey. The gullies were classified into homogeneous groups, using Cluster analysis method, and effective factors for categorizing were determined using factor analysis. Values, coefficient of determination and significance of the relationships were determined using regression method in Minitab 16 and SPSS 23 software. The results of this research indicated a strong relationship between ground slope and drainage area in both cases of gully development and incision at 1% significance level. The relationship for gully development had an exponent (b) equal to -0.365 and R^2 equal to 0.564, indicating the dominant impact of overland flow on the gully development. Although using values of ground slopes measured on DEM did not decrease the significance level, values of exponent (b), intercept (a) and R^2 decreased. Value of intercept (a) indicates low resistance of landscape to gully erosion in the studied environments.

Keywords

Gully erosion · Topographic threshold · Digital elevation model · Iran

M. Soufi (✉)
Department of Soil Conservation and Watershed Management, Fars Research and Education Center for Agriculture and Natural Resources, Shiraz, Iran

R. Bayat
Department of Soil Conservation Engineering, Institute of Soil Conservation and Watershed Management, Tehran, Iran

A. Davudirad
Department of Soil Conservation and Watershed Management, Markazi Research and Education Center for Agriculture and Natural Resources, Arak, Iran

M. Zanjanijam
Department of Soil Conservation and Watershed Management, Zanjan Research and Education Center for Agriculture and Natural Resources, Zanjan, Iran

H. Esaei
Department of Soil Conservation and Watershed Management, Golestan Research and Education Center for Agriculture and Natural Resources, Gorgan, Iran

25.1 Introduction

Soil erosion is recognized as the key factor for soil degradation in the world (Oldeman et al. 1990; Valentin et al. 2005; Liniger and Critchley 2007; Gutierez et al. 2009). Gully erosion as an important sediment source, conveyance of flow and sediment from upstream to downstream channel networks and contamination of water bodies by transported sediment and chemicals, needs to be better understood, managed and its effects mitigated (Poesen et al. 2003, 2011; Torri and Poesen 2014). Gully erosion as a geomorphic phenomenon has different thresholds including topographic, hydraulic, rainfall, soil and land use (Poesen et al. 2003; Phillips 2006). Different methods were used to define topographic conditions for gully erosion, such as stream power index (Kakembo et al. 2009), topographic wetness index (Gutierez et al. 2015), and topographic threshold (Monsiers et al. 2015; Torri and Poesen 2014). Topographic threshold is expressed by the relationship between drainage area (A) and slope of the soil surface (S), controlling the development and position of gully heads in various environments (Torri and Poesen 2014). This approach is applied to show the location of gully heads and threshold of the drainage area and slope gradient to initiate and/or retreat headcuts. The process of the retreat of gully heads at a global scale remains poorly understood (Vanmaercke et al. 2016). In spite of one century research on gully erosion, there is still some knowledge gap which needs to be addressed (Castillo and Gomez 2016).

Patton and Schumm (1975) and Begin and Schumm (1979) began modelling of gully erosion as a threshold process when flow shear stress (τ_F) overcomes the shear strength of the channels (τsoil). This is developed by Montgomery and Dietrich (1994). The topographic threshold conditions for gully erosion are reported as a logarithmic plots of the upslope drainage area (A) and slope gradient of the soil surface as $S = aA^b$, where A (ha) is the area of the catchment draining towards the gully head and S (tangent, m/m) is the slope of soil surface at the gully head. The threshold coefficient (a) reflects the resistance of the site to gully development. The lowest values were observed for cropland followed by values for rangeland, pasture and forest (Torri and Poesen 2014). Negative exponent ($-b$) showed the reverse

relationship between the slope gradient and drainage area which means surface runoff while positive exponent (b) reveals that subsurface runoff is a dominant hydrologic process acting on gully development. Montgomery and Dietrich (1994) suggested exponent b between 0.5 and 0.875 corresponding to supercritical laminar flow and turbulent flow, respectively. Although most of the exponent b represented laminar flow, they are rare for concentrated flow in the fields (Torri and borselli 2003). The results of recent studies about threshold conditions for a variety of land uses in some parts of the world revealed that cultivated croplands need less topographic threshold than noncultivated lands for gully development (Poesen et al. 2003; Fig. 25.1). Land use change reduces the topographic threshold due to reduction of soil resistance and biomass. Therefore, the range of drainage area and slope of the soil surface were between 0.2–10,000 ha and 0.01–1 m/m in noncultivated lands and between 0.02–100 ha and 0.003–0.8 m/m in the cultivated croplands (Table 25.1). The results of previous research indicated that the field survey of the drainage area and the slope of soil surface (Fig. 25.1, line 1) produce larger values than analysis of the aerial photos and topographic maps

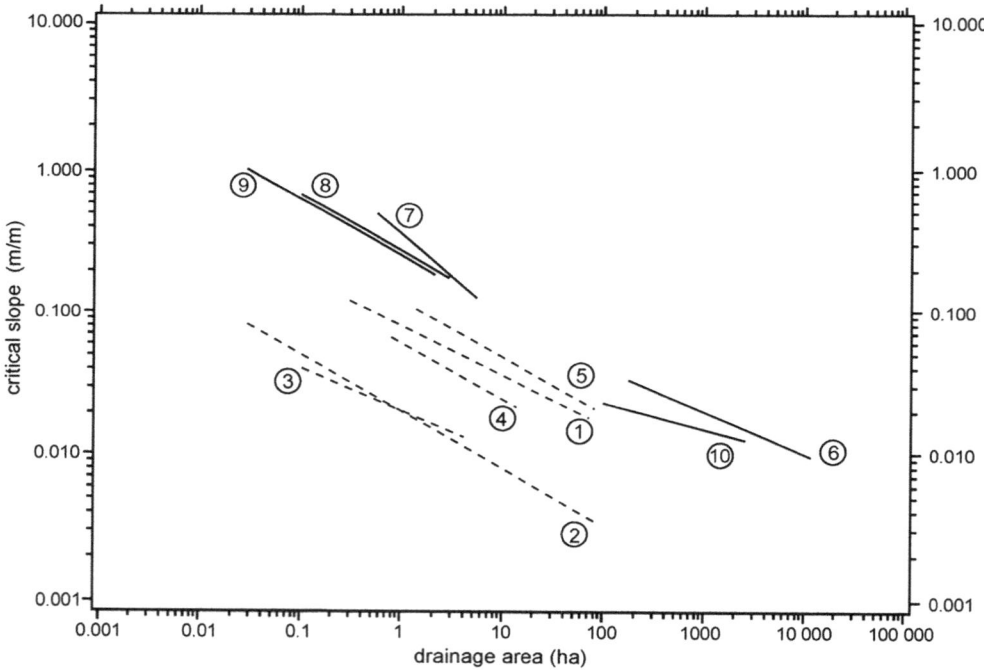

Fig. 25.1 The relationship between drainage area and critical slope of the soil surface for incipient gully development in some parts of the world (Poesen et al. 2003). Dotted lines indicate the threshold conditions for ephemeral gully development in cultivated cropland (1–5), and solid lines indicate the threshold conditions for gully head development in noncultivated land (6–10), (1) central Belgium, field survey (Poesen 1993); (2) central Belgium, measurement on aerial photos and topographic maps (Vandaele et al. 1996); (3) Portugal, analysis of aerial photos and topographic maps (Vandaele et al. 1996); (4) France, analysis of aerial photos and topographic maps (Vandaele et al. 1996);

(5) Southern Britain, field survey (Boardman 1992); (6) USA (Colorado), sage brush and scattered trees, analysis of photos and topographic maps (Patton and Schumm 1975); (7) USA (Nevada), field survey, open oak woodland and grassland (Montgomery and Dietrich 1988); (8) USA (California), coastal prairie, field survey (Montgomery and Dietrich 1988); (9) USA (Oregon), field survey, logged forest (Montgomery and Dietrich 1988); (10) Australia (New South Wales), swampy, reed covered valley floors, field survey (Nanson and Erskine 1988)

Table 25.1 The values of drainage area and slope of soil surface for threshold conditions in different land uses and methods of survey

Slope of soil surface (m/m)	Drainage area (ha)	Land use	Method	References
0.02–0.12	0.2–70	Cultivated land	Field survey	Poesen (1993)
0.003–0.8	0.02–100	Cultivated land	Analysis of aerial photos and topographic maps	Vandaele et al. (1996)
0.015–0.04	0.1–2	Cultivated land	Analysis of aerial photos and topographic maps	Vandaele et al. (1996)
0.025–0.07	0.5–10	Cultivated land	Analysis of aerial photos and topographic maps	Vandaele et al. (1996)
0.015–0.1	1–80	Cultivated land	Field survey	Boardman (1992)
0.01–0.04	200–10,000	Sage brush and scattered trees	Analysis of aerial photos and topographic maps	Patton and Schumm (1975)
0.02–0.5	0.2–4	Oak wood land and grassland	Field survey	Montgomery and Dietrich (1988)
0.3–0.7	0.2–2	Coastal prairie	Field survey	Montgomery and Dietrich (1988)
0.02–1	0.2–1	Logged forest	Field survey	Montgomery and Dietrich (1988)
0.015–0.025	100–2000	Swampy reed cover	Field survey	Nanson and Erskine (1988)

(Fig. 25.1, line 2). This means different methods of measurement in the same or similar locations yield different results.

The results of the research conducted in China (Sun et al. 2014) and in Australia (Monos-Robles et al. 2010) indicated the importance of slope gradient and drainage area on the rate of gully erosion. Vanmaercke et al. (2016) believed that the volumetric rates of gully head retreat (GHR) were significantly correlated to the runoff contributing area of the gully and rainy day normal. Topographic threshold could be changed based on the morphometric characteristics of the gullies such as gully depth in loess plateau of Belgium (Nachtergaele et al. 2002). Deeper gullies had a larger intercept (*a*) and exponent (*b*). Monsiers et al. (2015) presented the lowest intercept (*a*) values (0.078–0.090) for catchments treated with sloping drainage ditches and the highest ones (0.198–0.205) for stone bund catchments on the contour in Northern Ethiopia. The majority of studies indicated that exponent (*b*) varied between −0.2 and −0.7 (Vandaele et al. 1996; Desmet et al. 1999; Vandekerckhove et al. 2000; Nachtergaele et al. 2002; Hessel and Asch 2003; Vanwalleghem et al. 2005). Dietrich et al. (1993) believed a vast range of slope gradient was required to determine the relationship between the drainage area and slope of soil surface because of the effect of various natural and human factors on this relationship. Different methods of measurement of drainage area and slope gradient have been used by recent studies. Some of them (e.g. Morgan and Mngomezulu 2003) used the average slope gradient of catchment upslope the gully head; others used slope gradient of the valley bottom (e.g. Boardman 1992); some others used basin area per flow width (e.g. Prosser and Abernethy 1996); and also others used division of the difference of elevation to gully head area (e.g. Hancock and Evans 2006). Due to the vast

area of gully erosion in Iran about 14,000 km² (Soufi et al. 2017; Soufi and Bayat 2015, 2016) and huge damages to croplands and rangelands, roads, bridges and village and increased migration, it is necessary to study the topographic threshold of the gullies to manage and mitigate them. Therefore, four provinces from southwest (Fars), central (Markazi), northwest (Zanjan) and northeast (Golestan) were selected for the study of topographic conditions of gully erosion in Iran.

25.2 Materials and Methods

25.2.1 Study Area

This survey was carried out on 13 regions with massive gully erosion in four provinces including Fars, Golestan, Markazi and Zanjan in Iran (Fig. 25.2 and Table 25.2). Five regions including Lamerd, Alamarvdasht, Mishan, Baba Arab and Ghazian were studied in Fars province (Table 25.2). Gully erosion covers more than 300 km² in Lamerd and Alamarvdasht; the land type is alluvial which has originated from Miocene geologic formation. These regions are located 50 km far from Persian gulf with 150 and 200 m above the sea level; the average annual temperature and precipitation were 27 °C, 240 mm and 250 mm, respectively (Lamerd and Alamarvdasht stations, 20 years). Land use is rain-fed farms, poor rangeland and palm gardens. Most part of the area is poor rangeland. Mishan is located in the southwest of Fars province with semi-arid climate, 620 m above the sea level with hilly area; the average annual temperature and precipitation were 24 °C and 645 mm (Abdegah station, 18 years) and land use was rain-fed wheat and barley. Baba Arab is

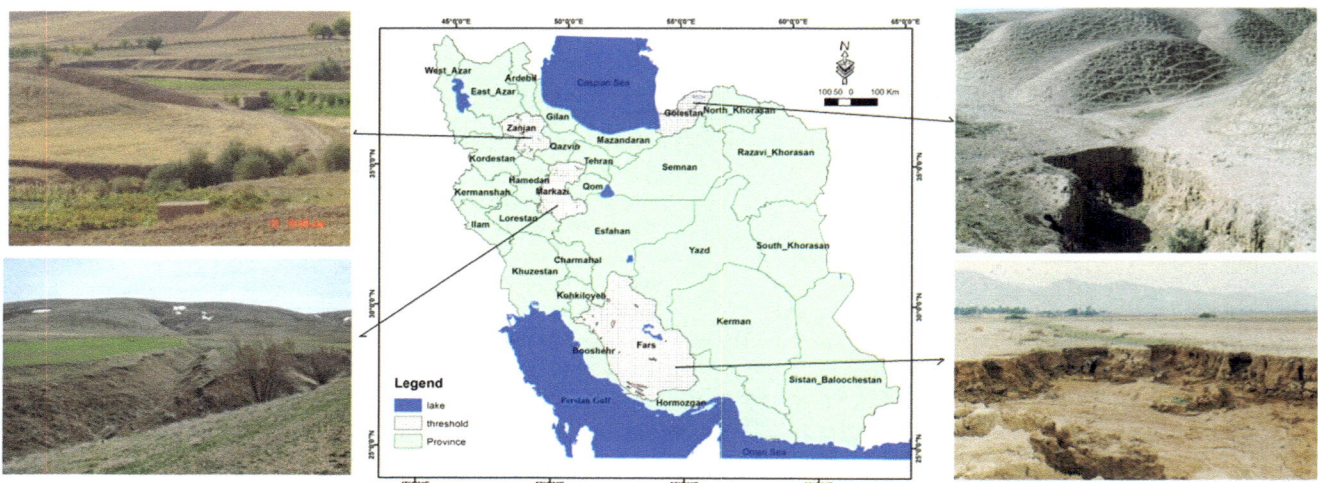

Fig. 25.2 The studied provinces for topographic threshold of gully erosion in Iran, Gully erosion in Bidgineh, Zanjan (above left), Zahirabad, Markazi (below left), Hajighoshan, Golestan (above right) and Lamerd, Fars (below right)

located in the east of Fars with hilly land type and mild arid desertic climate, and 1093 m above the sea level; the average temperature and precipitation were 23 °C and 245 mm (Baba Arab station, 31 years), respectively, with 90 mm maximum daily rainfall and rain-fed farms. Also, Ghazian is located in the north of Fars province with hilly land type and cold semi-arid climate; the average annual temperature and precipitation there were 12.5 °C and 239 mm, respectively, with 61 mm maximum daily rainfall (Dehbid station, 30 years), and land type was irrigated farms.

Three regions including Robatturk, Zahirabad and Peik were studied in Markazi province (Table 25.2). Robatturk is located in the alluvial river terrace near Delijan city with arid climate. It is located 1959 m above the sea level. Average annual temperature and precipitation were 13 °C and 235.7 mm, respectively (Moteh station, 20 years). Land type is irrigated farms and gardens on deep soil. Zahirabad (Gavar station, 25 years) is located 2237 m above the sea level near Shazand, with semi-humid climate and average annual temperature and precipitation equal 11.6 °C and 457 mm, respectively. Gullies are located in old alluvial terrace with irrigated and rain-fed farms on semi-deep soil. Peik region is located near Zarandieh city on a plateau land type and is 1146 m above the sea level with mild arid climate. The average annual temperature and precipitation were 13.4 °C and 235 mm, respectively (Peik station, 18 years). The gully area is located around the Shoor river on deep soil with irrigated farms and poor rangeland with gravel fragments on the soil surface.

Three regions including Alagol, Tamargharaghozi and Hajighoshan were studied in Golestan province (Table 25.2). Alagol is located on the hilly area 20 m above the sea level and mild desertic arid climate. Average annual temperature and

precipitation were 17.2 °C and 200 mm, respectively (Chaat station, 13 years). Land type is hilly with salt loess, and the land use is poor rangeland with loam soil texture. Tamargharaghozi region is located on a hilly old loess plateau 140 m above the sea level with mild semi-arid climate. The average annual temperature and precipitation were 17.4 °C and 521.6 mm, respectively (Kalaleh station, 11 years). Dominant land use is poor rangeland with rain-fed farms in some locations on the soils with salt and heavy texture. Hajighoshan region is located in the east of Ghonbad-kavoos city. It is located on loess hills 180 above the sea level with mild semi-arid climate. Average annual temperature and precipitation were 17.6 °C and 417 mm, respectively (Gonbad kavoos station, 29 years). Land use is dominantly very poor rangeland with rain-fed farms in some parts of the region. Its soil is similar to that of Alagol region with high silt and SAR and exchangeable sodium.

Two regions including Bidgineh and Chapchap were studied in Zanjan province (Table 25.2). Bidgineh region is located in the east of Zanjan province with cold semi-arid climate in 1802 m above the sea level. Average annual temperature and precipitation were 16 °C and 358.5 mm, respectively (Galtoog station, 30 years). Land type is hilly marl and sandstones. Soil texture varies between loam and loam clay. Land use is rain-fed farms and gardens resulting from rangeland change. Chapchap region is located on hilly land which is 1235 m above the sea level with extra cold semi-arid climate in the west of Zanjan province. The average annual temperature and precipitation were 20 °C and 250.7 mm, respectively (Filehkas station, 8 years). Soil texture varies between loam and loam sand. Land use is dominantly poor range land with rain-fed farms in some parts of the region.

Table 25.2 Characteristics of the studied regions in the selected provinces in Iran

Location	Climate	Annual precipitation (mm)	Land use	Land type	Coordinates
Fars (lamerd)	Arid	250	Cultivated cropland (dry farm)	Alluvial plain	27°15'; 2720'N 52°30'; 53°, 45'E
Fars (Alamarvdasht)	Arid	240	Cultivated dry farm and poor rangeland	Alluvial plain	27°28'; 2754'N 52°36'; 53°, 18'E
Fars (Mishan)	Mild semi-arid	620	Cultivated dry farm	Hill	30°00'; 30°02'N 50°53'; 53°, 57'E
Fars (Babaarab)	Mild arid	90	Cultivated cropland	Hill	28°30'; 30°38'N 53°40'; 53°59'E
Fars (Ghazian)	Cold semi-arid	239	Cultivated cropland	Hill	30°25'; 30°30'N 53°04'; 53°12'E
Zanjan (Bidgineh)	Cold semi-arid	358.5	Cultivated cropland and gardens	Hill	36°29'; 36°32'N 48°10'; 48°18'E
Zanjan (Chapchap)	Extra cold semi-arid	250.7	Rangeland and rain-fed farms	Hill	36°57'; 36°59'N 47°57'; 47°59'E
Golestan (Alagol)	Cold semi-arid	355.1	Rangeland + cultivated cropland	Hill	37°20'; 37°25'N 54°32'; 54°3'E
Golestan (Tamargharaghozi)	Mild semi-arid	521.6	Rain-fed wheat and barelt farms	Hill	37°29'; 37°31'N 55°30'; 54°32'E
Golestan (Hajighoshan)	Mild semi-arid	417	Poor rangeland and rain-fed farms	Hill	37°26'; 37°28'N 55°22'; 55°24'E
Markazi (Zahirabad-Shazand)	Cold semi-arid	457	Irrigated and rain-fed farm	Plateau (old alluvial and colluvial)	33°53'; 33°55'N 49°14'; 49°17'E
Markazi (Robatturk-Delijan)	Cold arid	235.7	Irrigated farm and garden	Hill	33°42'; 33°46'N 50°49'; 50°52'E
Markazi (Peik-Zarrandieh)	Mild arid	235	Poor rangeland + irrigated *Pistacia*	Plateau	35°17'; 35°20'N 50°39'; 50°50'E

25.2.2 Methodology

25.2.2.1 Measurement of Drainage Area and Slope of Soil Surface Upslope of Gully Heads or Point of Gully Incision

Measurement of ground slope with different methods yielded different threshold lines (Vandaele et al. 1996); therefore, to produce standard values for interpretation, it is necessary to use field survey and compare other methods in different landscapes (Poesen et al. 2003; Torri and Poesen 2014). Drainage area and slope of soil surface were measured upslope of the gully heads (for development) and the point of gully incision (for incision). Slope of soil surface was measured using digital elevation model (DEM) and field survey. Slope was measured few metres far in the upstream of headcuts (for gully development) and upstream of points that gullies formed using Sento clinometers in the field survey. Also, slope was measured on DEM at the so-called points. DEM was produced using topographic maps with a scale of 1:25,000 in the environment of Arc/view.

25.2.2.2 Statistical Methods

25.2.2.2.1 Cluster Analysis

The relationship of drainage area and slope of soil surface was established in three cases. In the first case, the relationship was estimated for the total 300 gullies. In the second case, the gullies were divided into homogeneous groups using cluster analysis by Minitab software, version 16. Homogeneous groups were determined using Ward and Average methods in the 60% similarity level. The result of Ward method was used for estimation of the relationship between the drainage area and slope because of the least variance and more groups. In the third case, the gullies were divided into different groups based on their class area.

25.2.2.2.2 Factor Analysis

Analysis of principle components was used to determine the degree of determination or dependence of different variables in order to omit independent variables. Recognition of data suitability was done using coefficient of KMO, using SASS software. Data are suitable for factor analysis if KMO is more than 0.5. Factor analysis used diagrams of loading plot and score plot. The method of Varimax rotation was used to produce smaller groups of variables with strong determination for simplicity of interpretation and also factors could be defined based on fewer variables.

25.2.2.3 Geometric Characteristics of Drainage Basin

Position of each gully including headcut and point of gully incision was recorded using GPS and transferred to DEM. The border of each gully basin was determined using Av. Swat software. In this method, the boundary of drainage basin for gully heads and points of gully incision was depicted. Geometric parameters of drainage basin including area, perimeter, and length were measured for each gully basin for development and incision stages in Arc/View environment, and then different indices were estimated for drainage basin of each gully.

25.2.2.4 Indices of Drainage Basin (Alizadeh 1998)

- Compactness coefficient of Gravelius: division of basin perimeter to perimeter of a circle with an area equal to basin area. Relationship of Gravelius is $C = 0.28P/\sqrt{A}$ in which, P is perimeter and A is area of drainage basin. This coefficient is near 1 for circle basins and more than 1 for elongated basins.

- Form factor: dividing average width by the length of the drainage basin or the length of the longest stream. Length and width of the drainage basin were obtained using topographic map of the basin and dividing the area by the basin length, respectively. Therefore, form factor could be shown as $FF = A/L^2$, in which, L is the length and A is the area of drainage basin. Form factor equal to 1 represents the circle basin and smaller than 1 indicates the elongated one.

- Elongation ratio: dividing the area by the width of drainage basin. Its relationship is shown as $Re = 2/L(A/\pi)^{0.5}$. Smaller ratio from 1 represent the elongated drainage basin.

- Circle ratio: dividing the drainage area by a circle area that has equal perimeter of drainage basin. Its relationship is $R_c = A/A_c \frac{A}{A_c}$ in which A is the drainage area and A_c circle area with equal perimeter to the drainage area. Ratio equal to 1 indicates the circle basin and elongated one has smaller ratio than 1.

25.2.2.5 Gully Dimensions

Dimensions of the first-order gullies including length, top width, bed width and depth were measured using a tape meter in the field. Depth and width of the gullies were measured in a distance from the gully head with a uniform cross-section shape.

25.2.2.6 Physical and Chemical Characteristics of Soil

Soil characteristics were measured from the collected soil samples from the gully heads and gully banks from the surface layer in which the dominant plant roots existed. Particle size and bulk density were determined using hydrometer and steel cores, respectively. The cores were dried up in an oven for 24 h with 105 °C temperature. Bulk density was calculated by dividing the dry weight by volume of cores. Organic carbon was measured using Walkley–Black method. Ec and pH were measured by Ec and pH meter device using saturated soil extract. Potassium and sodium were measured using flame photometer. Calcium and magnesium were measured using titration with EDTA (Handbook no. 467, Soil and Water Research Institute 2008).

25.2.2.7 Land Use and Characteristics of Soil Surface

Characteristics of the soil surface including vegetation cover, litter, surface gravel fragments and bare soil were measured in quadrates with 1 m per 1 m dimensions on the ground surface around the gullies. At least 10 quadrates were used for each gully head and also gully bank. Land use and its condition were determined by field survey.

25.2.2.8 Relationship Between Drainage Area and the Slope of Soil Surface

The relationship between the drainage area (independent variable) and slope of soil surface (dependent variable) was determined for three cases including total gullies (300), homogeneous groups by cluster analysis, and the area class of the drainage basin with specific interval for two stages of development and incision. The relationship between the drainage area and slope was determined using Regression method in SPSS version 21. The type of relationship was power. The unit of the drainage area and slope was hectare and meter/meter, respectively. Data were plotted on the logarithmic scale. Negative exponent shows the surface runoff and positive one indicates the dominant role of subsurface runoff as the most important hydrologic process accounting for gully development or/and incision.

25.3 Results and Discussion

25.3.1 Values of the Measured Parameters for Development Stage

Values of 26 parameters were compared for gullies in 4 provinces, as shown in Table 25.3. Maximum, minimum and average values were compared. Seventy three percent of the studied gullies had a drainage area lower than 1 ha and

Table 25.3 Comparison of statistical values of 26 parameters measured for upslope of the gully head of 300 gullies in Fars, Markazi, Golestan and Zanjan provinces

Variable	Average	Minimum	Maximum
Drainage area (ha)	115.79	0.01	5138.34
Ground slope (%)	6.45	0.01	38
Basin Perimeter (m)	2374.95	26.1	41,179.0
Basin length (m)	894.99	11.04	15,508.0
Form factor	0.30	0.01	0.86
Compactness coefficient	1.54	0.20	15.48
Elongation ratio	0.50	0	1.11
Circle ratio	0.78	0.06	9.8
Clay (%)	27	0	66
Sand (%)	32	5	75
Silt (%)	39.5	11	72.2
EC (ds/m)	5.36	0.10	68
pH	7.92	5.6	17.2
Na (meq/l)	22.99	0.27	310.46
Ca + Mg (meq/l)	24.98	1.16	344.4
Bulk density (g/cm^3)	1.31	1.04	1.55
Organic carbon (%)	0.53	0	3.9
Land use	Poor rangeland	Rain-fed farm	Irrigated farm and garden
Top width (m)	4.10	0.3	29.8
Bed width(m)	1.58	0.2	50
Depth (m)	1.80	0.07	15
Top width/depth	3.87	0.27	42.31
Vegetation cover (%)	19.5	0	80
Litter (%)	4.3	0	20
Gravel fragment (%)	17.4	0	81
Bare soil (%)	58.9	2.1	100

16% of them had a drainage area between 1 and 10 ha. Results indicated that 99.3% of the studied gullies had a slope of soil surface lower than 15°.

The average data for 300 gullies indicates that gully erosion occurred in basins with overgrazed poor rangelands with 20% vegetation cover versus 58.9% bare soil and 17% gravel fragments on the semi-salt soil with loam texture, and rain-fed farms with low organic carbon (0.5%). Medium size gullies were formed in semi-elongated basins with an average depth of 1.8 m and width/depth ratio of 3.87 (Table 25.3).

25.3.2 Relationship Between Drainage Area and Ground Slope with Field Measurement

25.3.2.1 Clustering Gullies into Homogeneous Groups

Clustering was done using 26 parameters for the development stage and 22 parameters for the incision stage, as shown in Table 25.3. Parameters related to gully dimensions were not used for clustering in the incision stage. Four

homogeneous groups of gullies were created in the ward method (Fig. 25.3 and Table 25.4). Groups 1 and 2 included gullies of Fars province. Group 1 (Table 25.4) included 7 gullies with drainage areas between 5138 ha (gully no. 1) and 1907 ha (gully no 20). Group 2 (Table 25.4) included

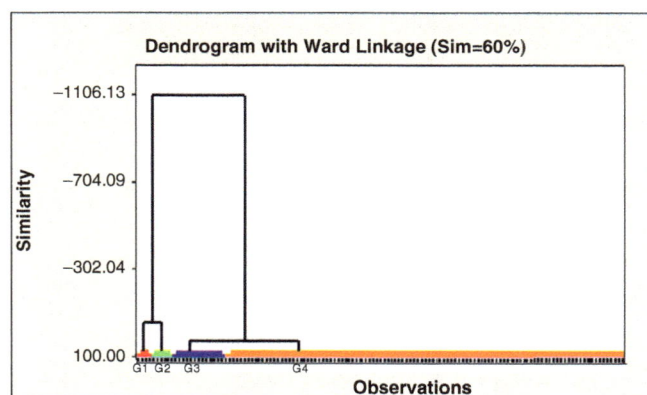

Fig. 25.3 Dendrogram of Ward method for categorizing of gullies into four groups in the development stage (G1 = group 1, red line with 96.3% similarity, G2 = group 2, green line with 93% similarity, G3 = group 3, blue line with 91% similarity and, G4 = group 4, yellow line with 90% similarity)

Table 25.4 Homogeneous groups of gullies using Ward method in the development stage

Gully group	Method	Case	Gully numbers	No. of gully
G1	Field survey	Development	20-11-9-6-4-2-1	7
G2	Field survey	Development	-42-30-29-25-22-21 -15-10-7-5-3	11
G3	Field survey	Development	8-13-14-16-18-19-23- 24-26-27-28-40-45- 61-536-G28	16
G4	Field survey	Development	Fars (117 gullies), Zanjan (total), Markazi (total-G28) and Golestan (total-536)	266

11 gullies with drainage areas between 1318 ha (gully no. 25) and 245 ha (gully no. 5) from Fars province. Group 3 included 16 gullies, 15 gullies from Fars with drainage area between 339 ha (gully no.16) and 59 ha (gully no. 23) and 1 gully from Golestan province. The remainder included 117 gullies of Fars province, total gullies in Zanjan, Golestan (minus 1) and Markazi (minus 1) provinces which had less than 10 ha drainage area.

25.3.2.2 Effective Factors for Categorizing Gullies

Adequacy of data was determined for factor analysis using the average value of KMO. For 300 gullies, it was equal 0.6762; therefore, effective factors for categorizing gullies were classified into seven categories based on the values obtained (Table 25.5). Effective factors for the stage of gully incision and for the method of DEM were the same. Data shown in Table 25.5 indicated that the most important group of effective variables belonged to the drainage basin of gully heads (factor 1) including the area, length and perimeter, and particle size was the least effective variable (factors 6 and 7) for categorizing the gullies into homogeneous groups (Table 25.5). Other factors such as soil characteristics and gully dimensions had an intermediate importance (Table 25.5). The same results are shown in Fig. 25.4; as shown, more important variables such as drainage area, length and perimeter of the basin had larger values than other variables (Fig. 25.4, above left); concentrations of

gullies are shown in groups (Fig. 25.4, above right), and the importance of each variable is shown by Eigen (specific) values (Fig. 25.4, below left).

25.3.3 Relationship Between Drainage Area and Slope of the Soil Surface for the Gully Head

25.3.3.1 Relationship for Total 300 Gullies Without Categorizing

The results indicated that power relationship between the drainage area and slope of the soil surface existed for 300 selected gullies from four provinces of Iran (Table 25.6). Table 25.6 shows that the relationship between the drainage area and slope was significant at 1% level and had negative exponents (b) for both development and incision stages for both methods of field surveying and digital elevation model (DEM). Data in Table 25.6 indicate that the field measurement of slope yielded values for exponent (a), intercept (a) and coefficient of determination bigger than DEM measurement in the development stage. In other words, in development stage, the values of b and a for field measurement of the slope, were 0.365 and 0.025 versus 0.326 and 0.017 for DEM (Table 25.6, rows 1 and 2). Negative exponent indicates the reverse relationship between the drainage area and slope of the soil surface and implies that surface runoff is the dominant hydrologic process for gully head development in the four provinces studied. The coefficient of determination (R^2) in the gully development with field surveying is equal to 0.564 (Table 25.6, row 1). This means that 75.5% of variation in the slope of the soil surface upslope in the gully heads could be interpreted by the variations in the drainage area. For gully incision, coefficients of the Eq. (25.3) (0.012 in row 3 in Table 25.6) were smaller than those (0.025 and 0.017) for gully development (Eqs. 25.1 and 25.2 in Table 25.6).

25.3.3.2 Relationship Between the Drainage Area and Slope for the Gully Groups Using Cluster and Factor Analysis

For development stage, cluster analysis divided the gullies into four groups (Table 25.4). Relationships for groups 1 and

Table 25.5 The factors affecting the categorizing of gullies in the development stage

Factor	1	2	3	4	5	6	7
Variable	X1 = basin area X3 = basin perimeter X5 = basin length	X14 = Na X12 = Hydraulic conductivity X15 = Ca + Mg	X19 = Ave. width X21 = Ave. depth	X6 = form factor X7 = Re	X23 = vegetation cover (%) X26 = Bare soil (%)	X11 = silt X10 = sand	X9 = clay
Value	X1 = +0.956 X3 = +0.950 X5 = +0.936	X14 = −0.920 X12 = −0.915 X15 = −0.832	X19 = +0.920 X21 = +0.902	X6 = −0.899 X7 = −0.874	X23 = +0.901 X26 = −0.670	X11 = +0.828 X10 = −0.822	X9 = +0.939

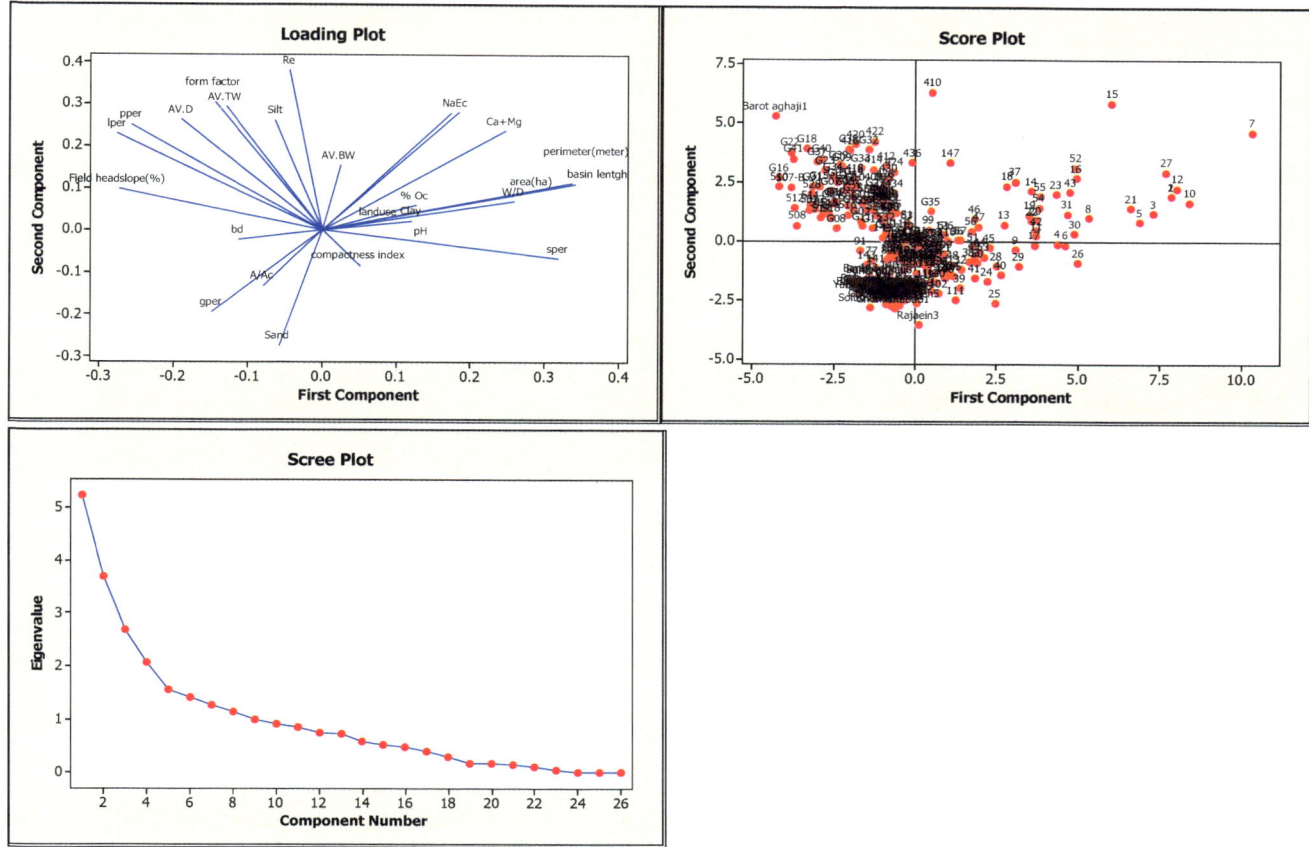

Fig. 25.4 The loading (above, left) score (above, right) and scree (below, left) plots of effective variables for categorizing gullies in development stage and field survey

Table 25.6 Comparison of the relationships between the drainage area and slope of the soil surface upslope gully head and point of gully incision in Iran

Eq.	Method	Case	Equations	R^2	P
25.1	Field survey	Development	$Slope(\%) = 0.025(Area)^{-0.365}$	0.564	≤ 0.000
25.2	DEM	Development	$Slope(\%) = 0.017(Area)^{-0.326}$	0.435	≤ 0.000
25.3	Field survey	Incision	$Slope(\%) = 0.012(Area)^{-0.309}$	0.189	≤ 0.000

2 were not significant, but they were significant for groups 3 and 4 with a significance level of 5% and 1%, respectively (Table 25.7). Group 3 with 16 gullies from Fars (14), Markazi (1) and Golestan (1) had a determination coefficient equal to 0.266. The relationship with $a = 0.037$ and $b = -0.502$ implies the dominant impact of surface runoff on the gully development (row 1, Table 25.7). Group 4 with 266 gullies (Table 25.4) had a higher level of significance (1%), but smaller intercept (a), exponent (b) and determination coefficient (r_2) than group 3 (rows 1 and 2 on Table 25.7).

For the incision stage, groups 3 [with 50 gullies from Fars (45), and Golestan (5)] and 4 [with 223 gullies from Fars (81), Zanjan (total), Markazi (total) and Golestan (total-5) provinces] were significant at the level of 5% (rows 3 and

4, Table 25.7). These groups had a smaller intercept (a) and exponent (b) (rows 3 and 4 in Table 25.7) than group 3 and 4 for gully development (rows 1 and 2 in Table 25.7).

Comparison of slope values between the two methods of measurement (field and DEM) using paired samples t-test indicated that there was a significant difference between them at 5% level; therefore, the relationship between the drainage area and slope of soil surface was estimated for DEM groups. Cluster analysis divided the gullies into four groups in the case of DEM. Relationships for groups 1, 2 and 3 were not significant, but there was a significant association only for group 4 with a significance level of 1%. Determination coefficient and intercept (a) were smaller (row 5 in Table 25.7) than group 4 with field survey (row 2 in

Table 25.7 Comparison of the relationships between the drainage area and slope of soil surface upslope gully head and point of gully incision for different homogenous groups of gullies in Iran

Group no. and no. of gullies	Method	Case	Equations	R^2	P
3(16)	Field survey	Development	$S = 0.037(A)^{-0.502}$	0.266	≤ 0.041
4(266)	Field survey	Development	$S = 0.031(A)^{-0.263}$	0.243	≤ 0.000
3(50)	Field survey	Incision	$S = 0.018(A)^{-0.428}$	0.089	≤ 0.035
4(223)	Field survey	Incision	$S = 0.014(A)^{-0.194}$	0.025	≤ 0.017
4(266)	Digital elevation model (DEM)	Development	$S = 0.019(A)^{-0.285}$	0.196	≤ 0.000

Table 25.7), but its exponent (b) was a little bigger than that for field survey (row 5 and 2 in Table 25.7).

25.3.3.3 Categorizing Gullies Based on Drainage Area

Torri and Poesen (2014) did not recommend using of gullies with big drainage area because of the occurrence of rainfall in a small portion of basin. Therefore, using larger drainage areas might yield wrong results due to smaller exponent (b) than reality for topographic threshold. Based on this theory, the selected gullies were divided into five classes (0.006–0.1, 0.11–0.99, 1–9.99, 10–1000 and > 1000 ha). Results in Table 25.8 indicated that the relationship between the drainage area and slope of soil surface was significant at the significance levels of 1% and 5%, respectively, for gullies at 6 out of 12 class areas. With increasing drainage area, the exponent (b) decreased in development and incision stages with field surveying method (Table 25.8, rows 1–4). The results revealed that result reported by Torri and Poesen (2014) was proved for the method of field surveying in gully development and incision stages. It means that with increasing the class area from 1–9.99 ha to 10–1000 ha in the development stage and from 0.1–0.99 ha to >1000 ha in the incision stage, exponent (b) decreased from 0.688 to 0.288 and from 1.37 to 0.766, respectively. However, for the method of DEM, the exponent (b) did not decrease with increasing drainage area. For example, exponent (b) increased from 0.359 to 1.070 with increase of the class area from 0.1–0.99 to 1–9.99 class area (rows 5 and 6 in Table 25.8).

Also, intercept (a), determination coefficient and significance level in the development case for both methods of filed survey and DEM decreased in the line of increasing drainage area (Table 25.8, rows 1–2, 5–6), but in the case of incision, the intercept (a) increased with increasing drainage area

(Table 25.8, rows 3–4). The first class area (0.1–0.99 ha) had a positive exponent; this means that the subsurface flow acted as the dominant process for gully incision in basins smaller than 1 ha. Results of this research indicated that exponent (b) of the significant relationships for the total gullies were between −0.309 and −0.365 (Table 25.6); it was between −0.194 and −0.502 for four gully groups by cluster analysis (Table 25.7) and between −0.288 and 1.37 for gullies grouped by the class area. These values are comparable with those presented by Nazari Samani et al. (2005) for southwest of Iran, Hessel and Asch (2003), Cheng et al. (2007) for loess plateau of China, Vandekerckhove et al. (2000) for Portugal and Spain and Vandaele et al. (1996) for loess belt of Belgium and Gutierez et al. (2009) for SW Spain.

Montgomery and Dietrich (1988) presented $b = 0.5$ for laminar overland flow and 0.8 for turbulent flow, so the values of exponent for the relationships between the drainage area and slope of the soil surface for group 3 in the development and incision stages with field survey were 0.502 and 0.428 that implies the existence of laminar flow. Some exponents (rows 1, 4 and 5 in Table 25.8) were between −0.688 (row 1, Table 25.8) and −1.070 (row 5, Table 25.8), revealing the turbulent flow. Comparison of intercept (a) in Tables 25.6, 25.7 and 25.8 indicated low resistance of the sites to erosion because most cases had an intercept near 0.03 that is similar to that for cultivated land in recent studies, such as that of Vandaele et al. (1996).

25.4 Conclusion

The results of this research indicated that power relationship existed between the drainage area and slope of the soil surface for 300 selected gullies from four provinces of Iran.

Table 25.8 Categorization of the gullies based on drainage area for the development case using field measurement of slope

Class area (ha)	Method	Case	Equations	R^2	P
1–9.99	Field survey	Development	$S = 0.033(A)^{-0.688}$	0.169	≤ 0.006
10–1000	Field survey	Development	$S = 0.015(A)^{-0.288}$	0.16	≤ 0.017
0.1–0.99	Field survey	Incision	$S = 0.004(A)^{1.37}$	0.209	≤ 0.000
>1000	Field survey	Incision	$S = 0.292(A)^{-0.766}$	0.441	≤ 0.004
1–9.99	Digital elevation model (DEM)	Development	$S = 0.025(A)^{-1.070}$	0.343	≤ 0.000
0.1–0.99	Digital elevation model (DEM)	Development	$S = 0.032(A)^{-0.359}$	0.050	≤ 0.022

This sort of relationship existed for different stages of development and incision and different methods of slope measurement. The sign of power in most cases was negative that implies the dominant impact of overland flow for gully erosion. The value of b, a and coefficient of determination decreased with changing the method of slope measurement from field to DEM and from development to incision stage.

The relationship between the drainage area and slope of the soil surface was significant for the basins area 1–9.99 ha and 10–1000 ha in the development stage and 0.1–0.99 ha in the incision stage. Exponent b was negative for both categories of area in the development stage (1–9.9 ha and 10–1000 ha), but it was positive for basin area 0.1–0.99 ha that shows the impact of the subsurface flow in the incision stage. The results are in the same line with those of the research conducted by Vandaele et al. (1996), Cheng et al. (2007), Hessel and Asch (2003) and Vandekerckhove et al. (2000).

Acknowledgements This project is sponsored by Institute of Soil Conservation and Watershed Management in Iran. We acknowledge their support and also the Research and Education Centers of Fars, Markazi, Golestan and Zanjan Provinces in Iran.

References

Alizadeh A (1998) Principles of Applied Hydrology. University of Imam Reza, Mashhad (In Persian).

Begin Z B, Schumm S A (1979) Instability of alluvial floors: A method for its assessment. Transactions of the ASAE. 22(2):347-350.

Boardman J (1992) Current erosion on the South Downs: implications for the past. In: Bell M, Boardman J. (Eds.), Past and Present Soil Erosion. Oxbow, Oxford, 9 –19.

Castillo C, Gomez J A (2016) A century of gully erosion research: Urgency, complexity and study approaches, Earth Science Reviews 160; 300-319

Cheng H, Zou X, Wu Y, Zhang CH, Zheng Q, Jiang Zh (2007) Morphology parameters of ephemeral gully in characteristics hill slopes on the Loess Plateau of China. Soil & Tillage Research, 4-14.

Desmet P, Poasen J, Govers G, Vandaele K (1999) Importance of slope gradient and contributing area for optimal prediction of incision and trajectory of ephemeral gullies. Catena 37:377-392.

Dietrich W E, Willson C J, Montgomery D R, Mckean J (1993) Analysis of Erosion Model. Journal of Geological, 101:259-278.

Gutierez A G, Conoscenti C, Angileri S E, Rotigliano E, Schnabel S (2015) Using topographical attributes to evaluate gully erosion proneness (susceptibility) in two mediterranean basins: advantages and limitations, Natural Hazards 79; 291-314.

Gutierez A G, Schnabel S, Contador F L (2009) Gully erosion, land use and topographical thresholds during the last 60 years in a small rangeland catchment in SW Spain, Land Degradation and Development 20(5);535-550.

Hancock G R, Evans K G (2006) Gully position, characteristics and geomorphic thresholds in an undisturbed catchment in northern Australia, Hydrological Processes 20: 2935-2951.

Hessel R, Asch T V (2003) Modelling gully erosion for a small catchment on the Chinese Loess Plateau Catena 54:131-146.

Kakembo V, Xanga W W, Rowntree K (2009) Topographic thresholds in gully development on the hillslopes of communal areas in Ngqushwa Local Municipality, Eastern Cape, South Africa, Geomorphology 110(3-4); 188-194.

Liniger H, Critchley W (2007) Where the Land is Greener: case studies and analysis of soil and water conservation initiatives worldwide, WOCAT, FAO and CDE, 364.

Monos-Robles C; Reid N; Fraizer P; Tighe M; Briggs S V; Wilson B (2010) Factors related to gully erosion in woody encroachment in south-eastern Australia, CATENA 83(2-3);148-157.

Monsiers E, Poesen J, Dessie M, Adgo E, Verhoest N E C, Deckers J and Nyssen J (2015) Effects of drainage ditches and stone bunds on topographical thresholds for gully head development in North Ethiopia, Geomorphology 234; 193-203.

Montgomery D R, Dietrich W E (1988) Where do channels begin. *Nature 336*:232- 234.

Montgomery D R, Dietrich W E (1994) A physical based model for the topographic control on shallow landsliding, Water Resources Research, 30(4):1153-117.

Morgan R P C, Mngomezulu D (2003) Threshold conditions for initiation of valley-side gullies in The Middle Veld of Swaziland, Catena *50*:401-414.

Nachtergaele J, Poesen J, Wijdenes D O, Vandekerckhove L (2002) Medium-term evolution of a gully developed in a loess-derived soil, Geomorphology, *46(3)*:223-239.

Nanson G C, Erskine W D (1988) Episodic changes in channels and floodplains on coastal rivers in New South Wales. In: Warner, R.F. (Ed.), IN: Fluvial Geomorphology of Australia. Academic Press Australia, Marrickville, *NSW*, 201–221.

Nazari Samani A, Ahmadi H, Jafari M, Boggs G, Ghoddousi J, Malekian A (2005) Geomorphic threshold condition for gully erosion in Southwestern Iran (Boushehr- Samal watershed), Journal of Asian Earth Sciences *35*:180-189.

Oldeman L R, Hakkeling R U, Sombroek W G (1990) World map of the status of human-induced soil degradation: an explanatory note. International Soil Reference and Incision Centre.

Patton P C, Schumm S A (1975) Gully Erosion, Northwestern Colorado: A Threshold Phenomenon, Geology 56:88-90.

Phillips J D (2006) Evolutionary geomorphology: thresholds and non-linearity in landform response to environmental change, Tobacco road team department of geography, university of Kentucky, Lexington.KY 40506-0027 USA.

Poesen J (1993) Gully typology and gully control measures in the European loess belt. In: Wicherek S (ed.), Farm Land Erosion in Temperate Plains Environment and Hills, Elsevier, Amsterdam, 221– 239.

Poesen J, Nachtergaele J, Verstraeten G, Valentin C (2003) Gully erosion and environmental change: importance and research needs, Catena, 50:91-93.

Poesen J, Torri D, Vanwalleghem T (2011) Gully erosion: procedures to adopt when modeling soil erosion in landscape affected by gullying. Chapt. 19, in Morgan R P C and Nearing M A (Eds) Handbook of Erosion Modelling. ISBN: 978-1-4051-9010-7, Wiley – Blackwell, 360-386.

Prosser I P, Abernethy B (1996) Predicting the topographic limits to a gully network using a digital terrain model and process thresholds. Water Resources Research, 32: 2289-2298.

Soil and Water Research Institute (2008) Manual for laboratory analysis of soil and water samples, no. 467 (In Persian).

Soufi M, Bayat R (2015) Morphoclimatic Classification of Gullies in Different Climates of I.R.Iran (phase 2). Organization of Research

Education and Extension for Agriculture. Ministry of Jihad –E-Agriculture. Iran, SBN no. 48474 (In Persian).

Soufi M, Bayat R (2016) Morphoclimatic Classification of Gullies in Different Climates of I.R.Iran (phase 3). Organization of Research Education and Extension for Agriculture. Ministry of Jihad –E-Agriculture. Iran, SBN no. 50689 (In Persian).

Soufi M, Bayat R, Charkhabi A H (2017) Morphoclimatic Classification of Gullies in Different Climates of I.R.Iran (phase 1). Organization of Research Education and Extension for Agriculture. Ministry of Jihad –E- Agriculture. Iran, SBN no. 51544 (In Persian).

Sun W; Shao Q; Liu J; Zhai J (2014) Assessing the effects of land use and topography on soil erosion on the Loess Plateau in China, CATENA 121;151-163.

Torri D, Borselli L (2003) Equation for high rate gully erosion. Catena 50: 449-467.

Torri D, Poesen J (2014) A review of topographic threshold conditions for gully head development in different environments, Earth Science Reviews, 130:73-85.

Valentin C, Poesen J, Li Y (2005) Gully erosion: Impacts, factors and control, Catena 63:132-153.

Vandaele K, Govers G, Wesemael B (1996) Geomorphic threshold conditions for ephemeral gully incision, Geomorphology 16 (2):161-173.

Vandekerckhove L, Poesen J, Oostwoud Wijdenes D, Nachtergaele J, Kosmas C, Roxo M J, De Figueiredo T (2000) Thresholds for gully incision and sedimentation in Mediterranean Europe, Earth Surface Processes and Landforms 25:1201–1220.

Vanmaercke M, Poesen J, Van Mele B, Demuzere M, Bruynseels A, Golosov V, Rodrigues Bezerra JF, Bolysov S, Dvinskih A, Frankl A, Fuseina Y, Teixeira Guerra AJ, Haregeweyn N, Ionita I, Imwangana F M, Moeyersons J, Moshe I, Nazari Samani A, Yermolaev O (2016) How fast do gully headcuts retreat? Earth Science Reviews 154;336-355.

Vanwalleghem T, Poesen J, Nachtergaele J, Verstraeten G (2005) Characteristics, controlling factors and importance of deep gullies under cropland on loess -derived soils. Geomorphology 69:76-91

Majid Soufi is an Associate Professor at the Department of Watershed Management and Soil Conservation, Fars Research and Education Center for Agriculture and Natural Resources, Shiraz, Iran. He got his Ph.D. in 1997 from the University of New South Wales in Sydney, Australia. He did his master's in watershed management in 1990 and B.S. in range and watershed management engineering in 1985 from Tehran University. He has professional experience of more than 30 years in the field of natural resources management and soil conservation, especially in gully erosion research and mitigation. He has presented 250 research papers in the national and international conferences and has published 55 research papers in scientific national and international journals. He was consultant of many Iranian projects in watershed and drought management which cooperated with UNDP, UNESCO, and FAO.

Reza Bayat is an Assistant Professor in the Department of Water and Soil Conservation Engineering at Soil Conservation and Watershed Management Research Institute, where has been a faculty member since 2001. He is Ph.D. candidate at Lorestan University and completed his M.Sc. studies at Tehran University and undergraduate studies at Gilan University. His research interests lie in the area of soil management, especially in soil erosion and conservation, and GIS application in this area. In recent years, he has focused on gully erosion and rainfall simulation, especially works with some mulches. He has collaborated actively with researchers in several other disciplines of environmental science, particularly watershed management and pollution.

Aliakbar Davudirad is an Assistant Professor in the Department of Soil Conservation and Watershed Management at Markazi Agricultural and Natural Resources and Education Center, Agricultural Research, Education and Extension Organization (AREEO). He has been a faculty member since 2004, holding a Ph.D. in Watershed Management Sciences and Engineering (Tarbiat Modares University, Iran). His research interests lie in the fields of watershed management, especially adaptive management and land degradation, zero net land degradation (ZNLD), strategic adaptive management in river engineering, and gully erosion. He has collaborated actively with researchers in several other disciplines of environmental science, particularly river engineering and hydrology.

Majid Zanjanijam is an Assistant Professor in the Department of Water and Soil Conservation Engineering at Soil Conservation and Watershed Management Research Institute, where has been a faculty member since 2001. He completed his M.Sc. and undergraduate studies at Tehran University. His research interests lie in the area of soil management, especially in soil erosion and conservation, and GIS application in this area. In recent years, he has focused on gully erosion. He has collaborated actively with researchers in several other disciplines of environmental science, particularly watershed management and pollution.

Hossein Esaei is a Research Scientist at Gorgan Research and Education Center for Agriculture and Natural Resources. He has received his B.Sc. from Gorgan University. His interest is soil and water conservation.

A Review on the Gully Erosion and Land Degradation in Iran

Mohsen Hosseinalizadeh, Mohammad Alinejad, Ali Mohammadian Behbahani, Farhad Khormali, Narges Kariminejad, and Hamid Reza Pourghasemi

Abstract

Water-induced soil erosion is one of the main causes of land degradation. In this category, gully erosion is the most important type of erosion which causes different problems and occurs in various pedo-climatic regions of Iran. For this study, around 60 relevant gully erosion case studies in the last decades with acceptable spatial patterns were considered. The most important control factors and analysis methods were considered. Most of the studies were carried out in semiarid and arid climatic regions. Recently, data mining and high techniques are commonly used in gully erosion susceptibility studies. Gully erosion was recorded in different climatic regions with annual rainfall range of 81–1200 mm. It also occurred in different topographic conditions (mean slope 2–27%) and different lithological units. Gully headcut retreatment varied from 0.99 to 1.4 m year^{-1} in various pedo-climatic regions, and ephemeral gully erosion in semiarid climate of Iranian Loess Plateau has resulted in a soil loss of 4 t ha^{-1} year^{-1}. Based on the obtained results, enough studies paid to gully erosion and various controlling factors have been determined. So its remediation and control should be regarded in future studies.

Keywords

Land degradation · Gully erosion · Iran

26.1 Introduction

Iran is exposed to various types of land degradation which is caused by multiple forces such as climate change and desertification. Researches reveal that the degradation of lands are mostly accelerated by combined pressures of agricultural activities, overgrazing of livestock, urbanization, population growth, deforestation, droughts, land salinization and lack of proper land management. Soil erosion, as a land degradation process is a potential source of damage to environment and human properties. Not only it degrades the soil quality on-site, but also increases sediment-related problems off-site (Poesen 2018). As stated by Ahmadi (2004), around 1.2 million km^2 of the land area of Iran is affected by different types of soil erosion. For instance, the maximum annual soil loss due to water erosion from farmlands in Iran is around 32 tonnes ha^{-1}. The total cost of soil and water degradation from cropland is estimated around 157,000 billion IR Rials (US\$3742 million) which is nearly 4% of the total gross domestic product (GDP) of the country and also one-third of the agricultural GDP annually. Gully erosion (GE) is one of the few geomorphological evidences of a past soil erosion which reflects the impacts of environmental changes on the landscape. It is among the major drivers of land degradation (Poesen 2011) which rapidly increases the runoff and sediment connectivity in landscapes and intensifies the off-site effects of water erosion and sediment production (Castillo and Gómez 2016; Pourghasemi et al. 2017; Rahmati et al. 2016). It often occupies less than 5% of the total catchment area but generates between 10 and 95% of the total sediment mass at the same scale (Poesen et al. 2003). According to the Iranian Ministry of Agriculture (Jahad-e-Keshavarzi), around 0.2 billion cubic metres of storage capacity of dams is

M. Hosseinalizadeh · A. Mohammadian Behbahani · N. Kariminejad
Department of Watershed and Arid Zone Management, Gorgan University of Agricultural Sciences and Natural Resources, Gorgan, Iran

M. Alinejad
Young Researchers Club, Gorgan Branch, Islamic Azad University, Gorgan, Iran

F. Khormali
Department of Soil Sciences, Gorgan University of Agricultural Sciences and Natural Resources, Gorgan, Iran

H. R. Pourghasemi (✉)
Department of Natural Resources and Environmental Engineering, College of Agriculture, Shiraz University, Shiraz, Iran

© Springer Nature Switzerland AG 2020
P. K. Shit et al. (eds.), *Gully Erosion Studies from India and Surrounding Regions*, Advances in Science, Technology & Innovation,
https://doi.org/10.1007/978-3-030-23243-6_26

reduced by annual sedimentation. So, 0.5% of the annual potential dam capacity is lost due to sedimentation (modified after World Bank 2005) which must be considered in prediction of water-induced soil erosion at watershed scale (Shruthi et al. 2015). The problems by GE have been well-documented in different parts of the world in the last decades. Globally, there is a comparable spatial variability for gully retreat rates (Vanmaercke et al. 2016). For example in Iran, gully head retreat has been measured up to 1.4 m year^{-1} during 1968–1994 and 1.2 m year^{-1} during 1994–2009 in Boushehr province with annual sediment mobilization of around 26.8 m^3 ha^{-1} (Nazari Samani et al. 2018) and 0.99 m year^{-1} in Iranian Loess Plateau (ILP) from 2001 to 2014 (Ghezelsofloo et al. 2018). Once this kind of soil erosion happens, the local topography and landscape change drastically. It also takes part in significant interactions with other soil erosion processes. For instance, gully channel formation has been shown to enhance the drainage of landscape leading to a reduction of water tables, a decrease in base flow and an increase of stormflow (Costa and Prado Bacellar 2007). Piping and gully headcut erosion have interaction as well. Piping may trigger gully headcut and conversely, gully headcut may induce piping erosion. Based on a point pattern analysis, gully headcuts were positively related to piping collapses and piping collapses were positively correlated with each other. Generally, multi-piping collapses induced gully erosion (Hosseinalizadeh et al. 2018). GE also reduces biomass production, particularly in the vicinity of the gully networks (Frankl et al. 2016). Apart from the unsustainable land use management, the agricultural activities were intensified over the last five decades in Iran. For instance, from the 1950s until 2008, around five million hectares of forest area was reduced (Emadodin et al. 2012) and the number of wells increased around tenfold in the last four decades. Globally, in terms of soil erosion, Iran holds the second rank with 2–2.5 billion tonnes annual amount of soil losses, which is equivalent to 8% of the global soil erosion (Najafi 2005). More than half the area of Iran (88 million ha) is in a critical state of specific erosion (Najafi 2005). In some provinces of Iran, gullies are initiated in different ecological and climatological conditions (Soufi 2002). Despite the various studies of GE in different parts of Iran, there is still no documented data which considers all different viewpoints across the whole country. Therefore, most of the gully relevant studies from different agro-climatic regions were considered to remove this deficiency.

26.2 Materials and Methods

This study consisted of three phases: (1) selection of some GE studies with suitable spatial patterns across Iran which consider all agro-pedo-climatic conditions; (2) determining the most important factors which induce GE occurrence; (3) considering all models including GE susceptibility and monitoring.

26.2.1 Description of the Study Area

About 70% of the total precipitation of Iran falls in only 40% of the land area. So there are significant spatio-temporal variations of rainfall in the country (Ahmadi-Givi and Parhizkar 2008). The average annual potential evaporation has high variation as well which varies from approximately 700 mm near the Caspian Sea to more than 4000 mm in central parts of Iran (Qadir et al. 2008).

26.2.2 Methodology

For this study, around 60 GE relevant case studies of the last decades with almost suitable spatial patterns were considered (Fig. 26.1). Controlling factors of GE and methods of these studies are listed in Table 26.1. The amount of annual precipitation for these studies changes from 81 to 1200 mm for Isfahan and Guilan provinces, respectively. These GE studies involve all climate regions in Iran (Fig. 26.2), and their slope is between 2 and 27%. They are located in different lithological formations as well (Fig. 26.3).

Golestan province has different climate regions with around 20% of the area covered by highly fertile and degradable loess deposits. Some photos of GE of this province are shown in Fig. 26.4.

26.3 Results and Discussion

Our survey indicated that in most case studies, in addition to topographic and edaphic factors (Abedini 2013; Balandeh et al. 2013; Khojeh et al. 2012; Rahi 1998; Shadfar and Sobhzahedi 2007; Balandeh et al. 2013), land use and lithology have also been taken into consideration (Gholami et al. 2017; Khaje 2017; Mohammad Ebrahimi et al. 2016; Nikpour

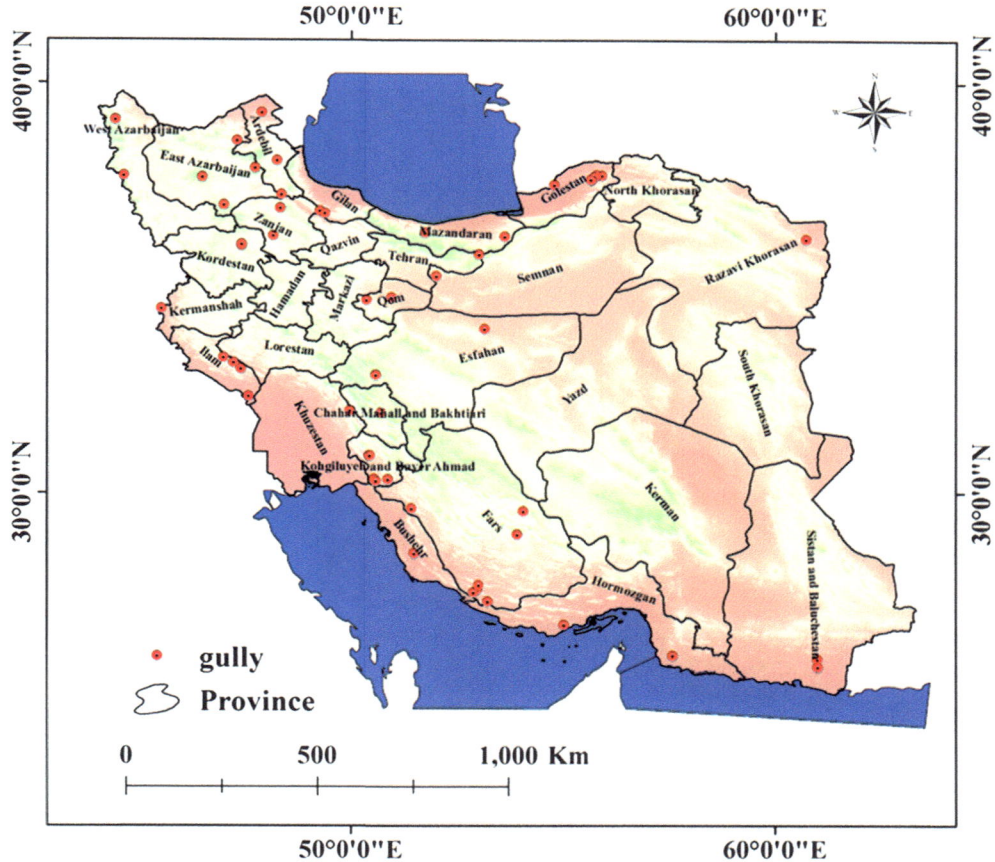

Fig. 26.1 Location of some gully relevant studies

et al. 2017; Shahbazi et al. 2016; Soleimanpour et al. 2009; Zanjani jam et al. 2012). The hydrological and climatic factors have been mentioned just in some studies (Rahmani et al. 2009; Shahab et al. 2017; Zarei et al. 2017). Only in one GE study, tectonic activity was applied as controlling factor of GE (Nikpour et al. 2017). Less attention paid to climatic and hydrological factors in GE studies is probably due to not having enough precipitation data of all the areas of study. So, in this case, only lumped model could be applied (constant climatic and hydrological factors across almost whole the study area). In this case, more information is needed to get spatial distribution of precipitation and runoff in modelling gully erosion. GE is usually studied by statistical analysis (Aghrazi et al. 2013; Ghoduosi and Davoudi Rad 2005; Ghasemi dehnavi et al. 2016; Khazaei et al. 2012; Noormohammadi et al. 2009; Yamani and Akbarian 2013; Jahantigh and Tabe 2016), Remote Sensing (Asghari Saraskanroud 2017; Bayati Khatibi et al. 2011; Ghoduosi and Davoudi Rad 2005; Khojeh et al. 2012; Neysi et al. 2017 Servati et al. 2013; Shafeie et al. 2011; Sufi 2005), Data mining (Pourghasemi et al. 2017 Hosseinalizadeh et al.

2018; Rahmati et al. 2017; Shahab et al. 2017) and Field Research (Arab ghashghaei et al. 2011; Asghari Saraskanroud 2017; Esmaeili et al. 2013; Ghasemi dehnavi et al. 2016; Gorji et al. 2015; Maghsudi et al. 2012; Nikpour et al. 2017; Servati et al. 2013; Shadfar and Sobhzahedi 2007; Shahbazi et al. 2016; Zanjani jam et al. 2012). Just a few studies have considered models to estimate gully erosion (Ahmadi et al. 2007; Nazari Samani et al. 2018; Zarei et al. 2017). Although GE occurs in most climatic regions of Iran, most of GE studies are devoted to semiarid and arid condition due to their intense rainfall and sparse vegetation. As stated by Poesen et al. (2003), the area which GE occupies is low, but its on/off-site side effects should be considered in natural resources management. It is unclear why no prediction study of gully development has been performed in Iran so far. Most of the GE studies have focused on GE monitoring, as well as empirical and statistical modelling. So in this regard, there remains the lack of considering different future scenarios which contribute in gully development. Models like CAEZAR Lisflood (Coulthard et al. 2013) and Dynamic SedNet (Wilkinson et al. 2014) should be applied to predict gully volumetric, areal and

Table 26.1 GE relevant case studies in different provinces

Researcher	Province study	Factor	Study method	Climate	Rainfall (m)	Height (m)	Slope (%)
Rahi (1998)	Boushehr	Soil-S	Zoning	Warm	250	300–1950	0–5
Bayati Khatibi (2004)	East Azerbaijan	Domain length-S-Domain shape	RS	Semiarid	350		1.9–4
Ghoduosi and Davoudi Rad (2005)	Zanjan	soil texture-(silt, clay)-EC-pH-Saturation of soil	RS-statistical tests	Arid to semiarid cold	406	1110–2197	3.82
Sufi (2005)	Fars	The area of the bare grounds-along the road	RS	Warm and arid	263		
Shahini and Charkhabi (2007)	Golestan	S-EC-Vegetation	RS-statistical analysis	Semiarid	270	1–25	
Shadfar and Sobhzahedi (2007)	Guilan	Petrological units-S-aspect-LU	FR	Semiarid Mediterranean	1200		5–20
Ahmadi et al. (2007)	Semnan	Model inputs	FAO, Thompson, SCS I and II	Semiarid			
Soleimanpour et al. (2009)	Fars	Soil texture-vegetation-sub-surface hydrologic-LU-L	RS-regression	Warm and arid	906		8
Soleimanpour et al. (2009)	Fars	Sub-surface hydrologic-vegetation-soil texture-LU-L	Regression	Semiarid	906	540	8
Noormohammadi et al. (2009)	Ilam	Insignificant variables	Regression-statistical tests	Semiarid	428.7		0–10
Rahmani et al. (2009)	Fars	Groundwater	Comparative	Arid and moderate		1560	1–2
Sadeghi et al. (2010)	Isfahan	LU	RS-Regression	Warm and arid	420	2100–3642	13.3
	Golestan	S-Average height, R, L, soil type, density and vegetation and LU	RS	Semiarid	470	124–250	0–5
Rahmanad Rad et al. (2010)	Sistan and Baluchestan	CEC-EC-SAR-ESP-LU	comparative	Warm and arid	114.2	0–64	0–0.5
Bayati Khatibi et al. (2011)	East Azerbaijan	S-L-soil texture-LU	Zoning-RS	Semiarid	322	1520–2950	
Shafeie et al. (2011)	Kohgiluyeh and Boyerahmad (Dehdasht)	SAR-ESP-EC-silt	RS-Comparative	Arid	534	620–1000	20–30
Fattahi and Javidkia (2011)	Qom	Soil-climate	RS	Dry and desert moderate	145	800–3141	
Arab ghashghaei et al. (2011)	Tehran	EC-plaster-SAR	Zoning-FR-RS	Mediterranean-semi-humid ultracold		2120–3360	0–5
Shafeie et al. (2011)	Kohgiluyeh and Boyerahmad (Gachsaran)	SAR-ESP-EC-Silt					
Novhegar and Heydarzadeh (2011)	Hormozgan	Silt-R	FR-statistical analysis	Arid	98.1		0–6
Hosseinzadeh et al. (2011)	Mazandaran	LU-drainage density-S-height	Regression	Warm and temperate	1096	2740–3569	20–30
Mahdavi et al. (2011)	Ilam	S-R	Zoning and regression	Semiarid cold	423.2	971–1584.68	2–5
Zanjani jam et al. (2012)	Zanjan	Soil erosion-destruction of vegetation-LU	FR	Semiarid cold	365	300–3330	
Khazaei et al. (2012)	Kohgiluyeh and Boyerahmad	Physico-chemical properties of soil-vegetation-R	FR-statistical analysis	Cold Mediterranean-subtropical	534	397–3403	
Khojeh et al. (2012)	Golestan	Silt, soluble salts, soil saturation moisture	RS-regression	Semiarid and arid	363		5.95
Esmaeelnezhad et al. (2012)	Guilan	Clay-exchangeable sodium-Na	FR-RS	Arid	245		

(continued)

Table 26.1 (continued)

Researcher	Province study	Factor	Study method	Climate	Rainfall (m)	Height (m)	Slope (%)
Yasrebi et al. (2013)	Ilam	S-L-silt-sand-SAR	RS-comparative	Semiarid	428	500–2790	6–12
Servati et al. (2013)	Golestan	S-distance from drainage network LU-R-L-soil-distance from gully-PC	FR -RS	Cold and semiarid	482	594–1270	3–8
Karami et al. (2013)	Fars	Mapping gully using object-oriented analysis methods	RS	Semiarid and moderate	250	485	
Abedini (2013)	East Azerbaijan	Topography-S	Regression	Semiarid	310	1878–3750	12.92
Balandeh et al. (2013)	West Azerbaijan	Height-area-S-aspect	RS	Semiarid and moderate	351		6.5–43.5
Aghrazi et al. (2013)	Markazi	S-LU-vegetation	FR-statistical analysis	Cold and semiarid	475	2237	15.94
Esmaeili et al. (2013)	Kordestan	L-LU-S-aspect-domain length-moisture-topography	FR-RS	Cold and arid	300	1600–2153	5–10
Yamani and Akbarian (2013)	Hormozgan	EC-silt-sand-lime-the plaster-magnesium-calcium-sodium-potassium	FR-RS-statistical tests	Warm and arid	–	–	–
Farajzadeh et al. (2012)	Mazandaran	L-soil-TMI-LU	RS-Multiple regression	Sub-humid	631	9.8–15.63	
Esmaeili and Shovkati (2015)	Kordestan	L-LU-S-aspect-domain length	Logistic regression	Cold and arid	300	1600–2153	
Karimi Sangchini and Ownegh (2015)	Chaharmahal and Bakhtiari	LU	Logistic regression-zoning	Semi-humid and cold	608.6	1456–2565	
Gorji et al. (2015)	Isfahan	Local conditions (change water level)	RS-FR	Warm and dry desert	81.5		
Jahantigh and Tabe (2016)	Sistan and Baluchestan	ESP-SAR-Mg-Ca-So4	FR-statistical analysis	Hyper arid	100	25–62	2
Mohammad Ebrahimi et al. (2016)	Golestan	Silt-clay-EC-pH-OC-CEC-SAR-ESP-L	Comparative	Semiarid	400	350–400	5–12
Shahbazi et al. (2016)	Kermanshah	LU	FR	Moderate and semiarid	370	80–300	
Ghasemi dehnavi et al. (2016)	Tehran	S-EC-SAR	FR-statistical analysis	Semiarid	274	830–1501	5–50
Mokarram and mahmudi (2017)	Fars	OM-deep ditch	Multiple linear regression-FR	Warm and arid	343	1252–3185	
Shahab et al. (2017)	Ardebil (OrtaDagh)	S-AS-surface runoff-properties of soil	RS-regression (PCA)	Semiarid	271.2		20–30
Shahab et al. (2017)	Ardebil (Mola Ahmad)	S-AS-surface runoff- properties of soil	RS-regression (PCA)	Semiarid	303.9		20–30
Shahab et al. (2017)	Ardebil (Sarcham)	AS-subsurface flow- properties of soil	RS-regression (PCA)	Semiarid	384.6		20–30
Nikpour et al. (2017)	Ilam	Climate factors-S-aspect-L-LU-tectonic activity	FR-RS	Arid and semiarid	165	50–1200	
Mohammad ebrahimi (2017)	Golestan	Soil texture-saturation moisture-EC-neutralizing agents-CEC-SAR-ESP	Comparative	Semiarid, semi-wet	552	430–882	13.46
Shahab et al. (2017)	Ardebil	S-surface and groundwater flow-AS-clay-runoff	Data mining-regression	Semiarid	303	1400	20–30
Asghari Saraskanroud (2017)	East Azerbaijan	EC-SAR-temperature and R	RS-FR-laboratory studies	Cold and semiarid	246.6	1240–1282	

(continued)

Table 26.1 (continued)

Researcher	Province study	Factor	Study method	Climate	Rainfall (m)	Height (m)	Slope (%)
Gholami et al. (2017)	Fras	L-LU S-distance from the river-distance from the road	RS	Dry and desert climate	211.5	413–1127	
Zarei et al. (2017)	Golestan	Model inputs	FR-EGEM	Moderate and semiarid cold	424	229–566	27.6
Nazari Samani et al. (2018)	Boushehr	Model inputs	Thompson, SCS I and II	Desertic condition	180	29–1141	
Mehraban et al. (2018)	Khorasan Razavi	L and distance from the river	RS	Semiarid and arid	230	650	
Hosseinalizadeh et al. (2019a, b)	Golestan	Models inputs (soil texture, LU, L, altitude, TWI, S, PC, DD, and SA)	Machine learning-RS	Semiarid	370	336–548	27
Ahmadi et al. (2018)	West Azerbaijan	L-LU	RS	Semi-humid and cold	190	1531–2922	
Azareh et al. (2019)	Ilam	Model inputs (soil texture, LU, L, altitude, TWI, S, PC, DD, SA and distance from rivers)	Maximum entropy, data mining Machine learning	Semiarid	450	423–2795	

PCA principal component analysis, *RS* remote sensing, *FR* field research, *EC* electrical conductivity, *OM* organic matter, *AS* aggregate stability, *PC* plan curvature, *S* slope, *LU* land use, *L* lithology, *EGEM* ephemeral gully erosion model, *SCS* soil conservation service, *TWI* topographic wetness index, *PC* plan curvature, *DD* drainage density, *SA* slope aspect

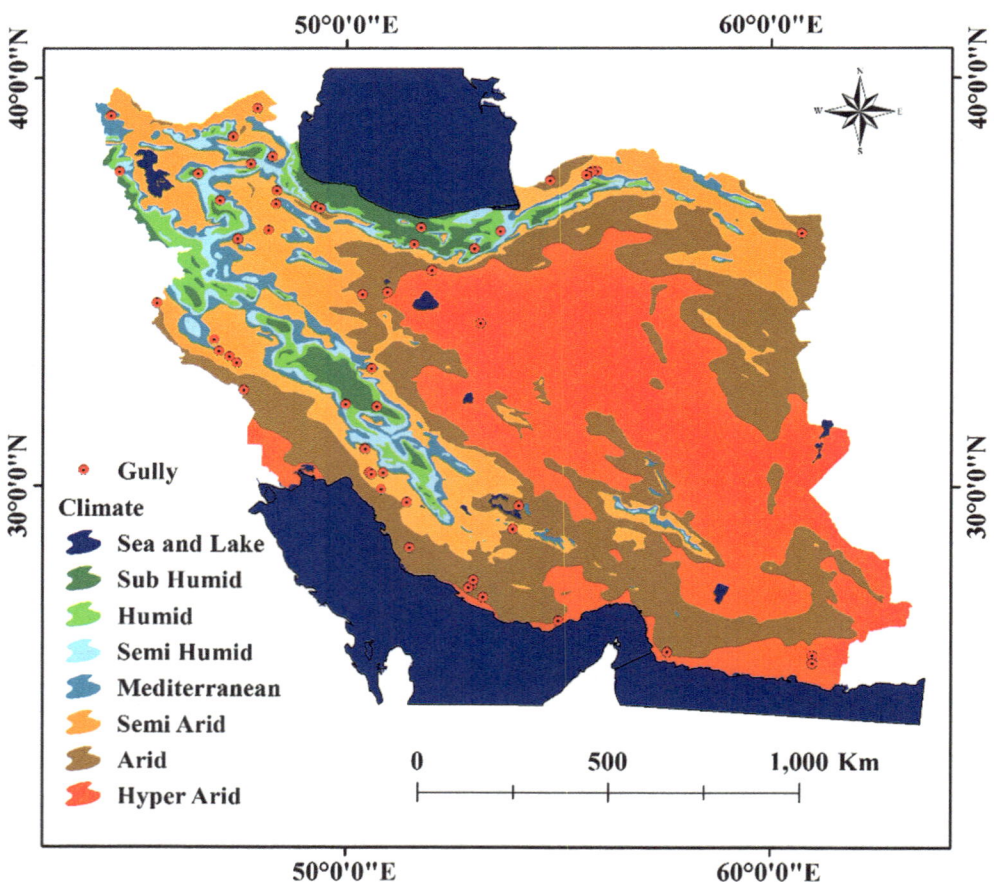

Fig. 26.2 Climatic pattern of Iran

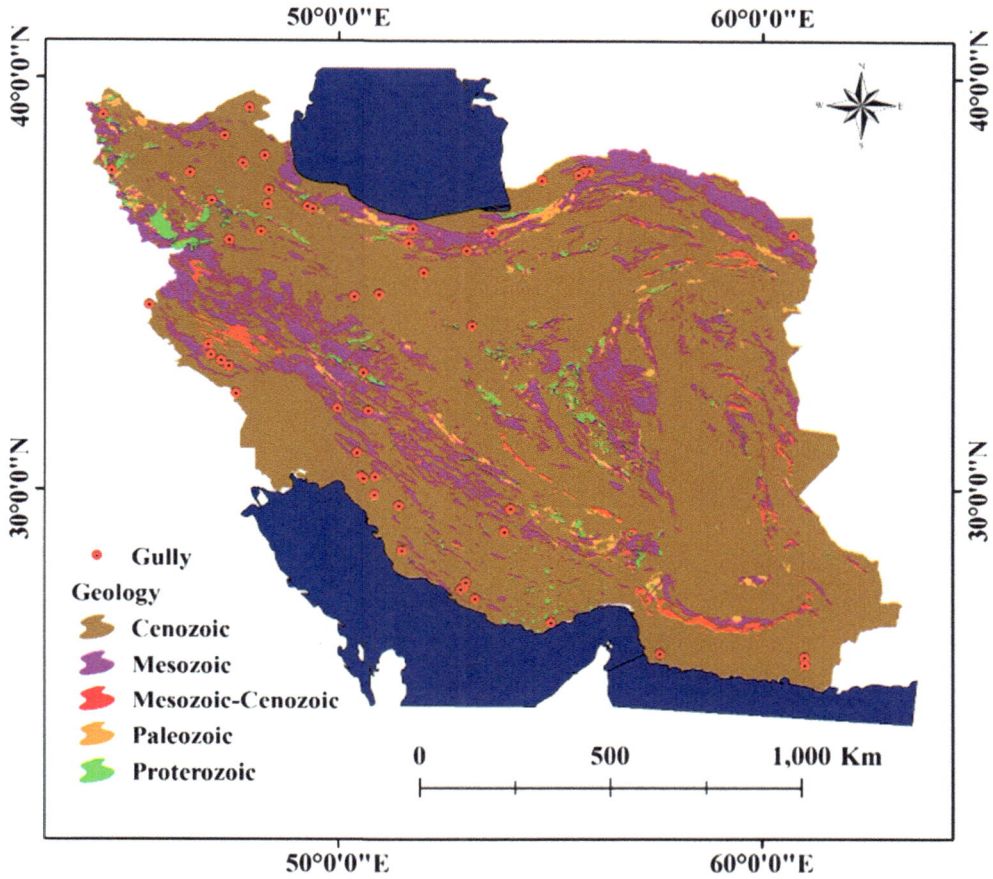

Fig. 26.3 Geological map of Iran

linear development in order to account for the effect of probable extreme precipitation on soil and water management.

26.4   Conclusion

Gully erosion is the main destructive type of water-induced soil erosion which causes land degradation and environmental and property damage in different agro-climatic regions of Iran. It decreases soil quality on-site, and increases significant sediment connectivity problems downstream. In this research, various factors were considered as the driving forces of gully erosion. For example, gully headcut retreat has been measured up to 1.4 m year^{-1} in coastal desert of Bushehr province (SW of Iran) and 0.99 m year^{-1} in ILP, and annual soil loss from ephemeral gully erosion for each gully in loess-derived soil has been measured up to 4 t ha^{-1} in Golestan with semiarid climate (Zarei et al. 2017) which needs much more attention in land use planning. Land use changes and agricultural intensification are very significant in Iran (Emadodin et al. 2012) which should be considered in future scenarios.

Fig. 26.4 Gully erosion in easternmost (A:C) and North (D) part of Golestan province

References

Abedini, M. (2013). Quantitative analysis of gully erosion in Kalghan-Chaj basin (east of Sahand). Geography and Urban Area Journal, 3 (7), 97–110. (In Persian)

Aghrazi, H. Davoudi Rad, A. Mordian, M. and Sufi, M. 2013. Investigating the threshold of gully erosion in Zahirabad Shazand watershed-Markazi Province. Journal of Watershed Engineering and Management. 6(1): 1–9. (In Persian)

Ahmadi, H.A.S.A.N., Mohammadi, A.A., Ghodousi, J.A.M.A.L. and Salajegheh, A., 2007. Testing four models for prediction of gully head advancement (case study: Hableh Rood basin-Iran). Desert, 12 (1): 61–68.

Ahmadi, H. 2004. The study of desertification in Iran. Journal of Forest and Rangeland, Teheran, Iran

Ahmadi, R. Nazarnejad, H. and Najafi, S. 2018. Distribution and erosion intensity in Omarabad, Orumieh. 13th National Conference on the Science and Engineering of Watersheds of Iran and the 3rd National

Conference on Natural and Environmental Preservation. Ardabil Researcher University. (In Persian)

Ahmadi-Givi, F., and Parhizkar, D. 2008. A study of the relationship between ENSO and the distribution of annual precipitation in Iran in the period 1971–2000. Iranian Geophysics Journal. 2: 25–37. (in Persian).

Arab Ghashghaei, Z. Nikkamo, D. Shadfar, S. and Moeini, A. 2011. Gully erosion zonation in Firoozkooh watershed. Geographical Quarterly. Scientific-Research. 8(31): 108–119. (In Persian)

Asghari Saraskanroud, S. 2017. Analysis of effective factors in gully formation and development. Journal of geographical space. 17 (58): 285–301. (In Persian)

Azareh, A., Rahmati, O., Rafiei-Sardooi, E., Sankey, J.B., Lee, S., Shahabi, H. and Ahmad, B.B., 2019. Modelling gully-erosion susceptibility in a semi-arid region, Iran: Investigation of applicability of certainty factor and maximum entropy models. Science of the Total Environment. 655: 684–696.

Balandeh, N. Ahmadi, A. Sokouti, R. and Darbandi, S. 2013. Determination of topographical threshold for initiation of gully erosion using GIS (Case Study: Reyhanlou-West Azerbaijan). Soil Applied Research. 1-15. (In Persian)

Bayati Khatibi, M., Rajabi, M., & Karami, F., 2011. Investigation of topographic thresholds and analysing role of surface materials on gully development in the hillslope of semiarid areas. Case study: shoorchay. Journal of Geography and Environmental Planning, 22 (1), 15–34. (In Persian)

Bayati Khatibi, M. 2004. Analysis of topographical and morphological factors on gully erosion, case Study: Northern slope of Gushe Dagh. Geographical research. 49: 53-70. (In Persian)

Castillo, C. and Gómez, J.A., 2016. A century of gully erosion research: Urgency, complexity and study approaches. Earth-Science Reviews. 160: 300–319.

Costa, F.M., Prado Bacellar, L.A., 2007. Analysis of the influence of gully erosion in the flow pattern of catchment streams, Southeastern Brazil. Catena 69, 230–238

Coulthard, T.J., Neal, J.C., Bates, P.D., Ramirez, J., de Almeida, G.A. and Hancock, G.R., 2013. Integrating the LISFLOOD-FP 2D hydrodynamic model with the CAESAR model: implications for modelling landscape evolution. Earth Surface Processes and Landforms, 38(15).1897–1906.

Emadodin, I., Narita, D., Bork, H.R., 2012. Soil degradation and agricultural sustainability: an overview from Iran. Environ. Dev. Sustain 14, 611–625.

Esmaeelnezhad, L. Seyedmohammadi Meresht, J. and Bakhshipour, R. 2012. Effects of chemical and mineralogical properties of marls on different erosion types in the south of Guilan province. Watershed Management Research. 97:7–16. (In Persian)

Esmaeili, R. and Shovkati, R. 2015. Sensitivity assessment of gully erosion using logistic regression, in Salavatabad basin of Kurdistan province. Geographical Studies of Arid Zones. 5(20): 91–104. (In Persian)

Esmaeili, R. Jokar Sarhangi, E. and Shovkati, R. 2013. Gully erosion susceptibility assessment in Bijar region, Kurdistan province. Earth science research. 3(11):1–14. (In Persian)

Farajzadeh, M., Afzali, A., Khalili, A. and Kalishi, A. 2012. Evaluation of the sensitivity gully erosion Using the Multivariate Regression Model, Case Study: Southeast of Mazandaran province; Kyasar. Journal of Environmental Erosion Research. 2(2):42–57. (In Persian)

Fattahi, M. M., Javidkia, H. R. 2011. Morpho-climatic study of gully erosion in arid areas, Case study: Qom province. Watershed Engineering and Management. 3(3): 131–139. (In Persian)

Frankl, A., Deckers, J., Moulaert, L., Van Damme, A., Haile, M., Poesen, J. and Nyssen, J., 2016. Integrated solutions for combating gully erosion in areas prone to soil piping: innovations from the drylands of Northern Ethiopia. Land Degradation & Development, 27(8).1797–1804.

Ghasemi Dehnavi, A. Sarikhani, R. Peyrawan, H. R. Shoaei, Z. and Kerimi Khaledi, M. 2016. Investigation of chemical and physical properties of neogen marls effects on the forms and intensity of erosion in Varamin area, southwest of Tehran province. Applied Sedimentology. 3(6): 57–69. (In Persian)

Ghezelsofloo, A.A. Maghrebi, M. and Daroughe, F. 2018. Identification of Expansion Rate in Active Gullies using Remote Sensing. Journal of Water and Sustainable Development. 5(1): 67–72. (in Persain)

Ghoduosi J., and Davoudi Rad E. 2005. Effect of physical and chemical soil properties on gully erosion occurrence and their morphology. P.7–8. Proceeding of the 2nd soil and sediment conference. 4-6 Sep. Tehran, Iran. (In Persian)

Gholami, M. Ahmadi, M., and Mahmoodi, M. 2017. Analysis of geomorphological limitations in physical development of city with an emphasis on gully erosion (case study: Mohr city in the southern of Fars province). Journal of natural environmental hazards. 6(12): 105–124. (In Persian)

Gorji, L. Shahzeydi, S. and Ramasht, M.H. 2015. Effective factors on development of gully erosion in Khur and Biabank area. Journal of geographic space. 16(56): 169–184. (In Persian)

Hosseinalizadeh, M., Kariminejad, N. and Alinejad, M., 2018. An application of different summary statistics for modelling piping collapses and gully headcuts to evaluate their geomorphological interactions in Golestan Province, Iran. Catena. 171: 613–621.

Hosseinzadeh, M. Esmaeili R., Kalhor S., and Nosrati, K. 2011. Local changes assessment of gully erosion under geomorphology factors. Environmental Erosion Research. 1: 57–66. (In Persian)

Hosseinalizadeh, M., Kariminejad, N., Chen, W., Pourghasemi, H. R., Alinejad, M., Mohammadian Behbahani, A., & Tiefenbacher, J. P. (2019a). Gully headcut susceptibility modeling using functional trees, naïve Bayes tree, and random forest models. Geoderma

Hosseinalizadeh, M., Kariminejad, N., Chen, W., Pourghasemi, H. R., Alinejad, M., Behbahani, A. M., & Tiefenbacher, J. P. (2019b). Spatial modelling of gully headcuts using UAV data and four best-first decision classifier ensembles (BFTree, Bag-BFTree, RS-BFTree, and RF-BFTree). Geomorphology, 329, 184-193.

Jahantigh, M and Tabe, M. 2016. Comparing soil physico-chemical characteristics and trapezoidal and v-shaped gully morphology with different land uses in dry areas, case study: Hossinzahi and Nalint regions of Chabahar. Journal of Watershed Engineering and Management. 9(3): 308–317. (In Persian)

Karami, A. Khorani, A. Fallah Shamsi, R. Musavi, V., and Khosravi, GH. R. 2013. Applications of object-oriented remote sensing in gully erosion. Geomatics Conference. (In Persian)

Karimi Sangchini, E. and Ownegh, M. 2015. Evaluation of gully erosion hazard by statistical models in Naghan Inter basin, Chaharmahal Va Bakhtiari province. Journal of Water and Soil Conservation. 22(5): 315–319. (In Persian)

Khaje, J. 2017. Investigation of the effect of earth environmental factors on initiation and expansion of gully erosion by using geographical information system (case study in Temer Ghareh Ghozi, Kalaleh, Golestan province). Journal of Watershed Management Research. 5 (12).202–212.(In Persian)

Khazaei, M., Shafiee, A. and Molaei, A. 2012. Assessment of the effective factors on gully development in Maroon watershed. Journal of soil researches, 26(2): 153–163. (In Persian)

Khojeh, N. Ghoddosi, J and Esmaeli, R. 2012. Investigation of the relation of soil physicochemical characteristics and initiation and expansion of gully erosion in Temer Ghareh Ghozi watershed, Golestan province. Watershed Management Research. 5: 27–41. (In Persian)

Maghsudi, M. Shadfar, P. and Abbasi, M. 2012. Zonation of gully erosion in Zwaryan watershed of Qom province. Quantitative geomorphology studies. 3: 35–52. (In Persian)

Mahdavi, Y. Kazemi, M. Rezaie, P. and Nor Mohamadi, F. 2011. Effects of area-slope threshold and rainfall components on gully development in Badreh watershed). Journal of RS & GIS for Natural Resources. 2 (3): 39–51. (In Persian)

Mehraban, M., Golkarian, A. and Khosravi, Kh. 2018. Gravity erosion sensitivity evaluation using maximum entropy model (Case study: Shoorzal area of Khorasan Razavi province). Third National Conference on Soil Conservation and Watershed Management. (In Persian)

Mohammad Ebrahimi., M. Javadi., M. R. and Vafakhah., M. 2016. Determination of Effective Factors on the Occurrence of Digitated Gully Erosion in the AghEmam Watershed. Journal of Water and Soil. 30 (6): 1978-1992. (In Persian)

Mohammad Ebrahimi, M. 2017. Identification of causes of the occurrence of gully erosion with emphasis on soil-related factors (case study: Agh Imam watershed). Journal of Soil and Water Conservation. 7(2). 1–13. (In Persian)

Mokarram, M, and Mahmudi, A.R. 2017. Investigation of morphometric characteristics of gullies and Relationship between morphometric parameters and soil characteristics. Quantitative Geomorphological Research. 5(3): 133–145. (In Persian)

Najafi, Q. 2005. Land and agricultural lands in Iran. Monthly Dehati Magazine,

Nazari Samani, A., Tavakoli Rad, F., Azarakhshi, M., Reza Rahdari, M., and Rodrigo-Comino, J. 2018. Assessment of the Sustainability of the Territories Affected by Gully Head Advancements through

Aerial Photography and Modeling Estimations: A Case Study on Samal Watershed, Iran. Sustainability, 10(8), p. 2909.

Neysi, S., Khalili Moghaddam, B. and Zoratipour, A. 2017. Modeling marl gully development in Khuzestan. Journal of Rangeland and Watershed Management. 70(2): 531–541. (In Persian)

Nikpour, N., Fotuhi, S., Negaresh, H. and Sistani, M. 2017. Morphometric erosion and effective factors in gully development. Journal spatial analysis of environmental hazards. 4(1): 97–112. (In Persian)

Noormohammadi, F., Sadeghi, S.H.R., Soufi, M., and Yasrebi, B. 2009. Assessment of relationship between important runoff parameters and gully erosion in Darehshahr rangeland watershed. Journal of Rangeland. 3(3): 533–545. (In Persian)

Novhegar, A. and Heydarzadeh, M. 2011. The study of physicochemical characteristics and morphometry of gullying area (case study: Gezir, Hormozgan province). Environmental Erosion Research. 2011; 1 (1): 29–44

Poesen, J., 2011. Challenges in gully erosion research. Landform analysis. 17: 5–9.

Poesen, J., 2018. Soil erosion in the Anthropocene: Research needs. Earth Surface Processes and Landforms. 43(1):64–84.

Poesen, J., Nachtergaele, J., Verstraeten, G. and Valentin, C., 2003. Gully erosion and environmental change: importance and research needs. Catena. 50(2–4): 91–133.

Pourghasemi, H.R., Yousefi, S., Kornejady, A. and Cerdà, A., 2017. Performance assessment of individual and ensemble data-mining techniques for gully erosion modeling. Science of the Total Environment. 609: 764–775.

Qadir, M., Qureshi, A. S. and Cheraghi, S. A. M. 2008. Extent and characterization of salt-affected soils in Iran and strategies for their amelioration and management. Land Degradation and Development. 19: 214–227.

Rahi, GH. 1998. Review mechanism and causes the formation of gullies in Genaveh city. Master thesis, Tarbiat Modarres University. 102. (In Persian)

Rahmanad Rad, J., Khosravi, F., Rigi Nezhad, Sh. 2010. Characteristics of soil chemistry in gully development in Dashtiari, Chabahar. Quarterly Journal of Applied Geology. 6(1): 1–9. (In Persian)

Rahmani, M., Mesbah, H., Hosseini Marandi, H. and Najafi Nejad, A. 2009. Investigating the effect of groundwater drop on the gully In the plain of Niriz Fars. Fifth National Conference on Watershed Management Science and Engineering Iran, (Sustainable management of natural disasters). Karaj. (In Persian)

Rahmati, O., Haghizadeh, A., Pourghasemi, H.R. and Noormohamadi, F. 2016. Gully erosion susceptibility mapping: the role of GIS-based bivariate statistical models and their comparison. Natural Hazards. 82(2): 1231–1258.

Rahmati, O., Tahmasebipour, N., Haghizadeh, A., Pourghasemi, H.R. and Feizizadeh, B. 2017. Evaluating the influence of geo-environmental factors on gully erosion in a semi-arid region of Iran: An integrated framework. Science of the Total Environment. 579: 913–927.

Sadeghi, H. R., Shojaee, GH. R. and Moradi, H. R. 2010. Relationship between land use and soil erosion in Manderijan Catchment in Zayandehrud Dam Basin. Journal of Watershed Engineering and Management. 3(2): 143–148. (In Persian)

Servati, M.R., Ghahrudi Tali, M., Gol Karami, A. and Najafi, E. 2013. Geomorphic thresholds of gully formation in Kechik basin, Golestan province. Applied Geographical Sciences Research. 4(32): 231–249. (In Persian)

Shadfar, S. and Sobhzahedi, Sh. 2007. Investigation of gully erosion using Analytical Hierarchy Process Model in Roudbar. National Conference on Watershed Management Science and Engineering Iran. (In Persian)

Shafeie, A., Khazayi, M., Molayi, A. and Sofi, M. 2011. Study and comparison of pedological and morphoclimatical characteristics of gullies. Journal of Irrigation and Water. 5: 26–38. (In Persian)

Shahab, H., Emami, H., Hagh Nia, Gh. H. Esmali, A. 2017. Determination the most important physical and mechanical soil properties on increasing cross sections in Ardebil province. Journal of Water and Soil. 30(6): 2060–2077. (In Persian)

Shahbazi, Kh. Salajegh, A., Jafari, M., Ahmadi, H., Nazari Samani, A. and Khosroshahi, M. 2016. Comparison of hydraulic thresholds of gully erosion in different land uses (case study: Qasarshirin region, Kermanshah province). Iranian Natural Resources Journal. 69(4): 931–947. (In Persian)

Shahini, Gh. R. and Charkhabi, A. H. 2007. Relationship between gully erosion and soil characteristics in the loess hills of Gorgan. 10th Iranian Soil Science Congress, Karaj. (In Persian)

Shruthi, R.B.V., Kerle, N., Jetten, V., Abdellah, L., Machmach, I., 2015. Quantifying temporal changes in gully erosion areas with object oriented analysis. Catena 128,262–277.

Soleimanpour, S.M., Soufi, M., Ahmadi, H., 2009. Determining Effective Factors on Gully Development in Konartakhte Region,Fars Province. Journal of Water and Soil 23, 131-141 (In Persian)

Soufi, M. 2002. Characteristics and causes of gully erosion in Iran. In 12th ISCO Conference, Beijing, China.

Sufi, M. 2005. Determine the effect of vegetation and urban development in the expansion of gullies in south of Fars province. Third National Conference On Erosion and Sediment. 35–350. (In Persian)

Vanmaercke, M., Poesen, J., Van Mele, B., Demuzere, M., Bruynseels, A., Golosov, V., Bezerra, J.F.R., Bolysov, S., Dvinskih, A., Frankl, A. and Fuseina, Y., 2016. How fast do gully headcuts retreat?. Earth-Science Reviews, 154, pp. 336–355.

Wilkinson, S.N., Dougall, C., Kinsey-Henderson, A.E., Searle, R.D., Ellis, R.J. and Bartley, R., 2014. Development of a time-stepping sediment budget model for assessing land use impacts in large river basins. Science of the Total Environment, 468, pp. 1210–1224.

Yamani, M. and Akbarian, M. 2013. Effect of sedimentology in the development of tunnel erosion IN the Flush Makran Formation (Case STUDY: Jask). Geography and environmental hazards. 7: 1–17. (In Persian)

Yasrebi, B., Noor Mohammadi, F., Sadeghi, H. and Sufi, M. 2013. Determining the role of topographic factors at the beginning of gully erosion (Case Study: Darre shahr, Ilam). Iranian Journal of Watershed Management Science. 21: 53–58. (In Persian)

Zanjani Jam, M., Soufi, M., Bayat, R. and Rasouli, M. 2012. Investigation on morpho - climatic characteristics of gullies in order to classify gully affected regions in Zanjan province. Journal of Watershed Management Research (Pajouhesh & Sazandegi), 99: 2-10. (In Persian)

Zarei, H., Najafinejad, A., Hosseinalizadeh, M. and Alipour, K. 2017. Efficiency assessment of the EGEM to estimate gully erosion in Iky-Aghzly watershed of Golestan province. Journal of Water and Soil Conservation. 24(5): 147–162. (In Persian)

Mohsen Hosseinalizadeh is received Bachelor degree in Range and Watershed Management in 2002 from Yazd University and graduated from Gorgan University of Agricultural Sciences and Natural Resources in Watershed Management (M.Sc). He is awarded Ph.D. degree from University of Tehran (June 2012). Presently he is working as Assistant Professor at Gorgan University of Agricultural Sciences and Natural Resources (GUASNR, IRAN). His main research interests are spatial statistic, bioengineering especially in Piping, Gully erosion, and landslide.

Mohammad Alinejad, born in Tabas city of South Khorasan province in Iran country. His Bachelor's degree is in the Ferdowsi University of Mashhad in Natural Resources Engineering-desert management and dryland. He is Master's Degree in Department of Watershed and Arid Zone Management of Gorgan University of Agricultural Sciences and

Natural Resources in desert management. He is currently working on various research projects in the Gorgan University of Agricultural Sciences and Natural Resources.

Ali Mohammadian Behbahani   is working as an Assistant Professor at GUASNR. He received Bachelor of Science in Range and Watershed Management from Gorgan University of Agricultural Sciences and M.Sc. from the University of Tehran. He received his Ph.D. degree from the University of Basel. His main research interests are geomorphological processes, soil erosion modeling and simulation, wind erosion and dust emission control.

Farhad Khormali   is a Professor of pedology at Gorgan University of Agricultural Sciences and Natural Resources, Iran. He is also a committee member of the Soil Science Society and Quaternary Society of Iran as well as editor in chief of the Journal of Soil Management and Sustainable Agriculture. His main research interests are soil genesis, loess–paleosols studies, micromorphology, clay mineral weathering, and soil quality and soil carbon storage.

Narges Kariminejad   is a Ph. D student of Desert Control and Management in Department of Watershed and Arid Zone Management,

Gorgan University of Agricultural Sciences and Natural Resources, Gorgan, Iran. She has received B.Sc. and M.Sc degrees in Environmental Science from Shiraz University, Iran. She has published some research articles in various international reputed journals.

Hamid Reza Pourghasemi   is an Associate Professor of Watershed Management Engineering in the College of Agriculture, Shiraz University, Iran. He has a B.Sc. in Watershed Management Engineering of the University of Gorgan (2004), Iran, an MSc in Watershed Management Engineering, from Tarbiat Modares University, Iran (2008), and a PhD in Watershed Management Engineering from the same University (Feb 2014). His main research interests are GIS-based spatial modelling using machine learning/data mining techniques in different fields such landslide, flood, gully erosion, forest fire, land subsidence, species distribution modelling, and groundwater/hydrology. Also, Hamid Reza works on Multi-Criteria Decision Making methods in Natural Resources and Environment.

He has published more of 90 peer reviewed papers in high-quality journals, with three chapters in Springer. Also, he published two books in Springer   (https://www.springer.com/gp/book/9783319733821)   and Elsevier (https://www.elsevier.com/books/spatial-modeling-in-gis-and-r-for-earth-and-environmental-science/pourghasemi/978-0-12-815226-3).

Mahdis Amiri and Hamid Reza Pourghasemi

Abstract

Preparing and mapping gully erosion (GE) is a basic instrumentation to land use projecting and reducing destruction of the land. The purpose of the current investigation was to assess gully erosion spatial modeling using multivariate adaptive regression spline (MARS) model in Maharlou watershed, Fars Province, Iran. The current study is consisted from two important parts including (1) recognizing dependent and variables, e.g., gully erosion inventory map (GEIM) and gully effective agents, and (2) running a famous machine learning algorithm named the MARS in order to gully erosion mapping. Gully erosion inventory map is randomly separated into two categories: training and validation datasets. Then, nine causative factors including land use, distance from rivers, clay percent, geology, pH, NDVI, drainage density, distance from roads, and slope direction are recognized, and their maps are classified in the ArcGIS. Also, the GESM was created using the MARS model in the R statistical environment. The outcomes of the MARS technique of the 30% of the unused gully points used in the modeling procedure based on the ROC curve. Results demonstrated that the ultimate gully erosion map had a top precision with AUC values 96.3% for accuracy data set.

Keywords

Gully erosion · MARS · GIS and R · Fars province

M. Amiri
Department of Watershed and Arid Zone Management, Gorgan University of Agricultural Sciences and Natural Resources, Gorgan, Iran

H. R. Pourghasemi (✉)
Department of Natural Resources and Environmental Engineering, College of Agriculture, Shiraz University, Shiraz, Iran

27.1 Introduction

Soil erosion as a worldwide issue has different hazards on soil and water resources (Swarnkar et al. 2018). Gully erosion is one of the great problems all over the universe, particularly in arid and semiarid areas, where vegetation cover is low to very low (Sankey and Draut 2014). Gullies are profound channels eroded by the condensation of the current water and the elimination of high soils in the land (USDA-SCS 1966). The outcomes of soil erosion could be extremely hazardous in each area (Boardman and Favis-Mortlock 1998). Gully erosion is controlled by several important agents, such as groundwater flow, groundwater movement, and soil piping (Poesen et al. 2018). However, gully erosion is a conclusion of normal activities; human operations can speed up gully organization and its expansion (Ionita et al. 2015). It is the most destructive kind of water erosion due to the soluble and alkaline organization in the forests, rangelands, and agricultural lands (Lesschen et al. 2007). In Iran, soil erosion is one of the most important problems, particularly in agriculture, natural resources, and the environment. Approximately, 125 million hectares of lands are subject to water erosion (Refahi 2009). So, considering different environmental agents affecting gully erosion, identifying the connection of among the above layers with the incidence of gullies, and predicting of gully erosion regions are essential strategies for managing water and soil resources (Shit et al. 2015). In this regard, the GIS tools is helpful for exploration of gullies and identifying the sensitivity of gully erosion (Zakerinejad and Maerker 2015). In recent years, researchers from all over the world have used different methods to study gully erosion. The models can be separated into three categories: (1) science according to models, e.g., the analytical hierarchy process (AHP) (Zakerinejad and Maerker 2014); (2) bivariate and multivariate statistical models including weight of evidence (WOE) (Zabihi et al. 2018), logistic regression (LR) (Dewitte et al. 2015), maximum entropy (ME), information value (IV) (Conforti et al. 2011),

P. K. Shit et al. (eds.), *Gully Erosion Studies from India and Surrounding Regions*, Advances in Science, Technology & Innovation, https://doi.org/10.1007/978-3-030-23243-6_27

conditional analysis (CA) (Conoscenti et al. 2013), and frequency ratio (FR) (Rahmati et al. 2016); and (3) data mining model, such as multivariate adaptive regression splines (MARS) (Gómez-Gutiérrez et al. 2015; Conoscenti et al. 2018), random forest (RF) (Arabameri et al. 2018), support vector machine (SVM) (Pourghasemi et al. 2017), classification and regression trees (CART) (Märker et al. 2011), and artificial neural networks (ANN) (Pourghasemi et al. 2017). In general, different statistical models including, logistic regression, information value, frequency ratio, index of entropy, and weight of evidence widely used to construct gully erosion maps (GESM) (Rahmati et al. 2016; Mararakanye and Sumner 2017; Selkimäki and González-Olabarria 2017; Al-Abadi and Al-Ali 2018; Arabameri et al. 2018; Meliho et al. 2018; Zabihi et al. 2018). According to Conforti et al. (2011), statistical models provide general benefits to different independent layers without any limitation.

Therefore, the aim of this study was to identify gully places in the Maharlou watershed and to determine the factors affecting the occurrence of gully erosion and eventually the GESM using the MARS machine learning model.

27.2 Material and Methods

27.2.1 Study Area

The Maharlou watershed (latitudes of 29° 1′ to 29° 58′ N and longitudes of 53° 12′ to 53° 28′ E) has a total area of 4274 km² and lies in an arid and semiarid climate (Fig. 27.1). It is located in the southwestern part of Iran, among the Zagros Mountains. In the study area, rainfall ranges from 150 mm on the flat to 650 mm on the high mountains (Sigaroodi et al. 2014). It has an average altitude of about 1500 m above sea level.

27.2.2 Methodology

As you can see in Fig. 27.2, the method used in this research contains of five main steps: (1) preparing fundamental thematic layers containing, geology map (1:100,000 scale), topography map (1:25,000 scale), and Landsat 8 satellite images, (2) definition and extraction of nine effective factors on gully erosion occurrence; (3) identification of the location of gullies and preparing the GEIM, (4) modeling of the GESM using the MARS model, and (5) validation of the Gully Erosion Inventory Mapping (GESM) based on the ROC curve.

27.2.2.1 Gully Erosion Inventory Mapping

In this search, 207 gullies were recognized using vast field of surveys recorded the coordination of each using gully polygons the global positioning system (GPS). From the 207 gullies in the Maharlou watershed, 70% (146 gully positions) were chosen for modeling (training) and 30% (61 gully positions) for credit goals (Rahmati et al. 2016; Conoscenti et al. 2018; Garosi et al. 2018). The points of training and validation gullies are placed in Fig. 27.1.

27.2.2.2 The GESM Predictor Variables

Gully erosion depends on various events depending on various factors such as geology, topography, hydrology, soil characteristics, climate, and human activities (Gómez-Gutiérrez et al. 2015; Arabameri et al. 2018). There is no global guideline to choose effective layers for GESM modeling. Prior researchers chosen different factors as independent variables (Conoscenti et al. 2013; Dube et al. 2014; Azareh et al. 2019) as following:

27.2.2.2.1 Slope Aspect

As a rule, the slope direction can control the evapotranspiration, vegetation cover, and solar radiation in an area (Sidle and Ochiai 2006; Wang et al. 2011). Therefore, it seems essential factor to detect the susceptibility of an area to gully erosion (Umar et al. 2014). The slope direction map of the Maharlou watershed was produced using the digital elevation model (DEM) by a 10 m × 10 m pixel size. It consist of nine classes: flat, north, northeast, east, southeast, south, southwest, west, and northwest.

27.2.2.2.2 Land Use

Land use map is a main driver for gully erosion occurrence and its subsequently land degradation in universal (Agnesi et al. 2011). The land use map was collected from the Natural Resources Office of Fars Province and then updated by Google Earth images. There are various types of land use in the Maharlou watershed including agricultural land, forest land, bare land, residential areas, wetland, grave and shrubbery, salty lands, and rangeland types.

27.2.2.2.3 Lithology

The lithology and physical properties of the geological materials affect the process of the surface of the land and causes degradation and expansion of gully erosion (Agnesi et al. 2011; El Maaoui et al. 2012). In this study, the geological map of the study area was prepared by Geology Organization of Iran in 1:100,000 scale and applied in this paper; the Maharlou watershed is covered up by different kinds of lithological and classified into seven groups.

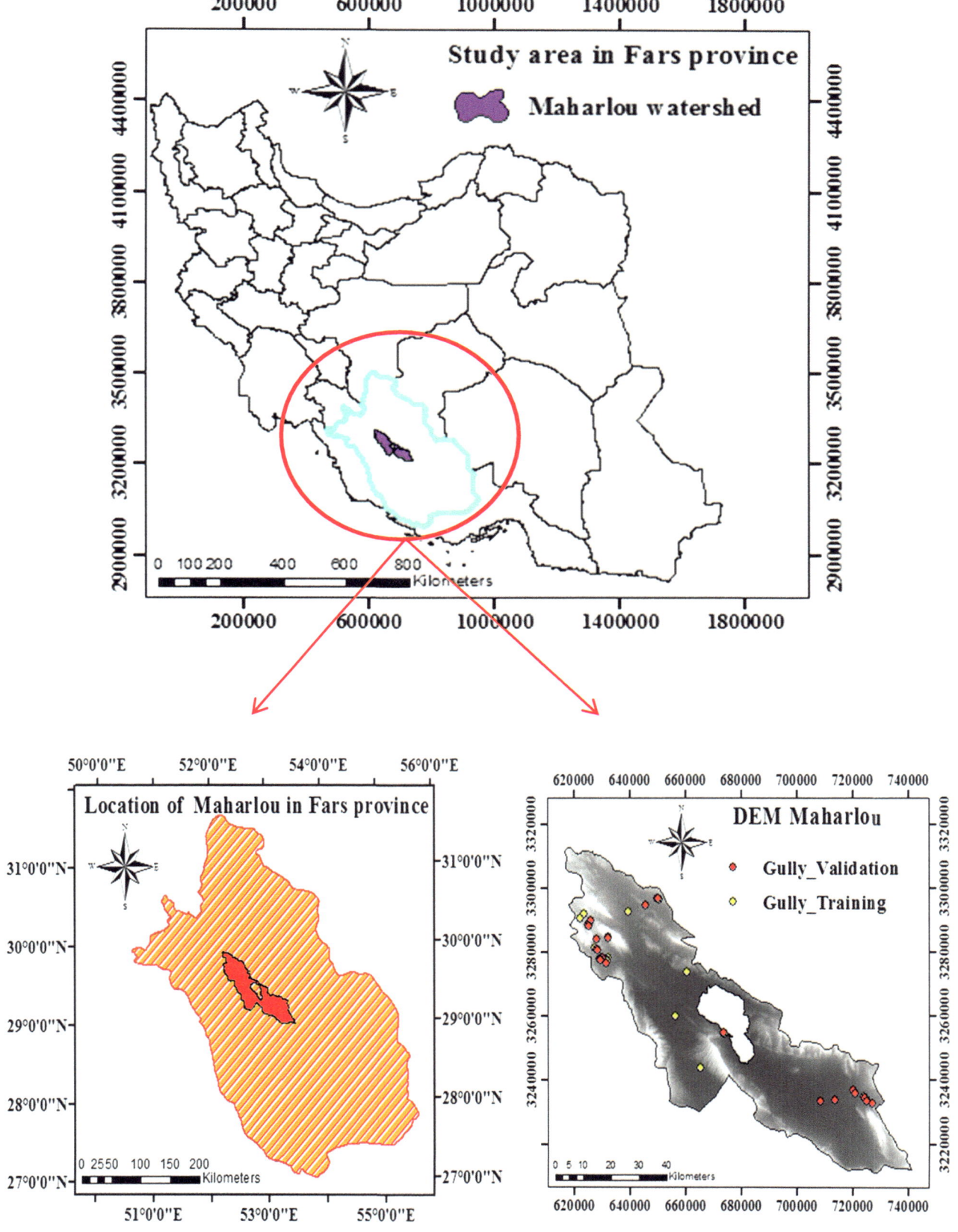

Fig. 27.1 Case study in Fars Province and gully erosion points is located on a digital elevation model of the Maharlou watershed, Iran

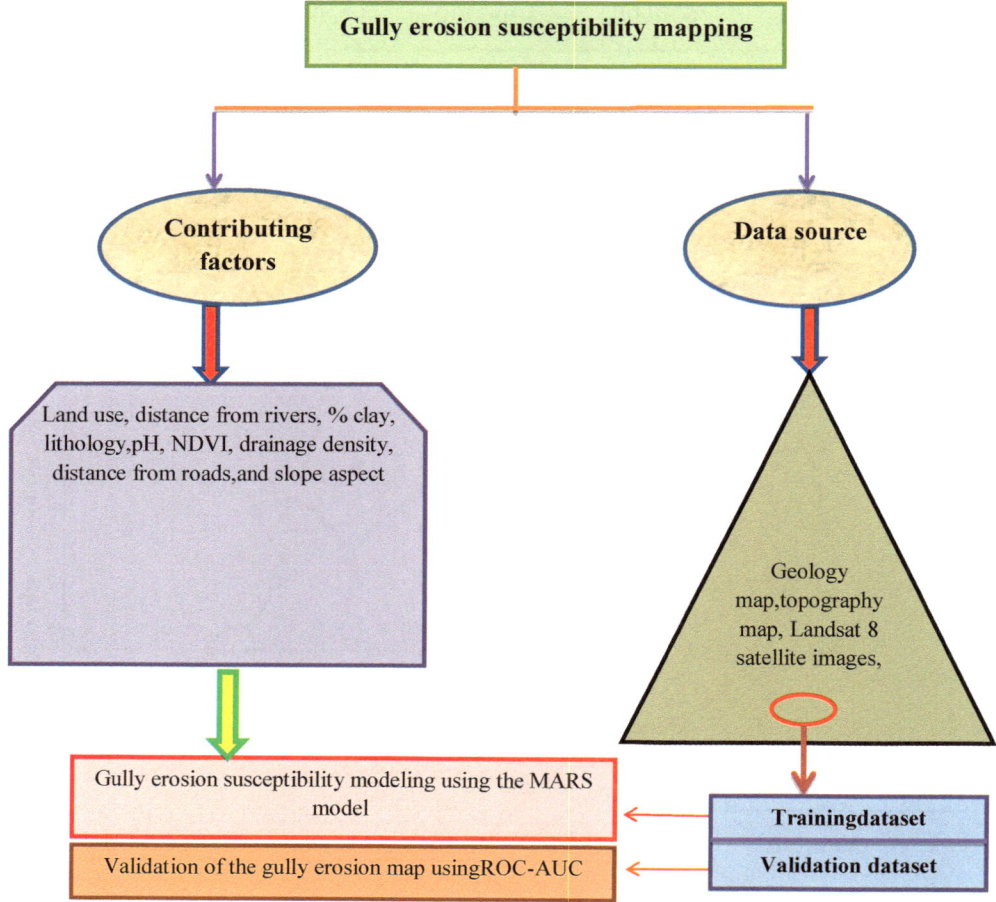

Fig. 27.2 Methodology of the current research applied in the study

27.2.2.2.4 Normalized Difference Vegetation Index

The normalized difference vegetation index (NDVI) was prepared in ArcGIS 10.2.2 (Eq. 27.1) using Landsat 8 satellite images downloaded from the United States Geological Survey (USGS) page. This map was separated into five categorize as follows (Pourghasemi et al. 2014): smaller or equal to zero (water, snow, and ice), 0–0.05 (bare soil), 0.05–0.1 (sparse vegetation), 0.1–0.5 (compact vegetation), and more than 0.5 (forest cover).

$$NDVI = NIR - Red/NIR + Red \qquad (27.1)$$

where NIR and Red values is the infrared and red portion of the electromagnetic spectrum, respectively. The NDVI values range from −0.17 to 1 in the study area.

27.2.2.2.5 Drainage Density

Drainage density is affected by the rainfall, lithology, vegetation type, slope, soil texture, and infiltration in the watershed. Drainage density causes to the distance between drainages or stream in a network (Rahmati and Pourghasemi 2017). High drainage density can represent locations on the landscape with low permeability scenarios, outcoming in higher runoff and higher levels of spatial distribution, and thus can be a helpful factor on gully erosion sensibility (Conoscenti et al. 2018). The drainage density map of the Maharlou watershed was made using Line Density tools in ArcGIS 10.2.2. According to drainage density map, minimum and maximum of values were 0.08 and 7.36 km/km^2.

27.2.2.2.6 Distance from Rivers

The potential impact of river networks on the sensitivity of gully erosion is another important factor (Azareh et al. 2019). In order to provide map of distance from rivers, a topography maps at 1:25,000 scales has been used. Distance from rivers in the study area varies from 0 to 1740.46 m.

27.2.2.2.7 Distance from Roads

The distance from roads map of the Maharlou watershed obtained from road lines using Euclidean distance tools in ArcGIS 10.2.2 Environment. A change of distance from roads was from 0 to 6549.34 m.

27.2.2.2.8 Soil Characteristics (pH and Clay Percent)

In order to make soil characteristics, 227 soil samples gathered from the field and clay percent using hydrometers were measured (Gee and Bauder 2002) in the lab. In continue, through sieve No. 10 (diameter 2 mm), 50 g of soil was transmissive from the sifter to plastic containers and added to 100 ml of 5% Calgon solution. The contents are mixed in a special container after 24 h and mixed for 5 min. Then, the contents of the container were added to the 1-l cylinder and added distilled water to the specimens in a 1-l cylinder. Then, this solution was mixed using a hand mixer until a homogeneous mixture is created. To read the hydrometer, the temperature of the mixture was taken 40 min, 1 h, and 2 h, respectively. After applying the modifications to read the temperature and hydrometer, then the percentage of silt was measured using a hydrometer method in the lab. The clay layer was produced according to inverse distance weighted (IDW) method (Setianto and Triandini 2013; Hong et al. 2016). Soil acidity was measured using pH meter (potentiometer) according to soil saturation method (Mclean 1982), and its map is made based on the IDW method (Eq. 27.2). In general, clay percent in the study area was from 5.61 to 62.99, whereas pH values are varying from 5.98 to 9.64.

$$\lambda_i = \frac{D_i^{-\alpha}}{\sum_{i=1}^{n} D_i^{-\alpha}} \quad (27.2)$$

where λ_i is the weight of the point i, D_i is the distance between the point i and the unknown point and α is equal to the weighing power (Setianto and Triandini 2013).

27.2.2.3 Multivariate Adaptive Regression Spline

In this study, the Multivariate Adaptive Regression Spline (MARS) model done using the "earth" package in R 3.4.3 (Milborrow 2009). The MARS is a nonparametric regression method and a linear extending the model provided by Friedman (1991). The MARS algorithm divides the range of interpretive variables into regions and produces for each of these regions a linear regression equation (Zabihi et al. 2016). The main aim of the MARS model is to solve regression equations problems and predictions of continuous affiliate values or the result of a variable of a set of independent or predictions variables (Felicísimo and Gómez-Muñoz 2004). The MARS combines classical linear regression with splines and dual recessive partitioning for preparing a linear or nonlinear based on relation among the dependent and independent variables (Felicísimo and Gómez-Muñoz 2004; Felicísimo et al. 2012). The general equation of MARS can be written as Eq. (27.3) (Felicísimo and Gómez-Muñoz 2004):

$$y = f(x) = \beta_0 + \sum_{m=1}^{m} \beta_m \, h_m(x) \quad (27.3)$$

where y is the value predicted by the function $f(x)$, β_0 is the constant, and M is a sum of terms, each of them are formed by coefficients β_m, and $h_m(x)$ is an individual basis function (BF).

27.2.2.4 Accuracy Assessment of the GESM

To GESM validation, 30% of gully erosion sites that were not available in the modeling proceeding were used. Receiver operating characteristics (ROC) are generally used to evaluate prediction precision (Gorsevski et al. 2006). Generally, the high-quality MARS model has the maximum AUC, and the values of AUC changing from 0.5 to 1 (Park 2010). As a rule, if the AUC values are 0.9–1, 0.8–0.9, 0.7–0.8, 0.6–0.7, and 0.5–0.6, respectively, then it indicates excellent, very good, good, moderate, and poor validation floors, respectively (Yesilnacar 2005).

27.3 Results and Discussion

27.3.1 Preparation of Gully Erosion Susceptibility Maps Using the MARS Model

In the present study, after identifying 207 gully polygons in Maharlou watershed in Fars Province, and also relationship of gully locations and effective factors, final GESM was prepared using the MARS model in R statistical software, and then it exported to the ArcGIS software. The ultimate GESM (gully erosion map) will be obtained from the MARS model. It is revealed in Fig. 27.3. The map that presented gully erosion susceptibility (Fig. 27.3) was an assortment of four floors containing low, moderate, high and very high according to natural break (NB) algorithm (Sezer et al. 2011; Zabihi et al. 2018; Pourghasemi et al. 2019). The results of this map indicated that 83.62%, 4.76%, 3.33%, and 8.27% of the Mahrlou watershed were located in low, moderate, high, and very high susceptibility, respectively (Fig. 27.3 and Table 27.1).

27.3.2 Validation of GESM

The outcomes of validation of the GESM using the ROC and its AUC are presented in (Fig. 27.4 and Table 27.2). The outcomes indicated an AUC value of 0.963 with forecast precision of 96.3% and S.E. of 0.015. Generally, the validity

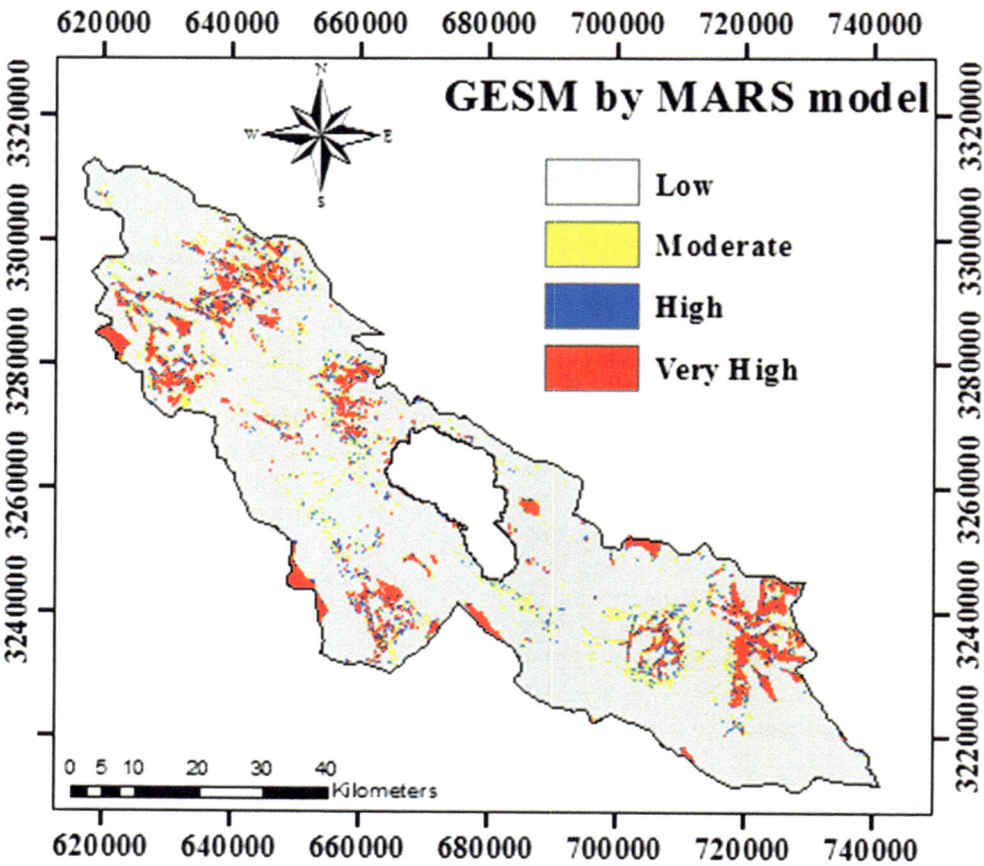

Fig. 27.3 The GESM produced by the MARS model in Maharlou watershed

outcomes of the model showed that the MARS model has an excellent ability. The above model is an automatic and transformative data mining which incorporate the benefits of recessive division (Breiman et al. 1984). Balashi et al. (2009) suggested that the description of developed models using the MARS is usually simpler. Gutiérrez et al. attempted to model the potential of gully erosion using a nonparametric MARS model. The results of the evaluation of gully erosion map using the MARS model in Maharlou watershed showed

a good efficiency of the mentioned model with the area under the curve (AUC) values of 0.963 for the validation data. The conclusion of the current paper is in line with the results of Gutiérrez et al. They stated that the MARS model had an excellent accuracy (AUC = 0.98). Also, results of another study in the investigation of groundwater sensitivity in western Sicily, Italy, using logistic regression and multivariate adaptive regression splines showed higher predictive function of the MARS (AUC = 0.881–0.912) model. The MARS technique has more positive points than the customary regression analysis. The MARS specifies only the most important expositive variables from a user-specified order (Zabihi et al. 2016). This means that the user may use several variables at the beginning of the analysis, and MARS will only select the most important ones, which will result in the ultimate consequence (Zabihi et al. 2016). The process of pruning the variables that are finite in the prognostication of measuring results is limited (Kennison and Cox 2013).

Table 27.1 Percentage of the susceptibility of gully erosion in each class

Susceptibility class	Pixel	Area%
Low susceptibility	33,648,038	83.62
Moderate susceptibility	1,916,528	4.76
High susceptibility	1,341,558	3.33
Very high susceptibility	3,330,065	8.27

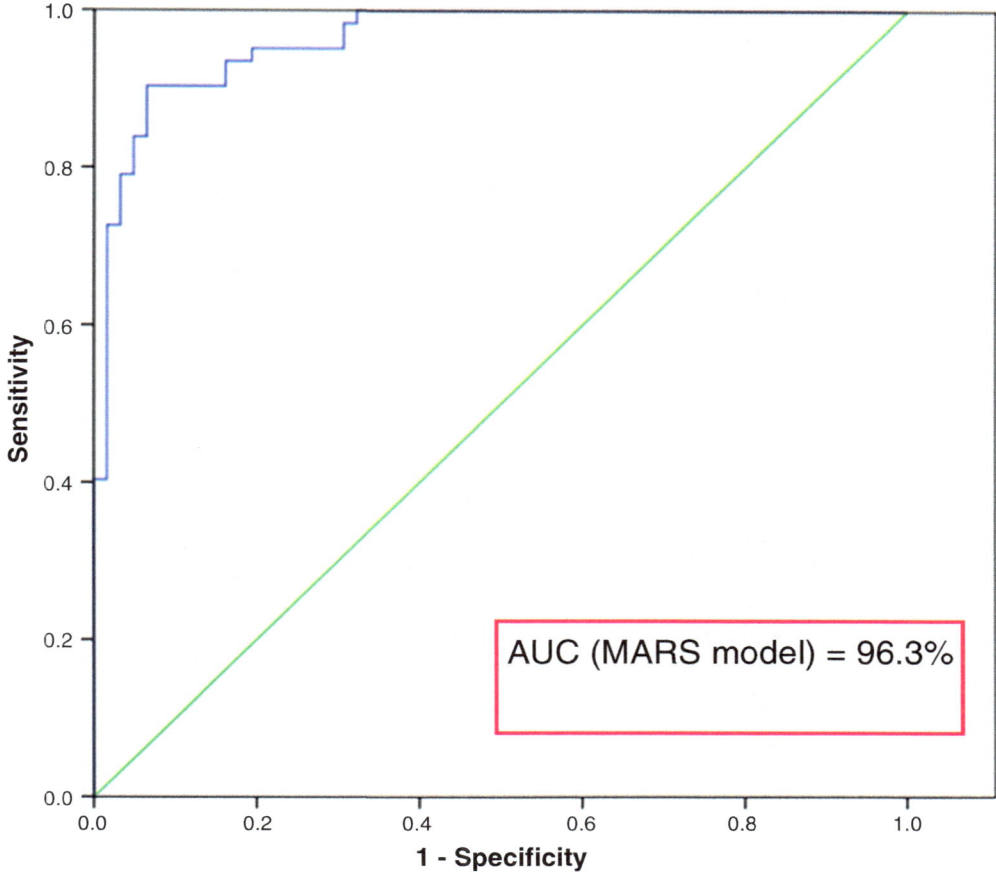

Fig. 27.4 Validation of model, the area under the curve

27.4   Conclusion

The purpose of the current research was to evaluate the spatial modeling of gully erosion using the MARS model in the Maharlou watershed, Fars province, Iran. Nine thematic layers as gully erosion effective layers are considered in this search. They were land use, distance from rivers, % clay, lithology, pH, NDVI, drainage density, distance from roads, and slope aspect. In order to spatial modeling, from total of total erosion locations, 70% and 30% was applied for modeling and evaluation. After preparing GESM, the built model was validated using the ROC-AUC. Confirmation of outcomes indicated that the MARS model with AUC values of 0.963, according to nine important agent's maps of the studied area, was highly accurate. Thus, to predict regions that are sensitive to gully erosion is helpful in the performance of protective measures and decreasing feasible losses.

Table 27.2   The ROC curve for susceptible gully erosion maps using the MARS model

Area under the curve				
			Asymptotic 95% confidence interval	
AUC	Standard error	Asymptotic significant	Lower bound	Upper bound
0.963	0.015	0.000	0.934	0.992

References

Agnesi, V., Angileri, S., Cappadonia, C., Conoscenti, C., Rotigliano, E., 2011. Multiparametric GIS analysis to assess gully erosion susceptibility: a test in southern Sicily, Italy. Landf. Anal. 7, 15–20.

Al-Abadi, A.M., Al-Ali, A.K., 2018. Susceptibility mapping of gully erosion using GIS-based statistical bivariate models: a case study from Ali Al-Gharbi District, Maysan Governorate, southern Iraq. Environ. Earth Sci. 77 (6), 249. https://doi.org/10.1007/s12665-018-7434-2.

Arabameri, A., Pradhan, B., Pourghasemi, H.R., Rezaei, K., Kerle, N., 2018. Spatial Modelling of Gully Erosion Using GIS and R Programing: A Comparison among Three Data Mining Algorithms. Appl. Sci. 8 (8), 1369. https://doi.org/10.3390/app8081369.

Azareh, A., Rahmati, O., Rafiei-Sardooi, E., Sankey, J. B., & Lee, S. 2019. Modelling gully-erosion susceptibility in a semi-arid region, Iran: Investigation of applicability of certainty factor and maximum entropy models. Science of the Total Environment, 655, 684–696.

Balashi MS, Mcguire AD, Duffy P, Flannigan M, Walsh J, Mellilo J (2009) Assessing the response of area burned to changing climate in western boreal North America using a multivariate adaptive regression splines (MARS) approach. Glob Chang Biol 15:578–600

Breiman L, Friedman JH, Olshen RA, Stone CJ (1984) Classification and regression trees. Wadsworth.

Boardman, J., Favis-Mortlock, D., 1998. Modelling Soil Erosion by Water, first ed. Springer-Verlag Berlin Heidelberg. https://doi.org/10.1007/978-3-642-58913-3.

Conoscenti, C., Agnesi, V., Angileri, S., Cappadonia, C., Rotigliano, E., Märker, M., 2013. AGIS-based approach for gully erosion susceptibility modelling: a test in Sicily, Italy. Environ. Earth Sci. 70 (3), 1179–1195.

Conoscenti, C., Agnesi, V., Cama, M., Caraballo-Arias, N.A., Rotigliano, E., 2018. Assessment of gully erosion susceptibility using multivariate adaptive regression splines and accounting for terrain connectivity. Land Degrad. Dev. 29 (3), 724–736.

Conforti, M., Aucelli, P.P.C., Robustelli, G., Scarciglia, F., 2011. Geomorphology and GIS analysis for mapping gully erosion susceptibility in the Turbolo stream catchment (Northern Calabria, Italy). Nat. Hazards 56, 881–898.

Dewitte, O., Daoudi, M., Bosco, C., Van Den Eeckhaut, M., 2015. Predicting the susceptibility to gully initiation in data-poor regions. Geomorphology 228, 101–115.

Dube, F., Nhapi, I., Murwira, A., Gumindoga, W., Goldin, J., Mashauri, D.A., 2014. Potential of weight of evidence modelling for gully erosion hazard assessment in Mbire District–Zimbabwe. Phys. Chem. Earth 67, 145–152. https://doi.org/10.1016/j.pce.2014.02.002 Pt. A/B/C.

El Maaoui, M.A., Sfar Felfoul, M., Boussema, M.R., Snane, M.H., 2012. Sediment yield from irregularly shaped gullies located on the Fortuna lithologic formation in semi-arid area of Tunisia. Catena 93, 97–104.

Friedman JH. (1991). Multivariate adaptive regression splines. Ann Statist 19(1):1–67.

Felicísimo AM, Gómez-Muñoz A (2004) GIS and predictive modelling: a comparison of methods applied to forestal management and decision-making. In: Geographical Information Systems Research UK. Proceedings of the GIS Research UK 12th Annual Conference:143–144. Norwich.

Felicísimo A, Cuartero A, Remondo J, Quirós E (2012) Mapping landslide susceptibility with logistic regression, multiple adaptive regression splines, classification and regression trees, and maximum entropy methods: a comparative study. Landslides. doi:https://doi.org/10.1007/s10346-012-0320.

Gorsevski, P.V., Gessler, P.E., Foltz, R.B., Elliot, W.J., 2006. Spatial prediction of landslide hazard using logistic regression and ROC analysis. Trans. GIS 10 (3), 395–415.

Garosi, Y., Sheklabadi, M., Pourghasemi, H.R., Besalatpour, A.A., Conoscenti, C., Van Oost, K., 2018. Comparison of differences in resolution and sources of controlling factors for gully erosion susceptibility mapping. Geoderma 330, 65–78.

Gómez-Gutiérrez, Á., Conoscenti, C., Angileri, S.E., Rotigliano, E., Schnabel, S., 2015. Using topographic attributes to evaluate gully erosion proneness (susceptibility) in two Mediterranean basins: advantages and limitations. Nat. Hazards 79 (1), 291–314.

Gee, G.W., Bauder, D. 2002. Particle size analysis. In: Dane JH, Topp GC, (eds). Methods of Soil Analysis. Part 4, Physical Methods. Soil Sci. Soc. Am. 5, 255-293.

Hong, H., Pourghasemi, H.R., Pourtaghi, Z.S. 2016. Landslide susceptibility assessment in Lianhua County (China): A comparison between a random forest data mining technique and bivariate and multivariate statistical. Geomorphology. 259, 105-118.

Ionita, I., Fullen, M.A., Zgłobicki, W., Poesen, J., 2015. Gully erosion as a natural and human-induced hazard. Nat. Hazards 79, 1–5.

Kennison RF, Cox J (2013) Health and functional limitations predict depression scores in the health and retirement study; results straight from MARS. Calif J Health Promot 11(1):97–108.

Lesschen, J.P., Kok, K., Verburg, P.H., Cammeraat, L.H., 2007. Identification of Vulnerable Areas for Gully Erosion under Different Scenarios of Land Abandonment in Southeast Spain. Catena 71 (1), 110–121. https://doi.org/10.1016/j.catena.2006.05.014.

Mclean, E.O. (1982) Soil pH and Lime Requirement. In: Page, A.L., Ed., Methods of Soil Analysis. Part 2. Chemical and Microbiological Properties, American Society of Agronomy, Soil Science Society of America, (pp. 199-224).

Mararakanye, N., Sumner, P.D., 2017. Gully erosion: a comparison of contributing factors in two catchments in South Africa. Geomorphology 288, 99–110.

Märker, M., Pelacani, S., Schröder, B., 2011. A functional entity approach to predict soil erosion processes in a small Plio-Pleistocene Mediterranean catchment in Northern Chianti, Italy. Geomorphology 125, 530–540. https://doi.org/10.1016/j.geomorph.2010.10.022.

Meliho, M., Khattabi, A., Mhammdi, N., 2018. A GIS-based approach for gully erosion susceptibility modelling using bivariate statistics methods in the Ourika watershed, Morocco. Environ. Earth Sci. 77 (18), 655. https://doi.org/10.1007/s12665-018-7844-1.

Milborrow S (2009) Derived from mda: mars by Trevor Hastie and RobTibshirani. earth: Multivariate Adaptive Regression Splines, 2009.R Package, http://CRAN.R-project.org/package=earth.

Park, N. W. (2010). Application of Dempster-Shafer theory of evidence to GIS-based landslide susceptibility analysis. Environmental Earth Science, 62(2), 367-376.

Poesen, J., Vanwalleghem, T., Deckers, J., 2018. Gullies and closed depressions in the Loess Belt: scars of human–environment interactions. Landscapes and Landforms of Belgium and Luxembourg. Springer, Cham, pp. 253–267.

Pourghasemi, H. R., Moradi, H. R., Fatemi Aghda, S. M., Gokceoglu, C., Pradhan, B. 2014. GIS-based landslide susceptibility mapping with probabilistic likelihood ratio and spatial multi-criteria evaluation models (North of Tehran, Iran), Arabian Journal of Geosciences. 7, 1857-1878.

Pourghasemi, H.R., Gayen, A., Panahi, M., Rezaie, F., Blaschke, T., 2019. Multi-hazard probability assessment and mapping in Iran. Science of the Total Environment, 692, 556–571.

Pourghasemi, H.R., Yousefi, S., Kornejady, A., Cerdà, A., 2017. Performance assessment of individual and ensemble data-mining techniques for gully erosion modeling. Sci. Total Environ. 609, 764–775.

Rahmati, O., Pourghasemi, H.R., 2017. Identification of critical flood prone areas in data scarce and ungauged regions: A comparison of three data mining models. Water Resour. Manag. 31 (5), 1473–1487.

Rahmati, O., Haghizadeh, A., Pourghasemi, H.R., Noormohamadi, F., 2016. Gully erosion susceptibility mapping: the role of GIS based bivariate statistical models and their comparison. Nat. Hazards 82 (2), 1231–1258. https://doi.org/10.1007/s11069-016-2239-7.

Refahi, H., 2009. Soil erosion by water & conservation. Tehran University Press, pp. 10–202 (In Farsi with English Summary).

Sankey, J.B., Draut, A.E., 2014. Gully annealing by aeolian sediment: field and remote sensing investigation of aeolian–hillslope–fluvial interactions, Colorado River corridor, Arizona, USA. Geomorphology 220, 68–80.

Selkimäki, M., González-Olabarria, J.R., 2017. Assessing gully erosion occurrence in forestlands in Catalonia (Spain). Land Degrad. Dev. 28 (2), 616–627.

Setianto, A., & Triandini, T. (2013). Comparison of Kriging and Inverse distance Weighted (IDW) Interpolation methods in Lineament extraction and Analysis. Journal of Southeast Asian Applied Geology. 5(1), 21-29.

Sezer, E.A., Pradhan, B., Gokceoglu, C., 2011. Manifestation of an adaptive neuro-fuzzy model on landslide susceptibility mapping: Klang valley, Malaysia. Expert Syst. Appl. 38, 8208–8219.

Swarnkar, S., Malini, A., Tripathi, S., Sinha, R., 2018. Assessment of uncertainties in soil erosion and sediment yield estimates at ungauged basins: an application to the Garra River basin, India. Hydrol. Earth Syst. Sci. 22, 2471–2485. https://doi.org/10.5194/hess-22-2471-2018.

Shit, P.K., Paira, R., Bhunia, G., Maiti, R., 2015. Modeling of potential gully erosion hazard using geo-spatial technology at Garbheta block, West Bengal in India. Model. Earth Syst. Environ. 1 (1–2), 1–16. https://doi.org/10.1007/s40808-015-0001-x.

Sidle, R.C., Ochiai, H., 2006. Landslides: Processes, Prediction, and Land Use, Water Res Monograph. vol. 18. American Geophysical Union, Washington, DC, p. 312.

Sigaroodi, S. K., Chen, Q., Ebrahimi, S., Nazari, A., & Choobin, B. 2014. Long-term precipitation forecast for drought relief using atmospheric circulation factors: a study on the Maharloo Basin in Iran. Hydrology. Earth System. Sciences, 18, 1-12.

United States Department of Agriculture, Soil Conservation Service (USDA-SCS), 1966.Procedure for determining rates of land damage, land depreciation, and volume of sediment produced by gully erosion. Technical Release No. 32. US GPO 1990-261-419:20727/SCS. US Government Printing Office, Washington, DC.

Umar, Z., Pradhan, B., Ahmad, A., Jebur, M.N., Tehrany, M.S., 2014. Earthquake induced landslide susceptibility mapping using an integrated ensemble frequency ratio and logistic regression models in West Sumatera Province, Indonesia. Catena 118, 124–135.

Wang, L., Wei, S., Horton, R., Shao, M.A., 2011. Effects of vegetation and slope aspect on water budget in the hill and gully region of the Loess Plateau of China. Catena 87(1), 90–100.

Yesilnacar, E. K. (2005). The application of computational intelligence to landslide susceptibility mapping in Turkey (Ph.D Thesis Department of Geomatics the University of Melbourne).

Zakerinejad, R., Maerker, M., 2014. Prediction of gully erosion susceptibilities using detailed terrain analysis and maximum entropy modeling: a case study in the Mazayejan Plain, Southwest Iran. Geogr. Fis. Din. Quaternaria 37 (1), 67–76. https://doi.org/10.4461/GFDQ.2014.37.7.

Zakerinejad, R., Maerker, M., 2015. An integrated assessment of soil erosion dynamics with special emphasis on gully erosion in the Mazayejan basin, southwestern Iran. Nat. Hazards 79 (1), 25–50.

Zabihi, M., Mirchooli, F., Motevalli, A., Darvishan, A.K., Pourghasemi, H.R., Zakeri, M. A., Sadighi, F., 2018. Spatial modelling of gully erosion in Mazandaran Province, northern Iran. Catena 161, 1–13.

Zabihi, M., Pourghasemi, H.R., Pourtaghi, Z., & Behzadfar, M. 2016. GIS-based multivariate adaptive regression spline and random forest models for groundwater potential mapping in Iran. Environment Earth Science, 75, 1-19.

Mahdis Amiri is a PhD student at Gorgan University in Iran. Her major interests are management and control of desert. She holds a Bachelor and Master of Science degrees from Shiraz University. Her master's thesis title was spatial modeling of gully erosion and she is currently working on multi-hazards spatial modeling using machine learning techniques.

Hamid Reza Pourghasemi is an Associate Professor of Watershed Management Engineering in the College of Agriculture, Shiraz University, Iran. He has a BSc in Watershed Management Engineering of the University of Gorgan (2004), Iran, an MSc in Watershed Management Engineering, from Tarbiat Modares University, Iran (2008), and a PhD in Watershed Management Engineering from the same University (Feb 2014). His main research interests are GIS-based spatial modelling using machine learning/data mining techniques in different fields such landslide, flood, gully erosion, forest fire, land subsidence, species distribution modelling, and groundwater/hydrology. Also, Hamid Reza works on Multi-Criteria Decision Making methods in Natural Resources and Environment.

He has published more of 90 peer reviewed papers in high-quality journals, with three chapters in Springer. Also, he published two books in Springer (https://www.springer.com/gp/book/9783319733821) and Elsevier (https://www.elsevier.com/books/spatial-modeling-in-gis-and-r-for-earth-and-environmental-science/pourghasemi/978-0-12-815226-3).

Hamid Reza Pourghasemi, Amiya Gayen, Sk. Mafizul Haque, and Shibiao Bai

Abstract

Gully erosion susceptibility mapping (GESM) is a valuable tool for sustainable land use management and reducing soil erosion. Gully erosion and its formation are a natural process; it greatly threatens agriculture, environment, ecosystem disruption, and natural resources. The objective of this present study is to develop a GESM by implementation of well acceptable SVM learning algorithm in Golestan Province, Kalaleh Township, Iran. Primarily, gully sites were obtained by comprehensive field observations. After that, 12 gully erosion predisposing factors were selected to assess the gully erosion susceptibility map. The 12 conditioning factors were aspect, altitude, drainage density, lithology, slope angle, slope length, distance from river, profile curvature, drainage density, TWI, distance from road, and plan curvature. Finally, gully erosion susceptibility map was prepared using the SVM model in "R" environment. In the final stage, assessment of the prediction accuracy of the susceptibility model with the help of training (70%) and validation datasets (30%) of gully location was done. The predicted susceptibility map was validated with the help of receiver operating characteristic (ROC) curve, true skill statistics (TSS), and deviance value. The results indicated that the areas under the curve (AUC) were calculated as 94.3% and 97.0% based on validation and training dataset, respectively. Furthermore, the TSS, deviance, and correlation values were 0.84, 0.50, and 0.85, respectively. So, the results of other indices including, sensitivity, specificity, and Cohen's Kappa (CK) showed that SVM model has reasonable prediction accuracy for the cases of gully erosion susceptibility assessment. As regards the SVM model, a total area of 11.66% was identified as the hazard prone area of the mentioned Town ship. So, it is concluded that the gully erosion map serves as an important tool for protective action and watershed management, specifically at the initiation of the gully to protect the development of land degradation.

Keywords

Gully erosion · True skill statistics · Area under the curve · Support vector machine · Machine learning

28.1 Introduction

Today, majority of country population are now facing problems from huge land degradation and soil erosion (Morgan 2009). Gully erosion is considered as an important mode of soil degradation process that influence on agricultural productivity, environmental degradation in downstream areas, and increasing sediment yields (Carey et al. 2001; Guerra et al. 2016). Sometimes, gully floor materials in a bad land landscape reveal some specific characteristics of a particular local environment rather its vast geological settings (Haque and Ghosh 2018). Gully is a narrow deep channel with a cross-sectional area that is greater than 929 cm^2 (Poesen 1993). The effect of gully development is found in ecosystems based on their role and function (Bernatek-Jakiel et al. 2017). So, it is important to predict spatial distribution of gully erosion and gully development in each area.

Several researchers tried to assess the impact of topographical and geo-environment factors on gully erosion (Amiri et al. 2019). A number of researchers study the impact of chemical and physical soil properties on gully erosion

H. R. Pourghasemi (✉)
Department of Natural Resources and Environmental Engineering, College of Agriculture, Shiraz University, Shiraz, Iran
e-mail: hr.pourghasemi@shirazu.ac.ir

A. Gayen · S. M. Haque
Department of Geography, University of Calcutta, Kolkata, West Bengal, India

S. Bai
College of Marine Sciences and Engineering, Nanjing Normal University, Nanjing, China

© Springer Nature Switzerland AG 2020
P. K. Shit et al. (eds.), *Gully Erosion Studies from India and Surrounding Regions*, Advances in Science, Technology & Innovation,
https://doi.org/10.1007/978-3-030-23243-6_28

(Wilson et al. 2015; Verachtert et al. 2013). In the field of GESM, a number of statistical and expert-based models have been implemented for more detailed assessment and evolution of these processes (Verachtert et al. 2011). These models are RUSLE (Shit et al. 2015; Gayen et al. 2019a, b), artificial neural-network (Rahmati et al. 2017), frequency ratio (Zabihi et al. 2018a, b), Analytical Hierarchy Process (Svoray et al. 2012), fuzzy logic (Metternicht and Gonzalez 2013), logistic regression (Conoscenti et al. 2014), and index of entropy (Rahmati et al. 2016).

But, still some machine learning models are not widely used for the assessment of GESM. Machine learning models are useful to assess the susceptibility zones for gullying (Mohsen Hosseinalizadeh et al. 2019). These models can trickle data from different measurement scales and easily work with various types of predisposing factors (Rahmati et al. 2017). In this research, the support vector machine (SVM) was implemented to predict gully erosion susceptibility map in area more probable to gully development.

The objectives of this work are (1) spatial modelling the occurrence of gully erosion in a hilly region of Kalaleh Township, Iran, and (2) validation of implemented model using different techniques including the ROC, TSS, correlation, deviance, sensitivity, specificity, and Cohen's Kappa index.

28.2 Materials and Methods

28.2.1 Description of the Study Area

Kalaleh Township is a part of Golestan Province in Northeast Iran which is situated on the southeast shore of the Caspian Sea. It is one of the most effected Townships of Golestan Province by the influence of gully erosion. The Kalaleh Township has an area covered by 4072 km^2. It is located between 37° 10′ to 37° 20′ Northern latitudes and 56° 00′ to 57° 00′ Eastern longitudes (Fig. 28.1). The altitude ranges from 100 to 2200 m from mean sea level. This study area falls under hilly rugged topography with silt-clay soil that is prone to gully erosion.

28.2.2 Gully Erosion Inventory Mapping

In general, existing gullies and their spatial distribution mapping are required to assess the relationship with distribution of gullies and the conditioning factors of the gully erosion.

Firstly, existing gullies were collected with the help of a Global Positioning System (GPS) and later the gully erosion shape file was drawn using Google Earth images (Fig. 28.1). Finally, the total gullies are subdivided into two subsets 70%

and 30% as training and validation purpose (Gayen and Pourghasemi 2019; Zabihi et al. 2018a, b).

28.2.3 Preparing Gully Erosion Effective Factors

To evaluate the gully erosion probability, it is most important to consider the influencing factors of gully erosion (Gayen et al. 2019b). Gully erosion is considered as threshold-dependent process (Rahmati et al. 2017). Twelve predisposing factors were considered based on a review of past studies (Conoscenti et al. 2008; Pourghasemi et al. 2017; Zabihi et al. 2018a, b), field experiments, and data existence in this region. These 12 identified factors were aspect, altitude, drainage density, lithology, slope angle, slope length, drainage density, TWI, distance from river, plan curvature, soil types, distance from road, and profile curvature (Fig. 28.2a–l). Finally, the GESM was developed using the SVM model in an R3.5.1environment.

Altitude digital elevation model (DEM) is selected to be an important topographical factor in order to assess gully erosion (Zabihi et al. 2018a, b). Various altitudes influence gully erosion processes (Conoscenti et al. 2014) due to the changes in climatic condition and vegetation types. Then, few morphometric variables including slope aspect, slope length, plan curvature, slope angle, and profile curvature were developed from the DEM with the help of ArcGIS 10.5. The slope aspect was also found to be a key factor for gully erosion (2a) (Agnesi et al. 2011). The slope aspect also influences climatic characteristics such as the drying winds, volume of rainfall, amount of sunlight, and the morphological structure of the watershed that affect the occurrence of gully erosion (Agnesi et al. 2011; Wang et al. 2011). The slope angle (Fig. 28.2i) is also considered as an important predisposing factor controlling surface runoff, detachment of soil particles, and the drainage intensity (Valentin et al. 2005). The LS is an important parameter in RUSEL model for the quantitative prediction of soil erosion (2j). The LS factor is considered as an important sediment-transporting factor due to its influence in the surface runoff (Gayen and Saha 2017). The LS factor was calculated based on Moore and Burch (1986) as:

$$LS = \left(\text{Flow accumulation} \times \frac{\text{Cell size}}{22.13} \right)^{0.4}$$
$$\times \sin \left(\frac{\text{Slope}}{0.0896} \right)^{1.3} \quad (28.1)$$

The technique for calculating LS factor requires a slope steepness and flow accumulation maps.

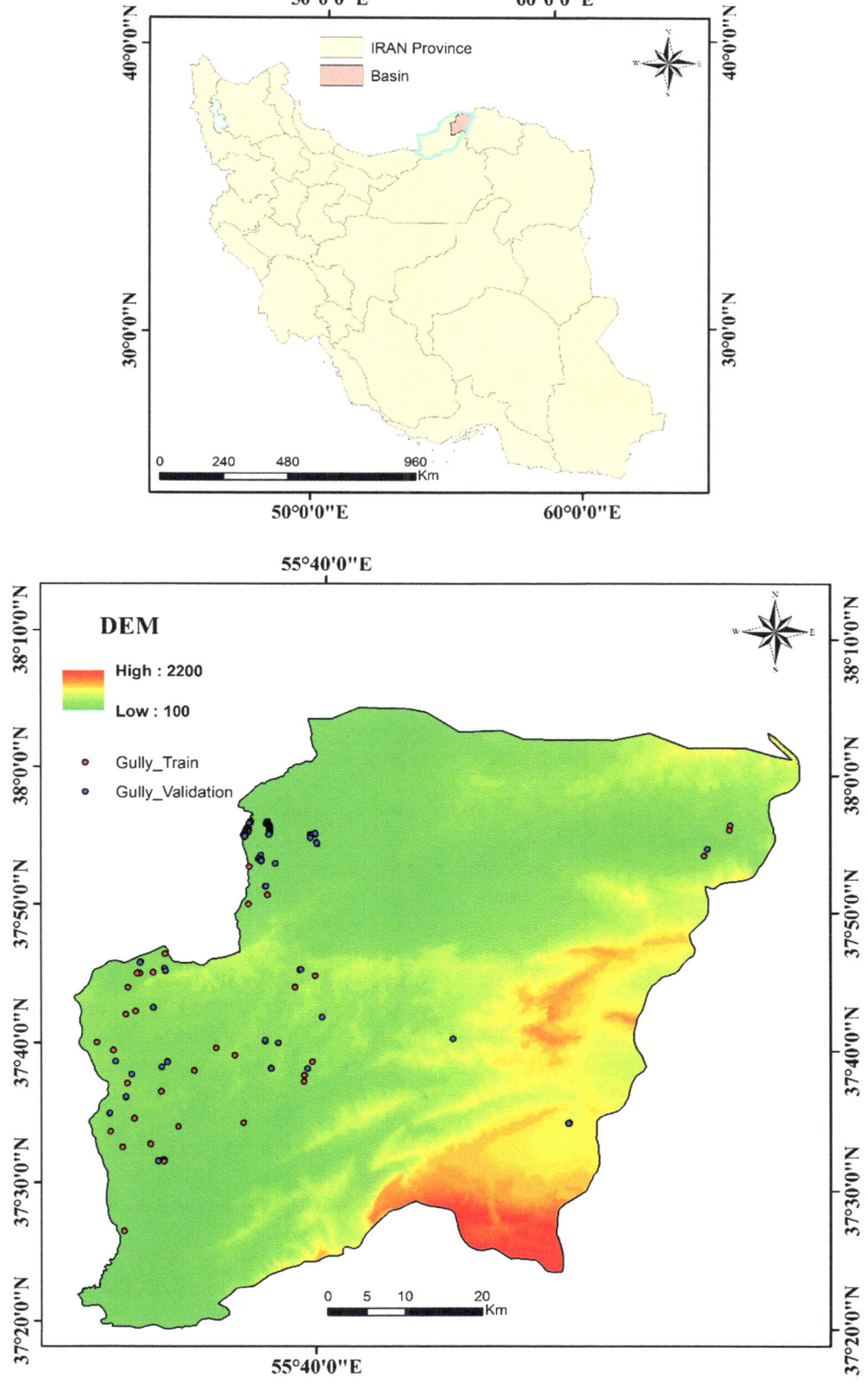

Fig. 28.1 Location of the study area in Golestan Province, Kalaleh Township, Iran

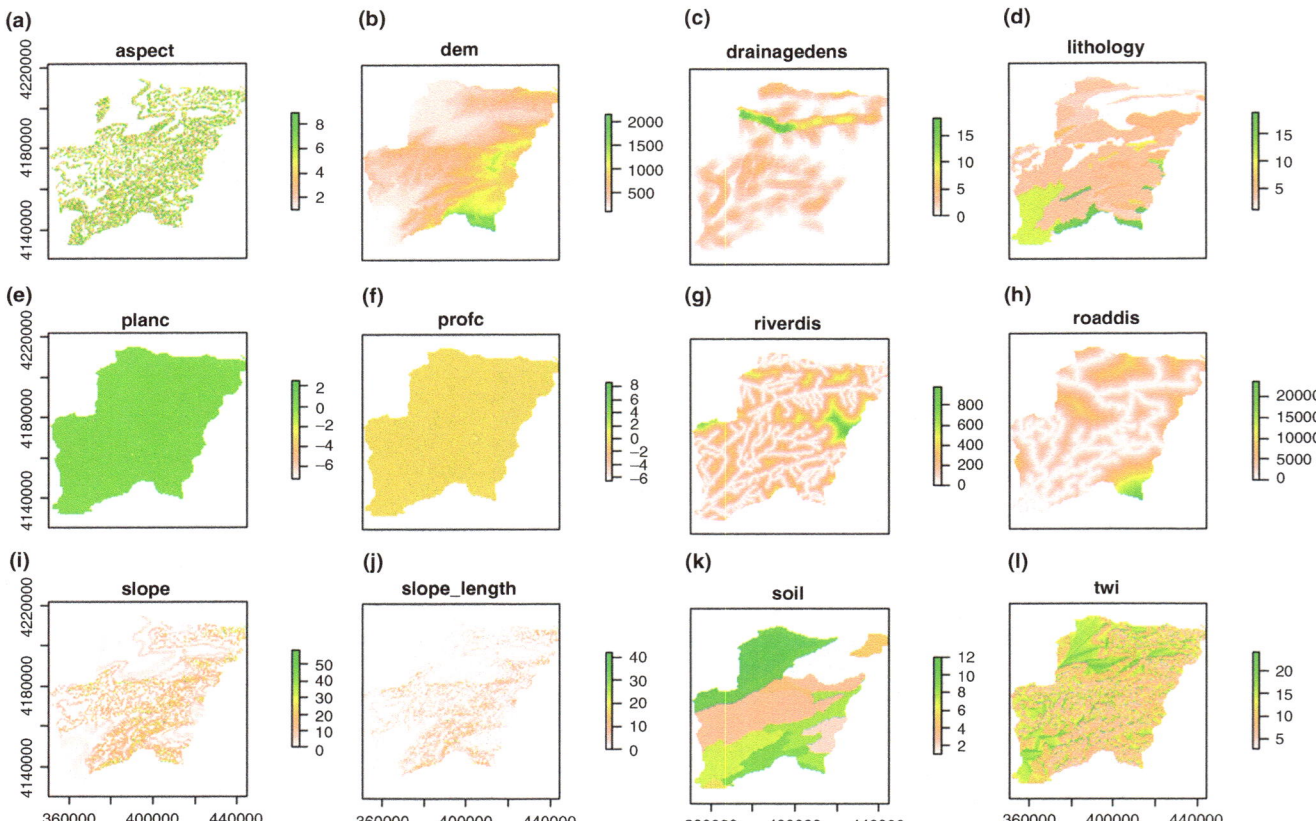

Fig. 28.2 Gully erosion conditioning factors: (**a**) slope aspect, (**b**) altitude, (**c**) drainage density, (**d**) lithology, (**e**) plan curvature, (**f**) profile curvature, (**g**) distance from river, (**h**) distance from road, (**i**) slope in degree, (**j**) slope length, (**k**) soil types (**l**) TWI

The Kalaleh Township was covered by four major soil orders, which were Entisols/Aridisols, Inceptisols, Rock Outcrops/Entisols, and Rock Outcrops/Inceptisols (Fig. 28.2k). Generally, the soil map of the Township was in 1:100,000 scale. Soil types are also considered as an important factor that influences the erodibility, organic matter, and soil mineralogy. Lithology is an influencing factor for gully erosion because of its great effect on rock types, mineral characteristics, surface runoff, and permeability (Hong et al. 2015). So, the impermeable and hard rock types decrease the gully formation, and soft rock type increases the gully formation process (Hong et al. 2015). In the Township, there are 18 different lithological units spread all over the study area according to the Geological Survey of Iran (Fig. 28.2d). These lithological units are Ksn, Qsd, Kat, Qsw, Ksr, Kad-ab, Qft2, Ekh, PeEck, Qm, Ktr, JKsj, Qft1, Kad, Jmz, J1, PlQc, and Jd.

Finally, the drainage density, distance from road, and distance from rivers were generated from topographical maps of the Kalaleh Township. The distance from river is another prime factor in gully erosion (Amiri et al. 2019). The

rivers influence the moisture condition and stability of the slope due to toe erosion and saturation of the low-lying areas of the slope (Cevik and Topal 2003). This map was produced based on the Euclidian Distance Tool in ArcGIS 10.5 using rivers and streams (Fig. 28.2g). Based on the vector lines of rivers, the drainage density map was produced using Line Density tool in ArcGIS (Fig. 28.2c). The distance from the roads plays an important role in gully erosion (Amiri et al. 2019). These newly colonized peoples are settled near the roads and also construct new roads to access the vegetation cover as subsistence land (Gayen and Saha 2018) that has influenced natural slope conditions as a result of road construction, and this may influence the susceptibility of gully erosion (Gayen et al. 2019a, b). The distance to road was calculated using the vector lines of roads according to Distance Function Tool in ArcGIS 10.5 (Fig. 28.2h). The Topographical Wetness Index (TWI) is an important secondary topographical factor (Rahmati et al. 2016) that indicates the amount of water accumulation at a site. The TWI is calculated by the applying Eq. (28.2) according to Moore and Wilson (1991).

$$\text{TWI} = \ln(\alpha/\tan\beta) \qquad (28.2)$$

where α denotes the upslope area drainage from a point and $\tan\beta$ denotes the degree of inclination of slope at the point.

The plan and profile curvatures are also considered to be important factors for gully erosion. The factor is calculated as the slope perpendicular to the slope gradient and it influences the divergence and convergence of water flow across the slope surface. The plan and profile curvatures are generated from the DEM in GIS (Fig. 28.2e–f).

28.2.4 Gully Erosion Spatial Modelling

28.2.4.1 Application of Support Vector Machine Model

The Support Vector Machine (SVM) is a statistical learning theory or supervised classifier model that has been widely used as a data mining skill to solve complex classification and regression problems (Jebur et al. 2014; Chen et al. 2016). It is more useful when the dataset are more complex linear separable error because of its producing efficiency for complex curved boundaries (Kalantar et al. 2017). The SVM is very useful for data processing of nonlinear relationships by the kernel function ($K(x_i, x) = \varphi(x_i). \varphi(x)$) (Hong et al. 2016). The SVM classifier attempts to find an optimal separation hyperplane for the positive and negative classes (Gayen et al. 2019b; Hong et al. 2016). When the training dataset considered n number of samples, it is denoted by $(x_i y_i)(i = 1, \ldots, n)$ where x_i denotes the input and y_i denotes the output and the problem denotes a function $f(x)$. For the cases of linear separable, the hyperplane can be defined by the following equation (Hong et al. 2016):

$$\frac{yi(w * xi + b)}{\geq 1} - \delta i \qquad (28.3)$$

where w indicates the coefficient vector that indicates the hyperplane tendency in the feature space, δi means the positive slack variables, and b indicates the offset of the hyperplane from the primary (Cortes and Vapnik 1995). The optimal hyperplane resolves an optimization problem as (Jebur et al. 2014):

$$\text{Minimize} \sum_{i=1}^{n} ai - \frac{1}{2} \sum_{i=1}^{n} \sum_{j=1}^{n} aiajyiyj(xixj) \qquad (28.4)$$

$$\text{Subject} \sum_{i=1}^{n} aiyi = 0, 0 \leq ai \leq C \qquad (28.5)$$

where ai is the lag range multiplier and C refers to the penalty.

In this study, the SVM model with the radial basic function was chosen due to its robustness as reported by number of researchers (Pourghasemi et al. 2013a, b).

28.2.4.2 Validation of Gully Erosion Susceptibility Map

28.2.4.2.1 The ROC–AUC Value

To validate gully erosion susceptibility model, the ROC curve has been implemented and calculated the AUC for qualitative evaluation of the SVM model (Hong et al. 2016). It is a well-known method for the quantitative assessment of the predictive probability model (Pourghasemi et al. 2012). The fittest model has a curve with the greater AUC; in most ideal models, the AUC values ranges from 0.5 to 1.0. If the AUC values are 0.9–1, represent the excellent prediction accuracy of an implemented model (Yesilnacar and Topal 2005). When the AUC value is between 0.5 and 0.6, it indicates the poor prediction accuracy of an applied model. The ROC curve was developed for validation of GESM using unused gully location data or validation datasets (prediction rate curve for future gully erosion occurrence).

28.2.4.2.2 True Skill Statistics

Generally, the True Skill Statistics (TSS) is a quantitative assessment technique for the assessment of predicative models. This methodology is widely used for analysis in the accuracy of weather forecasts (Accadia et al. 2005). For the cases of confusion matrix, it is defined as:

$$\text{TSS} = \text{Sei} + \text{Spi} - 1 \qquad (28.6)$$

where, Sei is sensitivity and Spi is specificity, respectively (Liu et al. 2009). Its values vary from +1 to −1; if the TSS value is less than 0.4, it indicates poor prediction accuracy and values close to +1 indicates excellent predictive capacity of the model (Allouche et al. 2006).

28.2.4.2.3 Calculation of Correlation (COR)

For the validation purpose, correlation static was used based on Karl Pearson's method. It is used for assessment of the relationship between two variables (i.e., dependent and independent variables) (Pearson 1897). The relation between two variables is defined as Eq. (28.7):

$$r = \frac{\sum (X - \overline{X})(Y - \overline{Y})}{\sqrt{\sum (X - \overline{X})^2} \sqrt{(Y - \overline{Y})^2}} \qquad (28.7)$$

where r denotes the coefficient of correlation. Y is the mean of Y variable, and X is the mean of X variable. Generally, the value of "r" varies from +1 to −1 where +1 represents highly positive relationship and −1 represents strongly negative relationship between dependent and independent variables.

28.2.4.2.4 Calculation Deviance Value

The deviance value was first introduced by Guisan and Zimmermann (2000) in the field of ecology. Now, it has become a widely acceptable method for model accuracy assessment in various fields of research (e.g., Engler et al. 2004; Liu et al. 2009). The proportion of D^2 was also implemented in species distribution models. Now, it is also adjusted based on the asymptotic X^2 distribution of the log-likelihood statistics. In this field of GESM, the deviance value was determined by the following equation (Mittlböck and Schemper 1996):

$$D^2 = 1 - \log L(\hat{\beta}) / \log L(\hat{\beta}_0) \qquad (28.8)$$

where

$$L(\hat{\beta}) = \sum_{i=1}^{n} [o_i \log p_i + (1 - o_i) \log (1 - p_i)] \qquad (28.9)$$

$$L(\hat{\beta}_0) = p \log p + (1 - p) \log (1 - p), p$$
$$= \frac{1}{n} \sum_{i=1}^{n} o_i \qquad (28.10)$$

28.2.4.2.5 Fitting Performance Measures (Sensitivity, Specificity, and Cohen's Kappa Index)

In this work, the final result was validated by specificity, sensitivity, and Cohen's Kappa (CK) Index (Eqs. 28.11–28.13) (Rossi and Reichenbach 2016). Specificity (1 − false positive rate) and sensitivity are given by Rossi et al. (2010):

$$\mathrm{CK} = (A - B)/(1 - B) \qquad (28.11)$$

where

$$A = (\mathrm{TP} + \mathrm{TN})/(\mathrm{TP} + \mathrm{TN} + \mathrm{FP} + \mathrm{FN}) \qquad (28.12)$$

and

$$B = ((((\mathrm{TN} + \mathrm{FN}) * (\mathrm{TN} + \mathrm{FP}))/(\mathrm{TP} + \mathrm{TN} + \mathrm{FP} + \mathrm{FN}))$$
$$+ (((\mathrm{TP} + \mathrm{FP}) * (\mathrm{TP} + \mathrm{FN}))/(\mathrm{TP} + \mathrm{TN} + \mathrm{FP} + \mathrm{FN})/(\mathrm{TP} + \mathrm{TN} + \mathrm{FP} + \mathrm{FN})$$
$$(28.13)$$

where A is agreement, B is chance, TN denotes true negative, TP denotes true positive, FN denotes false negative, and FP denotes false positive.

28.3 Results and Discussion

Preparation Gully Erosion Susceptibility Map Using the SVM Model

After exploration of spatial interaction between gully location and effective factors, the separated data was entered into the R 3.5.1 environment. Finally, the GESM was developed based on pixel level for the entire study area. Lastly, the GESM was classified using Jenk's natural break classification method (Naghibi et al. 2015; Zabihi et al. 2018a, b). Then GESM was classified into four classes, namely, low, very high, moderate, and high (Fig. 28.3).

The GESM which is produced based on the SVM model, 41.78% of the total area is found to have low gully erosion potential. High and very high zones make up 8.32% and 3.84% of the total Township (Table 28.1). The moderate susceptible area for gully erosion area is 46.56% of the total area of study (Fig. 28.4). Also, the AUC values of 0.943 with model prediction capacity are 94.3% with appropriate standard error of 0.014 but in case of success rate curve (using training dataset, i.e., 70% of the data), the AUC value is displayed as 0.97 (97%) (Fig. 28.5). Also, fourfold cross-validation of robustness method provides good results in respect of the fist trying (onefold) in model prediction accuracy process (Fig. 28.6).

The statistical evaluation criteria (TSS, deviance, and correlation) were calculated to compare the probabilistic models for validation datasets. The SVM model provides excellent prediction results in terms of TSS, correlation, and deviance (Table 28.2) for the cases of gully erosion susceptibility assessment in this Kalaleh Township. The TSS and correlation values are 0.85 and 0.84, respectively, and also results displayed lowest deviance value of 0.5. The lowest deviance represents better acceptability of this model (Table 28.2).

Finally, the model performance was measured by model evaluation plot [specificity (green line), sensitivity (red line), and Cohen's Kappa Index (blue line)] of SVM model in GESM process (Fig. 28.7). So, based on these rules we can conclude that the SVM model provides good or adequate results that means model has good prediction power for gully erosion susceptibility assessment in this study area.

Normally, the SVM model provides good results in different field of hazard susceptibility assessment. The SVM model has the advantages of handling complex objects, nonlinear relationship and over fitting problems, and is very robust to noise (Amiri et al. 2019). A number of researchers showed that the SVM model also provides better prediction results in different fields such as landslide assessment (Pourghasemi

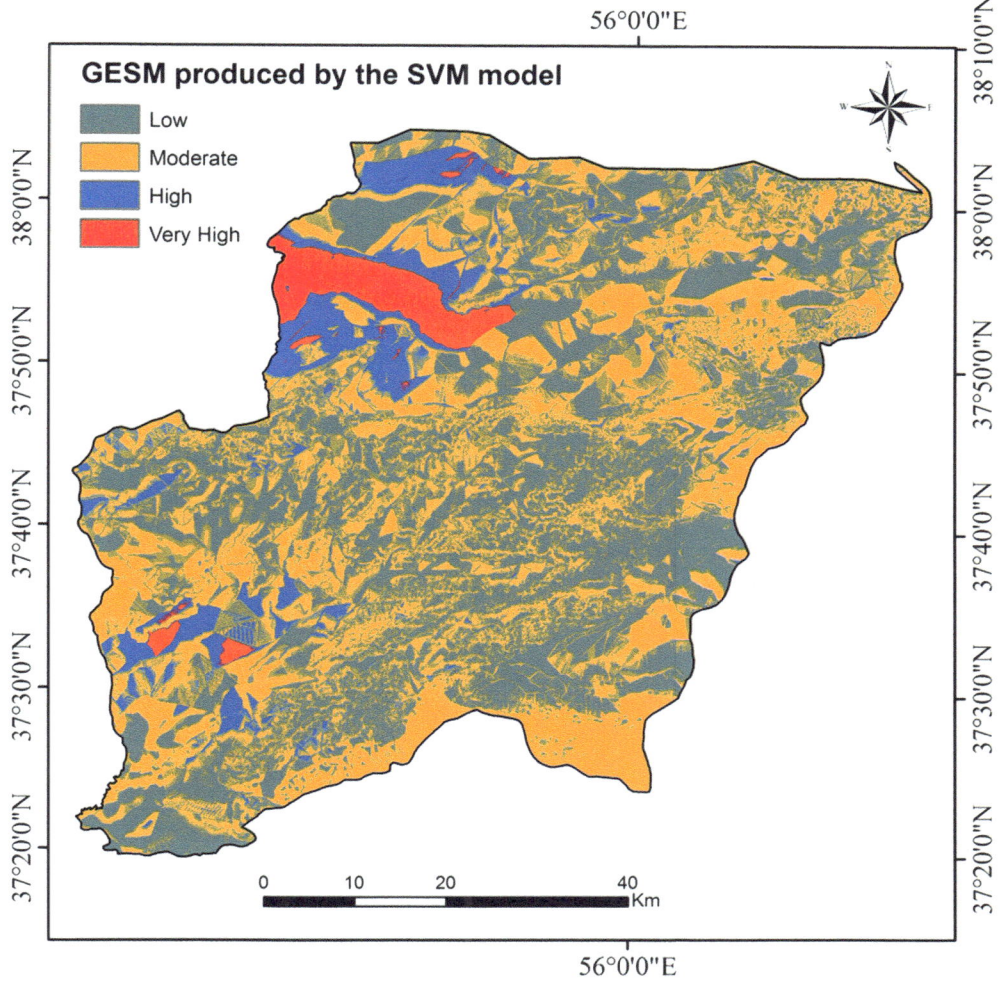

Fig. 28.3 Gully erosion susceptibility map produced based on SVM model

et al. 2013a, b) and flood susceptibility mapping (Tehrany et al. 2015). Tien Bui et al. (2016) developed five various machine learning models for shallow landslide susceptibility (SLS) assessment. The final results indicated that SVM model provide better result (AUC = 88.77) in the cases of SLS assessment. Rahmati et al. (2016) showed that the SVM model provides excellent accuracy with an AUC value more than 90%. So, finally we can conclude that this model has excellent prediction capacity in cases of gully erosion in this Township.

Table 28.1 Percentage of area of under different susceptibility classes

Gully erosion susceptibility class	Number of pixels in SVM model	Percentage of each class
Low	2,182,570	41.78
Moderate	2,432,682	46.56
High	434,724	8.32
Very high	174,562	3.34

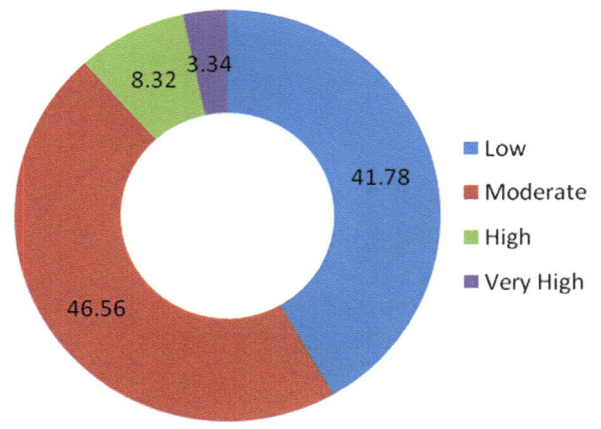

Fig. 28.4 Percentage of soil erosion susceptibility areas that fall into the various classes (i.e., low, moderate, high, and very high) of SVM Susceptibility map

28.4 Conclusion

The main objective of this work was to evaluate the spatial modeling of gully erosion by implementation of the SVM model in Kalaleh Township. Gully erosion is a very sensitive problem in the northwest part of Township. For a comprehensive regional development plan, a good evaluation of the gully erosion processes is essential. For the assessment of spatial distribution of gully erosion, some researchers tried to develop different methods by which evaluate gully erosion susceptible at the regional level. So, GESM is necessary for comprehensive regional development and solving these land degradation problems. In this study, the SVM machine learning model was applied according to 12 important factors, viz., slope, aspect, elevation, distance from road, LS, soil type, TWI, distance from river, plan curvature, drainage density, profile curvature, lithology, and distance from road for identifying the gully erosion probable areas. Therefore, GESM can be applied for sustainable land use

Fig. 28.5 Model validation of gully erosion susceptibility maps using validation dataset (30%) (Prediction rate curve)

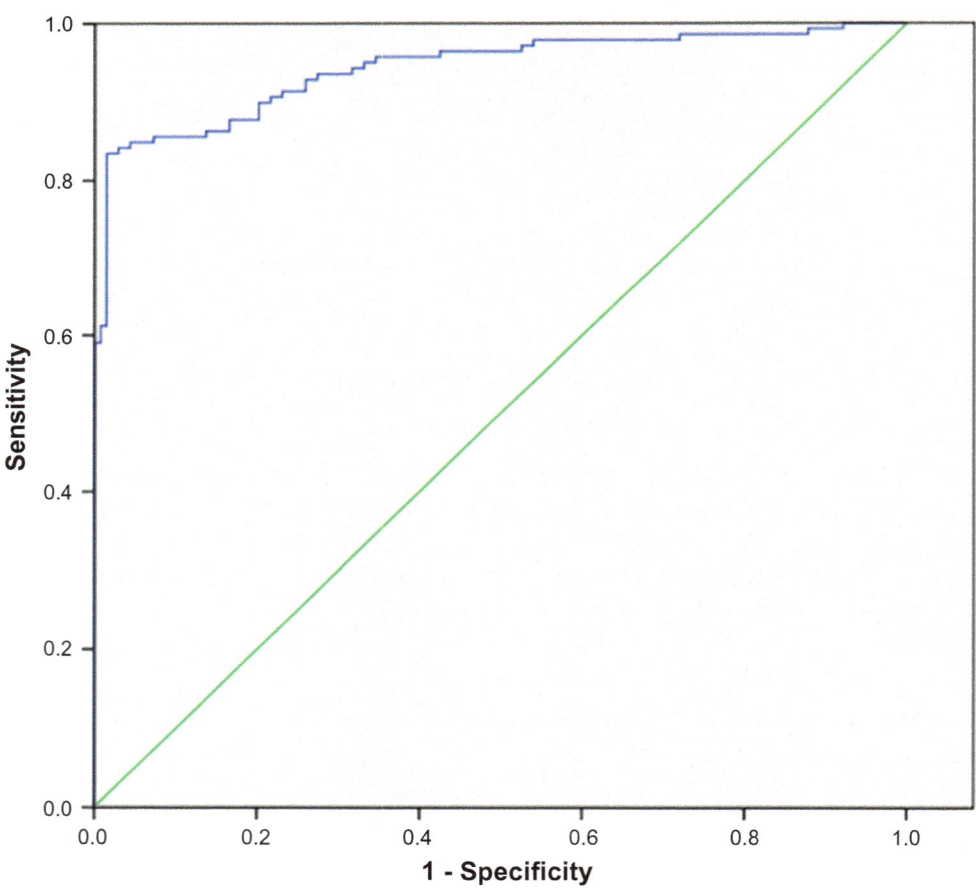

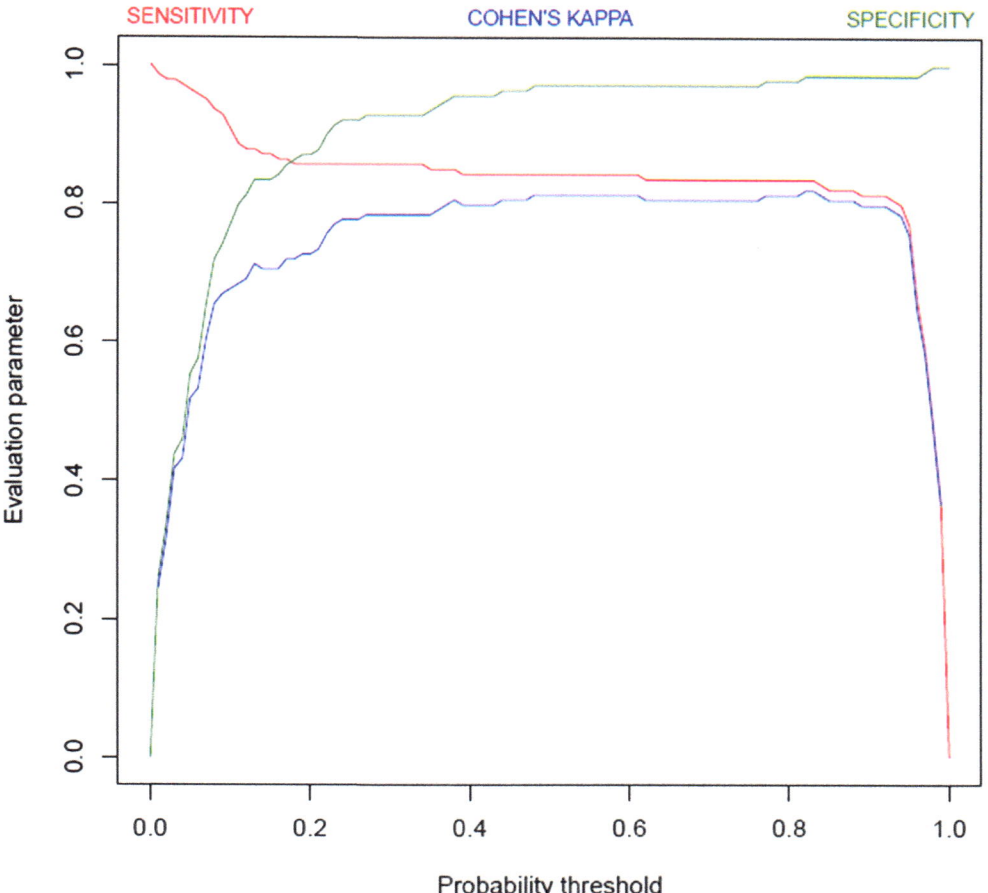

Fig. 28.6 The fitting performance measures, sensitivity (red line), specificity (green line), and Cohen's Kappa index (blue line) of SVM model

management in respect to plant cover, protection of soil, trickling the surface runoff, and structure of soil; it is urgent to reduce the surface runoff by the diversion of long-period floods in the upstream areas. To evaluate the predictive accuracy of the GESM, the shape of the ROC curves has been observed. According to the ROC curves, TSS, correlation, deviance, sensitivity, specificity, and Cohen's Kappa index of the forecasted map have the reasonably good predictors of future gully activity. Finally, the outcomes depict that the machine learning models are capable enough to demarcate the gully erosion areas. Therefore, the prepared gully erosion susceptible map can be used as an important tool for conservation and sustainable planning of gully erosion prone areas of Kalaleh Township.

Table 28.2 Various types of accuracy assessment tests and their results

Accuracy assessment methods	Results
Area under the curve	0.97
Correlation (COR)	0.85
True skill statistic (TSS)	0.84
Deviance	0.50

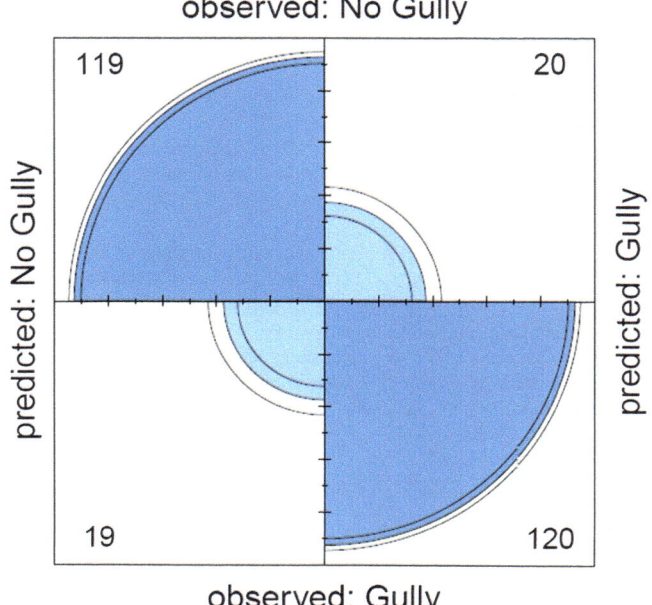

Fig. 28.7 Fourfold plot based on validation dataset

References

Accadia, C., Mariani, S., Casaioli, M., Lavaqnini, A., Speranza, A., 2005. Verification of precipitation forecasts from two limited-area models over Italy and comparison with ECMWF forecasts using a resampling technique. Weather and Forecasting, 20, 276-300.

Agnesi, V., Angileri, S., Cappadonia, C., Conoscenti, C., Rotigliano, E., 2011. Multi-parametric GIS analysis to assess gully erosion susceptibility: a test in southern Sicily, Italy. Landf Anal. 7, 15-20.

Allouche, O., Tsoar, A., Kadmon, R., 2006. Assessing the accuracy of species distribution models: prevalence, kappa and the true skill statistic (TSS). J. Appl. Ecol. 43, 1223–1232.

Amiri M, Pourghasemi HR, Ghanbarian GA, Afzali SF (2019) Assessment of the importance of gully erosion effective factors using Boruta algorithm and its spatial modelling and mapping using three machine learning algorithms. Geoderma 340:55-69.

Amor, D., and Pfaff, A., 2008. Early history of the impact of road investments on deforestation in the Mayan forest. In Working paper, Nicholas School of the Environment and Sanford School of Public Policy, Duke University, Durham, NC, USA.

Bernatek-Jakiel, A., Jakiel, M., Krzemień, K., 2017. Piping dynamics in mid-altitude mountains under a temperate climate: Bieszczady Mts., Eastern Carpathians. Earth Surf. Process. Landf. 42, 1419–1433.

Carey, B., Gray, J., Seagrave, C., 2001. Gully Erosion. Department of Natural Resources and Mines, The State of Queensland. http://www.gcenvironment.org.au/pdf/LM81w.pdf.

Cevik, E., Topal, T., 2003. GIS-based landslide susceptibility mapping for a problematic segment of the natural gas pipeline, Hendek (Turkey). Environ. Geol. 44, 949–962.

Chen, W., Chai, H., Zhao, Z., Wang, Q., Hong, H., 2016. Landslide susceptibility mapping based on GIS and support vector machine models for the Qianyang County, China. Environ Earth Sci 75(6), 1–13.

Conoscenti, C., Angileri, S., Cappadonia, C., Rotigliano, E., Agnesi, V., Maʾrker, M., 2014. Gully erosion susceptibility assessment by means of GIS-based logistic regression: a case of Sicily (Italy). Geomorphology 204 (1), 399–411.

Conoscenti, C., Maggio, C.D., Rotigliano, E., 2008. Soil erosion susceptibility assessment and validation using a geostatistical multivariate approach: a test in Southern Sicily. Natural Hazards 46 (3), 287-305.

Cortes, C., Vapnik, V., 1995. Support vector networks. Mach. Learn. 20 (3), 273-297.

Engler, R., A. Guisan, Rechsteiner, L., 2004. An improved approach for predicting the distribution of rare and endangered species from occurrence and pseudo-absence data. Journal of Applied Ecology 41,263–274.

Gayen, A., Pourghasemi, H.R., 2019. Spatial Modeling of Gully Erosion: A New Ensemble of CART and GLM Data-Mining Algorithms. Spatial Modeling in GIS and R for Earth and Environmental Science, pp 653-669.

Gayen, A., Saha, S., 2017. Application of weights-of-evidence (WoE) and evidential belief function (EBF) models for the delineation of soil erosion vulnerable zones: a study on Pathro river basin, Jharkhand, India. Modeling Earth Systems and Environment 3, 1123-1139.

Gayen, A., Saha, S., 2018. Deforestation probable area predicted by logistic regression in Pathro river basin: a tributary of Ajay River. Spatial Information Research 26 (1), 1-9.

Gayen, A., Saha, S., Pourghasemi, H.R., 2019a. Soil erosion Assessment using RUSLE model and its Validation by FR probability model. Geocarto International. DOI: https://doi.org/10.1080/10106049.2019.1581272

Gayen, A., Pourghasemi, H.R., Saha, S., Keesstra, S.D., Bai, S., 2019b. Gully erosion susceptibility assessment and management of hazard-prone areas in India using different machine learning algorithms. Science of the Total Environment, DOI: https://doi.org/10.1016/j.scitotenv.2019.02.436.

Guerra, C.A., Maes, J., Geijzendorffer, I., Metzger, M.J., 2016. An assessment of soil erosion

Guisan, A., and Zimmermann, N.E., 2000. Predictive habitat distribution models in ecology. Ecological Modelling 135, 147-186.

Haque, M., Ghosh, S., 2018. Microstructural Evidence of Palaeo-Coastal Landform from Westernmost Fringe of Lower Ganga-Brahmaputra Delth. Quaternary Geomorphology in India, pp. 61-78.

Hong, H., Pradhan, B., Bui, D.T., Xu, C., Youssef, A.M., Chen, W., 2016. Comparison of four kernel functions used in support vector machines for landslide susceptibility mapping: a case study at Suichuan area (China). Geomatics Natural Hazards and Risk 8 (2), 544-569.

Hong, H., Pradhan, B., Xu, C., Tien Bui, D., 2015. Spatial prediction of landslide hazard at the Yihuang area (China) using two-class kernel logistic regression, alternating decision tree and support vector machines. Catena 133, 266–281.

Jebur, M.N., Pradhan, B., Shafapour Tehrany, M., 2014. Optimization of landslide conditioning factors using very high-resolution airborne laser scanning (LiDAR) data at catchment scale. Remote Sens Environ 152, 150–165.

Kalantar, B., Pradhan, B., Naghibi, S.A., Motevalli, A., Mansor, S., 2017. Assessment of the effects of training data selection on the landslide susceptibility mapping: a comparison between support vector machine (SVM), logistic regression (LR) and artificial neural networks (ANN). Geomatics, Natural hazards and Risk 9 (1), 49-69.

Liu, C., White, M., Newell, G., 2009. Measuring the accuracy of species distribution models: a review. In: Andersen, R.S., Braddock, R.D., Newham, L.T.H. (Eds.), Proceedings 18th World IMACs/MODSIM Congress. Cairns, Australia, pp. 4241–4247.

Metternicht, G., Gonzalez, S., 2013. FUERO: Foundations of a fuzzy exploratory model for soil erosion hazard prediction. Environmental Modelling and Software 20(6), 715-728.

Mittlböck, M., and Schemper, M., 1996. Explained variation for logistic regression. Statistics in Medicine, 15, 1987-1997.

Mohsen Hosseinalizadeh, M., Kariminejad, N., Rahmati, O., Keesstra, S., Alinejad, M., Behbahani, A.M., 2019. How can statistical and artificial intelligence approaches predict piping erosion susceptibility? Science of the Total Environment 646, 1554–1566.

Moore, I. D., Wilson, J. P., 1991. Length-slope factors for the revised universal soil loss equation: simplified method of estimation. Journal of Soil and Water Conservation 47(5), 423–428.

Moore, I.D., Burch, G.J., 1986. Physical basis of the length-slope factor in the Universal Soil

Morgan, R.P., 2009. Soil Erosion and Conservation. John Wiley & Sons.

Naghibi, S.A., Pourghasemi, H.R., Dixon, B., 2015. GIS-based groundwater potential mapping using boosted regression tree, classification and regression tree, and random forest machine learning models in Iran. Environmental Monitoring and Assessment 188 (1).

Pearson, K., 1897. Mathematical contributions to the theory of evolution. On a form of spurious correlation which may arise when indices are used in the measurement of organs. Proceeding of the royal society of London 60, 489-498.

Poesen, J., 1993. Gully typology and gully control measures in the European loess belt. In: prevention by vegetation in Mediterranean Europe: current trends of ecosystem service provision. Ecol. Indic. 60, 213–222.

Pourghasemi, H. R, Pradhan, B., Gokceoglu, C., Moezzi, K.D., 2013b. A comparative assessment of prediction capabilities of Dempster–Shafer and Weights-of-evidence models in landslide susceptibility mapping using GIS. Geomatics Natural Hazards and Risk 4 (2), 93-118.

Pourghasemi, H.R., Jirandeh, A.G., Pradhan, B., Xu, C., Gokceoglu, C., 2013a. Landslide susceptibility mapping using support vector machine and GIS at the Golestan Province, Iran. Journal of Earth System Science 122 (2), 349-369.

Pourghasemi, H.R., Mohammady, M., Pradhan, B., 2012. Landslide susceptibility mapping using index of entropy and conditional probability models in GIS: Safarood Basin, Iran. Catena 97, 71-84.

Pourghasemi, H.R., Yousefi, S., Kornejady, A., Cerdà, A., 2017. Performance assessment of individual and ensemble data-mining techniques for gully erosion modeling. Science of The Total Environment 609, 764-775.

Rahmati, O., Haghizadeh, A., Pourghasemi, H.R., Noormohamadi, F., 2016. Gully erosion susceptibility mapping: the role of GIS-based bivariate statistical models and their comparison. Nat. Hazards 82 (2), 1231–1258.

Rahmati, O., Tahmasebipour, N., Haghizadeh, A., Pourghasemi, H.R., Feizizadeh, B., 2017. Evaluating the influence of geo-environmental factors on gully erosion in a semi-arid region of Iran: an integrated framework. Science of The Total Environment 579, 913-927.

Rossi, M., Guzzetti, F., Reichenbach, P., Cesare Mondini, A., Peruccacci, S., 2010. Optimal landslide susceptibility zonation based on multiple forecasts. Geomorphology 114, 129–142.

Rossi, M., Reichenbach, P., 2016. LAND-SE: a software for statistically based landslide susceptibility zonation, version 1.0. Geosci. Model Dev. 9, 3533–3543.

Shit, P.K., Nandi, A.S., Bhunia, G.S., 2015. Soil erosion risk mapping using RUSLE model on jhargram sub-division at West Bengal in India. Model. Earth Syst. Environ. 1: 28. https://doi.org/10.1007/s40808-015-0032-3

Svoray, T., Michailov, E., Cohen, A., Rokah, L., Sturm, A., 2012. Predicting gully initiation: comparing data mining techniques, analytical hierarchy processes and the topographic threshold. Earth Surf. Process. Landforms 37, 607-619.

Tehrany, M.S., Pradhan, B., Mansor, S., Ahmad, N., 2015. Flood susceptibility assessment using GIS-based support vector machine model with different kernel types. Catena 125, 91-101.

Tien Bui, D., Tuan, T.A., Klempe, H., Pradhan, B., Revhaug, I., 2016. Spatial prediction models for shallow landslide hazards: a comparative assessment of the efficacy of support vector machines, artificial neural networks, kernel logistic regression, and logistic model tree. Landslides 13, 361–378.

Valentin, C., Poesen, J., Li, Y., 2005. Gully erosion: Impacts, factors and control. Catena 63 (2-3), 132-153.

Verachtert, E., Devoldere, S., Van Den Eeckhaut, M., Poesen, J., Deckers, J., 2011. Impact of land use and soil properties on piping in Belgium. Landform Analysis. 17, pp. 215–218.

Verachtert, E., Van Den Eeckhaut, M., Martínez-Murillo, J.F., Nadal-Romero, E., Poesen, J., Devoldere, S., Wijnants, N., Deckers, J., 2013. Impact of soil characteristics and land use on pipe erosion in a temperate humid climate: field studies in Belgium. Geomorphology 192, 1–14.

Wang, L., Wei, S., Horton, R., Shao, M., 2011. Effects of vegetation and slope aspect on water budget in the hill and gully region of the Loess Plateau of China, Catena 87 (1), 90-100.

Wilson, G.V., Rigby, J.R., Dabney, S.M., 2015. Soil pipe collapses in a loess pasture of Goodwin Creek watershed, Mississippi: role of soil properties and past land use. Earth Surf. Process. Landf. 40 (11), 1448–1463.

Yesilnacar E, Topal T. 2005. Landslide susceptibility mapping: a comparison of logistic regression and neural networks methods in a medium scale study, Hendek region (Turkey). Eng. Geol. 79: 251–266.

Zabihi, M., Mirchooli, F., Motevalli, A., Darvishan, A.K., Pourghasemi, H.R., Zakeri, M.A., Sadighi, F., 2018a. Spatial modelling of gully erosion in Mazandaran Province, northern Iran. Catena 161, 1-13.

Zabihi, Z., Pourghasemi, H.R., Motevalli, A., Zakeri, M.A., 2018b. Gully erosion modeling using GIS-based data mining techniques in northern Iran: a comparison between boosted regression tree and multivariate adaptive regression spline. Natural Hazards GIS-Based Spatial Modeling Using Data Mining Techniques, 1-26.

Hamid Reza Pourghasemi is an Associate Professor of Watershed Management Engineering in the College of Agriculture, Shiraz University, Iran. He has a BSc in Watershed Management Engineering of the University of Gorgan (2004), Iran, an MSc in Watershed Management Engineering, from Tarbiat Modares University, Iran (2008), and a PhD in Watershed Management Engineering from the same University (Feb 2014). His main research interests are GIS-based spatial modelling using machine learning/data mining techniques in different fields such landslide, flood, gully erosion, forest fire, land subsidence, species distribution modelling, and groundwater/hydrology. Also, Hamid Reza works on Multi-Criteria Decision Making methods in Natural Resources and Environment.

He has published more of 90 peer reviewed papers in high-quality journals, with three chapters in Springer. Also, he published two books in Springer (https://www.springer.com/gp/book/9783319733821) and Elsevier (https://www.elsevier.com/books/spatial-modeling-in-gis-and-r-for-earth-and-environmental-science/pourghasemi/978-0-12-815226-3).

Amiya Gayen is a Ph.D. student, Department of Geography, University of Calcutta, Kolkata, India. He obtained his M.Sc. in Geography from Presidency University, Kolkata. His research field is soil geography, natural hazard, forest health assessment, and environmental geography including GIS and R techniques. Simultaneously, he is working as an Assistant Professor under RUSA in the Department of Geography, Midnapore College (Autonomous), Midnapore, West Bengal.

Sk. Mafizul Haque was awarded M.A. in Geography and Environment Management from Vidyasagar University, Midnapore in the year of 2004. He was an Assistant Professor in the Department of Geography, Aliah University, Kolkata from 2012 to 2015. After awarded the Ph.D. degree in 2014 from the University of Calcutta, he joined in the Department of Geography, University of Calcutta as an Assistant Professor in 2015. His area of inclination is Change Science—mainly the application of geo-spatial technology on urban landscape, micro-climate, habitat analysis, urban environment and palaeo-climate study. Dr. Haque has published more than 17 research papers in the reputed national and international journals. He was engaged in the training programme of SIPRD, Kalyani, Govt. of West Bengal during March to June, 2015. Currently he is coordinator of IIRS outreach programme at Department of Geography, University of Calcutta.

Shibiao Bai is a Professor in Nanjing Normal University. His current research interests: (1) Giant Landslide processes using OSL and TCN dating; (2) Landslide hazard and risk mapping. The research Projects: (1) 2015 Jiangsu provincial key R&D Program (Social Development) "The key Tech. research for the flood and geohards in city" (BE2015704). 2015-2018. (2) The European Space Agency (ESA) and Ministry of Science and Technology of China (MOST) Dragon 4 Cooperation Program "Spatio-temporal landslide identification and activity assessment for hazard and risk investigations in Longnan region, Northwest China" (ID: 32365_3). 2016–2020. (3) The Opening Fund of Key Laboratory of Mountain Surface Process and Hazards of Chinese Academy of Sciences (KLMHESP-17-07). 2017–2019.

29

Narges Javidan, Ataollah Kavian, Hamid Reza Pourghasemi, Christian Conoscenti, and Zeinab Jafarian

Abstract

Soil erosion is a serious problem affecting most of the countries. This study was carried out in Gorganrood Watershed (Iran), which extends for 10,197 km^2 and is severely affected by gully erosion. A gully headcut inventory map consisting of 307 gully headcut points was provided by Google Earth images, field surveys, and national reports. Gully conditioning factors including significant geo-environmental and morphometric variables were selected as predictors. Maximum entropy (ME) model was exploited to model gully susceptibility, whereas the area under the ROC curve (AUC) and drawing receiver operating characteristic (ROC) curves were employed to evaluate the performance of the model.

The highly acceptable predictive skill of the ME model confirms the reliability of the procedure adopted to using this model in other gully erosion studies, as they are qualified to rapidly producing accurate and robust GESMs (gully erosion susceptibility maps) for making decisions and management of soil and water. The result is useful for local administrators to recognize the areas that are most susceptible to gully erosion and to best allocate resources for soil conservation approaches.

Three different sample datasets including 70% for training and 30% for validation were randomly prepared to evaluate the robustness of the model for gully erosion. The accuracy of the predictive model was evaluated by drawing ROC curves and by calculating the area under the ROC curve (AUC). The ME model performed excellently both in the degree of fitting and in predictive performance (AUC values well above 0.8), which resulted in accurate predictions.

Keywords

Gully erosion · Susceptibility · Geographic information systems (GIS) · Maximum entropy (ME) model · Area under the ROC curve (AUC)

29.1 Introduction

Soil erosion by water constitutes a serious land degradation event affecting around one billion hectares area in the world (Lal 2003), causing the decrease of vegetation growth, filling of reservoirs and valleys, geo-environmental degradation, the loss of a great amount of soil, and siltation of water courses (Kosmas et al. 1997; Vandekerckhove et al. 2000; Vanwalleghem et al. 2005). Gully erosion is one of the erosive processes and the most intricate erosion phenomena (Chaplot et al. 2005), most often triggered or intensed by a combination of extreme rainstorms and inappropriate land use (Chaplot et al. 2005). Gully erosion contains an extensive range of subprocesses, such as headcut, fluting, piping, tension crack progress, and mass wasting (Imeson and Kwaad 1980; Angileri et al. 2016). Generally, the increasing interest in analyzing gully erosion reflects the need to enhance our knowledge on its impacts and condition factors that alter under a wide range of causes (Chaplot et al. 2005). Gullies involve intricate processes controlled by a variability of closely related variables such as soil texture, lithology, land use and vegetation cover, climate, and topography (Conforti et al. 2011). As gully erosion is threshold phenomena (Angileri et al. 2016), several studies have focused on defining the topographic and hydraulic conditions for predicting and assessing the initiation of gully susceptibility mapping (Govers 1985); to a threshold approach, features and

N. Javidan (✉) · A. Kavian · Z. Jafarian
Sari Agricultural Sciences and Natural Resources University, Sari, Iran

H. R. Pourghasemi
Department of Natural Resources and Environmental Engineering, College of Agriculture, Shiraz University, Shiraz, Iran

C. Conoscenti
Physical Geography and Geomorphology, University of Palermo, Palermo, Italy

locations of erosion processes may also be anticipated by using bivariate to multivariate statistical methods. These techniques allow a scholar to explain the occurrence of soil erosion processes, by crossing the spatial distribution of the gully erosion landforms (the outcome) with that of a set of the predictors (geo-environmental variables).

In the field of geomorphology, statistical and data mining methods have been widely adopted to evaluate and assess landslide susceptibility mapping (e.g., Ballabio and Sterlacchini 2012; Lombardo et al. 2015; Conoscenti et al. 2015). An increasing number of studies have also applied a stochastic approach for zoning erosion susceptibility. In addition to bivariate analyses (Conoscenti et al. 2013; Magliulo 2010, 2012), different multivariate statistical approaches have also been employed to this aim, such as logistic regression (Lucà et al. 2011; Conoscenti et al. 2014), classification and regression trees (Geissen et al. 2007; Märker et al. 2011), frequency ratio (Conforti et al. 2011; Lucà et al. 2011), weights of evidence (Dube et al. 2014; Tahmassebipoor et al. 2016), the analytical hierarchy process (Zakerinejad and Märker 2014), and multivariate adaptive regression splines (Gómez Gutiérrez et al. 2009, 2015; Conforti et al. 2011).

Angileri et al. (2016) created an integrated water erosion susceptibility map by heuristically combining the resulting rill-interrill erosion and gully erosion susceptibility maps in central-northern Sicily (Italy). In this study, the Stochastic Gradient Tree boost (SGT) was tested as a multivariate statistical tool. The results of validation, based on ROC curves, showed excellent to outstanding accuracies of the models (Conoscenti et al. 2014, 2016) and a high prediction skill. This result can be useful for local administrators to recognize the areas most susceptible to water erosion and best allocate resources for soil conservation strategies.

Svoray et al. (2012) used different machine learning models, such as SVM, decision tree (DT), SV, and ANN, for predicting gully initiation, and then, they compared their results with the AHP (analytic hierarchy process) and TT (topographic threshold) methods. The results of this study indicated that machine learning models provide a better predictive ability of gully susceptibility mapping than the use of both TT and AHP methods. Rahmati et al. (2018) investigated the capability and robustness of a novel hybrid model, namely, the logistic model tree (LMT), and compared it with state-of-the-art models such as the support vector machine and C4.5 models for groundwater spring potential. Three different sample datasets (S1, S2, and S3) were randomly prepared. Results showed that the LMT model had the highest accuracy performance for all three validation datasets although a small sensitivity to change in input data was occasionally observed for this model. Also Rahmati et al. (2016) evaluated different machine learning models (SVM with four kernel types, BP-ANN, RF, and BRT) for predicting the susceptibility of gully erosion. Three different sample datasets (S1, S2, and S3) were randomly prepared to evaluate the robustness of the models. In terms of accuracy, the RF, RBF-SVM, BRT, and P-SVM models performed excellently both in the degree of fitting and in predictive performance (AUC values well above 0.9), which resulted in accurate predictions. Additionally, it was found that the performance of RF and RBF-SVM for modeling gully erosion occurrence is quite stable when the learning and validation samples are changed. A review of the research indicated that only a few studies have employed maximum entropy (ME) model for assessing gully erosion susceptibility, and according to research background, there is no comprehensive study to compare and evaluate the capability of ME model for assessing gullying erosion susceptibility. To address the research gaps, in the present study, we adopted maximum entropy (ME) model as a data mining technique for analyzing, exploring, and predicting the spatial occurrence of gully erosion processes. ME is a very popular method that has been already demonstrated to provide credible models of landslide susceptibility (Kornejady et al. 2017a). This model was selected for the spatial prediction of gully erosion because of the following reasons: (1) It can model the nonlinear relationship between the conditioning factors and gully occurrence, (2) it can work using different types of independent variables and can handle data from various measurement scales, and (3) according to previous studies in this area that any studies have applied this model for evaluating the ability and robustness of it for susceptibility mapping.

Gorganrood has witnessed gully erosion which has made a big cause for concern in the Gorganrood Watershed and gave the organizations a pause to rethink the weirdness of their constant development strategies (CONRWMGP 2009). Astonishingly, cutting down trees located at the upstream of Gorganrood basin and land-use changes were reported as the main starting points to the erosion phenomenon (CONRWMGP 2009). From 1990 to 2005 due to the presence of loess soils in the northern Golestan Province, 430,000 ha of these areas are affected by erosion. Soil erosion in this province is 6–5 tons/hectare/year in forest areas (CONRWMGP 2009). Gully erosion in Marave Tape and Kalale leads to the loss of soil, the imposition of plenty of costs, and reduced agricultural potential and has caused the migration of people in the villages of this region, and intense soil erosion processes affect the productivity of agricultural systems. Consequently, it is of prominent importance to define a valid model to assess and evaluate the susceptibility of the territory to the development of gully processes, and Gorganrood Watershed as an area highly prone to gully erosion will be the focus of this research. The main scope of current study is exploring the ability and robustness of ME model to predict the spatial occurrence of gully erosion using different sample datasets and evaluation criteria.

29.2 Materials and Methods

29.2.1 Study Area

The Gorganrood Watershed is located in the Golestan Province which is situated in the northeastern part of Iran and covers an area which extends for 10,197 km² and is severely affected by gully erosion. The study area lies between the latitudes of 36° 34′ to 38° 15′ N and the longitudes of 54° 5′ to 56° 8′ E (Fig. 29.1). Topographically, this region has mountainous area and flat land. The central and western parts are generally characterized as plain and flat areas. Average elevation of the study area is between −95 and 3652 m. The mean annual rainfall is approximately between 231 and 848 mm. The southern section has a typical mountain climate, and the central and northern regions have a temperate Mediterranean climate. The average minimum and maximum temperature is 11 and 18.5 °C, respectively. In the last decade, this area has challenged with natural hazards and has faced with much intensive gully erosion, so this study area was selected as potential gully erosion-prone area in Golestan province. Figure 29.2 presents some photographs of gullies.

29.2.2 Methodology

Figure 29.3 illustrates the methodological flowchart of the approach that was used for gully erosion susceptibility mapping analysis using maximum entropy model. As shown, the flowchart consists of four steps: (1) preparing thematic layers or effective conditioning factor, (2) determination of effective factors step, (3) gully erosion susceptibility modeling using the ME machine learning techniques, and (4) validation of the susceptibility maps using the ROC-AUC curve.

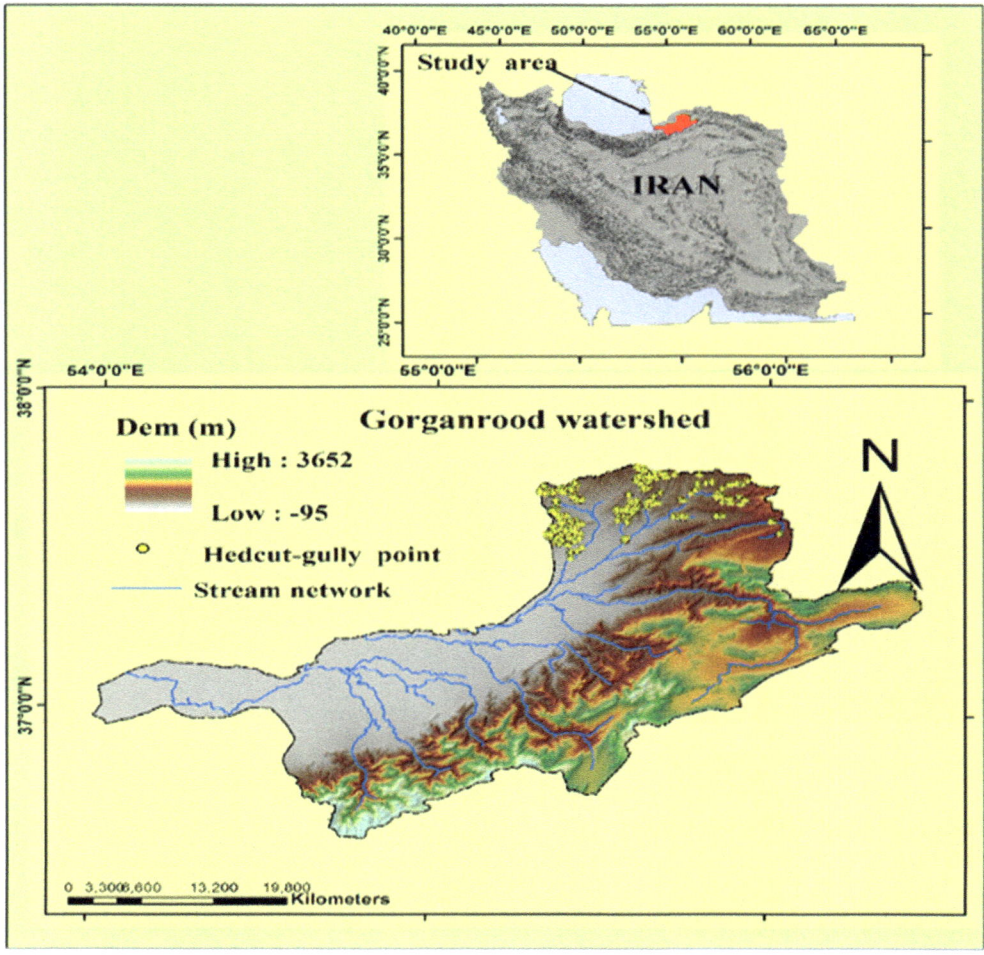

Fig. 29.1 Location map of the study area with the elevations of the study area

Fig. 29.2 Photographs of gully erosions in Golestan province

29.2.3 Gully Headcut Inventory Mapping Preparation

A key step for susceptibility mapping is the preparation of an inventory of hazard landforms (Conoscenti et al. 2014). The Gully headcut inventory for Gorganrood Watershed was a compiled field investigation. Considering that some hazard points located in the field investigation may be missed, "Google Earth" was used for gully interpretation. The inventory map for gully erosion is a collection of occurrences (307 headcut points). In this study, gully point inventory map was compiled using documentary sources in

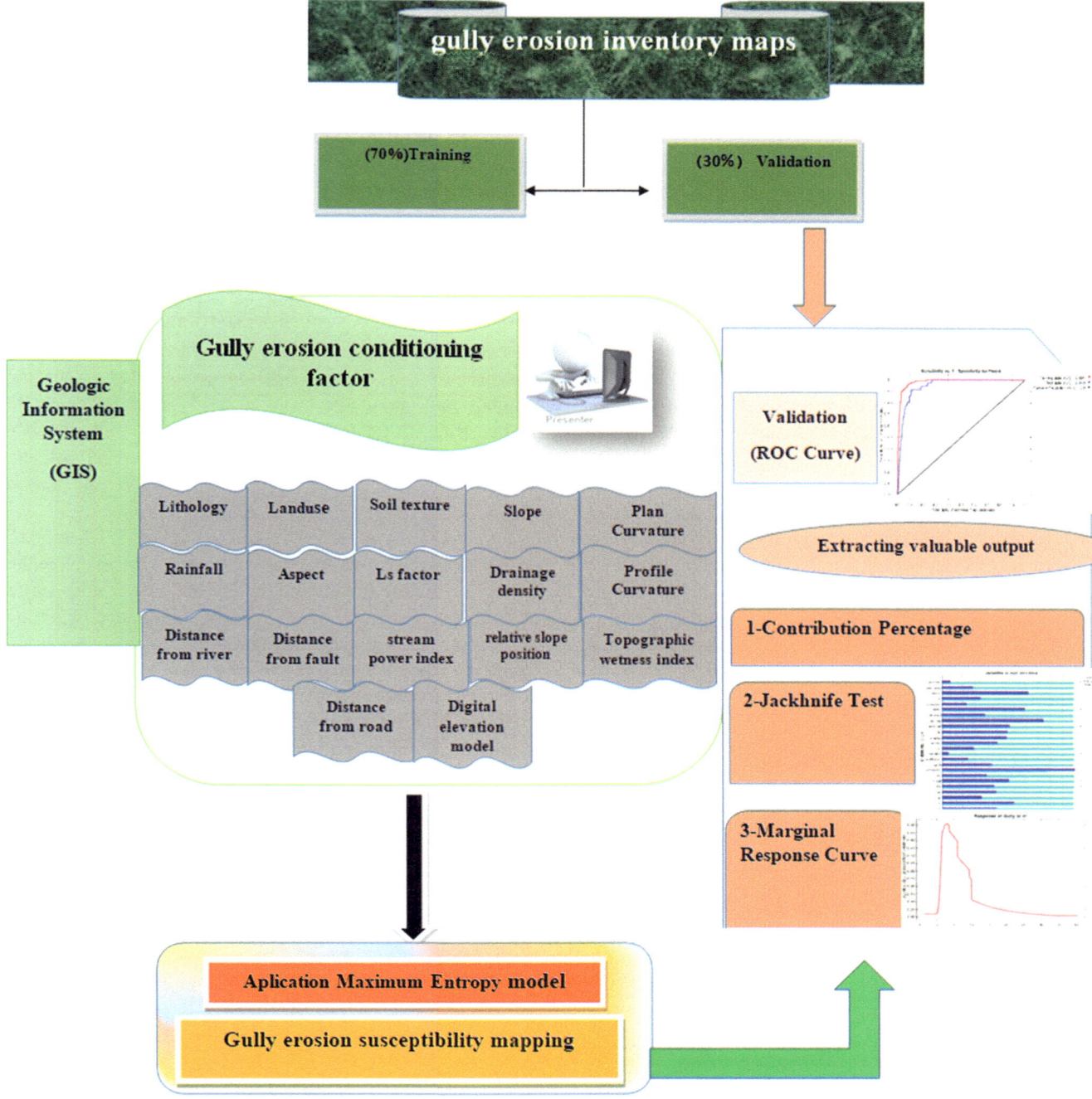

Fig. 29.3 Flowchart of methodology used for gully erosion susceptibility map spatial modeling in the Gorganrood Watershed, Golestan Province, Iran

Golestan Natural Resources and Watershed Administration and extensive field investigation, i.e., GPS points. When developing the machine learning model, the training data should be different from those used in the validation part (Lee et al. 2007). A randomly partition algorithm was used to separate training points from the validation points (Pradhan 2010; Pourtaghi and Pourghasemi 2014; Rahmati et al. 2016). From the total point, 70% were used in the model building (training), and the remaining 30% were used for the validation. To assess the robustness of the model data sensitivity (Conoscenti et al. 2014; Angileri et al. 2016; Cama et al. 2017), three sample datasets, S1, S2, and S3, were prepared. The sample datasets (S1) were shown in Fig. 29.1.

29.2.4 Gully Erosion Conditioning Factors

It is essential to determine the effective factors on different natural hazards and man-made fatalities in order to perform gully susceptibility maps (Kia et al. 2012). A good understanding of the main hazard-related factors is required to recognize the susceptible areas. Factors are usually used in the scientific literature to analyze the hazards. Therefore, the conditioning factors were selected from the literature review. In this study, ArcGIS 10.5 and System for Automated Geoscientific Analyses (SAGA) software were used to produce and display these data layers. For application of statistical and data mining models, all the mentioned factors were converted to a raster grid with 30 m × 30 m grid cells. All the conditioning factors were primarily continuous, and some of them were classified into different categories based on expert knowledge and literature review (Jaafari et al. 2014; Youssef 2015; Saponaro et al. 2015).

The predicting factors used in this work for gully erosion are (**a**) digital elevation model (m), (**b**) slope percent, (**c**) slope aspect, (**d**) land use, (**e**) profile curvature, (**f**) plane curvature, (**g**) TWI, (**h**) LS factor, (**i**) soil texture, (**j**) distance to fault (m), (**k**) distance to roads (m), (**l**) distance to streams (m), (**m**) drainage density, (**n**) annual rainfall (mm), (**o**) stream power index (SPI), (**p**) relative slope position (RSP), and (**q**) lithological formation.

In this study, most of the conditioning factors for these areas were selected from the literature review (Pradhan 2010; Lee et al. 2012). Then, the SPSS software was used to analyze data. These factors were used in consideration of the effect of correlation among independent variables.

When two independent variables are highly correlated, it is a problem. The problem is named multicollinearity. The VIF (variance inflation factor) and tolerance are two significant indexes for multicollinearity diagnosis. In fact, VIF is simply the reciprocal of tolerance; on the other hand, tolerance is $1 - R^2$ for the regression of that variable against all the other independents, deprived of the dependent variable. VIF measures the degree to which the interrelatedness of the variable with other conditioning factors inflates the variance of the estimated regression coefficient for the predictor. Consequently, the square root of the VIF is the degree to which the collinearity has increased the standard error for that predictor variable. A VIF of 5 or 10 and above and/or a tolerance of less than 0.20 or 0.10 indicates a multicollinearity problem (O'Brien 2007; Ozdemir 2011). According to this point among the variables used in current research, Terrain Ruggedness Index (TRI) with VIF >5 and tolerance <0.1 was eliminated. Table 29.1 shows there is not any multicollinearity between independent factors in present study.

Further, previous studies (Archer et al. 1997; Chen et al. 2010) have used SA as an exploratory technique to define the effect of variable variations on model outputs, allowing a quantitative evaluation of the relative importance of uncertainty sources. So, in this study, the contribution of each input factor to the uncertainty of the model outputs was investigated using SA (Convertino et al. 2014). In this study, to evaluate the uncertainty of predicted maps using the SA, a jackknife test was conducted for investigating the effects of removing any of the conditioning factors on the susceptibility map (Yost et al. 2008). The jackknife test can

Table 29.1 The multicollinearity diagnosis indexes for variables

Model		Unstandardized coefficients		Standardized coefficients			Collinearity statistics	
		B	Std. Error	Beta	t	Sig.	Tolerance	VIF
1	(Constant)	−0.773	0.445		−1.736	0.085		
	River density	3.458	0.289	0.944	11.978	0.000	0.386	2.589
	RSP	0.220	0.114	0.113	1.925	0.056	0.694	1.442
	Slope	−0.001	0.003	−0.053	−0.498	0.619	0.210	4.752
	Soil	−0.008	0.021	−0.021	−0.379	0.705	0.788	1.269
	SPI	2.104E-5	0.000	0.013	0.232	0.817	0.734	1.362
	TWI	0.006	0.012	0.034	0.561	0.575	0.656	1.525
	DEM	2.993E-5	0.000	0.034	0.474	0.636	0.454	2.205
	Distance from river	4.669E-5	0.000	0.232	3.105	0.002	0.428	2.335
	Land use	0.022	0.012	0.091	1.758	0.081	0.902	1.109
	Litho	0.001	0.003	0.022	0.380	0.704	0.687	1.455
	Melton	0.010	0.023	0.041	0.453	0.651	0.294	3.406
	Plan curvature	0.088	0.142	0.038	0.617	0.538	0.634	1.576
	Profile curvature	0.052	0.099	0.032	0.528	0.598	0.658	1.519
	Rain	0.001	0.001	0.068	1.021	0.309	0.547	1.828
	Aspect	−0.018	0.011	−0.085	−1.668	0.097	0.917	1.091

[a]Dependent Variable: Occurrence
Coefficients[a]

be considered to evaluate the relative strengths of each predictor variable (Phillips et al. 2006; Yost et al. 2008; Park 2015) According to the jackknife test results, variables with zero importance including SPI, LS factor, aspect, and profile curvature factor were eliminated; subsequently, the model was run with the remaining variables for all three datasets.

DEM of the study area with a 30 m pixel size was produced using digital contour data prepared from the Department of Natural Resources Management of Iran and was considered as a continuous factor. From this DEM, physiographical and geomorphological layers such as the slope aspect, slope percent, and curvature layers were extracted using ArcGIS 10.2 software. The once essential morphometric parameter that is important on gully erosion is slope percent. The slope percent comprises a great part of the natural landscape and is an important factor as it affects drainage density, surface runoff, influences, vegetation structure of the soil erosion, soil moisture, and geomorphological processes (Nagarajan et al. 2000; Gallardo-Cruz et al. 2009; Conforti et al. 2011; Geroy et al. 2011; Lucà et al. 2011; Kornejady et al. 2017a, b). Slope aspect is another important factor in preparing susceptibility maps and related to such factors as precipitation, snow meltwater, land cover, soil moisture patterns, and physiographic trends and therefore can be impressive on hydrologic conditions (Ercanoglu and Gokceoglu 2002; Sidle and Ochiai 2006; Yalcin 2008; Vahidnia et al. 2010; Poiraud 2014; Meinhardt et al. 2015). This GCF was classified into nine classes of main and submain categories in addition to a flat class. The suitable geomorphological information can be extracted through the analysis of curvature (Shafapour Tehrany et al. 2014a,b; Khosravi et al. 2016; Moghaddam et al. 2015). The slope curvature map was compiled with three categories: convex, concave, and flat. Positive curvature exhibits convex (> +0.1), negative curvature depicts concave (<−0.1), and zero curvature represents flat ((−0.1) – (+0.1)). Also profile and plan curvatures have a range of negative and positive values and reflect a different description in each of these indexes. Negative and positive values in profile curvature represent convexity (increasing flow velocity) and concavity (reducing flow velocity), respectively. In contrast, negative and positive values in the plan curvature represent concavity (flow convergence) and the convexity (flow divergence), respectively (Jenness 2013; Kornejady et al. 2017b). Values close to zero represent neutral curvature in both cases. Using tools extension in ArcGIS 10.5, the curvature map was produced and classified.

Land use plays a significant role in the operation of geomorphological and hydrological processes by directly or indirectly influencing on evapotranspiration, infiltration, runoff generation, and sediment dynamics (Maestre and Cortina 2002; Meinhardt et al. 2015). Settlement areas, which are typically made by impervious surfaces, increase the inundation (Shafapour Tehrany et al. 2013). On the other hand, vegetated and forest areas are less prone to erosion due to the positive relationship between vegetation density and infiltration capability. Also agricultural activities have an important impact on gully erosion development and nitiation (Zucca et al. 2006). The land-use map of the study area was prepared from the Natural Resources Office of Golestan Province and modified by Google Earth images. The land use of the study area consists of lake, residential area, forest, range, drying farming, irrigation farming, rock, and saline land.

Soil texture is commonly identified as a significant controlling mechanism of infiltration, runoff generation, and, consequently, the inundation and hazard occurrence (Cosby et al. 1984; Gyssels et al. 2002; Vandekerckhove et al. 2003). The water infiltration primarily depends upon soil texture. This layer was prepared by digitizing the soil texture map of Golestan Province (1:100,000 scale) obtained from the Agriculture Department, Iran. The soil texture in the study area consists of sandy loam, clay loam, sandy clay loam, silty clay, silty clay loam, and silty loam. This is the only soil map available for the study area.

Moore et al. (1991) and Grabs et al. (2009) mentioned topographic wetness index (TWI) that presents the spatial distribution of wetness conditions and the tendency of gravitational forces to move water downslope, which is defined according to the following equation:

$$\text{TWI} = \ln\left(\frac{\propto}{\tan\beta}\right) \qquad (29.1)$$

where α is the cumulative upslope area draining through a point (per unit contour length) and $\tan\beta$ is the slope angle at the point. It ranges from 1.20 to 22.92. In this study, TWI map was prepared in GIS 10.5 software.

Distance from river is one of the key conditioning factors due to its important impact (Glenn et al. 2012; Fernández and Lutz 2010). There is no doubt that road construction played a strong negative impact on slope stability where water flow concentrates may be suitable for hosting gullies (Conoscenti et al. 2014; Jungerius et al. 2002).

Layers of the proximity were produced using Euclidean distance function in ArcGIS 10.5 software and varying from 0 to 11,720 m from roads, 0 to 15,080 m from streams, and 0 to 55,212 m from faults. The roads and rivers were extracted from the national topographic map at the scale of 1:50,000.

The drainage density is also one of the main conditioning factors that strongly contribute in many hazard occurrences (Pourghasemi et al. 2013). According to Shafapour Tehrany et al. (2014b), a high drainage density causes larger surface runoff ratio. Drainage pattern of an area is affected by several factors such as the structure and nature of the soil characteristics, geological formation, infiltration rate, vegetation cover condition, and slope degree (Pourtaghi and Pourghasemi 2014). In order to convert the drainage network pattern to measurable quantity, the drainage density was determined using Line Density tools in ArcGIS software.

Rainfall-triggered landslides have brought great damages to communication substructures, properties, and pasture biomass production (Shimizu 1988; Lan et al. 2004; Duc 2012). The annual mean rainfall map of Gorganrood Watershed was prepared based on the rainfall data extracted from the Regional Water Organization of Golestan Province. This map was created using fifty-three stations and statistical period of 2009–2016 based on inverse distance weight (IDW) interpolation method (Bui et al. 2012) (Eq. 29.2). This map ranges from 384 to 810 mm/year. The precipitation map is prepared in a raster format of 30×30 m in ArcGIS 10.5 as an input layer for susceptibility assessment.

$$\lambda_i = \frac{D_i^{-\alpha}}{\sum_{i=1}^{n} D_i^{-\alpha}} \qquad (29.2)$$

where λ_i is weight of the point i, D_i is the distance between the point i and the point of unknown, and α is equal to the weighing power (Bui et al. 2012).

Relative slope position is the tool one calculates several terrain indices from digital elevation model. General information on the computational concept can be found in Böhner and Selige (2006). Relative slope position expresses the morphological characteristics of a cell: slope position or distance from the river and morphological setting (Lombardo et al. 2016).

Lithology indicators play a dominant role in determining gully erosion susceptibility (Yalcin 2008; Conforti et al. 2011; Song et al. 2012; Zhu et al. 2014; Meinhardt et al. 2015) because gully erosion is particularly dependent on the lithology properties and different lithological units display significant differences in landslide instability. Also lithology is considered as an essential factor to the temporal and spatial variations of drainage basin hydrology (Miller et al. 1990). Lithological units have different susceptibilities to active hydrological processes. In our study, the lithological map of the study area was produced according to the available geological maps on a scale of 1:100,000 obtained from the Geological Survey Department, Iran. The Gorganrood Watershed is covered by various types of lithological formations and classified into 24 groups (Table 29.2). Finally, the lithology map was classified into 24 groups.

The corresponding maps for condition factors are presented in Fig. 29.4.

29.2.5 Maximum Entropy Model

Performance of ME model generation gully erosion susceptibility map and prediction rate were done using the Maxent software (version 13.0.6.0) with the default settings. Maxent's default settings are a set of model parameters achieved as a result of a tuning approach using the NCEAS dataset (Elith et al. 2006). Phillips et al. (2006) proposed the maximum entropy (ME) model for predictive modeling of geographical species distribution on the basis of the most important environmental condition, when presence data are available for modeling (Phillips et al. 2004, 2006). Maximum entropy density estimation can also be explained from a decision theoretic perspective as robust Bayes estimation. Maxent is based on a machine learning response that makes predictions from incomplete data (Medley 2010; Moreno et al. 2011). Maxent produces in ASCII format a continuous prediction of specific presence that ranges from 0 to 100 (Boubli and De Lima 2009). The Maxent distribution is calculated over the set of pixels representing the study area that have data for all environmental conditions. However, if the number of pixels is very large, processing time increases without a significant development in modeling performance. For that reason, when the number of pixels with data is larger than 10,000, a random sample of 10,000 "background" pixels is used to represent the variety of environmental variables present in the data (Yost et al. 2008). The other parameters for model were tuned a convergence threshold of 0.00001 (the stoppage time), the number of replicates of 3 and 500 iterations. Also the model was tested by three replicated run types, i.e., cross-validate, bootstrap, and subsample. The tuned parameters for model resulted in a feature termed plateau effect which implicates that there will not be any important changes in model results, essentially AUC values.

An efficient algorithm (Douaik et al. 2004, 2005) based on the maximum entropy approach is deterministic to probability distribution function (Baldwin 2009; Phillips et al. 2006). Three random partitions of the presence records were made to assess the average behavior of Maxent, following Phillips et al. (2006). Training and validation datasets were prepared in Excel format, and the conditioning factors were converted from raster to ASCII format, which is required in Maxent software. During the model run, 70% of datasets are randomly chosen using random selection algorithm for model training in the calibration phase. This machine learning technique allows for investigation of the relationship between a dependent variable and several independent variables that in

Table 29.2 Lithology of Gorganrood Watershed

Group	Code	Explanation	Formation
1	Ksr	Ammonite-bearing shale with interaction of orbitolina limestone	Sarcheshmeh
2	Pz1a. bv	Andesitic basaltic volcanic	–
	Pz1av	Andesitic volcanic	
3	Jsc	Conglomerate	–
	E1c	Pale red, polygenic conglomerate and sandstone	
	Plc	Polymictic conglomerate and sandstone	
	Murc	Red conglomerate and sandstone	
4	K	Cretaceous rocks in general	–
	pC-C	Late Proterozoic-early Cambrian undifferentiated rocks	
	Kl	Lower Cretaceous undifferentiated rocks	
	Pz	Undifferentiated lower Paleozoic rocks	
	P	Undifferentiated Permian rocks	
	Ku	Undifferentiated Permian rocks	
5	Jch	Dark gray argillaceous limestone and marl	Chamanbid
6	Pr	Dark gray medium-bedded to massive limestone	Ruteh limestone
7	TRJs	Dark gray shale and sandstone	Shemshak
8	Cm	Dark gray to black fossiliferous limestone with subordinate black shale	Mobarak
9	Cl	Dark red medium-grained arkosic to subarkosic sandstone and micaceous siltstone	Lalun
10	PlQc	Fluvial conglomerate, piedmont conglomerate, and sandstone	–
11	Sn	Greenish gray, shale, sandstone, sandylime, coral limestone, and dolomite	Niur
12	Jmz	Gray thick-bedded limestone and dolomite	Mozduran
13	Ksn	Gray to black shale and thin layers of siltstone and sandstone	Sanganeh
14	Murm	Light red to brown marl and gypsiferous marl with sandstone intercalations	Dalichai
	Murmg	Gypsiferous marl	
	E1m	Marl, gypsiferous marl, and limestone	
	Mur	Red marl, gypsiferous marl, sandstone, and conglomerate	
	Kad-ab	Undifferentiated unit including argillaceous limestone, marl, and shale	
	Jd	Well-bedded to thin-bedded, greenish-gray argillaceous limestone with intercalations of calcareous shale	
15	Qft1	High-level piedmont fan and valley terrace deposits	–
	Qft2	Low-level piedmont fan and valley terrace deposits	
	Qal	Stream channel, braided channel, and floodplain deposits	
	Qs,d	Unconsolidated windblown sand deposit including sand dunes	
16	Jl	Light gray, thin-bedded to massive limestone	Lar
17	Dp	Light red to white, thick-bedded quartzarenite with dolomite intercalations and gypsum	Padeha
18	pCmt2	Low-grade, regional metamorphic rocks (greenschist facies)	–
19	Ekh	Olive green shale and sandstone	Khangiran
20	Kat	Olive green glauconitic sandstone and shale	Aitamir
21	Pd	Red sandstone and shale with subordinate sandy limestone	Dorud
22	Jbash	Shale with intercalations of sandstone	Bashcalateh
23	Qsw	Swamp	–
	Qm	Swamp and marsh	
24	TRe2	Thick-bedded dolomite	Elikah
	TRe	Thick-bedded gray oolitic limestone; thin-platy, yellow to pinkish shaly limestone with worm tracks and well- to thick-bedded dolomite and dolomitic limestone	

this study are gully occurrence and conditioning factors, respectively. In the main output of the ME model, each pixel is allocated a presence probability value.

For susceptibility mapping, the model starts with a uniform distribution and performs a number of repetitions based on the most important geo-environmental factors until no further improvements in the spatial prediction are made (Phillips et al. 2004, 2006). Maxent can fit complex relationships to environmental variables through the use of threshold and hinge features and can model interactions between environmental variables (Edrén et al. 2010). The goal of ME model is to estimate a target probability

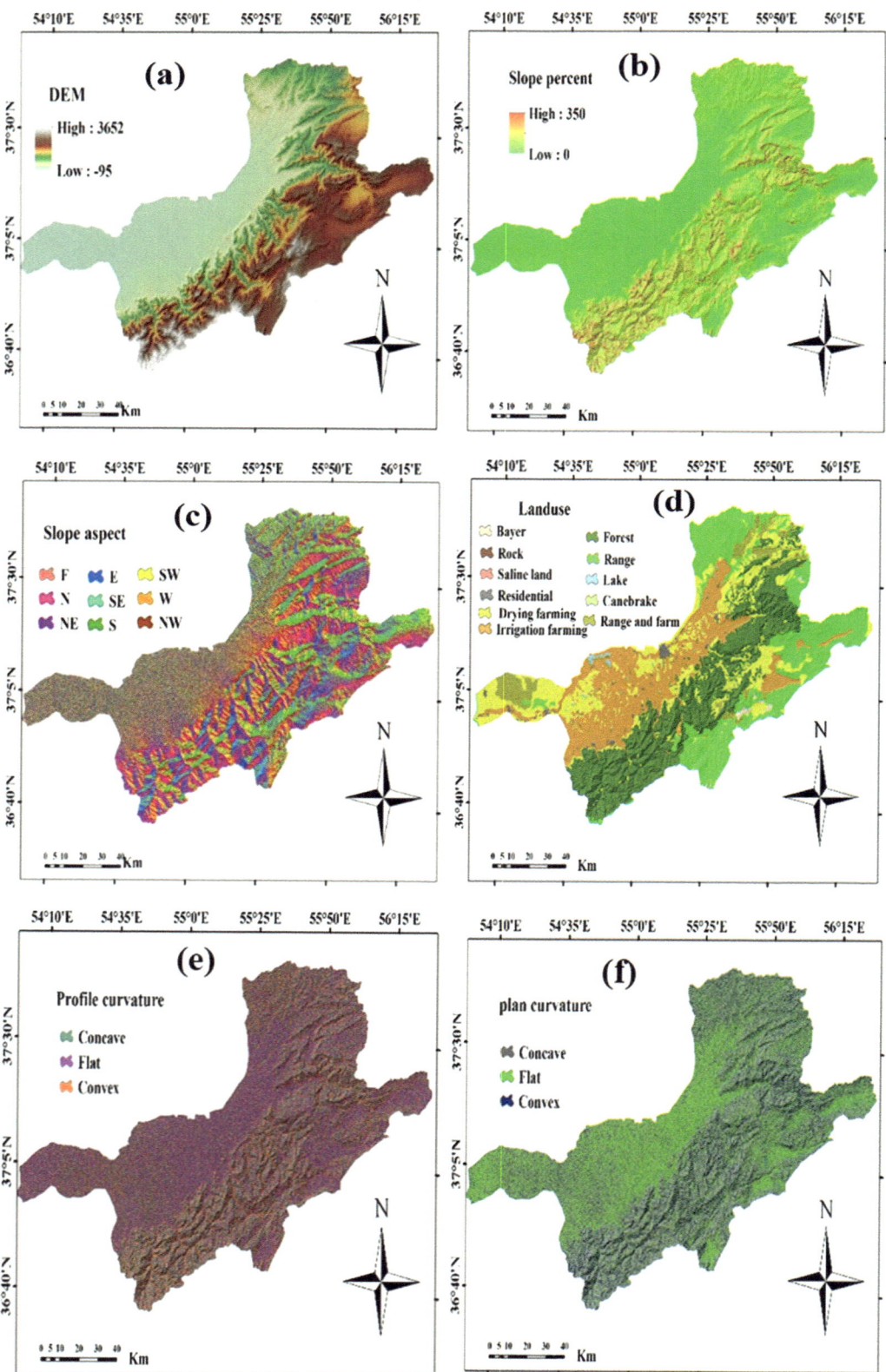

Fig. 29.4 Inset map displaying the flood, gully, and landslide conditioning factors (**a**) DEM (m), (**b**) slope percent, (**c**) slope acpect, (**d**) land use, (**e**) profile curvature, (**f**) plan curvature, (**g**) TWI, (**h**) LS factor, (**i**) soil texture, (**j**) distance to fault (m), (**k**) distance to roads (m), (**l**) distance to streams (m), (**m**) drainage density, (**n**) annual rainfall (mm), (**o**) stream power index, (**p**) relative slope position, (**q**) lithological formation

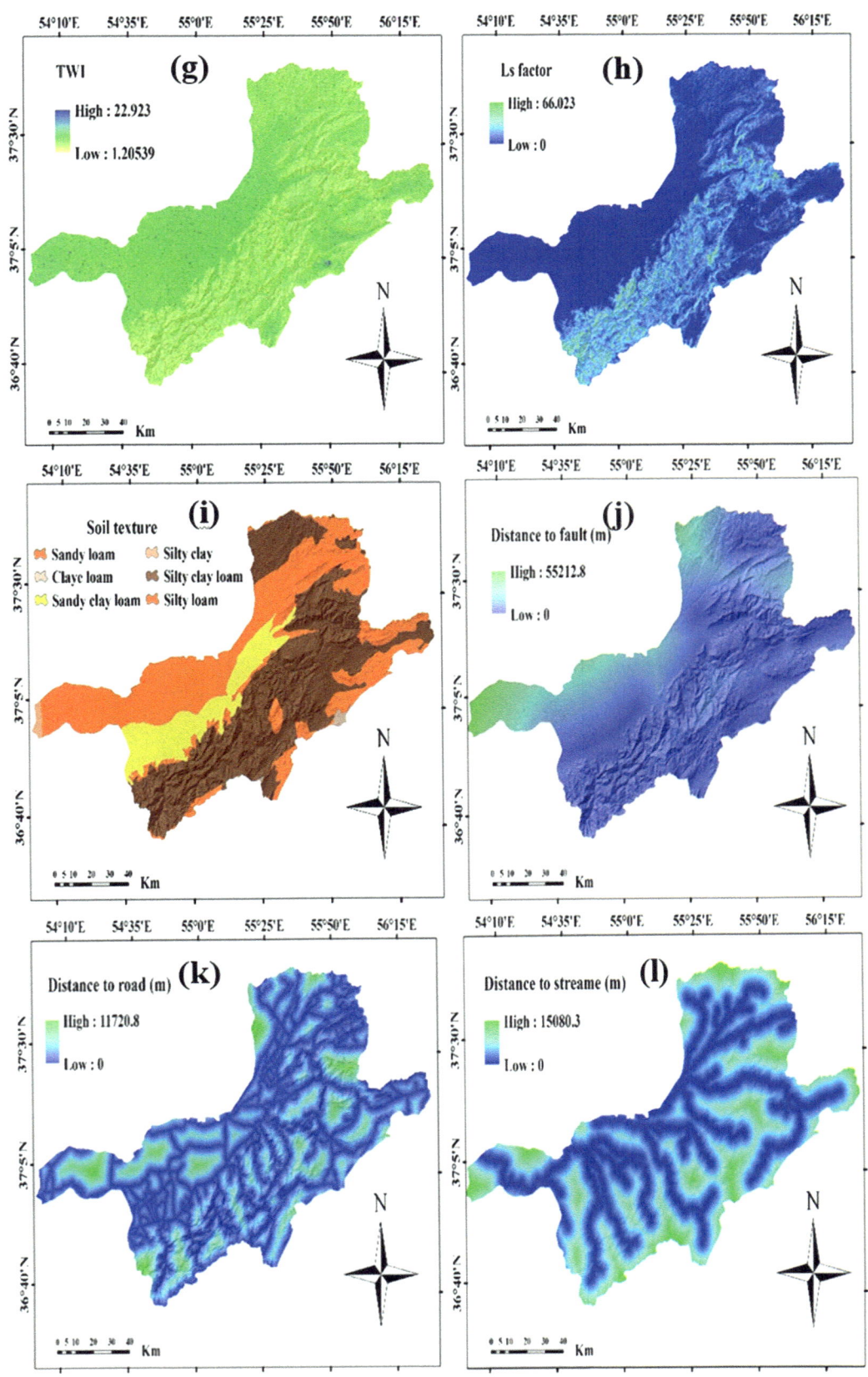

Fig. 29.4 (continued)

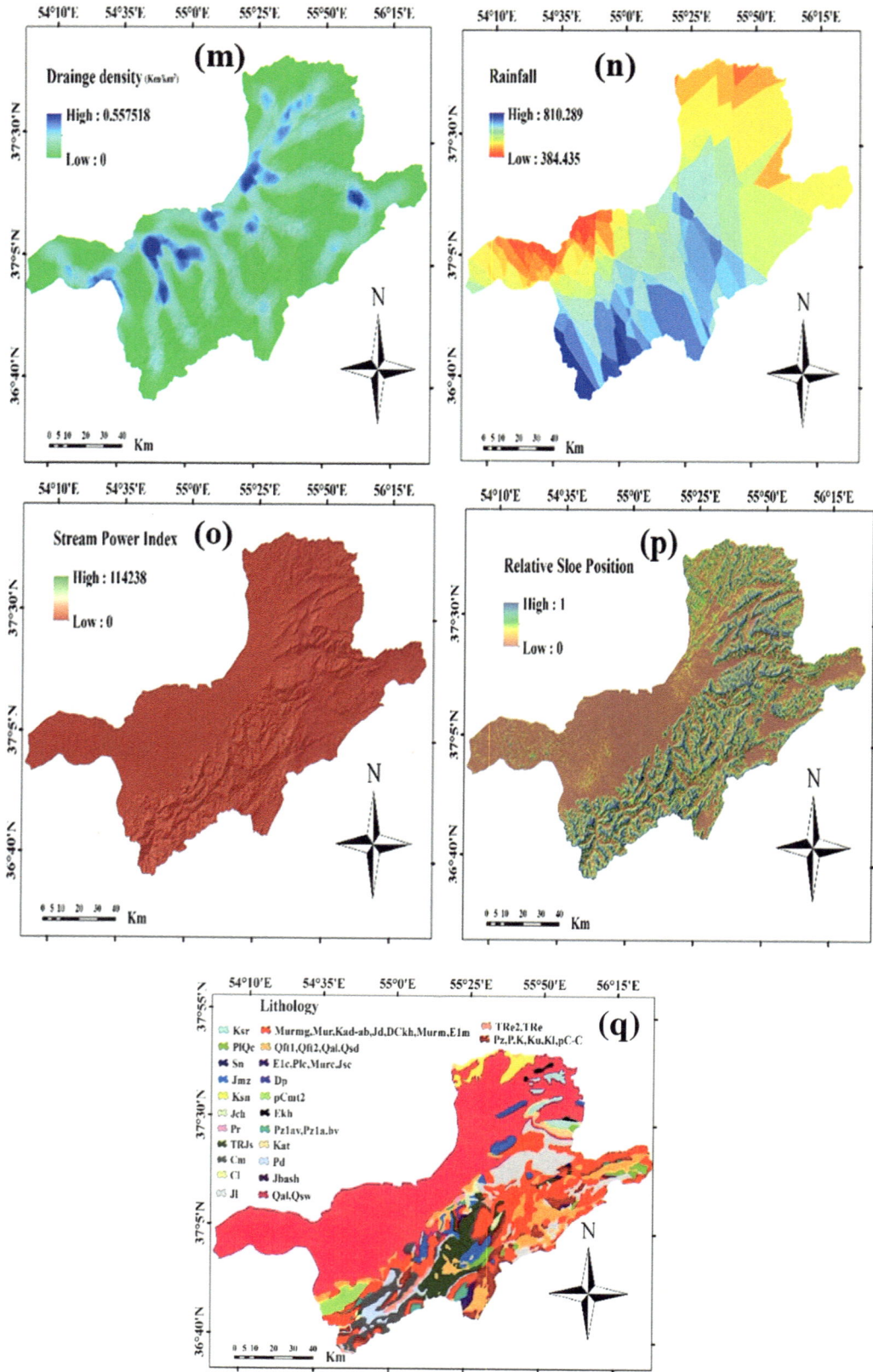

Fig. 29.4 (continued)

distribution (Φ) by finding the probability distribution of maximum entropy, subject to a set of constraints X that represent our incomplete information about the target distribution. Conditioning factors are used to explain the moment limitation on the probability distribution (Φ). The moment, such as the mean, is received from the values of the conditioning factors at all hazard point well located (with high productivity). By applying the ME algorithm, the most uniform distribution is selected from among many possible distributions (Phillips and Dudík 2008).

When using Maxent, a given space "x" represents the set of discrete grid cells covering the study area and $\Phi(x)$ (non-negative and sums to one) the value of the target probability distribution at location x. Each grid cell of "x" ($x_1, x_2, \ldots x_m$) is provided with the most important environmental variables defined, such as geographical and topographical ones. The probability distribution function in the study area, which the target is present at location x, denoted as $P(y = 1|x)$ is expressed as shown (Park 2015; Phillips et al. 2006):

$$P(y = 1|x) = \frac{P(y = 1)P(x|y = 1)}{P(x)} = \frac{P(y = 1)\Phi(x)}{1/|x|} \quad (29.3)$$

where $|X|$ and $P(y = 1)$ are the number pixels or locations and the overall prevalence of target occurrences, over the study area, respectively. $\Phi(x)$ estimated by the ME algorithm is equal to a Gibbs probability distribution (Phillips and Dudík 2008; Phillips et al. 2006). As discussed in Phillips et al. (2006), the Maxent distribution belongs to the family of Gibbs distributions.

All input conditioning factors are introduced as random variables of the model according to the ME algorithm described by Convertino et al. (2014) that represent their uncertainty. The detailed explanation of mathematical formulation of this model is shown in Phillips et al. (2006) and Phillips and Dudík (2008).

The ME model was built using all three training groups of the sample datasets (i.e., S1, S2, and S3) in the training step. The mentioned maps of every hazard were classified according to four classification techniques in GIS environment, Equal Interval, namely Natural Breaks, Geometrical Interval and Quintile and, into four different hazard potentiality zones, including low, medium, high, and very high. By comparing the results of each classification technique and the distribution of training and validation points on the high and very high potentiality zones, it was found that the quantile classification technique provided the most accurate distribution (Fig. 29.5). This agrees with Razandi et al. (2015) and Naghibi and Pourghasemi (2015), and in that, quantile classification technique is a good classifier in susceptibility mapping.

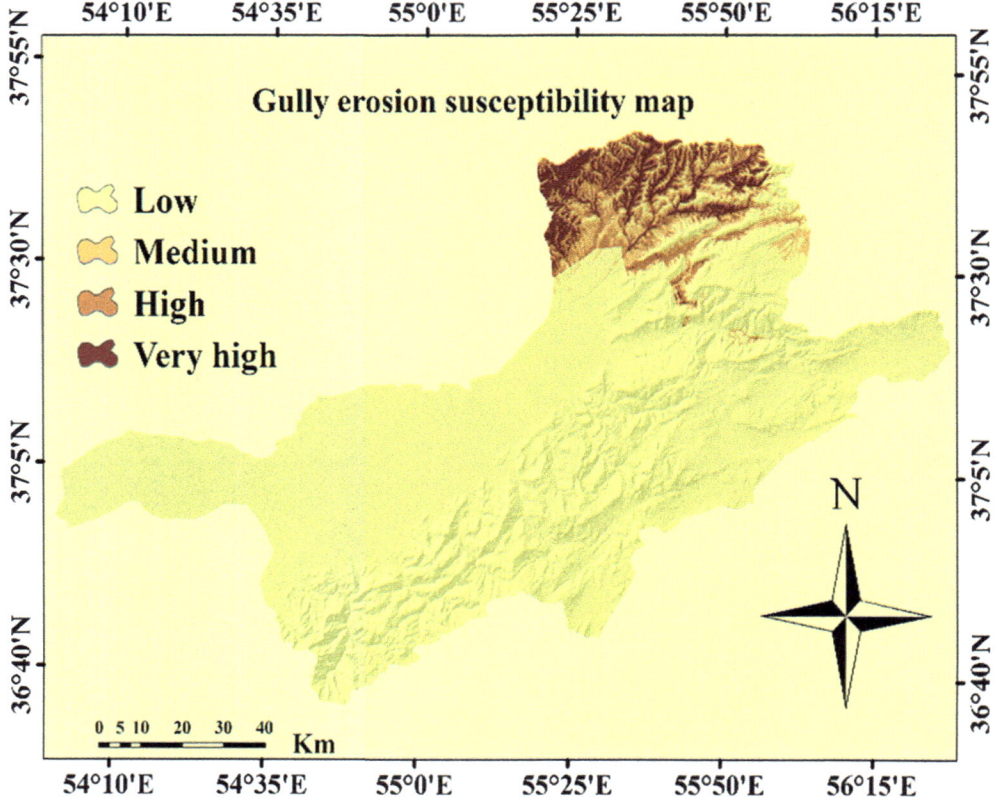

Fig. 29.5 Gully erosion susceptibility mapping of the study area using ME model

29.2.6 Evaluation of Predictive Performance

Validation step is the most important process of modeling (Chang-Jo and Fabbri 2003). In order to evaluate the prediction accuracy model, the receiver operating characteristic (ROC) curve can be used. In this method, the area under the ROC curves (AUC) can measure the prediction accuracy qualitatively (Maier and Dandy 2000; Bui et al. 2012). The receiver operating characteristic (ROC) curve has been generally applied in several researches to quantitatively assess the efficiency of susceptibility map (Nampak et al. 2014; Umar et al. 2014 Park et al. 2013), and it is part of the Maxent output (Elith et al. 2006). The ROC curve is a scientific technique of describing the proficiency of deterministic and probabilistic and prediction systems (Swets 1988). Zipkin et al. (2012) confirmed that area under the ROC curve (AUC) is useful in quantifying the uncertainty in model predictions while can account for recognition biases associated with estimation.

Here, the predictive performance for ME model in gully erosion susceptibility mapping has been investigated through the use of the AUC (Lin et al. 2015), and afterward, goodness of fit (i.e., degree of fitting) and prediction efficacy of the algorithms were studied, respectively (Oh and Pradhan 2011). Goodness of fit demonstrates the capability of the algorithm in estimating the training subset, while the predictive performance is a fundamental step for model accuracy in predicting a validation dataset (30% of points do not use the training process) (Bui et al. 2012; Umar et al. 2014). When the training and validation points are altered, the robustness of the predictive model is determined as the stability of the model's outputs in terms of accuracy models. Since the acceptance of a predictive model requires the evaluation of its robustness to small changes of the input data (i.e., data sensitivity), gully susceptibility model was prepared on three different samples of mapping units. So the robustness of the predictive models and their stability were furthermore evaluated when the training and validation samples are altered (S1, S2, and S3), i.e., a replicate method (Conoscenti et al. 2014; Angileri et al. 2016, Rahmati et al. 2017, 2018).

Model was employed to the mentioned datasets and was tested using the validation datasets. In the ROC curve, a comparison takes place between the computed event map with the validation dataset.

The robustness of the model was calculated by differentiating the maximum and minimum accuracy values based on each evaluation criteria (Conoscenti et al. 2014; Rahmati et al. 2018):

$$R_{AUC-ROC} = AUC - ROC_{max} - AUC - ROC_{min} \quad (29.4)$$

where $R_{AUC\text{-}ROC}$ is the robustness of a model based on AUC-ROC criteria and AUC-ROC$_{max}$ is the maximum

accuracy values among all datasets. In addition, minimum accuracy values are denoted as AUC-ROC$_{min}$ among all three datasets. These analyses were performed in both training and validation steps.

The use of ME model can be important because it allows to evaluate the uncertainties and mapping of the variance components (Convertino et al. 2014; Diniz-Filho et al. 2009), giving more information on where more research is needed to minimize the variance of relative decrease (PRD) of AUC values as a percentage was also calculated to investigate the dependency of model output on the influence of conditioning factors. A response curve is used to quantify the behavior of variables and to recognize relationships between each conditioning factor and the hazards modeling.

29.3 Results and Discussion

29.3.1 Application of Maximum Entropy Model

The susceptibility map for gully erosion and each dataset in the study area was produced using both continuous and categorical datasets with 10,000 background samples. Finally, the ME model was built using all three training groups of the sample datasets (i.e., S1, S2, S3) in the training step. Gully erosion susceptibility mapping of the study area using is presented in Fig. 29.5. Also Fig. 29.6 shows the relative distribution of the average of gully erosion susceptibility classes for three groups of the sample datasets. In the gully erosion susceptibility map, 8.22% of the pixels in the study area fell into high and very high susceptibility classes,

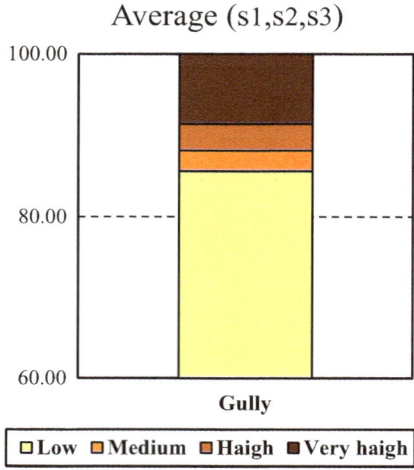

Fig. 29.6 Relative distributions of the average of four susceptibility classes for Gully susceptibility map

respectively. The highest percentage belongs to the low class. In addition, the statistical characteristics of the probabilistic prediction of the gully of three sample datasets are shown in Table 29.3.

29.3.2 Sensitivity Analysis

Model input layer sets unavoidably contain some uncertainties. A sensitivity analysis was applied, to investigate the conditioning factor with the strongest effect on the result of susceptibility prediction and to assess these uncertainties (Moreno et al. 2011; Park 2015). The jackknife test index was used in two procedures: (1) one-by-one factor removal (OOFR) that is a back calculation of Kappa values and (2) only one factor involved (OOFI) that follows a forward propagation. This was used to recognize the most influential factors for gully erosion. These results implied the conditioning factors with highest importance. The main variables for gully erosion include rainfall, DEM and lithology.

Therefore, as stated by Convertino et al. (2014), SA allows modelers and managers to identify the conditioning factors (i.e., input variables) that reduce the variance of the model output to the most, which is significantly vital in understanding the model structure.

29.3.3 Analyzing the Response Curves

Figure 29.7 illustrates the response curves of one dataset (S1) for some of the important conditioning factors used for gully erosion assessment. In response curves of DEM, it is obvious that gully erosion occurs in areas with low elevation and in plains, so with increasing elevation, the susceptibility values decreased.

The response curves of rainfall for gully show that the highest amount of gully erosion occurred in a rainfall of 450–500 mm. With increasing rainfall, gully erosion decreases, since the area with high slope has a lot of rainfall, with low probability of gully erosion occurrence. Also gully

occurred in group 13 and 1 with Sanganeh and Sarcheshmeh formation, respectively.

Consequently, a relatively higher contribution susceptibility prediction was obtained among the some categorical datasets. However, these lesser contributions of some categorical layers did not mean that the categorical data layers were unusable for susceptibility mapping. As discussed in Park (2015), all these categorical layers had affect on the final prediction result, and then simultaneously considered with continuous datasets.

29.3.4 ME Model Performance

The results of ME model (based on all three sample datasets) show different ranges of susceptibility values of hazards. The prediction accuracy of the models based on the AUC value can be classified into three classes of accuracy following the classification proposed by Hosmer and Lemeshow (2000): 0.7, 0.8, and 0.9 AUC thresholds were adopted to acceptable, excellent, and outstanding performance, respectively (Conoscenti et al. 2014, 2016). The goodness-of-fit results are shown in Table 29.4. It can be observed that performance values for the applied model based on AUC-ROC case of gully range from 0.916 to 0.926 (mean = 0.920). Therefore, it can be observed that model for all three hazards has the excellent performances. Figure 29.8 shows the AUC value and ROC curve for the prediction rate of the susceptibility maps for S1. However, as the training sample datasets were used to generate the model, they could not be used to assess and evaluate their prediction ability. The validation analysis shows how well the model can predict susceptibility mapping in a given area. The results of the applied model were verified using validation datasets (30% of gully headcut samples) and based on AUC-ROC evaluation criteria in the validation step (Table 29.4). On the other hand, the prediction rate curve used the validation hazard points determined how well the model and conditioning factors anticipate the hazards occurrence (Pradhan 2013). According to the ME results, AUC-ROC changes between 0.918 and 0.922 (mean = 0.920). The validation of the results established a strong agreement between the distribution of the existing gully erosion (validation dataset) and the predictive maps of the ME model. According to the accuracy classification (described in the methodology section), the ME model indicated an excellent predictive skill (AUC-ROC > 0.9), based on three sample datasets (Conoscenti et al. 2014, 2016). Therefore, the validation results proved that this applied model not only exhibits an excellent performance in terms of the AUC method, but its performance is also quite stable when the validation samples changed. The findings of robustness based on AUC are presented in Fig. 29.9. Since the accuracy values are very similar for the three sample

Table 29.3 Statistical characteristics of the probability values obtained from ME models

| Model | Dataset | Probabilistic prediction values | |
		Mean	SD
Gully	S1	0.0437	0.1213
	S2	0.0449	0.1234
	S3	0.0472	0.1284

SD Standard deviation

Fig. 29.7 The response curves of one dataset (S1) for some of the important conditioning factors used for gully erosion assessment

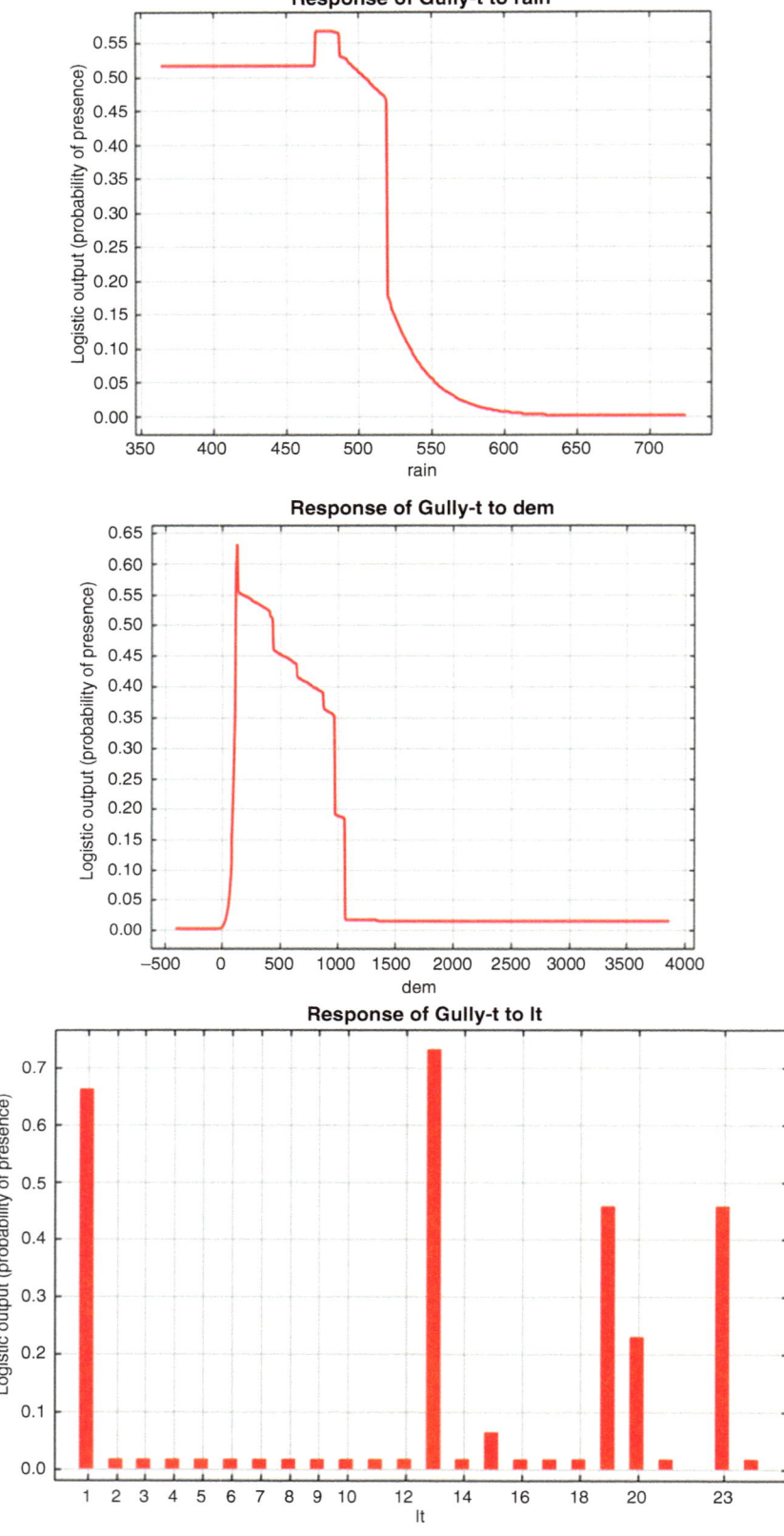

Table 29.4 Predictive performance of models based on three sample datasets (S1, S2, and S3) in the training and validation step

Evaluation criteria (AUC)	Dataset	ME model
		Gully
Training	S1	0.916
	S2	0.920
	S3	0.926
	Mean	0.920
Validation	S1	0.920
	S2	0.918
	S3	0.922
	Mean	0.92

datasets, model was robust when the validation group changes because there were only slight variations when changes of the training and validation datasets were performed. Furthermore, from a model stability viewpoint, the almost excellent agreement between training and validation *AUC* values for applied model demonstrates that this model is most stable and overfitting has also been avoided (Marmion et al. 2008). Based on these results, it is obvious that the ME model can be applied as efficient machine learning models to evaluate susceptibility gully erosion. Gully erosion occurs with nonlinear relationships. This machine learning model does not need prior elimination of outliers or data transformation and can fit complex nonlinear relationships between gully susceptibility and automatically analyzes interaction effects between conditioning factors (i.e., predictors).

Since almost all of the most controlling factors are not human-made and are of natural agents (not human-made) and changing natural factors are almost impossible, it can be conceived that the mitigation actions should mostly be built on a conformity technique (from training to performance mitigation/remedial measures and awareness) which requires harmony between public policies and individual actions (e.g., basin stakeholders) (Thomalla et al. 2006). Risk and hazard management should be taken into account prior to disaster management. To start off such activities and schematization for further land-use planning, the hazard-based susceptibility mapping can be valuable platforms.

29.4   Conclusion

The purpose of this study is evaluation of GESM (gully erosion susceptibility mapping) using data mining technique in the Gorganrood Watershed. It is essential for detecting the robust and accurate model for reducing errors in gully erosion susceptibility modeling and delineating gully-prone areas. The maximum entropy model was used to determine spatial modeling of gully erosion. Seventeen geo-environmental factors (e.g., elevation, slope percent, slope aspect, plan curvature, TWI, distance from rivers, distance from fault, distance from roads, drainage density, lithology, annual mean rainfall, land use, soil texture, TPI, and convergence index) were selected as predictors.

For spatial modeling, from the total each hazard locations, 70% were used for modeling and 30% for model evaluation. Validation of models was done based on the receiver

Fig. 29.8 The ROC curves of one dataset (S1) for gully erosion

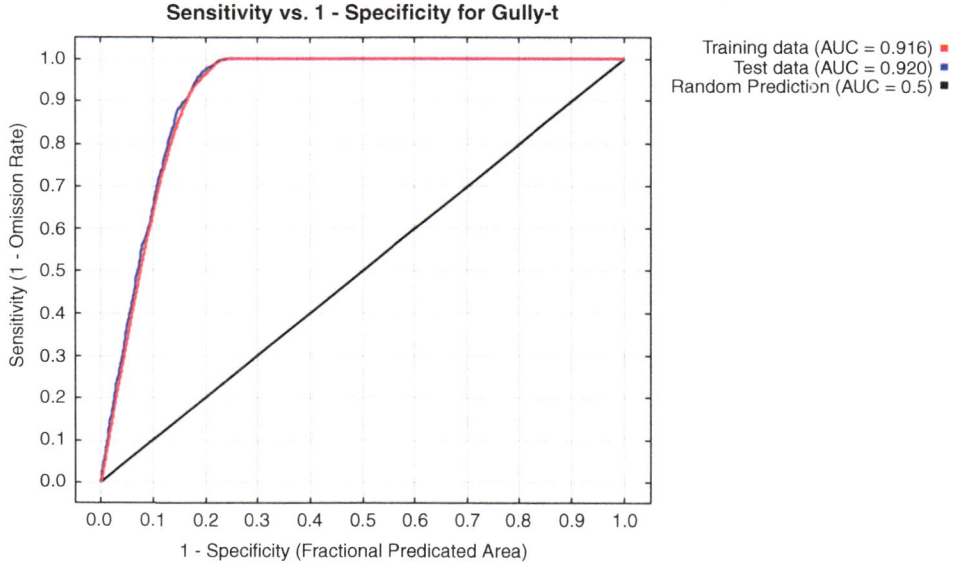

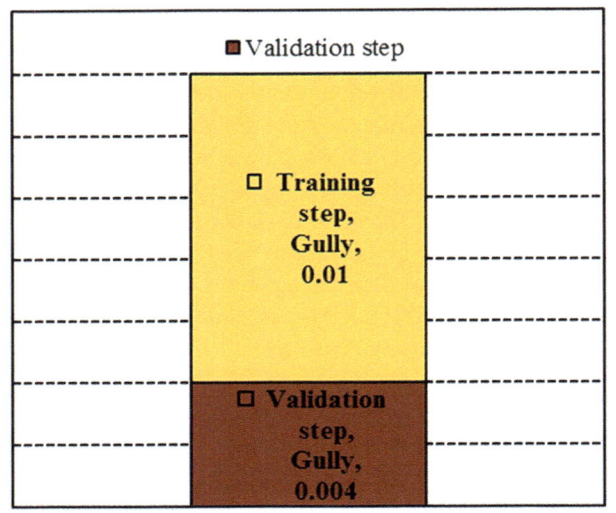

Fig. 29.9 Robustness of the ME model in training and validation steps based on AUC

operating characteristic (ROC) curve and the area under the curve. The ME model performed excellently both in the degree of fitting and in predictive performance (AUC values well above 0.8), which resulted in accurate predictions.

An integrated analysis was carried out in which the most important phenomenon (gullies) was incorporated and put into gully erosion susceptibility to help the decision makers to act most advisedly. This work demonstrates the usage of GESM in the Gorganrood Watershed, Iran, and may be utilized in land use and hazard mitigation planning, and it is a tool in planning a new facility and for insurance purpose.

Listed below are data obtained using data mining model:

- ME model accompanied by a robust pattern was identified as useful model, practicality, and correctly recognizing the presence locations of gullies locations in the study area.
- These results implied that the conditioning factors with highest importance when used in isolation are rain, DEM, and lithology for gully erosion, which therefore appear to have the most useful information by themselves. In other words, DEM was the main controlling factor among all, and the lithology was identified as the most important independent variable. Since changing the participated natural variables and reallocation of inhabitants from danger areas are inconceivable and economically unreasonable, the adaptive precautionary actions should be undertaken.
- Results show almost 80% of the area is located in the low to very low susceptibility zones, but 20% of the area is subjected to gully occurrence.
- The result of this study demonstrates the gully erosion susceptibility map over an area provides beneficial information for remediation strategies and establishes land-use plans.

Acknowledgments This research was supported by Regional Water Authority of Golestan province, and the authors would like to thank them for providing the discharge and meteorological data and the Forests, Ranges and Catchment Management Organization (FRWO) of Golestan for providing the data and maps.

References

Archer, G.E.B., Saltelli, A., Sobol, I. M., 1997. Sensitivity measures, ANOVA-like techniques and the use of bootstrap. Journal of Statistical Computation and Simulation, 58(2), 99-120

Angileri, S.E., Conoscenti, C., Hochschild, V., Märker, M., Rotigliano, E., & Agnesi, V., 2016. Water erosion susceptibility mapping by applying stochastic gradient treeboost to the imera Meridionale River basin (Sicily, Italy). *Geomorphology, 262*, 61-76.

Böhner, J., Selige, T., 2006. Spatial prediction of soil attributes using terrain analysis and climate regionalisation. Gottinger Geographische Abhandlungen, 115, 13-28

Boubli, J. P., De Lima, M.G., 2009. Modeling the geographical distribution and fundamental niches of Cacajao spp. and Chiropotes israelita in Northwestern Amazonia via a maximum entropy algorithm. International Journal of Primatology, 30(2), 217-228.

Baldwin, R.A., 2009. Use of maximum entropy modeling in wildlife research. Entropy 11, 854–866.

Bui, D., Pradhan, B., Lofman, O., Revhaug, I., Dick, O.B., 2012. Landslide susceptibility mapping at Hoa Binh province (Vietnam) using an adaptive neuro-fuzzy inference system and GIS. Computers & Geosciences, 45, 199–211.

Ballabio, C., & Sterlacchini, S., 2012. Support vector machines for landslide susceptibility mapping: the Staffora River Basin case study, Italy. *Mathematical geosciences, 44*(1), 47-70.

Cosby, B.J., Hornberger, G.M., Clapp, R.B., Ginn, T.R., 1984. A statistical exploration of the relationships of soil moisture characteristics to the physical properties of soils. Water Resour Res, 20, 682–690.

Chang-Jo, F.C., Fabbri, A.G., 2003. Validation of spatial prediction models for landslide hazard mapping. Natural Hazards, 30 (3), 451–472.

Chaplot, V., Giboire, G., Marchand, P., Valentin, C., 2005. Dynamic modelling for linear erosion initiation and development under climate and land-use changes in northern Laos. Catena, 63(2-3), 318-328. doi:https://doi.org/10.1016/j.catena.2005.06.008

[CONRWMGP] Central Office of Natural Resources and Watershed Management in Golestan Province., 2009. Detailed action plan, Iran, pp 230.

Chen, Y., Lawless, C., Gillespie, C.S., Wu, J., Boys, R.J., Wilkinson, D. J., 2010. CaliBayes and BASIS: integrated tools for the calibration, simulation and storage of biological simulation models. Briefings in bioinformatics, *11*(3), 278-289.

Convertino, M., Muñoz-Carpena, R., Chu-Agor, M.L., Kiker, G.L., Linkov, I., 2014. Untangling drivers of species distributions: global sensitivity and uncertainty analyses of MAXENT. Environmental Modelling & Software, 51, 296–309.

Conoscenti, C., Agnesi, V., Angileri, S., Cappadonia, C., Rotigliano, E., Märker, M., 2013. A GIS-based approach for gully erosion susceptibility modelling: a test in Sicily, Italy. Environ Earth Sci, 70,1179-1195.

Conoscenti, C., Angileri, S., Cappadonia, C., Rotigliano, E., Agnesi, V., Märker, M., 2014. Gully erosion susceptibility assessment by means of GIS-based logistic regression: A case of Sicily (Italy). Geomorphology, 204(1), 399–411.

Conoscenti, C., Ciaccio, M., Caraballo-Arias, N. A., Gómez-Gutiérrez, Á., Rotigliano, E., & Agnesi, V., 2015. Assessment of susceptibility to earth-flow landslide using logistic regression and multivariate adaptive regression splines: a case of the Belice River basin (western Sicily, Italy). *Geomorphology*, *242*, 49-64.

Conoscenti, C., Rotigliano, E., Cama, M., Caraballo-Arias, N.A., Lombardo, L. and Agnesi, V., 2016. Exploring the effect of absence selection on landslide susceptibility models: A case study in Sicily, Italy. Geomorphology, 261, 222–235.

Conforti, M., Aucelli, P. P., Robustelli, G., & Scarciglia, F., 2011. Geomorphology and GIS analysis for mapping gully erosion susceptibility in the Turbolo stream catchment (Northern Calabria, Italy). *Natural hazards*, *56*(3), 881-898.

Cama, M., Lombardo, L., Conoscenti, C., Rotigliano, E., 2017. Improving transferability strategies for debris flow susceptibility assessment: Application to the Saponara and Itala catchments (Messina, Italy). Geomorphology, 288, 52–65.

Douaik, M., Phillips, S., Schapire, R., 2004. Performance guarantees for regularized maximum entropy density estimation. Proceedings of the 17th Annual Conference on Computational Learning Theory. Springer, Berlin, Heidelberg, pp. 655–662.

Douaik, A., Meirvenne, M.V., Tóth, T., 2005. Soil salinity mapping using spatio-temporal Kriging and Bayesian maximum entropy with interval soft data. Geoderma, 128, 234–248.

Diniz-Filho, J.A.F., Bini, L.M., Rangel, T.F., Loyola, R.D., Hof, C., Nogués-Bravo, D., Araújo, M.B., 2009. Partitioning and mapping uncertainties in ensembles of forecasts of species turnover under climate change. Ecography, 32, 897–906.

Duc, D.M., 2012., Rainfall-triggered large landslides on 15 December 2005 in Van Canh District, Binh Dinh Province, Vietnam. Landslides, 10, 219–230.

Dube, F., Nhapi, I., Murwira, A., Gumindoga, W., Goldin, J., Mashauri, D.A., 2014. Potential of weight of evidence modelling for gully erosion hazard assessment in Mbire District-Zimbabwe. Physics and Chemistry of the Earth, Parts A/B/C, 67, 145–152.

Ercanoglu, M., Gokceoglu, C., 2002. Assessment of landslide susceptibility for a landslide prone area (north of Yenice, NW Turkey) by fuzzy approach. Environmental geology, 41, 720–730.

Elith, J., Graham, C.H., Anderson, R.P., Dudik, M., Lohmann, L.G., Loiselle, B.A., 2006. Novel methods improve prediction of species' distribution from occurrence data. Ecography, 29, 129–151.doi: https://doi.org/10.1111/j.2006.0906-7590.04596.x.

Edrén, S.M., Wisz, M.S., Teilmann, J., Dietz, R., Söderkvist, J., 2010. Modelling spatial patterns in harbour porpoise satellite telemetry data using maximum entropy. Ecography, 33, 698-708.

Fernández, D.S., Lutz, M.A., 2010. Urban flood hazard zoning in Tucumán Province, Argentina, using GIS and multicriteria decision analysis. Engineering Geology, 111, 90–98.

Jungerius, P.D., Matundura, J., Van de Ancker, J.a.M., 2002. Road construction and gully erosion in West Pokot, Kenya. Earth Surface Processes and Landforms, 27, 1237–1247.

Jenness, J., 2013. DEM Surface Tools for ArcGIS. Jenness Enterprises Available at. http://www.jennessent.com/arcgis/surface:area.htm.

Jaafari, A., Najafi, A., Pourghasemi, H.R., Rezaeian, J., Sattarian, A., 2014. GIS-based frequency ratio and index of entropy models for landslide susceptibility assessment in the Caspian forest, northern Iran. International Journal of Environmental Science and Technology, 11(4), 909-926.

Hosmer, D.W., Lemeshow, S., 2000. Applied Logistic Regression, Wiley Series in Probability and Statistics. Wiley.

Imeson, A. C., & Kwaad, F. J. P. M., 1980. Gully types and gully prediction. *Geografisch Tijdschrift*, *14*(5), 430-441.

Govers, G., 1985. Selectivity and transport capacity of thin flows in relation to rill erosion. Catena 12(1), 35-49. doi:https://doi.org/10.1016/S0341-8162(85)80003-5.

Gyssels, G., Poesen, J., Nachtergaele, J., Govers, G., 2002. The impact of sowing density of small grains on rill and ephemeral gully erosion in concentrated flow zones. Soil and Tillage Research, 64(3), 189–201.

Geissen, V., Kampichler, C., López-de Llergo-Juárez, J.J., Galindo-Acántara, A., 2007. Superficial and subterranean soil erosion in Tabasco, tropical Mexico: Development of a decision tree modeling approach. Geoderma 139(3-4), 277–287. doi:https://doi.org/10.1016/j.geoderma.2007.01.002

Grabs, T., Seibert, J., Bishop, K., Laudon, H., 2009. Modeling spatial patterns of saturated areas: a comparison of the topographic wetness index and a dynamic distributed model. Journal of Hydrology, 373, 15–23.

Gómez Gutiérrez, Á., Schnabel, S., Lavado Contador, J.F., 2009. Using and comparing two nonparametric methods (CART and MARS) to model the potential distribution of gullies. Ecological Modelling 220 (24), 3630-3637. doi:https://doi.org/10.1016/j.ecolmodel.2009.06.020.

Geroy, I.J., Gribb, M.M., Marshall, H.P., Chandler, D.G., Benner, S.G., McNamara, J.P., 2011. Aspect influences on soil water retention and storage. Hydrological Processes, 25(25), 3836–3842.

Glenn, E., Morino, K., Nagler, P., Murray, R., Pearlstein, S., Hultine, K., 2012. Roles of saltcedar (Tamarix spp.) and capillary rise in salinizing a non-flooding terrace on a flow-regulated desert river. *Journal of arid environments, 79*, 56-65.

Gómez Gutiérrez, Á., Conoscenti, C., Angileri, S.E., Rotigliano, E., Schnabel, S., 2015. Using Topographical attributes to model the spatial distribution of gullying from two Mediterranean basins: advantages and limitations. Natural Hazards. doi:https://doi.org/10.1007/s11069-015-1703-0.

Gallardo-Cruz, J.A., Pérez-García, E.A., Meave, J.A., 2009. β-Diversity and vegetation structure as influenced by slope aspect and altitude in a seasonally dry tropical landscape. Landscape Ecology. 24(4), 473–482.

Kosmas, C., Danalatos, N., Cammeraat, L.H., Chabart, M., Diamantopoulos, J., Farand, R., Gutierrez, L., Jacob, A., Marques, H., Martinez-Fernandez, J., Mizara, A., Moustakas, N., Nicolau, J. M., Oliveros, C., Pinna, G., Puddu, R., Puigdefabregas, J., Roxo, M., Simao, A., Stamou, G., Tomasi, N., Usai D., Vacca, A., 1997. The effect of land use on runoff and soil erosion rate under Mediterranean conditions. Catena 29, 45-59. doi:https://doi.org/10.1016/S0341-8162(96)00062-8.

Kia, M. B., Pirasteh, S., Pradhan, B., Mahmud, A.R., Sulaiman, W.N. A., Moradi, A., 2012. An artificial neural network model for flood simulation using GIS: Johor River Basin, Malaysia. Environmental Earth Sciences, 67(1), 251-264.

Khosravi, K., Nohani, E., Maroufinia, E., Pourghasemi, H.R., 2016. A GIS-based flood susceptibility assessment and its mapping in Iran: a comparison between frequency ratio and weights-of-evidence bivariate statistical models with multi-criteria decision-making technique. Natural Hazards, 83(2), 947-987.

Kornejady, A., Ownegh, M., Bahremand, A., 2017a. Landslide susceptibility assessment using maximum entropy model with two different data sampling methods. Catena, 152, 144-162.

Kornejady, A., Ownegh, M., Rahmati, O., Bahremand, A., 2017b. Landslide susceptibility assessment using three bivariate models considering the new topo-hydrological factor: HAND. Geocarto International, 32, 1–68

Lal, R., 2003. Offsetting global CO$_2$ emissions by restoration of degraded soils and intensification of world agriculture and forestry. Land Degradation & Development, vol. 14(3), 309-322. doi:https://doi.org/10.1002/ldr.562.

Lan, H.X., Zhou, C.H., Wang, L.J., Zhang, H.Y., Li, R.H., 2004. Landslide hazard spatial analysis and prediction using GIS in the Xiaojiang watershed, Yunnan, China. Engineering geology, 76, 109–128.

Lee, S., Ryu, J.H., Kim, I.S., 2007. Landslide susceptibility analysis and its verification using likelihood ratio, logistic regression, and artificial neural network models: case study of Youngin, Korea. Landslides, 4(4), 327–338.

Lee, M. J., Kang, J. E., & Jeon, S., 2012. Application of frequency ratio model and validation for predictive flooded area susceptibility mapping using GIS. In Geoscience and Remote Sensing Symposium (IGARSS), 2012 IEEE International, pp, 895-898.

Lucà, F., Conforti, M., Robustelli, G., 2011. Comparison of GIS-based gullying susceptibility mapping using bivariate and multivariate statistics: Northern Calabria, South Italy. Geomorphology, 134, 297–308.

Lin, Y.P., Deng, D., Lin, W.C., Lemmens, R., Crossman, N.D., Henle, K., Schmeller, D.S., 2015. Uncertainty analysis of crowd-sourced and professionally collected field data used in species distribution models of Taiwanese moths. Biological conservation, 181, 102-110.

Lombardo, L., Cama, M., Conoscenti, C., Märker, M., & Rotigliano, E., 2015. Binary logistic regression versus stochastic gradient boosted decision trees in assessing landslide susceptibility for multiple-occurring landslide events: application to the 2009 storm event in Messina (Sicily, southern Italy). *Natural Hazards*, *79*(3), 1621-1648.

Lombardo, L., Bachofer, F., Cama, M., Märker, M., & Rotigliano, E., 2016. Exploiting Maximum Entropy method and ASTER data for assessing debris flow and debris slide susceptibility for the Giampilieri catchment (north-eastern Sicily, Italy). *Earth Surface Processes and Landforms*, *41*(12), 1776-1789.

Miller, J.R., Ritter, D.F., Kochel, R.C., 1990. Morphometric assessment of lithologic controls on drainage basin evolution in the Crawford Upland, south-central Indiana. American Journal of Science, 290, 569–599.

Moore, I.D., Grayson, R.B., Ladson, A.R., 1991. Digital terrain modelling: a review of hydrological, geomorphological, and biological applications. Hydrological processes, 5(1), 3-30.

Maier, H.R., Dandy, G.C., 2000. Neural networks for the prediction and forecasting of water resources variables: a review of modelling issues and applications. Environmental modelling & software, 15, 101–124

Maestre, F.T., Cortina, J., 2002. Spatial patterns of surface soil properties and vegetation in a Mediterranean semiarid steppe. Plant Soil, 241(2), 279–291.

Marmion, M., Hjort, J., Thuiller, W., Luoto, M., 2008. A comparison of predictive methods in modelling the distribution of periglacial landforms in Finnish Lapland. Earth surface processes and landforms, 33(14), 2241-2254.

Medley, K.A., 2010. Niche shifts during the global invasion of the Asian tiger mosquito, Aedes albopictus Skuse (Culicidae), revealed by reciprocal distribution models. Global ecology and biogeography, 19(1), 122-133.

Magliulo, P., 2010. Soil erosion susceptibility maps of the Janare Torrent Basin (Southern Italy). J. Maps , 6, 435–447.

Magliulo, P., 2012. Assessing the susceptibility to water-induced soil erosion using a geomorphological, bivariate statistics-based approach. Environ. earth Sci, 67, 1801–1820.

Moreno, R., Zamora, R., Molina, J.R., Vasquez, A., Herrera, M.Á., 2011. Predictive modeling of microhabitats for endemic birds in South Chilean temperate forests using maximum entropy (Maxent). Ecological Informatics, 6(6), 364-370.

Märker, M., Pelacani, S., Schröder, B., 2011. A functional entity approach to predict soil erosion processes in a small Plio-Pleistocene Mediterranean catchment in Northern Chianti, Italy. Geomorphology 125(4), 530-540. doi:https://doi.org/10.1016/j.geomorph.2010.10.022

Meinhardt, M., Fink, M., Tunschel, H., 2015. Landslide susceptibility analysis in central Vietnam based on an incomplete landslide inventory: comparison of a new method to calculate weighting factors by means of bivariate statistics. Geomorphology, 234, 80–97.

Moghaddam, D.D., Rezaei, M., Pourghasemi, H.R., Pourtaghie, Z.S., Pradhan, B., 2015. Groundwater spring potential mapping using bivariate statistical model and GIS in the Taleghan watershed, Iran. Arabian Journal of Geosciences, 8(2), 913-929

Nagarajan, R., Roy, A., Kumar, R.V., Mukherjee, A., Khire, M.V. 2000. Landslide hazard susceptibility mapping based on terrain and climatic factors for tropical monsoon regions. Bulletin of Engineering Geology and the Environment, 58(4), 275-287.

Nampak, H., Pradhan, B., Manap, M.A., 2014. Application of GIS based data driven evidential belief function model to predict groundwater potential zonation. Journal of Hydrology, 513, 283-300.

Naghibi, S.A., Pourghasemi, H.R., 2015. A comparative assessment between three machine learning models and their performance comparison by bivariate and multivariate statistical methods in groundwater potential mapping. Water resources management, 29(14), 5217-5236.

O'brien, R. M., 2007. A caution regarding rules of thumb for variance inflation factors. Quality & quantity, 41(5), 673-690.

Ozdemir, A., 2011. Using a binary logistic regression method and GIS for evaluating and mapping the groundwater spring potential in the Sultan Mountains (Aksehir, Turkey). Journal of Hydrology, 405 (1-2), 123-136

Oh, H. J., Pradhan, B., 2011. Application of a neuro-fuzzy model to landslide-susceptibility mapping for shallow landslides in a tropical hilly area. Computers & Geosciences, 37(9), 1264-1276.

Phillips, S.J., Dudík, M., Schapire, R.E., 2004 A maximum entropy approach to species distribution modeling. In Proceedings of the twenty-first international conference on Machine learning (p. 83). ACM, Banff, Canada.

Phillips, S.J., Anderson, R.P., Schapire, R.E., 2006. Maximum entropy modeling of species geographic distributions. Ecological modelling, 190(3-4), 231-259.

Phillips, S.J., Dudík, M., 2008. Modeling of species distributions with maxent: new extensions and a comprehensive evaluation. Ecography 31, 161–175.

Pradhan, B., 2010. Flood susceptible mapping and risk area estimation using logistic regression, GIS and remote sensing. Journal of Spatial Hydrology, 9,1–18.

Pradhan, B., 2013. A comparative study on the predictive ability of the decision tree, support vector machine and neuro-fuzzy models in landslide susceptibility mapping using GIS. Computers & Geosciences, 51, 350-365.

Pourghasemi, H.R., Jirandeh, A.G., Pradhan, B., Xu, C., Gokceoglu, C., 2013. Landslide susceptibility mapping using support vector machine and GIS at the Golestan Province, Iran. Journal of Earth System Science, 122(2), 349-369.

Pourtaghi, Z.S., Pourghasemi, H.R., 2014. GIS-based groundwater spring potential assessment and mapping in the Birjand Township, southern Khorasan Province, Iran. Hydrogeology Journal, 22(3), 643-662.

Poiraud, A., 2014. Landslide susceptibility–certainty mapping by a multi-method approach: a case study in the tertiary basin of Puy-en-Velay (Massif central, France). Geomorphology, 216, 208–224.

Park, S., Choi, C., Kim, B., Kim, J., 2013. Landslide susceptibility mapping using frequency ratio, analytic hierarchy process, logistic regression, and artificial neural network methods at the Inje area, Korea. Environmental earth sciences, 68(5), 1443-1464

Park, N.W., 2015. Using maximum entropy modeling for landslide susceptibility mapping with multiple geoenvironmental data sets. Environmental Earth Sciences, 73(3), 937-949.

Razandi, Y., Pourghasemi, H.R., Neisani, N.S., Rahmati, O. 2015. Application of analytical hierarchy process, frequency ratio, and certainty factor models for groundwater potential mapping using GIS. Earth Science Informatics, 8(4), 867-883.

Rahmati, O., Pourghasemi, H. R., Melesse, A.M., 2016. Application of GIS-based data driven random forest and maximum entropy models for groundwater potential mapping: a case study at Mehran Region, Iran. Catena, 137, 360-372.

Rahmati, O., Tahmasebipour, N., Haghizadeh, A., Pourghasemi, H.R., Feizizadeh, B., 2017. Evaluation of different machine learning models for predicting and mapping the susceptibility of gully erosion. Geomorphology, 298, 118-137.

Rahmati, O., Naghibi, S.A., Shahabi, H., Bui, D.T., Pradhan, B., Azareh, A., Melesse, A.M., 2018. Groundwater spring potential modelling: comprising the capability and robustness of three different modeling approaches. Journal of hydrology, 565, 248-261.

Swets, J.A., 1988. Measuring the accuracy of diagnostic systems. Science, 240 (4857),1285–1293.

Shimizu, M., 1988. Prediction of slope failures due to heavy rain using the tank model. Proceedings of the 5th International Symposium on Landslides. Lausanne, 1, pp. 771–776.

Sidle, R.C., Ochiai, H., 2006. Landslides: processes, prediction, and land use. Water Research Monograph, 18. Washington, DC: American Geophysical Union; p 312.

Svoray, T., Michailov, E., Cohen, A., Rokah, L., & Sturm, A., 2012. Predicting gully initiation: comparing data mining techniques, analytical hierarchy processes and the topographic threshold. *Earth Surface Processes and Landforms, 37*(6), 607-619.

Song, Y., Gong, J., Gao, S., Wang, D., Cui, T., Li, Y., Wei, B., 2012. Susceptibility assessment of earthquake-induced landslides using Bayesian network: a case study in Beichuan, China. Computers & Geosciences, *42*, 189-199.

Shafapour Tehrany, M., Pradhan, B., Jebur, M.N., 2013. Spatial prediction of flood susceptible areas using rule based decision tree (DT) and a novel ensemble bivariate and multivariate statistical models in GIS. Journal of Hydrology, 504, 69–79.

Shafapour Tehrany, M., Lee, MJ., Pradhan, B., Jebur, M.N, Lee, S., 2014a. Flood susceptibility mapping using integrated bivariate and multivariate statistical models. Environmental earth sciences, 72, 4001–4015

Shafapour Tehrany, M., Pradhan, B., Jebur, M.N., 2014b. Flood susceptibility mapping using a novel ensemble weights-of-evidence and support vector machine models in GIS. Journal of Hydrology, 512:332–343

Saponaro, A., Pilz, M., Wieland, M., Bindi, D., Moldobekov, B., Parolai, S., 2015. Landslide susceptibility analysis in data-scarce regions: the case of Kyrgyzstan. Bulletin of Engineering Geology and the Environment, 74(4), 1117-1136.

Thomalla, F., Downing, T., Spanger-Siegfried, E., Han, G., Rockström, J., 2006. Reducing hazard vulnerability: towards a common approach between disaster risk reduction and climate adaptation. Disasters, 30(1), 39–48.

Tahmassebipoor, N., Rahmati, O., Noormohamadi, F., Lee, S., 2016. Spatial analysis of groundwater potential using weights-of-evidence and evidential belief function models and remote sensing. Arabian Journal of Geosciences, 9(1), 79.

Umar, Z., Pradhan, B., Ahmad, A., Jebur, M.N., Tehrany, M.S., 2014. Earthquake induced landslide susceptibility mapping using an integrated ensemble frequency ratio and logistic regression models in West Sumatera Province, Indonesia. Catena, 118,124–135.

Vandekerckhove, L., Poesen, J, Oostwoudwijdenes, D.J., Gyssels, G., Beuselinck, L., De Luna, E., 2000. Characteristics and controlling factors of bank gullies in two semi arid Mediterranean environments. Geomorphology, 33,37–58

Vandekerckhove, L., Poesen, J., Govers, G., 2003. Medium-term gully headcut retreat rates in Southeast Spain determined from aerial photographs and ground measurements. Catena, 50 (2-4), 329-352.

Vanwalleghem, T., Poesen, J., Nachtergaele, J., Verstraeten, G., 2005. Characteristics, controlling factors and importance of deep gullies under cropland on loess derived soils. Geomorphology 69:76–91.

Vahidnia, M.H., Alesheikh, A.A., Alimohammadi, A., Hosseinali, F., 2010. A GIS-based neuro-fuzzy procedure for integrating knowledge and data in landslide susceptibility mapping. Computers & Geosciences. 36, 1101–1114.

Yost, A. C., Petersen, S. L., Gregg, M., & Miller, R., 2008. Predictive modeling and mapping sage grouse (Centrocercus urophasianus) nesting habitat using Maximum Entropy and a long-term dataset from Southern Oregon. Ecological Informatics, *3*(6), 375-386

Yalcin, A., 2008. GIS-based landslide susceptibility mapping using analytical hierarchy process and bivariate statistics in Ardesen (Turkey): comparisons of results and confirmations. Catena, 72, 1–12.

Youssef, A.M., 2015. Landslide Susceptibility Delineation in the Ar-Rayth Area, Jizan, Kingdom of Saudi Arabia, by using analytical hierarchy process, frequency ratio, and logistic regression models. Environmental Earth Sciences, *73*(12), 8499-8518

Zucca, C., Canu, A., Della Peruta, R., 2006. Effects of land use and landscape on spatial distribution and morphological features of gullies in an agropastoral area in Sardinia (Italy). Catena, 68(2), 87–95.

Zipkin, E.F., Grant, E.H.C., Fagan, W.F., 2012. Evaluating the predictive abilities of community occupancy models using AUC while accounting for imperfect detection. Ecological Applications, 22(7), 1962-1972.

Zakerinejad, R., Märker, M., 2014. Prediction of Gully erosion susceptibilities using detailed terrain analysis and maximum entropy modeling: a case study in the Mazayejan Plain, Southwest Iran. *Geogr Fis Din Quat, 37*(1), 67-76.

Zhu, A.X., Wang, R.X., Qiao, J.P., Qin, C.Z., Chen, Y.B., Liu, J., Du, F., Lin, Y., Zhu, T.X., 2014. An expert knowledge-based approach to landslide susceptibility mapping using GIS and fuzzy logic. Geomorphology, 214, 128–138.

Narges Javidan is PhD student of Sari Agricultural Sciences and Natural Resources University. Her research activities are mainly focused on GIS and Modeling of soil erosion processes and Modeling of hydrology.she obtained a MSc degree in watershed engineering, in 2013 at the University of Gorgan, and was accepted in PhD degree in Watershed Science and Engineering, in 2015, at the University of Sari. She is author of seven scientific-research/scientific-promotional papers and six Conference paper related to her main research interests.

Ataollah Kavian is Associate Professor of watershed engineering at the Department of Natural Resources of the University of Sari Agricultural Sciences and Natural Resources. His research activities are mainly focused on GIS, RS and Modeling of soil erosion processes and landslide, flood and drought processes. He obtained an M.Sc. degree in watershed engineering, in 2002, and a PhD in Watershed Science and Engineering, in 2006, at the University of Tehran. He obtained the role of Associate Professor, in 2006. He is author of 52 ISI papers related to his main research interests.

Hamid Reza Pourghasemi is an Associate Professor of Watershed Management Engineering in the College of Agriculture. Shiraz University, Iran. He has a BSc in Watershed Management Engineering of the University of Gorgan (2004), Iran, an MSc in Watershed Management Engineering, from Tarbiat Modares University, Iran (2008), and a PhD in Watershed Management Engineering from the same University (Feb 2014). His main research interests are GIS-based spatial modelling using machine learning/data mining techniques in different fields such landslide, flood, gully erosion, forest fire, land subsidence, species distribution modelling, and groundwater/hydrology. Also, Hamid Reza works on Multi-Criteria Decision Making methods in Natural Resources and Environment.

He has published more of 90 peer reviewed papers in high-quality journals, with three chapters in Springer. Also, he published two books in Springer (https://www.springer.com/gp/book/9783319733821) and Elsevier (https://www.elsevier.com/books/spatial-modeling-in-gis-and-r-for-earth-and-environmental-science/pourghasemi/978-0-12-815226-3).

Christian Conoscenti is Associate Professor of Physical Geography and Geomorphology at the Department of Earth and Marine Sciences of the University of Palermo. His research activities are mainly focused on GIS and statistical analysis of soil erosion and landslide processes. He is member of the Academic Board of the PhD in Earth and Marine Sciences of the University of Palermo. He obtained a MSc degree in Geological Sciences, in 2000, and a PhD in Geology, in 2006, at the University of Palermo. He obtained from the Italian Ministry of Education, University and Research (MIUR), the National Scientific Qualification (ASN) in the field of Physical Geography and Geomorphology for the role of Associate Professor, in 2103, and for the role of Full Professor, in 2017. He is author of 34 ISI papers related to his main research interests.

Zeinab Jafarian is Associate Professor of Rangeland Management at the Department of Natural Resources of the University of Sari Agricultural Sciences and Natural Resources. Her research activities are mainly focused on GIS, RS and Modeling of soil erosion processes and statistical analysis. She obtained a M.Sc. degree in Rangeland Management, in 2002, and a PhD in Rangeland Management, in 2006, at the University of Tehran. She obtained the role of Associate Professor, in 2006. She is author of 15 ISI papers related to her main research interests.

Iwan Rudiarto, Isna Rahmawati, and Anang Wahyu Sejati

Abstract

Soil erosion has been a major threat in land degradation
processes around the world. High level of soil erosion in
particular area may influence community livelihood where
land resource as the main source of family income is being
threatened. This study was carried out in Dieng Plateau,
Central Java Province, Indonesia, with the aim to seek the
level of soil erosion as well as to find out how resilient
farm families are in the study area toward land degradation. Soil erosion assessment was performed using
RUSLE method with various spatial data such as Landsat
images, rainfall, soil erodibility, slope data, and conservation practice, while community resilience assessment was
performed by comparing community preparedness to its
vulnerability from 67 farm household samples. The results
show most of high level of soil erosion occurred in area
dominated by steep slope with less vegetation cover. It is
also confirmed that soil erosion has accelerated due to
deforestation indicated by the increasing area for soil
erosion level 61–180 tons/ha/year from 1871 ha in 2007
to 2174 ha in 2017 (+4.07%). While the highest soil
erosion level more than 180 tons/ha/year was increased
for about 1.45% from 226 ha in 2017 to 34 ha in 2017, it
was found as well that community resilience in the study
area is classified at low level (0.27–1.01) with score 0.56.
In general, the community in the study area is not resilient
toward land degradation processes and hence jeopardizes
livelihood sustainability.

Keywords

Land degradation · Soil erosion · Community resilience ·
Preparedness · Vulnerability · Central Java · Indonesia

30.1 Introduction

The evolution of soils, which is influenced by the environmental factors, is moving toward through a long period of
time. Once human being exploits soil for agricultural
purposes, these processes are being accelerated and consequently leading to the rapid changes in soil properties, where
land is degraded (Douglas 1994). According to Geist (2005),
the acceleration of land degradation may come from biophysical aspect (e.g., land management), socioeconomic aspect
(e.g., income and land tenure), and political aspect (e.g.,
incentives and political stability). Referring to the concept
above, therefore, land degradation can be defined as the loss
of a sustained economy, cultural, or ecological function due
to human activity in combination with natural processes
(Geist 2005). In terms of land capability, Douglas (1994)
defined land degradation as the reduction in the capability
of land to produce benefits from a particular land use under a
specified form of land management which includes vegetation degradation, water degradation, climate deterioration,
losses to urban and industrial development, and soil
degradation.

In agricultural production, land degradation, which
comprises a bunch of processes, has a direct impact on the
socioeconomic of people, such as productivity decline and
income loss. The arising problem following land degradation
has been the major issue in land management for agricultural
purposes and obviously reduced the capability of land in
terms of production. All the actions in regard to agriculture
activities are directly related to the sort of land degradation
and finally affect the human life socially, economically,
ecologically, and institutionally. The changing shape of

I. Rudiarto (✉) · A. W. Sejati
Department of Urban and Regional Planning, Diponegoro University,
Semarang, Indonesia
e-mail: iwan.rudiarto@pwk.undip.ac.id

I. Rahmawati
Postgraduate School in Environmental Science, Diponegoro University,
Semarang, Indonesia

P. K. Shit et al. (eds.), *Gully Erosion Studies from India and Surrounding Regions*, Advances in Science, Technology & Innovation,
https://doi.org/10.1007/978-3-030-23243-6_30

land degradation into different aspects of human life has challenged scholars to develop resilience indices for community living with land degradation issues. The notion of community resilience is being prevalent since Adger (2000) place the resilience concern at the community level. Ever since then, many studies have been conducted in assessing the community resilience in different related variables (Cote and Nightingale 2012). Resilience research is the key to assess the ability of the community to survive from natural disturbances (Wilson 2012). Research on community resilience has widely spread with various concerns, such as natural disaster, climate change, rural/urban resilience, collective action, globalization, but less concern on land degradation. Among those few less examples were done by Kelly et al. (2015) and Wilson et al. (2016). Their research connected land degradation process with community resilience from various aspects such as social, economy, natural, infrastructural, institutional, and cultural.

This chapter is then proposed to fulfill the lack of research on the connection of land degradation and community resilience with specific purpose is to assess land degradation and the level of community resilience in Dieng Plateau, one of mountainous area in Indonesia. Following the concept of land degradation, we concern specifically on soil erosion assessment which been the major issue and thread in our study area either socially, economically, institutionally, or physically.

30.2 Material and Methods

30.2.1 Study Area

The study was conducted in Kejajar Sub-District which is a rural mountain area located in Dieng Plateau, Wonosobo Regency, Central Java Province. Dieng Plateau is an area with a great environmental problem in particular land degradation and water deficiency as the result of overuse and misuse of land resources which may threaten the continuation of agricultural activities. It is also one of the mountain areas in Indonesia with high level of deforestation and high environmental risk with different kinds of off-farm activities existing and considered as a significant contribution to the family income. Typical physical environmental characteristic has made this area suitable for agricultural production particularly for vegetables and potato and therefore led to inappropriate extensive crop production which may accelerate soil erosion.

Topographically, Dieng Plateau located on a mountain area with three areas division; higher area, middle area, and lower area. Higher area ranges from 2000 to 3000 m above sea level (asl), middle area from 1500 m to 2000 m asl, and lower area from 500 m to 1500 m asl. According to Rudiarto and Doppler (2013), higher area is dominated by potato

agricultural area with subsistence-market oriented farming system and less infrastructure found. Middle area has more less similar condition as compared to high area but more developed with subsistence and market-oriented farming system. Lower area is likely dominated by market-oriented farming system with less effort of managing sloppy land as the area mostly plain area. Figures 30.1 and 30.2 show the topographical condition of the study area.

30.2.2 Data Collection

As this study wants to show the development of land degradation in terms of soil erosion and community resilience, data needs to be grouped into two types: spatial and non-spatial data. Spatial data in 2007 and 2017 such as land use, soil type, rainfall, and slope were elaborated to find soil erosion level while non-spatial data were used to identify the level of community resilience in the study area. Non-spatial data were collected from 67 farm household samples exaggerated by land degradation in Dieng Plateau. Standardized questionnaires were distributed with a random sampling technique. Table 30.1 shows the data needs for this study.

30.2.3 Assessment of Soil Erosion

The Revised Universal Soil Loss Estimation (RUSLE) was applied to assess the level of soil erosion in the study area. RUSLE (Renard et al. 1997) was developed from the Universal Soil Loss Estimation (USLE) founded by Wischmeier and Smith (1978) as the empirically based model. RUSLE produces the average rates of soil loss per unit area annually based on factors influencing soil erosion such as rainfall, soil type, slope, crop management, and control practice (Rudiarto and Doppler 2013). The modification of USLE into RUSLE is more on the site location of the model application where more slopes are available. RUSLE model gives more opportunity to calculate soil erosion with several slope lengths and the average results of soil erosion rates follows the particular slope length (Angima et al. 2003). Even within a large area, RUSLE enables to estimate soil erosion potential on a cell-by-cell basis, and it may produce a better spatial pattern of soil loss (Shinde et al. 2011). RUSLE is calculated based on the following equation:

$$A = R \times K \times LS \times C \times P \qquad (30.1)$$

where A is the average soil loss (tones/ha/year), R is the rainfall erosivity index, K is the soil erodibility factor, L is the length factor (m), S is the slope factor (%), C is the crop management factor, and P is the conservation practice.

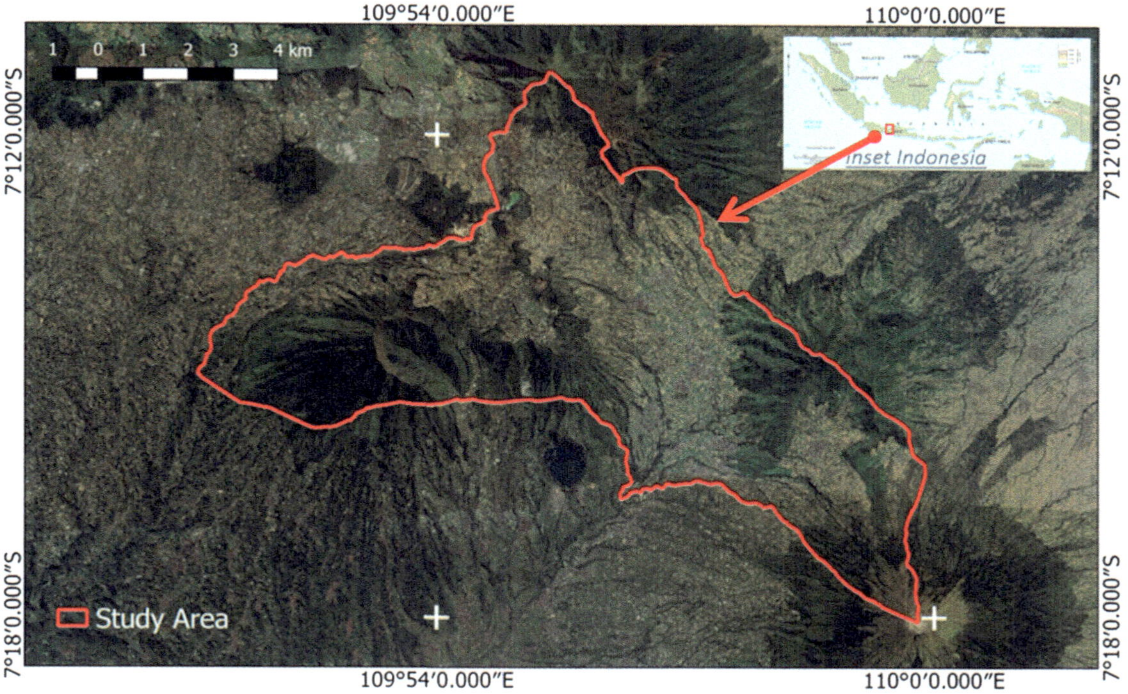

Fig. 30.1 Location of the study area

30.2.4 Assessment of Community Resilience

As this study is also aimed to discuss the resilience condition of the community living in the study area, a set of Resilience Index (RI) with its components were proposed. Based on previous study, the components of resilience index were divided into different dimensions or variables such as economic, social, physical/infrastructure, institutional, natural, and community capacity (Bruneau et al. 2003; Rose 2004;

Mayunga 2007; Cutter et al. 2008; Simpson 2006; Norris et al. 2008; Shaw 2009; Ainuddin and Routray 2012; Kusumastuti et al. 2014). Though many studies on resilience have been conducted, most of the scholars agreed that it is multifaceted which includes different aspects such as social, economic, ecology, institution, infrastructure, and community (Bruneau et al. 2003; Cutter et al. 2008; Norris et al. 2008).

Community resilience was calculated by comparing preparedness and vulnerability scores from dimensions

Fig. 30.2 (a) Agriculture expansion in forest area; (b) Settlement distribution in higher area

Table 30.1 Data needs

Data and information need	Sources	Data types
Non-spatial data for community resilience assessment (see Table 30.4 for details)		
Preparedness Economic, social, infrastructure, community capacity, and institution *Vulnerability* Economic, social, infrastructure, community capacity, institution, and hazard	Micro-level survey through household samples with standardized questionnaires distribution	Primary data
Spatial data for soil erosion assessment (year 2007 and 2017)		
Administrative/boundary Land use Soil type and quality Topography Climate/rainfall Other related and derived spatial data	Satellite images (landsat images TM 5 for 2007 and Landsat 8 for 2017), DEM data from IFSAR 5 m, analog/digital maps from Geospatial Information Agency, local planning and development board, local agriculture agency	Secondary and primary data

mentioned above. As stated by Simpson (2006), the comparison of preparedness and vulnerability results in community resilience. Preparedness describes the capacity of the community in coping with disaster, while vulnerability concerns on the disaster exposure. Scoring method was applied in order to obtain a more realistic resilience condition in the study area. This study used six dimensions (social, economy, institution, infrastructure, natural, and community capacity) as the basis for the assessment of community resilience. Natural dimension was named as hazard (Shaw 2009) which related to land degradation (Table 30.4 for details).

To obtain the final score of each component, both preparedness score and vulnerability score were determined as the amount of score from all dimensions, while the dimension scores were resulted from the average of its indicators. Score Indicator (SI) was calculated following this formula:

$$\text{SI}_x = \frac{T_x}{n} \qquad (30.2)$$

where SI_x = Score of indicator x, T_x = total score of indicator x, and n is the number of samples. When the values for each indicator were calculated, then they were summed up to obtain score of dimensions.

Once preparedness and vulnerability were calculated, the two contributing components are then combined using the equation:

$$\text{RS} = \frac{\text{PS}}{\text{VS}} \qquad (30.3)$$

where RS = resilience score was resulted from the ratio of PS = preparedness score to VS = vulnerability score. The resilience of community then produced by associating preparedness versus the vulnerability. Resilience score is divided into three classes: low resilience, moderate resilience,

and high resilience. To decide the length of each class interval, we use this formula:

$$\text{Length of class interval} = \frac{S_{\max} - S_{\min}}{\text{Number of classes}} \qquad (30.4)$$

The length of class interval is determined by number of classes and maximum – minimum scores. The maximum score is 3.00 while the minimum score is 1.00, and the number of classes is 3, then the length of class is 0.74. So, as the final classification, the resilience score *0.27–1.01 classified as low resilience, 1.02–1.76 classified as moderate resilience, and 1.77–2.50 classified as high resilience.*

30.3 Results and Discussion

30.3.1 Soil Erosion

The analysis of soil erosion was completed by comparing potential soil loss in 2007 and 2017. The purpose in comparing those two is to show the physical environmental change for 10 years as well as to see the changing of environmental risk due to agricultural development in the study area. As shown in Figs. 30.3 and 30.4, potential soil loss in 2007 and 2017 was classified into four classifications from less than 15 tons/ha/year up to more than 180 tons/ha/year. Potential soil loss is dominated by the range from 15 to 60 tons/ha/year with a total area of 2983 ha in 2007 and 2711 in 2017 followed by less than 15 tons/ha/year, covered the area about 2374 ha in 2007 and 2235 in 2017. Spatially, potential soil loss of less than 15 tons/ha/year was mostly found and distributed in the southern east part of the study area, while soil loss from 15 to 60 tons/ha/year concentrated more in the middle and western part. Soil loss from 61 to 180 tons/ha/year was found more on the steep slope area which covered area of 1871 ha in 2007 and 2174 ha in 2017. The highest

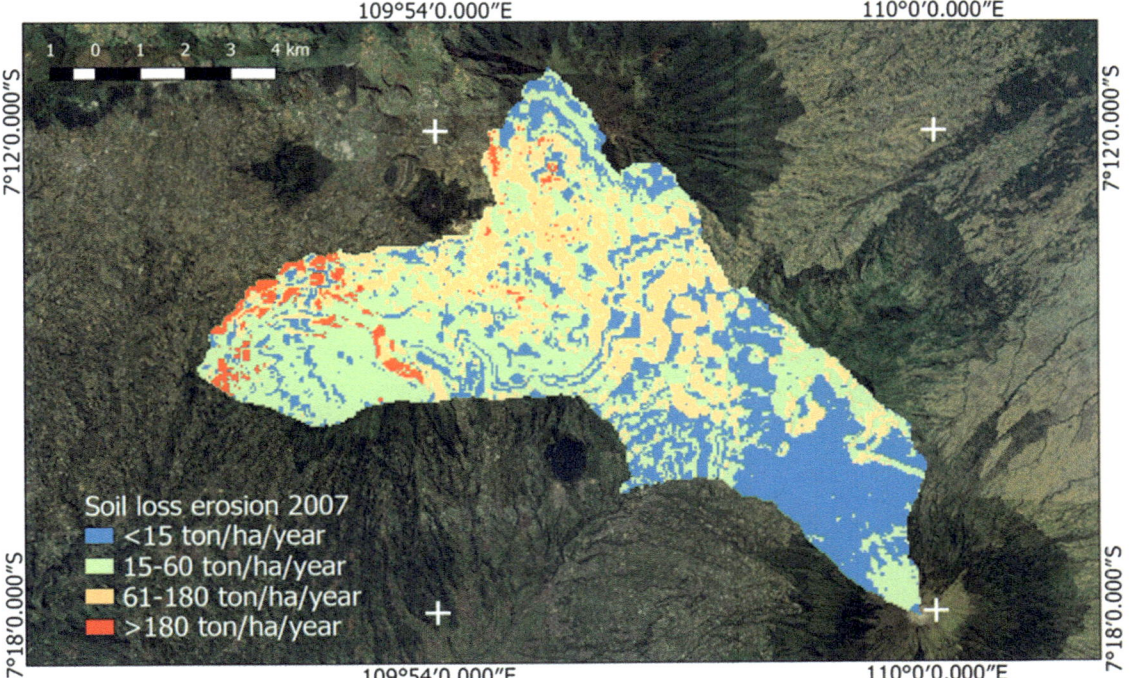

Fig. 30.3 Spatial distribution of soil loss in 2007

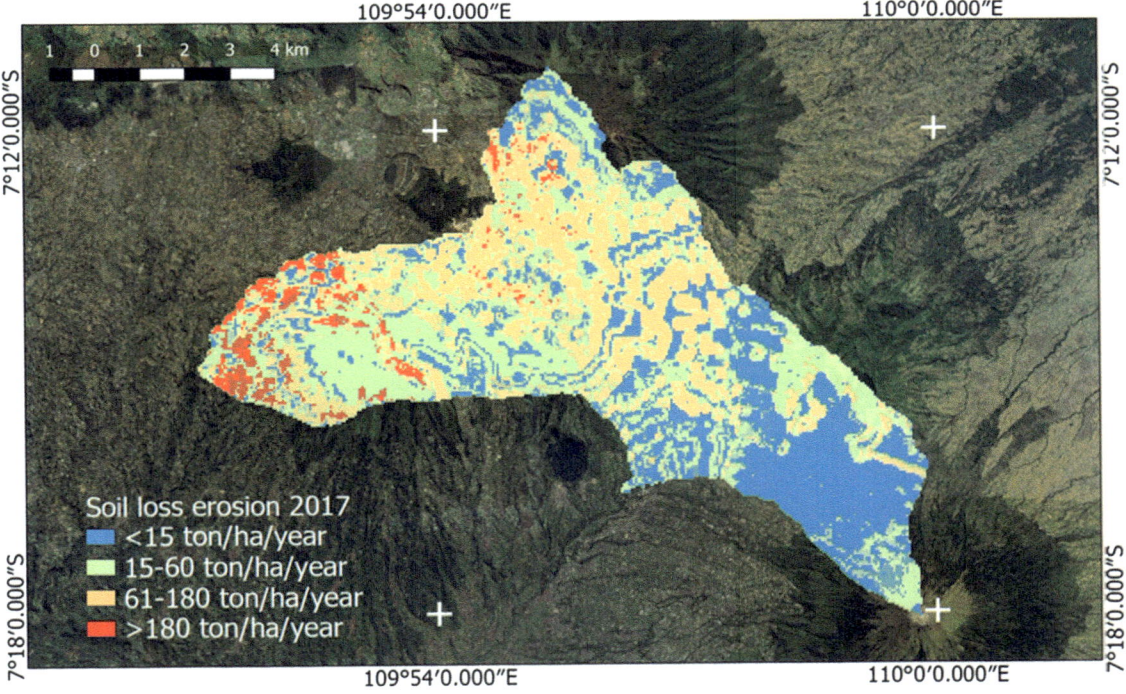

Fig. 30.4 Spatial distribution of soil loss in 2017

potential soil loss with more than 180 tons/ha/year is distributed more in very steep slope area particularly in the western part which covered the area of 226 ha in 2007 and 334 ha in 2017.

As the comparison (Table 30.2), during 10 years of development, soil loss has increased quite ominously for the rate of 61 to 180 tons/ha/year from 25.10% in 2007 to 29.17% in 2017. This increasing is also followed by the highest rate of

Table 30.2 Average rate of soil loss

Range of rate	Area (ha)				Difference	
	2007	%	2017	%	Area (ha)	%
<15 ton/ha/year	2374	31.85	2235	29.98	−139	−1.87
15–60 ton/ha/year	2983	40.02	2711	36.37	−272	−3.65
61–180 ton/ha/year	1871	25.10	2174	29.17	304	+4.07
>180 ton/ha/year	226	3.03	334	4.48	107	+1.45
Total	7454	100.00	7454	100.00		

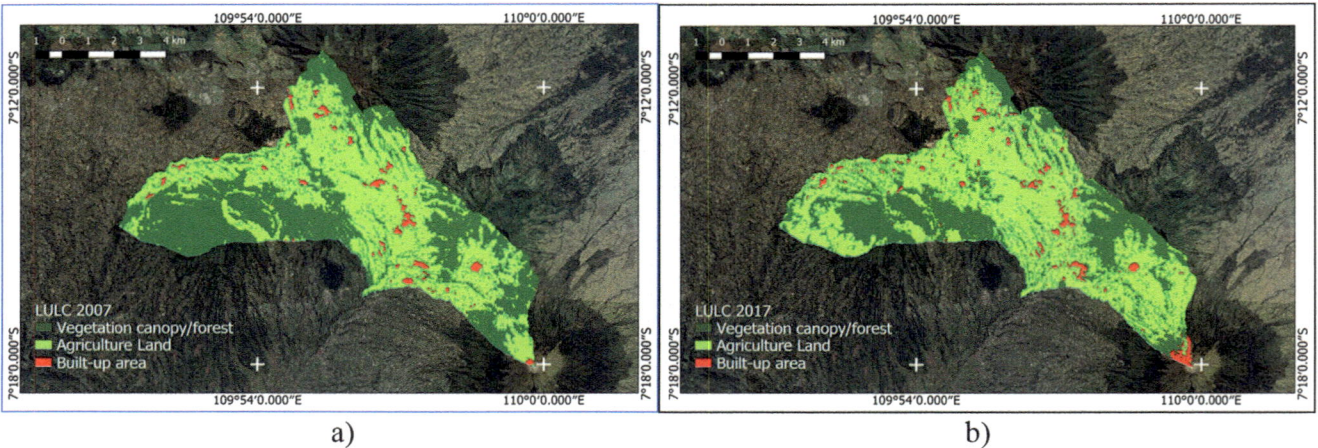

Fig. 30.5 Land use: (**a**) 2007 and (**b**) 2017

Table 30.3 Land use change

Land use	Area (ha)				Difference (ha)	
	2007	%	2017	%	Area (ha)	%
Vegetation canopy/forest	3273	43.91	2573	34.52	−700	−9.39
Agricultural	3865	51.85	4412	59.19	547	+7.34
Built-up area	316	4.24	469	6.29	153	+2.05
Total	7454	100.00	7454	100.00		

soil loss with more than 180 tons/ha/year with total difference 1.45% from 3.03% in 2007 to 4.48% in 2017. Simultaneously, two other rates of soil loss have been reduced from 31.85% in 2007 to 29.98% in 2017 for soil loss less than 15 tons/ha/year and from 40.02% in 2007 to 36.37% in 2017 for 15–60 tons/ha/year. Similar trend showed from previous study done by Rudiarto and Doppler (2013) from 1991 to 2006 where soil loss in Dieng Plateau increased for about 7.63% from 53.02% in 1991 to 60.65% in 2006 for the higher range of rates, whereas for the lower rates, the tendency showed declining rate for about 11.23% from 1991 to 2006.

As shown in Fig. 30.5a, b, the spatial distribution of land use has been changed significantly during 10 years' development at the higher area. This happened particularly to the forest land in the western and southern part of the study area. On the other hand, agricultural land seems more dominant in the area where forest land has been declined. Built up area is found more in 2017 distributed close to the current condition in 2007. In Table 30.3, forest land has been decreased for 9.39% from 3273 ha in 2007 to 2573 in 2017. Agricultural land increased 7.34% from 3865 ha in 2007 to 4412 ha in 2017, while built up area increased for 2.05% from 2007 to 2017. These results showed that deforestation in the study area is at critical phase, and as shown in Table 30.2, the soil erosion has been accelerated accordingly. The alteration of soil rates due to land use change significantly affects the amount and character of protection covering the surface materials (Park 2002). Therefore, the obstinacy of land from soil erosion risk is determined by a specific land use cover.

There are actually two ways of estimating erosion rates by using RUSLE model: potential erosion and actual erosion. Potential erosion implies only on the natural calculation (*R*,

K, L, and S factor) where no human interference has been considered. It means that under the natural conditions, the erosion still occurs. Accordingly, actual erosion is calculated by considering C and P factor as the human interference factor. The interference of human is heavily subjected for any kind of land use change activities, such as deforestation, land clearance, cultivation, and other forms of activities which obviously influences the soil erosion (Renschler et al. 1999; Ananda and Herath 2003; Rudiarto and Doppler 2013; Ganasri and Ramesh 2015). Therefore, the C-factor of vegetative cover as well as types of conservation practices play an important role in determining the actual rate of soil erosion in the study area. If the soil loss rates increased in a specific period, then it may be concluded that the practice of agricultural activities has influenced the CP factors' availability. This condition confirmed that the environmental risk has been subsequently increased due to shifting of human activities on land, such as agricultural expansion to the forest area (Zhao et al. 2013; Ganasri and Ramesh 2015).

30.3.2 Community Resilience

As mentioned before, community resilience is the proportion of preparedness to vulnerability in a specific location. Therefore, the community resilience is discussed in its components: preparedness and vulnerability. The complete scores for preparedness and vulnerability as well as its dimension and components are available in Table 30.4.

30.3.2.1 Preparedness Score

The dimension as well as the indicator scores of preparedness in the study area shows the total score of 7.48. The highest score is community capacity dimension, while the lowest is infrastructure dimension. The score of preparedness from the highest to the lowest are community capacity, social, institutional, economic and infrastructure. The higher score shows the better level of community preparedness. Figure 30.6a shows the level of preparedness in the study area, while the scores are available in Table 30.4.

The higher result of community capacity among other dimensions due to the effort applied by the farmers in soil management such as crop rotation and soil conservation. Farmers are willing to do crop rotation and grow additional commodities in order to give more economic value to the family. Since Dieng Plateau is located on various sloppy areas, ranged from plain to highly step slope areas, terrace system with additional soil conservation practice such as stone has been commonly found. Poor agricultural practice related to soil conservation has been one of the main issues

Table 30.4 Score of vulnerability and preparedness dimensions and indicators

Components of resilience	Dimensions	Indicators	Score	Mean	Total
Vulnerability	Social	Dependency ratio	1.49	1.73	13.34
		Level education of household head	2.33		
		Social conflict	1.51		
	Community Capacity	Knowledge on farming practices	2.60	1.90	
		Understanding of land degradation	1.54		
		Understanding of soil conservation	1.55		
	Economic	Source of income	2.76	2.59	
		Decrease of agricultural yield	2.24		
		Decrease of household income	2.78		
	Institutional	Number of government assistantship programs	1.66	2.01	
		Access to local government assistance programs	2.37		
	Infrastructure	The availability of irrigation facilities	2.46	2.46	
	Hazard	Crop failure	2.82	2.60	
		Soil erosion	2.72		
		Number of natural disaster in past 3 years	2.25		
Preparedness	Social	Farmer group activities	1.99	1.73	7.48
		Farmers' participation in farmer group	1.48		
	Community Capacity	Crop rotation on agricultural land	2.01	2.02	
		Land conservation on agricultural land	2.03		
	Economic	Saving ownership	1.31	1.36	
		Additional jobs available	1.40		
	Institutional	Socialization of sustainable agricultural	1.52	1.37	
		Soil conservation programs from governance	1.22		
	Infrastructure	The availability of cheek dam in agricultural land	1.00	1.00	
		The availability of diversion ditch in agricultural land	1.00		

Source: Simpson (2006), Cutter et al. (2008), Shaw (2009), Cutter et al. (2010), Kusumastuti et al. (2014)

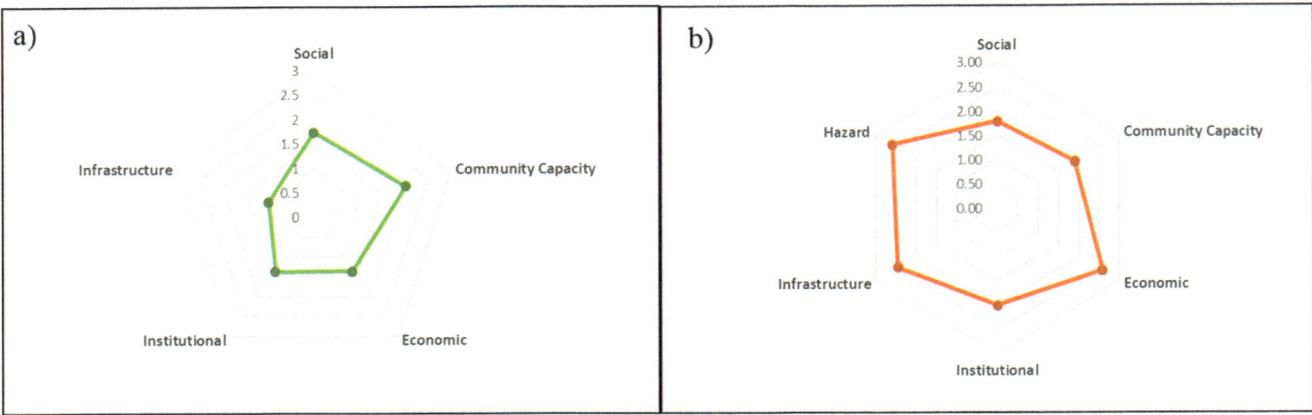

Fig. 30.6 (a) Preparedness and (b) vulnerability

among other causes in soil erosion (Kelly et al. 2015). Therefore, farmers have to develop their capacity to overcome the natural challenge related to soil and land conditions that exist in the study area in order to reduce soil loss and environmental risk to sustain agricultural activity.

Social dimension is the second main issue in terms of preparedness with a moderate score (1.73). Two components in this dimension were assessed, i.e., farmer group activities and farmers' participation in the farmer's group. The score for farmer group activities was higher (1.99) as compared to farmer's participation (1.48). The assessment of activities and participation of the farmers in social dimension is to show the level of social interaction and intervention. Farmer group can be used as an opportunity to build partnerships as well as to develop and to transfer knowledge and experience in regard to sustainable farming system.

In terms of institutional dimension, the preparedness score is relatively low (1.37) from two aspects, i.e., socialization of sustainable agricultural and land conservation programs from government. Based on the questionnaire results, socialization of sustainable agriculture was applied instead of land conservation programs. It is indicated by the score 1.52 for socialization of sustainable agriculture and 1.22 for land conservation programs. It shows that land conservations program is likely less followed by the community. However, although government program was not always followed by the community, government programs are actually there to encourage farmers in enhancing environmental awareness due to land degradation.

The economic dimension in preparedness shows also a relatively low score (1.36). This happens due to low saving ownership (1.31) and lack of additional job (1.40). Low-saving ownership indicates that farm families in the study area have less ability to organize and to manage their

income and expenses. The ability to saving money is also considered as an important part in dealing with uncertainty and inevitable disturbances. Additional job availability was found limited since most of the farmers were less educated which may lessen the opportunity to get diversified jobs. Additional job is very much related to the education level and skill owned by the farmers.

Among all dimension of preparedness, infrastructure was found as the lowest score with only 1.00. This lowest score was assessed from the availability of check dam and diversion ditch which showed score 1.00 for both. All respondents stated that there are no such agricultural infrastructures applied in order to reduce the effect of soil erosion in their farm field. Therefore, the farmers are not well prepared for soil erosion which makes them vulnerable.

30.3.2.2 Vulnerability Score

As shown in Table 30.4, the vulnerability score is 13.34 with hazard dimension as the highest score and social dimension as the lowest. The complete score of each dimension in vulnerability aspects leads by hazard (2.60), economic (2.59), infrastructure (2.46), institutional (2.01), community capacity (1.90), and social (1.78). The higher the score indicates, the more vulnerable the community. The vulnerability diagram from all dimensions is presented in Fig. 30.6b.

Hazard dimension identified as the most vulnerable among other dimension. The score of hazard dimension in vulnerability aspect was calculated from three components, i.e., crop failure (2.82), soil erosion (2.72), and a number of natural disasters for past 3 years (2.25). The higher score of hazard dimension is due to most of farm families experienced with crop failure and having trouble with unpredictable risk. High level of land degradation in the study area was not followed by proper land management which directly influenced soil fertility and increasing natural hazard. Hazard dimension is very close to different aspects such as soil quality, water quality and availability, type of vegetation, and terrain accessibility (Basso et al. 2010; Sendzimir et al. 2011). From field survey, there were about 20 landslides and 8 floods disaster that occurred in the study area for the last 3 years. According to TKPD (2013), the intensity of landslide and flood disaster annually occurred in most of the villages in Dieng Plateau and obviously becomes a major thread for livelihood sustainability in the future.

In terms of economic dimension, land degradation has been contributing to the vulnerability more on decreasing of household income (2.78) followed by source of income (2.76) and decreasing agricultural yield (2.24). Those three components were interconnected where land degradation may result in the destruction of farm field or decreasing of soil quality. Disturbance on farm filed due to land degradation may result to decreasing of agricultural yield and consequently less income gain. The farmers therefore need to find solutions and alternatives for future income possibilities to cope with economic vulnerability.

Infrastructure dimension was assessed from the availability of irrigation facilities which shows a high vulnerable score (2.46). Since the study area is a sloppy area with high different of altitudes, water distribution has become a major challenge. Irrigation facilities are one of the basic components of agriculture which can affect the soil fertility. Farmers build non-permanent irrigation channels in dry season and detach it in rainy season. Farmers with limited budget built small pond in their farm field to overcome water need but only for temporarily.

Institutional vulnerability is quite vulnerable with score 2.01 from two components, i.e., number and access to government assistantship program. Overall, farm families get access to local government assistance programs. There are a quite number of assistantship programs from local government in order to increase the quality of farm families' life. Similar to institutional vulnerability, community capacity also has a quite vulnerable score (1.90). Vulnerable score in community capacity was derived from three indicators, i.e., knowledge on farming practices (2.60), understanding of land degradation (1.54), and soil conservation (1.55).

Social dimension was found as the less vulnerable dimension with total score of 1.78 from three indicators: dependency ratio (1.49), social conflict (1.51), and the level of education (2.33). From the field survey, it was found that most of the respondents were low educated with elementary or primary school graduated. It is also relevance with the community capacity dimension particularly knowledge on farming practices which indicated vulnerable since the education level is low.

30.3.2.3 Resilience Score

The resilience score of farm families due to land degradation in the study area is 0.56 classified as low resilience (0.27–1.01) (Table 30.5). With that classification, farm families were not resilient concerning land degradation as the capacity to cope with the threats in terms of preparedness is lower than its vulnerability. Nevertheless, since the resilience is derived from the comparison of preparedness and vulnerability aspects, therefore, change of scores of each indicator in those two aspects is influential. More effort in enhancing preparedness components will subsequently reduce the vulnerability and at the same time, level of resilience improves. In simple term, people are more resilient if they are less vulnerable and more prepared within a specific condition.

The contribution of vulnerability aspect in hazard, economic, and infrastructure dimension was found less resilient as the score in vulnerability is high. Hazard dimension was identified as the most vulnerable while social dimension as the less vulnerable. The high level of vulnerability in hazard dimension is due to high-level soil erosion followed by other disasters such as flood that occurred in the study area. On the other hand, inappropriate agricultural practices such as intensification and over use of soil have been contributing to higher level of land degradation. Agricultural practices therefore have become one of the major contributions in soil erosion (Collins et al. 2001; Qian 2002; Nunes and Seixas 2003). Economic dimension has significant role in creating community resilience and land degradation processes and even previous study argued that both of them is the most important to the community (Gray and Moseley 2005; van Oudenhoven et al. 2011). Land degradation reduced soil quality where top soil as the main element washed out by the water. Therefore, the existence of land resource as the main source in agriculture sector can threat the livelihood sustainability. Infrastructure dimension in terms of the availability of irrigation facilities is one of the major determinants in farm production. A well-prepared and managed irrigation facility determines the sustainability of water supply.

Other dimensions of vulnerability aspect such as community capacity, social, and institutional were also significant in shaping community resilience. Social and community capacity showed a more resilience condition while institutional was

Table 30.5 Resilience score

Resilience score			
Preparedness	Vulnerability	Total score	Range of classification
7.48	13.34	**0.56**	**0.27–1.01**	**Low**
			1.02–1.76	Moderate
			1.77–2.50	High

The value of 0.56 is calculated from 7.48/13.34 while classification value is the range for each resilience category. The value 0.56 is in the range between 0.27 and 1.01, categorized as low resilience fairly resilience. The result also showed that farm families were more vulnerable in terms of knowledge of farming practices. Socialization and education in various types of soil conservation are imperative issues to fill out the lack of knowledge in farming practice as well as more assistantship delivery against land degradation and other natural hazard in the study area. However, it is very often found that ideas and suggestion toward more resilience trajectories were hardly applied. Stakeholders are sometimes reluctant to break through the dependencies on current systems or traditions which called as cultural resistance (Burton et al. 2008). Therefore, a more convincing approach followed by simple practices is more workable for the community since they like to see more simple and applicable approaches.

Concerning the preparedness aspect, it was found that infrastructure dimension as the less resilience and community capacity was the most resilience dimension. Agricultural infrastructure is noteworthy to reduce the risk of land degradation. This issue is also related to farmers' knowledge in farming practice where the existence of agriculture infrastructures was not really necessary. In their opinion, agricultural infrastructure will reduce the farm size and having these infrastructures mean more money to spend. They have to expend more money to build it. It is also found that farmer's involvement in government program and social interaction are relatively low which makes them less resilience both in institutional and social dimension. A limited number of farm family heads reported to have a strong social network in farmers' group. To be the member in farmer's group is of importance to the farmer, as it may increase the opportunity to discuss similar circumstances and access the relevance issues in improving farm productivity. Social dimension further gives the instance of farmers into planning process through decision or policy making within the community (Kelly et al. 2015; Wilson et al. 2016). Economic dimension was found less resilient where economically the farmers have only small money left and, on the other hand, their education level is not adequate to access more jobs.

30.4 Conclusion

Land degradation in the study area is being escalated most by forest land conversion into agricultural land and conservation practices which were very limited. This escalation was indicated by the level of soil loss in 2007 and 2017. Agricultural expansion to the forest area with steep slope condition and poor land management are the major combinations in accelerating soil loss. It is confirmed by the analysis that most of soil loss increment to higher level is due to those reasons. Land expansion for agricultural purpose and low level of soil

conservation techniques indicate the needs of developing oriented strategies. However, the function of land is not only as the production factor but also as other important factors that support human life, like recreation, cultural, habitat, and regulation function. Hence, to have a good erosion control, we need to understand the historical changes of the environment in a particular area.

Vulnerability aspect is more dominant than preparedness aspect in determining the level of resilience in the study area. In principal, community very much depends on the higher level of preparedness and lower level of vulnerability to achieve more community resilience. Dimensions and indicators in community resilience calculations are able to inform household, farm families, and even stakeholders on what kind of variables should be addressed for more resilience community in facing land degradation issues. It is a decisive method entirely based on the condition of the community. High level of community resilience can be further improved by enhancing community preparedness concerning the availability of agricultural infrastructures, government assistance programs in soil conservation, and socioeconomic characteristics of farm families. Therefore, micro-level approaches in order to portray more real condition on how land degradation impact livelihood development is a great challenge for future research.

References

Adger, W. N. (2000). Social and ecological resilience: are they related? *Progress in human geography, 24*(3), 347-364.

Ainuddin, S., & Routray, J. K. (2012). Community resilience framework for an earthquake prone area in Baluchistan. *International Journal of Disaster Risk Reduction, 2*, 25-36.

Ananda, J., & Herath, G. (2003). Soil erosion in developing countries: A socio-economic appraisal. *Journal of Environmental Management, 68*(4), 343–353. https://doi.org/10.1016/S0301-4797(03)00082-3

Angima, S. D., Stott, D. E., O'Neill, M. K., Ong, C. K., & Weesies, G. A. (2003). Soil erosion prediction using RUSLE for central Kenyan highland conditions. *Agriculture, Ecosystems and Environment, 97*(1–3), 295–308. https://doi.org/10.1016/S0167-8809(03)00011-2.

Basso, B., De Simone, L., Ferrara, A., Cammarano, D., Cafiero, G., Yeh, M. L., & Chou, T. Y. (2010). Analysis of contributing factors to desertification and mitigation measures in Basilicata region. Italian Journal of Agronomy, 5(3S), 33-44.

Bruneau, M., Chang, S. E., Eguchi, R. T., Lee, G. C., O'Rourke, T. D., Reinhorn, A. M., ... & Von Winterfeldt, D. (2003). A framework to quantitatively assess and enhance the seismic resilience of communities. *Earthquake spectra, 19*(4), 733-752.

Burton, R. J., Kuczera, C., & Schwarz, G. (2008). Exploring farmers' cultural resistance to voluntary agri-environmental schemes. *Sociologia ruralis, 48*(1), 16-37.

Collins, A. L., Walling, D. E., Sichingabula, H. M., & Leeks, G. J. L. (2001). Using 137Cs measurements to quantify soil erosion and redistribution rates for areas under different land use in the Upper Kaleya River basin, southern Zambia. *Geoderma, 104*(3-4), 299-323.

Cote, M., & Nightingale, A. J. (2012). Resilience thinking meets social theory: situating social change in socio-ecological systems (SES) research. *Progress in Human Geography, 36*(4), 475-489.

Cutter, S. L., Barnes, L., Berry, M., Burton, C., Evans, E., Tate, E., & Webb, J. (2008). A place-based model for understanding community resilience to natural disasters. *Global Environmental Change, 18*(4), 598–606. https://doi.org/10.1016/j.gloenvcha.2008.07.013.

Cutter, S. L., Burton, C. G., & Emrich, C. T. (2010). Disaster Resilience Indicators for Benchmarking Baseline Conditions. *Journal Of Homeland Security and Emergency Management, 7*(1), 1–52. https://doi.org/10.2202/1547-7355.1732.

Douglas, M. (1994). *Sustainable use of agricultural soils: a review of the prerequisites for success or failure.* Berne, Switzerland.

Ganasri, B. P., & Ramesh, H. (2015). Assessment of soil erosion by RUSLE model using remote sensing and GIS - A case study of Nethravathi Basin. *Geoscience Frontiers, 7*(6), 1–9. https://doi.org/10.1016/j.gsf.2015.10.007.

Geist, H. (Ed.). (2005). Our earth's changing land: an encyclopedia of land-use and land-cover change. Greenwood Publishing Group.

Gray, L. C., & Moseley, W. G. (2005). A geographical perspective on poverty–environment interactions. *Geographical Journal, 171*(1), 9-23.

Kelly, C., Ferrara, A., Wilson, G. A., Ripullone, F., Nolè, A., Harmer, N., & Salvati, L. (2015). Community resilience and land degradation in forest and shrubland socio-ecological systems: Evidence from Gorgoglione, Basilicata, Italy. *Land Use Policy, 46*, 11–20. https://doi.org/10.1016/j.landusepol.2015.01.026.

Kusumastuti, R. D., Husodo, Z. A., Suardi, L., & Danarsari, D. N. (2014). Developing a resilience index towards natural disasters in Indonesia. *International journal of disaster risk reduction, 10*, 327-340.

Mayunga, J. S. (2007). *Understanding and Applying the Concept of Community Disaster Resilience: A Capital-based Approach. Megacities as Hotspots of Risk: Social Vulnerability and Resilience Building.* https://doi.org/10.1146/annurev.energy.32.051807.090348.

Norris, F. H., Stevens, S. P., Pfefferbaum, B., Wyche, K. F., & Pfefferbaum, R. L. (2008). Community resilience as a metaphor, theory, set of capacities, and strategy for disaster readiness. *American journal of community psychology, 41*(1-2), 127-150.

Nunes, J. P. C., & Seixas, J. (2003). Impacts of extreme rainfall events on hydrological soil erosion patterns: application to a Mediterranean watershed. *World Resource Review, 15*(3), 336-351.

Park, C. (2002). *The environment: principles and applications.* Routledge. London.

Qian, C. (2002). Soil Erosion and Ecological Reestablishment in the Nianchu River Valley in Tibet. *Chinese Journal of Ecology, 21*(5), 74-77.

Renard, K. G., Foster, G. R., Weesies, G. A., McCool, D. K., & Yoder, D. C. (1997). Predicting soil erosion by water: a guide to conservation planning with the Revised Universal Soil Loss Equation (RUSLE) (Vol. 703). United States Department of Agriculture, Washington, DC.

Renschler, C. S., Mannaerts, C., & Diekkrüger, B. (1999). Evaluating spatial and temporal variability in soil erosion risk - Rainfall erosivity and soil loss ratios in Andalusia, Spain. *Catena, 34*(3–4), 209–225. https://doi.org/10.1016/S0341-8162(98)00117-9.

Rose, A. (2004). Defining and measuring economic resilience to disasters. *Disaster Prevention and Management: An International Journal, 13*(4), 307-314.

Rudiarto, I., & Doppler, W. (2013). Impact of land use change in accelerating soil erosion in Indonesian upland area : A case of Dieng Plateau, Central Java - Indonesia. *International Journal of AgriScience, 3*(July), 558–576.

Sendzimir, J., Reij, C. P., & Magnuszewski, P. (2011). Rebuilding resilience in the Sahel: regreening in the Maradi and Zinder regions of Niger. *Ecology and Society, 16*(3).

Shaw, R. (2009). Climate disaster resilience: Focus on coastal urban studies in Asia. *Asian Journal of Environment and Disaster Management, 1*(June), 1–15.

Shinde, V., Tiwari, K. N., & Singh, M. (2011). Prioritization of micro watersheds on the basis of soil erosion hazard using remote sensing and geographic information system, 2(July), 1–7. Retrieved from papers3://publication/uuid/74DFB22B-3598-48E8-A48C-C19AE6DEECCF

Simpson, D. M. (2006). Indicator Issues and Proposed Framework for a Disaster Preparedness Index (DPi). *Center for Hazard Research and Policy Development University of Louisville*, 1–18. Retrieved from http://www.fritzinstitute.org/PDFs/WhitePaper/DaveSimpson IndicatorsRepor.pdf.

TKPD (Tim Kerja Pemulihan Dieng/Local Working Group for Dieng Restoration). (2013). Penyusunan Analisis Sosial Budaya dan Mitigasi Bencana untuk Mendukung Rehabilitasi dan Konservasi Lahan di Kawasan Dieng-Wonosobo. Yogyakarta: Wanamukti Mandiri.

van Oudenhoven, F. J., Mijatović, D., & Eyzaguirre, P. B. (2011). Social-ecological indicators of resilience in agrarian and natural landscapes. *Management of Environmental Quality: An International Journal, 22*(2), 154-173.

Wilson, G.A. (2012). Community resilience and environmental transitions. London: Routledge/Earthscan.

Wilson, G., Quaranta, G., Kelly, C., & Salvia, R. (2016). Community resilience, land degradation and endogenous lock-in effects: evidence from the Alento region, Campania, Italy. *Journal of Environmental Planning and Management, 59*(3), 518-537.

Wischmeier, W.H., Smith, D.D. (1978). Predicting rainfall erosion losses. A Guide to conservation planning. United States Department of Agriculture, Agricultural Research Service (USDA-ARS) Handbook No. 537. United States Government Printing Office, Washington, DC.

Zhao, G., Mu, X., Wen, Z., Wang, F., & Gao, P. (2013). Soil erosion, conservation, and eco-environment changes in the Loess Plateau of China. Land Degradation & Development, 24(5), 499-510.

Iwan Rudiarto obtained his bachelor's degree in urban and regional planning from Diponegoro University, Indonesia, in 1998. He continued his master's degree in Germany and pursued his Master of Science (M. Sc.) in Land Management and Land Tenure from Technical University of Munich in 2005. In 2010, he graduated with Dr. sc. agr. from Hohenheim University, Germany. Currently, he is working as a lecturer and researcher at the same department where he obtained his bachelor's degree, for more about 20 years. His expertise is majorly in rural planning and development, land management, and GIS modeling. His current research, which is mostly funded by the Ministry of Research, Technology and Higher Education and Diponegoro University, is related to disaster management, urban and rural resilience, and rural development. He is also active in *Geoplanning Journal* as one of the editors.

Isna Rahmawati completed her bachelor's degree in geography from State University of Semarang, Indonesia. She has recently graduated from Master's Program in Environmental Studies, Diponegoro

University, Indonesia, and entitled with Master of Science (M.Sc.) degree. Her interest is mostly on environmental and socioeconomic issues. She had joined several environmental volunteer programs such as planting mangrove trees and cleaning up of river and coastal areas to preserve environment in various institutions. Currently, she is doing research on environment and community resilience in a mountain area of Java, Indonesia.

Anang Wahyu Sejati is working as a lecturer and researcher in the Department of Urban and Regional Planning, Diponegoro University, Indonesia. He received his bachelor's degree and Master of Urban and Regional Planning (MURP) from Diponegoro University, Indonesia. His research interest is spatial modeling for urban and regional planning with GIS and remote sensing.

Taofeeq Sholagberu Abdulkadir, Raza Ul Mustafa Muhammad, Olayinka Gafar Okeola, Wan Yusof Khamaruzaman, Bashir Adelodun, and Saheed Adeniyi Aremu

Abstract

Land degradation in the form of erosion is a serious geo-hazard threatening land and water resources sustainability. Its socioeconomic and ecological impacts necessitated its geospatial prediction via susceptibility analysis. Accuracy of susceptibility mapping depends largely on modeling techniques and causative factors (CFs) considered. The study implements scarcely used CFs and support vector machine (SVM) technique for geospatial prediction of soil erosion for low and peak rainfall cycles in a complex watershed of Cameron Highlands Malaysia. The CFs considered are non-redundant static (drainage density, length slope, lineament density, and erodibility) along with some dynamic (surface temperature, soil moisture index, rainfall erosivity, and vegetation index) CFs. Four kernel functions of SVM with optimized kernel parameters were used for spatial prediction of erosion for both rainfall cycles. The model performances were validated using the commonest evaluation criteria. The results indicated that polynomial SVM outperformed other models and was used in developing soil erosion susceptibility maps. The distribution pattern of erosion showed that majority of the area are "low and moderately" susceptible. The analysis shows that most of the highly susceptible locations are within the urban and agriculture land-use. The study provides information on erosional land forms that could be advanced to gullies. This medium-scale susceptibility map produced could be useful for sustainable watershed management for mitigating erosion.

T. S. Abdulkadir (✉)
Department of Water Resources and Environmental Engineering, University of Ilorin, Ilorin, Kwara State, Nigeria

Department of Civil and Environmental Engineering, UniversitiTeknologi PETRONAS, Bandar Seri Iskandar, Perak, Malaysia

R. U. M. Muhammad
Department of Civil and Environmental Engineering, UniversitiTeknologi PETRONAS, Bandar Seri Iskandar, Perak, Malaysia

Centre for Urban Resource Sustainability, Institute of Self-Sustainable Building, UniversitiTeknologi PETRONAS, Seri Iskandar, Perak, Malaysia

O. G. Okeola
Department of Water Resources and Environmental Engineering, University of Ilorin, Ilorin, Kwara State, Nigeria

W. Y. Khamaruzaman
Department of Civil and Environmental Engineering, UniversitiTeknologi PETRONAS, Bandar Seri Iskandar, Perak, Malaysia

B. Adelodun
Department of Agricultural and Biosystems Engineering, University of Ilorin, Ilorin, Kwara State, Nigeria

S. A. Aremu
Department of Water Resources and Environmental Engineering, University of Ilorin, Ilorin, Kwara State, Nigeria

Lower Niger River Basin Development Authority, Ilorin, Kwara State, Nigeria

Keywords

SVM · Rainfall · Dynamic factors · Gully erosion · Susceptibility

31.1 Introduction

The key basic human needs are intricately linked to water-food nexus, and un-abating world population increase is mounting pressure on land exploitation with resultant impacts on soil and water. Soil erosion, one of the major causes of land degradation (Agnesi et al. 2011; George and Anu 2018; Nampak et al. 2018; Rizeei et al. 2016), is a key factor that endangers water and soil (Arabameri et al. 2018). Decisions made relating to water and food resources are often politically contested (Allouche et al. 2014; Williams et al. 2014). The consequential role of piping in landslides and gully erosion cannot be overemphasized. Piping, the movement of fine particles out of the soil layer by seepage forces, is a key suspect in the commencement of quite process leading to gully erosion, landslides, and other direct land degradation processes (George and Anu 2018). Asia is ranked second globally with average erosion rate of 16.6 Mg/ha/year followed by South America with a rate of 22.1 Mg/ha/year (Walling and Webb 1983). It results in economic loss, changes of rural livelihood pattern (Kerr 1998), natural resource degradation, and increasing sediment deposition (Kelley and Nater 2000) and has impact on sustainable development and ecosystem services (Pimentel and Burgess 2013; Tinker 1997). Due to these, several studies are ongoing particularly on the predictability of this phenomenon before unleashing havocs on lives and economy. Gully erosion develops in watercourses or other locations where runoff concentrates. It started from rill erosion which may later advances to gullies. It dissects lands, reducing its productivity and values via destruction of farm, harboring vermin, sediment generation, etc. Thus, gully erosion is arguably a huge threat in the attainment of five UN sustainable development goals (SDGs) for this region: SDSs 1 (No poverty), 2 (Zero hunger), 3 (Good health and well-being), 6 (Clean water and sanitation), and 11 (sustainable cities and communities). The SDGs are follow-up of the MDGs established as a response to world poverty, inequality, and insecurity, which have developed into the drivers of resource management. It is imperative to intensify research efforts for the prediction of erosion susceptible locations and mitigate its impacts.

Several studies have been carried out in the last decades on developing models for the evaluation of erosion rates and its susceptibility analysis and mapping at different scales and objectives (Agnesi et al. 2011; Dewitte et al. 2015; Gómez-Gutiérrez et al. 2015; Khatun 2017; Lucà et al. 2011; Mararakanye and Le Roux 2012; Pournader et al. 2018; Sujatha and Sridhar 2018; Wang et al. 2016). Most methods quantify eroded soil volumes by empirically or physically based approach linking soil loss rates to the values of a set of environmental variables or mechanical properties of terrains. The common method is Universal Soil Loss Equation (USLE), and its revised version, RUSLE model (Pham et al. 2018), is widely used in the developing countries because of their simplicity and flexibility and it needs less data than most of the other erosion models (Duarte et al. 2016; Thlakma et al. 2018). Other models that find applicability in soil erosion susceptibility assessment for their compatible integration with GIS are Soil and Water Assessment Tool (SWAT), European Soil Erosion Model (EUROSEM), and Annualized Agricultural Non-Point Source (AnnAGNPS) (Kouli et al. 2009; Thlakma et al. 2018). Other models for assessing soil erosion and gully rate determination quantitatively and qualitatively include Erosion Potential Method; Modified Pacific Southwest Interagency

Committee Model (MPSIAC); Chemicals, Runoff, and Erosion from Agricultural Management Systems (CREAMS); Water Erosion Prediction Project (WEEP); and Ephemeral Gully Erosion Model (EGEM) (Althuwaynee et al. 2014; Barber and Mahler 2010; Pournader et al. 2018).

In measuring, mapping, and monitoring gully erosion, several approaches coupled with remote sensing and GIS have been developed and implemented to obtain spatiotemporal continuous gully information (Wang et al. 2016). For examples, high spatial resolution of classical aerial photography and terrestrial laser scanner (TLS) find applicability in the ability to measure gullies. However, advance use of unmanned aerial vehicles (UAVs) with integrated autopilot technology image acquisition are finding relevancy in the study of gully erosion (d'Oleire-Oltmanns et al. 2012; Hugenholtz et al. 2013; Stöcker et al. 2015). Also, automatic 3D photo-reconstruction techniques enhance image processing for those images are derived from uncalibrated and nonmetric cameras. The 3D photo-reconstruction techniques from the images obtained from UAV to generate high-resolution topographical data and ortho-images are being used for monitoring gully erosion (Stöcker et al. 2015; Wang et al. 2016).

Recently, statistical techniques have been employed for the evaluation of erosion susceptibility by defining the geostatistical relationships between the geographical variability of selected physical attributes and the spatial distribution of the evidence for the water-induced erosion processes. Thus, allowing the generation of maps defining the susceptibility levels and expressing the relative probability of future occurrence of erosion landforms (Conoscenti et al. 2008). Subsequently, the susceptibility analyses advanced with various statistical and machine learning methods such as logistic regression, analytical hierarchy processes, bivariate statistics, boosted regression trees, artificial neural network, random forest, weights-of-evidence, index of entropy, information value, maximum entropy, and frequency ratio for spatial prediction of susceptibility to various erosion types (Akgün and Türk 2011; Conforti et al. 2011; Conoscenti et al. 2014; Lucà et al. 2011; Pourghasemi et al. 2017; Rahmati et al. 2016, 2017). However, some machine learning techniques such as SVM are scarcely used in erosion susceptibility mapping despite its predictive capability. Pradhan (2013) emphasized that both the choice and quality of CFs and the effectiveness of modeling techniques could impact on the susceptibility accuracy. Analysis of the existing literatures indicated lack of specific guidelines in selecting CFs for susceptibility studies (Lee and Pradhan 2007; Tehrany et al. 2015). Therefore, various researchers have applied different CFs (with varying number and types) and modeling techniques for developing susceptibility maps. Extensive review by Prosdocimi et al. (2016) identified CFs as one of the present knowledge gaps in soil erosion modeling. Researchers believed that the accuracy of susceptibility mapping increases with increase in the number of CFs (Donati and Turrini 2002; Tehrany et al. 2015). Conversely, Remondo et al. (2003) opined that increasing number of CFs does not necessarily increase the susceptibility accuracy if the CFs involved are redundant. Survey of previous literature revealed dominate consideration of static CFs out of which many are redundant that could cause overweighing of the model results (Magliulo 2012). These factors do not respond to change in rainfall cycle and often remain unchanged for a relatively long period of time. However, previous studies are devoid of some crucial dynamic CFs associated with rainfall cycles despite roles in water-induced soil erosion. Therefore, this study aimed to develop geospatial prediction model for erosion susceptibility analysis using SVM technique for low and peak rainfall cycles in a complex watershed of Cameron Highlands, Malaysia. The region, located at $4°19'–4°37'$N and $101°21'–101°30'$E, is a mountainous landscape with an average elevation of approximately 1180 m, highest at Gunung Brinchang, of about 2032 m above mean sea level (IEA 2006). About 25% of the watershed have elevation less than 1000 m, while two-thirds fall within the range of 1100–1600 m (Gasim et al. 2009) and many of the other peaks are above 1524 m. Cameron Highlands is classified as a complex landscape which is predominantly steep with about 60% of land areas steeper than 20° (Fortuin 2006). This, coupled with high rain, makes it highly susceptible to gully erosion and landslides (Pradhan 2010). Eighty-one percent of the area has high erosion risk due to increasing human activities and nature of the watershed (World Wildlife Fund Malaysia 2002). The adverse impacts currently experienced include deterioration of water quality (Jaafar et al. 2010; Jamil et al. 2014; Khalik et al. 2013), landslide on steep slope terrains (Basith 2011; Conoscenti et al. 2008), and siltation of rivers and built water bodies, i.e., Ringlet Dam, which sometimes triggers flood reducing hydropower generation (IEA 2006; Jansen et al. 2012). Sustainable land-use management in a complex and fragile landscape like Cameron Highlands remains challenging to all the stakeholders (Aminuddin et al. 2005). In mitigating this menace, susceptibility assessment is required to geospatially predict active/potential erosion distribution and identify critical locations.

31.2 Materials and Methods

31.2.1 Causative Factors for Gully Erosion Process

Gully erosion process is majorly caused by surface runoff which is channeled across an unprotected land and washes away the available soil particles along the drainage lines. This process is often influenced by different watershed characteristics such as land-use/cover, soil type, rainfall characteristics, etc. Thus, several CFs have been implemented by different researchers for susceptibility analyses and mapping. This study considered topographic, land-use/cover, geology, hydrology, and climatic and soil factors which are broadly grouped into static and dynamic causative factors (CFs). These were obtained from different sources as presented in Table 31.1. The CFs were selected due to their impacts on soil erosion processes and previous implementation in erosion susceptibility analysis. The non-redundant static CFs considered are drainage density (Kavzoglu et al. 2015; Pal 2016; Vijith et al. 2012), length slope (Conoscenti et al. 2013; Lucà et al. 2011), lineament density (Pradeep et al. 2015; Vijith et al. 2012), and soil type (erodibility) (Conoscenti et al. 2013; Dube et al. 2014), and their detailed descriptions could be found in the aforementioned literatures. Similarly, dynamic CFs such as R-factor (Magliulo 2012), LST (Xue et al. 2011), and SMI (Jamali 2004; Vahabi and Nikkami 2008) have been highlighted by researchers to influence erosion. However, these dynamic factors are rarely considered in the susceptibility analysis despite their roles in erosion processes. Hence, they were introduced to achieve a valid and accurate erosion susceptibility mapping. The preprocessing and processing procedures for the extraction of both static and dynamic CFs are described in the following sections.

Table 31.1 Spatial data source and their purpose

S/no.	Spatial data	Format	Source	Purpose used for
1	DEM (5 m resolution)	Digital	Global Pixel Solutions, Malaysia	For extraction of static factors such as length slope (LS) factor, drainage density
2	Soil map	Digital	Malaysian Department of Agriculture, US Food and Agriculture Organization (FAO)	For extraction of soil erodibility factor
3	Multispectral Landsat-8 data (30 m resolution)	Digital	United States Geological Survey (USGS) database	For extraction of factors: land surface temperature (LST), soil moisture index (SMI), lineament, and NDVI
4	Rainfall data (daily data)	Excel format	Malaysian Department of Irrigation and Drainage	For computation and development of rainfall erosivity (R-factor)

31.2.1.1 Extraction of Lineament Density

Due to the roles of lineament density in soil erosion, it was considered in erosion susceptibility modeling. Lineaments were derived from atmospherically corrected Landsat-8 data using ArcMap® 10, ENVI 5.3, and Geomatica 2016 software. The following are the procedures for the extraction:

1. The principal component image (PCI) of Landsat-8-pan-sharpened reflected bands was used, as this carries most information suitable for lineament extraction. The pan-sharpened layer of Landsat-8 OLI bands was processed with Gram-Schmidt sharpened surface reflectance (GSSR). This serves as the input file into ENVI 5.3 to obtain PC1 (8-bit grayscale 15 m spatial resolution). The GSSR layer was loaded into ENVI 5.3 Classic in this sequence: *transform >principal_component>forward_PC_ rotation >compute_new_statistics_and_rotation*to obtain PCA. The output PCI was saved as an image file with a resolution of *"8-bit grayscale" in TIFF/GeoTIFF*format.

2. For the "automatic lineament" extraction, "LINE Module Control Panel" of Geomatica 2016 software was used to further process PC1 to extract the lines. PC1 was launched into Geomatica platform in this sequence: *Tool>Algorithm_Liberian>Line: Lineament_Extraction>LINE_Module_Control_Panel*. LINE module control panel settings were adjusted appropriately before running the program. The extracted lineaments were saved as ArcView Shapefile (.shp) as "lineaments" for subsequent processing in the ArcMap® software. ThePCI-LINE module extracts lineaments from the single 8-bit image through three basic steps: edge detection, thresholding, and curve extraction.

3. The output from Step 2 was processed in ArcMap® with split-line model to split the compounded lines into simple lines and edit the lineament attributes. The respective results from Steps 1 and 2 which are PC1 and extracted lineaments were both loaded onto ArcMap® platform. The lines were checked for the presence of compounded lines using editor command in ArcMap®. However, compounded lines are inevitable in such a large study area. In the context of lineament, compounded lines are the ones that composed of segmented lines which need to be splitted at the respective vertices. This was done by using "Split Line At Vertices" from data management on python modeler of ArcMap® to produce simple straight lines with two endpoints. Then, the file was renamed as "lineaments_split". Attribute Table of the "lineaments_split" was edited to add coordinates (x_1, y_1) and (x_2, y_2) to the endpoints of each line. Subsequently, lineament density was estimated around each grid cells within a specific search area using "Line Density" command of Spatial Analyst Tool to display the output raster file (line_density) using "lineaments_splitted" as input polyline features. Area unit and output cell size were set to square kilometers and 30, respectively, before running the program for line_density computation. The output (line_density) map was re-sampled using cubic convolution method during display.

31.2.1.2 Extraction of Length Slope Factor

Length slope (LS) is a topographic factor comprised of slope gradient and slope length factors derivable from digital elevation model (DEM). A relatively high-resolution DEM of 5 m was used for its derivation following the procedures presented below:

1. Prior to application of raw DEM, it was first assigned a Universal Transverse Mercator (UTM) projection, Northern zone 47 datum WGS 1984 coordinate system applicable to the study area.
2. The projected DEM was hydrologically corrected by filling the sinks. Then, the flow direction and flow accumulation were determined using flow direction tool in "ArcHydro tool" extension of ArcMap® 10.
3. The resulting map obtained was then used for the evaluation of slope map. Finally, LS-factor map was developed using Moore and Burch's Equation defined in Eq. (31.1) (Parveen and Kumar 2012).

$$LS = \left(FlowAcc \times \frac{resolution}{22.1} \right)^{0.6} \times \left(Sin(SlopeDem) \times \frac{0.0174}{0.09} \right)^{1.3} \quad (31.1)$$

where flowAcc is the flow accumulation and SlopeDem is the slope derived from DEM.

31.2.1.3 Extraction of Drainage Density

Drainage density map was extracted from hydrologically corrected DEM following the procedures described as follows:

1. The "logical option" of "math tool" of ArcMap was used to select the predefined stream lengths to be captured in the stream extraction. All stream features greater than 1000 m were selected using flow accumulation as the input file.
2. In the "hydrology tool," flow lengths were estimated using "stream lengths" greater than 1000 m as input file.
3. Using the "flow lengths" and "flow direction" as the input files, "stream links" and "stream order" were evaluated.
4. "Stream order" was then converted to "stream feature" for the development of drainage density map for the watershed.

31.2.1.4 Development of Rainfall Erosivity Factor

Daily rainfall data for the weather stations within and around the Cameron Highlands were obtained from Malaysia Department of Irrigation and Drainage (DID) for a period of 11 years (2006–2016). The missing rainfall data were fixed using nearest neighborhood interpolation technique. A widely applied mathematical model (in Eq. 31.2) proposed by Wischmeier and Smith (1958) for the estimation of R-factor from monthly rainfall records was adopted. The model was applied for each of the stations within and around the study area to evaluate the corresponding erosivity values. The estimated R-factor values for the stations with known weather stations' coordinates were imported into ArcMap®. Interpolation was carried out for other locations within the watershed using simple kriging interpolation technique to develop R-factor map.

$$R = \sum_{i=1}^{12} 1.735 \times 10^{1.5 \log_{10}\left[\frac{P_i}{P} - 0.08188\right]} \quad (31.2)$$

where R is the rainfall erosivity factor (MJ mm/ha/year), P_i is the monthly average rainfall for ith month (mm), and P is the annual average rainfall (mm) (Khosrokhani and Pradhan 2014).

31.2.1.5 Development of Soil Erodibility Factor

Soil erodibility is one of the key parameters to be considered in soil erosion susceptibility analysis (Wang et al. 2013). It measures the proneness of a particular soil type to erosion by actions of rainfall and runoff (Ali and Hagos 2016). Digital soil map for Cameron Highlands was extracted from the map obtained from the Malaysia Department of Agriculture and digital soil map of the world (DSMW) for K-factor estimation and development of its map. William's approach for the computation of K-factor (Eq. 31.3) was adopted in this study. The detailed equations and descriptions for estimating component factors are present in Eqs. (31.4)–(31.8) (Wawer et al. 2005). The required corresponding soil properties for applying William's method were extracted from the DSMW database. The estimated K-factor was then exported to ArcMap environment for the development of erodibility map for the entire Cameron Highlands watershed.

$$K_{\text{factor}} = f_{\text{coarse}_{\text{sand}}} \times f_{\text{clay}_{\text{silt}}} \times f_{\text{org}} \times f_{\text{hi.sand}} \quad (31.3)$$

where $f_{\text{coarse}_{\text{sand}}}$ is the factor for high coarse-sand content, $f_{\text{clay}_{\text{silt}}}$ is the factor for high clay-to-silt ratios, f_{org} is the factor for high organic carbon content, and $f_{\text{hi. sand}}$ is for soils with extremely high sand content.

$$f_{\text{coarse}_{\text{sand}}} = \left[0.2 + 0.3 \times \exp\left(-0.256 \times m_s \left(\frac{100 - m_{\text{silt}}}{100}\right)\right)\right] \quad (31.4)$$

$$f_{\text{clay}_{\text{silt}}} = \left[\frac{m_{\text{silt}}}{(m_c + m_{\text{silt}})}\right]^{0.3} \quad (31.5)$$

$$f_{\text{org}} = \left[1 - \frac{0.25 \times \text{OC}}{\text{OC} + \exp(3.72 - 2.95 \times \text{OC})}\right] \quad (31.6)$$

$$f_{\text{hi.sand}} = \left[1 - \frac{0.7 \times A}{A + \exp(-5.51 + 22.9A)}\right] \quad (31.7)$$

$$A = \left[\frac{(100 - m_s)}{100}\right] \quad (31.8)$$

where m_s is the percentage of sand fraction defined to be in the range 0.05–2.00 mm in diameter, m_c is the percentage of clay fraction defined to be <0.002 mm in diameter, m_{silt} is the percentage of silt fraction defined to be in the range 0.002–0.05 mm in diameter, and OC is the percentage of organic carbon present in the soil sample.

31.2.1.6 Extraction of Normalized Difference Vegetation Index

Normalized difference vegetation index (NDVI) map was extracted from Bands 4 (RED) and 5 (near infrared) of Landsat-8 image. In Band 4, vegetation has very high reflectance within the spectral band, while built-up areas have a very high reflectance in Band 5. Prior to application of these Bands, they were treated for atmospheric and sun angle corrections by converting band digital number (DN) to "top of atmospheric (ToA) reflectance" using Eq. (31.9). This was subsequently corrected for sun angle using Eq. (31.10) and the NDVI map evaluated using Eq. (31.11).

$$L_\lambda = M_L Q_{\text{cal}} + A_L \quad (31.9)$$

$$\text{DN}_{\text{corr.}} = \frac{\text{DN}_{\text{atm}}}{\text{Sin}(\varnothing)} \quad (31.10)$$

$$\text{NDVI} = \frac{\text{NIR} - \text{RED}}{\text{NIR} + \text{RED}} + 1 \quad (31.11)$$

where L_λ is ToA spectral radiance with SI unit watt/$(\text{m}^2 \text{ s rad } \mu\text{m}^{-1})$, M_L is the radiance multiplicative band number, A_L is the radiance additive band number, Q_{cal} is the quantized and calibrated digital number (DN), and ϕ is the sun angle.

31.2.1.7 Development of Land Surface Temperature

A relatively clear Landsat-8 data was acquired from USGS archive for the study which is located on Row/Path: 57/127. The thermal Band 10 was used for retrieving LST for the Cameron Highlands. The brightness temperatures are measured from the ToA radiance (L_λ) of satellite's thermal infrared (TIR) sensors using Plank's law (Friedl and Davis 1994). TIR band digital number was converted to ToA spectral radiance using the algorithms in Eq. (31.9). The equations required for the computation of land surface emissivity (e), brightness temperature (BT), proportion vegetation (Pv), and land surface temperature (LST) were given, respectively, in Eqs. (31.12)–(31.15). The detailed procedures can be found in Abdulkadir et al. (2017).

$$BT = K_2 \Big/ \ln\left(\frac{K_1}{L_\lambda}+1\right) \tag{31.12}$$

$$e = 0.004\text{Pv} + 0.986 \tag{31.13}$$

$$\text{Pv} = \left[\frac{(\text{NDVI} - \text{NDVI}_{\min})}{\text{NDVI}_{\max} - \text{NDVI}_{\min}}\right]^2 \tag{31.14}$$

$$\text{LST} = \frac{\text{BT}}{\left(1 + w \times \frac{\text{BT}}{p} \times \ln(e)\right)} \tag{31.15}$$

where K_1 and K_2 are the band specific thermal conversion constants, p is the Plank's constant, $\text{NDVI}_{\max}$ and $\text{NDVI}_{\min}$ are maximum and minimum NDVI values obtainable from NDVI map, M_L is the radiance multiplicative band number (rescaling factor), A_L is the radiance additive band number (rescaling factor), and Q_{cal} is the quantized and calibrated digital number (DN).

31.2.1.8 Development of Soil Moisture Index

Having developed NDVI and LST maps as previously described, pixel-based values for each of the parameters were extracted and plotted against each other. The soil moisture index (SMI) was evaluated by defining parameters $\text{LST}_{i,\max}$ and $\text{LST}_{i,\max}$ $\text{LST}_{i,\min}$ for a given NDVI plot. From LST-NDVI space diagram, linear regression equations for the dry and wet edges were obtained. These are typically represented as in Eqs. (31.16) and (31.17). The LST_i empirical parameters for dry and wet edges, when modeled as a linear fit to the observation data (i.e., constants a, b, a', and b'), were determined from a two-dimensional scatter plot of LST against NDVI. Hence, SMI on pixel basis was evaluated with Eq. (31.18).

$$\text{LST}_{i,\min} = a + b \times \text{NDVI}_i \tag{31.16}$$

$$\text{LST}_{i,\max} = a' + b' \times \text{NDVI}_i \tag{31.17}$$

$$\text{SMI} = \frac{\text{LST}_{i,\max} - \text{LST}_i}{\text{LST}_{i,\max} - \text{LST}_{i,\min}} \tag{31.18}$$

where $\text{LST}_{i,\max}$, and $\text{LST}_{i,\min}$ are maximum and minimum surface temperatures for a given NDVI. LST_i is the observed instantaneous surface temperature at the specific pixel. The constants a, b, a', and b' are empirical parameters evaluated from the dry and wet edges of linear fits (Ridd 1995).

31.2.2 Support Vector Machine Technique for Susceptibility Modeling

Support vector machine (SVM) is a statistical learning technique proposed by Sujay and Paresh (2014) for solving classification and regression problems by constructing n-dimensional hyperplanes that optimally separate data values into categories (Hearst et al. 1998). It is one of the most sophisticated supervised techniques in artificial intelligence derived from statistical learning theory and structural risk minimization principle (Tehrany et al. 2015; Vapnik 1998) whose accuracy exceeds many other data mining, bivariate and multivariate techniques (Sujay and Paresh 2014). It uses kernel tricks on training dataset such as linear, sigmoid, polynomial, or radial basis function (RBF) with their parameters (tolerance, γ; polynomial degree, d; and cost function, C) (Bui et al. 2012; Tehrany et al. 2015) to reshape nonlinear problems into linear and process-able classes by generating separating hyperplanes (Jebur et al. 2014). The flowchart for the implementation of SVM for soil erosion susceptibility modeling is depicted in Fig. 31.1, and detailed procedures are defined in the following sections:

31.2.2.1 Data Preparation and Partitioning for SVM Application

The first stage in performing soil erosion susceptibility mapping is to develop erosion inventory map followed by development of geodatabase of erosion-related CFs. Some of the surveyed gully and rill erosion locations in the study area are shown in Fig. 31.2a–d. These were documented during multiple field survey for the development of inventory map for the area under consideration. In a watershed, there is possibility of occurrence (denoted with 1) or non-occurrence (denoted with 0) of soil erosion at different locations. This makes the response variable to be binary or dichotomous in nature. It is recommended to have

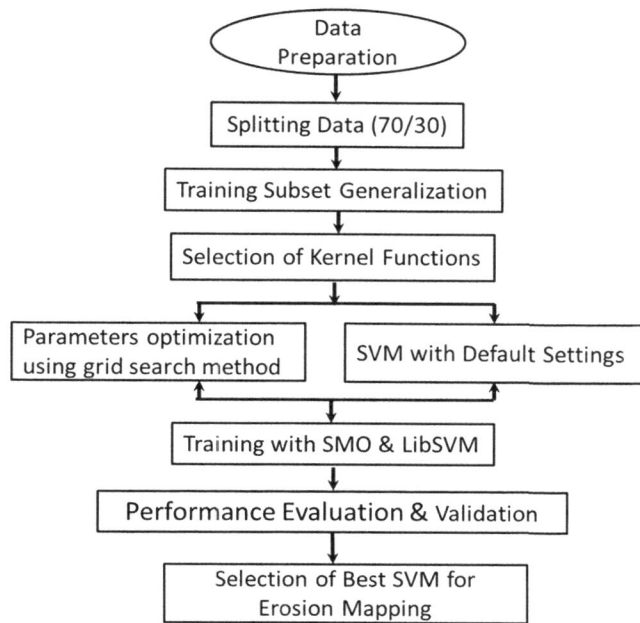

Fig. 31.1 Flowchart for application of SVM

approximately equal number of erosion-present points (1) and erosion-absent (0) points (Ayalew and Yamagishi 2005). It is mandatory that all the CFs have the same projection for easier acquisition of data. Moreover, Conoscenti et al. (2014) recommended that all CFs should have uniform scale/ extent (number of columns and rows) and range as the dependent variable. From the developed erosion inventory map, equal number of erosion-absent points were randomly created using slope map as a guide. All the CFs were projected to UTM 47 North, WGS 84, and have similar grid sizes of 30 by 30 m and column by row of 1025 × 1112. Based on the response variable, the corresponding erosion CFs' (predictor variables) data values were generated for SVM modeling.

In preprocessing stage, the generated dataset was partitioned into training and testing datasets using unsupervised filtering algorithms to avoid duplication. Each of these datasets is composed of both response and predictor variables. Trend in the previous studies on application of machine learning showed that there is no specific rule of thumb for the proportion of training and testing datasets. However, Guyon (1996) recommended that the proportion for validation (testing) set should be inversely proportional to the square root of the number of free adjustable parameters. The testing set (v) to training set (t) size ratio, v/t, scales yielded 33.33% proportion of the dataset for testing, due to the consensus of the researchers to have more training dataset than testing. Thus, the dataset was partitioned into 70% training and 30% testing datasets. This conforms with the proportion of dataset considered in similar studies (Jebur et al. 2014; Tehrany et al. 2014, 2015).

31.2.2.2 Selection of SVM Kernel Functions, Kernel Parameters, and Its Optimization

The nature of environmental problems like soil erosion is often nonlinear due to complexity of the watershed characteristics. Thus, it is imperative to apply different kernel functions such as linear, polynomial, radial basis function (RBF), and sigmoid. Selection of kernel type is very vital in SVM modeling due to their impacts on the successful training and classification precision (Tehrany et al. 2015). After the preparation of dataset as described in Sect. 31.2.2.1, the kernel functions were selected and implemented one after the other. For each of the kernel function, the corresponding kernel parameters described in Table 31.2 were defined. The C controls the trade-off between training errors and margin that often prevent over-fitting of the model (Marjanović et al. 2011), while γ controls the degree of nonlinearity of the SVM model (Bui et al. 2012). It is recommended that these parameters be predefined for specific studies to obtain accurate modeling results (Pourghasemi et al. 2013). In most studies, it's often difficult to select kernel parameters that will produce optimum model results particularly when modeling nonlinear problems. Hence, parameter optimization is indispensable to select optimum parameters. In this study, grid-search method was adopted for selecting optimal cost parameter, C; tolerance, γ; and degree of polynomial, d.

31.2.2.3 SVM Susceptibility Modeling and Performance Evaluation

The training dataset (x_i, y_i) are used in mapping the input dataset into high dimensional feature space. The pair of training dataset (x_i, y_i) for $x_i \in R^n$, $y_i \in (1, 0)$ for $i = 1, 2, \ldots$ m. The elements x are the array of erosion CFs consisting of LS-factor, R-factor, K-factor, lineament density, drainage density, NDVI, LST, and SMI. The separating hyperplane is generated in the original space of n coordinates (x_i parameters in vector x) between the points of two distinct classes (Marjanović et al. 2011). The two classes are erosion-present and erosion-absent pixels represented by 1 and 0, respectively, for the training dataset. If the point is located over the hyperplane, it will be classified as 1, otherwise, 0. Hence, SVM is implemented to establish optimal separating hyperplanes that would classify the training dataset into the two pixel classes. The training topology for this study is depicted in Fig. 31.3.

The most popular WEKA packages, LibSVM and SMOreg algorithms, were used for SVM classification or regression. The packages can handle both linear and nonlinear SVM analyses. Sequential minimal optimization (SMO) is one of the optimization algorithms for the training of a given dataset. It uses heuristics to partition the training dataset into smaller units that can be solved analytically. The training of dataset was applied to find Lagrange multipliers (α_i) by maximizing a dual quadratic program

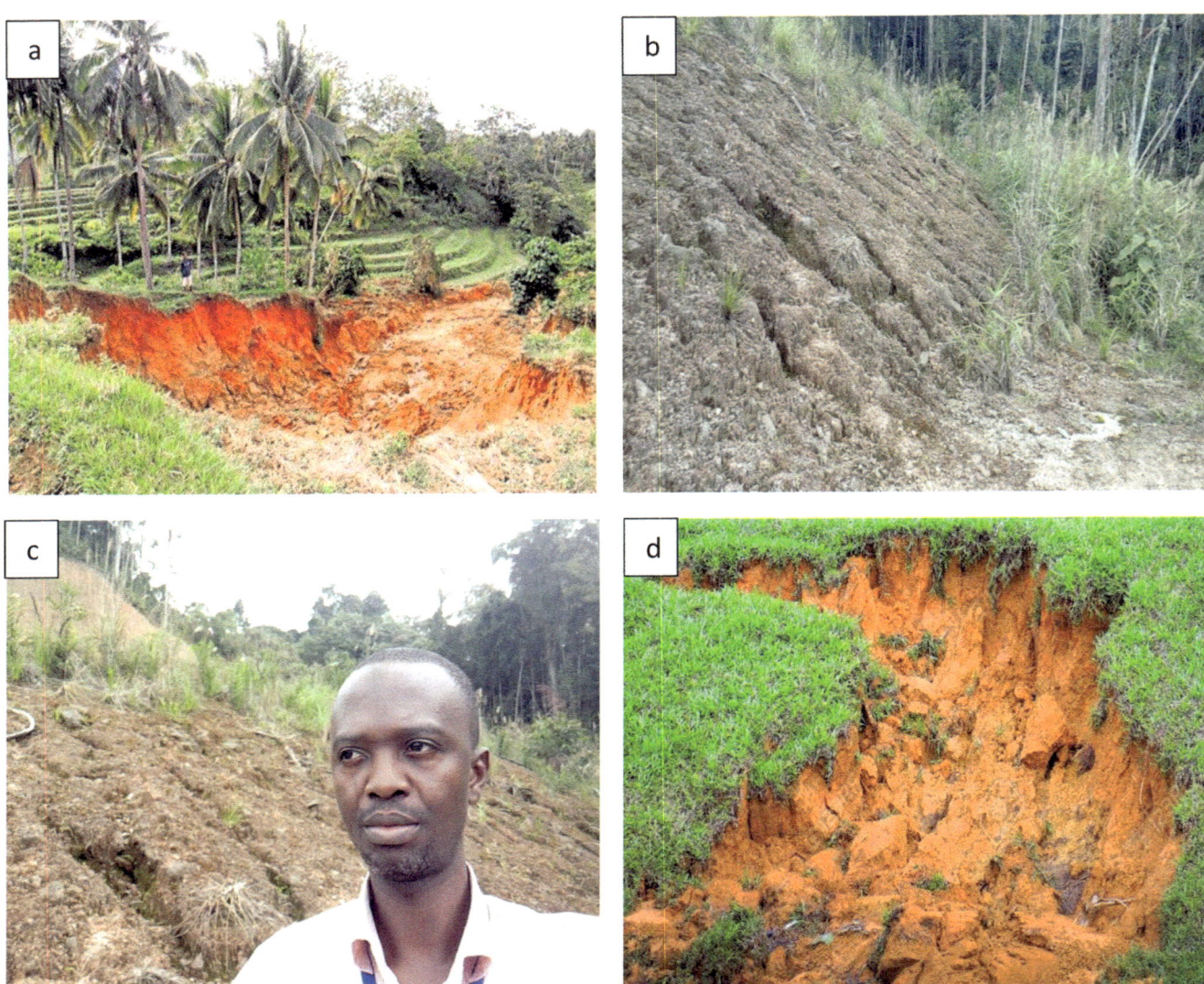

Fig. 31.2 Some erosional landforms in the study area

Table 31.2 Different kernel functions (Bui et al. 2012; Tehrany et al. 2015)

Kernel function $K(x_i, x_j)$	Kernel representation	Kernel parameter
Linear, LNR	x_i^T, x_j	–
Polynomial, Poly	$(-\gamma . x_i^T . x + C)^d$, C is often given as 1	γ, d
Sigmoid, Sigmd	$\tanh((-\gamma . x_i^T . x + C)^d$,	γ
Radial basis function, RBF	$\exp(-\gamma x_i - x_j^2)$, for $\gamma > 0$	γ

(QP). The SMO algorithms search through the feasible region of the dual quadratic problems and maximize Eq. (31.19) by decomposing it into fixed size QP subproblems and solving the smallest possible optimization problem at each step (Hearst et al. 1998; Platt 1998).

$$\max L_D(\alpha) = \sum_{i=1}^{n} \alpha_i - \tfrac{1}{2} \sum_{i=1}^{n}$$

$$\times \sum_{i=1}^{n} \alpha_i \alpha_j y_i y_j K(x_i, x_j) \qquad (31.19)$$

Subject to $0 \leq \alpha_i \leq C$

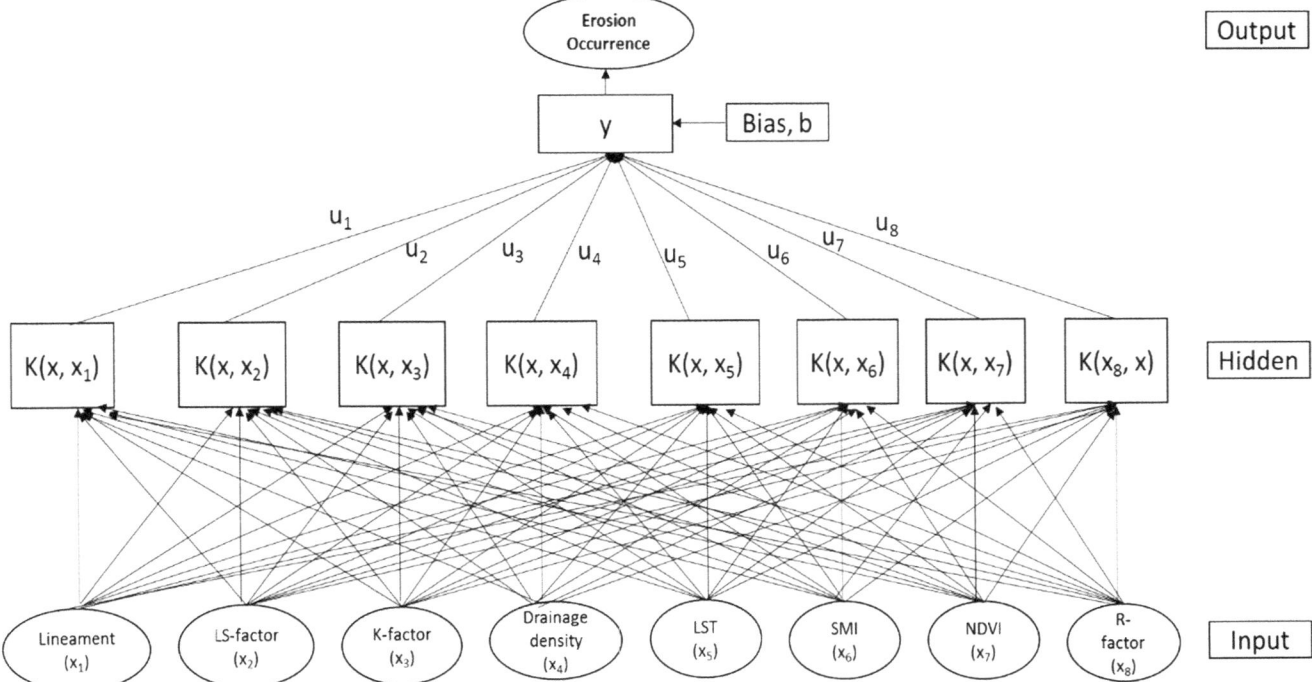

Fig. 31.3 Training architecture for erosion susceptibility

$$\sum_{i=1}^{n} \propto_i y_i = 0, \forall_i$$

where C is the SVM hyperparameter and $K(x_i, x_j)$ is kernel functions.

The optimized kernel parameters were used for the respective kernel functions for the classification purposes. SVM in WEKA requires that one of the classification training approaches must be selected for the training of the model. Hence, model performance in terms of success and prediction rates can be evaluated for the training and testing datasets, respectively. There are four available training techniques in WEKA machine learning workbench which include training dataset, supplied test data, percentage split, and cross validation techniques. In cross validation technique, the dataset was partitioned into n-folds ($n = 10$ for this study). The training will be conducted on all the partitions except one that is held out as the test set. The repetition of this process will create n-number of the models and avail each fold the chance of being held out as test set. The cross validation was selected in this study for the training of the model because of its accuracy over the others. Then, the success rate of the model was evaluated. In the same manner, test dataset was introduced through "supplied test data" option of the work bench to evaluate the prediction rate of the model with the test dataset.

Finally, performance of the models were evaluated with the indexes: area under the curve (AUC), kappa index (k),

prediction rate (PR), success rate (SR), and classification errors (Althuwaynee et al. 2012; Bui et al. 2012; Jebur et al. 2014; Shit et al. 2015; Tehrany et al. 2014, 2015). Comparative analysis among all the SVM kernel functions was conducted to evaluate their relative performances. In the end, SVM classification results were used for the development of erosion susceptibility mapping for Cameron Highlands.

31.2.3 Erosion Susceptibility Modeling

Susceptibility technique is used to assess the relative probability of occurrence of erosion at a particular location compared to other locations under the influence of triggering factors known as CFs. Eight CFs were considered as predictor variables and a response variable (which is a binary class composed of eroded and non-eroded pixels). The CFs were grouped into static and dynamic factors. In order to evaluate the seasonality impact of dynamic CFs on the erosion susceptibility, the low and peak rainfall cycles were considered since the set of CFs respond to change in rainfall cycles. Hence, Landsat images for February 26, 2016, and November 8, 2016, were downloaded representing the respective low and peak rainfall cycles. NDVI, LST, R-factor, and SMI were extracted for these two periods for multitemporal analysis of erosion susceptibility mapping. The erosion base map in Fig. 31.4a that consists of

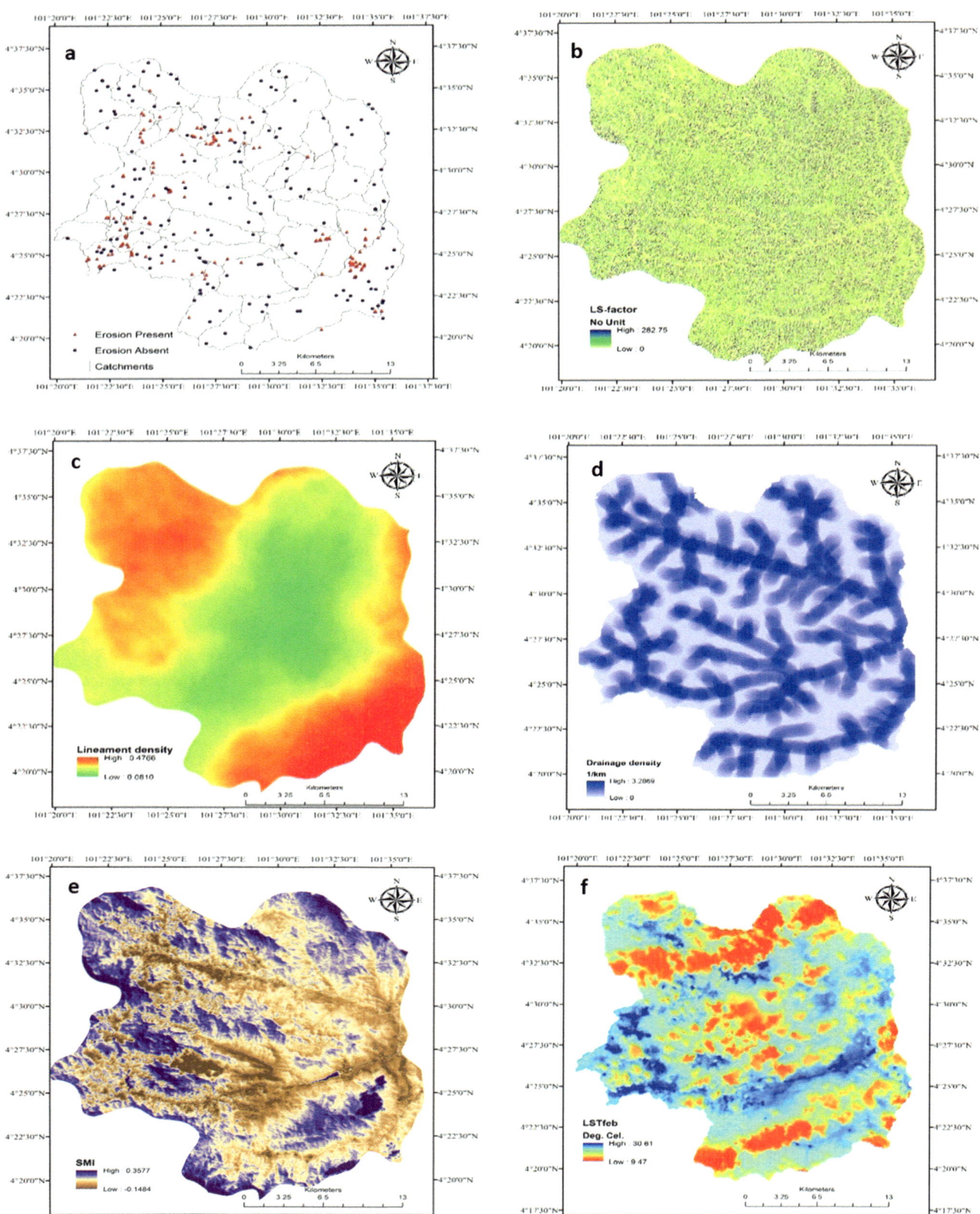

Fig. 31.4 (**a**) Inventory map. (**b**) Length slope factor. (**c**) Lineament density. (**d**) Drainage density. (**e**) Soil moisture index. (**f**) Land surface temperature. (**g**) NDVI. (**h**) Rainfall erosivity. (**i**) Soil erodibility factor

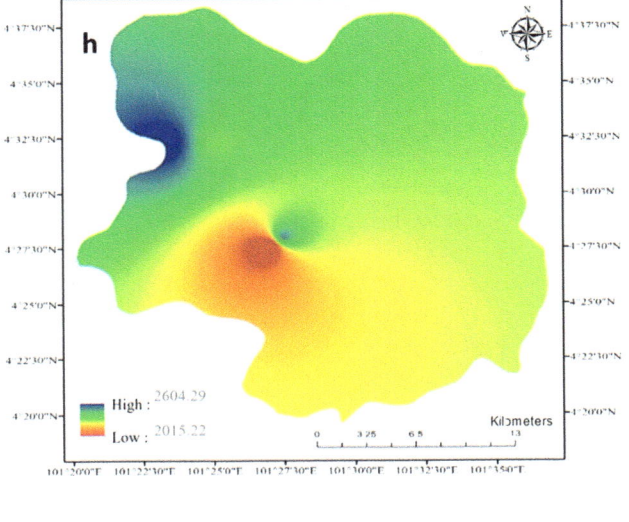

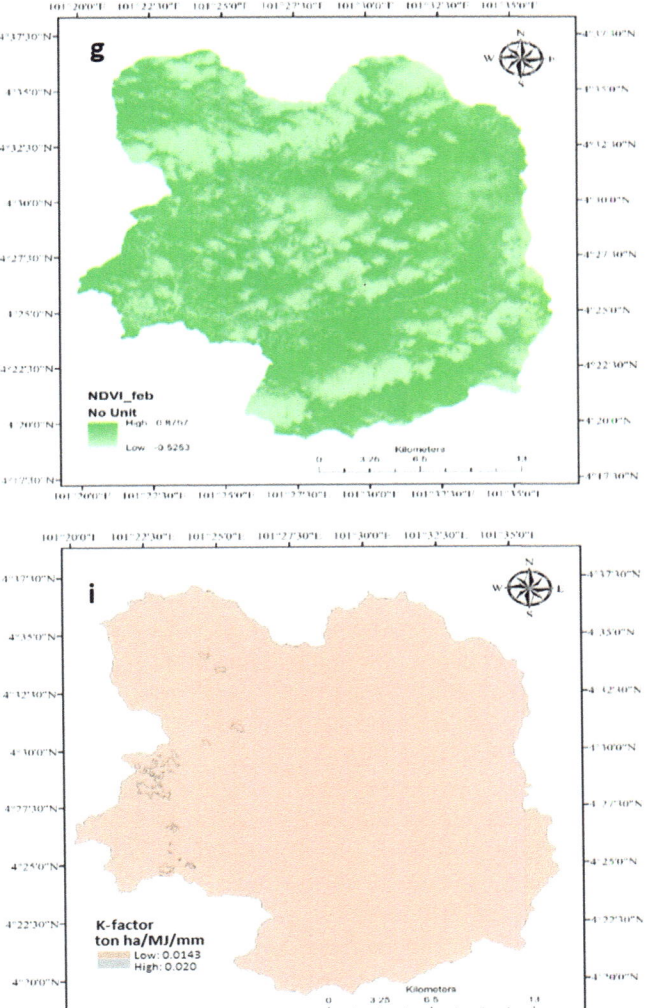

Fig. 31.4 (continued)

159 erosion-present and equal numbers of erosion-absent locations was randomly created across the watershed. The erosion-present locations are made up of both gullies and rill erosional landforms. The weights for the CFs of the model having the best performance were applied for development of erosion susceptibility maps for the study area.

31.3 Results and Discussion

The analysis involved nine attributes (i.e., 8 CFs as the predictor and the response variables, binary classes of absence or presence) with total of three hundred and eighteen cases. The cases comprised of erosion-present and erosion-absent cases with their corresponding values of the erosion CFs. Figure 31.4b–i shows the geospatial distribution of all the CFs within the study area. The dataset is composed of static and dynamic CFs for the month of low rainfall cycle. In order to achieve accurate modeling results, kernel parameters for each of the kernel functions were optimized. The cost function (C) and tolerance (γ) were optimized for the SVM-RBF and sigmoid SVM, C, γ, and the polynomial degree (d) for polynomial SVM and only C for linear SVM using grid search technique. Table 31.3 shows the optimized kernel parameter obtained which were used for SVM modeling. The non-zero positive C values obtained indicate that the polynomial kernel is inhomogeneous in nature. This is a free parameter trading off the influence of higher-order versus lower-order terms in the polynomial.

These parameters were applied on low rainfall cycle dataset for erosion susceptibility mapping and the results are presented in Table 31.4. The results therein comprised

Table 31.3 Optimized kernel parameters

Kernel	Cost parameter, C	Tolerance, γ	Polynomial degree, d
Linear SVM	10	–	–
Polynomial SVM	14	5.2	4
RBF SVM	12	1.6	–
Sigmoid SVM	16	2.0	–

Table 31.4 Kernel functions for low rainfall cycle

Kernel	SR	PR	Kappa index		AUC		MAE		RMSE	
			Training	Testing	Training	Testing	Training	Testing	Training	Testing
Linear SVM	0.9258	0.8778	0.8477	0.7552	0.828	0.838	0.1732	0.1214	0.2014	0.3091
Polynomial SVM	0.9287	0.8844	0.8512	0.7911	0.896	0.885	0.1001	0.1012	0.3029	0.5511
RBF SVM	0.8022	0.7104	0.6872	0.6324	0.804	0.791	0.3003	0.3321	0.4130	0.4014
Sigmoid SVM	0.6286	0.5527	0.1601	0.1840	0.510	0.501	0.4148	0.4011	0.6181	0.7047

of SR, PR, k, AUC, mean absolute error (MAE), and root mean squared error (RMSE) obtained for the training and testing for model performance evaluation. However, AUC and kappa index are the commonest validation criteria in susceptibility studies. The analysis shows an SR of 0.9258 representing the training cases that were correctly classified with respective MAE and RMSE of 0.1732 and 0.2014 for training dataset. However, the SR only is insufficient to valid the efficacy of the model because it gives model performance on the training dataset. Hence, PR was evaluated on the test dataset that were not included in the training exercise to ascertain their performance. Implementing the results on test dataset indicated a prediction rate of 0.8778. The respective classification errors MAE and RMSE were 0.1214 and 0.3091.

Using kappa index and AUC as validation criteria showed that polynomial SVM outperformed other models during testing followed by linear SVM. The performance of polynomial SVM suggests the nonlinearity and complexity of watershed characteristics with respect to soil erosion. This is slightly similar to the results obtained by Tehrany et al. (2015) who evaluated the capability of SVM kernel function in flood susceptibility mapping. Furthermore, sigmoid SVM had the least performance and the highest classification errors. The closeness of its AUC value to 0.5 and the k value to 2.0 underscores its weakness in erosion susceptibility analysis and mapping (Althuwaynee et al. 2012; Tehrany et al. 2014).

Table 31.5 presents the weights obtained for each of the CFs for different kernel functions. These underscore the strength of each CFs which appear slightly similar. However,

their magnitudes differ from one another. The results showed that all the CFs had positive relationships with occurrence of erosion except NDVI. From the three kernel functions, topographic landscape appeared to be most significant of all the CFs. Specifically, in polynomial SVM, LS-factor's weight of 2.632 was the highest followed by lineament density. Being the best model, the corresponding weights of polynomial SVM were considered for the development of soil erosion susceptibility map (Fig. 31.4).

From the developed erosion susceptibility map (Fig. 31.5), the classes are very low, low, moderate, high, and very high susceptible zones constituting 3.6%, 68.8%, 25.1%, 1.8%, and 0.7% of the total area of Cameron Highlands watershed, respectively. The output map showed most of the areas are in the categories of "low and moderate" susceptibilities. It was observed that the areas within "high and very high" susceptibilities fell in the urban and agriculture areas. This underscores the results of human disturbance in the watershed thereby causing land degradation.

Similar analyses were performed for the peak rainfall cycle dataset. The previous dynamic CFs for the low rainfall cycle were replaced with that of the month of November (i.e., peak rainfall cycle) for another round of SVM susceptibility modeling for multitemporal analysis. However, static CFs remain unchanged because they do not respond to rainfall cycles. With optimized kernel parameters, SVM classification was carried out for all the kernel functions. The results of the analyses are presented in Table 31.6. The results also showed that polynomial SVM had the best performance among all the SVM models.

Table 31.5 Variable of importance of each CFs for low rainfall cycle

Weights	K-factor	LS-factor	Lineament density	Drainage density	LST	SMI	NDVI	R-factor
Linear SVM	0.027	2.572	1.520	0.214	1.041	1.423	−3.133	0.175
Polynomial SVM	0.021	2.632	1.641	0.201	1.442	1.376	−3.820	0.153
RBF SVM	0.043	2.632	1.492	0.234	1.984	1.403	−2.811	0.148

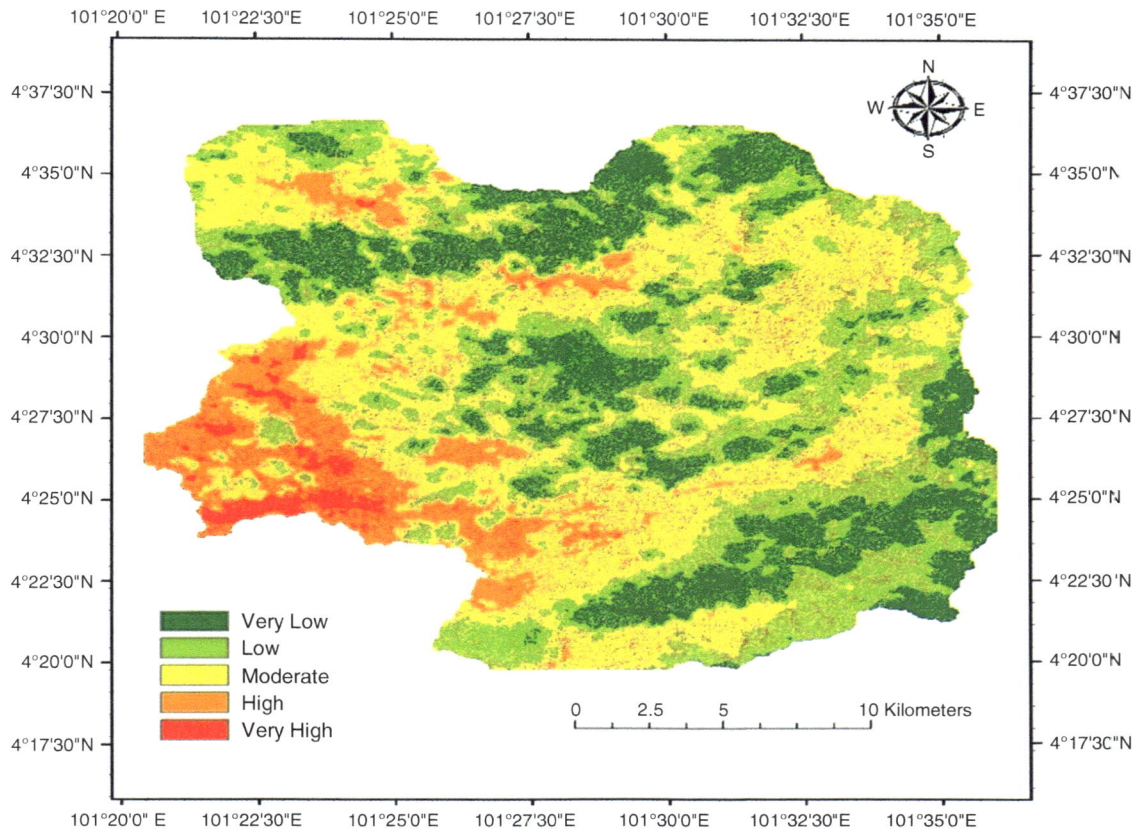

Fig. 31.5 Soil erosion susceptibility map for polynomial SVM-Poly for low rainfall cycle

Table 31.6 Kernel functions peak rainfall cycle

Kernel	SR	PR	Kappa index		AUC		MAE		RMSE	
			Training	Testing	Training	Testing	Training	Testing	Training	Testing
Linear SVM	0.8901	0.8687	0.7828	0.7568	0.7924	0.8001	0.1661	0.1003	0.1988	0.2145
Polynomial SVM	0.9226	0.8830	0.8642	0.8012	0.8226	0.8605	0.0913	0.1001	0.1093	0.2111
RBF SVM	0.8837	0.8855	0.7011	0.7112	0.7320	0.7522	0.2002	0.2046	0.2210	0.3441
Sigmoid SVM	0.5843	0.5088	0.1982	0.2002	0.5013	0.5116	0.2133	0.2822	0.4008	0.5145

Table 31.7 shows the CFs' weights obtained for each of the SVM models highlighting the strength and types of the relationships existing between CFs and the occurrence of soil erosion. Comparing these results with that of low rainfall cycle showed relatively similar behaviors in terms of the strengths and nature of the relationships of the CFs with soil erosion. Since polynomial SVM outperformed others, thus, their weights were used in developing erosion susceptibility map (Fig. 31.6) for Cameron Highlands. The only difference observed when compared with the previous analysis was in the percentage area susceptible to soil erosion. The percentage area susceptible to erosion during the peak rainfall cycle was observed to be 1.8%, 66.5%, 28.1%, 2.5%, and 1.1% for the very low, low, moderate, high, and very high

Table 31.7 Weights for CFs for each kernel functions for peak rainfall cycle

Weights	K-factor	LS-factor	Lineament density	Drainage density	LST	SMI	NDVI	R-factor
SVM-LNR	0.021	2.554	1.600	0.301	1.116	1.394	−3.164	0.224
SVM-Poly	0.027	2.661	1.688	0.202	1.443	1.365	−3.843	0.302
SVM-RBF	0.050	2.645	1.498	0.244	1.984	1.355	−2.755	0.146

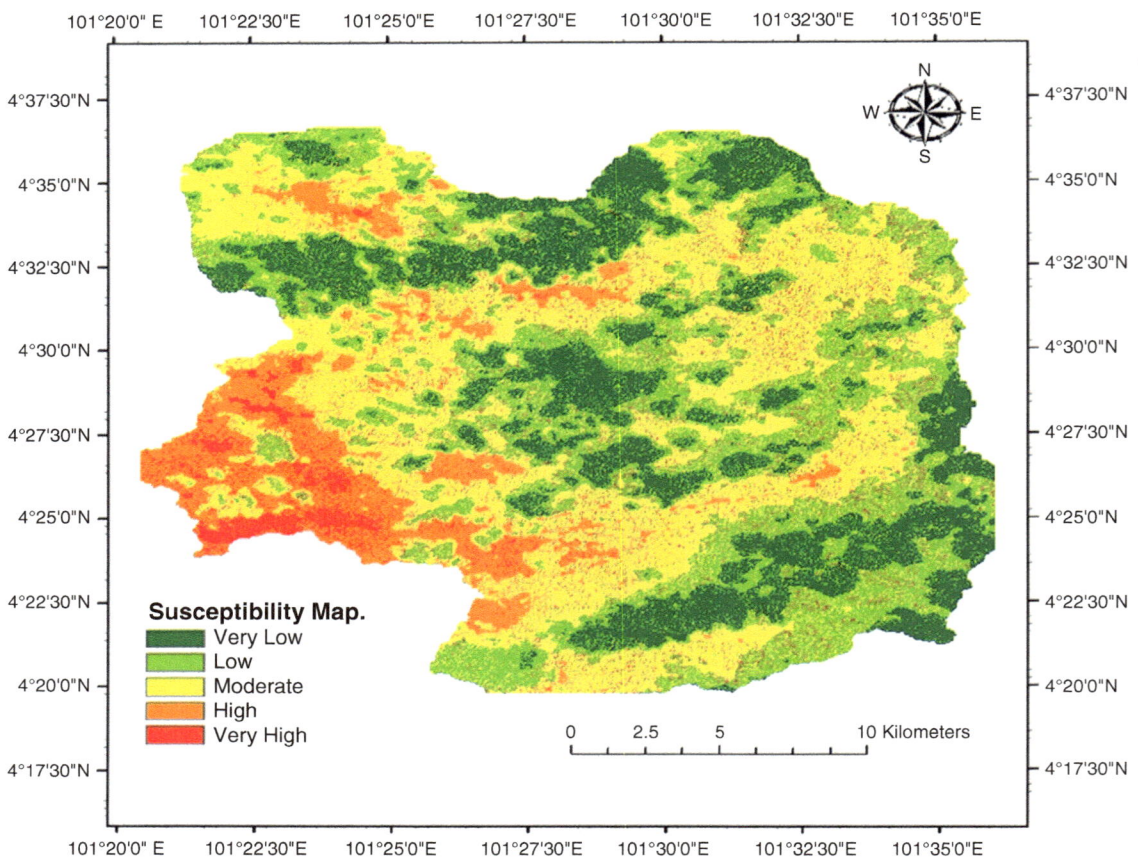

Fig. 31.6 Soil erosion susceptibility map for SVM-Poly for peak rainfall cycle

susceptible zones. This shows a slight increase in the area "high and very high" susceptible regions compare to low rainfall cycle period.

The distribution pattern of erosion susceptibility showed that majority of the areas are under "very low, low, and moderate" category of susceptibility. The analysis shows that most of the highly eroded locations are within the urban and agriculture land-use. This is because erosion rates are often enhanced at locations that are denuded and reshaped for urban and agricultural development. These susceptibility maps would provide early warning to the watershed users and other stakeholders on the proneness of some certain locations to soil erosion. This would greatly assist in sustainable management of the watershed that will alleviate soil erosion and its related hazards.

31.4 Conclusion

The study implements SVM for geospatial prediction of soil erosion under both static and dynamic CFs for low and peak rainfall cycles. This machine learning technique was applied to the complex watershed of Cameron Highlands to model its

erosion susceptibility. A total of eight CFs non-redundant CFs were selected including the unpopular dynamic CFs that are often neglected in many related studies. Four SVM kernel functions such as linear, polynomial, sigmoid, and RBF were implemented with optimized kernel parameters. The performances of SVM models were evaluated using AUC, SR, PR, and kappa index. The results indicated that polynomial SVM had the highest performance with sigmoid SVM having the lowest accuracy with relatively large classification errors. The performance of polynomial SVM on the other hand proved its suitability and superiority in modeling the complex and nonlinear soil erosion processes. Due to its outstanding predictive capability of this model, its corresponding weights for all CFs were used for the development of soil erosion susceptibility map for the study area. The analysis of the CFs considered in this study indicated that all had positive relationships with the occurrence of soil erosion except NDVI. The distribution patterns of erosion showed that majority of the area are "low and moderately susceptible." The analysis shows that most of the highly susceptible locations are within the urban and agriculture land-use. This is because erosion rates are often prevalent where land are denuded and reshaped for anthropogenic activities.

Susceptibility analyses of both low and peak rainfall cycles yielded similar results except that there was increase in susceptible areas during the peak rainfall cycle. This study could provide information on the erosional landforms that could advance to gullies especially during the peak rainfall cycle. This medium-scale erosion susceptibility map produced would be of great benefits to the watershed managers in mitigating the environmental and socioeconomic-related challenges of soil erosion and thus contribute to achieving six SDGs earlier mentioned.

Acknowledgment This research was funded by Universiti Teknologi PETRONAS via 2016 URIF grant 0153AA-G04.

References

Abdulkadir TS, Muhammad MR, Khamaruzaman WY, Ahmad HM (2017) Assessing the influence of terrain characteristics on spatial distribution of satellite derived land surface parameters in mountainous areas. In: 37th IAHR World Congress, Kuala Lumpur Malaysia, 13-18th August 2017. pp 2897-2906

Agnesi V, Angileri S, Cappadonia C, Conoscenti C, Rotigliano E (2011) Multi-parametric GIS analysis to assess gully erosion susceptibility: a test in southern Sicily, Italy Landform Analysis 17:15-20

Akgün A, Türk N (2011) Mapping erosion susceptibility by a multivariate statistical method: a case study from the Ayvalık region, NW Turkey Computers & Geosciences 37:1515-1524

Ali SA, Hagos H (2016) Estimation of soil erosion using USLE and GIS in Awassa catchment, Rift valley, Central Ethiopia Geoderma Regional 7:159-166

Allouche J, Middleton C, Gyawali D (2014) Nexus Nirvana or Nexus Nullity? A dynamic approach to security and sustainability in the water-energy-food nexus. In: STEPS Working Paper no. 63, STEPS Centre, Institute of Development Studies. University of Sussex, Brington,

Althuwaynee OF, Pradhan B, Lee S (2012) Application of an evidential belief function model in landslide susceptibility mapping Computers & Geosciences 44:120-135

Althuwaynee OF, Pradhan B, Park H-J, Lee JH (2014) A novel ensemble bivariate statistical evidential belief function with knowledge-based analytical hierarchy process and multivariate statistical logistic regression for landslide susceptibility mapping Catena 114:21-36

Aminuddin B, Ghulam M, Abdullah WYW, Zulkefli M, Salama R (2005) Sustainability of current agricultural practices in the Cameron Highlands, Malaysia Water, Air, & Soil Pollution: Focus 5:89-101

Arabameri A, Pradhan, B., Pourghasemi, H.R., Rezaei, K. and Kerle, N. (2018) Spatial Modelling of Gully Erosion Using GIS and R Programing: A Comparison among Three Data Mining Algorithms. Applied Sciences 8:1369

Ayalew L, Yamagishi H (2005) The application of GIS-based logistic regression for landslide susceptibility mapping in the Kakuda-Yahiko Mountains, Central Japan Geomorphology 65:15-31

Barber M, Mahler R (2010) Ephemeral gully erosion from agricultural regions in the Pacific Northwest, USA. Ann. Wars. Univ. Life Sci Land Reclam 42:23-29

Basith A (2011) Landslide susceptibility modelling under environmental changes: A case study of Cameron Highlands, Malaysia., Universiti Teknologi Petronas

Bui DT, Pradhan B, Lofman O, Revhaug I, Dick OB (2012) Landslide susceptibility mapping at Hoa Binh province (Vietnam) using an adaptive neuro-fuzzy inference system and GIS Computers & Geosciences 45:199-211

Conforti M, Aucelli PP, Robustelli G, Scarciglia F (2011) Geomorphology and GIS analysis for mapping gully erosion susceptibility in the Turbolo stream catchment (Northern Calabria, Italy) Natural hazards 56:881-898

Conoscenti C, Agnesi V, Angileri S, Cappadonia C, Rotigliano E, Marker M (2013) A GIS-based approach for gully erosion susceptibility modeling: A test in Sicily, Italy Environmental Earth Sciences 70:1179-1195

Conoscenti C, Angileri S, Cappadonia C, Rotigliano E, Agnesi V, Märker M (2014) Gully erosion susceptibility assessment by means of GIS-based logistic regression: a case of Sicily (Italy) Geomorphology 204:399-411

Conoscenti C, Di Maggio C, Rotigliano E (2008) Soil erosion susceptibility assessment and validation using a geostatistical multivariate approach: a test in Southern Sicily Natural Hazards 46:287-305

d'Oleire-Oltmanns S, Marzolff I, Peter KD, Ries JB (2012) Unmanned aerial vehicle (UAV) for monitoring soil erosion in Morocco Remote Sensing 4:3390-3416

Dewitte O, Daoudi M, Bosco C, Eeckhaut MVD (2015) Predicting the susceptibility to gully initiation in data-poor regions Geomorphology 228:101-115

Donati L, Turrini M (2002) An objective method to rank the importance of the factors predisposing to landslides with the GIS methodology: application to an area of the Apennines (Valnerina; Perugia, Italy) Engineering Geology 63:277-289

Duarte L, Teodoro A, Gonçalves J, Soares D, Cunha M (2016) Assessing soil erosion risk using RUSLE through a GIS open source desktop and web application Environ Monit Assess 188:1-16

Dube F, Nhapi I, Murwira A, Gumindoga W, Goldin J, Mashauri D (2014) Potential of weight of evidence modelling for gully erosion hazard assessment in Mbire District–Zimbabwe Physics and Chemistry of the Earth, Parts A/B/C 67:145-152

Fortuin R (2006) Soil erosion in Cameron Highlands: An erosion rate study in a highland area Saxion University, Deventer, the Netherlands, Regional Environmental Awareness Cameron Highlands:1-83

Friedl M, Davis F (1994) Sources of variation in radiometric surface temperature over a tallgrass prairie Remote Sensing of Environment 48:1-17

Gasim MB, Surif S, Toriman ME, Rahim SA, Elfithri R, Lun PI (2009) Land-use change and climate-change patterns of the Cameron Highlands, Pahang, Malaysia The Arab World Geographer 12:51-61

George CM, Anu VV (2018) Predicting piping erosion susceptibility by statistical and artificial intelligence approaches-a review International Research Journal of Engineering and Technology 5:239-243

Gómez-Gutiérrez Á, Conoscenti C, Angileri SE, Rotigliano E, Schnabel S (2015) Using topographical attributes to evaluate gully erosion proneness (susceptibility) in two Mediterranean basins: advantages and limitations Natural Hazards 79:291-314

Guyon I (1996) A scaling law for the validation-set training-set ratio. Unpublished Technical Report, AT&T Bell Laboratories

Hearst MA, Dumais ST, Osman E, Platt J, Scholkopf B (1998) Support vector machines. Intelligent Systems and their Applications IEEE Geoscience and remote sensing letters 13:18-28

Hugenholtz CH et al. (2013) Geomorphological mapping with a small unmanned aircraft system (sUAS): Feature detection and accuracy assessment of a photogrammetrically-derived digital terrain model Geomorphology 194:16-24

IEA (2006) Hydropower Good Practices: Environmental mitigation measures and benefits case study 04-03: Reservoir sedimentation-cameron highlands hydroelectric scheme, Malaysia. Hydropower Implementing Agreement Annex VIII, International Energy Agency

Jaafar O et al. (2010) Modeling the impacts of ringlet reservoir on downstream hydraulic capacity of bertain river using XPSWMM in cameron highlands, Malaysia Research Journal of Applied Sciences 5:47-53

Jamali HMB (2004) Study of soil moisture in relation to soil erosion in the proposed Tamcitaro Geopark, Central Mexico: A case of the Zacandaro sub-watershed., International Institute for Geo-information Science and Earth Observation

Jamil NR, Ruslan MS, Toriman ME, Idris M, Razad AA (2014) Impact of Landuse on Seasonal Water Quality at Highland Lake: A Case Study of Ringlet Lake, Cameron Highlands, Pahang. In: From Sources to Solution. Springer, pp 409-413

Jansen L, Lariyah MS, Mohamed NMD, Pierre YJ (2012) Challenge in running hydropower as source of clean energy: Ringlet reservoir, Cameron Highlands case study. Paper presented at the Proceedings National Graduate Conference, Universiti Tenaga Nasional, Malaysia, November 8-10

Jebur MN, Pradhan B, Tehrany MS (2014) Optimization of landslide conditioning factors using very high-resolution airborne laser scanning (LiDAR) data at catchment scale Remote Sensing of Environment 152:150-165

Kavzoglu T, Sahin EK, Colkesen I (2015) Selecting optimal conditioning factors in shallow translational landslide susceptibility mapping using genetic algorithm Engineering Geology 192:101-112

Kelley DW, Nater EA (2000) Historical sedimentflux from three watersheds into Lake Pepin, Minnesota, USA J Environ Qual 29:561–568

Kerr J (1998) The economics of soil degradation: from national policy to farmers' fields Soil Erosion at Multiple Scales: Principals and Methods for Assessing Causes and Impacts International Board for Soil Research and Management (IBSRAM), CABI Publishers, Wallingford, UK:21-38

Khalik WMAWM, Abdullah MP, Amerudin NA, Padli N (2013) Physicochemical analysis on water quality status of Bertam River in Cameron Highlands, Malaysia Journal of Materials and Environmental Science 4 488-495

Khatun S (2017) Detection of Soil Erosion Potential Zones and Estimation of Soil Loss in Kushkarani River Basin of Eastern India Journal of Geography, Environment and Earth Science International 12:1-17

Khosrokhani M, Pradhan B (2014) Spatio-temporal assessment of soil erosion at Kuala Lumpur metropolitan city using remote sensing data and GIS Geomatics, Natural Hazards and Risk 5:252-270

Kouli M, Soupios P, Vallianatos F (2009) Soil erosion prediction using the revised universal soil loss equation (RUSLE) in a GIS framework, Chania, Northwestern Crete, Greece Environmental Geology 57:483-497

Lee S, Pradhan B (2007) Landslide hazard mapping at Selangor, Malaysia using frequency ratio and logistic regression models Landslides 4:33-41

Lucà F, Conforti M, Robustelli G (2011) Comparison of GIS-based gullying susceptibility mapping using bivariate and multivariate statistics: Northern Calabria, South Italy Geomorphology 134:297-308

Magliulo P (2012) Assessing the susceptibility to water-induced soil erosion using a geomorphological, bivariate statistics-based approach Environmental Earth Sciences 67:1801-1820

Mararakanye N, Le Roux J (2012) Gully location mapping at a national scale for South Africa South African Geographical Journal 94:208-218

Marjanović M, Kovačević M, Bajat B, Voženílek V (2011) Landslide susceptibility assessment using SVM machine learning algorithm Engineering Geology 123:225-234

Nampak H, Pradhan B, Mojaddadi Rizeei H, Park H-J (2018) Assessment of Land Cover and Land Use Change Impact on Soil Loss in a Tropical Catchment by Using Multi-Temporal SPOT-5 Satellite Images and RUSLE model Land Degrad Dev 29:3440-3455

Pal S (2016) Identification of soil erosion vulnerable areas in Chandrabhaga river basin: a multi-criteria decision approach Modeling Earth Systems and Environment 2:5

Parveen R, Kumar U (2012) Integrated approach of universal soil loss equation (USLE) and geographical information system (GIS) for soil

loss risk assessment in Upper South Koel Basin, Jharkhand Journal of Geographic Information System 4:588

Pham TG, Degener J, Kappas M (2018) Integrated universal soil loss equation (USLE) and Geographical Information System (GIS) for soil erosion estimation in A Sap basin: Central Vietnam Int Soil Water Conserv Res 6:99-110

Pimentel D, Burgess M (2013) Soil erosion threatens food production Agriculture 3:443-463

Platt J (1998) Sequential minimal optimization: A fast algorithm for training support vector machines

Pourghasemi HR, Jirandeh AG, Pradhan B, Xu C, Gokceoglu C (2013) Landslide susceptibility mapping using support vector machine and GIS at the Golestan Province, Iran J Earth Syst Sci 122:349-369

Pourghasemi HR, Yousefi S, Kornejady A, Cerdà A (2017) Performance assessment of individual andensemble data-mining techniques for gully erosion modeling Science of Total Environment 609:764-775

Pournader M, Ahmadi H, Feiznia S, Karimi H, Peirovan HR (2018) Spatial prediction of soil erosion susceptibility: an evaluation of the maximum entropy model Earth Science Informatics 11:389-401 doi: https://doi.org/10.1007/s12145-018-0338-6

Pradeep GS, Krishnan MN, Vijith H (2015) Identification of critical soil erosion prone areas and annual average soil loss in an upland agricultural watershed of Western Ghats, using analytical hierarchy process (AHP) and RUSLE techniques Arabian Journal of Geosciences 8:3697-3711

Pradhan B (2010) Application of an advanced fuzzy logic model for landslide susceptibility analysis International Journal of Computational Intelligence Systems 3:370-381

Pradhan B (2013) A comparative study on the predictive ability of the decision tree, support vector machine and neuro-fuzzy models in landslide susceptibility mapping using GIS Computers & Geosciences, 51:350-365

Prosdocimi M, Cerdà A, Tarolli P (2016) Soil water erosion on Mediterranean vineyards: A review Catena 141:1-21

Rahmati O, Pourghasemi HR, Zeinivand H (2016) Flood susceptibility mapping using frequency ratio and weights-of-evidence models in the Golastan Province, Iran Geocarto International 31:42-70

Rahmati O, Tahmasebipour N, Haghizadeh A, Pourghasemi HR, Feizizadeh B (2017) Evaluating the influence of geo-environmental factors on gully erosion in a semi-arid region of Iran: An integrated framework. Sci Total Environ 579:913-927

Remondo J, González A, De Terán JRD, Cendrero A, Fabbri A, Chung C-JF (2003) Validation of landslide susceptibility maps; examples and applications from a case study in Northern Spain Natural Hazards 30:437-449

Ridd MK (1995) Exploring a VIS (vegetation-impervious surface-soil) model for urban ecosystem analysis through remote sensing: comparative anatomy for cities International journal of remote sensing 16:2165-2185

Rizeei HM, Maryam AS, Pradhan B, Ahmad N (2016) Soil erosion prediction based on land cover dynamics at the Semenyih watershed in Malaysia using LTM and USLE models Geocarto international 31:1158-1177

Shit PK, Paira R, Bhunia G, Maiti R (2015) Modeling of potential gully erosion hazard using geo-spatial technology at Garbheta block, West Bengal in India Modeling Earth Systems and Environment 1:1-16

Stöcker C, Eltner A, Karrasch P (2015) Measuring gullies by synergetic application of UAV and close range photogrammetry - A case study from Andalusia, Spain Catena 132:1-11

Sujatha ER, Sridhar V (2018) Spatial Prediction of Erosion Risk of a Small Mountainous Watershed Using RUSLE: A Case-Study of the Palar Sub-Watershed in Kodaikanal South India Water 10:1608

Sujay RN, Paresh CD (2014) Support vector machine applications in the field of hydrology: a review Applied Soft Computing 19:372-386

Tehrany MS, Pradhan B, Jebur MN (2014) Flood susceptibility mapping using a novel ensemble weights-of-evidence and support vector machine models in GIS Journal of hydrology 512:332-343

Tehrany MS, Pradhan B, Mansor S, Ahmad N (2015) Flood susceptibility assessment using GIS-based support vector machine model with different kernel types Catena 125:91-101

Thlakma SR, Iguisi E, Odunze AC, Jeb DN (2018) Estimation of Soil Erosion Risk in Mubi South Watershed, Adamawa State, Nigeria. J Remote Sensing & GIS 7:226

Tinker PB (1997) The environmental implications of intensified land use in developing countries Philos Trans R Soc Lond B 352:1023–1033

Vahabi J, Nikkami D (2008) Assessing dominant factors affecting soil erosion using a portable rainfall simulator International Journal of Sediment Research 23:376-386

Vapnik V (1998) Statistical Learning Theory. Wiley, New York, NY

Vijith H, Suma M, Rekha V, Shiju C, Rejith P (2012) An assessment of soil erosion probability and erosion rate in a tropical mountainous watershed using remote sensing and GIS Arabian Journal of Geosciences 5:797-805

Walling DE, Webb BW (1983) Patterns of sediment yield. Gregory KJ (Ed) Background to Paleohydrology. Wiley, Chichester

Wang B, Zheng F, Römkens MJ, Darboux F (2013) Soil erodibility for water erosion: A perspective and Chinese experiences Geomorphology 187:1-10

Wang R et al. (2016) Gully Erosion Mapping and Monitoring at Multiple Scales Based on Multi-Source Remote Sensing Data of the Sancha River Catchment, Northeast China ISPRS Int J Geo-Inf 5:200

Wawer R, Nowocien E, Podolski B (2005) Eal and calculated Kusle erodibility factor for selected Polish soils Polish Journal of Environmental Studies 14:655-658

Williams J, Bouzarovski S, Swyngedouw E (2014) Politicising the Nexus: Nexus Technologies, Urban Circulation, and the Coproduction of Water-energy. Nexus Network Think Piece Series, Paper 001,

Wischmeier WH, Smith DD (1958) Rainfall energy and its relationship to soil loss Eos, Transactions American Geophysical Union 39:285-291

World Wildlife Fund Malaysia (2002) Community and non-governmental organisation (NGO) partnership in highland catchment management in Malaysia.

Xue X, Luo Y, Zhou X, Sherry R, Jia X (2011) Climate warming increases soil erosion, carbon and nitrogen loss with biofuel feedstock harvest in tallgrass prairie Gcb Bioenergy 3:198-207

Taofeeq Sholagberu Abdulkadir is a Lecturer in the Department of Water Resources and Environmental Engineering (Formerly in Civil Engineering Department), University of Ilorin, Nigeria. He received his Bachelor and Master of Civil Engineering degrees from University of Ilorin, Nigeria and his PhD in Civil and Environmental Engineering from Universiti Teknologi Petronas, Malaysia. His research interests include water resources modeling, watershed management and conservation, artificial neural network, machine learning, GIS and remote sensing. He has published few research articles in reputable national and international journals and conference proceedings.

Raza Ul Mustafa Muhammad is a Senior Lecturer in the Department of Civil and Environmental Engineering, Universiti Teknologi Petronas, Malaysia. He completed his Bachelor of Agricultural Engineering from the University of Agriculture Faisalabad, Pakistan; master's degree in Water Resources Engineering, University of Engineering and Technology, Pakistan; and PhD degree from Universiti Teknologi Petronas, Malaysia. His research interest includes artificial neural network, machine learning applications to water resources, and sediment modeling, among others.

Olayinka Gafar Okeola is a Civil Engineer in the Department of Civil Engineering, University of Ilorin. He received his PhD degree from University of Ilorin in 2010. He has been involved in teaching, mentoring, and research. His key research interest includes decision support systems, watershed hydrology, policy, urban water supply, hydro-economics, and GIS. He has publications in reputable journals and conference proceedings. He received the Pillars of Nation Building Award in 2012 as a Distinguished Academic ICON from the Strategic Institute for Natural Resources and Human Development of Nigeria (SINRHD).

Wan Yusof Khamaruzaman is an Associate Professor, Department of Civil and Environmental Engineering, Universiti Teknologi Petronas, (UTP) Malaysia. He obtained his Bachelor of Science in Civil Engineering from Loughborough University of Technology, United Kingdom, Master of Science in Irrigation Engineering from University of Southampton, Southampton, and PhD in Water Resources and GIS from Coventry University, Coventry, United Kingdom. He has over 30 years of teaching and research experience during which he has graduated several research students. He is currently the cluster leader of Water and Environmental Engineering, UTP. He has won several grants and gold medals in academic exhibitions. His research interests include storm water management, flood modeling, hydrological modeling, coastal erosion, watershed management, computational fluid dynamics, and GIS modeling.

Bashir Adelodun is a Lecturer in the Department of Agricultural and Biosystems Engineering, University of Ilorin, Nigeria. He received both his Bachelor of Engineering and Master of Engineering degrees from the University of Ilorin. He is currently doing his PhD research program in land and water engineering in the Department of Agricultural Civil Engineering, Kyungpook National University, South Korea. He has over 8 years of experience in both teaching and research. His research includes water resources management, water quality and wastewater treatment, soil remediation and environmental pollution, and hydrologic and water resources modeling and simulation.

Saheed Adeniyi Aremu is an Associate Professor in the Department of Water Resources and Environmental Engineering, University of Ilorin, Nigeria. Presently, he is the Managing Director, Lower Niger River Basin, Ilorin, Nigeria. He bagged his Master of Civil Engineering from University of Ibadan and Doctor of Philosophy from University of Ilorin, Nigeria. He has been teaching and researching for over two decades and has published several research papers in reputable national and international Journals including conference proceedings books and chapters. His research interests include water resources management, environmental modeling, pollution, waste management, GIS, and remote sensing.

Index

A

Afforestation, 13, 18, 160, 236, 241
Aggradation and degradation theory, 6
Akarsa watershed, 187–203
Analytical hierarchy process (AHP), 93–106, 405, 416, 428, 462
Arc-GIS, 51, 118, 188, 189, 199, 203

B

Badlands, ix, 46, 48, 65, 69, 105, 109–112, 114–116, 118, 122–125, 137, 155, 208, 212, 214, 216, 222, 231
Bamboo based technology, 307–318
Bamboo plantation, 13, 308, 310–312, 315–317
Bed width, 11, 12, 254, 386, 387
Below ground biomass, 295
Bio-engineering, 13, 14
Bivariate analyses, 370, 371, 373, 376, 377, 379, 405
Buffer strips, 222, 224, 226, 228–230
Bulk density, 22, 25, 120, 222, 295–297, 308, 327, 360, 386, 387

C

Chambal valley, 3, 6, 12, 307
Check dams, 14, 15, 17, 158, 160, 238, 248, 292, 310, 313, 322, 337–343, 456
Chotanagpur plateau, 36, 47, 148, 152
Cluster analysis, 49, 385, 386, 388–390
Community resilience, 449–458
Concentrated flow, 2, 6, 65, 87, 93, 169, 170, 221–230, 235, 295–297, 300, 302, 304, 382
Cross-section area, 256–258

D

Darcy-Weisbach friction factor, 223, 224, 227, 229, 230
Dendrogram, 387
Denudational processes, 70, 79, 134, 198
Detritus layer, 21, 22, 24, 26, 27, 29
Disintegration index, 22
Drainage density, 37, 41, 94, 96, 98, 99, 101, 104, 106, 109, 153, 163, 170, 188, 189, 191–193, 197, 199, 236, 240, 255, 256, 259, 260, 340, 398, 408, 411, 416, 418, 422, 432, 433, 436, 443, 463, 464, 467, 470, 472, 473
Drainage network, 69, 70, 189, 240, 253, 259, 341, 346, 378, 434

E

Ephemeral channel, 163–183
Erosion pin, 110, 111, 115, 118–120, 123, 124

F

Factor analysis, 101, 386, 388–390
Fish resources, 350
Fourfold plot, 423
Froude Number (Fr), 223, 227–230, 249

G

Gangani, 22–24, 27, 28, 69, 208, 211, 212, 214, 218, 346, 348–350
Garbheta badland, 70
Geo-environmental factors, 134, 137, 144, 187, 188, 203, 415, 435, 443
Geomorphic threshold, ix, 45–66
Gorganrood watershed, 427–444
Groundwater fluctuation, 268–271, 273
Gully beds erosion, 221–230
Gully classification, 10, 11, 238
Gully collapse, ix, 21–32
Gully head, 6, 7, 10, 12–14, 21, 22, 27, 28, 46, 47, 49, 50, 53, 56, 57, 59, 63, 66, 115, 136, 208, 211, 214, 222, 238, 258, 280, 284, 290, 307–309, 317, 327, 358, 360, 362, 369–379, 381–383, 385–390, 394
Gully headcut, 14, 51, 63, 66, 142, 208, 372, 386, 394, 399
Gully morphology, 85, 87, 134, 148, 150, 207–218
Gully reclamability, 10, 11
Gully stabilize, 14, 231, 317
Gully susceptibility, 99, 101, 142–143, 428, 432, 440

H

Hydraulic share stress, 223
HydroCAD, 331–334
Hydrogeomorphic, 46, 53, 341

I

Index of diversity, 347
Initiation of gully, 53, 56, 251–261, 279, 427
In-situ scouring experiment, 309

K

Kangsabati basin, 164–166, 172, 346, 348–350, 354
Kappa index, 39, 416, 420, 423, 469, 472–474
Kernel functions, 419, 467–469, 471–474
Khoai badland, 147–160

L

Landslides, 40, 57, 133, 134, 238–240, 249, 321, 323, 324, 378, 421, 428, 434, 436, 457, 461, 463
Lateritic region, 148
Lateritic tract, 22, 208
Lateritic uplands, ix, 208

© Springer Nature Switzerland AG 2020
P. K. Shit et al. (eds.), *Gully Erosion Studies from India and Surrounding Regions*, Advances in Science, Technology & Innovation,
https://doi.org/10.1007/978-3-030-23243-6

Livelihood improvement, 318
Logistic regression (LR), 397, 405, 410, 416, 428, 462

M
Machine learning, ix, 39–41, 398, 406, 415–423, 428, 429, 431, 434,
        443, 462, 467, 469, 474
Maharlou watershed, 406–411
Mahi ravine, 8, 12
Manning roughness coefficient, 223, 224, 230
Maximum entropy model (ME), 398, 405, 427–444, 462
Micro-watershed, 309, 337–343
Multicollinearity, 432
Multi-influencing factor (MIF), 187–203
Multivariate adaptive regression splines (MARS) model, ix, 405–411,
        428

N
Normalized difference vegetation index (NDVI), 189, 192, 408, 411,
        463, 465–467, 469, 470, 472–474

O
Oceanic upwelling theory, 7
Orthophoto, 360, 369, 374

P
Palaeogenesis, 47
Particle size distribution (PSD), 22, 24, 26, 27, 31, 32, 370, 372
Pedogeomorphology, 46
Photogrammetry, 93, 207–218
Pielou's evenness index, 347, 348
Potential land degradation zone (PLDZ), 187–203
Profilometer, 110, 111, 114–118, 123

Q
QGIS, 254

R
Rainfall simulator, 110, 111, 119, 121–122, 325, 332
Random forest (RF) model, ix, 35–43
Rangamati badland, 222, 231
Ravine area restoration, 11, 12
R environment, 416, 420
Resource conservation, 307–318
Reynolds number (Re), 223, 226–230
River water quality, 345–354
Run-off, 2, 39, 46, 69, 98, 118, 140, 147, 163, 199, 222, 235, 252, 280,
        296, 307, 322, 337, 345, 358, 369, 382, 393, 408, 416, 433, 462

S
Sediment yield (SY), ix, 54, 104, 109–111, 118–120, 122, 125, 147,
        151, 157, 163–183, 222, 236, 240–242, 248, 279–292, 302, 321,
        415
Shannon diversity index, 347, 348

Siddheswari river basin, 279–292
Silabati river basin, 346, 349, 350, 353, 354
Simpson's index, 347
Slope aspect, 37, 41, 141, 198, 398, 406, 416, 418, 433, 443
Slope instability, 322
Soil and Water Assessment Tool (SWAT) model, 238, 240, 279–292,
        462
Soil anti-scouribility, 295, 296, 300, 303–305
Soil detachment rate, 296, 300–303
Soil erodibility, 55, 112, 166, 190, 191, 194, 197, 199, 281, 283, 295,
        296, 300, 301, 354, 370, 450, 463, 465, 470
Soil erosion, 2, 21, 35, 45, 69, 93, 109, 133, 148, 163, 188, 211, 222,
        236, 251, 279, 295, 307, 321, 337, 345, 357, 378, 381, 393, 405,
        415, 427, 450, 461
Soil moisture index (SMI), 463, 466, 467, 469, 470, 472, 473
Soil organic matter (SOM), 6, 7, 22, 24–27, 29, 31, 32
Spatial analysis, 370–372, 461–475
Stream length-gradient index, 148, 150
Stream power index (SPI), 134, 137, 141, 144, 381, 432, 436
Support vector machine (SVM), 406, 415–423, 428, 462, 463, 466–469,
        471–474
Surface runoff, ix, 14, 46, 53, 55, 105, 147, 151, 163, 168, 222, 228,
        280, 281, 283, 284, 287–289, 296, 312, 337, 354, 358, 360, 365,
        382, 388, 389, 416, 418, 423, 433, 434, 463
Susceptibility map, 35, 36, 41, 133, 145, 405–410, 427–430, 435,
        439–441, 443, 462, 463, 466, 469, 471, 472

T
Tension cracks, 8, 427
Terai region, 251–261
Topographical factors, 175, 416, 418
Top width, 10, 12, 366, 386, 387
Tropical plateau, 163–181, 202
True skill statistics (TSS), 416, 419, 420, 423
Tunneling, 7–9, 358

U
Univariate, 370, 372, 373, 375, 377, 379
Upliftment theory, 6
Upstream drainage system, 256

V
Vegetative cover, 109, 231, 312, 318, 322, 323, 455
Vetiver root, 325–328, 331–334
Vulnerability, 26, 57, 59, 66, 94, 198, 316, 451, 452, 455–458

W
Weighted linear combination (WLC), 96, 99, 101, 105
Weight of evidence model (WOE), 133, 145, 405
Wetness index (WI), 37, 134, 137, 140–142, 144, 381, 398, 416, 418,
        422, 433, 436, 443
Width-depth ratio, 74, 76, 77, 256, 258, 387

Y
Yamuna valley, 3, 6, 12, 17, 309, 311, 312, 315

Printed by Printforce, the Netherlands